OXFORD READINGS IN PHILOSOPHY

THE PHILOSOPHY OF BIOLOGY

THE PHILOSOPHY OF BIOLOGY

Edited by

DAVID L. HULL

and

MICHAEL RUSE

OXFORD UNIVERSITY PRESS
1998

Oxford University Press, Great Clarendon Street, Oxford OX2 6DP

Oxford New York

Athens Auckland Bangkok Bogota Bombay Buenos Aires
Calcutta Cape Town Dar es Salaam Delhi Florence Hong Kong Istanbul
Karachi Kuala Lumpur Madras Madrid Melbourne Mexico City
Nairobi Paris Singapore Taipei Tokyo Toronto Warsaw

and associated companies in
Berlin Ibadan

Oxford is a trade mark of Oxford University Press

Published in the United States by
Oxford University Press Inc., New York

Introduction and selection © Oxford University Press 1998

British Library Cataloguing in Publication Data
Data available

Library of Congress Cataloging in Publication Data
The philosophy of biology / edited by David L. Hull and Michael Ruse.
(Oxford readings in philosophy)
Includes bibliographical references (p.   ) and index.
1. Biology—Philosophy. I. Hull, David L. II. Ruse, Michael. III. Series.
QH331.P468   1997 570'.1—dc21 97-36921
ISBN 0-19-875213-X.
ISBN 0-19-875212-1 (pbk.)

1 3 5 7 9 10 8 6 4 2

Typeset by Best-set Typesetter Ltd., Hong Kong
Printed in Great Britain by
Biddles Ltd., Guildford and King's Lynn

CONTENTS

CONTENTS

INTRODUCTION

DAVID L. HULL AND MICHAEL RUSE

During the past three decades, philosophy of biology has come into its own. It is now a mature discipline. In fact, the classic papers in this discipline are so familiar by now that they need not be reproduced again. Instead, more recent papers by some of the newer members of the profession are included in this anthology. Of the thirty-six papers reproduced here, all but three appeared in the last decade. And members of our profession are not limited to professional philosophers. Rather, a third of the papers in this anthology were authored or co-authored by scientists, primarily biologists. In no other area of philosophy of science have philosophers and scientists co-operated to the extent that they have in philosophy of biology.

Of the traditional issues in philosophy of biology, we have included four—adaptation, units of selection, function, and species. As central as adaptation is to evolutionary biology, problems arise with respect to its application. Is it as slippery a notion as Gould and Lewontin in their classic paper claim, or is it no different in kind from other fundamental concepts in biology? Fitness has played such an extensive role in philosophy of biology that some critics refer to our discipline derisively as 'the philosophy of fitness'. In this anthology an equally important issue—the levels at which selection occurs—is discussed at some length. On no other issue have philosophers and biologists co-operated to greater mutual benefit.

Over the years two distinct senses of 'function' have emerged in the philosophical literature—Cummins functions and Wright functions. Can they be merged into a single, unambiguous usage, as Philip Kitcher suggests? As in the case of function, one would think that nothing new could possibly be said about the species problem. However, the papers included here show that this is not the case. There *is* something new under the sun. Phylogenetic species concepts are nothing if not novel. Finally, although development surely deserves to be a central issue in the philosophy of biology, it has been all but ignored until quite recently. As the two papers on the subject included in this anthology indicate, it is likely to become a major topic for future research.

The papers mentioned so far represent the sort of issues that professional philosophers have addressed in the past. But younger members of our profession have begun to participate in discussions of more socially relevant problems. For example, those people who know the least about biological evolution are the most certain that it is progressive. Evolutionary biologists are not so sure. As intuitively obvious as the notion of evolutionary progress seems to be, it is very difficult to show explicitly that it is anything but an illusion. In the past few years, students of science have repeatedly emphasized (and occasionally documented) that forces and factors other than reason, argument, and evidence influence the course of science. Is a belief in progress one of these external forces?

Homo sapiens is a biological species like any other. Lots of species are sexually dimorphic. So are we. But the human species is peculiarly social. We live in societies. In these societies there is more to sexual dimorphism than just sex. Gender also matters. So does sexual preference. Peculiarly human notions of altruism play important roles in society, roles that on the surface seem to conflict with a notion of altruism developed by biologists. One would think that biology would have something to contribute to these emotionally charged issues.

Until recently, physics has had a corner on big science. Partially in response to this uneven allotment of resources, biologists launched a massive effort to map the entire human genome base pair by base pair. This led to controversy centred on finances. The fear was that money normally spent on a variety of biological programmes would be redirected to this one huge project, and these fears have proved to be well-founded. But more general fears arose concerning the social and ethical implications of the Human Genome Project. For the first time, money was set aside in a government-supported project to study these issues and implications. As a result, the philosophical literature on the Human Genome Project is huge.

Religious fundamentalism is on the rise again, from Algeria to Arkansas. One of the effects of this resurgence that is peculiar to the United States is an increasing effort by Creationists to require that Bible stories be taught in high school biology classes. To some extent this controversy has pitted strongly religious people against those of us who are at most indifferent to religion. In addition, the scholarly credentials of the two sides have been decidedly different. Creationists either lack higher degrees or have degrees in areas unrelated to evolutionary biology. In this anthology we have included two sides of the issue argued by scholars who are not only religious themselves but also respected scholars.

PART I

ADAPTATION

INTRODUCTION TO PART I

MICHAEL RUSE

In his *Dialogues Concerning Natural Religion*, David Hume drove a skewer through the Argument from Design—that argument for God's existence which claims that organisms are so well put together that their features (their 'adaptations' for survival and reproduction) necessitate the supposition of a divine artificer. Yet, in the absence of an alternative, the argument continued to hold sway right through the first half of the nineteenth century. Small wonder, then, that when Charles Darwin had become convinced of the truth of evolution, he should have laboured to find a mechanism which would explain not simply change but change in an adaptive fashion. The hand and the eye simply could not have come about through chance. But, argued Darwin, they could have come about through 'natural selection' or the 'survival of the fittest'. Since so many more organisms are born than can survive and reproduce, there is a consequent struggle for existence, and success in this struggle comes through the special features possessed by the winners alone. Given enough time, this winnowing, or selecting, of the successful or 'fit' leads to full-blown adaptation.

From its introduction in 1859 in the *Origin of Species*, adaptation (and Darwin's mechanism of selection) has been controversial. Not everyone was convinced that the organic world is so very adapted. German natural theology had always stressed the primacy of the isomorphisms between different organisms—the analogy between the bones of the arm and hand of the human, the wing of the bird, the forelimb of the horse, the flipper of the whale, and so on. Although, clearly, Darwin's evolution in itself was explaining the fact of these isomorphisms (known now as 'homologies')— they are the legacies of common ancestors—attention was still directed away from selection and adaptation. Even evolutionists continued to think it just too improbable that any non-directed (by God) force could account for the organic intricacies that were stressed by enthusiasts for the Argument from Design. Either the argument must be wrong in its premises— things are not so very well adapted—or there must be an unknown mechanism that can explain the full range of adapted life. In the first part of this century, many saw the notion of randomness introduced by the theoretician Sewall Wright as speaking to both sides of this supposition. He argued that 'genetic drift'—the chance matings of organisms in small populations—can be significant in micro-evolutionary situations, even though the changes produced are not adaptive.

In fact, English evolutionists always felt sympathy for the Darwinian position and its promotion of adaptation and natural selection. Then, about the time of the Second World War—thanks particularly to discoveries made on fruit-flies by the Russian-American population geneticist Theodosius Dobzhansky—general opinion in North America became more of this opinion also. A consensus emerged that held until about twenty-five years ago, when things started to unravel again. The anomalies and exceptions began to loom larger, and before long the whole question of adaptation again became a matter of focus and discussion.

The survey discussion on the meaning(s) of adaptation, by the biologist Mary Jane West-Eberhard (Ch. 1), sets the scene. She provides evidence of the ambivalence that biologists today feel, not so much about adaptation as such—no one denies that there are standard cases of adaptation and that selection was involved in their production—but about how ubiquitous it is. Is it the case that most organic features are adaptive, and that the exceptions are just that: exceptions? Or is adaptation just one of a range of states in which we find organic features, in which case one could just as easily find features that are the results of non-selective mechanisms and so are not adaptive at all? In any case, however one answers these questions, is it heuristically useful to assume adaptiveness unless forced to conclude otherwise? Is the 'adaptationist programme' a good scientific strategy?

Richard Dawkins (Ch. 2) and Daniel Dennett (Ch. 3) have little doubt on where they stand or on the best line of action for the evolutionist. Dawkins, one of today's most popular writers on evolution, is an ardent Darwinian, and believes that natural selection is by far the most important mechanism of biological change. For him it is, if not positively unthinkable, then biologically highly implausible that any other mechanism could even approach the effectiveness of selection. He stresses not just the adaptiveness of organisms, but their adaptive complexity—the fact that things like the hand and the eye really are very subtle and organized entities. Dawkins is right with the natural theologians—especially the most noted of all, Archdeacon Paley, who authored the highly influential *Natural Theology* (1802)—in thinking that blind chance simply could not have brought such phenomena into existence. However, after running through the alternatives, Dawkins argues that it is impossible that other proposed evolutionary mechanisms (like, for instance, the Lamarckian inheritance of acquired characteristics) could have done any better. This leaves the field to natural selection. It, and it alone, can account for the nature of the organic world.

Much in agreement with Dawkins, Dennett (Ch. 3) takes the argument further, claiming that adaptationism—supposing that organic features are as if designed—is indeed a crucial heuristic principle in evolutionary argumentation. Deciding whether or not to take an adaptationist stance is not an optional extra. To fail to do so is to fail to do evolutionary biology as we understand it. Without assumptions about design (more precisely, assumptions about design-like features brought about by natural selection), one simply does not have appropriate questions to ask about organic features. And without questions, there will be no answers.

Next, we have the paleontologists Stephen Jay Gould and Elizabeth Vrba. Gould, both a professional scientist and (like Dawkins) a popular writer on matters evolutionary, is today's greatest critic of the adaptationist stance. He argues that it is often little more than a carry-over of a long discarded natural theology and is quite inappropriate—positively dangerous—in evolutionary biology today. Here, Gould joins forces with Vrba to show that opposition to adaptationism is not blind or indiscriminate. They appreciate with other biologists the importance of adaptation as produced by natural selection. It is rather that they think that uncritical adaptationism is only going to lead to trouble, and through the neologism of 'exaptation'—a term referring to a feature produced for one purpose and then put to use for another—they hope to save what is of value in the adaptationist stance. At the same time, they want to move on to highlight and understand aspects of evolutionary processes which demand more subtle treatment.

The philosopher of biology Elliott Sober (Ch. 5) has the final word, offering both a synthesis and a critique of much that has been said recently about adaptation. He has incisive things to say about the kinds of arguments seen already in this section, although his conclusion is far from negative regarding adaptationist thinking. He argues that much of the controversy is the result of confusing a claim about nature and a methodological directive about the best way to do biology.

We suspect that neither Sober nor the other contributors will have the last word on the subject of adaptation; but they give a sense of why the debate about adaptation has been so intense and why it is so important for an understanding of evolutionary biology.

1

ADAPTATION: CURRENT USAGES

MARY JANE WEST-EBERHARD

In contemporary evolutionary biology an 'adaptation' is a characteristic of an organism whose form is the result of selection in a particular functional context (see Williams 1966, Futuyma 1986). Accordingly, the process of 'adaptation' is the evolutionary modification of a character under selection for efficient or advantageous (fitness-enhancing) functioning in a particular context or set of contexts. The word is sometimes also applied to individual organisms to denote the 'propensity to survive and reproduce' in a particular environment (general adaptation) (see Mayr 1988). Ernst Mayr (1986) suggests substituting the term 'adaptedness' for this usage.

The use of 'adaptation' by evolutionary biologists thus differs from that in some other areas of biology, where the term can refer to short-term physiological adjustments by phenotypically plastic individuals (adaptability) or to a change in the responsiveness of muscle/nerve tissue upon repeated stimulation.

According to strict usage in evolutionary biology, it is correct to consider a character an 'adaptation' for a particular task only if there is some evidence that it has evolved (been modified during its evolutionary history) in specific ways to make it more effective in the performance of that task, and that the change has occurred due to the increased fitness that results. Incidental ability to perform a task effectively is not sufficient; nor is mere existence of a good fit between organism and environment. To be considered an adaptation, a trait must be shown to be a consequence of selection for that trait, whether natural selection or sexual and social selection—whether the selective context involves what Darwin called 'the struggle for existence', or competitive interactions with conspecifics.

Several kinds of evidence can contribute to determining whether or not a characteristic of an organism is an adaptation (after Curio 1973, elaborating on suggestions of Tinbergen 1967). The first is correlation between

First published in E. Fox Keller and E. Lloyd (eds.), *Keywords in Evolutionary Biology* (Cambridge, Mass.: Harvard University Press, 1992), 13–18. Reprinted by permission.

character and environment or use. A character shows evidence of being an adaptation if (a) the same form or similar forms occur in similar environments in a number of different species, especially in unrelated species (due to convergence); (b) variant forms of a character in a number of related species (e.g. of a single genus) accord with differences in the environments of the respective species, or with the mode of usage of the character in different species; (c) variant forms appearing in different life stages during ontogeny accord with differences in the environment or behaviour of the respective life stages; or (d) for complex characters in a particular context, the more their component aspects can be related point by point to function in that context (the goodness of 'design' of Williams 1966: 12 ff.).

The second kind of evidence used in determining whether a characteristic is an adaptation is that which results from altering a character. An organ or behaviour is experimentally altered or eliminated, in order to see how this affects its efficiency in a particular function or environmental condition.

A third kind of evidence is obtained through comparison of naturally occurring variants (individual differences). The efficiency or reproductive success of different forms or morphs within a species are compared in the situation(s) where they are hypothesized to function as adaptations.

All of these approaches provide evidence for or against the hypothesis that the structural peculiarities of a trait owe their existence (spread and persistence) in a population to their contribution to fitness via performance of a particular task.

An example can serve to illustrate some of the difficulties in applying the adaptation hypothesis to particular cases. The elaborately sculptured and species-specific forms of the head and thoracic horns of male beetles have been imagined to be adaptations for fighting, for digging, and for influencing female choice of mates. Observations of behaviour, however, demonstrate that the structural details of beetle horns and the differences between related species correspond to inter-specific differences in the particular ways they are wielded during battles between males; their special features are not used in special ways during courtship or digging, although they are occasionally used to hold females or to enlarge holes occupied by beetles (Eberhard 1979, 1980). Thus the available evidence supports the hypothesis that beetle horns are adaptations for fighting, and that they are only incidentally or secondarily used during mating and digging. It could be argued, however, that the structural peculiarities observed are developmental or pleiotropic results of traits evolved in other contexts (the 'exaptations' of Gould and Vrba 1982), and that the high degree of correlation with behaviour (which is difficult to consider merely coincidental) has been produced by selection to use these incidentally common structures to the individual's advantage in fights; by this interpretation, horn morphology would be a non-adaptation, and the form of behaviour an adaptation.

It is not always easy to apply the distinction between adaptation and incidental use, even given information on present employment and evolutionary history. Suppose an incidental use or secondary function were to persist, while the original, evolved function disappeared (e.g. horns came to be used exclusively for digging, even though they had not been modified in that context). Strict adherence to the above definition would not permit horns to be considered an adaptation for digging, even though digging had become the exclusive context for their use, and even though they might be maintained (rather than lost) under selection in that context. The concept of 'pre-adaptation' has been applied to such cases, in which a trait has evolved in one context and has come to be used (function) in another.

Suppose a horn used secondarily but exclusively in digging undergoes some small modification enhancing the digging function. Can it then be considered an 'adaptation' for digging? Evidently it can, although this points up another difficulty in the distinction: how much modification is necessary to consider a character an adaptation in a particular context? What, indeed, is a 'character', as opposed to a feature or modification of a character? The designation of an aspect of the phenotype as a character (whether an adaptation or not) is always somewhat arbitrary: is digging behaviour, along with horn morphology, part of a single co-selected trait? This would classify the pre-adapted horn as part of a new 'adaptation'.

Curio (1973) argues that when exactly the same character is employed in more than one context and contributes to fitness in all contexts, it should be regarded as an adaptation only for that context where it makes the greatest contribution to fitness. Such an argument can lead to contradictions in applying the above criteria: for example, if the form of a character has been shaped in the past primarily by a function presently of less importance (in terms of fitness) than another use (which by Curio's criterion would be the primary adaptive context even if not effecting evolutionary modification of the character). In most discussions, the historical criterion (rather than fitness difference) would predominate: the character would be considered an 'adaptation for' the function in which it was originally or primarily shaped by selection. Even when multiple uses are completely contemporaneous in their fitness effects, Curio's criterion seems difficult to apply, given that, in so far as the same form can serve multiple functions, the sum of all (even minor) contributions to fitness could influence form in the face of counter-selection (in other contexts) favouring alternative forms. These considerations regarding multiple functions apply as well to questions of selection at different levels of organization, whereby the same trait may simultaneously affect, for example, the survival or replication rate of individuals and groups, and hence the population frequencies of their constituent genotypes.

Given current usage of the word 'adaptation', it is clear that not all observable evolved characteristics of organisms are properly regarded as adaptations. In their efforts to explain peculiarities of form, biologists often attempt to apply a hypothesis of adaptation with insufficient empirical support. Several authors have argued in favour of parsimony in the use of this term (e.g. Williams 1966, Curio 1973, Gould and Lewontin 1979). They stress the importance of considering alternative explanations for particular and even complex characters, especially the hypotheses that form can be vestigial (the product of selective forces no longer operating) or the incidental result of developmental processes evolved under selection for other aspects of the phenotype.

Stephen Jay Gould (1984) has proposed that covariance of characters could be accepted as 'positive evidence' of non-adaptation, and has erected a dichotomy of 'automatic sequelae' (non-adaptations) versus selected traits (adaptations). This criterion of non-adaptation tacitly requires some analysis of adaptation, however, because it is impossible to tell from covariance alone which of several developmentally associated traits has been most important in the spread and/or maintenance of the set. Furthermore, one cannot assume that covariant aspects have not been modified independently of each other. For example, Gould (1981) interpreted the male-like female display morphology and behaviour of the genitalic displays of female hyenas as a non-adaptation, evolved by selection in males and only incidentally or secondarily expressed in females. However, female genital displays are known to function as appeasement gestures (Wickler 1966, Eibl-Eibesfeldt 1970), and if modified or somewhat specialized due to selection on females, they would qualify as adaptations. This would be true even if a set of characters used in this way originated via a regulatory mutation that allowed them to be expressed in females as well as in males (where the original set had been formed by selection). Indeed, new adaptations may sometimes originate as co-adapted character sets, whose expression has been shifted between sexes or life stages (via heterochrony) and then modified in the new context (see West-Eberhard 1989).

Gould (1984) also argued that 'ecophenotypic responses' to environmental conditions cannot be regarded as adaptations, because they are not 'genetically mediated'; but this criterion for non-adaptation (environmental influence in phenotype determination) cannot hold unequivocally: plasticity itself can be seen as an adaptation. Furthermore, ecophenotypic responses are always products of gene–environment interaction, and thus are genetically mediated (see West-Eberhard 1989). By Gould's criterion, all environmentally cued, facultatively expressed phenotypes would presumably be classified as 'non-adaptations', including the winter pelage of hibernating mammals, the restive walking behaviour of the swarming phase of migratory locusts, and the ability of chameleons to match the background colouration of their resting-places.

Developmental mechanism *per se* does not provide enough information to determine whether or not a trait is an adaptation, though it might provide information on how non-adaptive traits are maintained (e.g. via covariance with adaptive traits), and even on how adaptive traits originate. An aspect of the phenotype that is a secondary 'by-product' of selection for another aspect (in the sense of being either completely covariant with it or a less commonly expressed product of the same genotype) may have the following relationships to adaptation and selection.

(a) The observed frequency and form of the secondary aspect of the phenotype may be completely owing to characteristics evolved under selection for a covariant aspect, in which case the character would not be regarded as an adaptation.

(b) More than one covariant aspect of the phenotype may contribute simultaneously to fitness in different functional contexts (e.g. pleiotropic effects of a single gene) from the time of their (simultaneous) origin and be concurrently favoured by selection. I would call both positively selected traits adaptations, even if one of them made a greater contribution to the fitness and spread of the covariant set and its underlying genes, because both aspects contribute to the rate of spread of the set in competition with alternatives; Curio (1973) would term only the greater contributor to fitness an adaptation.

(c) The initial spread or frequency of the secondary aspect of the phenotype in the population may have been entirely due to selection for a covariant aspect, but its form and/or frequency of expression may have been modified in the context in which it is expressed. In this case a phenotype not originally an adaptation has become an adaptation by evolution in its own context.

To classify a pleiotropic or secondary effect as a non-adaptation requires showing not only that it is (a) only expressed together with a developmentally related trait that is a proved adaptation, but also evidence that (b) concurrent positive selection and (c) independent modification do not apply.

Overly facile application of the term 'adaptation' encourages the assumption that all characters are adaptive; for this reason, some authors have urged restraint on use of the term. It remains the case, however, that persistent attempts to discern the adaptive significance of phenotypic traits—to apply an adaptation hypothesis—have been a primary and fruitful occupation of evolutionary biologists since before Darwin. There is still controversy over the importance of selection and adaptation versus non-adaptation in the evolution of phenotypes. Although adaptation cannot be assumed, some authors argue that it should be regarded as the most important (commonly supported) hypothesis for the spread and persistence of organismic traits: 'The experimental study of adaptation has unravelled adaptive values in such unobtrusive and inconspicuous details of organismic organization that one should think of a character as having survival value until the contrary has been demonstrated' (Curio 1973: 1046). Richard Lewontin (1978: 230) gave the following compelling reason for continuing to pursue the 'adaptationist' programme that seeks to explain characters in terms of their evolved functions, in spite of its difficulties:

Even if the assertion of universal adaptation is difficult to test because simplifying assumptions and ingenious explanations can almost always result in an ad hoc adaptive explanation, at least in principle some of the assumptions can be tested in some cases. A weaker form of evolutionary explanation that explained some proportion of the cases by adaptation and left the rest to allometry, pleiotropy, random gene fixations, linkage and indirect selection would be utterly impervious to test. It would leave the biologist free to pursue the adaptationist program in the easy cases and leave the difficult ones on the scrap heap of chance. In a sense, then, biologists are forced to the extreme adaptationist program because the alternatives, although they are undoubtedly operative in many cases, are untestable in particular cases.

REFERENCES

Curio, E. (1973), 'Towards a Methodology of Teleonomy', *Experientia*, 29: 1045–58.
Eberhard, W. G. (1979), 'The Function of Horns in *Podischnus agenor* (Dynastinae) and Other Beetles', in M. S. Blum and N. A. Blum (eds.), *Sexual Selection and Reproductive Competition* (New York: Academic Press), 231–58.
—— (1980), 'Horned Beetles', *Scientific American*, 242/3: 166–81.
Eibl-Eibesfeldt, I. (1970), *Ethology: The Biology of Behaviour* (New York: Holt, Rinehart and Winston).
Futuyma, D. J. (1986), *Evolutionary Biology*, 2nd edn. (Sunderland, Mass.: Sinauer).
Gould, S. J. (1981), 'Hyena Myths and Realities', *Natural History*, 90: 16–24.

Gould, S. J. (1984), 'Covariance Sets and Ordered Geographic Variation in *Cerion* from Aruba, Bonaire and Curacao: A Way of Studying Nonadaptation', *Systematic Zoology*, 33/2: 217–37.

—— and Lewontin, R. C. (1979), 'The Spandrels of San Marco and the Panglossian Paradigm', *Proceedings of the Royal Society of London*, B 205: 581–98.

—— and Vrba, E. (1982), 'Exaptation: A Missing Term in the Science of Form', *Paleobiology*, 8: 4–15; reproduced as Ch. 4.

Lewontin, R. C. (1978), 'Adaptation', *Scientific American*, 239/3: 212–30.

Mayr, E. (1986), 'Natural Selection: The Philosopher and the Biologist', *Paleobiology*, 12/2: 233–9.

—— (1988), *Toward a New Philosophy of Biology* (Cambridge, Mass.: Harvard University Press).

Tinbergen, N. (1967), 'Adaptive Features of the Black-Headed Gull *Larus ridibandus*', *Proceedings of the 14th Interantional Ornithological Congress*, 43–59.

West-Eberhard, M. J. (1989), 'Phenotypic Plasticity and the Origins of Diversity', *Annual Review of Ecology and Systematics*, 20: 249–78.

Wickler, W. (1966), 'Ursprung und biologische Deutung des Genitalprasentierens mannlicher Primaten', *Tierpsychologie*, 23: 422–37.

Williams, G. C. (1966), *Adaptation and Natural Selection* (Princeton: Princeton University Press).

2

UNIVERSAL DARWINISM

RICHARD DAWKINS

It is widely believed on statistical grounds that life has arisen many times all around the universe (Asimov 1979, Billingham 1981). However varied in detail alien forms of life may be, there will probably be certain principles that are fundamental to all life, everywhere. I suggest that prominent among these will be the principles of Darwinism. Darwin's theory of evolution by natural selection is more than a local theory to account for the existence and form of life on Earth. It is probably the only theory that *can* adequately account for the phenomena that we associate with life.

My concern is not with the details of other planets. I shall not speculate about alien biochemistries based on silicon chains, or alien neurophysiologies based on silicon chips. The universal perspective is my way of dramatizing the importance of Darwinism for our own biology here on Earth, and my examples will be mostly taken from Earthly biology. I do, however, also think that 'exobiologists' speculating about extraterrestrial life should make more use of evolutionary reasoning. Their writings have been rich in speculation about how extraterrestrial life might work, but poor in discussion about how it might *evolve*. This essay should, therefore, be seen firstly as an argument for the general importance of Darwin's theory of natural selection; secondly as a preliminary contribution to a new discipline of 'evolutionary exobiology'.

The 'growth of biological thought' (Mayr 1982) is largely the story of Darwinism's triumph over alternative explanations of existence. The chief weapon of this triumph is usually portrayed as *evidence*. The thing that is said to be wrong with Lamarck's theory is that its assumptions are factually wrong. In Mayr's words: 'Accepting his premises, Lamarck's theory was as legitimate a theory of adaptation as that of Darwin. Unfortunately, these premises turned out to be invalid.' But I think we can say something stronger: *even accepting his premisses*, Lamarck's theory is *not* as legitimate a theory of adaptation as that of Darwin because, unlike Darwin's, it

First published in D. S. Bendall (ed.), *Evolution from Molecules to Man* (Cambridge: Cambridge University Press, 1983), 403–25. Reprinted by permission.

is *in principle* incapable of doing the job we ask of it—explaining the evolution of organized, adaptive complexity. I believe this is so for all theories that have ever been suggested for the mechanism of evolution except Darwinian natural selection, in which case Darwinism rests on a securer pedestal than that provided by facts alone.

Now, I have made reference to theories of evolution 'doing the job we ask of them'. Everything turns on the question of what that job is. The answer may be different for different people. Some biologists, for instance, get excited about 'the species problem', while I have never mustered much enthusiasm for it as a 'mystery of mysteries'. For some, the main thing that any theory of evolution has to explain is the diversity of life— cladogenesis. Others may require of their theory an explanation of the observed changes in the molecular constitution of the genome. I would not presume to try to convert any of these people to my point of view. All I can do is to make my point of view clear, so that the rest of my argument is clear.

I agree with Maynard Smith (1969) that 'The main task of any theory of evolution is to explain adaptive complexity, i.e. to explain the same set of facts which Paley used as evidence of a Creator'. I suppose people like me might be labelled neo-Paleyists, or perhaps 'transformed Paleyists'. We concur with Paley that adaptive complexity demands a very special kind of explanation: either a Designer, as Paley taught, or something such as natural selection that does the job of a designer. Indeed, adaptive complexity is probably the best diagnostic of the presence of life itself.

ADAPTIVE COMPLEXITY AS A DIAGNOSTIC CHARACTER OF LIFE

If you find something, anywhere in the universe, whose structure is complex and gives the strong appearance of having been designed for a purpose, then that something either is alive, or was once alive, or is an artefact created by something alive. It is fair to include fossils and artefacts, since their discovery on any planet would certainly be taken as evidence for life there.

Complexity is a statistical concept (Pringle 1951). A complex thing is a statistically improbable thing, something with a very low a priori likelihood of coming into being. The number of possible ways of arranging the 10^{27} atoms of a human body is obviously inconceivably large. Of these possible ways, only very few would be recognized as a human body. But this is not, by itself, the point. Any existing configuration of atoms is, a posteriori, unique, as 'improbable', with hindsight, as any other. The point is that, of all possible ways of arranging those 10^{27} atoms, only a tiny minority would constitute anything remotely resembling a machine that worked to keep itself in being, and to reproduce its kind. Living things are not just statistically improbable in the trivial sense of hindsight: their statistical improbability is limited by the a priori constraints of design. They are *adaptively* complex.

The term 'adaptationist' has been coined as a pejorative name for one who assumes 'without further proof that all aspects of the morphology, physiology and behavior of organisms are adaptive optimal solutions to problems' (Lewontin 1979, 1983). I have responded to this elsewhere (R. Dawkins 1982*a*: ch. 3). Here, I shall be an adaptationist in the much weaker sense that I shall only be *concerned* with those aspects of the morphology, physiology, and behaviour of organisms that are undisputedly adaptive solutions to problems. In the same way, a zoologist may specialize on vertebrates without denying the existence of invertebrates. I shall be preoccupied with undisputed adaptations because I have defined them as my working diagnostic characteristic of all life, anywhere in the universe, in the same way as the vertebrate zoologist might be preoccupied with backbones because backbones are the diagnostic character of all vertebrates. From time to time I shall need an example of an undisputed adaptation, and the time-honoured eye will serve the purpose as well as ever (Paley 1828, Darwin 1859, any fundamentalist tract). 'As far as the examination of the instrument goes, there is precisely the same proof that the eye was made for vision, as there is that the telescope was made for assisting it. They are made upon the same principles; both being adjusted to the laws by which the transmission and refraction of rays of light are regulated' (Paley 1828: i. 17).

If a similar instrument were found upon another planet, some special explanation would be called for. Either there is a God, or, if we are going to explain the universe in terms of blind physical forces, those blind physical forces are going to have to be deployed in a very peculiar way. The same is not true of non-living objects, such as the moon or the solar system (see below). Paley's instincts here were right.

My opinion of Astronomy has always been, that it is *not* the best medium through which to prove the agency of an intelligent Creator . . . The very simplicity of [the heavenly bodies'] appearance is against them . . . Now we deduce design from relation, aptitude, and correspondence of *parts*. Some degree therefore of *complexity* is necessary to render a subject fit for this species of argument. But the heavenly bodies do not, except perhaps in the instance of Saturn's rings, present themselves to our observation as compounded of parts at all. (Paley 1828: ii. 146–7)

A transparent pebble, polished by the sea, might act as a lens, focusing a real image. The fact that it is an efficient optical device is not particularly interesting because, unlike an eye or a telescope, it is too simple. We do not feel the need to invoke anything remotely resembling the concept of design. The eye and the telescope have many parts, all co-adapted and working together to achieve the same functional end. The polished pebble has far fewer co-adapted features: the coincidence of transparency, high refractive index, and mechanical forces that polish the surface in a curved shape. The odds against such a threefold coincidence are not particularly great. No special explanation is called for.

Compare how a statistician decides what P value to accept as evidence for an effect in an experiment. It is a matter of judgement and dispute, almost of taste, exactly when a coincidence becomes too great to stomach. But, no matter whether you are a cautious statistician or a daring statistician, there are some complex adaptations whose 'P value', whose coincidence rating, is so impressive that nobody would hesitate to diagnose life (or an artefact designed by a living thing). My definition of living complexity is, in effect, 'that complexity which is too great to have come about through a single coincidence'. For the purposes of this essay, the problem that any theory of evolution has to solve is how living adaptive complexity comes about.

In the book referred to above, Mayr (1982) helpfully lists what he sees as the six clearly distinct theories of evolution that have ever been proposed in the history of biology. I shall use this list to provide me with my main headings. For each of the six, instead of asking what the evidence is, for or against, I shall ask whether the theory is *in principle* capable of doing the job of explaining the existence of adaptive complexity. I shall take the six theories in order, and will conclude that only Theory 6, Darwinian selection, matches up to the task.

Theory 1. Built-in Capacity For, or Drive Toward, Increasing Perfection

To the modern mind this is not really a theory at all, and I shall not bother to discuss it. It is obviously mystical, and does not explain anything that it does not assume to start with.

Theory 2. Use and Disuse Plus Inheritance of Acquired Characters

It is convenient to discuss this in two parts.

Use and disuse   It is an observed fact that on this planet living bodies sometimes become better adapted as a result of use. Muscles that are exercised tend to grow bigger. Necks that reach eagerly towards the tree-tops may lengthen in all their parts. Conceivably, if on some planet such acquired improvements could be incorporated into the hereditary information, adaptive evolution could result. This is the theory often associated with Lamarck, although there was more to what Lamarck said. Crick (1982: 59) says of the idea: 'As far as I know, no one has given *general* theoretical reasons why such a mechanism must be less efficient than natural selection. 'In this section and the next I shall give two general theoretical objections to Lamarckism of the sort which, I suspect, Crick was calling for. I have discussed both before (R. Dawkins 1982*b*), so will be brief here. First, the shortcomings of the principle of use and disuse.

The problem is the crudity and imprecision of the adaptation that the principle of use and disuse is capable of providing. Consider the evolution-ary improvements that must have occurred during the evolution of an organ such as an eye, and ask which of them could conceivably have come about through use and disuse. Does 'use' increase the transparency of a lens? No, photons do not wash it clean as they pour through it. The lens and other optical parts must have reduced, over evolutionary time, their spherical and chromatic aberration; could this come about through in-creased use? Surely not. Exercise might have strengthened the muscles of the iris, but it could not have built up the fine feedback control system which controls those muscles. The mere bombardment of a retina with coloured light cannot call colour-sensitive cones into existence, or connect up their outputs so as to provide colour vision.

Darwinian types of theory, of course, have no trouble in explaining all these improvements. Any improvement in visual accuracy could signifi-cantly affect survival. Any tiny reduction in spherical aberration may save a fast-flying bird from fatally misjudging the position of an obstacle. Any minute improvement in an eye's resolution of acute coloured detail may crucially improve its detection of camouflaged prey. The genetic basis of any improvement, however slight, will come to predominate in the gene pool. The relationship between selection and adaptation is a direct and close-coupled one. The Lamarckian theory, on the other hand, relies on a much cruder coupling: the rule that the more an animal uses a certain bit of itself, the bigger that bit ought to be. The rule occasionally might have some validity, but not generally, and, as a sculptor of adaptation, it is a blunt hatchet in comparison to the fine chisels of natural selection. This point is universal. It does not depend on detailed facts about life on this particular planet. The same goes for my misgivings about the inheritance of acquired characters.

Inheritance of acquired characters The problem here is that acquired characters are not always improvements. There is no reason why they should be, and indeed the vast majority of them are injuries. This is not just a fact about life on earth. It has a universal rationale. If you have a complex and reasonably well-adapted system, the number of things you can do to it that will make it perform less well is vastly greater than the number of things you can do to it that will improve it (Fisher 1958). Lamarckian evolution will move in adaptive directions only if some mechanism—selection—exists for distinguishing those acquired characters that are improvements from those that are not. Only the improvements should be imprinted into the germ line.

Although he was not talking about Lamarckism, Lorenz (1966) emphasized a related point for the case of learned behaviour, which is perhaps the most important kind of acquired adaptation. An animal learns to be a better animal during its own lifetime. It learns to eat sweet foods, say, thereby increasing its survival chances. But there is nothing inherently nutritious about a sweet taste. Something, presumably natural selection, has to have built into the nervous system the arbitrary rule: 'treat sweet taste as reward', and this works because saccharine does not occur in nature, whereas sugar does.

Similarly, most animals learn to avoid situations that have, in the past, led to pain. The stimuli that animals treat as painful tend, in nature, to be associated with injury and increased chance of death. But again, the connection must ultimately be built into the nervous system by natural selection, for it is not an obvious, necessary connection (M. Dawkins 1980). It is easy to imagine artificially selecting a breed of animals that enjoyed being injured, and felt pain whenever their physiological welfare was being improved. If learning is adaptive *improvement*, there has to be, in Lorenz's phrase, an innate teaching mechanism, or 'innate schoolmarm'. The principle holds even where the reinforcers are 'secondary', learned by association with primary reinforcers (Bateson 1983).

It holds, too, for morphological characters. Feet that are subjected to wear and tear grow tougher and more thick-skinned. The thickening of the skin is an acquired adaptation, but it is not obvious why the change went in this direction. In man-made machines, parts that are subjected to wear get thinner, not thicker, for obvious reasons. Why does the skin on the feet do the opposite? Because, fundamentally, natural selection has worked in the past to ensure an adaptive, rather than a maladaptive, response to wear and tear.

The relevance of this for would-be Lamarckian evolution is that there has to be a deep Darwinian underpinning even if there is a Lamarckian surface structure: a Darwinian choice of which potentially acquirable characters shall in fact be acquired and inherited. As I have argued before (R. Dawkins 1982a: 164–77), this is true of a recent, highly publicized immunological theory of Lamarckian adaptation (Steele 1979). Lamarckian mechanisms cannot be fundamentally responsible for adaptive evolution. Even if acquired characters are inherited on some planet, evolution there will still rely on a Darwinian guide for its adaptive direction.

Theory 3. Direct Induction by the Environment

Adaptation, as we have seen, is a fit between organism and environment. The set of conceivable organisms is wider than the actual set. And there is a set of conceivable environments wider than the actual set. These two subsets match each other to some extent, and the matching is adaptation. We can re-express the point by saying that information from the environment is present in the organism. In a few cases this is vividly literal—a frog carries a picture of its environment around on its back. Such information is usually carried by an animal in the less literal sense that a trained observer, dissecting a new animal, can reconstruct many details of its natural environment.

Now, how could the information get from the environment into the animal? Lorenz (1966) argues that there are two ways, natural selection and reinforcement learning, but that these are both *selective* processes in the broad sense (Pringle 1951). There is, in theory, an alternative method for the environment to imprint its information on the organism, and that is by direct 'instruction' (Danchin 1979). Some theories of how the immune system works are 'instructive': antibody molecules are thought to be shaped directly by moulding themselves around antigen molecules. The currently favoured theory is, by contrast, selective (Burnet 1969). I take 'instruction' to be synonymous with the 'direct induction by the environment' of Mayr's Theory 3. It is not always clearly distinct from Theory 2.

Instruction is the process whereby information flows directly from its environment into an animal. A case could be made for treating imitation learning, latent learning, and imprinting (Thorpe 1963) as instructive, but for clarity it is safer to use a hypothetical example. Think of an animal on some planet, deriving camouflage from its tiger-like stripes. It lives in long dry 'grass', and its stripes closely match the typical thickness and spacing of local grass blades. On our own planet such adaptation would come about through the selection of random genetic variation, but on the imaginary planet it comes about through direct instruction. The animals go brown except where their skin is shaded from the 'sun' by blades of grass. Their stripes are therefore adapted with great precision, not just to any old habitat, but to the precise habitat in which they have sunbathed, and it is this same habitat in which they are going to have to survive. Local populations are automatically camouflaged against local grasses. Information about the habitat, in this case about the spacing patterns of the grass blades, has flowed into the animals, and is embodied in the spacing pattern of their skin pigment.

Instructive adaptation demands the inheritance of acquired characters if it is to give rise to permanent or progressive evolutionary change. 'Instruction' received in one generation must be 'remembered' in the genetic (or equivalent) information. This process is in principle cumulative and progressive. However, if the genetic store is not to become overloaded by the accumulations of generations, some mechanism must exist for discarding unwanted 'instructions' and retaining desirable ones. I suspect that this must lead us, once again, to the need for some kind of selective process.

Imagine, for instance, a form of mammal-like life in which a stout 'umbilical nerve' enabled a mother to 'dump' the entire contents of her memory in the brain of her foetus. The technology is available even to our nervous systems: the corpus callosum can shunt large quantities of information from right hemisphere to left. An umbilical nerve could make the experience and wisdom of each generation automatically available to the next, and this might seem very desirable. But without a selective filter, it would take few generations for the load of information to become unmanageably large. Once again we come up against the need for a selective underpinning. I will leave this now, and make one more point about instructive adaptation (which applies equally to all Lamarckian types of theory).

The point is that there is a logical link-up between the two major theories of adaptive evolution—selection and instruction—and the two major theories of embryonic development—epigenesis and preformationism. Instructive evolution can work only if embryology is preformationistic. If embryology is epigenetic, as it is on our planet, instructive evolution cannot work. I have expounded the argument before (R. Dawkins 1982a: 174–6), so I will abbreviate it here.

If acquired characters are to be inherited, embryonic processes must be reversible: phenotypic change has to be read back into the genes (or equivalent). If embryology is preformationistic—the genes are a true blueprint—then it may indeed be reversible. You can translate a house back into its blueprint. But if embryonic development is epigenetic: if, as on this planet, the genetic information is more like a recipe for a cake (Bateson 1976) than a blueprint for a house, it is irreversible. There is no one-to-one mapping between bits of genome and bits of phenotype, any more than there is mapping between crumbs of cake and words of recipe. The recipe is not a blueprint that can be reconstructed from the cake. The transformation of recipe into cake cannot be put into reverse, and nor can the process of making a body. Therefore acquired adaptations cannot be read back into the 'genes', on any planet where embryology is epigenetic.

This is not to say that there could not, on some planet, be a form of life whose embryology was preformationistic. That is a separate question. How likely is it? The form of life would have to be very different from ours, so much so that it is hard to visualize how it might work. As for reversible embryology itself, it is even harder to visualize. Some mechanism would have to scan the detailed form of the adult body, carefully noting down, for instance, the exact location of brown pigment in a sun-striped skin, perhaps turning it into a linear stream of code numbers, as in a television camera. Embryonic development would read the scan out again, like a television receiver. I have an intuitive hunch that there is an objection in principle to this kind of embryology, but I cannot at present formulate it clearly. All I am saying here is that, if planets are divided into those where embryology is preformationistic and those, like Earth, where embryology is epigenetic, Darwinian evolution could be supported on both kinds of planet, but Lamarckian evolution, even if there were no other reasons for doubting its existence, could be supported only on the preformationistic planets—if there are any.

The close theoretical link that I have demonstrated between Lamarckian evolution and preformationistic embryology gives rise to a mildly entertaining irony. Those with ideological reasons for hankering after a neo-Lamarckian view of evolution are often especially militant partisans of epigenetic, 'interactionist', ideas of development, possibly—and here is the irony—for the very same ideological reasons (Koestler 1967, Ho and Saunders 1982).

Theory 4. Saltationism

The great virtue of the idea of evolution is that it explains, in terms of blind physical forces, the existence of undisputed adaptations whose statistical improbability is enormous, without recourse to the supernatural or the mystical. Since we *define* an undisputed adaptation as an adaptation that is too complex to have come about by chance, how is it possible for a theory to invoke only blind physical forces in explanation? The answer—Darwin's answer—is astonishingly simple when we consider how self-evident Paley's Divine Watchmaker must have seemed to his contemporaries. The key is that the co-adapted parts do not have to be assembled *all at once*. They can be put together in small stages. But they really do have to be *small* stages. Otherwise we are back again with the problem we started with: the creation by chance of complexity that is too great to have been created by chance!

Take the eye again, as an example of an organ that contains a large number of independent co-adapted parts, say N. The a priori probability of any one of these N features coming into existence by chance is low, but not incredibly low. It is comparable to the chance of a crystal pebble being washed by the sea so that it acts as a lens. Any one adaptation on its own could, plausibly, have come into existence through blind physical forces. If each of the N co-adapted features confers some slight advantage on its own, then the whole many-parted organ can be put together over a long period of time. This is particularly plausible for the eye—ironically in view of that organ's niche of honour in the Creationist pantheon. The eye is, *par excellence*, a case where a fraction of an organ is better than no organ at all; an eye without a lens or even a pupil, for instance, could still detect the looming shadow of a predator.

To repeat, the key to the Darwinian explanation of adaptive complexity is the replacement of instantaneous, coincidental, multi-dimensional luck by gradual, inch by inch, smeared-out luck. Luck is involved, to be sure. But a theory that bunches the luck up into major steps is more incredible than a theory that spreads the luck out in small stages. This leads to the following general principle of universal biology. Wherever in the universe adaptive complexity shall be found, it will have come into being gradually through a series of small alterations, never through large and sudden increments in adaptive complexity. We must reject Mayr's fourth theory, saltationism, as a candidate for explanation of the evolution of complexity.

It is almost impossible to dispute this rejection. It is implicit in the definition of adaptive complexity that the only alternative to gradualistic evolution is supernatural magic. This is not to say that the argument in favour of gradualism is a worthless tautology, an unfalsifiable dogma of the sort that creationists and philosophers are so found of jumping about on. It is not *logically* impossible for a full-fashioned eye to spring *de novo* from virgin bare skin. It is just that the possibility is statistically negligible.

Now it has recently been widely and repeatedly publicized that some modern evolutionists reject 'gradualism', and espouse what Turner (1982) has called 'theories of evolution by jerks'. Since these are reasonable people without mystical leanings, they must be gradualists in the sense in which I am here using the term: the 'gradualism' that they oppose must be defined differently. There are actually two confusions of language here, and I intend to clear them up in turn. The first is the common confusion between 'punctuated equilibrium' (Eldredge and Gould 1972) and true saltationism. The second is a confusion between two theoretically distinct kinds of saltation.

Punctuated equilibrium is not macro-mutation, not saltation at all in the traditional sense of the term. It is, however, necessary to discuss it here, because it is popularly regarded as a theory of saltation, and its partisans quote, with approval, Huxley's criticism of Darwin for upholding the principle of *Natura non facit saltum* (Gould 1980). The punctuationist theory is portrayed as radical and revolutionary and at variance with the 'gradualistic' assumptions of both Darwin and the neo-Darwinian synthesis (e.g. Lewin 1980). Punctuated equilibrium, however, was originally conceived as what the orthodox neo-Darwinian synthetic theory should truly predict, on a palaeontological time-scale, if we take its embedded ideas of allopatric speciation seriously (Eldredge and Gould 1972). It derives its 'jerks' by taking the 'stately unfolding' of the neo-Darwinian synthesis, and *inserting* long periods of stasis separating brief bursts of gradual, albeit rapid, evolution.

The plausibility of such 'rapid gradualism' is dramatized by a thought experiment of Stebbins (1982). He imagines a species of mouse evolving larger body size at such an imperceptibly slow rate that the differences between the means of successive generations would be utterly swamped by sampling error. Yet even at this slow rate Stebbins's mouse lineage would attain the body size of a large elephant in about 60,000 years, a time-span so short that it would be regarded as instantaneous by palaeontologists. Evolutionary change too *slow* to be detected by micro-evolutionists can nevertheless be too *fast* to be detected by macro-evolutionists. What a palaeontologist sees as a 'saltation' can in fact be a smooth and gradual change so slow as to be undetectable to the micro-evolutionist. This kind of palaeontological 'saltation' has nothing to do with the one-generation macro-mutations that, I suspect, Huxley and Darwin had in mind when they debated *Natura non facit saltum*. Confusion has arisen here, possibly because some individual champions of punctuated equilibrium have also, incidentally, championed macro-mutation (Gould 1982). Other 'punctuationists' have either confused their theory with macro-mutationism, or have explicitly invoked macro-mutation as one of the mechanisms of punctuation (e.g. Stanley 1981).

Turning to macro-mutation, or true saltation itself, the second confusion that I want to clear up is between two kinds of macro-mutation that we might conceive of. I could name them, unmemorably, saltation (1) and saltation (2), but instead I shall pursue an earlier fancy for airliners as metaphors, and label them 'Boeing 747' and 'Stretched DC-8' saltation. 747 saltation is the inconceivable kind. It gets its name from Sir Fred Hoyle's much quoted metaphor for his own cosmic misunderstanding of Darwinism (Hoyle and Wickramasinghe 1981). Hoyle compared Darwinian selection to a tornado, blowing through a junkyard and assembling a Boeing 747 (what he overlooked, of course, was the point about luck being 'smeared-out' in small steps—see above). Stretched DC-8 saltation is quite different. It is not in principle hard to believe in at all. It refers to large and sudden changes in *magnitude* of some biological measure, without an accompanying large increase in adaptive information. It is named after an airliner that was made by elongating the fuselage of an existing design, not adding significant new complexity. The change from DC-8 to Stretched DC-8 is a big change in magnitude—a saltation, not a gradualistic series of tiny changes. But, unlike the change from junk-heap to 747, it is not a big increase in information content or complexity, and that is the point I am emphasizing by the analogy.

An example of DC-8 saltation would be the following. Suppose the giraffe's neck shot out in one spectacular mutational step. Two parents had necks of standard antelope length. They had a freak child with a neck of modern giraffe length, and all giraffes are descended from this freak. This is unlikely to be true on Earth, but something like it may happen elsewhere in the universe. There is no objection to it in principle, in the sense that there is a profound objection to the (747) idea that a complex organ like an eye could arise from bare skin by a single mutation. The crucial difference is one of complexity.

I am assuming that the change from short antelope's neck to long giraffe's neck is *not* an increase in complexity. To be sure, both necks are exceedingly complex structures. You couldn't go from *no* neck to either kind of neck in one step: that would be 747 saltation. But once the complex organization of the antelope's neck already exists, the step to the giraffe's neck is just an elongation: various things have to grow faster at some stage in embryonic development; existing complexity is preserved. In practice, of course, such a drastic change in magnitude would be highly likely to have deleterious repercussions which would render the macro-mutant unlikely to survive. The existing antelope heart probably could not pump the blood up to the newly elevated giraffe head. Such practical objections to evolution by 'DC-8 saltation' can only help my case in favour of gradualism, but I still want to make a separate, and more universal, case against 747 saltation.

It may be argued that the distinction between 747 and DC-8 saltation is impossible to draw in practice. After all, DC-8 saltations, such as the proposed macro-mutational elongation of the giraffe's neck, may appear very complex: myotomes, vertebrae, nerves, blood vessels, all have to elongate together. Why does this not make it a 747 saltation, and therefore rule it out? But although this type of 'co-adaptation' has indeed often been thought of as a problem for any evolutionary theory, not just macro-mutational ones (see Ridley 1982 for a history), it is so only if we take an impoverished view of developmental mechanisms. We know that single mutations can orchestrate changes in growth rates of many diverse parts of organs, and, when we think about developmental processes, it is not in the least surprising that this should be so. When a single mutation causes a drosophila to grow a leg where an antenna ought to be, the leg grows in all its formidable complexity. But this is not mysterious or surprising, not a 747 saltation, because the organization of a leg is already present in the body before the mutation. Wherever, as in embryogenesis, we have a hierarchically branching tree of causal relationships, a small alteration at a senior node of the tree can have large and complex ramified effects on the tips of the twigs. But although the change may be large in magnitude, there can be no large and sudden increments in adaptive information. If you think you have found a particular example of a large and sudden increment in adaptively complex information in practice, you can be certain the adaptive information was already there, even if it is an atavistic 'throw-back' to an earlier ancestor.

There is not, then, any objection in principle to theories of evolution by jerks, even the theory of hopeful monsters (Goldschmidt 1940), provided that it is DC-8 saltation, not 747 saltation, that is meant. Gould (1982) would clearly agree: 'I regard forms of macromutation which include the sudden origin of new species with all their multifarious adaptations intact *ab initio*, as illegitimate.' No educated biologist actually believes in 747 saltation, but not all have been sufficiently explicit about the distinction between DC-8 and 747 saltation. An unfortunate consequence is that Creationists and their journalistic fellow-travellers have been able to exploit saltationist-sounding statements of respected biologists. The biologist's intended meaning may have been what I am calling DC-8 saltation, or even non-saltatory punctuation; but the Creationist *assumes* saltation in the sense that I have dubbed 747, and 747 saltation would, indeed, be a blessed miracle.

I also wonder whether an injustice is not being done to Darwin, owing to this same failure to come to grips with the distinction between DC-8 and 747 saltation. It is frequently alleged that Darwin was wedded to gradualism, and therefore that, if some form of evolution by jerks is proved, Darwin will have been shown to be wrong. This is undoubtedly the reason for the ballyhoo and publicity that has attended the theory of punctuated equilibrium. But was Darwin really opposed to all jerks? Or was he, as I suspect, strongly opposed only to 747 saltation?

As we have already seen, punctuated equilibrium has nothing to do with saltation; but anyway, I think it is not at all clear that, as is often alleged, Darwin would have been discomfited by punctuationist interpretations of the fossil record. The following passage, from later editions of the *Origin*, sounds like something from a current issue of *Paleobiology*: 'the periods during which species have been undergoing modification, though very long as measured by years, have probably been short in comparison with the periods during which these same species remained without undergoing any change'.

Gould (1982: 84) shrugs this off as somehow anomalous and away from the mainstream of Darwin's thought. As he correctly says: 'You cannot do history by selective quotation and search for qualifying footnotes. General tenor and historical impact are the proper criteria. Did his contemporaries or descendants ever read Darwin as a saltationist?' Certainly nobody ever accused Darwin of being a saltationist. But to most people saltation means macro-mutation, and, as Gould himself stresses, 'Punctuated equilibrium is not a theory of macromutation'. More importantly, I believe we can reach a better understanding of Darwin's general gradualistic bias if we invoke the distinction between 747 and DC-8 saltation.

Perhaps part of the problem is that Darwin himself did not have the distinction. In some anti-saltation passages it seems to be DC-8 saltation that he has in mind. But on those occasions he does not seem to feel very strongly about it: 'About sudden jumps', he wrote in a letter in 1860, 'I have no objection to them—they would aid me in some cases. All I can say is, that I went into the subject and found no evidence to make me believe in jumps [as a source of new species] and a good deal pointing in the other direction' (quoted in Gillespie 1979: 119). This does not sound like a man fervently opposed, in principle, to sudden jumps. And of course there is no reason why he *should* have been fervently opposed, if he only had DC-8 saltations in mind.

But at other times he really is pretty fervent, and on those occasions, I suggest, he is thinking of 747 saltation: 'it is impossible to imagine so many co-adaptations being formed all by a chance blow' (quoted in Ridley 1982: 52, 67). As the historian Neal Gillespie puts it:

For Darwin, monstrous births, a doctrine favored by Chambers, Owen, Argyll, Mivart, and others, from clear theological as well as scientific motives, as an explanation of how new species, or even higher taxa, had developed, was no better than a miracle: 'it leaves the case of the co-adaptation of organic beings to each other and to their physical conditions of life, untouched and unexplained'. It was 'no explanation' at all, of no more scientific value than creation 'from the dust of the earth'. (Gillespie 1979: 118)

As Ridley (1982: 67) says of the 'religious tradition of idealist thinkers [who] were committed to the explanation of complex adaptive contrivances by intelligent design': 'The greatest concession they could make to Darwin was that the Designer operated by tinkering with the generation of diversity, designing the variation.' Darwin's response was: 'If I were convinced that I required such additions to the theory of natural selection, I would reject it as rubbish . . . I would give nothing for the theory of Natural selection, if it requires miraculous additions at any one stage of descent.'

Darwin's hostility to monstrous saltation, then, makes sense if we assume that he was thinking in terms of 747 saltation—the sudden invention of new adaptive complexity. It is highly likely that that is what he was thinking of, because that is exactly what many of his opponents had in mind. Saltationists such as the Duke of Argyll (though presumably not Huxley!) wanted to believe in 747 saltation, precisely because it *did* demand supernatural intervention. Darwin did not believe in it, for exactly the same reason. To quote Gillespie again (p. 120): 'for Dawrin, designed evolution, whether manifested in saltation, monstrous births, or manipulated variations, was but a disguised form of special creation'.

I think this approach provides us with the only sensible reading of Darwin's well-known remark that 'If it could be demonstrated that any complex organ existed, which could not possibly have been formed by numerous, successive, slight modifications, my theory would absolutely break down'. That is not a plea for gradualism, as a modern palaeobiologist uses the term. Darwin's theory is falsifiable, but he was much too wise to make his theory *that* easy to falsify! Why on earth *should* Darwin have committed himself to such an arbitrarily restrictive version of evolution, a version that positively invites falsification? I think it is clear that he didn't. His use of the term 'complex' seems to me to be clinching. Gould (1982: 84) describes this passage from Darwin as 'clearly invalid'. So it is invalid if the alternative to slight modifications is seen as DC-8 saltation. But if the alternative is seen as 747 saltation, Darwin's remark is valid and very wise. Notwithstanding those whom Miller (1982) has unkindly called Darwin's more foolish critics, his theory is indeed falsifiable, and in the passage quoted he puts his finger on one way in which it might be falsified.

There are two kinds of imaginable saltation, then, DC-8 saltation and 747 saltation. DC-8 saltation is perfectly possible, undoubtedly happens in the laboratory and the farmyard, and may have made important contributions to evolution. 747 saltation is statistically ruled out unless there is supernatural intervention. In Darwin's own time, proponents and opponents of saltation often had 747 saltation in mind, because they believed in—or were arguing against—divine intervention. Darwin was hostile to (747) saltation, because he correctly saw natural selection as an *alternative* to the miraculous as an explanation for adaptive complexity. Nowadays saltation means either punctuation (which isn't saltation at all) or DC-8 saltation, neither of which Darwin would have had strong objections to in principle, merely doubts about the facts. In the modern context, therefore, I do not think Darwin should be labelled a strong gradualist. In the modern context, I suspect that he would be rather open-minded.

It is in the anti-747 sense that Darwin was a passionate gradualist, and it is in the same sense that we must all be gradualists, not just with respect to life on Earth, but with respect to life all over the universe. Gradualism in this sense is essentially synonymous with evolution. The sense in which we may be non-gradualists is a much less radical, although still quite interesting, sense. The theory of evolution by jerks has been hailed on television and elsewhere as radical and revolutionary, a paradigm shift. There is, indeed, an interpretation of it which is revolutionary, but that interpretation (the 747 macro-mutation version) is certainly wrong, and is apparently not held by its original proponents. The sense in which the theory might be right is not particularly revolutionary. In this field you may choose your jerks so as to be revolutionary, *or* so as to be correct, but not both.

Theory 5. Random Evolution

Various members of this family of theories have been in vogue at various times. The 'mutationists' of the early part of this century—De Vries, W. Bateson, and their colleagues—believed that selection served only to weed out deleterious freaks, and that the real driving force in evolution was mutation pressure. Unless you believe mutations are directed by some mysterious life force, it is sufficiently obvious that you can be a mutationist only if you forget about adaptive complexity—forget, in other words, most of the consequences of evolution that are of any interest! For historians there remains the baffling enigma of how such distinguished biologists as De Vries, W. Bateson, and T. H. Morgan could rest satisfied with such a crassly inadequate theory. It is not enough to say that De Vries's view was blinkered by his working only on the evening primrose. He only had to look at the adaptive complexity in his own body to see that 'mutationism' was not just a wrong theory: it was an obvious non-starter.

There post-Darwinian mutationists were also saltationists and anti-gradualists, and Mayr treats them under that heading, but the aspect of their view that I am criticizing here is more fundamental. It appears that they actually thought that mutation, on its own without selection, was sufficient to explain evolution. This *could* not be so on any non-mystical view of mutation, whether gradualist or saltationist. If mutation is undirected, it is clearly unable to explain the adaptive directions of evolution. If mutation is directed in adaptive ways, we are entitled to ask how this comes about. At least Lamarck's principle of use and disuse makes a valiant attempt at explaining how variation might be directed. The 'mutationists' didn't even seem to see that there was a problem, possibly because they underrated the importance of adaptation—and they were not the last to do so. The irony with which we must now read W. Bateson's dismissal of Darwin is almost painful: 'the transformation of masses of populations by imperceptible steps guided by selection is, as most of us now see, so inapplicable to the fact that we can only marvel . . . at the want of penetration displayed by the advocates of such a proposition' (1913, quoted in Mayr 1982: 884).

Nowadays some population geneticists describe themselves as supporters of 'non-Darwinian evolution'. They believe that a substantial number of the gene replacements that occur in evolution are non-adaptive substitutions of alleles whose effects are indifferent relative to one another (Kimura 1968). This may well be true, if not in Israel (Nevo 1983) maybe somewhere in the universe. But it obviously has nothing whatever to contribute to solving the problem of the evolution of adaptive complexity. Modern advocates of neutralism admit that their theory cannot account for adaptation, but that doesn't seem to stop them regarding the theory as interesting. Different people are interested in different things.

The phrase 'random genetic drift' is often associated with the name of Sewall Wright, but Wright's conception of the relationship between random drift and adaptation is altogether subtler than the others I have mentioned (Wright 1980). Wright does not belong in Mayr's fifth category, for he clearly sees selection as the driving force of adaptive evolution. Random drift may make it easier for selection to do its job by assisting the escape from local optima (R. Dawkins 1982a: 40), but it is still selection that is determining the rise of adaptive complexity.

Recently palaeontologists have come up with fascinating results when they perform computer simulations of 'random phylogenies' (e.g. Raup 1977). These random walks through evolutionary time produce trends that look uncannily like real ones, and it is disquietingly easy, and tempting, to read into the random phylogenies apparently adaptive trends which, however, are not there. But this does not mean that we can admit random drift as an explanation of real adaptive trends. What it might mean is that some

are copied. For instance, the extent to which they are 'particulate' as opposed to 'blending' probably has a more important bearing on evolution than their detailed molecular or physical nature. Similarly, a universe-wide classification of replicators might make more reference to their dimensionality and coding principles than to their size and structure. DNA is a digitally coded one-dimensional array. A 'genetic' code in the form of a two-dimensional matrix is conceivable. Even a three-dimensional code is imaginable, although students of Universal Darwinism will probably worry about how such a code could be 'read'. (DNA is, of course, a molecule whose three-dimensional structure determines how it is replicated and transcribed, but that doesn't make it a three-dimensional code. DNA's meaning depends upon the one-dimensional sequential arrangement of its symbols, not upon their three-dimensional position relative to one another in the cell.) There might also be theoretical problems with analogue, as opposed to digital codes, similar to the theoretical problems that would be raised by a purely analogue nervous system (Rushton 1961).

As for the phenotypic levers of power by which replicators influence their survival, we are so used to their being bound up into discrete organisms or 'vehicles' that we forget the possibility of a more diffuse extracorporeal or 'extended' phenotype. Even on this Earth a large amount of interesting adaptation can be interpreted as part of the extended phenotype (R. Dawkins 1982a: chs. 11, 12, and 13). There is, however, a general theoretical case that can be made in favour of the discrete organismal body, with its own recurrent life cycle, as a necessity in any process of evolution of advanced adaptive complexity (ibid. ch. 14), and this topic might have a place in a full account of Universal Darwinism.

Another candidate for full discussion might be what I shall call divergence, and convergence or recombination of replicator lineages. In the case of Earth-bound DNA, 'convergence' is provided by sex and related processes. Here the DNA 'converges' within the species after having very recently 'diverged'. But suggestions are now being made that a different kind of convergence can occur among lineages that originally diverged an exceedingly long time ago. For instance, there is evidence of gene transfer between fish and bacteria (Jacob 1983). The replicating lineages on other planets may permit very varied kinds of recombination, on very different time-scales. On Earth the rivers of phylogeny are almost entirely divergent: if main tributaries ever recontact each other after branching apart, it is only through the tiniest of trickling cross-streamlets, as in the fish/bacteria case. There is, of course, a richly anastomosing delta of divergence and convergence due to sexual recombination *within* the species, but only within the species. There may be planets on which the 'genetic' system permits much more cross-talk at all levels of the branching hierarchy, one huge fertile delta.

I have not thought enough about the fantasies of the previous paragraphs to evaluate their plausibility. My general point is that there is one limiting constraint upon all speculations about life in the universe. If a life-form displays adaptive complexity, it must possess an evolutionary mechanism capable of generating adaptive complexity. However diverse evolutionary mechanisms may be, if there is no other generalization that can be made about life all around the universe, I am betting it will always be recognizable as Darwinian life. The Darwinian law (Eigen 1983) may be as universal as the great laws of physics.[1]

REFERENCES

Asimov, I. (1979), *Extraterrestrial Civilizations* (London: Pan).

Atkins, P. W. (1981), *The Creation* (Oxford: W. H. Freeman).

Bateson, P. P. G. (1976), 'Specificity and the Origins of Behavior', *Advances in the Study of Behavior*, 6: 1–20.

——(1983), 'Rules for Changing the Rules', in O. S. Bendall (ed.), *Evolution from Molecules to Man* (Cambridge: Cambridge University Press), 483–507.

Billingham, J. (1981), *Life in the Universe* (Cambridge, Mass.: MIT Press).

Burnet, F. M. (1969), *Cellular Immunology* (Melbourne: Melbourne University Press).

Cairns-Smith, A. G. (1982), *Genetic Takeover* (Cambridge: Cambridge University Press).

Crick, F. H. C. (1982), *Life Itself* (London: Macdonald).

Danchin, A. (1979), 'Thèmes de la biologie: théories instructives et théories selectives', *Revue des questions scientifiques*, 150: 151–64.

Darwin, C. R. (1859), *The Origin of Species*, 1st edn., repr. 1968 (Harmondsworth: Penguin).

Dawkins, M. (1980), *Animal Suffering: The Science of Animal Welfare* (London: Chapman and Hall).

Dawkins, R. (1982a), *The Extended Phenotype* (Oxford: W. H. Freeman).

——(1982b), 'The Necessity of Darwinism', *New Scientist*, 94: 130–2; repr. in J. Cherfas (ed.), *Darwin Up to Date* (London: New Scientist, 1982), 61–3.

Eigen, M. (1983), 'Self-Replication and Molecular Evolution', in D. S. Bendall (ed.), *Evolution from Molecules to Man* (Cambridge: Cambridge Univerity Press), 105–30.

Eldredge, N., and Gould, S. J. (1972), 'Punctuated Equilibria: An Alternative to Phyletic Gradualism', in T. J. M. Schopf (ed.), *Models in Paleobiology* (San Francisco: Freeman Cooper), 82–115.

[1] As usual I have benefited from discussions with many people, including especially Mark Ridley, who also criticized the manuscript, and Alan Grafen. Dr F. J. Ayala called attention to an important error in the original spoken version of the paper.

Fisher, R. A. (1958), *The Genetical Theory of Natural Selection* (New York: Dover).

Gillespie, N. C. (1979), *Charles Darwin and the Problem of Creation* (Chicago: University of Chicago Press).

Goldschmidt, R. (1940), *The Material Basis of Evolution* (New Haven: Yale University Press).

Gould, S. J. (1980), *The Panda's Thumb* (New York: W. W. Norton).

——(1982), 'The Meaning of Punctuated Equilibrium and its Role in Validating a Hierarchical Approach to Macroevolution', in R. Milkman (ed.), *Perspectives on Evolution* (Sunderland, Mass.: Sinauer), 83–104.

——(1983), 'Irrelevance, Submission, and Partnership: The Changing Role of Palaeontology in Darwin's Three Centennials and a Modest Proposal for Macroevolution, in D. S. Bendall (ed.), *Evolution from Molecules to Man* (Cambridge: Cambridge University Press), 347–66.

Ho, M.-W., and Saunders, P. T. (1982), 'Adaptation and Natural Selection: Mechanism and Teleology', in *Towards a Liberatory Biology* (Dialectics of Biology Group, general ed. S. Rose: London: Allison and Busby), 85–102.

Hoyle, F., and Wickramasinghe, N. C. (1981), *Evolution from Space* (London: J. M. Dent).

Jacob, F. (1983), 'Molecular Tinkering in Evolution', in D. S. Bendall (ed.), *Evolution from Molecules to Man* (Cambridge: Cambridge University Press), 131–44.

Kimura, M. (1968), 'Evolutionary Rate at the Molecular Level', *Nature*, 217: 624–6.

Koestler, A. (1967), *The Ghost in the Machine* (London: Hutchinson).

Lewin, R. (1980), 'Evolutionary Theory under Fire', *Science*, 210: 883–7.

Lewontin, R. C. (1970), 'The Units of Selection', *Annual Review of Ecology and Systematics*, 1: 1–18.

——(1979), 'Sociobiology as an Adaptationist Program', *Behavioral Science*, 24: 5–14.

——(1983), 'Gene, Organism, and Environment', in D. S. Bendall (ed.), *Evolution from Molecules to Man* (Cambridge: Cambridge University Press), 273–85.

Lorenz, K. (1966), *Evolution and Modification of Behaviour* (London: Methuen).

Maynard Smith, J. (1969), 'The Status of Neo-Darwinism', in C. H. Waddington (ed.), *Towards a Theoretical Biology* (Edinburgh: Edinburgh University Press), 82–9.

Mayr, E. (1982), *The Growth of Biological Thought* (Cambridge, Mass.: Harvard University Press).

Miller, J. (1982), *Darwin for Beginners* (London: Writers and Readers).

Nevo, E. (1983), 'Population Genetics and Ecology: The Interface', in D. S. Bendall (ed.), *Evolution from Molecules to Man* (Cambridge: Cambridge University Press), 287–321.

Paley, W. (1828), *Natural Theology*, 2nd edn. (Oxford: J. Vincent).

Pringle, J. W. S. (1951), 'On the Parallel between Learning and Evolution', *Behaviour*, 3: 90–110.

Raup, D. M. (1977), 'Stochastic Models in Evolutionary Palaeontology', in A. Hallam (ed.), *Patterns of Evolution* (Amsterdam: Elsevier), 59–78.

Ridley, M. (1982), 'Coadaptation and the Inadequacy of Natural Selection', *British Journal for the History of Science*, 15: 45–68.

Rushton, W. A. H. (1961), 'Peripheral Coding in the Nervous System', in W. A. Rosenblith (ed.), *Sensory Communication* (Cambridge, Mass.: MIT Press), 169–88.

Stanley, S. M. (1981), *The New Evolutionary Timetable* (New York: Basic Books).

Stebbins, G. L. (1982), *Darwin to DNA, Molecules to Humanity* (San Francisco: W. H. Freeman).

Steele, E. J. (1979), *Somatic Selection and Adaptive Evolution* (Toronto: Williams and Wallace).

Thorpe, W. H. (1963), *Learning and Instinct in Animals*, 2nd edn. (London: Methuen).

Turner, J. R. G. (1982), review of R. J. Berry, *Neo-Darwinism*, *New Scientist*, 94: 160–2.

Wright, S. (1980), 'Genic and Organismic Selection', *Evolution*, 34: 825–43.

3

THE LEIBNIZIAN PARADIGM

DANIEL C. DENNETT

> If, among all the possible worlds, none had been better than the rest, then God would never have created one.
>
> Leibniz (1710)

> The study of adaptation is not an optional preoccupation with fascinating fragments of natural history, it is the core of biological study.
>
> Colin Pittendrigh (1958)

Leibniz, notoriously, said that this was the best of all possible worlds, a striking suggestion that might seem preposterous from a distance, but turns out to throw an interesting light on the deep questions of what it is to be a possible world, and on what we can infer about the actual world from the fact of its actuality. In *Candide*, Voltaire created a famous caricature of Leibniz, Dr Pangloss, the learned fool who could rationalize any calamity or deformity—from the Lisbon earthquake to venereal disease—and show how, no doubt, it was all for the best. Nothing *in principle* could prove that this was not the best of all possible worlds.

Gould and Lewontin memorably dubbed the *excesses* of adaptationism the 'Panglossian Paradigm', and strove to ridicule it off the stage of serious science. They were not the first to use 'Panglossian' as a term of criticism in evolutionary theory. The evolutionary biologist J. B. S. Haldane had a famous list of three 'theorems' of bad scientific argument: the Bellman's Theorem ('What I tell you three times is true', from 'The Hunting of the Snark' by Lewis Carroll), Aunt Jobisca's Theorem ('It's a fact the whole world knows', from Edward Lear, 'The Pobble Who Had No Toes'), and Pangloss's Theorem ('All is for the best in this best of all possible worlds', from *Candide*). John Maynard Smith then used the last of these more particularly to name 'the old Panglossian fallacy that natural selection favours adaptations that are good for the species as a whole, rather than

First published in D. C. Dennett, *Darwin's Dangerous Idea* (New York: Simon and Schuster, 1995), 238–51. Reprinted by permission.

acting at the level of the individual'. As he later commented, 'It is ironic that the phrase "Pangloss's theorem" was first used in the debate about evolution (in print, I think, by myself, but borrowed from a remark of Haldane's), not as a criticism of adaptive explanations, but specifically as a criticism of "group-selectionist", mean-fitness-maximising arguments' (Maynard Smith 1988: 88). But Maynard Smith is wrong, apparently. Gould has recently drawn attention to a still earlier use of the term by a biologist, William Bateson (1909), of which he, Gould, had been unaware when he chose to use the term. As Gould (1993b: 312) says, 'The convergence is hardly surprising, as Dr Pangloss is a standard synecdoche for this form of ridicule.' For the more apt or fitting a brain-child is, the more likely it is to be born (or borrowed) independently in more than one brain.

Voltaire created Pangloss as a parody of Leibniz, and it is exaggerated and unfair to Leibniz—as all good parody is. Gould and Lewontin similarly caricatured adaptationism in their article attacking it, so parity of reasoning suggests that, if we wanted to undo the damage of that caricature, and describe adaptationism in an accurate and constructive way, we would have a title ready-made: we could call adaptationism, fairly considered, the 'Leibnizian Paradigm'.

The Gould and Lewontin article has had a curious effect on the academic world. It is widely regarded by philosophers and other humanists who have heard of it or even read it as some sort of *refutation of adaptationism*. Indeed, I first learned of it from the philosopher/psychologist Jerry Fodor, a lifelong critic of my account of the intentional stance, who pointed out that what I was saying was pure adaptationism (he was right about that), and went on to let me in on what the *cognoscenti* all knew: Gould and Lewontin's article had shown adaptationism 'to be completely bankrupt'. (For an instance of Fodor's view in print, see Fodor 1990: 70). When I looked into it, I found out otherwise. In 1983, I published a paper in *Behavioral and Brain Sciences*, 'Intentional Systems in Cognitive Ethology', and since it was unabashedly adaptationist in its reasoning, I included a coda, 'The "Panglossian Paradigm" Defended', which criticized both Gould and Lewontin's paper and—more particularly—the bizarre myth that had grown up around it.

The results were fascinating. Every article that appears in *BBS* is accompanied by several dozen commentaries by experts in the relevant fields, and my piece drew fire from evolutionary biologists, psychologists, ethologists, and philosophers, most of it friendly but some remarkably hostile. One thing was clear: it was not just some philosophers and psychologists who were uncomfortable with adaptationist reasoning. In addition to the evolutionary theorists who weighed in enthusiastically on my side

(Dawkins 1983, Maynard Smith 1983), and those who fought back (Lewontin 1983), there were those who, though they agreed with me that Gould and Lewontin had not refuted adaptationism, were eager to downplay the standard use of optimality assumptions that I claimed to be an essential ingredient in all evolutionary thinking.

Niles Eldredge (1983: 361) discussed the reverse engineering of functional morphologists: 'You will find sober analyses of fulcra, force vectors and so forth: the understanding of anatomy as a living machine. Some of this stuff is very good. Some of it is absolutely dreadful.' He went on to cite, as an example of good reverse engineering, the work of Dan Fisher (1975) comparing modern horseshoe crabs with their Jurassic ancestors:

Assuming only that Jurassic horseshoe crabs also swam on their backs, Fisher showed they must have swum at an angle of 0–10 degrees (flat on their backs) and at the somewhat greater speed of 15–20 cm/sec. Thus the 'adaptive significance' of the slight differences in anatomy between modern horseshoe crabs and their 150-million-year-old relatives is translated into an understanding of their slightly different swimming capabilities. (In all honesty, I must also report that Fisher does use optimality in his arguments: He sees the differences between the two species as a sort of trade-off, where the slightly more efficient Jurassic swimmers appear to have used the same pieces of anatomy to burrow somewhat less efficiently than their modern-day relatives.) In any case, Fisher's work stands as a really good example of functional morphological analysis. The notion of adaptation is naught but conceptual filigree—one that may have played a role in motivating the research, but one that was not vital to the research itself. (Eldredge 1983: 362)

But in fact the role of optimality assumptions in Fisher's work—beyond the explicit role that Eldredge conceded—is so 'vital', and indeed omnipresent, that Eldredge entirely overlooked it. For instance, Fisher's inference that the Jurassic crabs swam at 15–20 cm/sec has as a tacit premiss that those crabs *swam at the optimal speed for their design*. (How does he know they swam at all? Perhaps they just lay there, oblivious of the excess functionality of their body shapes.) Without this tacit (and, of course, dead obvious) premiss, no conclusion at all could be drawn about what the *actual* swimming speed of the Jurassic variety was.

Michael Ghiselin (1983: 363) was even more forthright in denying this unobvious obvious dependence:

Panglossianism is bad because it asks the wrong question, namely, What is good? . . . The alternative is to reject such teleology altogether. Instead of asking, What is good? we ask, What has happened? The new question does everything we could expect the old one to do, and a lot more besides.

He was fooling himself. There is hardly a single answer to the question 'What has happened (in the biosphere)?' that doesn't depend crucially on assumptions about what is good.[1] As we just noted, you can't even avail yourself of the concept of a homology without taking on adaptationism, without taking the intentional stance.

So now what is the problem? It is the problem of how to tell good—irreplaceable—adaptationism from bad adaptationism, how to tell Leibniz from Pangloss.[2] Surely one reason for the extraordinary influence of Gould and Lewontin's paper (among non-evolutionists) is that it expressed, with many fine rhetorical flourishes, what Eldredge called the 'backlash' against the concept of adaptationism among biologists. What were they reacting against? In the main, they were reacting against a certain sort of laziness: the adaptationist who hits upon a truly nifty explanation for why a particular circumstance should prevail, and then never bothers to test it—because it is too good a story, presumably, not to be true. Adopting another literary label, this time from Rudyard Kipling (1912), Gould and Lewontin call such explanations 'Just So Stories'. It is an enticing historical curiosity that Kipling wrote his *Just So Stories* at a time when this objection to Darwinian explanation had already been swirling around for decades;[3] forms of it were raised by some of Darwin's earliest critics (Kitcher 1985: 156). Was Kipling inspired by the controversy? In any case, calling the adaptationists' flights of imagination 'Just So Stories' hardly does them credit; as delightful as I have always found Kipling's fantasies about how the elephant got its trunk, and the leopard got its spots, they are quite simple and unsurprising tales compared with the amazing hypotheses that have been concocted by adaptationists.

[1] Doesn't my assertion fly in the face of the claims of those cladists who purport to deduce history from a statistical analysis of shared and unshared 'characters'? (For a philosophical survey and discussion, see Sober 1988.) Yes, I guess it does, and my review of their arguments (largely via Sober's analyses) shows me that the difficulties they create for themselves are largely, if not entirely, due to their trying so hard to find non-adaptationist ways of drawing the sound inferences that are dead obvious to adaptationists. For instance, those cladists who abstain from adaptation talk cannot just help themselves to the obvious fact that having webbed feet is a pretty good 'character' and having dirty feet (when examined) is not. Like the behaviourists who pretended to be able to explain and predict 'behaviour' defined in the starkly uninterpreted language of geographical trajectory of body parts, instead of using the richly functionalistic language of searching, eating, hiding, chasing, and so forth, the abstemious cladists create majestic edifices of intricate theory, which is amazing, considering they do it with one hand tied behind their backs, but strange, considering that they wouldn't have to do it at all if they didn't insist on tying one hand behind their backs. (See also Dawkins 1986a: ch. 10, and Ridley 1985: ch. 6.)

[2] The myth that the point of the Gould and Lewontin paper was to destroy adaptationism, not correct its excesses, was fostered by the paper's rhetoric, but in some quarters it backfired on Gould and Lewontin, since adaptationists themselves tended to pay more attention to the rhetoric than the arguments: 'The critique by Gould and Lewontin has had little impact on practitioners, perhaps because they were seen as hostile to the whole enterprise, and not merely to careless practise of it' (Maynard Smith 1988: 89).

[3] Kipling began publishing the individual stories in 1897.

Consider the greater honey guide, *Indicator indicator*, an African bird that owes its name to its talent for leading human beings to wild beehives hidden in the forest. When the Boran people of Kenya want to find honey, they call for the bird by blowing on whistles made of sculpted snail shells. When a bird arrives, it flies around them, singing a special song—its 'follow-me' call. They follow as the bird darts ahead and waits for them to catch up, always making sure they can see where it's heading. When the bird reaches the hive, it changes its tune, giving the 'here-we-are' call. When the Boran locate the beehive in the tree and break into it, they take the honey, leaving wax and larvae for the honey guide. Now, don't you ache to believe that this wonderful partnership actually exists, and has the clever functional properties described? Don't you want to believe that such a marvel could have evolved under some imagined series of selection pressures and opportunities? I certainly do. And, happily, in this case, the follow-up research is confirming the story, and even adding nifty touches as it does so. Recent controlled tests, for instance, showed that the Boran honey-hunters took much longer to find hives without the help of the birds, and 96 per cent of the 186 hives found during the study were encased in trees in ways that would have made them inaccessible to the birds without human assistance (Isack and Reyer 1989).

Another fascinating story, which strikes closer to home, is the hypothesis that our species, *Homo sapiens*, descended from earlier primates via an intermediate species that was aquatic (Hardy 1960, Morgan 1982, 1990)! These aquatic apes purportedly lived on the shores of an island formed by the flooding of the area that is now in Ethiopia, during the late Miocene, about seven million years ago. Cut off by the flooding from their cousins on the African continent, and challenged by a relatively sudden change in their climate and food sources, they developed a taste for shellfish, and over a period of a million years or so they began the evolutionary process of returning to the sea that we know was undergone earlier by whales, dolphins, seals, and otters, for instance. The process was well under way, leading to the fixation of many curious characteristics that are otherwise found *only* in aquatic mammals—not in any other primate, for example— when circumstances changed once again, and these semi-seagoing apes returned to a life on the land (but typically on the shore of sea, lake, or river). There, they found that many of the adaptations they had developed for good reasons in their shell-diving days were not only not valuable but a positive hindrance. They soon turned these handicaps to good uses, however, or at least made compensations for them: their upright, bipedal posture, their subcutaneous layer of fat, their hairlessness, perspiration, tears, inability to respond to salt deprivation in standard mammalian ways, and, of course, the diving reflex—which permits even new-born human infants to survive sudden submersion in water for long periods with no ill effects. The details—and there are many, many more—are so ingenious, and the whole aquatic ape theory is so shockingly anti-establishment, that I for one would *love* to see it vindicated. That does not make it true, of course.

The fact that its principal exponent these days is not only a woman, Elaine Morgan, but an amateur, a science writer without proper official credentials in spite of her substantial researches, makes the prospect of vindication all the more enticing.[4] The establishment has responded quite ferociously to her challenges, mostly treating them as beneath notice, but occasionally subjecting them to withering rebuttal.[5] This is not necessarily a pathological reaction. Most uncredentialled proponents of scientific 'revolutions' are kooks who really are not worth paying any attention to. There really are a lot of them besieging us, and life is too short to give each uninvited hypothesis its proper day in court. But in this case, I wonder; many of the counter-arguments seem awfully thin and *ad hoc*. During the last few years, when I have found myself in the company of distinguished biologists, evolutionary theorists, palaeo-anthropologists, and other experts, I have often asked them just to tell me, please, exactly why Elaine Morgan must be wrong about the aquatic ape theory. I haven't yet had a reply worth mentioning, aside from those who admit, with a twinkle in their eyes, that they have often wondered the same thing. There seems to be nothing *inherently* impossible about the idea; other mammals have made the plunge, after all. Why couldn't our ancestors have started back into the ocean and then retreated, bearing some tell-tale scars of this history?

Morgan may be 'accused' of telling a good story—she certainly has—but not of declining to try to test it. On the contrary, she has used the story as leverage to coax a host of surprising predictions out of a variety of fields, and has been willing to adjust her theory when the results have demanded

---

[4] Sir Alister Hardy, the Linacre Professor of Zoology at Oxford, who originally proposed the theory, could hardly have been a more secure member of the scientific establishment, however.

[5] For instance, there is no mention at all of the aquatic ape theory, not even to dismiss it, in two recent coffee-table books that include chapters on human evolution. Philip Whitfield's *From So Simple a Beginning: The Book of Evolution* (1993) offers a few paragraphs on the standard savanna theory of bipedalism. 'The Primates' Progress', by Peter Andrews and Christopher Stringer, is a much longer essay on hominid evolution, in *The Book of Life* (Gould 1993a), but it, too, ignores the aquatic ape theory—the AAT. And, adding insult to oblivion, there has also been a wickedly funny parody of it by Donald Symons (1983), exploring the radical hypothesis that our ancestors used to *fly*—'The *flying on air theory*—FLOAT, as it is acronymously (acrimoniously, among the reactionary human evolution "establishment") known.' For an overview of the reactions, see Richards 1991.

it. Otherwise, she has stuck to her guns, and, in fact, invited attack on her views through the vehemence of her partisanship. As so often happens in such a confrontation, the intransigence and defensiveness, on both sides, have begun to take their toll, creating one of those spectacles that then discourage anyone who just wants to know the truth from having anything more to do with the subject. Morgan's latest book on the topic (1990) responded with admirable clarity, however, to the objections that had been lodged to date, and usefully contrasted the strengths and weaknesses of the aquatic ape theory to those of the establishment's history. And, more recently still, a book has appeared that collects essays by a variety of experts, for and against the aquatic ape theory: Roede *et al.* 1991. The tentative verdict of the organizers of the 1987 conference from which that book sprang is that, 'while there are a number of arguments favoring the AAT, they are not sufficiently convincing to counteract the arguments against it' (p. 324). That judicious note of mild disparagement helps ensure that the argument will continue, perhaps even with less rancour; it will be interesting to see where it all comes out.

My point in raising the aquatic ape theory is not to defend it against the establishment view, but to use it as an illustration of a deeper worry. Many biologists would like to say, 'A pox on both your houses!' Morgan (1990) deftly exposes the hand-waving and wishful thinking that have gone into the establishment's tale about how—and *why*—*Homo sapiens* developed bipedalism, sweating, and hairlessness on the savanna, not the sea-shore. Their stories may not be literally as fishy as hers, but some of them are pretty far-fetched; they are every bit as speculative, and (I venture to say) no better confirmed. What they mainly have going for them, so far as I can see, is that they occupied the high ground in the textbooks before Hardy and Morgan tried to dislodge them. Both sides are indulging in adaptationist 'Just So Stories', and since *some story or other* must be true, we must not conclude we have found *the* story just because we have come up with *a* story that seems to fit the facts. To the extent that adaptationists have been less than energetic in seeking further confirmation (or dreaded disconfirmation) of their stories, this is certainly an excess that deserves criticism.[6]

---

[6] The geneticist Steve Jones (1993: 20) gives us another case in point. There are more than 300 strikingly different species of cichlid fish in Lake Victoria. They are so different; how did they get there? 'The conventional view is that Lake Victoria must once have dried up into many small lakes to allow each species to evolve. Apart from the fish themselves, there is no evidence that this ever happened.' Adaptationist stories *do* get disconfirmed and abandoned, however. My favourite example is the now-discredited explanation of why certain sea turtles migrate all the way across the Atlantic between Africa and South America, spawning on one side, feeding on the other. According to this all-too-reasonable story, the habit started when Africa and South America were first beginning to split apart; at that time, the turtles were just going across the bay to spawn; the distance grew imperceptibly longer over the aeons, until their descendants dutiful-

But before leaving it at that, I want to point out that there are many adaptationist stories that *everybody* is happy to accept even though they have never been 'properly tested', just because they are too obviously true to be worth further testing. Does anybody seriously doubt that eyelids evolved to protect the eye? But that very obviousness may hide good research questions from us. George Williams points out that concealed behind such obvious facts may lie others that are well worth further investigation:

A human eye blink takes about 50 milliseconds. That means that we are blind about 5% of the time when we are using our eyes normally. Many events of importance can happen in 50 milliseconds, so that we might miss them entirely. A rock or spear thrown by a powerful adversary can travel more than a meter in 50 milliseconds, and it could be important to perceive such motion as accurately as possible. Why do we blink with both eyes simultaneously? Why not alternate and replace 95% visual attentiveness with 100%? I can imagine an answer in some sort of trade-off balance. A blink mechanism for both eyes at once may be much simpler and cheaper than one that regularly alternates. (Williams 1992: 152–3)

Williams has not himself yet attempted to confirm or disconfirm any hypothesis growing out of this exemplary piece of adaptationist problem-setting, but he has called for the research by asking the question. It would be as pure an exercise in reverse engineering as can be imagined.

Serious consideration of why natural selection permits simultaneous blinking might yield otherwise elusive insights. What change in the machinery would be needed to produce the first step towards my envisioned adaptive alternation or simple independent timing? How might the change be achieved developmentally? What other changes would be expected from a mutation that produced a slight lag in the blinking of one eye? How would selection act on such a mutation? (ibid. 153)

Gould himself has endorsed some of the most daring and delicious of adaptationist 'Just So Stories', such as the argument by Lloyd and Dybas (1966) explaining why cicadas (such as 'seventeen-year locusts') have reproductive cycles that are prime-numbered years long—thirteen years, or seventeen, but never fifteen or sixteen, for instance. 'As evolutionists', Gould says, 'we seek answers to the question, why. Why, in particular, should such striking synchroneity evolve, and why should the period between episodes of sexual reproduction be so long?' (Gould 1977: 99).[7] The answer—which makes beautiful sense, in retrospect—is that, by having a

ly cross an ocean to get to where their instinct still tells them to spawn. I gather that the timing of the breakup of Gondwanaland turns out not to match the evolutionary timetable for the turtles, sad to say, but wasn't it a cute idea?

[7] Gould has recently (1993*b*: 318) described his anti-adaptationism as the 'zeal of the convert', and elsewhere (1991: 13) confesses, 'I sometimes wish that all copies of *Ever Since Darwin* would self-destruct', so perhaps he would recant these words today, which would be a pity, since

large prime number of years between appearances, the cicadas minimize the likelihood of being discovered and later tracked as a predictable feast by predators who themselves show up every two years, or three years, or five years. If the cicadas had a periodicity of, say, sixteen years, then they would be a rare treat for predators who showed up every year, but a more reliable source of food for predators who showed up every two or four years, and an even-money gamble for predators that got in phase with them on an eight-year schedule. If their period is not a multiple of any lower number, however, they are a rare treat—not worth 'trying' to track—for any species that isn't lucky enough to have exactly their periodicity (or some multiple of it—the mythical thirty-four-year locust-muncher would be in fat city). I don't know whether Lloyd and Dybas's 'Just So Story' has been properly confirmed yet, but I don't think Gould is guilty of Panglossianism in treating it as established until proved otherwise. And if he really wants to ask and answer 'why' questions, he has no choice but to be an adaptationist.

The problem he and Lewontin perceive is that there are no standards for when a particular bit of adaptationist reasoning is too much of a good thing. How serious, really, is this problem even if it has no principled 'solution'? Darwin has taught us not to look for essences, for dividing lines between *genuine* function or *genuine* intentionality and mere *on-its-way-to-being* function or intentionality. We commit a fundamental error if we think that if we want to indulge in adaptationist thinking we need a license, and the only license could be the possession of a strict definition of, or criterion for, a genuine adaptation. There are good rules of thumb to be followed by the prospective reverse engineer, made explicit years ago by George Williams (1966). (1) Don't invoke adaptation when other, lower-level explanations are available (such as physics). We don't have to ask what advantage accrues to maple trees that explains the tendency of their leaves to fall *down*, any more than the reverse engineers at Raytheon need to hunt for a reason why GE made their widgets so that they would melt readily in blast furnaces. (2) Don't invoke adaptation when a feature is the outcome of some general developmental requirement. We don't need a special reason of increased fitness to explain the fact that heads are attached to bodies, or limbs come in pairs, any more than the people at Raytheon need to explain why the parts in GE's widget have so many edges and corners with right angles. (3) Don't invoke adaptation when a feature is a by-product of another adaptation. We don't need to give an

they eloquently express the rationale of adaptationism. Gould's attitude towards adaptationism is not so easily discerned, however. *The Book of Life* (1993a) is packed with adaptationist reasoning that made it past his red pencil, and thus presumably has his endorsement.

adaptationist explanation of the capacity of a bird's beak to groom its feathers (since the features of the bird's beak are there for more pressing reasons), any more than we need a special explanation of the capacity of the GE widget's casing to shield the innards from ultraviolct rays.

But you will already have noticed that in each case these rules of thumb can be overridden by a more ambitious enquiry. Suppose someone marvelling at the brilliant autumn foliage in New England asks *why* the maple leaves are so vividly coloured in October. Isn't this adaptationism run amok? Shades of Dr Pangloss! The leaves are the colours they are simply because once the summer energy-harvest season is over, the chlorophyll vanishes from the leaves, and the residual molecules have reflective properties that happen to determine the bright colours—an explanation at the level of chemistry or physics, not biological purpose. But wait. Although this may have been the only explanation that was true up until now, today it is true that human beings so prize the autumn foliage (it brings millions of tourist dollars to northern New England each year) that they protect the trees that are brightest in autumn. You can be sure that if you are a tree competing for life in New England, there is now a selective advantage to having bright autumn foliage. It may be tiny, and in the long run it may never amount to much (in the long run, there may be no trees at all in New England, for one reason or another), but this is how all adaptations get their start, after all, as fortuitous effects that get opportunistically picked up by selective forces in the environment. And of course there is also an adaptationist explanation for why right angles predominate in manufactured goods, and why symmetry predominates in organic limb-manufacturing. These may become utterly fixed traditions, which would be almost impossible to dislodge by innovation, but the reasons why *these* are the traditions are not hard to find, or controversial.

Adaptationist research always leaves unanswered questions open for the next round. Consider the leatherback sea turtle and her eggs:

Near the end of egg laying, a variable number of small, sometimes misshapen eggs, containing neither embryo nor yolk (just albumin) are deposited. Their purpose is not well understood, but they become desiccated over the course of incubation and may moderate humidity or air volume in the incubation chamber. (It is also possible that they have no function or are a vestige of some past mechanisms not apparent to us today.) (Eckert 1992: 30)

But where does it all end? Such open-endedness of adaptationist curiosity is unnerving to many theorists, apparently, who wish there could be stricter codes of conduct for this part of science. Many who have hoped to

contribute to clearing up the controversy over adaptationism and its backlash have despaired of finding such codes, after much energy has been
expended in drawing up and criticizing various legislative regimes. They
are just not being Darwinian enough in their thinking. Better adaptationist
thinking soon drives out its rivals by normal channels, just as second-rate
reverse engineering betrays itself sooner or later.

The eskimo face, once depicted as 'cold engineered' (Coon *et al.* 1950), becomes an
adaptation to generate and withstand large masticatory forces (Shea 1977). We do
not attack these newer interpretations; they may all be right. We do wonder, though,
whether the failure of one adaptive explanation should always simply inspire a
search for another of the same general form, rather than a consideration of alternatives to the proposition that each part is 'for' some specific purpose. (Gould and
Lewontin 1979: 586)

Is the rise and fall of successive adaptive explanations of various things
a sign of healthy science constantly improving its vision, or is it like the
pathological story-shifting of the compulsive fibber? If Gould and Lewontin had a serious alternative to adaptationism to offer, their case for the
latter verdict would be more persuasive, but although they and others have
hunted around energetically, and promoted their alternatives boldly, none
has yet taken root.

Adaptationism, the paradigm that views organisms as complex adaptive machines
whose parts have adaptive functions subsidiary to the fitness-promoting function of
the whole, is today about as basic to biology as the atomic theory is to chemistry.
And about as controversial. Explicitly adaptationist approaches are ascendant in
the sciences of ecology, ethology, and evolution because they have proven essential
to discovery; if you doubt this claim, look at the journals. Gould and Lewontin's
call for an alternative paradigm has failed to impress practicing biologists
both because adaptationism is successful and well-founded, and because its critics
have no alternative research program to offer. Each year sees the establishment
of such new journals as *Functional Biology* and *Behavioral Ecology*. Sufficient
research to fill a first issue of *Dialectical Biology* has yet to materialize. (Daly
1991: 219)

What particularly infuriates Gould and Lewontin, as the passage
about the Eskimo face suggests, is the blithe confidence with which adaptationists go about their reverse engineering, always sure that sooner or
later they will find *the reason* why things are as they are, even if it so
far eludes them. Here is an instance, drawn from Richard Dawkins's
discussion of the curious case of the flatfish (flounders and soles, for
instance) who when they are born are vertical fish, like herring or sunfish,
but whose skulls undergo a weird twisting transformation, moving one eye
to the other side, which then becomes the top of the bottom-dwelling
fish. Why didn't they evolve like those other bottom-dwellers, skates,

which are not on their side but on their belly, 'like sharks that have passed under a steam roller' (Dawkins 1986: 91)? Dawkins *imagines* a scenario (ibid. 92–3):

... even though the skate way of being a flat fish might *ultimately* have been the best design for bony fish too, the world-be intermediates that set out along this evolutionary pathway apparently did less well in the short term than their rivals lying on their side. The rivals lying on their side were so much better, in the short term, at hugging the bottom. In genetic hyperspace, there is a smooth trajectory connecting free-swimming ancestral bony fish to flatfish lying on their side with twisted skulls. There is not a smooth trajectory connecting these bony fish ancestors to flatfish lying on their belly. There is such a trajectory in theory, but it passes through intermediates that would have been—in the short term, which is all that matters—unsuccessful if they had ever been called into existence.

Does Dawkins *know* this? Does he know that the postulated inter-mediates were less fit? Not because he has seen any data drawn from the fossil record. This is a purely theory-driven explanation, argued a priori from the assumption that natural selection tells us the true story—some true story or other—about every curious feature of the biosphere. Is that objectionable? It does 'beg the question'—but what a question it begs! It assumes that Darwinism is basically on the right track. (Is it objection-able when meteorologists say, begging the question against supernatural forces, that there must be a purely physical explanation for the birth of hurricanes, even if many of the details so far elude them?) Notice that in this instance, Dawkins's explanation is almost certainly right— there is nothing especially daring about that particular speculation. More-over, it is, of course, exactly the sort of thinking a good reverse engineer should do. 'It seems so obvious that this General Electric widget casing ought to be made of two pieces, not three, but it's made of three pieces, which is wasteful and more apt to leak, so we can be damn sure that three pieces was seen as better than two in somebody's eyes, shortsighted though they may have been. Keep looking!' The philosopher of biology Kim Sterelny, in a review of *The Blind Watchmaker*, made the point this way:

Dawkins is admittedly giving only scenarios: showing that it's *conceivable* that (e.g.) wings could evolve gradually under natural selection. Even so, one could quibble. Is it really true that natural selection is so fine-grained that, for a protostick insect, looking 5% like a stick is better than looking 4% like one? (pp. 82–83). A worry like this is especially pressing because Dawkins's adaptive scenarios make no mention of the costs of allegedly adaptive changes. Mimicry might deceive potential mates as well as potential predators. ... Still, I do think this objection is something of a quibble because essentially I agree that natural selection is the only possible explanation of complex adaptation. So something like Dawkins' stories have got to be right. (Sterelny 1988: 424)[8]

REFERENCES

Bateson, W. (1909), 'Heredity and Variation in Modern Lights', in A. C. Seward (ed.), *Darwin and Modern Science* (Cambridge: Cambridge University Press), 85–101.
Coon, C. S., Garn, S. M., and Birdsell, J. B. (1950), *Races* (Springfield, Oh.: C. Thomas).
Daly, M. (1991), 'Natural Selection Doesn't Have Goals, but it's the Reason Organisms Do' (commentary on P. J. H. Schoemaker, 'The Quest for Optimality: A Positive Heuristic of Science?'), *Behavioral and Brain Sciences*, 14: 219–20.
Dawkins, R. (1983), 'Adaptationism Was Always Predictive and Needed No Defence' (commentary on Dennett 1983), *Behavioral and Brain Sciences*, 6: 360–1.
——(1986), *The Blind Watchmaker* (London: Longmans).
Dennett, D. C. (1983), 'Intentional Systems in Cognitive Ethology: The "Panglossian Paradigm" Defended', *Behavioral and Brain Sciences*, 6: 343–90.
Eckert, S. A. (1992), 'Bound for Deep Water', *Natural History*, 101 (March): 28–35.
Eldredge, N. (1983), 'A la Recherche du Docteur Pangloss' (commentary on Dennett 1983), *Behavioral and Brain Sciences*, 6: 361–2.
Fisher, D. (1975), 'Swimming and Burrowing in *Limulus* and *Mesolimulus*', *Fossils and Strata*, 4: 281–90.
Fodor, J. (1990), *A Theory of Content and Other Essays* (Cambridge, Mass.: MIT Press).
Ghiselin, M. (1983), 'Lloyd Morgan's Canon in Evolutionary Context', *Behavioral and Brain Sciences*, 6: 362–3.
Gould, S. J. (1977), *Ever Since Darwin* (New York: Norton).
——(1991), *Bully for Brontosaurus* (New York: Norton).
——(1993a) (ed.), *The Book of Life* (New York: Norton).
——(1993b), 'Fulfilling the Spandrels of World and Mind', in Selzer (1993), 310–36.
——and Lewontin, R. C. (1979), 'The Spandrels of San Marco and the Panglossian Paradigm: A Critique of the Adaptationist Programme', *Proceedings of the Royal Society*, B205: 581–98.
Hardy, A. (1960), 'Was Man More Aquatic in the Past?' *New Scientist*, 16: 642–5.
Isack, H. A., and Reyer, H-U. (1989), 'Honeyguides and Honey Gatherers: Interspecific Communications in a Symbiotic Relationship', *Science*, 243: 1343–6.

---

[8] Dawkins is not content to rest with Sterelny's dismissal of his own objections as 'quibbles', since, he points out (personal communication), they raise an important point often misunderstood: 'It is not up to individual humans like Sterelny to express their own commonsense scepticism of the proposition that 5% like a stick is significantly better than 4%. It is an easy rhetorical point to make: "Come on, are you really trying to tell me that 5% like a stick really matters when compared to 4%?" This rhetoric will often convince laymen, but the population-genetic calculations (e.g. by Haldane) belie common sense in a fascinating and illuminating way: because natural selection works on genes distributed over many individuals and over many millions of years, human actuarial intuitions are over-ruled.'

Jones, S. (1993), 'A Slower Kind of Bang' (review of E. O. Wilson, *The Diversity of Life*), *London Review of Books* 15(April): 20.

Kipling, R. (1912), *Just So Stories*, repr. 1952 (Garden City, NY: Doubleday).

Kitcher, P. (1985), 'Darwin's Achievement', in N. Rescher (ed.), *Reason and Rationality in Science* (Lanham, Md.: University Press of America), 127–89.

Leibniz, G. W. (1710), trans. G. M. Duncan (1958), *Theodicy (Essais de Théodicée sur la bonté de Dieu, la liberté de l'homme et l'origine du mal)* (Peru, Ill.: Open Court).

Lewontin, R. C. (1983), 'Elementary Errors about Evolution' (commentary on Dennett 1983), *Behavioral and Brain Sciences*, 6: 367–8.

Lloyd, M., and Dybas, H. S. (1966), 'The Periodical Cicada Problem', *Evolution*, 20: 132–49.

Maynard Smith, J. (1983), 'Adaptation and Satisficing' (commentary on Dennett 1983), *Behavioral and Brain Sciences*, 6: 70–1.

——(1988), *Did Darwin Go It Right?: Essays on Games, Sex and Evolution* (New York: Chapman and Hall).

Morgan, E. (1982), *The Aquatic Ape* (London: Souvenir).

——(1990), *The Scars of Evolution: What Our Bodies Tell Us about Human Origins* (London: Souvenir).

Pittendrigh, C. (1958), 'Adaptation, Natural Selection and Behavior', in A. Roe and G. G. Simpson (eds.), *Behavior and Evolution* (New Haven: Yale University Press), 390–416.

Richards, G. (1991), 'The Refutation That Never Was: The Reception of the Aquatic Ape Theory 1972–1986', in Roede *et al.* (1991), 115–26.

Ridley, M. (1985), *The Problems of Evolution* (Oxford: Oxford University Press).

Roede, M., Wind, J., Patrick, J. M., and Reynolds, V. (1991) (eds.), *The Aquatic Ape: Fact or Fiction?* (London: Souvenir Press).

Selzer, J. (1993), *Understanding Scientific Prose* (Madison: University of Wisconsin Press).

Shea, B. T. (1977), 'Eskimo Cranofacial Morphology, Cold Stress and the Maxillary Sinus', *American Journal of Physical Anthropology*, 47: 289–300.

Sober, E. (1988), *Reconstructing the Past* (Cambridge, Mass.: MIT Press).

Sterelny, K. (1988), review of Dawkins 1986, *Australasian Journal of Philosophy*, 66: 421–6.

Symons, D. (1983), 'FLOAT: A New Paradigm for Human Evolution', in G. M. Scherr (ed.), *The Best of the Journal of Irreproducible Results* (New York: Workman), 27–8.

Whitfield, P. (1993), *From So Simple a Beginning: The Book of Evolution* (New York: Macmillan).

Williams, G. C. (1966), *Adaptation and Natural Selection* (Princeton: Princeton University Press).

——(1992), *Natural Selection: Domains, Levels, and Challenges* (Oxford: Oxford University Press).

4

EXAPTATION—A MISSING TERM IN THE SCIENCE OF FORM

STEPHEN JAY GOULD AND ELISABETH S. VRBA

I. INTRODUCTION

We wish to propose a term for a missing item in the taxonomy of evolutionary morphology. Terms in themselves are trivial, but taxonomies revised for a different ordering of thought are not without interest. Taxonomies are not neutral or arbitrary hat-racks for a set of unvarying concepts; they reflect (or even create) different theories about the structure of the world. As Michel Foucault has shown in several elegant books (1965 and 1970, for example), when you know why people classify in a certain way, you understand how they think.

Successive taxonomies are the fossil traces of substantial changes in human culture. In the mid-seventeenth century, madmen were confined in institutions along with the indigent and unemployed, thus ending a long tradition of exile or toleration for the insane. But what is the common ground for a taxonomy that mixes the mad with the unemployed—an arrangement that strikes us as absurd. The 'key character' for the 'higher taxon', Foucault argues, was idleness, the cardinal sin and danger in an age on the brink of universal commerce and industry (Foucault's interpretation has been challenged by British historian of science Roy Porter, 1982). In other systems of thought, what seems peripheral to us becomes central, and distinctions essential to us do not matter (whether idleness is internally inevitable, as in insanity, or externally imposed, as in unemployment).

II. TWO MEANINGS OF ADAPTATION

In the vernacular, and in sciences other than evolutionary biology, the word 'adaptation' has several meanings all consistent with the etymology

First published in *Paleobiology*, 8/1 (1982): 4–15. Reprinted by permission. An equal time production; order of authorship was determined by a trans-oceanic coin flip.

of *ad* + *aptus*, or towards a fit (for a particular role). When we adapt a tool for a new role, we change its design consciously so that it will work well in its appointed task. When creationists before Darwin spoke of adaptation—for the term long precedes evolutionary thought—they referred to God's intelligent action in designing organisms *for* definite roles. When physiologists claim that larger lungs of Andean mountain peoples are adapted to local climates, they specify directed change for better function. In short, all these meanings refer to historical processes of change or creation for definite functions. The 'adaptation' is designed specifically for the task it performs.

In evolutionary biology, however, we encounter two different meanings—and a possible conflation of concepts—for features called adaptations. The first is consistent with the vernacular usages cited above: a feature is an adaptation only if it was built by natural selection for the function it now performs. The second defines adaptation in a static, or immediate way as any feature that enhances current fitness, regardless of its historical origin. (As a further confusion, adaptation refers both to a process and a state of being. We are only discussing state of being here—that is, features contributing to fitness. We include some comments about this further problem in section VI E.)

Williams, in his classic book on adaptation, recognized this dilemma, and restricted the term to its first, or narrower, meaning. We should speak of adaptation, he argues, only when we can 'attribute the origin and perfection of this design to a long period of selection for effectiveness in this particular role' (1966: 6). In his terminology, 'function' refers only to the operation of adaptations. Williams further argues that we must distinguish adaptations and their functions from fortuitous effects. He uses 'effect' in its vernacular sense—something caused or produced, a result or consequence. Williams's concept of 'effect' may be applied to a character, or to its usage, or to a potential (or process), arising as a consequence of true adaptation. Fortuitous effect always connotes a consequence following 'accidentally', and not arising directly from construction by natural selection. Others have adopted various aspects of this terminology for 'effects' *sensu* Williams (Paterson 1985, Vrba 1980, Lambert, MS). However, Williams and others usually invoke the term 'effect' to designate the *operation* of a useful character *not* built by selection for its current role—and we shall follow this restriction here (Table 4.1). Williams also recognizes that much haggling about adaptation has been 'encouraged by imperfections of terminology' (1966: 8), a situation that we hope to alleviate slightly.

Bock, on the other hand, champions the second, or broader, meaning in

TABLE 4.1 *A taxonomy of fitness*

Process	Character		Usage
Natural selection shapes the character for a current use—adaptation	adaptation		function
A character, previously shaped by natural selection for a particular function (an adaptation), is co-opted for a new use—co-optation		aptation	
A character whose origin cannot be ascribed to the direct action of natural selection (a non-aptation), is co-opted for a current use—co-optation	exaptation		effect

the other most widely cited analysis of adaptation from the 1960s (Bock and von Wahlert 1965, Bock 1967, 1979, 1980). 'An adaptation is, thus, a feature of the organism, which interacts operationally with some factor of its environment so that the individual survives and reproduces' (1979: 39).

The dilemma of subsuming different criteria of historical genesis and current utility under a single term may be illustrated with a neglected example from a famous source. In his chapter devoted to 'difficulties on theory', Darwin wrote (1859: 197):

The sutures in the skulls of young mammals have been advanced as a beautiful adaptation for aiding parturition, and no doubt they facilitate, or may be indispensable for this act; but as sutures occur in the skulls of young birds and reptiles, which have only to escape from a broken egg, we may infer that this structure has arisen from the laws of growth, and has been taken advantage of in the parturition of the higher animals.

Darwin asserts the utility, indeed the necessity, of unfused sutures, but explicitly declines to label them an adaptation because they were not built by selection to function as they now do in mammals. Williams follows Darwin, and would decline to call this feature an adaptation; he would designate its role in aiding the survival of mammals as a fortuitous effect. But Bock would call the sutures and the timing of their fusion an adaptation, and a vital one at that.

As an example of unrecognized confusion, consider this definition of adaptation from a biological dictionary (Abercrombie *et al*. 1951: 10): 'Any characteristic of living organisms which, in the environment they inhabit, improves their chances of survival and ultimately leaving descendants, in comparison with the chances of similar organisms without the characteristic; natural selection therefore tends to establish adaptations in a population.' This definition conflates current utility with historical genesis. What is to be done with useful structures not built by natural selection for their current role?

III. A DEFINITION OF EXAPTATION

We have identified confusion surrounding one of the central concepts in evolutionary theory. This confusion arises, in part, because the taxonomy of form in relation to fitness lacks a term. Following Williams (see Table 4.1), we may designate as an *adaptation* any feature that promotes fitness and was built by selection for its current role (criterion of *historical genesis*). The operation of an adaptation is its *function*. (Bock uses the term 'function' somewhat differently, but we believe we are following the biological vernacular here.) We may also follow Williams in labelling the operation of a useful character not built by selection for its current role as an *effect*. (We designate as an effect only the usage of such a character, not the character itself. But what is the unselected, but useful, character itself to be called? Indeed, it has no recognized name (unless we accept Bock's broad definition of adaptation—the criterion of current utility alone—and reject both Darwin and Williams). Its space on the logical chart is currently blank.

We suggest that such characters, evolved for other usages (or for no function at all), and later 'co-opted' for their current role, be called *exaptations*. (See VI A on the related concept of 'pre-adaptation'.) They are fit for their current role, hence *aptus*; but they were not designed for it, and are therefore not *ad aptus*, or pushed towards fitness. They owe their fitness to features present for other reasons, and are therefore *fit* (*aptus*) *by reason of* (*ex*) their form, or *ex aptus*. Mammalian sutures are an exaptation for parturition. Adaptations have functions; exaptations have effects. The general, static phenomenon of being fit should be called 'aptation', not 'adaptation'. (The set of aptations existing at any one time consists of two partially overlapping subsets: the subset of adaptations and the subset of exaptations. This also applies to the more inclusive set of aptations existing through time; see Table 4.1.)

IV. THE CURRENT NEED FOR A CONCEPT OF EXAPTATION

Why has this conflation of historical genesis with current utility attracted so little attention heretofore? Every biologist surely recognizes that some useful characters did not arise by selection for their current roles; why have we not honoured that knowledge with a name? Does our failure to do so simply underscore the unimportance of the subject? Or might this absent term, in Foucault's sense, reflect a conceptual structure that excluded it? And, finally, does the potential need for such a term at this time indicate that the conceptual structure itself may be altering?

Why did Williams not suggest a term, since he clearly recognized the problem and did separate usages into functions and effects (corresponding respectively to adaptations and to the unnamed features that we call exaptations)? Why did Bock fail to specify the problem at all? We suspect that the conceptual framework of modern evolutionary thought, by continually emphasizing the supreme importance and continuity of adaptation and natural selection at all levels, subtly relegated the issue of exaptation to a periphery of unimportance. How could non-adaptive aspects of form gain a proper hearing under Bock's definition (1979: 63): 'On theoretical grounds, all existing features of animals are adaptive. If they were not adaptive, then they would be eliminated by selection and would disappear.' Williams recognized the phenomenon of exaptation and even granted it some importance (in assessing the capacities of the human mind, for example), but he retained a pre-eminent role for adaptation, and often designated effects as fortuitous or peripheral—'merely an incidental consequence' he states in one passage (1966: 8).

We believe that the adaptationist programme of modern evolutionary thought (Gould and Lewontin 1979) has been weakening as a result of challenges from all levels, molecules to macro-evolution. At the biochemical level, we have theories of neutralism and suggestions that substantial amounts of DNA may be non-adaptive at the level of the phenotype (Orgel and Crick 1980, Doolittle and Sapienza 1980). Students of macro-evolution have argued that adaptations in populations translate as effects to yield the patterns of differential species diversification that may result in evolutionary trends (Vrba's effect hypothesis, 1980). If non-adaptation (or what should be called 'non-aptation') is about to assume an important role in a revised evolutionary theory, then our terminology of form must recognize its cardinal evolutionary significance—co-optability for fitness (see Seilacher 1972 on important effects of a non-aptive pattern in the structure and colouration of molluscs).

Some colleagues have said that they prefer Bock's broad definition because it is more easily operational. We can observe and experiment to determine what good a feature does for an organism now. To reconstruct the historical pathway of its origin is always more difficult, and often (when crucial evidence is missing) intractable.

To this we reply that we are not trying to dismantle Bock's concept. We merely argue that it should be called 'aptation' (with adaptation and exaptation as its modes). As aptation, it retains all the favourable properties for testing enumerated above.

Historical genesis is, undoubtedly, a more difficult problem, but we cannot therefore ignore it. As evolutionists, we are charged, almost by definition, to regard historical pathways as the essence of our subject. We cannot be indifferent to the fact that similar results can arise by different historical routes. Moreover, the distinction between ad- and ex-aptation, however difficult, is not unresolvable. If we ever find a small running dinosaur, ancestral to birds and clothed with feathers, we will know that early feathers were exaptations, not adaptations, for flight.

V. EXAMPLES OF EXAPTATION

A. Feathers and Flight-Sequential Exaptation in the Evolution of Birds

Consider a common scenario from the evolution of birds. (We do not assert its correctness, but only wish to examine appropriate terminology for a common set of hypotheses.) Skeletal features, including the sternum, rib basket, and shoulder joint, in late Jurassic fossils of *Archaeopteryx* indicate that this earliest known bird was probably capable of only the simplest feats of flight. Yet it was quite thoroughly feathered. This has suggested to many authors that selection for the initial development of feathers in an ancestor was for the function of insulation and not for flight (Ostrom 1974, 1979, Bakker, 1975). Such a fundamental innovation would, of course, have many small, as well as far-reaching, incidental consequences. For example, along no descendant lineage of this first feathered species did (so far as we know) a furry covering of the body evolve. The fixation, early in the life of the embryo, of cellular changes that lead on the one hand to hair, and on the other to feathers, constrained the subsequent course of evolution in body covering (Oster 1980).

Archaeopteryx already had large contour-type feathers, arranged along its arms in a pattern very much as in the wings of modern birds. Ostrom (1979: 55) asks: 'Is it possible that the initial (pre-*Archaeopteryx*) enlargement of feathers on those narrow hands might have been to increase the hand surface area, thereby making it more effective in catching insects?' He concludes (ibid. 56): 'I do believe that the predatory design of the wing skeleton in *Archaeopteryx* is strong evidence of a prior predatory function of the proto-wing in a cursorial proto-*Archaeopteryx*.' Later selection for changes in skeletal features and feathers, and for specific neuro-motor patterns, resulted in the evolution of flight.

The black heron (or black egret, *Egretta ardesiaca*) of Africa, like most modern birds, uses its wings in flight. But it also uses them in an interesting way to prey on small fish: 'Its fishing is performed standing in shallow water with wings stretched out and forward, forming an umbrella-like canopy which casts a shadow on the water. In this way its food can be seen' (McLachlan and Liversidge 1978: 39, Plate 6). This 'mantling' of the wings appears to be a characteristic behaviour pattern, with a genetic basis. The wing and feather structures themselves do not seem to be modified in comparison with those of closely related species, the individuals of which do not hunt in this way (A. C. Kemp, personal communication).

We see, in this scenario, a sequential set of adaptations, each converted to an exaptation of different effect that sets the basis for a subsequent adaptation. By this interplay, a major evolutionary transformation occurs that probably could not have arisen by purely increasing adaptation. Thus, the basic design of feathers is an adaptation for thermo-regulation and, later, an exaptation for catching insects. The development of large-contour feathers and their arrangement on the arm arise as adaptations for insect catching and become exaptations for flight. Mantling behaviour uses wings that arose as an adaptation for flight. The neuro-motor modifications governing mantling behaviour, and therefore the mantling posture, are adaptations for fishing. The wing *per se* is an exaptation in its current effect of shading, just as the feathers covering it also arose in different adaptive contexts but have provided much evolutionary flexibility for other uses during the evolution of birds.

B. Bone as Storage and Support

The development of bone was an event of major significance in the evolution of vertebrates. Without bone, vertebrates could not have later taken up life on land. Halstead (1969) has investigated the question: granting its subsequent importance as body support in the later evolution of vertebrates, why did bone evolve at such an early stage in vertebrate history? Some authors have hypothesized that bone initially arose as an osmo-regulatory response to life in freshwater. Others, like Romer (1963), postulate initial adaptation of bony 'armour' for a protective function. Pautard (1961, 1962) pointed out that any organism with much muscular activity needs a conveniently accessible store of phosphate. Following Pautard, and noting the seasonal cycle of phosphate availability in the sea,

Halstead (1969) suggested the following scenario. Calcium phosphates, laid down in the skin of the earliest vertebrates, evolved initially as an adaptation for storing phosphates needed for metabolic activity. Only considerably later in evolution did bone replace the cartilaginous endoskeleton and adopt the function of support for which it is now most noted.

Thus, bone has two major uses in extant vertebrates: support/protection and storage/homeostasis (as a storehouse for certain mineral ions, including phosphate ions). The ions in vertebrate bone are in equilibrium with those in tissue fluids and blood, and function in certain metabolic activities (Scott and Symons 1977). For instance, in humans, 90 per cent of body phosphorus is present in the inorganic phase of bone (Duthie and Ferguson 1973).

Following Halstead's analysis, the deposition of phosphate in body tissues originally evolved as an adaptation for a storage/metabolic function. The metabolic mechanism for producing bone *per se* can thus be interpreted as an exaptation for support. The metabolic mechanisms for depositing an increased quantity of phosphates and for mineralization, as well as the arrangement of bony elements in an internal skeleton, are then adaptations for support.

C. The Evolution of Mammalian Lactation

Dickerson and Geis (1969) recount how Alexander Fleming, in 1922, discovered the enzyme lysozyme. He had a cold and, for interest's sake, added a few drops of nasal mucus to a bacterial culture. To his surprise he found, after a few days, that something in the mucus was killing the bacteria: the enzyme lysozyme, since found in most bodily secretions and in large quantities in the whites of eggs. Lysozyme destroys many bacteria by lysing, or dissolving, the mucopolysaccharide structure of the cell wall. The amino acid sequence of α-lactalbumin, a milk protein of previously unknown function, was then found to be so close to that of lysozyme, that some relationship of close homology must be involved. Dickerson and Geis (1969: 77–8) write:

α-Lactalbumin by itself is not an enzyme but was found to be one component of a two-protein lactose synthetase system, present only in mammary glands during lactation.... The other component (the 'A' protein) had been discovered in the liver and other organs as an enzyme for the synthesis of N-acetyllactosamine from galactose and NAG. But the combination of the A protein and α-lactalbumin synthesizes the milk sugar lactose from galactose and glucose instead. The noncatalytic α-lactalbumin evidently acts as a control device to switch its partner from one potential synthesis to another. . . . It appears that when a milk-producing-system was being developed during the evolution of mammals, and when a need for a polysaccharide-synthesizing enzyme arose, a suitable one was found in part by modifying a pre-existing polysaccharide-cutting enzyme.

Thus, lysozyme, in all vertebrates in which it occurs, is probably an adaptation for the function of killing bacteria. Further evolution in mammals (alteration of a duplicated gene according to Dickerson and Geis 1969) resulted in α-lactalbumin, an adaptation (together with the A protein) for the function of lactose synthesis and lactation. Human lysozyme, in this scenario, is an adaptation for lysing the cell walls of bacteria, and an exaptation with respect to the lactose synthetase system.

D. Sexual 'Mimicry' in Hyenas

Females of the spotted hyena, *Crocuta crocuta*, are larger than males and dominant over them. Pliny, and other ancient writers, had already recognized a related and unusual feature of their biology in calling them 'hermaphrodites' (falsely, as Aristotle showed). The external genitalia of females are virtually indistinguishable from the sexual organs of males by sight. The clitoris is enlarged and extended to form a cylindrical structure with a narrow slit at its distal end; it is no smaller than the male's penis, and can also be erected. The *labia majora* are folded over and fused along the mid-line to form a false scrotal sac (though without testicles of course), virtually identical in form and position with the male's scrotum (Harrison Matthews 1939).

The literature on this sexual 'mimicry' is full of speculations about adaptive meaning. Most of these arguments have conflated current utility and historical genesis in assuming that the demonstration of modern use (Bockian adaptation) specifies the path of origin (adaptation as used by Williams and Darwin, and as advocated here). We suggest that the absence of an articulated concept of exaptation has unconsciously forced previous authors into this erroneous conceptual bind.

Kruuk (1972), the leading student of spotted hyenas, for example, notes that the enlarged sexual organs of females are used in an important behaviour known as the meeting ceremony. Hyenas spend long periods as solitary wanderers searching for carrion, but they also live in well-integrated clans that defend territory and engage in communal hunting. A mechanism for reintegrating solitary wanderers into their proper clan must be developed. In the meeting ceremony, two hyenas stand side by side, facing in opposite directions. Each lifts the inside hind leg, exposing an erect penis or clitoris to its partner's teeth. They sniff and lick each other's genitals for 10 to 15 seconds, largely at the base of the penis or clitoris and in front of the scrotum or false scrotum.

Having discovered a current utility for the prominent external genitalia of females, Kruuk (1972: 229–30) infers that they must have evolved for this purpose:

> It is impossible to think of any other purpose for this special female feature than for use in the meeting ceremony.... It may also be, then, that an individual with a familiar but relatively complex and conspicuous structure sniffed at during the meeting has an advantage over others; the structure would often facilitate this reestablishment of social bonds by keeping partners together over a longer meeting period. This could be the selective advantage that has caused the evolution of the females' and cubs' genital structure.

Yet another hypothesis, based upon facts known to every first-year biology student, virtually cries out for recognition. The penis and clitoris are homologous organs, as are the scrotum and *labia majora*. We know that high levels of androgen induce the enlargement of the clitoris and the folding over and fusion of the labia until they resemble penis and scrotal sac respectively. (In fact, in an important sense, they *are* then a penis and scrotal sac, given the homologies.) Human baby girls with unusually enlarged adrenals secrete high levels of androgen, and are born with a peniform clitoris and an empty scrotal sac formed of the fused labia.

Female hyenas are larger than males, and dominant over them. Since these features are often hormonally mediated in mammals, should we not conjecture that females attain their status by secreting androgens, and that the peniform clitoris and false scrotal sac are automatic, secondary by-products. Since they are formed anyway, a later and secondary utility might ensue; they may be co-opted to enhance fitness in the meeting ceremony, and then secondarily modified for this new role. We suggest that the peniform clitoris and false scrotal sac arose as non-aptive consequences of high androgen levels (a primary adaptation related to the unusual behavioural role of females). They are, therefore, exaptations for the meeting ceremony, and their effect in enhancing fitness through that ceremony does not specify the historical pathway of their origin.

Yet this obvious hypothesis, with its easily testable cardinal premiss, was not explicitly examined until 1979, after, literally, more than 2,000 years of speculation in the adaptive mode (both ancient authors and medieval bestiaries tried to infer God's intent in creating such an odd beast). Racey and Skinner (1979) found no differences in levels of androgen in blood plasma of male and female spotted hyenas. Female foetuses contained the same high level of testosterone as adult females. In the other two species of the family *Hyaenidae*, however, androgen levels in blood plasma are much lower for females than for males. Females of these species are not dominant over males, and do not develop peniform clitorises or false scrotal sacs.

We do not assert that our alternative hypothesis of exaptation must be correct. One could run the scenario in reverse (with a bit of forcing in our judgement): females 'need' prominent genitalia for the meeting ceremony; they build them by selection for high androgen levels; large size and dominance are a secondary by-product of the androgen. We raise, rather, a different issue: why was this evident alternative not considered, especially by Kruuk in his excellent exhaustive book on the species? We suggest that the absence of an explicitly articulated concept of exaptation has constrained the range of our hypotheses in subtle and unexamined ways.

E. The Uses of Repetitive DNA

For a few years after Watson and Crick elucidated the structure of DNA, many evolutionists hoped that the architecture of genetic material might fit all their presuppositions about evolutionary processes. The linear order of nucleotides might be the beads on a string of classical genetics: one gene, one enyzme; one nucleotide substitution, one minute alteration for natural selection to scrutinize. We are now, not even twenty years later, faced with genes in pieces, complex hierarchies of regulation, and, above all, vast amounts of repetitive DNA. Highly repetitive, or satellite, DNA can exist in millions of copies; middle-repetitive DNA, with its tens to hundreds of copies, forms about one-quarter of the genome in both *Drosophila* and *Homo*. What is all the repetitive DNA for (if anything)? How did it get there?

A survey of previous literature (Doolittle and Sapienza 1980, Gould 1983) reveals two emerging traditions of argument, both based on the selectionist assumption that repetitive DNA must be good for something if so much of it exists. One tradition (see Britten and Davidson 1971 for its *locus classicus*) holds that repeated copies are conventional adaptations, selected for an immediate role in regulation (by bringing previously isolated parts of the genome into new and favourable combinations, for example, when repeated copies disperse among several chromosomes). We do not doubt that conventional adaptation explains the preservation of much repeated DNA in this manner.

But many molecular evolutionists now strongly suspect that direct adaptation cannot explain the existence of all repetitive DNA: there is simply too much of it. The second tradition therefore holds that repetitive DNA must exist because evolution needs it so badly for a flexible future—as in the favoured argument that 'unemployed', redundant copies are free to alter because their necessary product is still being generated by the original copy (see Cohen 1976, Lewin 1975, and Kleckner 1977, all of whom also follow the first tradition and argue both sides). While we do not doubt that such future uses are vitally important consequences of repeated DNA, they simply cannot be the cause of its existence, unless we return to certain theistic views that permit the control of present events by future needs.

This second tradition expresses a correct intuition in a patently nonsensical (in its non-pejorative meaning) manner. The missing thought that supplies sense is a well-articulated concept of exaptation. Defenders of the second tradition understand how important repetitive DNA is to evolution, but only know the conventional language of adaptation for expressing this conviction. But since utility is a future condition (when the redundant copy assumes a different function or undergoes secondary adaptation for a new role), an impasse in expression develops. To break this impasse, we might suggest that repeated copies are non-apted features, available for co-optation later, but not serving any direct function at the moment. When co-opted, they will be exaptations in their new role (with secondary adaptive modifications if altered).

What, then, is the source of these exaptations? According to the first tradition, they arise as true adaptations, and later assume their different function. The second tradition, we have argued, must be abandoned. A third possibility has recently been proposed (or, rather, better codified after previous hints): perhaps repeated copies can originate for no adaptive reason that concerns the traditional Darwinian level of phenotypic advantage (Orgel and Crick 1980, Doolittle and Sapienza 1980). Some DNA elements are transposable; if these can duplicate and move, what is to stop their accumulation as long as they remain invisible to the phenotype (if they become so numerous that they begin to exert an energetic constraint upon the phenotype, then natural selection will eliminate them)? Such 'selfish DNA' may be playing its own Darwinian game at a genic level, but it represents a true non-aptation at the level of the phenotype. Thus, repeated DNA may often arise as a non-aptation. Such a statement in no way argues against its vital importance for evolutionary futures. When used to great advantage in that future, these repeated copies are exaptations.

VI. SIGNIFICANCE OF EXAPTATION

A. A Solution to the Problem of Pre-adaptation

The concept of pre-adaptation has always been troubling to evolutionists. We acknowledge its necessity as the only Darwinian solution to Mivart's (1871) old taunt that 'incipient stages of useful structures' could not function as the perfected forms do (what good is 5 per cent of a wing?). The incipient stages, we argue, must have performed in a different way (thermo-regulation for feathers, for example). Yet we traditionally apologize for 'pre-adaptation' in our textbooks, and laboriously point out to students that we do not mean to imply foreordination, and that the word is somehow wrong (though the concept is secure). Frazzetta (1975: 212), for example, writes: 'The association between the word "pre-adaptation" and dubious teleology still lingers, and I can often produce a wave of nausea in some evolutionary biologists when I use the word unless I am quick to say what I mean by it.'

Indeed, the word is wrong, and our long-standing intuitive discomfort is justified (see Lambert, MS). For if we divide the class of features contributing to fitness into adaptations and exaptations, and if adaptations were constructed (and exaptations co-opted) for their current use, then features working in one way cannot be pre-*ada*ptations to a different and subsequent usage: the term makes no sense at all.

The recognition of exaptation solves the dilemma neatly, for what we now incorrectly call 'pre-adaptation' is merely a category of exaptation considered before the fact. If feathers evolved for thermo-regulation, they become exaptations for flight once birds take off. If, however, with the hindsight of history, we choose to look at feathers while they still encase the running, dinosaurian ancestors of birds, then they are only potential exaptations for flight, or *pre-aptations* (that is, *aptus*—or fit—before their actual co-optation). The term 'pre-adaptation' should be dropped in favour of 'pre-aptation'. Pre-aptations are potential, but unrealized, exaptations; they resolve Mivart's major challenge to Darwin.

B. Primary Exaptations and Secondary Adaptations

Feathers, in their basic design, are exaptations for flight, but once this new effect was added to the function of thermo-regulation as an important source of fitness, feathers underwent a suite of secondary adaptations (sometimes called 'post-adaptations') to enhance their utility in flight. The order and arrangement of tetrapod limb bones is an exaptation for walking on land; many modifications of shape and musculature are secondary adaptations for terrestrial life.

The evolutionary history of any complex feature will probably include a sequential mixture of adaptations, primary exaptations, and secondary adaptations. Just as any feature is plesiomorphic at one taxonomic level and apomorphic at another (torsion in the class *Gastropoda* and in the phylum *Mollusca*), we are not disturbed that complex features are a mixture of exaptations and adaptations. Any co-opted structure (an exaptation) will probably not arise perfected for its new effect. It will therefore develop secondary adaptations for the new role. The primary exaptations and secondary adaptations can, in principle, be distinguished.

C. The Sources of Exaptation

Features co-opted as exaptations have two possible previous statuses. They may have been adaptations for another function, or they may have been non-aptive structures. The first has long been recognized as important, the second underplayed. Yet the enormous pool of non-aptations must be the wellspring and reservoir of most evolutionary flexibility. We need to recognize the central role of 'co-optability for fitness' as the primary evolutionary significance of ubiquitous non-aptation in organisms. In this sense, and at its level of the phenotype, this non-aptive pool is an analogue of mutation—a source of raw material for further selection.

Both adaptations and non-aptations, while they may have non-random proximate causes, can be regarded as randomly produced with respect to any potential co-optation by further regimes of selection. Simply put: all exaptations *originate* randomly with respect to their effects. Together, these two classes of characters, adaptations and non-aptations, provide an enormous pool of variability, at a level higher than mutations, for co-optation as exaptations. (Lambert, MS, has discussed this with respect to pre-adaptations only—pre-aptations in our terminology. He explored the evolutionary implications of the notion that for any function, resulting directly from natural selection at any one time, there may be multiple effects.)

If all exaptations began as adaptations for another function in ancestors, we would not have written this essay. For the concept would be covered by the principle of 'pre-adaptation'—and we would only need to point out that 'pre-aptation' would be a better term, and that etymology requires a different name for pre-aptations after they are established. Exaptations that began as non-aptations represent the missing concept. They are not covered by the principle of pre-aptation, for they were not adaptations in ancestors. They truly have no name, and concepts without names cannot be properly incorporated in thought. The great confusions of historical genesis and current utility primarily involve useful features that were not adaptations in ancestors—as in our examples of sexual 'mimicry' in hyenas and the uses of middle-repetitive DNA.

D. The Irony of our Terminology for Non-aptation

It seems odd to define an important thing by what it is not. Students of early geology are rightly offended that we refer to five-sixths of Earth history as Precambrian. Features not now contributing to fitness are usually called 'non-adaptations'. (In our terminology they are 'non-aptations'.) This curious negative definition can only record a feeling that the subject is 'lesser' than the thing it is not. We believe that this feeling is wrong, and that the size of the pool of non-aptations is a central phenomenon in evolution. The term 'non-adaptive' is but another indication of previous—and in our view false—convictions about the supremacy of adaptation. The burden of nomenclature is already great enough, and we do not propose a new term for features without current fitness. But we do wish to record the irony.

E. Process and State-of-being

Evolutionary biologists use the term adaptation to describe both a current state-of-being (as discussed here) and the process leading to it. This duality presents no problem in cases of true adaptation, where a process of selection directly produces the state of fitness. Exaptations, on the other hand, are not fashioned for their current role, and reflect no attendant process beyond co-optation (Table 4.1); they were built in the past either as non-aptive by-products or as adaptations for different roles.

Perhaps we should begin our analysis of process with a descriptive approach, and simply focus upon the set of features that increase their relative or absolute abundance within populations, species, or clades by the only general processes that can yield such 'plurifaction', or 'more making': differential branching or persistence (see Arnold and Fristrup 1982). This descriptive process of plurifaction has two basic causes. First, features may increase their representation actively by contributing to branching or persistence either as adaptations evolved by selection for their current function, or exaptations evolved by another route and co-opted for their useful effect. Secondly, and particularly at the higher level of species within clades, features may increase their own representation for a host of non-aptive reasons, including causal correlation with features contributing to fitness and fortuitous correlation found at such surprisingly high frequency in random simulations by Raup and Gould (1974). These non-aptive features establish an enormous pool for potential exaptation.

VII. CONCLUSION

The ultimate decision about whether we have written a trivial essay on terminology or made a potentially interesting statement about evolution must hinge upon the importance of exaptation, both in frequency and in role. We believe that the failure of evolutionists to codify such a concept must record an inarticulated belief in its relative insignificance.

We suspect, however, that the subjects of non-aptation and co-optability are of paramount importance in evolution. (When co-optability has been recognized—in the principle of 'pre-adaptation'—we have focused upon shift in role for features previously adapted for something else, not on the potential for exaptation in non-apted structures.) The flexibility of evolution lies in the range of raw material presented to processes of selection. We all recognize this in discussing the conventional sources of genetic variation—mutation, recombination, and so forth—presented to natural selection from the genetic level below. But we have not adequately appreciated that features of the phenotype themselves (with their usually complex genetic bases) can also act as variants to enhance and restrict future evolutionary change. Thus the important statement of Fisher's fundamental theorem considers only genetic variance in relation to fitness: 'The rate of increase in fitness of any organism at any time is equal to its genetic variance in fitness at that time' (Fisher 1958: 37). In an analogous way, we might consider the flexibility of phenotypic characters as a primary enhancer of, or damper upon, future evolutionary change. Flexibility lies in the pool of features available for co-optation (either as adaptations to something else that has ceased to be important in new selective regimes, as adaptations whose original function continues but which may be co-opted for an additional role, or as non-aptations always potentially available). The paths of evolution—both the constraints and the opportunities—must be largely set by the size and nature of this pool of potential exaptations. Exaptive possibilities define the 'internal' contribution that organisms make to their own evolutionary future.

A. R. Wallace, a strict adaptationist if ever there was one, none the less denied that natural selection had built the human brain. 'Savages' (living primitives), he argued, have mental equipment equal to ours, but maintain only a rude and primitive culture—that is, they do not use most of their mental capacities, and natural selection can only build for immediate use. Darwin, who was not a strict adaptationist, was both bemused and angered. He recognized the hidden fallacy in Wallace's argument: that the brain, though undoubtedly built by selection for some complex set of functions, can, as a result of its intricate structure, work in an unlimited number of ways quite unrelated to the selective pressure that constructed it. Many of these ways might become important, if not indispensable, for future survival in later social contexts (like afternoon tea for Wallace's contemporaries). But current utility carries no automatic implication about historical origin. Most of what the brain now does to enhance our survival lies in the domain of exaptation—and does not allow us to make hypotheses about the selective paths of human history. How much of the evolutionary literature on human behaviour would collapse if we incorporated the principle of exaptation into the core of our evolutionary thinking? This collapse would be constructive, because it would vastly broaden our range of hypotheses, and focus attention on current function and development (all testable propositions) instead of leading us to unprovable reveries about primal fratricide on the African savannah or dispatching mammoths at the edge of great ice sheets—a valid subject, but one better treated in novels that can be quite enlightening scientifically (Kurtén 1980).

Consider also the apparently crucial role that repeated DNA has played in the evolution of phenotypic complexity in organisms. If each gene codes for an indispensable enzyme (or performs any necessary function), asks Ohno (1970) in his seminal book, how does evolution transcend mere tinkering along established lines and achieve the flexibility to build new types of organization? Ohno argues that this flexibility must arise as the incidental result of gene duplication, with its production of redundant genetic material: 'Had evolution been entirely dependent upon natural selection, from a bacterium only numerous forms of bacteria would have emerged. . . . Only the cistron which became redundant was able to escape from the relentless pressure of natural selection, and by escaping, it accumulated formerly forbidden mutations to emerge as a new gene locus' (from the preface to Ohno 1970).

We argued in section V E that much of this repetitive DNA may arise for non-aptive reasons at the level of the individual phenotype (as in the 'selfish DNA' hypothesis). The repeated copies are then exaptations, co-opted for fitness and secondarily adapted for new roles. And they are exaptations in the interesting category of structures that arose as non-aptations, when the 'selfish DNA' hypothesis applies.

Thus, the two evolutionary phenomena that may have been most crucial to the development of complexity with consciousness on our planet (if readers will pardon some dripping anthropocentrism for the moment)— the process of creating genetic redundancy in the first place, and the myriad and inescapable consequences of building any computing device as complex as the human brain—may both represent exaptations that began as non-aptations, the concept previously missing in our evolutionary terminology. With examples such as these, the subject cannot be deemed unimportant!

In short, the codification of exaptation not only identifies a common flaw in much evolutionary reasoning—the inference of historical genesis from current utility. It also focuses attention upon the neglected but paramount role of non-aptive features in both constraining and facilitating the path of evolution. The argument is not anti-selectionist, and we view this essay as a contribution to Darwinism, not as a skirmish in a nihilistic vendetta. The main theme is, after all, co-optability for *fitness*. Exaptations are vital components of any organism's success.[1]

REFERENCES

Abercrombie, M., Hickman, C. H., and Johnson, M. L. (1951), *A Dictionary of Biology*, 5th edn. 1966 (Aylesbury: Hunt Bernard and Co. Ltd.).
Arnold, A. J., and Fristrup, K. (1982), 'The Hierarchical Basis for a Unified Theory of Evolution', *Paleobiology*, 8: 113–29.
Bakker, R. T. (1975), 'Dinosaur Renaissance', *Scientific American*, 232/4: 58–78.
Bock, W. (1967), 'The Use of Adaptive Characters in Avian Classification', in *Proceedings of the 14th International Ornithological Congress* (Pittsburgh: Carnegie Museum of Natural History), 66–74.
——(1979), 'A Synthetic Explanation of Macroevolutionary Change—A Reductionistic Approach', *Bulletin of the Carnegie Museum of Natural History*, 13: 20–69.
——(1980), 'The Definition and Recognition of Biological Adaptation', *American Zoologist*, 20: 217–27.
——and von Wahlert, G. (1965), 'Adaptation and the Form–Function Complex', *Evolution*, 10: 269–99.

[1] The following have commented on the manuscript: C. K. Brain, C. A. Green, A. C. Kemp, H. E. H. Paterson. One of us (E.S.V.) owes a debt to Hugh Paterson for an introduction, during extensive discussions, to the terminology of effects (*sensu* Williams). We both thank him for referring us to the examples of mantling behaviour in the black heron and lysozyme/α-lactalbumin evolution. D. M. Lambert has given us access to an unpublished manuscript, and has discussed with us the ubiquitous presence, and enormous importance, in evolution of what he and others call 'pre-adaptation'.

Britten, R. J., and Davidson, E. H. (1971), 'Repetitive and Non-Repetitive DNA Sequences and a Speculation on the Origins of Evolutionary Novelty', *Quarterly Review of Biology*, 46: 111–31.

Cohen, S. N. (1976), 'Transposable Genetic Elements and Plasmid Evolution', *Nature*, 263: 731–8.

Darwin, C. (1859), *On the Origin of Species* (London: J. Murray).

Dickerson, R. E., and Geis, I. (1969), *The Structure and Action of Proteins* (New York: Harper & Row).

Doolittle, W. F., and Sapienza, C. (1980), 'Selfish Genes, the Phenotype Paradigm, and Genome Evolution', *Nature*, 284: 601–3.

Duthie, R. B., and Ferguson, A. B. (1973), *Mercer's Orthopaedic Surgery*, 7th edn. (London: Edward Arnold).

Fisher, R. A. (1958), *Genetical Theory of Natural Selection*, 2nd rev. edn. (New York: Dover).

Foucault, M. (1965), *Madness and Civilization* (New York: Random House).

——(1970), *The Order of Things* (New York: Random House).

Frazzetta, T. H. (1975), *Complex Adaptations in Evolving Populations* (Sunderland, Mass.: Sinauer Associates).

Gould, S. J. (1983), 'What Happens to Bodies if Genes Act for Themselves?', in *Hen's Teeth and Horse's Toes* (New York: Norton), 166–76.

—— and Lewontin, R. C. (1979), 'The Spandrels of San Marco and the Panglossian Paradigm: A Critique of the Adaptationist Programme', in J. Maynard Smith and R. Holliday (eds.), *The Evolution of Adaptation by Natural Selection* (London: Royal Society), 147–64.

Halstead, L. B. (1969), *The Pattern of Vertebrate Evolution* (Edinburgh: Oliver & Boyd).

Harrison Matthews, L. (1939), 'Reproduction in the Spotted Hyena *Crocuta crocuta* (Erxleben)', *Philosophical Transactions of the Royal Society*, B230: 1–78.

Kleckner, N. (1977), 'Translocatable Elements in Procaryotes', *Cell*, 11: 11–23.

Kruuk, H. (1972), *The Spotted Hyena, A Study of Predation and Social Behavior* (Chicago: University of Chicago Press).

Kurten, B. (1980), *Dance of the Tiger* (New York: Pantheon).

Lewin, B. (1975), 'Units of Transcription and Translation', *Cell*, 4: 77–93.

McLachlan, G. R., and Liversidge, R. (1978), *Roberts' Birds of South Africa*, 4th edn. (Cape Town: John Voelcker Bird Book Fund, first pub. 1940).

Mivart, St. G. (1871), *On the Genesis of Species* (London: Macmillan).

Ohno, S. (1970), *Evolution by Gene Duplication* (New York: Springer).

Orgel, L. E., and Crick, F. H. C. (1980), 'Selfish DNA: The Ultimate Parasite', *Nature*, 284: 604–7.

Oster, G. (1980), 'Mechanics, Morphogenesis and Evolution', address to Conference on Macroevolution, Oct. 1980, Chicago.

Ostrom, J. H. (1974), '*Archaeopteryx* and the Origin of Flight', *Quarterly Review of Biology*, 49: 27–47.

——(1979), 'Bird Flight: How Did It Begin?', *American Scientist*, 67: 46–56.

Paterson, H. E. H. (1985), 'The Recognition Concept of Species', in E. S. Vrba (ed.), *Species and Speciation*, Transvaal Museum Monograph, 4 (Pretoria: Transvaal Museum), 21–9.

Pautard, F. G. E. (1961), 'Calcium, Phosphorus, and the Origin of Backbones', *New Scientist*, 12: 364–6.

——(1962), 'The Molecular-Biologic Background to the Evolution of Bone', *Clinical Orthopaedics*, 24: 230–44.

Porter, R. (1982), 'Shutting People Up', *Social Studies of Science*, 12: 467–76.

——(MS), 'Problems in the Treatment of "Madness" in English Science, Medicine and Literature in the Eighteenth Century'.

Racey, P. E., and Skinner, J. C. (1979), 'Endocrine Aspects of Sexual Mimicry in Spotted Hyenas *Crocuta crocuta*', *Journal of Zoology* (London), 187: 315–26.

Raup, D. M., and Gould, S. J. (1974), 'Stochastic Simulation and Evolution of Morphology—Towards a Nomothetic Paleontology', *Systematic Zoology*, 23: 305–22.

Romer, A. S. (1963), 'The "Ancient History" of Bone', *Annals of the New York Academy of Science*, 109: 168–76.

Scott, J. D., and Symons, N. B. B. (1977), *Introduction to Dental Anatomy* (London: Churchill Livingstone).

Seilacher, A. (1970), 'Arbeitskonzept zur Konstruktionsmorphologie', *Lethaia*, 3: 393–6.

——(1972), 'Divariate Patterns in Pelecypod Shells', *Lethaia*, 5: 325–43.

Vrba, E. S. (1980), 'Evolution, Species and Fossils: How Does Life Evolve?', *South African Journal of Science*, 76: 61–84.

Williams, G. C. (1966), *Adaptation and Natural Selection* (Princeton: Princeton University Press).

SIX SAYINGS ABOUT ADAPTATIONISM

ELLIOTT SOBER

Adaptationism is a doctrine that has meant different things to different people. Some define adaptationism so that it is obviously true; others define it so that it is obviously false. I prefer to do neither. In this essay, I want to isolate and discuss a reading of adaptationism that makes it a non-trivial empirical thesis about the history of life. I'll take adaptationism to be the following claim:

Natural selection has been the only important cause of most of the phenotypic traits found in most species.

I won't try to determine whether adaptationism, so defined, is true. That's something that biologists working on different traits in different species will be able to decide only in the long run. Rather, my task will be one of clarification. What does this statement mean, and how is it related to various familiar remarks that biologists have made about adaptationism, *pro* and *con*?

EVOLUTIONARY FORCES AND A NEWTONIAN ANALOGY

Objects released above the surface of the earth accelerate downward at a rate of 32 feet/second^2. Or rather, they do so unless some force other than the earth's gravitational attraction is at work. A similar principle can be stated for the process of natural selection. *In a population subject to natural selection, fitter traits become more common and less fit traits become more rare, unless some other force prevents this from happening.*[1] With sufficient time, this transformation rule means that the fittest of the available pheno-

---

First published as part of 'Evolution and Optimality—Feathers, Bowling Balls, and the Thesis of Adaptationism', *Philosophical Exchange*, 26 (1996): 41–57. Reprinted by permission.

[1] This description of evolutionary theory as a 'theory of forces' is drawn from Sober 1984. Selection can produce evolution only if the traits under selection are heritable. It makes no sense to talk of selection 'alone' producing an evolutionary outcome if this means that it does so

types (if there is just one) will be the only one that remains in the population. This resulting phenotype is said to be *optimal*, not in the sense that it is the best *conceivable* trait, but in the sense that it is the best of the traits *available*.

What might prevent the optimal phenotype from evolving? There are several possibilities (Maynard Smith 1978). First, random events induced by small population size can prevent fitter traits from increasing in frequency. Another possible preventer is the underlying genetics—the pattern by which phenotypes are coded by genotypes. The simplest example of this is the genetic arrangement known as heterozygote superiority; if there are three phenotypes, each coded by the diploid genotype found at a single locus, then the fittest phenotype will not evolve to fixation if it is encoded by the heterozygote genotype. Other more complicated genetic arrangements can lead to the same result. A third factor that can prevent the optimal phenotype from evolving is time. If a population begins with a range of phenotypes, it will take time for natural selection to transform this population into one in which the optimal phenotype has gone to fixation. If biologists start studying this population before sufficient time has elapsed, they will discover that the population is polymorphic. Here again, it is a contingent matter whether the best phenotype among the range of variants has attained 100 per cent representation.

The list of possible preventers could be continued (cf. e.g. Reeve and Sherman 1993), but I think the pattern is already clear. When selection is the only force guiding a population's evolution, the fittest of the available phenotype evolves. However, when other forces intrude, other outcomes are possible.

'Pure' natural selection has predictable results, but the world is never pure. For example, populations are never infinitely large, which means that random drift always plays some role, however small. Still, the question remains of how closely nature approximates the pure case. It is an empirical matter whether natural selection was the only important influence on the evolution of a particular trait in a particular population, or if nonselective forces also played an important role. A Newtonian analogy is useful. The Earth's gravitational force induces a component acceleration on objects released at its surface. However, since the Earth is not surrounded by a vacuum, falling objects always encounter air resistance. It is therefore an empirical matter whether gravitation is the only important

---

without heredity. Rather, the right way to understand the principle I describe in the text is that selection can be expected to lead to the evolution of fitter traits when like phenotype produces like phenotype. Departures from this simple rule of heredity can impede the ability of natural selection to lead fitter phenotypes to evolve, as explained below.

influence on the trajectory of a falling body or if other forces also play an important role. We know that objects are not the same in this respect; the trajectory of a bowling ball differs markedly from the trajectory of a feather. In physics, we are quite accustomed to this pluralistic view of the relative importance of different forces; adaptationism raises the question of whether pluralism is also the right view to take in evolution. If organisms are almost always like bowling balls, then forces other than natural selection can be ignored if one wishes to explain and predict which of the available phenotypes found in a population evolves. If organisms are often like feathers, this idealization will be a mistake. Perhaps selection was important without its being the *only* important cause.

With this conception of what adaptationism means, I now want to consider several remarks one commonly hears—from biologists, philosophers, cognitive scientists, and others—on both sides of this controversy.

Saying number 1: 'Natural selection is the only natural process that can produce adaptive complexity.'

In his essay 'Universal Darwinism', Richard Dawkins (1983 and Ch. 2 above) updates the design argument for the existence of God. If one examines the vertebrate eye, for example, and wants to explain its complexity, its organization, and why its parts conspire so artfully to allow the organism to see, the only naturalistic explanation one can think of is natural selection. Rather than conclude that adaptive complexity points to the existence of an intelligent designer, Dawkins argues that it points to the existence of a 'blind watchmaker'—that is, to the process of natural selection, which is not only blind, but mindless.

Richard Lewontin (1990) has pointed out that there are complex and orderly phenomena in nature that do not demand explanation in terms of natural selection. The turbulent flow of a waterfall is mathematically complex, but it is not the result of a selection process. The lattice structure of a crystal is highly ordered, but this is not the result of natural selection.

Dawkins might reply that waterfalls and crystals have not evolved; they are not the result of descent with modification. In addition, the complexity of waterfalls and the orderliness of crystals confer no advantage on the waterfalls or the crystals themselves. Dawkins's design argument could be formulated as the thesis that when evolution leads a trait to be found in all the organisms in a population, and that trait is complex, orderly, and benefits the organisms possessing it, the only plausible explanation of the trait's ubiquity is natural selection.

This argument leaves open a serious issue that Lewontin's response suggests. Is it possible to be more precise about the concepts of 'complexity' and 'order' so that the special features of traits that require selective explanation are made clear? I do not have an answer to this question, but in the present context I think we may set it to one side. In my opinion, we should grant that natural selection provides a plausible explanation of the vertebrate eye, and that no alternative explanation is now available. Adaptationism does not have to claim that none will ever be conceivable. Even though waterfalls and crystals attained their complexity and their orderliness by non-selective means, it is entirely unclear how non-selective processes could explain the structure of the vertebrate eye.

Dawkins takes this point to establish the correctness of adaptationism. However, the arch 'anti-adaptationists' Gould and Lewontin (1979), in their influential paper 'The Spandrels of San Marco and the Panglossian Paradigm: A Critique of the Adaptationist Programme', assert that 'Darwin regarded selection as the most important of evolutionary mechanisms (as do we).' What, then, is all the shouting about, if both sides agree that natural selection is important, indeed indispensable, as an explanatory principle?

Dawkins's argument provides a good reason to think that natural selection is an important part of the explanation of why the vertebrate eye evolved. However, this does not tell us whether the traits exhibited by the eye are optimal. Perhaps when we anatomize the organ into traits, we will discover that some of its features are optimal whereas others are not. As noted before, selection can be part of the explanation of a trait's evolution without that trait's being the best of the phenotypes available. The issue of adaptationism concerns not just the *pervasiveness* of natural selection, but its *power*.

Saying number 2: 'Adaptationism is incompatible with the existence of traits that initially evolve for one adaptive reason but then evolve to take over a new adaptive function.'

One of the main points of the spandrels paper is that it is important not to confuse the current utility of a trait with the reasons that the trait evolved in the first place. Natural history is filled with examples of *opportunistic switching*; traits that evolve because they perform one function are often appropriated to perform another. Sea turtles use their forelimbs to dig nests in the sand, but these forelimbs evolved long before turtles came out of the sea to build nests (Lewontin 1978). Insect wings evidently began to evolve because they facilitated thermal regulation, and only later helped organisms to fly (Kingsolver and Koehl 1985; for further discussion, see Reeve and Sherman 1993).

If adaptationism embodied a commitment to the view that there is little or no opportunistic switching in nature, the pervasiveness of this pattern would undermine adaptationism. However, most self-proclaimed adaptationists have no trouble with this idea. To be sure, some adaptationists have made the mistake of assuming that the current utility of a trait is the reason that the trait initially evolved. But this appears to be a mistake on the part of adaptationists, not a thesis that is intrinsic to the idea of adaptationism. It is useful to separate the proposition of adaptationism from the people who happen to espouse it (Sober 1993).

The idea of opportunistic switching places natural selection in the driver's seat. Selection governs the initial evolution of the trait, and selection governs its subsequent retention and possible modification. If adaptationism is a thesis about the power of natural selection, the existence of opportunistic switching is not central to the dispute.

Saying number 3: 'Adaptationism is incompatible with the existence and importance of constraints that limit the power of natural selection.'

The word 'constraint' has been used in many different ways; biologists talk about mechanical constraints, developmental constraints, phylogenetic constraints, genetic constraints, etc., etc. Underlying this diversity, however, is the idea that constraints limit the ability of natural selection to produce certain outcomes. To the degree that adaptationism emphasizes the power of natural selection, it apparently must minimize the importance of constraints (Reeve and Sherman 1993). However, as we will now see, this is correct for some so-called constraints, but not for others.

I mentioned earlier that the manner in which genotypes code phenotypes can prevent the fittest phenotype from evolving. If this pattern of coding is fixed during the duration of the selection process and does not itself evolve, then it is properly called a constraint on natural selection. Adaptationism as a research programme is committed to the unimportance of such constraints. The supposition is that a simplifying assumption about heredity—that like phenotype produces like phenotype—is usually close enough to the truth; the details of the underlying genetics would not materially alter one's predictions about which phenotypes will evolve.

I now want to consider two examples of a constraint of a different sort. Maynard Smith (1978) points out, in his discussion of running speed, that an animal's running speed increases as its leg bones get longer, but that lengthening the leg bone makes it more vulnerable to breaking. This means that running speed does not evolve in isolation from the effect that leg structure has on vulnerability to injury. The optimality modeller responds to this consideration by thinking about which bone shape is best, given the competing requirements of speed and strength. The existence of constraints does not refute the optimality approach, but gives it shape.

The second example I want to consider is the work on 'antagonistic pleiotropy' of Rose and Charlesworth (1981). They found that female drosophila have high fecundity early in life and low fecundity late, or have low fecundity early and high fecundity late. Females do not have high fecundity both early and late. For the sake of an example, imagine that this finding is due to the fact that all females have the same number of eggs. They vary in how they apportion these eggs to different stages of the life cycle. The fixed number of eggs thus serves as a constraint on the distribution of reproductive effort. Once again, the biologist need not take this result to show that an optimality model is inappropriate. Rather, the adaptationist will want to take account of the constraint: given that all females have the same number of eggs, what is the optimal distribution of eggs to different phases of the life cycle? If two distribution patterns are represented in the population, the optimality modeller will want to explore the possibility that this is a polymorphism created by natural selection.

The example described by Rose and Charlesworth might be termed a 'developmental constraint'. The reason is that if a fruit-fly lays lots of eggs early in life, this has consequences for what she will be able to do later. The example from Maynard Smith is less happily subsumed under this label, since leg length and leg strength are established simultaneously, not sequentially. Perhaps it should be called a 'mechanical' constraint instead.

Notice that in both these examples, a naïve analysis of the problem might suggest that there are four possible combinations of traits, whereas the reality of the situation is that there are just two. For example, we might naïvely suppose that zebras can have long leg bones or short ones, and that, as a quite separate matter, they can have strong leg bones or weak ones (see Fig. 5.1). The entries in this 2-by-2 table represent the fitnesses of the four combinations of traits; w is the highest value and z the lowest. If selection operated on all four of these variants, the optimal outcome would be the evolution of legs that are long and strong. However, given the fact that long legs tend to be weak, there are just two variants, whose fitnesses are x and y. What will evolve is either long and weak or short and strong, depending on which trade-off is better.

In this type of example, talk of constraints is really a way to describe the variation that natural selection has to act upon (Reeve and Sherman 1993). The question is not whether the fittest of the available phenotypes will evolve, but what the available phenotypes in fact are. If adaptationism is limited to a claim about the power of natural selection to ensure that the fittest of the available phenotypes will evolve, then the existence of

ELLIOTT SOBER

Leg strength

		Strong	Weak
Leg length	Long	*w*	*x*
	Short	*y*	*z*

FIG. 5.1. A 2-by-2 table of fitness.

constraints of this type is irrelevant. On the other hand, it must be admitted that some self-described adaptationists often hold that the range of variation available for selection to act upon is quite rich; for example, see Dawkins 1982: 32. This thesis about variation sometimes surfaces in debates about adaptationism in a manner that may be illustrated by an example suggested to me by Paul Bloom. Consider two hypotheses about how the human language faculty evolved:

(A) An ancestral human population contained a vast number of language structures; natural selection eliminated all but one of these. Thus, the present language faculty is the fittest of the alternatives that were available.

(B) Due to constraints on the physical form of human beings and their ancestors, there were just two phenotypes represented in the ancestral population: no language faculty at all and the language faculty that human beings now possess. The latter was fitter than the former in the evolution of our species, and natural selection ensured that this fitter phenotype was the one that evolved.

Under both hypotheses, natural selection caused the fittest available phenotype to evolve. However, natural selection seems to be 'doing more work' in (A) than in (B). Adaptationists such as Pinker and Bloom (1990) tend to favour hypotheses that resemble (A), whereas anti-adaptationists such as Chomsky (1988) advance claims that resemble (B).

Does the difference between (A) and (B) represent a disagreement about the 'power' of natural selection? Consider the following type of question: Why does this population now have phenotype Pa rather than phenotype Pc? Here Pa is the population's *actual* present phenotype and Pc is a *conceivable* phenotype that the population now does not possess. Selection will be the answer to more of these questions if (A) is true than it will if (B) is true. And constraints on variation will be the answer to more of these questions if (B) is true than it will if (A) is true (on the assumption that there are finitely many conceivable variants). However, neither of these judgements allows one to compare the power that selection and constraints *actually* exercised. I see no way to answer the following question: If (A) is true, which was the more important cause of the phenotype that evolved—selection or constraints? And the same holds for the parallel question about (B).

Consider the following two-stage process:

Conceivable variation   Actual variation   Variant that evolves

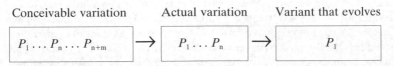

$$P_1 \ldots P_n \ldots P_{n+m} \rightarrow P_1 \ldots P_n \rightarrow P_1$$

Selection is the process that is responsible for what happens in the second stage of this process. Constraints on variation, on the other hand, determine which of the conceivable variants actually are represented in the ancestral population. Presumably m is a large number; there are many variants that one can conceive of that are not actually represented in ancestral populations. If so, selection effects a reduction from n variants to a single trait, whereas constraints explain why only n of the $n + m$ conceivable variants are actually represented. However, it would be a mistake to compare the 'power' of selection and of constraints by comparing the magnitudes of these two reductions. It is impossible to be very precise about how large m is; and a little imagination will make m so big that constraints always turn out to be more 'important' than selection. This is a hollow victory for anti-adaptationism, since it turns on no empirical fact. See Wright *et al.* 1992: 147–51 for further discussion of 'limits and selections'.

In the 'spandrels' paper, Gould and Lewontin (1979) emphasize the importance of the concept of evolutionary spin-off; a trait can evolve because it is correlated with another trait that is selected, rather than being directly selected itself.[2] The chin is apparently such a trait, and the architectural selection idea of a spandrel was used as a metaphor for this general category. Chins do not evolve independently of jaw structure; it is a misconception to think that chins evolved because they conferred some adaptive advantage of their own. However, if jaw structure evolved under the guidance of natural selection, and chins evolved as spin-off from selection on jaw structure, then it may still be true that natural selection has caused

---

[2] In Sober 1984 I discuss the difference between *selection-of* and *selection-for* in this connection.

the best available phenotype to evolve. The overarching category of *correlation of characters* subsumes mechanical constraints, developmental constraints, and evolutionary spin-off.

Let us now consider the idea of 'phylogenetic constraint'. When selection causes a trait to evolve, the trait evolves against a background of other traits that are already present in the population. Gould's (1980) example of the panda's 'thumb' illustrates this point; for ancestral pandas to evolve devices for stripping bamboo, these devices had to be modifications of traits that were already present. The spur of bone in the panda's wrist was a variant that was able to arise against this ancestral background biology; the panda was not going to evolve from scratch an efficient implement for stripping bamboo. Similar remarks apply to the skeletal structure that allows human beings to have upright gait. Phylogeny 'constrains' subsequent evolution in the sense that it provides the background of traits, whose modifications constitute the novelties that natural selection gets to act upon (Reeve and Sherman 1993). I hope it is clear that the recognition of phylogenetic constraints is not at all inconsistent with the claim that the optimal available phenotype evolves. Naïve adaptationists may forget about the importance of background biology; however, sophisticated adaptationists are still adaptationists.

In summary, if adaptationism asserts that natural selection ensures that the fittest available phenotype evolves, its relation to the concept of 'constraint' is less than straightforward. The view does deny that genetic constraints are important and pervasive, but it does not deny the existence and importance of mechanical, developmental, or phylogenetic constraints.

Saying number 4: 'Adaptationism is untestable; it involves the uncritical formulation of Just So Stories.'

It is possible to formulate an adaptationist thesis about all phenotypic traits, about most of them, or about some particular phenotype found in a particular population. Let us start with the last of these.

The trait I want do consider is sex ratio—the mix of males and females found in a population. R. A. Fisher (1930) analysed sex ratio by formulating a quantitative optimality problem: what mix of sons and daughters should a parent produce, if the goal is to maximize the number of grandchildren? Fisher showed that with certain assumptions about the population, the sex ratio strategy that will evolve is one in which parents invest equally in sons and daughters.[3] Given that human males have a slightly higher mortality rate than females, Fisher's model predicts that slightly

---

[3] For a simple exposition of this idea, see Sober 1993: 17.

more males than females will be conceived, that slightly more males than females will be born, and that the sex ratio among children will become even at the age when their parents stop taking care of them.

This adaptationist model is an instructive example with which to evaluate the charge that adaptationism is untestable. Fisher's explanation of sex ratio in human beings is testable. The obvious thing to check is whether its quantitative predictions about sex ratio are correct. In addition, Fisher's model rests on certain assumptions (e.g. that there is random mating), which can also be tested.

A further property of sex ratio theory is worth noting. Hamilton (1967) discovered that Fisher's argument is a special case of a more general pattern. If there is random mating, equal investment is the strategy that will evolve. But if there is inbreeding, a female-biased sex ratio will evolve. We can apply this body of theory to numerous species that exhibit different sex ratios, in each case checking whether the patterns of parental investment, mating system, and sex ratio are as the theory predicts. From the point of view of testing an optimality model, the sex ratio found in a single species is, so to speak, a single data point. To properly test a theory, several data points are needed. It is for this reason that a comparative perspective on testing adaptationist hypotheses is extremely important.

One often hears it said that adaptationist explanations are too 'easy' to invent. If one fails, it is easy to invent another. This is sometimes true, but it is not always so. What other explanation can we construct for the slightly male-biased sex ratio in human beings at conception that slowly changes to an even sex ratio later on? And how easy is it to invent a new and unified explanation of the pattern of variation in sex ratio that is found across different species? I'm not saying that no alternative explanation could exist, just that it is not so easy to invent one. The truth of the matter is that *some* adaptationist explanations are *difficult* to test. It is a double exaggeration to say that *all* adaptationist explanations are *impossible* to test.

The charge of untestability is often formulated by saying that if one adaptationist hypothesis turns out to be wrong, another can be invented to take its place. This comment does not assert that specific adaptive explanations are untestable; in fact, the complaint suggests that specific models can turn out to be wrong, which is why the need for new models arises. Rather, the criticism is levelled, not at a specific adaptationist explanation, but at an adaptationist claim that is more abstract. The claim that *there exists* an adaptive explanation of a specific trait is hard to prove wrong; such *existence claims* are harder to refute than specific concrete proposals.

It is important to recognize that the difficulty posed by existence claims is not limited to adaptationism. For example, consider the ongoing debate about whether the human language faculty is an adaptation to facilitate communication.[4] An alternative proposal that has been discussed is that the abilities that permit language use evolved for a quite different reason, and only subsequently were co-opted to facilitate communication. This is an existence claim; it says that a spin-off explanation exists, but does not provide the details of what the explanation is supposed to be. This type of conjecture is just as hard to test as existence claims that say that a trait was directly selected for some reason we-know-not-what.

Popper (1959) advanced a falsifiability criterion for distinguishing scientific propositions from unscientific ones. This criterion entails that existence claims of the kinds just described are not just *difficult* to refute, but *impossible* to refute, and therefore are not scientific statements at all. Shall we therefore conclude that adaptationism and anti-adaptationism are both unscientific—a plague on both their houses? Not at all—existence claims are testable, though they are not falsifiable in Popper's overly restrictive sense. If an adaptationist model about a specific trait is confirmed by data, then the anti-adaptationist existence claim about that trait is disconfirmed. Symmetrically, if an anti-adaptationist model about a specific trait is confirmed, then the adaptationist existence claim about that trait is disconfirmed. This is the pathway by which the existence claims advanced both by adaptationism and by anti-adaptationism as well can be tested. They do not inhabit a no man's land beyond scientific scrutiny (Reeve and Sherman 1993).

Adaptationist Just So Stories are sometimes easy to make up. The same is true of anti-adaptationist just-so stories. Adaptationism as a general thesis about all or most phenotypic traits is difficult to test. The same is true of pluralism, which views selection as one of several important causes of trait evolution. Specific adaptationist proposals are sometimes weakly supported by flimsy evidence, but the same can be said of some specific anti-adaptationist proposals. If adaptationism is a thesis about what has happened in nature, one cannot reject that thesis because biologists have not always tested the thesis with perfect rigour.

Saying number 5: 'Populations of organisms are always finite, always experience mutation, and frequently experience migration and assortative mating. Optimality models fail to represent these non-selective factors and therefore are false.'

It is true that optimality models ignore non-selective factors that frequently or always play a role in influencing trait evolution. However, the debate

---

[4] Pinker and Bloom (1990) and the accompanying commentaries on their target article provide an indication of current division of opinion on this issue.

about adaptationism does not concern the *existence* of such factors, but their *importance*. An optimality model predicts that a trait will evolve to a certain frequency. A perfectly realistic model, which accurately describes both selective and non-selective forces, also makes a prediction about what will evolve. Adaptationism asserts that these predictions will be the same or nearly the same.

Because adaptationism is a relatively monistic position, an adaptationist model will always fit the data less well than a pluralistic model. This is because an optimality model can be regarded as nested within a pluralistic model. Roughly speaking, they are related in the way the following two equations are related:

$$H_1: y = ax$$
$$H_2: y = bx + cw + dz$$

In these hypotheses, y is the dependent variable, x, w, and z are independent variables, and a, b, c, and d are adjustable parameters whose values must be estimated from the data. Because H_1 is nested within H_2, H_2 will always fit the available data better than H_1.[5]

Hypothesis choice in science is not guided exclusively by a concern for fitting the data. Scientists do not always prefer the more complex H_2 over the simpler H_1. Simplicity also plays a role in model selection, although the rationale for the weight given to simplicity is not completely understood.[6] Typically, scientists will see how well the simpler model H_1 fits the data; only if goodness-of-fit improves *significantly* by moving to H_2 will H_1 be rejected. A pluralistic model will always fit the data better than a relatively monistic model that is nested within it, but how much of an improvement pluralism provides depends on the data.

Saying number 6: 'Adaptationist thinking is an indispensable research tool. The only way to find out whether an organism is imperfectly adapted is to describe what it would be like if it were perfectly adapted.'

I think this last saying is exactly right. Optimality models are important even if they turn out to be false (Reeve and Sherman 1993, Sober 1993, Orzack and Sober 1994). To find out whether natural selection has controlled the evolution of a particular phenotypic trait, one must discover

---

[5] The two models will fit the data equally well in a case of zero dimensionality—when the best estimate of values for the parameters c and d is that $c = d = 0$. Note also that H_2 is a pluralistic model in which the independent variables combine additively. This is not the mathematical form that pluralistic models of evolution must inevitably take.

[6] Forster and Sober (1994) argue that H. Akaike's (1973) approach to the problem of model selection helps explain why simplicity matters in scientific inference.

Lewontin, R. (1980), 'Adaptation', in *The Encyclopedia Einaudi* (Milan: Einaudi); repr. in E. Sober (ed.), *Conceptual Issues in Evolutionary Biology* (Cambridge, Mass.: MIT Press, 1st edn. 1984), 235–51.

—— (1990), 'How Much Did the Brain Have to Change for Speech?', *Behavior and Brain Sciences*, 13: 740–1.

Maynard Smith, J. (1978), 'Optimization Theory in Evolution', *Annual Review of Ecology and Systematics*, 9: 31–56; repr. in E. Sober (ed.), *Conceptual Issues in Evolutionary Biology* (Cambridge, Mass.: MIT Press, 2nd edn. 1994), 91–118.

Orzack, S., and Sober, E. (1994), 'Optimality Models and the Test of Adaptationism', *American Naturalist*, 143: 361–80.

Pinker, S., and Bloom, P. (1990), 'Natural Language and Natural Selection', *Behavior and Brain Sciences*, 13: 707–84.

Popper, K. (1959), *The Logic of Scientific Discovery* (London: Hutchinson).

Reeve, H., and Sherman, P. (1993), 'Adaptation and the Goals of Evolutionary Research', *Quarterly Review of Biology*, 68: 1–32.

Rose, M., and Charlesworth, B. (1981), 'Genetics of Life History in *Drosophila melanogaster*', *Genetics*, 97: 173–96.

Sober, E. (1984), *The Nature of Selection* (Cambridge, Mass.: MIT Press; 2nd edn., Chicago: University of Chicago Press, 1994).

—— (1993), *Philosophy of Biology* (Boulder, Colo.: Westview Press; Oxford: Oxford University Press).

—— and Wilson, D. (1997), *Unto Others—The Evolution of Altruism* (Cambridge Mass.: Harvard University Press).

Williams, G. C. (1966), *Adaptation and Natural Selection* (Princeton: Princeton University Press).

Wright, E., Levine, A., and Sober, E. (1992), *Reconstructing Marxism: Essays on Explanation and the Theory of History* (London: Verso).

PART II

DEVELOPMENT

INTRODUCTION TO PART II

DAVID L. HULL

One of the great advantages of Mendelian genetics in the early decades of this century was that it bypassed the uncharted swamp of development. All Mendelian geneticists had to do was to work out the gene combinations that would explain the character distributions that their series of crosses revealed. They did not have to discover the actual developmental pathways that connected these postulated genes to their phenotypic effects. They assumed that such pathways actually exist, but they did not have to work them out in order to fulfil the goals of Mendelian genetics.

A parallel story can be told for the modern synthesis in evolutionary biology that took place between 1936 and 1947 (Mayr and Provine 1980). Biologists from all sorts of fields contributed to the modern synthesis with one major exception—development. Although *Entwicklungsmechanik* formed a major research programme from the 1880s to the present, it produced very little that evolutionary biologists could use in their work. If anything, it provided a whole series of hurdles to be overcome. The putative correlation between ontogeny and phylogeny is a case in point. Experimental embryologists amassed a huge backlog of observations on a variety of species, but did not come up with much in the way of developmental regularities that evolutionary biologists could incorporate into their synthesis (Hamburger 1980).

Early on in his career one of T. H. Morgan's professors tried to dissuade him from pursuing what came to be known as genetics because all the great discoveries in the next few years were going to be made in embryology. Generation after generation, biologists have thought that at long last developmental embryology was about to reach a stage in its development that would permit a major theoretical synthesis. Little by little, links between embryology and genetics have been forged to produce developmental genetics. However, a synthesis with evolutionary biology has been much slower in making itself felt. Present-day embryologists once again are predicting a major new synthesis. This time they may be right.

Embryology has entered into two recent disputes in evolutionary biology—adaptationism and punctuated equilibria. As the articles in the preceding section indicate, adaptationism is as central to evolutionary biology as it is problematic. If evolutionary biology cannot explain the myriad apparent adaptations that we see all around us, it will lose one of its claims to fame. But hypotheses about particular adaptations are extremely difficult to test, and one adaptive scenario can too easily replace another (Gould and Lewontin 1979). In addition, other forces in the evolutionary process should not be ignored. Chief among these other forces are developmental constraints.

When Eldredge and Gould (1972) first argued that the most typical pattern of evolutionary development is stasis punctuated by short bursts of rapid change, they had to explain stasis. What causes species to remain unchanged during long periods of time? Once again, one answer is developmental constraints. Unfortunately, no one has been able to say very much about how these developmental constraints actually work. In Chapter 6 Ron Amundson investigates the notion of developmental constraint to facilitate the integration of developmental and evolutionary biology.

Advocates of developmental systems theory are even more ambitious. They are not content to integrate current views on development into a larger evolutionary framework. Instead, they propose to replace current versions of evolutionary theory, including the replicator–interactor approach discussed in Part III of this volume, with a more general developmental view. One problem with more traditional versions of evolutionary theory is their dependence on the inadequately explicated notion of 'information'. One hint that traditional views of the evolutionary process may need fundamental reformulation is the continued difficulty of saying anything both succinct and accurate about the interplay between nature and nurture (see Oyama in Part VI).

Building on the work of Oyama (1985), Griffiths and Gray (Ch. 7) sketch their own developmental process approach. As they see it, evolution is best construed as the differential replication of total developmental processes or life cycles. On their view, genes have no special status. They are just one sort of developmental resource out of many. Phenotypic traits are another sort of developmental resource, and, like Dawkins (1982), Griffiths and Gray extend the 'phenotype' to cover all sorts of nonstandard phenomena, such as nests and spider webs. They also extend the notion of developmental resource to cover everything from a system that depends on the existence of past generations of this system to those that do not. For example, the replication of bush fires depends in part on the success of the plants for which they are the resource. But they argue that even those features that only persist (such as sunlight and gravity) also count as developmental resources.

One consequence of the developmental process approach to evolution is that it breaks down the sharp distinction between biological and cultural evolution. As usually described, biological evolution is gene-based, while cultural evolution is not. We can transmit knowledge to our children, but we can also transmit this same knowledge to other people in our culture— and genes play no determinant role in these processes. Perhaps people must have certain genes in order to be able to learn plane geometry, but the basic axioms of geometry are in no significant sense 'programmed' into our genes. They are transmitted culturally. However, on the developmental process view, genes are just one developmental resource out of many. Genetic transmission has no privileged role. Some may find this consequence of the developmental processes view a positive feature; some may find it decidedly negative.

Finally, Creationists have used the disputes among the scientists mentioned above to show that their objections to evolutionary theory are supported by genuine scientists. If embryologists from Darwin until well into this century can claim that evolutionary theory is false, if Mendelian geneticists at the turn of the century concurred in this judgement, if Gould (1980), Gould and Lewontin (1979), Webster and Goodwin (1996), and present-day developmentalists have such harsh things to say about Darwinian versions of the evolutionary process, then Creationists have been right all along in rejecting Darwinism (see Part X of this volume).

REFERENCES

Dawkins, R. (1982), *The Extended Phenotype* (San Francisco: Freeman, Cooper and Co.).
Eldredge, N., and Gould, S. J. (1972), 'Punctuated Equilibria: An Alternative to Phyletic Gradualism', in T. J. M. Schopf (ed.), *Models in Paleobiology* (San Francisco: Freeman, Cooper and Co.), 82–115.
Gould, S. J. (1980), 'Is a New and General Theory of Evolution Emerging?', *Paleobiology*, 6: 119–30.
——and Lewontin, R. C. (1979), 'The Spandrels of San Marco and the Panglossian Paradigm: A Critique of the Adaptationist Programme', *Proceedings of the Royal Society of London*, B205: 581–98.
Hamburger, V. (1980), 'Embryology and the Modern Synthesis in Evolutionary Theory', in Mayr and Provine (1980), 97–112.
Mayr, E., and Provine, W. B. (1980) (eds.), *The Evolutionary Synthesis: Perspectives on the Unification of Biology* (Cambridge, Mass.: Harvard University Press).

Oyama, S. (1985), *The Ontogeny of Information* (Cambridge: Cambridge University Press).

Webster, G., and Goodwin, B. (1996), *Form and Transformation: Generative and Relational Principles in Biology* (Cambridge: Cambridge University Press).

6

TWO CONCEPTS OF CONSTRAINT: ADAPTATIONISM AND THE CHALLENGE FROM DEVELOPMENTAL BIOLOGY

RON AMUNDSON

I. INTRODUCTION

Controversy has surrounded the so-called adaptationism of mainstream neo-Darwinian evolutionary theory during the past two decades. It has been argued that mainstream adaptationists systematically exaggerate the prevalence of adaptations in biology and are insensitive to possible non-adaptational explanations of biological phenomena. One alleged flaw in adaptationism is the failure to adequately recognize *developmental constraints*. This paper addresses the nature of the debate between adaptationists and advocates of constraint.

Most philosophers have learned of the adaptationism disputes from Gould and Lewontin (1979). While this article has attracted much discussion, for various reasons it does not focus philosophical attention on the issue of developmental constraints. It proposes a variety of grounds for distrusting adaptationism, including general methodological flaws. Developmental constraints are among the topics, but are not dealt with in particular depth. Philosophers are familiar with the methodological topics (e.g. falsifiability), and many are familiar with the topics from mainstream population genetics (e.g. genetic drift and pleiotropy) cited by Gould and Lewontin. Their article has been interpreted to claim that adaptationism is unfalsifiable. Various responses by both pro- and anti-adaptationists point out that unfalsifiability is an inappropriate criticism of a research programme, and that individual adaptationist hypotheses are indeed frequently falsified. (Actually Gould and Lewontin have accused neither adaptationism *qua* research programme nor individual adaptationist hypotheses of unfalsifiability. They claim, rather, that when such hypotheses

First published in *Philosophy of Science*, 61 (1994): 556–78. Reprinted by permission.

are falsified, other adaptationist hypotheses take their place. What seems never to be falsified is the belief that the trait is an adaptation of some kind.)

The philosophical discussions of falsification and drift were of unquestionable value, but left the core of the developmental constraint/adaptation conflict virtually unnoticed. Many people question why this conflict exists since, on many descriptions, the processes of natural selection and the processes of embryological development are perfectly compatible, indeed complementary.

From one perspective, the three alternatives of developmental constraint, adaptation, and drift form an ordered sequence. Developmental constraint tends to restrain selective adaptation, and adaptation tends to restrain drift. To believe that (almost) all biological traits are adaptations is to believe that natural selection is powerful enough both to *overcome* constraint and to *resist* random drift. In this sense, natural selection is a conservative force, a constraint, with respect to drift. Antonovics and van Tienderen (1991) are perplexed by talk of 'selective constraints' on drift, but the concept is natural, given the dynamics of the situation. We must only keep in mind that selection is not a *developmental* constraint.

From another perspective, these three alternatives are not so smoothly ordered. Mathematical population genetics is at the core of the modern synthesis. Genetic drift is perfectly possible, given the formulas of population genetics. Experiments and field studies are required to determine the relative importance in the natural world of selected and drifted traits. So the modern synthesis has not been uniformly adaptationist from its birth. Drift is a theoretical option, and its advocates have worked within the synthesis (Gould 1983, Beatty 1986, Burian 1986). Embryology and developmental biology are a different story. Embryology has never been an integrated part of the modern synthesis. This explains its unfamiliarity to most philosophers of evolutionary biology, and why an advocate of developmental constraint would see philosophers' emphasis on the population-genetic alternatives to adaptationism as a symptom of the problem. The neatly ordered series 'developmental constraint, adaptation, drift' includes two phenomena (adaptation and drift) that share a common scientific vocabulary, history, and mathematical formalism. The third phenomenon belongs to a field of study that has been isolated from the others during the entire history of the synthesis. V. Hamburger (1980), an embryologist whose career began in the 1930s, described the synthesis as having treated the processes of ontological development as a 'black box', the contents of which can be safely ignored. B. Wallace (1986), a major synthesis biologist and student of Dobzhansky, recently asked, 'Can embryologists contribute to an understanding of evolutionary mechanisms?' (p. 149). His answer was, not much.

In one of the few philosophical papers dealing with the tension between neo-Darwinism and embryology, K. Smith (1992) has discussed two degrees of developmental criticisms of neo-Darwinism. The radical 'process structuralists' believe that little of the modern synthesis is worth saving. The moderate 'general structuralists' believe that a new developmental synthesis is needed to integrate embryology and development on the one hand with the neo-Darwinian modern synthesis on the other. To its advocates, this new synthesis would be as far-reaching as the synthesis of Mendelian genetics and Darwinian natural history which originally formed the modern synthesis (Horder 1989, Gilbert 1991: ch. 23). The present essay will not be fine-grained enough to discriminate among developmental critics according to how harshly they view the modern synthesis, but will concentrate on the contrasts between 'general structuralist' approaches and the synthesis.

Developmental constraints are one of the principle topics on which developmental biologists have criticized the adaptationism of neo-Darwinism. An influential and accessible introduction to developmental constraints is Maynard Smith *et al.* 1985. R. Burian's contributions to that article are an exception to philosophers' lack of interest in development. The paper is a multiple-authored co-operative catalogue of various kinds of constraint (not all of them developmental), along with guidelines on how to classify them. It states the now-standard definition: 'A developmental constraint is a bias on the production of variant phenotypes or a limitation on phenotypic variability caused by the structure, character, composition, or dynamics of the developmental system' (ibid. 266). A problem with this paper, for philosophical purposes at least, is that it is too co-operative. The reader gets no feeling for the contentiousness of the issue. Why should the significance of these constraints be questioned by neo-Darwinian adaptationists? To understand, one must look to other sources which more openly express criticisms of neo-Darwinism from the developmental biologist's viewpoint. (Representative works are Goodwin *et al.* 1983, Bonner 1982, Thomson 1988, and Gould 1980. For more radical critiques, see Løvtrup 1987 and Goodwin 1984.)

The constraint/adaptation dispute is unlikely to find a quick resolution, due to a deep contrast in explanatory strategies between the adversaries. As a step toward explicating the complexity of issues which play a role, this essay will explore two distinct versions of the central concept in the dispute—the concept of *constraint*. The close attention I give to divergent 'meanings' of the term comes not from a hope of reducing the debate to a semantic one—this is not at all a semantic pseudo-problem. A better model for the present study can be seen in E. Mayr's (1980: 20ff., 1982: 742ff.) discussion of the various meanings attached to the term 'mutation' in the years preceding the modern synthesis. Mayr shows that the Mendelian geneticists and their Darwinian naturalist adversaries used the term with distinct meanings which now illustrate the deep theoretical differences. Understanding one's adversary's theoretical approach was impossible with this mismatch. Achieving a synthesis required overcoming this difference (as well as many others). As in the mutation case, the divergent meanings of 'constraint' fit neatly into divergent theoretical interests and commitments. The dispute is, at bottom, a clash of explanatory strategies, of approaches to explaining the nature of organic life. Charting the two meanings of constraint is not merely a semantic exercise, but an attempt to explicate the structure of the constraint/adaptation dispute. If a developmental synthesis actually occurs, future historians may comment on the divergent concepts of constraint just as Mayr has discussed the pre-synthesis differences on the term 'mutation'.

II. CONSTRAINTS AS ACTING ON ADAPTATION

The term 'constraint' implies some sort of restriction on variety or on change. In the adaptationism debates, what is being constrained? This question has two answers. The most common is that adaptation is being constrained. Developmental constraints, on this view, are restrictions placed by the mechanisms of embryology (for example) on the adaptive optimality of the adult organism. Natural selection simply cannot overcome the conservative forces of development, and sub-optimally adapted traits and organisms are the result. The view of constraints as restrictions on adaptation is expressed in Stephens and Krebs's (1986) discussion of optimality models in foraging theory. Optimality models have three elements (ibid. 5):

1. *Decision assumptions*: Which of the forager's problems (or choices) are to be analysed?
2. *Currency assumptions*: How are various choices to be evaluated?
3. *Constraint assumptions*: What limits the animal's feasible choices, and what limits the pay-off (currency) that may be obtained?

The 'currency' chosen for a model is some property presumed to contribute to fitness, but which can be directly measured. In foraging models the currency might be 'maximization of long-term average rate of energy intake'. The model builder constructs a set of external and internal constraints, and then makes an a priori calculation of the foraging behaviour which would optimize the currency given the constraints. External constraints are features of the environment which might limit the currency, such as the availability and distribution of the food resource. Internal constraints are features of the organism itself, such as the nature of its perceptual system. If the organism is discovered not to behave according to the optimal foraging model, there is a search for other unnoticed constraints which could account for the sub-optimality of the actual behaviour. It might be discovered, for example, that a bird which chooses a poor food source when a rich one is available is simply unable visually to discriminate the two food sources. When this new (internal) constraint is introduced, the behaviour may become optimal—that is, optimal *within the constraints* (ibid. 180).

From this line of thought many adaptationists conclude that the advocates of constraint have no argument. Other practitioners of neo-Darwinian adaptationism are less explicit than optimality theorists in how they specify constraints, but none of them believes that organisms can just evolve whatever they happen to need, at the drop of a hat. Thus, it is said, constraints are already openly recognized by adaptationists.

Using this conception, developmental constraints are simply one sort of internal constraint. Developmental constraints are *constraints on adaptation*. On this reading, the grounds for conflict between developmentalists and adaptationists is clear. Advocates of developmental constraints believe that adaptationists overlook some factors which limit adaptive optimality. Testing for optimality is more difficult when one is dealing with morphological traits than with behaviour patterns, of course. But in principle the resolution of the case is the same. First, prove that a morphological trait is less than optimally adaptive. Then trace the sub-optimality's source. If the source is unchangeable in the developmental system, we have discovered a developmental constraint. Moreover, we have shown that the trait is adaptively optimal (within that constraint). Far from refuting adaptationism, this example shows that adaptationist hypotheses are necessary *even for the discovery* of constraints. Stephens and Krebs reject Gould and Lewontin's criticism of adaptationism on precisely these grounds: 'Even if they serve no other purpose, well-formulated [adaptationist] design models are needed to identify constraints: without a design hypothesis there would be no basis for postulating any kind of constraint!' (ibid. 212).

III. CONSTRAINTS AS ACTING ON ORGANIC FORM

As plausible as the above interpretation of constraint may sound, constraint-on-adaptation does not accurately express the challenge to adaptationism which comes from developmental biology. Phylogenetically evolved adaptations *qua* adaptations are the primary *explananda* of natural selection, the central mechanism of neo-Darwinian theory. By contrast, developmental biology does not identify phylogenetic adaptations *or any derivative* of adaptations (e.g. constraints on them) as its primary *explananda*. Advocates of developmental constraint have a different notion than constraint-on-adaptation in mind. This can be seen in P. Alberch's (1982) proposed thought-experiment:

> [L]et us assume that the morphology of an organism can be described by two variables, x and y. If one plots all the observed forms, a distribution of the kind shown in Figure [6.1] is observed. That is, the observed forms are a subset of all possible forms. Furthermore, the observed forms are arranged in clusters, each cluster corresponding to a distinct species (e.g., Drosophila melanogaster) or to a class of pheno-deviants (e.g., D. melanogaster wingless mutant). How do we explain the empty spaces and the ordered pattern in morphology-space? There are basically two extreme explanations: (a) empty spaces represent nonadaptive forms that have been eliminated by natural selection; and (b) they are a reflection of the developmental constraints operating on the system, i.e., there are morphologies that cannot be produced by the developmental program. (p. 317)

To contrast the two hypotheses, Alberch proposes a hypothetical experiment in which all of the members in one of the clusters in the real world (Fig. 6.1) are allowed to reproduce for many generations while the forces of selection are reduced to a minimum (random mating imposed, no competitive interactions) and variation is increased as much as possible by mutagens. Lethal teratologies would also be logged to keep track of selection at embryonic stages. According to hypotheses (a) and (b) above, what patterns of descendent morphologies would one expect to find? Figure 6.2 represents the two possible outcomes: H_1 is (a), the hypothesis that selection explains all gaps—without selection, morphologies would no longer be clustered; morphospace would be smoothly filled; and H_2 represents (b), the developmental constraints hypothesis. Most of the clusters of H_2 are similar to those which had existed in Figure 6.1, one cluster of which (cluster A) had formed the ancestors of the organisms in Figure 6.2. The additional smaller clusters are (according to Alberch) those which would be unfit in the real world, and so presumably had been removed by selection from Figure 6.1. Most of the empty space in Figure 6.1 still remains in H_2. According to H_2, relatively little of the empty area outside the clusters in the morphospace of the real biological world was cleared (or blocked)

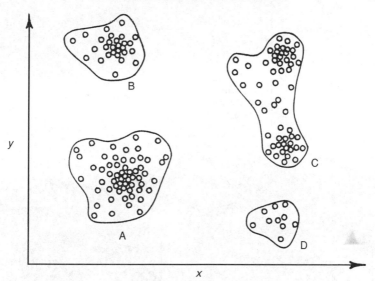

FIG. 6.1. The clustering of organisms in morphospace. From Alberch 1982: 316.
Copyright 1982 by Springer-Verlag. Reprinted by permission.

by natural selection. These hypotheses are acknowledged to be extremes,
but Alberch clearly leans toward H_2: 'In the second case the role of selection is basically stabilizing, being responsible for "pruning out" the non-
functional morphologies, and for determining the differential survival of
morphological types (states A, B, C, and D in Fig. [6.1]). However, the
realm of possible morphologies is basically determined by the internal
structure of the system' (ibid. 319). Morphospace is generally recognized
as clumpy at all levels of the genealogical hierarchy. Birds and mammals
cluster separately with open space between the clusters; so do felines and
canines, and plants and animals.

   This is a dramatic statement of the constraint advocate's position. However, let us note the following about the diagrams. Compare Alberch's
drawings with the adaptive landscapes introduced by S. Wright, among the
most familiar of evolutionary diagrams. Figure 6.3 is a sample. Two of the
dimensions of an adaptive landscape represent abstract genome space, so
to speak, just as Alberch's x and y represent abstract morphospace. The
third dimension is represented by the contours, which connect genome
points (or gene combinations) of equal adaptive values. The peaks are

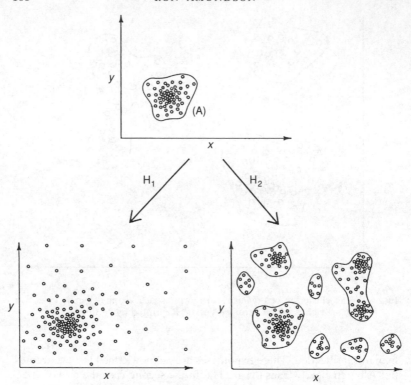

FIG. 6.2. Two hypotheses on the effects of removing natural selection from a population. From Alberch 1982: 318. Copyright 1982 by Springer-Verlag. Reprinted by permission.

areas of high fitness or adaptive value; the valleys, areas of low fitness. One thinks of a population being driven up an adaptive slope as natural selection increases the frequencies of alleles of high fitness.

At first glance, the clusters in figure 6.1 may have been interpreted as the familiar peaks of an adaptive landscape, and the empty areas as the valleys. This is not Alberch's intention. To 'see' adaptive peaks in the morphological clusters of Figure 6.1 is to *assume* H_1, the adaptationist explanation of the clustering of existing morphotypes. Alberch's drawing includes no dimension to represent the adaptiveness of morphotypes. It is purely a diagram of organic form. Indeed, H_2 specifically denies that adaptivity is

FIG. 6.3. An example of an adaptive landscape in the sense of S. Wright. The *x*-and *y*-axes would represent genetic space. Contour lines connect points of equal adaptive value; pluses and minuses are areas of high and low adaptivity.

responsible for the clumpiness of the morphospace. Another way of understanding this point is to think of the adaptive surfaces in Figure 6.2 as absolutely flat, having been flattened by Alberch's removal of selective forces. Hypothesis H_1 exhibits the pattern of variation which an adaptationist would expect to evolve on a flat adaptive landscape.

Unlike the main currents of neo-Darwinism, developmental biology does not focus its explanatory attention on adaptations *or on their absence*. Rather, developmental biology aims to explain *organic form* and its origins in the embryo. The *explanandum* is not adaptation, but form. Constraints thus proposed by developmental theorists are not constraints on adaptation, but *constraints on form*. Many other pictorial representations of the ranges of possible morphology can be found in the literature of developmental biology. The drawings contain no dimension representing the relative adaptivity of the 'permitted' and 'forbidden' forms. Examples are representations of the morphospace of coiled shells (Maynard Smith *et al.* 1985: 278, Schindel 1990), J. D. Murray's (1981) reaction–diffusion model of mammal coat colour patterning, and the drawings of permitted versus forbidden digit patterns frequently cited by students of the vertebrate limb (e.g. Alberch 1982, Holder 1983). Not natural selection, but rather the embryological mechanisms of growth are believed to permit or forbid these forms. Adaptive values are not the evidential basis from which the constraints are inferred.

To be sure, constraints on form (call them constraints$_r$) may result in constraints on adaptation (constraints$_A$). But this is not always or necessarily the case. The relation between constraints$_r$ and constraints$_A$ is not one of entailment, and to mistake the former for the latter is to measure developmental biology using an adaptationist's yardstick. If the developmentalists' contribution to the explanation of biological traits is limited to traits which are known (or asserted) to be non- or maladaptive, then the developmentalist has no business discussing traits believed to be adaptive. But in fact no developmentalist would abandon the field in this way. Developmentalists would claim that their contributions are a proper part of the full explanation of even the most wonderfully adapted trait.

IV. TWO APPROACHES TO PHYLLOTAXY

An example illustrates the distinction between constraints$_A$ and constraints$_r$. Phyllotaxy, a simple developmental system, is the pattern, usually spiralling, of the growth of leaves, bracts, or florets on plants. Examples are helical patterns of leaves on stems, seed covers on pine-cones, spiral patterns of seeds on sunflower heads, and florets on cauliflower and broccoli stems. An interesting feature of much phyllotaxy is that various particular patterns can be correlated with the Fibonacci number series. (The Fibonacci series is 1, 1, 2, 3, 5, 8, . . . , 55, 89, 144 . . . —each number the sum of the preceding two.) Particular phyllotactic patterns are associated with fractions in which the numerator and denominator are successive numbers in the Fibonacci series. The denominator indicates the number of leaves between successive exact overlaps as the leaves spiral along the stem. The numerator indicates the number of circuits around the stem before that overlap occurs. For pine-cones and sunflowers, it is easier to count the numbers of observable left-hand versus right-hand spirals; the fraction arrived at is equivalent to the circuit-counting method in stems and leaves. The Fibonacci number multiplied by 360 degrees gives the angular deflection from one leaf (bract, and such) to the next in a particular pattern. Plants often start early growth with a low-numbered Fibonacci pattern (e.g. 3/5) and transfer to higher numbers (34/55, 55/89) in later and larger stages. The absence of intermediate or divergent patterns, either within or between species, strongly hints at constraint.

Phyllotactic spirals have been discussed at least since Leonardo da Vinci. D. Thompson (1942: ch. 14) sketched that history. One notable attempt to give an adaptationist explanation of Fibonacci phyllotaxy was by C. Wright, the American mathematician and early supporter of Darwin. Wright pointed out that the Fibonacci series converges on an angle called the 'golden section', measuring approximately 137.5 degrees. This angle is an irrational portion of the full circle. If there were successive layerings of radial vectors about an axis distanced by the golden section, no vector would ever exactly overlap a lower vector. So if the phyllotactic angle of divergence were the golden section, no leaf would ever exactly shade any lower leaf from an overhead sun.

Thompson listed five reasons to doubt Wright's adaptationist explanation (ibid. 932). For example, the higher numbers in the Fibonacci series are close approximations to the golden section, but they are much rarer among plants than the low-numbered ratios which allow frequent overlaps. Furthermore, the golden section has no special adaptive significance—any angle of divergence which is irrational with respect to the full circle will do the job, and there are infinitely many such angles. So, Wright notwithstanding, Fibonacci phyllotaxy seems a good candidate for a developmental constraint.

K. Niklas (1988) studied the influence of Fibonacci phyllotaxy on adaptation. He first cited evidence that the phyllotactic pattern is developmentally conservative, including evidence that within an individual plant or a species it is insensitive to environmental variables, and that it varies among species in a discontinuous manner. He investigated the effects of the various patterns on the amount of sunlight striking the leaves. Using computer simulations, Niklas showed that the photosynthetic potentials of different patterns did indeed vary. This raises the question of why plants would develop according to a less-than-optimal phyllotactic pattern:

Model plants constructed with equal total leaf area and number differ significantly in flux, even when [phyllotactic patterns] are very similar.... Nonetheless, computer simulations indicate that a variety of morphological features can be varied, either individually or in concert, to compensate for the negative aspects of leaf crowding resulting from 'inefficient' phyllotactic patterns. Internodal distance and the deflection ('tilt') angle of leaves can be adjusted in simulations with different phyllotactic patterns to achieve equivalent light interception capacities. (Ibid. 12)

Evidence shows that these and at least some other possible non-phyllotactic traits (e.g. leaf opacity, spectral sensitivity) are more amenable to selection than is the phyllotactic pattern itself, and are not developmentally linked to it in such a way as to block their ability to compensate for the 'inefficiencies' of a specific pattern.

Niklas concludes that phyllotaxy, while a 'candidate' for a developmental constraint, is *not* a 'developmental constraint sensu stricto'. It is, rather, a 'limiting factor'. The difference is important: 'In the first case, the morphological domain is "constrained" by the internal structure of the developmental system. In the second case, the system provokes and defines changes in other facets of the organism's development. . . . The distinction between a "constraint" and a limiting factor is important, because it reflects a measure of plasticity within the developmental repertoire' (ibid. 9). But in the cases being discussed, the morphological domain is indeed constrained as to its possible phyllotactic patterns—only certain patterns are ever available, and they are without variation within a species. The plasticity which exists is not in the 'domain' of phyllotactic pattern *at all*— it occurs only in the compensating factors such as stem distance, leaf angle, and tilt. When Niklas denied that the 'morphological domain is "constrained" by the internal structure of the developmental system', the domain referred to could not be the positioning of leaves on a stem—*that* domain *is* so constrained. The intended domain must have included that pattern *together with* the set of traits which compensate for phyllotactically imposed limitations. And why is that group of traits bundled into one domain? Certainly not because they are developmentally integrated—by Niklas's hypothesis they cannot be. Either the entire morphology of the plant is the morphological domain, or the domain just happens to include phyllotaxy, stem distance, leaf shape, leaf angle, leaf tilt, surface opacity— that is, the 'limiting factor' together with its compensators. Such a domain is defined *post hoc* by what is needed in order to achieve adaptation. Niklas is measuring development with an adaptationist's yardstick, and in phyllotaxy he finds no constraint.

Niklas's notion of constraint is clearly constraint$_A$, constraint on adaptation. An unchangeable developmental pattern can count as a constraint$_A$ only if it irremediably reduces adaptation. Since 'limiting factors' are those which can be compensated for, they are not constraints$_A$: '[P]hyllotaxy may operate as a limiting factor, provoking compensatory adjustments in other morphological features, but, from the perspective of photobiology, it is not a developmental constraint *on performance*' (ibid. 14; emphasis added). On this concept, two equally canalized traits may differ on whether they count as constraints. A trait which can be compensated for is not a constraint$_A$ no matter how deeply it is entrenched in the developmental programme.

Another stance can be taken with respect to phyllotaxy. A developmentalist, with eyes on other phenomena than adaptation, takes an apparent constraint on organic form as *itself* a target for explanation—but of developmental rather than adaptational explanation. G. J. Mitchison (1977) is an example. Mitchison explains the Fibonacci series as a mathematical consequence of certain known or plausible features of stem growth and leaf placement. Positioning of a newly developing leaf is influenced by the positions of the leaves just below it; new leaves cannot originate too close to their predecessors or to the apex of growth. Mitchison develops a close-packing or 'touching-circle' model which assumes that leaf positioning is governed by something like an inhibitor mechanism:

This assumes that the leaves and apical tip of a plant produce an inhibitor which prevents new leaves from forming in their proximity. I shall assume that this inhibitor diffuses or is transported away from its sources, and that the new leaf is formed at the first site to appear beneath the growing apex where the inhibitor concentration falls below a fixed threshold. (Ibid. 273)

Mitchison shows that Fibonacci patterns will result from this mechanism, and from many other leaf-positioning mechanisms. He explains the Fibonacci number size of the pattern (i.e. 3/5 vs. 89/144) as a function of the rate of growth of the apex of the stem. On this model, the head of a sunflower which shows a dramatic 89/144 Fibonacci number at its perimeter would result from a 30-fold increase in the size of the apex (the growth zone for new bracts) during its growth. The rapid expansion of the sunflower's 'apex' can be seen from the fact that the growth zone is the circumference about the centre of the flower. Lower Fibonacci numbers correspond to plants whose stems increase in diameter only slightly during growth.

Let us consider Niklas's and Mitchison's results from both the adaptationist's and the developmentalist's point of view. To an adaptationist, Mitchison's conclusions are of little consequence. There is an obvious adaptive reason for leaves not sprouting too closely together, and some mechanism has evolved to keep them apart. Mitchison shows (only) how a broad range of possible leaf-spacing mechanisms would produce Fibonacci patterns. But Fibonacci patterns are not the *explananda* of adaptationist explanation, since they are (apparently) not adaptations. The patterns may have turned out to be obstacles to adaptation, constraints$_A$, but Niklas shows that such obstacles can be overcome. So Niklas's work has the important adaptationist effect of showing that even these non-adaptive, universal patterns need produce no reduction of overall adaptation. What may have been a constraint$_A$ (a constraint *sensu* Stephens and Krebs) turns out to be potentially innocuous (*vis-à-vis* adaptation).

The scene changes from a developmentalist perspective. Niklas produces no explanation of the forms of plants. He takes the existing phyllotactic patterns as a given. In contrast, Mitchison explains how a certain organic pattern comes to exist, given what we know about plant growth.

He points out the features of the growth of plants which generate the Fibonacci patterns in all their variety, and explains why non-Fibonacci patterns are rare. The adaptive relation of Fibonacci patterns to the photosynthetic potential of plants is irrelevant to Mitchison's enterprise. The biological functions of leaves or of seeds play no role in the analysis. Whether Fibonacci patterns contribute to adaptation, whether they are 'limiting factors' or even absolute barriers to optimality, is inconsequential to the correctness of Mitchison's developmentalist explanation. Mitchison intended to explain organic form, not adaptation.

The work of these two biologists is consistent. Mitchison addresses organic form, Niklas, the adaptive effects of that form. In present terminology, Mitchison gives a developmental explanation of a phyllotactic constraint$_F$, while Niklas shows that the same constraint$_F$ is not (necessarily) a constraint$_A$. Being primarily interested in adaptation, Niklas expresses his conclusion as the discovery that phyllotactic pattern is not a developmental constraint 'sensu stricto'. He is correct in that it is not a constraint 'sensu accommodationis' (in the sense of adaptation). But it is still a constraint$_F$ *sensu stricto*, a genuine constraint on organic form.

The distinction between constraints$_A$ and constraints$_F$ is only implicit in developmental biologists' discussions of constraint. Nevertheless, Niklas misreads his developmentalist sources. He cites Alberch (1982) in support of his conclusion that phyllotaxy is not a constraint. But neither in the Alberch nor in the Maynard Smith *et al.* (1985) definition quoted above are developmental constraints defined as *reductions of* adaptation. Adaptation is mentioned in neither of the definitions. It is Niklas's own adaptationist orientation which fills in the missing reference to adaptation.

So developmental constraints as seen by practitioners of developmental biology are defined by their effects on organic form rather than on adaptation. Such constraints$_F$ surely influence adaptation; the 'versus' in 'adaptation versus constraint' is not meaningless. But the effects of a constraint on adaptation are secondary consequences of its effects on form, at least from a developmentalist's perspective. The primary *explanandum* of developmental biology is the origin of form.

V. HOW FORM RELATES TO ADAPTATION

The assumption that constraint on form entails constraint on adaptation seems natural, but I will explore its grounds. Under what conditions do constraints$_F$ create constraints$_A$? A constraint on potential adaptation will only occur when the variant which is prohibited by a constraint$_F$ *would be* selectively favoured *if* that variation were to exist. That is, whether a constraint$_F$ gives rise to a constraint$_A$ depends on whether the environment would selectively favour forms *forbidden* by the constraint$_F$ over forms *permitted* by the constraint$_F$.

In the real world, which traits fit this description? Which traits are such that their prohibited variants would be selectively favoured, were they only allowed to exist? Naturalistic observation obviously will not answer this question—the variation required for observable differential fitness is absent by hypothesis. Some sort of hypothetical reasoning must be invoked. In some cases it would be simple; immunity to a juvenile lethal disease would presumably always be selectively favoured over the lack of such immunity. But for interesting morphological cases the assessment might be difficult. Consider the task of comparing the adaptivity of the single proximal bone in the tetrapod limb (humerus, femur) with the adaptivity of a probably prohibited double proximal bone. The prospects of a well-founded empirical assessment, for the entire tetrapod group or any subgroup, seem dim indeed.

So empirical proofs that specific constraints$_F$ yield specific constraints$_A$ may be elusive. None the less, general theoretical orientations have implications for the issue. What sort of theoretical commitment leads to an expectation that development constrains adaptation? Let us re-examine our trio of theoretical positions in the adaptationism dispute: constraint, adaptation, and drift. Adaptedness is a relation between organic form (or other phenotypic trait) and environment. Adaptedness is a relational, ecological concept. Neo-Darwinism explains states of adaptedness as resulting from natural selection. Natural selection is a two-stage process involving (1) the production of heritable variation; and (2) the winnowing of that variation by environmental demands, with these two stages repeating themselves in each generation. The debates between adaptationists and advocates of drift concern phenomena at the second stage of this process, the ecological level. The core of the dispute is whether each trait is (and has been) under selective pressure during its history. Adaptationists consider the world a demanding place, and believe that even small differences between traits have selective consequences. Advocates of drift consider the world far less demanding. They believe that many variations are effectively neutral in their selective value. (Neutralism and drift are not synonymous, as I will show.) In the absence of selection, the statistical principles of population genetics imply that traits will drift, with the likelihood of a random variant becoming fixed in the population depending on effective population size. The judgements of each side on the adaptive status of current traits (adaptive or selectively neutral) generally match the historical explanations offered for the existence of the traits (natural selection or products of drift).

Developmental constraints function not at the ecological second stage of the process of natural selection, but the first stage, the production of heritable variation. They bias that production. In the course of studying how organic form is (ontogenetically) produced, developmental biologists believe they have discovered embryological processes which can produce only a certain range of phenotypic variation. The issue is not the amount of variation which is possible, but the range of that variation. A comparison with generative linguistics is helpful here. Just as a hypothesized universal grammar of human languages can generate infinitely many different potential human languages, the generative processes of embryology have an unlimited number of possible variants. But, just as all languages generated by a universal grammar are governed by certain constraints, so are all of the possible outcomes of the embryological processes of a given phylum. The similarity here is not accidental—developmental theories are generative theories.

What would an advocate of constraint$_F$ say about the adaptive status of constrained$_F$ traits? Adaptation is a topic at the second level of natural selection, where the winnowing of the less well adapted forms occurs. The discovered facts concerning the embryological development of form imply nothing about the fitness relations between that form and its eventual environment. The existence of strong constraints$_F$ have no immediate implications for the existence of constraints$_A$. Information about the environment, not just form, is needed before judgements of adaptation are possible. An embryologist *qua* embryologist has no more to say about the adaptedness of particular organic forms than, say, a climatologist, a scientist on the opposite, environmental side of the relational field of ecology. To be sure, some embryologists (or climatologists) might be interested in how their subject-matter ties in with adaptation, but the major research programmes of developmental biology and climatology can be conducted in isolation from questions of adaptation. The existence (or not) of constraints$_F$ requires no assessment of the adaptedness of the resulting forms.

Advocates of constraints$_F$ may choose to take a position on the disputes about the second level of natural selection. Justifying any such position would require evidence from ecology, of course (adaptation being an ecological concept). Let us consider the options. An advocate of strong constraints$_F$ may be a neutralist regarding the ubiquity of selective forces. If so, he would expect that many constraints$_F$ would have no effect on adaptedness. After all, the world is a non-demanding and an open place to a neutralist. There would presumably be room for 'purposeless' conservatism of pattern, just as there is room for 'purposeless' variation and drift. For such a *constraint$_F$ neutralist*, constraints$_F$ may well not result in constraints$_A$.

It was noted above that neutralism was not synonymous with drift. This is why. Some selectively neutral traits might be present in a species because they drifted to fixation, others because they are the products of a developmental constraint. Neutralism is true of both sorts of traits, but drift explains only the first. The almost universal identification of neutralism with drift would seem to be another effect of the isolation of developmental biology from mainstream evolutionary theory.

A different advocate of constraints$_F$ might not be a neutralist at the ecological level of discussion. She might indeed believe that all organic traits have adaptive importance which will be strongly tested by the process of winnowing. But, since this person also believes that strong constraints$_F$ exist, she would believe that almost any constraint$_F$ would be likely to result in a constraint$_A$, a reduction of potential adaptation. We might label this person a *constraint$_F$ adaptationist*.

But this sounds paradoxical. Does this argument not pit adaptationists against the advocates of constraint? Well, yes and no. We must tease apart two aspects of the position called 'adaptationism'. I propose the following terminology:

Soft adaptationism: All organic traits have adaptive values on which the winnowing process of natural selection operates (or would operate if there were a variant). For this reason a constraint on form is (probably) a constraint on adaptation. Contradicts neutralism.

Neutralism: Many organic traits are adaptively neutral, so a constrained trait might well be adaptively neutral. Contradicts soft adaptationism.

It can be seen that soft adaptationism does not (in itself) deny the current existence of constraints$_F$; the 'constraint$_F$ adaptationist' is a soft adaptationist who believes in constraints$_F$. An adaptationist who denies the existence of developmental constraints does so by claiming that natural selection can (and has) overcome any such constraints. This position can be called:

Hard adaptationism: All organic traits have adaptive values, and those adaptive values, via the principle of natural selection, provide the proper historical explanation of the existence of those traits. Any developmental constraints can be (and have been) overcome by the forces of natural selection.

The assumption that a constraint$_F$ must result in a constraint$_A$ is soft adaptationism. Hard adaptationism adds to soft adaptationism the claim that selection can conquer any constraint$_A$. Since selection has conquered any constraints$_A$ and (almost) all constraints$_F$ are constraints$_A$, selection must have conquered (almost) all constraints$_F$. (Exceptions would be constrained$_F$ traits which are fortuitously adaptive.) Other soft adaptationists may believe in the continued existence of constraints$_F$, and believe that the existing constraints$_F$ impose an adaptive disadvantage to the organisms so constrained (as compared with hypothetically similar organisms which lack the constraint). But such a conclusion does not follow from the existence of constraints$_F$. It requires soft adaptationism—that is, the denial of neutralism with respect to constrained$_F$ traits.

Let us relate these distinctions to Alberch's and Wright's diagrams. The point of contention between neutralism and soft adaptationism is the shape of the adaptive surface. Soft adaptationists believe that (almost) all points on the surface lie on a relatively steep slope. Neutralists believe that large areas on the surface are flat, reflecting the lack of selection on the range of traits associated with those areas. An advocate of constraints$_F$ (such as Alberch) believes that the distribution of morphologies in the two-dimensional morphospace is to be explained by the processes of embryology, and that even on an adaptively flat landscape (like Fig. 6.2) most of the pattern would remain. So Alberch would not expect the clusters in morphospace to match the contours of the adaptive landscape. But notice that the clusters may fail to match the adaptive contours in two different ways. First, there may not *be* many contour lines in this landscape, reflecting the neutralist opinion that adaptive landscapes have large, flat plains. Second, there may be many steep contours, but only a weak correlation between the pattern of the clusters and the shape of the adaptive landscape. This second possibility is the 'constraint$_F$ adaptationist' belief that constraints$_F$ usually do yield constraints$_A$; distributions of morphotypes would only partially correlate with contours of high adaptivity. Only the hard adaptationist, who denies both neutralism and the strength of constraints$_F$, would expect Alberch's morphological clusters to perch precisely atop adaptive peaks.

This is why the notion that development constrains adaptation arises from adaptationist biology. One must have soft adaptationist leanings even to worry about developmental constraints reducing adaptation.

I do not mean to suggest that soft adaptationism is controversial, or that each of the above positions is equally justified. The modern synthesis has successfully and justifiably brought most modern thinkers to the view that adaptation is an extremely prominent feature in the organic world. But it is important to understand the programme of developmental biology in its own terms, and not simply in terms of its sometime and oblique opposition to neo-Darwinism. It is mistaken to infer the lack of a constraint$_r$ from a high degree of adaptation in an organism, and it is mistaken to infer a reduction in adaptation from the existence of a constraint$_r$.

VI. CONCLUSIONS

The recognition of developmental constraints is only a small part of the revisions to neo-Darwinian evolutionary theory being urged by developmental biologists. Constraints are the point at which the two traditions come closest. As we have seen, even here is a gap; the two sides mean different things by the word 'constraint'. Among other topics for which developmental theorists claim superior explanatory resoures over neo-Darwinians are long-term evolutionary trends, rapid evolutionary change, parallel and convergent evolution, and the origins of higher taxa. Some constraints$_r$ are even seen as enhancing the possibilities for adaptive changes. It is argued, for example, that the plasticity of certain developmental mechanisms allows for correlated changes in form without the requirement that each correlated part be the target of independent selection (Rachootin and Thomson 1981). Such correlations are still constraints$_r$, since the correlated features must change synchronously. This is incoherent if constraints are defined as restricting adaptation, as they are by adaptationists like Niklas, Stephens, and Krebs. Developmentalists sometimes recognize that an overemphasis of the term 'constraint' gives a false picture of their intended contributions to evolutionary theory. From a group report on a 1981 conference: 'Every time that someone mentioned a "constraint", someone else reinterpreted it as an "evolutionary opportunity" for a switch to a new mode of life, and a third person would bring up the subject of the complementary "flexibility"' (Horn *et al.* 1982: 217). It is beyond the scope of the present essay to describe all of these ideas, let alone to evaluate them in comparison to mainstream neo-Darwinian explanations of the same phenomena.

Hamburger claimed that the modern synthesis treated ontogenetic development as a black box which could safely be ignored by evolutionary biology. Mayr (1991: 8), on the other hand, attributed the non-participation of embryologists in the synthesis to the embryologists' own disinterest. Both are probably correct. Blame need not be assessed here, especially since the bracketing of the complexities of development was probably a necessary condition for the remarkable achievements of the modern synthesis. Fifty years after the synthesis, the role of developmental biology may need reappraisal. The developmental biologists' arguments should be seen as assertions that the bracketing of development should end, that the insides of the black box of development have causal relevance to evolutionary biology.

Even though the constraints issue is not the most exciting aspect of developmentalists' theoretical ambitions, the semantic confusions exposed above strongly prejudice the argument. For example, recall the statement by Stephens and Krebs that claims of constraint presuppose adaptationist research, that 'without a design hypothesis there would be no reason to postulate a constraint!' Taken as a claim about constraints on form, this statement is blatantly false. The patterns in Alberch's morphospace diagrams are based on a knowledge of form alone, not on a discovery of sub-optimal adaptivity. The forbidden morphologies of digit patterns are determined not from surveys of the digit patterns which actually occur in nature, but from a knowledge of the developmental processes which build those digit patterns. Murray's constraints on colour patterns are proposed on generative, developmental bases alone—adaptationist design hypotheses are not consulted. It is false to claim that constraints on form are discovered by embryologists in the same way as constraints on optimal foraging are discovered by ethologists. Developmental biology is a source of knowledge independent of adaptationism.

Classifying developmental constraints as constraints$_A$ has a second pernicious effect. It trivializes the detailed causal understanding which developmentalists believe is essential to evolutionary biology. An example can be seen in Dawkins 1982. Here are listed a number of explanations for the imperfection of adaptation. They include time-lags (the environment might have changed too recently for natural selection to have operated), the variability of environments (an organism cannot be perfectly adapted to every micro-environment), costs and materials (birds cannot grow wings of titanium alloys), and 'available genetic variation'. Developmental constraints fit in this category. As Dawkins puts it elsewhere, 'no mammal will ever sprout wings like an angel unless mammalian embryological patterns are susceptible to this kind of change' (1986: 311). These are indeed factors accepted by neo-Darwinians which would explain imperfection of adaptation. The operation of these constraints$_A$ in any given case would presumably be determined in the manner of Stephens and Krebs—by discovering imperfect adaptation—rather than by a prior causal understanding of the process which *produced* the constraint. Even though here and elsewhere Dawkins acknowledges the complexity of embryology and the limits it places on the available genetic variation, the actual insides of the embryological black box remain irrelevant to his discussions. Something is in that box, it is complicated, and it reduces available genetic variation. But its exact details do not matter to evolutionary biology. Variation which is lacking because of details of the developmental process is no more important than variation which by chance just has not occurred.

In other words, the consequence of treating developmental constraints as constraints$_A$ is that the black box can remain closed. The box can be alluded to as the source of an identifiable constraint on adaptive perfection, not unlike the changing environment. The detailed causal accounts which fill texts and journals of developmental biology need have no more relevance to evolutionary biology than the theoretical details of what causes earthquakes, hurricanes, or ice ages. In contrast, advocates of a developmental synthesis are asking for much more than a mere acknowledgement of adaptive imperfection. They want to integrate the complex and internal details of embryology into the study of evolution. The significance of developmental constraints cannot be reduced to the language of adaptive imperfection.

In this way the debate between the modern synthesis and its developmentalist critics is similar to another great black box debate in twentieth-century science, between behaviourists and their opponents who favoured cognitive and neurological theories. There are many dissimilarities, of course; issues like intentionality and consciousness are (fortunately) not central to evolutionary theory. But just as synthesis adaptationists deny the causal importance of embryology to evolutionary theory, so behaviourists deny the causal importance of internal states of the psychological organism, either cognitive or neurological states (see Amundson and Smith 1984 and Amundson 1989, 1990, on similarities in debates within psychology and biology). Neither adaptationists nor behaviourists actually deny complex goings-on inside the embryo or inside the brain. They claim that the important scientific issues are understandable without the need for a detailed knowledge of the intervening processes. Neo-Darwinian evolutionary theorists know that genes somehow build phenotypes and then get winnowed and passed on as a result of phenotype/environment interactions. Likewise, behaviourists know that somehow an organism's stimuli (including reinforcing stimuli and such) connect with responses, and that the connecting involves lots of complicated interactions among neurons. But just as the details of neurological or cognitive processes are seen as irrelevant to the explanation of behaviour, the details of development are seen as irrelevant to evolution. All that matters are the input–output characteristics of the black boxes. Genotypes determine phenotypes, and stimuli are connected to responses. That is all that needs to be known about the insides of the processes by behaviourists or neo-Darwinians. Developmental biologists, like cognitivists and neuropsychologists before them, face the challenge of arguing for the causal relevance of the insides of a black box.

The above paragraph is intended as an explication, not a vindication, of developmentalist critiques of neo-Darwinism. I do not share the common philosophical prejudice that behaviourism had obvious methodological flaws (see L. D. Smith 1986). Furthermore, the bracketing of problematic domains is scientifically respectable. Evolutionary biology was built on a huge black box—Darwin could never have written the *Origin of Species* if he had not wisely bracketed the mechanism of inheritance. All that was required of inheritance for Darwin was that somehow some of the phenotypic variation seen in natural populations is passed on to descendants— the detailed insides of the black box of inheritance could (and did) remain unknown for decades. The modern synthesis depended on the surprising realization that Mendelian genetics *was* the inside of Darwin's black box.

The proponents of the developmental synthesis have a difficult task. Pre-synthesis Darwinians at least realized the need for a theory of inheritance, although they doubted that Mendelism was that theory. Most post-synthesis neo-Darwinians do not require developmental biological contributions to evolution theory. Developmentalists may or may not be able to demonstrate that a knowledge of the processes of ontogenetic development is essential for the explanation of evolutionary phenomena. If they can demonstrate this, and provide well-founded developmental/ evolutionary explanations, the result will be a dramatic synthesis of divergent explanatory and theoretical traditions.[1]

REFERENCES

Alberch, P. (1982), 'Developmental Constraints in Evolutionary Processes', in J. T. Bonner (ed.), *Evolution and Development* (New York: Springer-Verlag), 313–32.
Amundson, R. (1989), 'The Trials and Tribulations of Selectionist Explanations', in K. Hahlweg and C. A. Hooker (eds.), *Issues in Evolutionary Epistemology* (New York: State University of New York Press), 413–32.
——(1990), 'Doctor Dennett and Doctor Pangloss: Perfection and Selection in Psychology and Biology', *Behavioral and Brain Sciences*, 13: 577–84.

[1] The ideas expressed in this paper were stimulated by discussions with Stephen Jay Gould and Pere Alberch. Versions have received valuable commentary from many people, including Daniel Dennett, Kim Sterelny, Scott Gilbert, Kelly Smith, and especially Elliott Sober. The work was supported by NSF grant SBE-9122646.

——and Smith, L. D. (1984), 'Clark Hull, Robert Cummins, and Functional Analysis', *Philosophy of Science*, 51: 657–66.

Antonovics, J., and van Tienderen, P. H. (1991), 'Ontoecogenophyloconstraints? The Chaos of Constraint Terminology', *Trends in Ecology and Evolution*, 6: 166–9.

Beatty, J. (1986), 'The Synthesis and the Synthetic Theory', in W. Bechtel (ed.), *Integrating Scientific Disciplines* (Dordrecht: M. Nijhoff), 125–35.

Bonner, J. T. (1982) (ed.), *Evolution and Development* (New York: Springer-Verlag).

Burian, R. M. (1986), 'On Integrating the Study of Evolution and of Development', in W. Bechtel (ed.), *Integrating Scientific Disciplines* (Dordrecht: M. Nijhoff), 209–28.

Dawkins, R. (1982), *The Extended Phenotype* (Oxford: Oxford University Press).

——(1986), *The Blind Watchmaker* (New York: Norton).

Gilbert, S. F. (1991), *Developmental Biology*, 3rd edn. (Sunderland, Mass.: Sinauer).

Goodwin, B. C. (1984), 'Changing from an Evolutionary to a Generative Paradigm in Biology', in J. W. Pollard (ed.), *Evolutionary Theory* (New York: Wiley & Sons), 99–120.

——Holder, N., and Wylie, C. C. (1983), *Development and Evolution* (Cambridge: Cambridge University Press).

Gould, S. J. (1980), 'The Evolutionary Biology of Constraint', *Daedalus*, 109: 39–52.

——(1983), 'The Hardening of the Modern Synthesis', in M. Grene (ed.), *Dimensions of Darwinism* (Cambridge: Cambridge University Press), 71–93.

——and Lewontin, R. C. (1979), 'The Spandrels of San Marco and the Panglossian Paradigm: A Critique of the Adaptationist Programme', *Proceedings of the Royal Society of London*, B205: 581–98.

Hamburger, V. (1980), 'Embryology and the Modern Synthesis in Evolutionary Theory', in E. Mayr and W. Provine (eds.), *The Evolutionary Synthesis* (Cambridge: Cambridge University Press), 97–112.

Holder, N. (1983), 'Developmental Constraints and the Evolution of Vertebrate Digit Patterns', *Journal of Theoretical Biology*, 104: 451–71.

Horder, T. J. (1989), 'Syllabus for an Embryological Synthesis', in D. B. Wake and G. Roth (eds.), *Complex Organismal Functions: Integration and Evolution in Vertebrates* (Chichester: Wiley & Sons), 315–48.

Horn, H. S., Bonner, J. T., Dohle, W., Katz, J. J., Koehl, M. A. R., Meinhardt, H., Raff, R. A., Reif, E. E., Stearns, S. C., and Strathmann, R. (1982), 'Adaptive Aspects of Development', in J. T. Bonner (ed.), *Evolution and Development* (New York: Springer-Verlag), 215–35.

Løvtrup, S. (1987), *Darwinism: The Refutation of a Myth* (London: Croom Helm).

Maynard Smith, J., Burian, R., Kauffman, S., Alberch, P., Campbell, J., Goodwin, B., Lande, R., Raup, D., and Wolpert, L. (1985), 'Developmental Constraints and Evolution', *Quarterly Review of Biology*, 60: 265–87.

Mayr, E. (1980), 'Prologue: Some Thoughts on the History of the Evolutionary Synthesis', in E. Mayr and W. Provine (eds.), *The Evolutionary Synthesis* (Cambridge: Cambridge University Press), 1–48.

——(1982), *The Growth of Biological Thought* (Cambridge, Mass.: Harvard University Press).

——(1991), 'An Overview of Current Evolutionary Biology', in L. Warren and H. Koprowski (eds.), *New Perspectives on Evolution* (New York: Wiley & Sons), 1–14.

Mitchison, G. J. (1977), 'Phyllotaxis and the Fibonacci Series', *Science*, 196: 270–5.

Murray, J. D. (1981), 'A Pre-Pattern Formation Mechanism for Animal Coat Markings', *Journal of Theoretical Biology*, 88: 161–99.

Niklas, K. J. (1988), 'The Role of Phyllotactic Pattern as a "Developmental Constraint" on the Interception of Light by Leaf Surfaces', *Evolution*, 42: 1–16.

Rachootin, S. P., and Thomson, K. S. (1981), 'Epigenetics, Paleontology, and Evolution', in G. Scudder and J. Reveal (eds.), *Evolution Today* (Pittsburgh: Hunt Institute), 181–93.

Schindel, D. E. (1990), 'Unoccupied Morphospace and the Coiled Geometry of Gastropods: Architectural Constraint or Geometric Covariation?', in R. M. Ross and W. D. Allmon (eds.), *Causes of Evolution* (Chicago: University of Chicago Press), 270–304.

Smith, K. C. (1992), 'Neo-Rationalism versus Neo-Darwinism: Integrating Development and Evolution', *Biology and Philosophy*, 7: 431–51.

Smith, L. D. (1986), *Behaviorism and Logical Positivism* (Stanford, Calif.: Stanford University Press).

Stephens, D. W., and Krebs, J. R. (1986), *Foraging Theory* (Princeton: Princeton University Press).

Thompson, D. W. (1942), *On Growth and Form*, 2nd edn. (Cambridge: Cambridge University Press).

Thomson, K. S. (1988), *Morphogenesis and Evolution* (New York: Oxford University Press).

Wallace, B. (1986), 'Can Embryologists Contribute to an Understanding of Evolutionary Mechanisms?', in W. Bechtel (ed.), *Integrating Scientific Disciplines* (Dordrecht: M. Nijhoff), 149–63.

7

DEVELOPMENTAL SYSTEMS AND EVOLUTIONARY EXPLANATION

P. E. GRIFFITHS AND R. D. GRAY

Few scientific ideas are so well embedded in popular culture as the idea that certain features of an organism are genetically determined, while others are acquired by interaction with the environment. There have been many attempts to recast the special role of the genes in an attempt to do justice to our knowledge of developmental processes. The division of traits into innate and acquired has been replaced by attempts to determine the relative influence of genetic and evironmental factors on each trait. The idea of a genetically specified outcome has been replaced by a genetic blueprint and then by a genetic programme. But all these accounts presume that the key to understanding development is to understand the interaction of two classes of developmental resources—genes and the rest. They are all *dichotomous* accounts of development.

Developmental systems theory rejects the dichotomous approach to development. The genes are just one resource that is available to the developmental process. There is a fundamental symmetry between the role of the genes and that of the maternal cytoplasm, or of childhood exposure to language. The full range of developmental resources represents a complex system that is replicated in development. There is much to be said about the different roles of particular resources. But there is nothing that divides the resources into two fundamental kinds. The role of the genes is no more unique than the role of many other factors.

Many authors have contributed to the developmental systems, or constructionist, tradition in the study of development.[1] We have drawn on this tradition, and particularly on the work of Susan Oyama, to produce a general account of development and evolution. We have tried to confront one major weakness of previous presentations of the developmental

First published in *Journal of Philosophy*, 91/6 (1994): 277–304. Reprinted by permission.

[1] Lehrman 1953, 1970; Stent 1981; Lewontin 1982, 1983; Oyama 1985, 1989; Ho 1986; Johnston 1987; Johnston and Gottlieb 1990; Nijhout 1990; Gray 1992; Moss 1992.

systems idea—the lack of any way of delimiting and individuating developmental systems. We suggest an aetiological solution: the developmental system consists of the resources that produce the developmental outcomes that are stably replicated in that lineage. By adopting this definition, we bring out the radical implications of the new approach to development for the theory of evolution. The developmental system goes far beyond the traditional phenotype, yet all its elements are parts of the evolutionary process. We argue that a reformulation of evolution in developmental systems terms maximizes the explanatory power of evolutionary theory.

The implications of the developmental systems approach are enormous. In the later part of the essay we try to sketch some of these. We argue that evolution is best construed as differential replication of total developmental processes or life cycles. We show that the well-known distinction between replicators and interactors is no longer of any great use in clarifying thought about evolution. Finally, we suggest that the developmental systems view makes it impossible to maintain the distinction between biological and cultural evolution. Both traditional processes are rejected in favour of a single, richer account of the replication of total developmental systems.

I. INNATENESS, GENETIC INFORMATION, OTHER CONFUSIONS

An early contribution to the development of developmental systems theory was Daniel S. Lehrman's (1953) attack on Konrad Lorenz's 1930s conception of innateness. The collapse of this conception, and of the idea of genetic information with which Lorenz replaced it, show the fundamental problems with dichotomous views of development.

Lorenz had described an innate trait as one whose origins are to be understood in terms of adaptation during evolution, and whose emergence is insensitive to environmental variation. Learned traits, on the other hand, are to be understood in terms of the organism's adjustment to its local environment, and are sensitive to variation in that environment. Lehrman pointed out that there is no conceptual link between the evolutionary and developmental elements of Lorenz's innateness concept, between the fact that a trait is an evolutionary adaptation and the fact that it is insensitive to environmental variation. It is of no evolutionary consequence whether a trait is sensitive to environmental variation, as long as the actual historical environment regularly provides the required input. 'Nature selects for outcomes' (Lehrman 1970: 28), and is indifferent to how they are achieved.

Lehrman supplemented this conceptual point with a host of examples of the role of environmental input in the production of evolved traits. The female rat abstains from eating her young, for example, only if she licks her genitalia during pregnancy. She will construct a nest and retrieve the young only if she has been exposed to temperature variations earlier in her life and had the chance to carry other objects around in her mouth (Lehrman 1953: 342–3).[2] A later example of the same kind, due to Gilbert Gottlieb (1981), makes the same point. Under normal developmental conditions, young ducklings develop a preference for the maternal call of their own species. Gottlieb discovered that they fail to develop this preference when devocalized in the egg. Exposure to their own prenatal call is required for the development of their preference for the (quite different) maternal call. Lehrman was at pains to point out that these sorts of facts do not show that all traits are 'learned', as opposed to innate. They show that reliable developmental outcomes occur because of reliable interactions between the developing organism and its environment. The fact that a trait has an evolutionary history has no implications about the role of environmental factors in the process by which it develops, except that the process is sufficiently reliable to produce similar outcomes in each generation.

In his later work, and partly in response to this critique, Lorenz eschewed the idea that some phenotypic traits are innate, while others are learned. We have found it hard to convince some philosophical devotees of Lorenz that he ever held this 'naïve' view, but we can hardly do better than to quote his own words. Lorenz noted that his earlier 'atomistic attitude' of conceiving complex behaviours as chains of elements, some of which were innate and some acquired, 'was a serious obstacle to the understanding of the relations between phylogenetic adaptation and adaptive modifications of behaviour. It was Lehrman's (1953) critique which, by a somewhat devious route, brought the full realisation of these relations to me' (1965: 80). Lorenz replaced his distinction between innate and acquired traits with a distinction between two sources of developmental information. Some of the information manifested in an organism's adaptation to its environment is phylogenetic, as opposed to ontogenetic. Phylogenetic information is transmitted in the genes, whereas ontogenetic information is gathered from the environment during development. Lorenz's classic experimental paradigm, the deprivation experiment, which was originally intended to reveal innate traits, was interpreted in this later work as revealing the presence of phylogenetic information. A rat reared in isolation and given no opportunity to practice maternal skills nevertheless constructs a species-typical nest, and retrieves its young in the species-

---

[2] The last two points are disputed by Lorenz in his (1965).

typical manner. Lorenz argues that this can only be explained by the genetic transmission of phylogenetic information: 'certain parts of the information which underly the adaptedness of the whole, and which can be ascertained by the deprivation experiment, are innate' (ibid. 40).

Lorenz admits that the deprivation experiment does not remove all sources of environmental input. No trait can develop without input from the environment. Trivially, the organism must eat if it is to grow. Less trivially, the rat must have experienced temperature variation and carrying things. The rationale of the deprivation experiment therefore requires a distinction between two sorts of environmental input, those which provide ontogenetic information for 'learning', and those which provide mere physical 'support' for the reading of phylogenetic information.

No biologist in his right senses will forget that the blueprint contained in the genome requires innumerable environmental factors in order to be realised. . . . During his individual growth the male stickleback may need water of sufficient oxygen content, copepods for food, light, detailed pictures on his retina, and millions of other conditions in order to enable him, as an adult, to respond selectively to the red belly of a rival. Whatever wonders phenogeny [sic] can perform, however, it cannot extract from these factors information that simply is not contained in them, namely the information that a rival is red underneath. (Ibid. p. 37)

This information, therefore, must be contained in the genome, which 'rules ontogeny'. Lorenz compares the roles of genome and environment in ontogeny to an architect's plan and the bricks and mortar in a building project (ibid. 42). Of all the resources that are utilized in the development of traits that represent phylogenetic adaptations, only one, the genome, provides information. The others merely provide raw materials.

Unfortunately for Lorenz, no suitable notion of information exists which will allow him to draw this distinction between the role of the genome and the role of other developmental resources. Timothy Johnston (1987) makes this point very clearly. We have a well-understood, mathematical notion of information, derived from communication theory. An event carries information about another event to the extent that it is correlated with that event. The 'transmission' of mathematical information is a matter of the systematic dependence of one system on another. In a classic example of learning, such as a rat finding out which foods are poisonous, there is just such a systematic dependence between the state of the environment and the later state of the organism. After learning, the internal state of the rat carries information about the state of the world. Information about the food has been transmitted to the rat. But in what Lorenz characterizes as the 'maturation' of an innate trait, there is an exactly similar dependence. Lehrman's original examples documented the ways in which developmental outcomes are contingent on the occurrence of interactions with the environment. The development of maternal care in the rat requires interaction with temperature variations, and with material that can be transported. Removal of these factors is reflected in changes in the phenotype, so they must be transmitting information to the phenotype. At the molecular level, cellular differentiation is dependent on a host of extragenomic factors. Induction of lactation in mammary cells in mice, for example, depends on the shape of the cells, which is in turn a function of the substrate to which they are attached (Moss 1992).

This symmetry between different causal factors in development is intrinsic to the concept of mathematical information. In the Lorenzian picture of 'maturation', the non-genetic developmental factors constitute the channel conditions under which the organism carries information about its genes, whereas in Lorenzian 'learning' the intrinsic organization of the organism constitutes channel conditions under which the state of the organism carries information about environmental factors. But it is always possible to reverse the roles of the sender and channel conditions. So it is equally open to us to interpret the maturation case as one in which the genes constitute channel conditions under which the organism carries information about some non-genetic developmental factor. We could also interpret the learning case as one in which the environmental factors are channel conditions under which the state of the organism tells us about its genes.[3]

Lorenz's failure to appreciate this symmetry shows us that he did not conceive of genetic information in terms of systematic dependence. Instead, he relied on some intentional or semantic conception of information. When the channel conditions are altered, the genes do not carry different information about the phenotype; they are just misinterpreted. Under abnormal developmental conditions, the phenotype misrepresents its genes. In fact, there are only two ways to make sense of the notion of information in development. First, the entire set of developmental resources, plus its spatio-temporal structure, may be said to contain information about evolved developmental outcomes in the unproblematic, mathematical sense of systematic dependence. But as long as we confine ourselves to this notion of information, there is no causal asymmetry in the role of different resources which makes it legitimate to regard some of them as carrying the information and the others as merely providing conditions in which it can be read. The second, more practical way to make sense of the notion of information in development is to embed the information in one resource by holding the state of the other resources fixed as

---

[3] A perfectly practical proposal given the extensive literature on species-specific patterns of associative learning. See e.g. Seligman and Hager 1972.

channel conditions under which that information is transmitted. But this move can be used to interpret any of the resources as the 'seat' of the information guiding development, and so it, too, fails to generate the traditional asymmetry between genetic and other factors.

Our critique of Lorenz can be applied to even the most sophisticated reconstruction of the idea that genes 'code for' phenotypic characteristics. Kim Sterelny and Philip Kitcher (1988) claim that a stretch of DNA codes for a trait relative to a 'standard' background of other genes and a 'standard' environment. Given these background conditions, changes in the gene are systematically linked to changes in the phenotypic trait. But consider the DNA in an acorn. If this codes for anything, it is for an oak tree. But the vast majority of acorns simply rot. So 'standard environment' cannot be interpreted statistically. The only interpretation of 'standard' that will work is 'such as to produce evolved developmental outcomes' or 'of the sort possessed by successful ancestors'. With this interpretation of 'standard environment', however, we can talk with equal legitimacy of cytoplasmic or landscape features coding for traits in standard genic backgrounds. No basis has been provided for privileging the genes over other developmental resources.

II. TAKING DEVELOPMENT SERIOUSLY

Developmental systems theory provides an alternative explanation of trans-generational stability of form. As Oyama argued in *The Ontogeny of Information* (1985), species-typical traits are constructed by a structured set of species-typical developmental resources in a self-organizing process that does not need a central source of information. Some of these developmental resources are genetic, others, from the cytoplasmic machinery of the zygote to the social events required for human psychological development, are non-genetic. The spatio-temporal disposition of the resources is itself a critical resource, as it helps induce self-organization. The fact that appropriately structured resources are available can receive an evolutionary explanation. The processes which effectively replicate themselves are those which find appropriately structured resources in each generation. An extended notion of inheritance, which stresses the role of past generations in structuring the developmental context of their successors, is thus a critical part of the theory.

The theory does not deny that there are distinctions among developmental processes. For example, Gottlieb (1976) suggests that different kinds of interactions may either facilitate, induce, or maintain developmental differences (Patrick Bateson (1983) notes that these distinctions are applicable indifferently to the roles of genetic and non-genetic factors). But the theory does deny that there are two fundamental kinds of developmental resources, genes and the rest, and that these two types play fundamentally different roles in development. It makes no more and no less sense to say that the other resources 'read off' what is 'written' in the genes than that the genes read off what is written in the other resources. The reading of the genes is a metaphor which has been of some historical utility, but which now retards the study of development, and, we shall argue, of evolution.

Perhaps the best metaphor for development is that of Stent (1981), who compares development to an idealized model of ecological succession. When an area of ground is denuded of its biota, the characteristic landscape of that region is re-established in a series of stages. Adventitious first colonizers, able to survive in the barren conditions, take advantage of the lack of competition to occupy the area. Their presence modifies factors such as the soil and micro-climate, making the area hospitable to the next phase of vegetation, and so forth. In this process, as in development, an outcome is replicated without any blueprint or programme, as a consequence of the presence of the same developmental resources. There is no room for any distinction between some resources that contain the information and others that 'read it' or 'provide the material conditions for its realization'. Nor is it possible to bypass a detailed analysis of the developmental process by going straight to the sources of the 'information' that is 'expressed' in the outcome.

The differences between the notion of information that is legitimate in this context and the everyday notion based on our experience of language is so great that it is very hard not to revert to the later notion, with all its inappropriate implications. It is perhaps for this reason that developmental systems theorists, and especially Oyama (1985), have eschewed the traditional metaphor of evolved traits being 'transmitted' from one generation to another. The picture that we have tried to convey with the metaphor of ecological succession is much better conveyed by saying that species-typical traits are reconstructed in the next generation by the interaction of the same sorts of developmental resources that were present in earlier generations. Oyama has also suggested that it is misleading to talk of the information used to reconstruct the phenotype being 'transmitted'. The resources that construct later stages of the developmental process are constructed by earlier stages. In Oyama's preferred terminology, information is itself the product of an ontogeny.

Figure 7.1 shows the developmental systems conception in

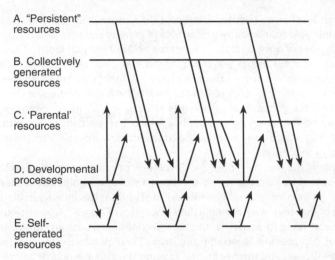

A. "Persistent" resources

B. Collectively generated resources

C. 'Parental' resources

D. Developmental processes

E. Self-generated resources

FIG. 7.1. Causal influences in four asexual generations of a lineage of developmental processes. Each arrow represents multiple inputs. Influence of each resource is contingent on the presence of the others. The effects of temporal order of interaction have been overlooked. The broad categories of resources are not intended to be exhaustive, and are made largely for convenience of exposition.

diagrammatic form. A developmental process is reconstructed through the interaction of suitably structured resources. Some, including the genes, are created by the immediate precursors of the generation in question. Others are generated over different periods of time by the collective activities of the population. Others, to be discussed below, persist without reference to these activities. A developmental process 'inherits' all these resources. Finally, many vital resources in development are generated by earlier stages of the developmental process itself.

III. INDIVIDUATING DEVELOPMENTAL SYSTEMS

For obvious reasons, even the most systematic presentations of developmental systems theory (e.g. Oyama 1985 and Gray 1992) have been more concerned to incorporate neglected elements into the developmental process than to exclude elements from it.[4] Little attention has been paid to setting out the limits of developmental systems, and to individuating one

[4] In writing this section, we have been influenced by Millikan 1984. In this area, as in many others, Millikan has broken the ground for those interested in biological teleology/teleonomy.

from another. Sterelny has criticized earlier versions of the theory on these grounds: 'Elvis Presley is part of my developmental system, being as he was causally relevant to the development of my musical sensibilities, such as they are. Yet surely there is no system, no sequence, no biologically meaningful unit, that includes me and Elvis' (personal communication). The 'Elvis Presley' problem helps us clarify the claims of developmental systems theory. The theory aims to provide an explanation of trans-generational stability of form which does not attribute it to the transmission of a blueprint or programme in the genome—a pseudo-explanation that inhibits work on the real mechanisms of development. So the theory is interested in those developmental resources whose presence in each generation is responsible for the characteristics that are stably replicated in that lineage. For example, we might contrast two influences on a new-born bird. The interaction between the new-born bird and the song of its own species, which occurs in each generation and helps explain how the characteristic song is produced, is part of the bird's developmental process. The interaction between the new-born bird and the noise that ruptures its eardrums plays no such role, and so is not part of the process.

Another way to draw this distinction is by distinguishing developmental outcomes that have evolutionary explanations from those which do not. The interactions that produce outcomes with evolutionary explanations are part of the developmental system. There is an evolutionary explanation of the fact that the authors of this paper have a thumb on each hand. We have thumbs because of the replication of thumbed ancestors. The thumb is an evolved trait. But the fact that one of us has a scar on his left hand has no such explanation. The scar is an individual trait (we are referring, of course, to the trait of having a scar just thus and so, not the general ability to scar). The resources that produced the thumbs are part of the developmental system. Some of those which produced the scar, such as the surgeon's knife, are not.

Various issues need to be clarified about this historical or aetiological characterization of the developmental system. First, the distinction between 'evolved' and 'individual' outcomes is not another version of the innate/acquired distinction. It is not a distinction between types of developmental processes. The fact that a developmental outcome has an evolutionary history is not an intrinsic property that can be determined by inspection of the outcome, or of the process that constructs it. By calling it an evolved outcome, we are merely indicating that it fits into a particular pattern of explanation. Similarly, when we privilege certain of the resources that go to construct an organism as 'the developmental system', we do so to point to the explanatory connection between the trans-generational stability of these resources and the trans-generational stability of certain developmental outcomes. For other explanatory purposes, such as the study of developmental abnormalities, a different system must be delineated.

The fact that the evolved/individual distinction is not a distinction between different types of developmental process cannot be too much stressed. Past interactions between evolutionary theory and developmental theory have not had happy outcomes. Evolutionary theorists like Lorenz used the category of innateness to substitute an evolutionary explanation for a genuine developmental explanation. We hope that a developmental systems account of evolution can avoid this mistake, because it makes the developmental mechanisms themselves the prime focus of evolutionary explanation.

A second important point of clarification is that the claim that all features of the developmental system can be given evolutionary explanations does not commit us to any form of adaptationism. Evolutionary explanation is 'adaptive-historical' explanation. The organism's response to any particular adaptive phase is determined in part by the historical resources and historical constraints accumulated in the lineage in response to past phases. The outcome of the process is affected by these resources and constraints, and they themselves are altered by the outcome of the process. The outcome is also influenced by the availability and order of variants and by the sheer stochasticity of the differential replication process. So even in cases where adaptation plays a role in the explanation of a particular trait, that explanation is very far from adaptationist. Furthermore, the developmental system is not a collection of separately evolved features. The system of interdependences that it represents is itself an evolutionary product. Vestiges and features produced because of developmental correlations are as much evolved features of the developmental system as features that offer some adaptive advantage. They, too, are subject to adaptive-historical explanation.[5]

IV. DEVELOPMENTAL SYSTEMS AND EXTENDED PHENOTYPES

The idea of a developmental system has certain parallels with Richard Dawkins's (1982) notion of the extended phenotype. We believe that

[5] The word 'adaptationism' was introduced by Gould and Lewontin (1979). The idea of adaptive-historical explanation is discussed in Griffiths (forthcoming). For examples of non-adaptationist evolutionary explanation in a developmental systems account, see Gray 1988. For an integration of the idea of developmental constraint with evolutionary explanation, see Smith 1992.

Dawkins's central insight was that many elements outside the traditional organism can be given an evolutionary explanation. Nests vary through evolution, and the number and form of current nests has been influenced by the relative effectiveness of these variations.

Dawkins claims that such explanations are possible because of the selection of genes associated with the production of extended phenotypic features, such as nests. But the central insight just described is completely independent of Dawkins's gene-selectionist view of evolution. The phenomena of habitat imprinting demonstrates very nicely how the association of an organism with an environmental feature could have an evolutionary explanation without the genes having an interesting role in the production of that trait. Klaus Immelmann (1975) cites a study of European mistle thrushes which clearly illusrates this. The expansion of this species' range from forest to parkland in France and Germany was shown to proceed, not by the spread of several local populations, but by the spread of a single population that had become habitat-imprinted on parkland rather than forest. The fate of different thrush lineages will depend on their interaction with the particular habitat with which they are reliably associated, and the fate of that habitat. The habitat is something they have acquired through evolution, as much as any other element of the phenotype. Yet the genetic variation between the two populations can be presumed to be random with respect to which habitat they have imprinted on. No difference in the mechanism in the two lineages is needed to sustain their association with two very different habitats.

We have argued against Dawkins's interpretation of his extended phenotype cases in genic terms, but we do not want to reject the cases themselves. We think that there are many valid evolutionary explanations of extra-genic developmental resources. The forms of nests, webs, and so forth do change over evolutionary time in a way that can be explained by the differential replication of lineages. In an earlier paper, Gray (1992) drew attention to the co-evolution of certain eucalypts and the bush fires that play such a role in their development. Developmental systems theory makes all developmental interactions subject to evolutionary explanation. Any species-typical occurrence that contributes to development has a history, and its continued occurrence has an adaptive-historical explanation. It is because of this feature that we claim that developmental systems theory maximizes the explanatory power of evolution. It allows the formulation in a single theoretical framework of all natural-historical narratives that are genuinely explanatory. This is a simple consequence of the way that we have defined the developmental system. It is precisely by having such an explanation that an item gets to be part of the system.

The developmental systems theorist's version of the 'extended pheno-type' is not subject to the sort of deflationary reinterpretation that Sterelny and Kitcher use to attack Dawkins (1976, 1982). According to Sterelny and Kitcher (1988), Dawkins's extended phenotypic features can be reduced to traditional behavioural phenotypic features. His talk of genes for webs or nests can be replaced by talk of genes for web- or nest-building behaviour. Thus, while they admit that Dawkins's picture of evolution can be illuminating, they deny that it explains anything that could not be effectively explained already. A traditional evolutionary theory can explain the evolution of the behaviour, and simply note the effects of this behaviour on the environment.

This deflationary strategy cannot be applied to many of the features that count as part of our 'extended phenotype'. Developmental systems theory claims to give evolutionary explanations of all the developmental interactions. Among them are those like the thrush case just described, in which the organism interacts with a persistent environmental feature. The interaction may have an evolutionary explanation. It may be that some lineages have survived because they were imprinted on an advantageous habitat. But the interaction cannot be reduced to a feature of the traditional behavioural phenotype. This is clearly shown by another example cited by Immelmann (1975). Cuckoo-style parasitic viduine finches have developed morphological subspecies and species on the basis of historic associations with different parasitized species. These associations are sustained by host imprinting. It is highly plausible that being associated with a successful host species, and one that has not developed anti-parasitic adaptations is a critical factor in success for the parasitic species. Developmental systems theory can give an evolutionary explanation of the developmental interaction between parasite lineage *A* and host lineage *B*. An account that confines itself to the traditional behavioural phenotype can only explain the general trait of host imprinting, which is common to all the species, and then state that this particular parasitic species has been historically associated with this particular host species. It cannot encompass the fact that association of the particular parasite *A* with the particular host *B* has itself evolved by the differential replication of this and other associations.

One difficulty arises, however, from our enthusiastic extension of the phenotype. Genes are a developmental resource, and their differential replication depends on the success of the system of which they are a part. More surprisingly, bush fires are a developmental resource, and their replication depends in part on the success of the plants for which they are a resource. But sunlight and gravity are also developmental resources, and play a critical role in determining evolved developmental outcomes. Surely, we are not proposing that features of this kind can be given evolutionary explanations? It is to this issue that we now turn.

V. WHAT IS REPLICATED IN DEVELOPMENT?

The evolutionary account of the limits of the developmental system given in Section II makes no distinction between developmental resources that owe their existence to the past generations of the developmental system, and those which exist independently of it. Some elements of the system are actively replicated by the parent organism (genes, cytoplasm, language traditions), and some are present because of the collective activities of the population (libraries, landscape features), but some merely persist from generation to generation (sunlight, gravity, parkland habitats for thrushes). Even if we have succeeded in showing that many elements of what is traditionally conceived of as the environment have an evolutionary history, surely we have gone too far by including merely persistent features? Surely there is no interesting sense in which persistent features are part of the evolutionary process? Instead, the objection goes, we should treat them as passive features of the environment, in the traditional fashion.

We are not impressed by this objection, and think that it overlooks an important sense in which persistent environmental features are part of the evolutionary process. Although the sun persists without reference to the evolution of a developmental system, its interaction with the rest of the system is highly contingent. A change in other developmental interactants which results in the organism behaving differently may substantially modify its interaction with the sun. If the organism becomes cave-dwelling, the interaction may cease completely. The phenomenon of habitat imprinting, in which an organism's choice of environment is a function of an earlier developmental interaction, shows the interaction with 'persistent' habitat features being actively replicated. There is a fundamental similarity between building a nest, maintaining one built by an earlier generation, and occupying a habitat in which nests simply occur (for example, as holes in trees). In all three cases, there may be an evolutionary explanation of the interaction of the nest with the rest of the developmental system.

The objection is useful, however, in that it forces us to consider more closely the ontology of developmental systems. We suggest that the primary focus of a constructionist account of development should be developmental processes, rather than developmental systems. The developmental process is a series of events which initiates new cycles of itself. We conceive of an evolving lineage as a series of cycles of a developmental process, where tokens of the cycle are connected by the fact that one cycle is initiated as a causal consequence of one or more previous cycles, and where small changes are introduced into the characteristic cycle as ancestral cycles initiate descendant cycles. The events which make up the developmental process are developmental interactions—events in which something causally impinges on the current state of the organism in such a way as to assist production of evolved developmental outcomes. The things that interact with the organism in developmental interactions are developmental resources. Some of the resources are products of earlier stages of the process, others are products of earlier cycles of the process, others exist independently of the process. These distinctions, while real, do not bear on the type of role which the entity plays in the developmental process.

The limits of the developmental process are set using the historical scheme of individuation which we applied to developmental systems in Section II. An interaction is part of the developmental process if it is of a type that has played a role in the evolution of the process. In the light of this revision, we might define a developmental system as the sum of the objects that participate in the developmental process, or, alternatively, as the sum of the developmental resources. We can now fix the limits of evolutionary explanation a little more precisely. All developmental interactions (as defined above) have evolutionary explanations, and some resources do. The distinction between explaining an interaction and explaining the resource that interacts is not just an *ad hoc* distinction invented to get around this problem. In a previous paper, Griffiths (1993) worried over the fact that *objets trouvés*, such as the shells occupied by hermit crabs, are clearly adaptations of those organisms, but do not owe their existence, either ontogenetically or phylogenetically, to that organism. We can now see that the interaction between the organism and this resource has an evolutionary explanation, while the resource itself has a quite separate explanation as part of the evolutionary history of another lineage of organisms.

VI. INDIVIDUATING DEVELOPMENTAL PROCESSES

The reformulation of developmental systems theory in terms of developmental processes allows us to resolve some outstanding puzzles for the theory. First, it allows us to confront the obvious objection that developmental systems do not form discrete generations, and so cannot provide the ancestor–descendant sequences required for evolution. Sterelny and others have suggested to us that once we lose sight of the sequence of individual genomes, separated from one another by the bottle-neck of the zygote, we have no univocal basis for dividing up a 'lineage' of developmental systems into discrete generations. It would be all very well if developmental systems theory only extended inheritance to include the non-genetic element of what we have labelled 'parental resources'. These resources, such as the maternal cytoplasm, cycle in synchrony with the genes. But, in fact, the developmental system includes the persistent resources, such as sunlight, and the population-generated resources, such as speech communities. The full range of developmental resources exhibits a bewildering variety of periodicities.

The objection can be put as a dilemma. On the one hand, does the developmental systems theory concede a privileged role to the genes in defining the temporal boundaries of an individual? That seems inconsistent with the whole thrust of the approach. On the other hand, if the theory rejects this privileged role for the genes, what account can it offer of the individuals in a lineage of developmental systems? The view developed in the last section allows us to slip between the horns of this dilemma. The central theoretical entity in our account is the developmental *process*, rather than the developmental *system*. The developmental process is a series of interactions with developmental resources which exhibits a suitably stable recurrence in the lineage. Its periodicity is unrelated to that of the resources themselves.

A simple thought-experiment can help to clarify how the move to developmental processes helps with the present objection. Imagine a lineage of asexuals in which each individual succeeds in begetting only a single viable offspring, and in which the individual dies with the birth of this offspring. What we have is a continuous series of developmental interactions. The problem is to find some way of chopping it up into generations (or individuals). Our proposal is to look for a particular sequence of interactions which is substantially repeated throughout the lineage. One repetition of this sequence of interactions constitutes a generation. Each repetition is an individual.

It is now possible to introduce some complications. First, suppose that the previous generation does not die at the inception of the next, and that an individual can give rise to more than one offspring, and to more than one batch of offspring. In that case, the lineage has a more complex topology, an irregular bush rather than a straight line, but the same empirical investigation can be carried out. Its aim might now be more clearly expressed as finding an atomic unit of which this topology is composed. In a further complication, we can suppose that reproduction is sexual. The topology of the lineage becomes reticulate, but there seems to be no additional obstacle to the search for an atomic unit, which might be more intuitively described as a *life cycle*. This may help to prevent confusions engendered by taking 'developmental process' to rcfer to development to adulthood.

It might be asked why we are so confident that the series of developmental interactions will have this cyclical structure. In reply, we are able to take over a well-known argument. The evolution of complex, functional structures requires a repeated life cycle during which structures are repeatedly reassembled. In Dawkins's version of this argument, he contrasts an organism that grows ever larger, with variations in its constituent cells merely giving it a mosaic structure, with one that grows to a finite size, and then begets descendants (1982: 256–64). Only in the latter case do variations have the opportunity to create a major reorganization of the overall structure of the organism, or of any of its complex subsystems. Thus, we can expect complex, functional systems to be produced by the repetition with variation of a developmental sequence. Dawkins construes this argument as providing support for the central importance of the Weissmanian bottleneck in evolution. He moves from the argument just outlined to a definition of an individual as a segment of a lineage isolated at each end by a single-cell bottle-neck. But this genophilic conception is not supported by the argument. What the argument actually supports is the view that the evolution of functional complexity will be favoured by the repeated reconstruction of the functional structures. This is entirely compatible with our view of development, in which these structures are produced by a developmental process/life cycle that draws on a wide range of inherited resources. On the developmental systems view, what separates individuals is not the existence of a developmental bottle-neck, but the fact that substantially similar functional structures are reconstructed anew from the developmental resources. We therefore help ourselves to the argument to explain our confidence that lineages of developmental processes/life cycles will exhibit the repetitive structure that we require.

We have tried to show that lineages will exhibit enough cyclical structure to support our proposal for individuating developmental systems. The next problem is that lineages may contain too much cyclical structure. In many lineages, larger developmental cycles exist that embed several traditional phenotypic life cycles. We may be faced, therefore, with an embarrassment of riches as regards repeated sequences of developmental events. In aphids, for example, the cycle of birth and death of traditional phenotypic individuals is nested within a larger seasonal cycle. A cycle of sexual reproduction is followed by a series of asexual reproductive cycles, terminated by a further sexual phase. D. H. Janzen (1977) and others have suggested that the whole asexual clone be regarded as the genuine 'evolutionary individual'. In our terms, this amounts to the suggestion that the development of individual aphids in the asexual phase be regarded as repeated components of a developmental process, like the development of individual leaves on a single plant. Clearly, we need to place extra conditions on the sorts of repeated developmental processes that constitute an 'evolutionary individual', as opposed to an iterated sequence in the development of an individual.

As a first step, we can make use of the evolutionary rationale that we have suggested for the cyclical nature of developmental processes. This leads us to reject the suggestion that the developmental process that produces an individual aphid is not in itself a life cycle in our sense. Variations in the resources that feed into the asexual production of an individual aphid can restructure this process in ways that are reflected in descendant processes. This process is therefore a life cycle of the sort that forms evolving lineages. This is not to suggest, of course, that the longer cycle is not also an evolutionary life cycle. Like many other accounts of evolution, the developmental systems view allows evolutionary units to embed one another. The key to identifying a new unit of self-replication will be to find new events and entities whose numbers, proportions, and properties can be explained as the result of the differential replication of the larger life cycles in which they are involved. Developmental systems accounts of intra-genomic evolution, as in the evolution of meiotic drive mechanisms, could be constructed. Developmental systems accounts of group selection are also possible.

The developmental systems position on the unit of selection debate is thus a form of pluralism. There may be several 'levels' of life cycles, accounting for different features of evolved systems. We are suspicious of the term 'level' here, however, since investigation has not yet proceeded far enough to determine to what extent processes at one 'level' can be considered independent of those at other 'levels'. There is also no real basis for making 'horizontal comparisons' among processes, so as to give a definite meaning to the statement that two processes are at the 'same level'. More investigation of these topics is clearly called for.

It might appear that this interpretation of the aphid case commits us to the view that, for example, metamorphosis in insects constitutes the end of one life cycle and the beginning of another.[6] Variation in a developmental

---

[6] Dawkins's discussion of this case is interestingly unresolved. He would presumably avoid the idea that metamorphic stages are individuals by pointing to the absence of a single-celled bottle-neck.

resource at this point certainly has the potential to cause major heritable alterations to the life cycle. This conclusion does not follow, however, as can be seen by considering the modified life cycle that would result from such a variation. Variation in a developmental resource that caused a different outcome to metamorphosis would give rise to a variant life cycle that recapitulated the phase before the metamorphosis. It would thus be a variant on the larger life cycle of which the metamorphosis is a phase, not a variant in a life cycle with the metamorphosis as its beginning. Hence metamorphosis is a stage in a single life cycle, not the end of one and the beginning of another.[7] Similar considerations explain why the growth of a leaf is an iterated component of a plant's life cycle, not a life cycle in itself. The descendants of variant leaves, if they have any, are variant plants.

A final clarification of our view can be obtained by considering the standard question of the status of vegetative clones. Consider a rhododendron bush that develops where the branch of another bush touches the ground. For Dawkins, its claim to be a new evolutionary individual is fatally undermined by its failure to pass through a single-cell bottle-neck. In his view, any organism that reproduces in this way, via a multicellular propagule, must eventually see its functional organization break down because of the divergent genetic interests of the various cell lineages of which it is composed. We reject this conclusion, because the single-cell bottle-neck is only one way of making a complex system function as a single evolutionary unit. Leo Buss (1987), for example, has drawn attention to the role of the rigidity of plant cell walls, and consequent restrictions on movement, in restricting the potential for conflict between cell lines and allowing the retention of vegetative reproduction as a major mode of propagation. On our view, the individual rhododendrons may well be genuine individuals. The growth of the plant from the initial rooted branch involves the reconstruction of its functional structures from a range of developmental resources, and gives ample opportunities for the development of variant forms as the result of alterations in one or more of those resources.

In summary, we claim that the individual, from a developmental systems perspective, is a process—the life cycle. It is a series of developmental events which forms an atomic unit of repetition in a lineage. Each life cycle is initiated by a period in which the functional structures characteristic of

---

[7] If the effect of a variation at metamorphosis was not as we have envisaged it, and the modified organism gave rise to descendants that bypassed the phases before the metamorphosis, we would have to say that what was previously a phase of a developmental process was now a developmental process in its own right. But in this extraordinary case that would be the right thing to say.

the lineage must be reconstructed from relatively simple resources. At this point there must be potential for variations in the developmental resources to restructure the life cycle in a way that is reflected in descendant cycles.

VII. TYPE AND TOKEN IN DEVELOPMENTAL PROCESSES

In reply to our treatment of vegetative reproduction in the last section, it might be objected that vegetative reproduction omits certain of the early stages of development seen in sexual reproduction in the same lineage, and so is not a repetition of the same developmental process. This objection picks up on the important fact that life cycles may have a disjunctive form, with different individuals having different characteristics. A developmental system can proliferate by producing a range of outcomes on different occasions. This accounts for much of the graded individual variation between organisms. Different humans develop a range of heights. In some circumstances one height is advantageous, in others a different height. The system that is replicated as a result of these individual successes and failures is one that produces a range of heights. All heights in the range are evolved outcomes.

A very similar conceptualization will allow developmental systems theory to encompass the idea of 'alternative life-history strategies'. The successful developmental systems in certain beetle lineages have been those which produce one outcome in response to one sort of interaction, and another in response to a different interaction. The first produces a large, well-armed morph, the second a smaller morph that avoids conflict. Morphs of one type regularly give rise to the other morph. Both morphs are expressions of the same developmental system. The two life cycles of individual rhododendrons, sexual or asexual, are alternative life-history strategies. This is perhaps obscured by the fact that they converge over time, rather than diverging, as in more stereotypical cases. Life cycles of one type regularly give rise to life cycles of the other, so both are segments of the same lineage.

We have shown that very different token developmental processes may be of the same type. This fact also allows us to capture Lorenz's insight that many of what we have termed 'individual traits'—those lacking an evolutionary explanation, may be seen as evolved traits, with full evolutionary explanation, if typed under a more general classification scheme. As Lorenz made famous, it is an evolved developmental outcome in certain waterfowl that they imprint on the first suitable object they see. So the thing they interact with is part of their developmental system. But it is not an evolved developmental outcome for them to imprint on any particular individual, like Lorenz's greylag goose Martina. Although Martina is part of her offspring's developmental system, it would be misleading to describe the situation this way. The general point here is that resources are parts of developmental systems because of generalizations about their role in producing evolved outcomes. In describing the system, we should use descriptions with sufficient generality to enter into these generalizations. There are evolutionary generalizations about the importance of imprinting on parents, and of imprinting on the first largish moving thing, but not one about the importance of imprinting on Martina, or on Lorenz.

This account of how to type-classify elements of the developmental system is the key to how the theory handles learning, and other cases where Lorenz would have invoked the interplay of 'ontogenetic information' and 'phylogenetic information'. Electric light sockets have as yet played little role in human evolutionary history, yet my fear of them (Griffiths 1990) has an evolutionary explanation. The key lies in choosing the right description. Fear of objects associated with injury, or with fear displays in conspecifics, is an evolved developmental outcome. There are evolutionary explanations of my acquiring a fear of any such object. So the resources that produce an organism with such fears are parts of the developmental system. My trait of being afraid of light sockets is an evolved developmental outcome, but only under a general description of the form: 'being afraid of objects with such and such a role in my past learning history'. The light sockets are part of my developmental system, but only under the general description of objects that play that role.

These considerations allow us to describe adequately the case of Sterelny and Elvis Presley raised in Section III. Perhaps there is an explanation of the ability to conform to the preferences of whatever group we find ourselves in at a certain age. In that case, Elvis is part of Sterelny's developmental system, but only under the description 'the preferred object of local aesthetic preference'. But perhaps the lineages that prefer Elvis are on a separate evolutionary trajectory! If an Elvis-filled upbringing makes its recipients likely to prefer Elvis, and if this preference makes them unlikely to achieve successful mating with anyone not similarly inclined, we have the potential for speciation! The ability of the developmental systems approach to explain relationships to individual objects, as well as to types of objects, comes to the fore here. In the extreme version of the Elvis case, there would be a lineage for whom Elvis was part of their developmental system, just as the scent of the home river is part of the developmental system of a lineage of salmon. The relationship between the individual lineage and this particular object is a key part of the evolutionary history of the lineage.

VIII. WHAT IS REPLICATED IN EVOLUTION?

Current mainstream accounts of evolution distinguish two sorts of entities that play distinct roles in the evolutionary process—replicators and interactors. The prototype replicators are genes. According to Dawkins (1976), the genes replicate themselves and exhibit continuity over the generations. They exhibit 'longevity, fecundity, and fidelity' and are potentially 'immortal' (ibid. 37–8). Other features of the organism are mere devices of the genes, whose role is to interact with the environment in the genes' interests. These phenotypic and extended phenotypic features he calls 'vehicles' (Hull (1988) has replaced Dawkins's loaded term 'vehicle' with the term 'interactor'). *Pace* Dawkins, we believe that the replicator/interactor distinction is not driven by considerations of evolutionary theory. It is the projection into evolution of the dichotomous views of development that we have criticized above. A developmental systems account of evolution has no use for the replicator/interactor distinction.

Dawkins has tried to insulate his gene-selectionist view of evolution from views about the role of the genes in development. He argues that 'when we are talking about development it is appropriate to emphasise nongenetic as well as genetic factors. But when we are talking about units of selection a different emphasis is called for, an emphasis on the properties of replicators' (Dawkins 1982: 98). But the two issues cannot be kept apart in this way, because the claim that only genes are replicators is based on an analysis of their role in development. To quote Dawkins himself: 'The special status of genetic factors is deserved for one reason only: genetic factors replicate themselves, blemishes and all, but non-genetic factors do not' (ibid. 99).

But what exactly is it that has the power to replicate itself? A segment of DNA isolated from the cytoplasmic machinery of ribosomes and proteins has no such power. Suppose we enumerate the whole cellular machinery needed to copy a strand of DNA, including the independently inherited centrioles, mitochondria, etc. This is very far from Dawkins's original vision of the immortal gene. Furthermore, under natural conditions this system only replicates itself because of the presence of all the other developmental resources. As Richard Lewontin has remarked, 'if anything in the world can be said to be self-replicating, it is not the gene, but the entire organism as a complex system' (1993: 48).[8]

Once again, the supposed asymmetry between the role of the genes and the role of other developmental resources evaporates when closely analysed. The genes replicate themselves by making a contribution to a developmental process that can initiate new cycles of itself. Other developmental resources do just the same. In one of the earliest responses to Dawkins, Bateson (1978) observed that, if we say a nest is a gene's way of making another gene, we may as well say that a gene is a nest's way of making another nest. The rhetoric of 'self-replicating' genes, while no doubt always intended as an ellipsis for replication in a broader organismic context, has distracted attention from this symmetry between the replication of genes and the replication of many other developmental resources.

According to developmental systems theory, all develomental interactions are replicated, as part of the replication of the developmental process/life cycle. Many of the elements of the developmental system—the developmental resources—are also replicated, as a consequence of the process. Some of these serve as resources for later stages of the process, others as resources for future generations. If we insist that a replicator have the intrinsic causal power to replicate itself, there will be only one replicator, the life cycle. But if we allow the status of 'replicator' to anything that is reliably replicated in development, there will be many replicators. In the terminology of Figure 7.1, the replication of developmental processes or life cycles (D) gives rise to the replication of all the developmental interactions that make up the process (represented by the arrows in Fig. 7.1) and of all the developmental resources that are not merely persistent (B, C, E). The theory of evolution is the theory of the change over time of the numbers, proportions, and properties of all these things.

IX. SELECTION AND COMPETITION IN DEVELOPMENTAL PROCESS EVOLUTION

Taking developmental processes, rather than genes or traditional phenotypes, to be the units of evolution requires a substantial reformulation of evolutionary theory. Yet the fundamental pattern of explanation—the development of complex form through variation and differential replication—is preserved. Perhaps the most radical departure is that the separation

---

[8] For an extended critique of the replicator/interactor distinction, see Greisemer, in press.

of organism and environment is called into question.[9] Evolution occurs because there are variations during the replication of life cycles, and some variations are more successful than others. Traditionally, variants are said to be exposed to independently existing selective forces, expressions of an independently existing environment. In the developmental systems representation, the variants differ in their capacity to replicate themselves. One variant does better than another, not because of a correspondence between it and some pre-existing environmental feature, but because the life cycle that includes interaction with that feature has a greater capacity to replicate itself than the life cycle that lacks that interaction. This perspective is appropriate, because many of the features of the traditional environment have evolutionary explanations. Organism and environment are both evolving as an effect of the evolution of differentially self-replicating life cycles. Life cycles still have fitness values, but these are interpreted, not as a measure of correspondence between the organism and its environment, but as measures of the self-replicating power of the system. Fitness is no longer a matter of 'fittedness' to an independent environment.

Our reinterpretation of natural selection as differential replication draws attention to a frequently neglected class of evolutionary events. There are many variations in developmental processes which are hard to interpret as improvements in the organisms fit to pre-existing selective forces. The cases of birds varying and differentially reproducing in virtue of different habitat associations (discussed in Section IV) provide one example. Another would be cases in which the organism's activity modifies its environment, as when a change in the habits of a eucalypt increases the probability of bush fires. These cases are more easily understood as the incorporation into the developmental process of elements that increase its self-replicating power.

One traditional notion that remains very little changed is that of competition. Competition occurs when two or more developmental processes utilize the same resources and there is a limit to these resources. This may occur because persistent features in the environment are developmental resources for both systems, as when members of different lineages occupy the same limited number of nest sites. It may also occur because resources produced by one process are utilized by another in a way that denies them to future cycles of the first process. Brood parasitism in birds and insects is one example. So competition occurs when a single developmental resource is part of more than one developmental system. Not all interpenetration of

[9] The dissolution of the organism/environment distinction has been urged by other proponents of the developmental systems perspective. See Lewontin 1983 and Oyama 1988.

developmental systems constitutes competition, however, Mutualisms are a positive form of interpenetration, and there are also neutral forms, such as hermit crabs occupying discarded whelk shells.

X. IMPLICATIONS FOR 'CULTURAL EVOLUTION'

The developmental process view changes the relationship between biological and cultural evolution. This distinction rests on a distinction between genetically transmitted and environmentally acquired traits. For example, Elliott Sober (1992) defines cultural evolution as the process in which traits are passed on by learning, rather than by the transmission of genes, and where fitness is measured by how many people learn them, not by how many copies of genes are passed on. Current discussions of evolution often give the impression that cultural evolution began when biological evolution left off! Humans, it is suggested, derived a set of 'biologically based' characters before and during a Pleistocene hunter–gatherer phase.[10] These traits are genetically based, and have been passed down largely unchanged. During this period, however, they acquired an enhanced capacity for learning and for the transmission of information. Cultural structures began to be passed down which are not genetically based. Most change since that period is the result of this latter process.

The developmental systems view implies that it is not possible to divide the traits of organisms into those with a genetic base, which can be explained by biological evolution, and those which are environmentally acquired and are the domain of cultural evolution. The means by which traits are reconstructed in the next generation are varied, and do not admit of any simple twofold division of the sort just described. Instead, all traits that are typical of a lineage are subject to a form of evolutionary explanation that describes how developmental processes replicate and differentiate into lineages as part of an adaptive-historical process. Many elements of the developmental systems associated with these processes can be given evolutionary explanations. Some of these will be elements of the traditional organisms, such as genes. Others will be elements of culture, such as the social structures that are required for the replication of evolved psychological traits in humans.

The developmental systems view emphasizes the currently marginalized fact that humans have had a culture since before they were human. This

---

[10] This view is clearly dominant in an important recent collection of papers on human evolution: Barkow *et al.* 1992.

culture is one of the developmental resources that feeds into the development of evolved traits. It has a history of development and differentiation among lineages as old as that of many other elements in the developmental system. Many species-typical features of human psychology may depend critically on stably replicated features of human culture. Many psychological features that are specific to certain human cultures may nevertheless have evolutionary explanations, since this variation may reflect differentiation among lineages of developmental systems. An obvious research programme within developmental systems theory is an attempt to locate critical developmental resources in human culture(s), and to study their influence on development, and how they themselves are replicated.

Two objections are commonly urged to the idea that cultural evolution can be accommodated in the same theoretical framework as the evolution of traditional biological traits. First, it is often remarked that culture changes much more rapidly that any biological trait. But how rapidly something changes depends on how it is taxonomized. The forms of relationship between the sexes in European society has changed greatly in the last thousand years, but it has remained fundamentally patriarchal. Developmental systems theory suggests an attempt to locate the fundamental developmental resources that account for the stability of this feature. These will be classified in such a way as to allow them to be identified across the whole range of such societies. The second common objection to evolutionary approaches to culture is that cultural traits are transmitted horizontally, rather than vertically, and that this gives cultural evolution a fundamentally different structure from biological evolution, in which traits are transmitted vertically. In such a process, it is suggested, the idea of lineages as the fundamental units of evolution is inappropriate. One response to problems of this kind would be to enlarge the size of the lineage groups studied so as to reduce the incidence of such transmission between the units of study (see O'Hara 1994: 12–22). But this may not be necessary, as the traditional contrast between cultural and biological is overdrawn on both sides. On the biological side, plant evolution and bacterial evolution involve a good deal of horizontal transmission (via hybridization and plasmid exchange). This calls for some revision of traditional methods in studying bacterial evolution, but not enough to render them unrecognizable.[11] On the cultural side, it is plausible that transmission is 'vertical' to a remarkable extent. Languages exchange items of vocabulary, but do not merge wholesale. This form of horizontal transmission is closely akin to

---

[11] For the implications of bacterial plasmid exchange for taxonomy, see Maynard Smith 1990. For implications of hybridization in plants, see McDade 1990, 1992.

plasmid exchange. Some studies (Cavalli-Sforza *et al.* 1988, Penny *et al.* 1993) have claimed a substantial parallelism between trees for language and genetic trees for human lineages. All elements of these comparisons are currently poorly empirically based, and should not be relied upon, but it is not inconceivable that Dr Johnson spoke more truly than he knew when he said that 'languages are the pedigrees of nations'.

XI. CONCLUSION

The developmental systems tradition in biology reflects a continued dissatisfaction among many workers with conventional, gene-centred accounts of development and evolution. Several authors have tried to replace this dichotomy with the idea of a developmental system. Our main aim in this essay has been to make this idea precise. We have shown how to define the system in terms of a developmental process. The developmental process or life cycle is a series of developmental events which forms a unit of repetition in a lineage. Each life cycle is initiated by a period in which the functional structures characteristic of the lineage must be reconstructed from relatively simple resources. At this point there must be potential for variations in the developmental resources to restructure the life cycle in a way that is reflected in descendant cycles. The developmental system is the structured set of resources from which the life cycle is reconstructed in each generation.

Developmental systems theory offers to free biology and the social sciences from the grip of dichotomous accounts of development. Traits need not be either genetic or environmental, either evolved or socially constructed. While there has been a general agreement that these dichotomies are mistaken, attempts to replace them have generally reproduced the same problem in a subtler form. For example, the insistence that all traits depend on both genic and non-genic factors is followed by an attempt to separate the contribution of the two and evaluate which is the more important in a particular case.[12] To take another case, the admission that a trait covaries with changes in the environment is explained by postulating several genetic programmes with environmental 'triggers' to choose among them.

We have also sketched the implications of developmental systems theory for the study of evolution. We argued that the prime unit of evolution (unit of self-replication) is the developmental process, or life cycle. Many

---

[12] For a critique of this attempt, see Lewontin 1974.

developmental resources interact with this process, and these have very different characters, ranging from the genes to persistent features of the environment, such as sunlight. But the interaction of all these features is subject to evolutionary explanation. Furthermore, when a feature is replicated, it is due to the replication of the whole process for which it is a resource. Conceiving evolution as the differential replication of developmental processes/life cycles therefore gives us maximum explanatory power, allowing us to explain everything that can be explained in terms of differential replication. As the last section has shown, this scope may be remarkable.[13]

REFERENCES

Barkow, J. H., Cosmides, L., and Tooby, J. (1992) (eds.), *The Adapted Mind* (New York: Oxford University Press).
Bateson, P. (1978), review of Dawkins 1976, *Animal Behaviour*, 78: 316–18.
——(1983), 'Genes, Embryology and the Development of Behaviour', in P. Slater and T. Halliday (eds.), *Animal Behaviour: Genes, Development and Learning* (Cambridge, Mass.: Blackwell), 52–81.
Buss, L. (1987), *The Evolution of Individuality* (Princeton: Princeton University Press).
Cavalli-Sforza, L. L., *et al.* (1988), 'Reconstruction of Human Evolution: Bringing Together Genetic, Archeological and Linguistic Data', *Proceedings of the National Academy of Sciences*, 85: 6002–6.
Dawkins, R. (1976), *The Selfish Gene* (New York: Oxford University Press).
——(1982), *The Extended Phenotype* (New York: Freeman).
Gottlieb, G. (1976), 'Conceptions of Prenatal Development: Behavioural Embryology', *Psychological Review*, 83: 215–34.
——(1981), 'Roles of Early Experience in Species-Specific Perceptual Development', in R. N. Aslin, J. R. Alberts, and M. P. Petersen (eds.), *Development of Perception* (New York: Academic Press), 5–44.
Gould, S. J., and Lewontin, R. C. (1979), 'The Spandrels of San Marco and the Panglossian Paradigm: A Critique of the Adaptationist Programme', *Proceedings of the Royal Society of London*, B205: 581–98.
Gray, R. D. (1988), 'Metaphors and Methods: Behavioral Ecology, Panbiogeography and the Evolving Synthesis', in M. W. Ho and S. W. Fox (eds.), *Evolutionary Processes and Metaphors* (New Yrok: Wiley), 209–42.
——(1992), 'Death of the Gene: Developmental Systems Strike Back', in P. E. Griffiths (ed.), *Trees of Life: Essays in Philosophy of Biology* (Boston: Kluwer), 165–209.

[13] In preparing this essay, we have benefited greatly from discussions with Susan Oyama, Kim Sterelny, and Patrick Bateson. Earlier drafts have been improved by suggestions from Robert Brandon, David Hull, Timothy Johnston, and Martyn Kennedy.

Greisemer, J. R. (in press), 'The Informational Gene and the Substantial Body: On the Generalisation of Evolutionary Theory by Abstraction', in N. Cartwright and M. Jones (eds.), *Varieties of Idealisation* (Amsterdam: Rodopi).

Griffiths, P. E. (1990), 'Modularity and the Psychoevolutionary Theory of Emotion', *Biology and Philosophy*, 2: 175–96.

——(1993), 'Functional Analysis and Proper Function', *British Journal for Philosophy of Science*, 44: 409–22.

——(1994), 'Cladistic Classifications and Functional Explanations', *Philosophy of Science*, 61: 206–27.

Ho, M. W. (1986), 'Heredity as Process', *Rivista di Biologica-Biology Forum*, 79: 407–47.

Hull, D. (1988), *Science as a Process* (Chicago: University of Chicago Press).

Immelmann, K. (1975), 'Ecological Significance of Imprinting and Early Learning', *Annual Review of Ecology and Systematics*, 6: 15–37.

Janzen, D. H. (1977), 'What are Dandelions and Aphids?', *American Naturalist*, 111: 586–9.

Johnston, T. (1987), 'The Persistence of Dichotomies in the Study of Behavioral Development', *Developmental Review*, 7: 149–82.

——and Gottlieb, G. (1990), 'Neophenogenesis: A Developmental Theory of Phenotypic Evolution', *Journal of Theoretical Biology*, 147: 471–95.

Lehrman, D. S. (1953), 'Critique of Konrad Lorenz's Theory of Instinctive Behaviour', *Quarterly Review of Biology*, 28: 337–63.

——(1970), 'Semantic and Conceptual Issues in the Nature–Nurture Problem', in his *Development and the Evolution of Behavior* (San Francisco: Freeman), 17–52.

Lewontin, R. C. (1974), 'The Analysis of Variance and the Analysis of Causes', *American Journal of Human Genetics*, 26: 400–11.

——(1982), *Human Diversity* (New York: Scientific American).

——(1983), 'The Organism as the Subject and Object of Evolution', *Scientia*, 118: 65–82.

——(1993), *Biology as Ideology: The Doctrine of DNA* (New York: HarperCollins).

Lorenz, K. (1965), *Evolution and the Modification of Behavior* (Chicago: University of Chicago Press).

McDade, L. (1990), 'Hybrids and Phylogenetic Systematics I. Patterns of Character Expression in Hybrids and their Implications for Cladistic Analysis,' *Evolution*, 44: 1685–1700.

——(1992), 'Hybrids and Phylogenetic Systematics II. The Impact of Hybrids on Cladistic Analysis', *Evolution*, 46: 1329–46.

Maynard Smith, J. (1990), 'The Evolution of Prokaryotes: Does Sex Matter?', *Annual Review of Ecology and Systematics*, 21: 1–12.

Millikan, R. G. (1984), *Language, Thought, and Other Biological Categories* (Cambridge, Mass.: MIT Press).

Moss, L. (1992), 'A Kernel of Truth? On the Reality of the Genetic Program', in D. Hull, M. Forbes, and K. Okruhlik (eds.), *Philosophy of Science Association Proceedings*, 1: 335–48.

Nijhout, H. F. (1990), 'Metaphors and the Role of Genes in Development', *Bioessays*, 12: 4410–46.

O'Hara, R. J. (1994), 'Evolutionary History and the Species Problem', *American Zoologist*, 34: 12–22.

Oyama, S. (1985), *The Ontogeny of Information* (New York: Cambridge University Press).

—— (1988), 'Stasis, Development and Heredity', in M. W. Ho and S. W. Fox (eds.), *Evolutionary Processes and Metaphors* (New York: Wiley), 255–74.

—— (1989), 'Ontogeny and the Central Dogma', in M. R. Gunnar and E. Thalen (eds.), *Systems and Development* (Hillsdale, NJ: Erlbaum), 1–34.

Penny, D., Watson, E. E., and Steel, M. A. (1993), 'Trees from Languages and Genes Are Very Similar', *Systematic Zoology*, 42: 382–4.

Seligman, M. E. P., and Hager, J. L. (1972) (eds.), *Biological Boundaries of Learning* (New York: Appleton, Century, Crofts).

Smith, K. C. (1992), 'Neo-rationalism versus Neo-Darwinism: Integrating Development and Evolution', *Biology and Philosophy*, 7: 431–52.

Sober, E. (1992), 'Models of Cultural Evolution', in P. E. Griffiths (ed.), *Trees of Life: Essays in Philosophy of Biology* (Boston: Kluwer), 17–39.

Stent, G. (1981), 'Strength and Weakness of the Genetic Approach to the Development of the Nervous System', in W. M. Cowan (ed.), *Studies in Developmental Neurobiology* (New York: Oxford University Press), 288–320.

Sterelny, K., and Kitcher, P. (1988), 'The Return of the Gene', *Journal of Philosophy*, 85/7: 339–61; reproduced as Ch. 8.

PART III

UNITS OF SELECTION

INTRODUCTION TO PART III

DAVID L. HULL

Periodically a book or paper will focus attention on a particular problem that has not been given sufficient attention in the past. G. C. Williams's *Adaptation and Natural Selection* (1966) is a case in point. Although previous authors had discussed group selection and its problems, Williams forced his fellow biologists to face up to some very hard issues. Another example of a book that has been the focus of immense attention is Richard Dawkins's *The Selfish Gene* (1976). Just as Williams raised serious doubts about the prevalence of the selection of 'groups', so Dawkins raised a series of objections to selection occurring even at the level of 'individuals'—that is, particular organisms. Dawkins argued that selection occurs almost exclusively at the level of individual genes or, more generally, replicators. In its most extreme form, gene selectionism is the view that ultimately the only things that compete with each other are alleles at the same locus. Genes at different loci cannot compete with each other in the relevant sense. In this sense, gene selectionism is an extremely monistic view of evolution.

These two books generated vast literatures both defending and attacking their general messages. Initially, defences of group selection tended to be quite muted. Group selection is not impossible. However, the conditions necessary for it to occur are quite stringent. As a result, group selection is quite rare. Later authors have argued that group selection, once properly defined, is much more common than critics suppose (Wilson and Sober 1994). Defences of organisms as units of selection have been anything but muted. Dawkins's gene selectionism obviously struck a nerve for both biologists and philosophers of biology. The major objection to gene selectionism has been that it is too 'reductionistic', as if all the activities at higher levels of organization can be reduced to changes in gene frequencies.

Dawkins does not ignore the role in selection of entities more inclusive than alternative alleles at a single locus. After all, Dawkins is one of the chief targets of those biologists who object to adaptationism, and organisms are the primary example of entities exhibiting adaptations (see Part I). They are the things that exhibit protective colouration and engage in arms races. However, Dawkins sees organisms as playing a definitely derivative role in selection. In fact, organisms are not necessary for selection at all, only phenotypic traits, and Dawkins (1982) extends the notion of phenotypic traits to include such non-standard examples as nests and spider webs (see Part II).

Although Dawkins models his system on the relation between genes and phenotypic traits, he defines it in the more general terms of replicators and vehicles. Replicators—in particular, germ-line replicators—are those entities that replicate themselves indefinitely through time. In addition, replicators construct vehicles that aid them in achieving their differential replication. The notion of replicators has itself spread quite extensively in the literature on selection since Dawkins first began promoting it. His complementary notion of vehicles has not been quite as successful. The relationship between Dawkins's replicators and vehicles is a matter of embryological development, and development continues to pose serious problems for evolutionary biologists (see Part II). As a result, Hull (1980) introduced an evolutionary alternative to vehicles—interactors (see also Brandon 1982). Interactors are entities that interact as cohesive wholes with their environments in such a way as to *cause* replication to be differential. The nature of these causes is left open. They might be embryological, but they need not be.

In addition to being a gene selectionist, Dawkins has been one of the most effective advocates of sociobiology—the biological explanation of behaviour and social organization, particularly human behaviour and social organization. Sociobiologists have tended to be reductionists in the sense that psychological and sociological phenomena are to be explained, not it terms of psychological or sociological theories, but in terms of biological theories. However, 'biologically' does not entail 'genetically'. Early attempts at explaining human behaviour biologically relied on selection occurring at the level of groups, and many sociobiologists continue to explain behaviour at the level of organisms, never proceeding to the genetic level.

Still, in part because of the influence of Dawkins, sociobiologists have also tended to be gene selectionists, attracting criticisms because of these related but independent positions. A second objection to sociobiology is that it is committed to genetic determinism. Once again, gene selectionism and gene determinism are related, but independent, positions. Gene selectionism concerns evolution, while genetic determinism relates to embryological development. According to gene selectionists, all higher-level phenomena can be explained ultimately by reference to genes and nothing but genes. Gene determinists argue that genes play a determinant role in embryology, while environments only 'influence' developmental pathways. We are constantly hearing about the discovery of a gene for schizophrenia, or baldness, or homosexuality. We rarely hear of the discovery of an environment for schizophrenia, or baldness, or homosexuality. Particular environments are necessary for the development of any trait, but gene determinists do not find their specification all that significant (but see Griffiths and Gray in Ch. 7).

Philip Kitcher (1985) has been one of the most effective critics of both gene selectionism and sociobiology. Thus, his paper with Kim Sterelny (Stevelny and Kitcher 1988, Ch. 8 below) defending gene selectionism caused quite a stir. Although Sterelny and Kitcher are far from disciples of Dawkins, they argue that many of the objections that have been raised to his gene selectionist position do not stand up to careful investigation. The two most important areas of dispute are the relationships between genes at different loci and the relationships between genes and traits. All that different alleles at the same locus do is compete with each other, but as the authors in Part III argue, genes at different loci can co-operate with each other. In addition, the embryological relationships between genes and the traits which they influence are extremely complicated and variable.

These authors also address the issue of the hierarchical organization of selection. If the living world is organized into two hierarchies, one of increasingly inclusive replicators, the other of increasingly inclusive interactors, what are the relationships between these two hierarchies and between the various levels in any one hierarchy? One answer is that entities in the replication hierarchy alternate with entities in the interactor hierarchy. Selection is a function of replication alternating with interaction alternating with replication, and so on. As far as the replication hierarchy is concerned, replication is concentrated at the lower levels, primarily at the level of the genetic material, while interaction occurs at a wide variety of levels, from single genes to organisms and possibly various sorts of groups. As Lloyd (1988) argues, the levels of selection controversy has always been about the levels at which interaction takes place. Because the preceding view of selection involves two hierarchies and interaction occurs at a wide variety of levels in its hierarchy, it is pluralistic.

Central to the problems surrounding selection is the more general notion of causation. For example, Brandon (1982) opts for a screening-off analysis of causation, while Sober (1984) adopts a Pareto-style conception. One of the benefits that Brandon sees for his screening-off analysis of causation is that it reveals an important asymmetry in selection:

152 DAVID L. HULL

phenotypes screen off genotypes, whereas genotypes do not screen off phenotypes. Hence, contrary to Dawkins, phenotypes are more fundamental than genotypes. As might be expected from Sober having adopted a different notion of causation, he disagrees (for the ensuing dispute see Sober 1992, Brandon *et al.* 1994, Sober 1994, and Brandon 1996).

REFERENCES

Brandon, R. N. (1982), 'The Levels of Selection', in P. Asquith and T. Nickles (eds.), *PSA 1982* (East Lansing, Mich.: Philosophy of Science Association), i. 315–22; reproduced as Ch. 9.
——(1996), *Concepts and Methods in Evolutionary Biology* (Cambridge: Cambridge University Press).
——Antonovics, J., Burian, R., Carson, S., Cooper, G., Davies, P. S., Horvath, C., Mishler, B. D., Richardson, R. C., Smith, K., and Thrall, P. (1994), 'Discussion: Sober on Brandon on Screening-Off and the Levels of Selection', *Philosophy of Science*, 61: 475–86.
Dawkins, R. (1976), *The Selfish Gene* (Oxford: Oxford University Press).
——(1982), *The Extended Phenotype* (San Francisco: Freeman, Cooper and Co.).
Hull, D. L. (1980), 'Individuality and Selection', *Annual Review of Ecology and Systematics*, 11: 311–32.
Kitcher, P. (1985), *Vaulting Ambition: Sociobiology and the Quest for Human Nature* (Cambridge, Mass.: MIT Press).
Lloyd, E. A. (1988), *The Structure and Confirmation of Evolutionary Theory* (New York: Greenwood Press).
Sober, E. (1984), *The Nature of Selection* (Cambridge, Mass.: MIT Press).
——(1992), 'Screening-Off and the Units of Selection', *Philosophy of Science*, 59: 142–52.
——(1994), *From a Biological Point of View: Essays in Evolutionary Biology* (Cambridge: Cambridge University Press).
Sterelny, K., and Kitcher, P. (1988), 'The Return of the Gene', *Journal of Philosophy*, 85: 339–61; reproduced as Ch. 8.
Williams, G. C. (1966), *Adaptation and Natural Selection* (Princeton: Princeton University Press).
Wilson, D. S., and Sober, E. (1994), 'Re-introducing Group Selection to the Human Behavioral Sciences', *Behavioral and Brain Sciences*, 17: 585–654.

8

THE RETURN OF THE GENE

KIM STERELNY AND PHILIP KITCHER

We have two images of natural selection. The orthodox story is told in terms of individuals. More organisms of any given kind are produced than can survive and reproduce to their full potential. Although these organisms are of a kind, they are not identical. Some of the differences among them make a difference to their prospects for survival or reproduction, and hence, on the average, to their actual reproduction. Some of the differences which are relevant to survival and reproduction are (at least partly) heritable. The result is evolution under natural selection, a process in which, barring complications, the average fitness of the organisms within a kind can be expected to increase with time.

There is an alternative story. Richard Dawkins[1] claims that the 'unit of selection' is the gene. By this he means not just that the result of selection is (almost always) an increase in frequency of some gene in the gene pool. That is uncontroversial. On Dawkins's conception, we should think of genes as differing with respect to properties that affect their abilities to leave copies of themselves. More genes appear in each generation than can copy themselves up to their full potential. Some of the differences among them make a difference to their prospects for successful copying, and hence to the number of actual copies that appear in the next generation. Evolution under natural selection is thus a process in which, barring complication, the average ability of the genes in the gene pool to leave copies of themselves increases with time.

Dawkins's story can be formulated succinctly by introducing some of his terminology. Genes are *replicators*, and selection is the struggle among

First published in *Journal of Philosophy*, 85/7 (1988): 339–61. Reprinted by permission.

[1] The claim is made in Dawkins 1976 and, in a somewhat modified form, in Dawkins 1982. We shall discuss the difference between the two versions in the final section of this essay, and our reconstruction will be primarily concerned with the later version of Dawkins's thesis. To forestall any possible confusion, our reconstruction of Dawkins's position does not commit us to the provocative claims about altruism and selfishness on which many early critics of Dawkins 1976 fastened.

active germ-line replicators. Replicators are entities that can be copied. Active replicators are those whose properties influence their chances of being copied. Germ-line replicators are those which have the potential to leave infinitely many descendants. Early in the history of life, coalitions of replicators began to construct *vehicles* through which they spread copies of themselves. Better replicators build better vehicles, and hence are copied more often. Derivatively, the vehicles associated with them become more common too. The orthodox story focuses on the successes of prominent vehicles—individual organisms. Dawkins claims to expose an underlying struggle among the replicators.

We believe that a lot of unnecessary dust has been kicked up in discussing the merits of the two stories. Philosophers have suggested that there are important connections to certain issues in the philosophy of science: reductionism, views on causation and natural kinds, the role of appeals to parsimony. We are unconvinced. Nor do we think that a willingness to talk about selection in Dawkinspeak brings any commitment to the adaptationist claims which Dawkins also holds. After all, adopting a particular perspective on selection is logically independent of claiming that selection is omnipresent in evolution.

In our judgement, the relative worth of the two images turns on two theoretical claims in evolutionary biology.

1. Candidate units of selection must have systematic causal consequences. If *X*s are selected for, the *X* must have a systematic effect on its expected representation in future generations.
2. Dawkins's gene selectionism offers a *more general theory* of evolution. It can also handle those phenomena which are grist to the mill of individual selection, but there are evolutionary phenomena which fit the picture of individual selection ill or not at all, yet which can be accommodated naturally by the gene selection model.

Those sceptical of Dawkins's picture—in particular, Elliott Sober, Richard Lewontin, and Stephen Jay Gould—doubt whether genes can meet the condition demanded in (1). In their view, the phenomena of epigenesis and the extreme sensitivity of the phenotype to gene combinations and environmental effects undercut genic selectionism. Although we believe that these critics have offered valuable insights into the character of sophisticated evolutionary modelling, we shall try to show that these insights do not conflict with Dawkins's story of the workings of natural selection. We shall endeavour to free the thesis of genic selectionism from some of the troublesome excrescences which have attached themselves to an interesting story.

I. GENE SELECTION AND BEAN-BAG GENETICS

Sober and Lewontin (1982) argue against the thesis that all selection is genic selection by contending that many instances of selection do not involve selection for properties of individual alleles. Stated rather loosely, the claim is that, in some populations, properties of individual alleles are not positive causal factors in the survival and reproductive success of the relevant organisms. Instead of simply resting this claim on an appeal to our intuitive ideas about causality, Sober has recently provided an account of causal discourse which is intended to yield the conclusion he favours, thus rebutting the proposals of those (like Dawkins) who think that properties of individual alleles can be causally efficacious (Sober 1984: chs. 7–9, esp. pp. 302–14).

The general problem arises because replicators (genes) combine to build vehicles (organisms), and the effect of a gene is critically dependent on the company it keeps. However, recognizing the general problem, Dawkins seeks to disentangle the various contributions of the members of the coalition of replicators (the genome). To this end, he offers an analogy with a process of competition among rowers for seats in a boat. The coach may scrutinize the relative times of different teams, but the competition can be analysed by investigating the contributions of individual rowers in different contexts (Dawkins 1976: 40–1, 91–2; 1982: 239).

Sober's Case

At the general level, we are left trading general intuitions and persuasive analogies. But Sober (and, earlier, Sober and Lewontin) attempted to clarify the case through a particular example. Sober argues that *heterozygote superiority* is a phenomenon that cannot be understood from Dawkins's standpoint. We shall discuss Sober's example in detail; our strategy is as follows. We first set out Sober's case: heterozygote superiority cannot be understood as a gene-level phenomenon, because only pairs of genes can be, or fail to be, heterozygous. Yet, being heterozygous can be causally salient in the selective process. Against Sober, we first offer an analogy to show that there must be something wrong with his line of thought: from the gene's-eye view, heterozygote superiority is an instance of a standard selective phenomenon: namely, *frequency-dependent* selection. The advantage (or disadvantage) of a trait can depend on the frequency of that trait in other members of the relevant population.

Having claimed that there is something wrong with Sober's argument, we then try to say what is wrong. We identify two principles on which the reasoning depends. First is a general claim about causal uniformity. Sober thinks that there can be selection for a property only if that property has a positive uniform effect on reproductive success. Second, and more specifically, in cases where the heterozygote is fitter, the individuals have no uniform causal effect. We shall try to undermine both principles, but the bulk of our criticism will be directed against the first.

Heterozygote superiority occurs when a heterozygote (with genotype Aa, say) is fitter than either homozygote (AA or aa). The classic example is human sickle-cell anaemia: homozygotes for the normal allele in African populations produce functional haemoglobin, but are vulnerable to malaria; homozygotes for the mutant ('sickling') allele suffer anaemia (usually fatal); and heterozygotes avoid anaemia, while also having resistance to malaria. The effect of each allele varies with context, and the contexts across which variation occurs are causally relevant. Sober writes:

In this case, the a allele does not have a unique causal role. Whether the gene a will be a positive or a negative causal factor in the survival and reproductive success of an organism depends on the genetic context. If it is placed next to a copy of A, a will mean an increase in fitness. If it is placed next to a copy of itself, the gene will mean a decrement in fitness. (Sober 1984: 303)

The argument against Dawkins expressed here seems to come in two parts. Sober relies on the principle

(A) There is selection for property P only if in all causally relevant background conditions P has a positive effect on survival and reproduction.

He also adduces a claim about the particular case of heterozygote superiority.

(B) Although we can understand the situation by noting that the heterozygote has a uniform effect on survival and reproduction, the property of having the A allele and the property of having the a allele cannot be seen as having uniform effects on survival and reproduction.

We shall argue that both (A) and (B) are problematic.

Let us start with the obvious reply to Sober's argument. It seems that the heterozygote superiority case is akin to a familiar type of frequency-dependent selection. If the population consists just of AAs and a mutation arises, the a allele, then, initially, a is favoured by selection. Even though it is very bad to be aa, a alleles are initially likely to turn up in the company of A alleles. So they are likely to spread, and as they spread, they find themselves alongside other a alleles, with the consequence that selection tells against them. The scenario is very similar to a story we might tell about interactions among individual organisms. If some animals resolve conflicts by playing hawk and others play dove, then, if a population is initially composed of hawks (and if the costs of bloody battle outweigh the benefits of gaining a single resource), doves will initially be favoured by selection.[2] For they will typically interact with hawks, and, despite the fact that their expected gains from these interactions are zero, they will still fare better than their rivals, whose expected gains from interactions are negative. But as doves spread in the population, hawks will meet them more frequently, with the result that the expected pay-offs to hawks from interactions will increase. Because they increase more rapidly than the expected pay-offs to the doves, there will be a point at which hawks become favoured by selection, so that the incursion of doves into the population is halted.

We believe that the analogy between the case of heterozygote superiority and the hawk–dove case reveals that there is something troublesome about Sober's argument. The challenge is to say exactly what has gone wrong.

Causal Uniformity

Start with principle (A). Sober conceives of selection as a *force*, and he is concerned to make plain the effects of component forces in situations where different forces combine. Thus, he invites us to think of the heterozygote superiority case by analogy with situations in which a physical object remains at rest because equal and opposite forces are exerted on it. Considering the situation only in terms of net forces will conceal the causal structure of the situation. Hence, Sober concludes, our ideas about units of selection should penetrate beyond what occurs on the average, and we should attempt to isolate those properties which positively affect survival and reproduction in every causally relevant context.

Although Sober rejects determinism, principle (A) seems to hanker after something like the uniform association of effects with causes that deterministic accounts of causality provide. We believe that the principle cannot be satisfied without doing violence to ordinary ways of thinking about natural selection, and, once the violence has been exposed, it is not obvious that there is any way to reconstruct ideas about selection that will fit Sober's requirement.

Consider *the* example of natural selection, the case of industrial melanism.[3] We are inclined to say that the moths in a Cheshire wood, where

---

[2] For details, see Maynard Smith 1982, and for a capsule presentation, Kitcher 1985: 88–97.
[3] The *locus classicus* for discussion of this example is Kettlewell 1973.

lichens on many trees have been destroyed by industrial pollutants, have been subjected to selection pressure, and that there has been selection for the property of being melanic. But a moment's reflection should reveal that this description is at odds with Sober's principle. For the wood is divisible into patches, among which are clumps of trees that have been shielded from the effects of industrialization. Moths who spend most of their lives in these areas are at a disadvantage if they are melanic. Hence, in the population comprising all the moths in the wood, there is no uniform effect on survival and reproduction: in some causally relevant contexts (for moths who have the property of living in regions where most of the trees are contaminated), the trait of being melanic has a positive effect on survival and reproduction, but there are other contexts in which the effect of the trait is negative.

The obvious way to defend principle (A) is to split the population into sub-populations and identify different selection processes as operative in different subgroups. This is a revisionary proposal, for our usual approach to examples of industrial melanism is to take a coarse-grained perspective on the environments, regarding the existence of isolated clumps of uncon-taminated trees as a perturbation of the overall selective process. None the less, we might be led to make the revision, not in the interest of honouring a philosophical prejudice, but simply because our general views about selection are consonant with principle (A), so that the reform would bring our treatment of examples into line with our most fundamental beliefs about selection.

In our judgement, a defence of this kind fails for two connected reasons. First, the process of splitting populations may have to continue much further—perhaps even to the extent that we ultimately conceive of individ-ual organisms as making up populations in which a particular type of selection occurs. For, even in contaminated patches, there may be varia-tions in the camouflaging properties of the tree trunks, and these variations may combine with propensities of the moths to cause local disadvantages for melanic moths. Second, as many writers have emphasized, evolutionary theory is a statistical theory, not only in its recognition of drift as a factor in evolution, but also in its use of fitness coefficients to represent the expected survivorship and reproductive success of organisms. The envis-aged splitting of populations to discover some partition in which principle (A) can be maintained is at odds with the strategy of abstracting from the thousand natural shocks that organisms in natural populations are heir to. In principle, we could relate the biography of each organism in the popu-lation, explaining in full detail how it developed, reproduced, and survived, just as we could track the motion of each molecule of a sample of gas. But evolutionary theory, like statistical mechanics, has no use for such a fine grain of description: the aim is to make clear the central tendencies in the history of evolving populations, and, to this end, the strategy of averaging, which Sober decries, is entirely appropriate. We conclude that there is no basis for any revision that would eliminate those descriptions which run counter to principle (A).

At this point, we can respond to the complaints about the gene's-eye view representation of cases of heterozygote superiority. Just as we can give sense to the idea that the trait of being melanic has a unique environment-dependent effect on survival and reproduction, so too we can explicate the view that a property of alleles—to wit, the property of directing the formation of a particular kind of haemoglobin—has a unique environment-dependent effect on survival and reproduction. The alleles form parts of one another's environments, and, in an environment in which a copy of the A allele is present, the typical trait of the S allele (namely, directing the formation of deviant haemoglobin) will usually have a positive effect on the chances that copies of that allele will be left in the next generation. (Notice that the effect will not be invariable, for there are other parts of the genomic environment which could wreak havoc with it.) If someone protests that the incorporation of alleles as themselves part of the environment is suspect, then the immediate rejoinder is that, in cases of behavioural interactions, we are compelled to treat organisms as parts of one another's environments.[4] The effects of playing hawk depend on the nature of the environment, specifically on the frequency of doves in the vicinity.[5]

[4] In the spirit of Sober's original argument, one might press further. Genic selectionists contend that an A allele can find itself in two different environments, one in which the effect of directing the formation of a normal globin chain is positive and one in which that effect is negative. Should we not be alarmed by the fact that the distribution of environments in which alleles are selected is itself a function of the frequency of the alleles whose selection we are following? No. The phenomenon is thoroughly familiar from studies of behavioural interactions—in the hawk–dove case we treat the frequency of hawks both as the variable we are tracking and as a facet of the environment in which selection occurs. Maynard Smith makes the parallel fully explicit in his 1987, esp. 125–6.

[5] Moreover, we can explicitly recognize the co-evolution of alleles with allelic environments. A fully detailed general approach to population genetics from the Dawkinsian point of view will involve equations that represent the functional dependence of the distribution of environments on the frequency of alleles and equations that represent the fitnesses of individual alleles in different environments. In fact, this is just another way of looking at the standard population-genetics equations. Instead of thinking of W_{AA} as the expected contribution to survival and reproduction of (an organism with) an allelic pair, we think of it as the expected contribution of copies of itself: i.e. of the allele A in the environment of another copy of the A allele. We now see W_{AS} as the expected contribution of A in environment S and also as the expected contribution of S in environment A. The frequencies p, q are not only the frequencies of the alleles, but also the frequencies with which certain environments occur. The standard definitions of the overall (net) fitnesses of the alleles are obtained by weighting

The Causal Powers of Alleles

We have tried to develop our complaints about principle (A) into a positive account of how cases of heterozygote superiority might look from the gene's-eye view. We now want to focus more briefly on (B). Is it impossible to reinterpret the examples of heterozygote superiority so as to ascribe uniform effects on survival and reproduction to allelic properties? The first point to note is that Sober's approach formulates the Dawkinsian point of view in the wrong way: the emphasis should be on the effects of properties of alleles, not on allelic properties of organisms (like the property of having an A allele), and the accounting ought to be done in terms of allele copies. Second, although we argued above that the strategy of splitting populations was at odds with the character of evolutionary theory, it is worth noting that the same strategy will be available in the heterozygote superiority case.

Consider the following division of the original population: let P_1 be the collection of all those allele copies which occur next to an S allele, and let P_2 consist of all those allele copies which occur next to an A allele. Then the property of being A (or of directing the production of normal haemoglobin) has a positive effect on the production of copies in the next generation in P_1, and conversely in P_2. In this way, we are able to partition the population and to achieve a Dawkinsian redescription that meets Sober's principle (A)—just in the way that we might try to do so if we wanted to satisfy (A) in understanding the operation of selection on melanism in a Cheshire wood or on fighting strategies in a population containing a mixture of hawks and doves.

Objection: the 'populations' just defined are highly unnatural, and this can be seen once we recognize that, in some cases, allele copies in the same organisms (the heterozygotes) belong to different 'populations'. Reply: so what? From the allele's point of view, the copy next door is just a critical part of the environment. The populations P_1 and P_2 simply pick out the alleles that share the same environment. There would be an analogous partition of a population of competing organisms which occurred locally in pairs such that some organisms played dove and some hawk. (Here, mixed pairs would correspond to heterozygotes.)

the fitnesses in the different environments by the frequencies with which the environments occur.

Lewontin has suggested to us that problems may arise with this scheme of interpretation if the population should suddenly start to reproduce asexually. But this hypothetical change could be handled from the genic point of view by recognizing an alteration of the co-evolutionary process between alleles and their environments: whereas certain alleles used to have descendants that would encounter a variety of environments, their descendants are now found only in one allelic environment. Once the algebra has been formulated, it is relatively straightforward to extend the reinterpretation to this case.

So the genic picture survives an important initial challenge. The moral of our story so far is that the picture must be applied consistently. Just as paradoxical conclusions will result if one offers a partial translation of geometry into arithmetic, it is possible to generate perplexities by failing to recognize that the Dawkinsian *Weltanschauung* leads to new conceptions of environment and of population. We now turn to a different worry, the objection that genes are not 'visible' to selection.

II. EPIGENESIS AND VISIBILITY

In a lucid discussion of Dawkins's early views, Gould claims to find a 'fatal flaw' in the genic approach to selection. According to Gould, Dawkins is unable to give genes 'direct visibility to natural selection' (1980: 90).[6] Bodies must play intermediary roles in the process of selection, and since the properties of genes do not map in one–one fashion on to the properties of bodies, we cannot attribute selective advantages to individual alleles. We believe that Gould's concerns raise two important kinds of issues for the genic picture: (i) Can Dawkins sensibly talk of the effect of an individual allele on its expected copying frequency? (ii) Can Dawkins meet the charge that it is the phenotype that makes the difference to the copying of the underlying alleles, so that, whatever the causal basis of an advantageous trait, the associated allele copies will have enhanced chances of being replicated? We shall take up these questions in order.

Do Alleles Have Effects?

Dawkins and Gould agree on the facts of embryology which subvert the simple Mendelian association of one gene with one character. But the salience of these facts to the debate is up for grabs. Dawkins regards Gould as conflating the demands of embryology with the demands of the theory of evolution. While genes' effects blend in embryological development, and while they have phenotypic effects only in concert with their gene-mates, genes 'do not blend as they replicate and recombine down the generations. It is this that matters for the geneticist, and it is also this that matters for the student of units of selection' (Dawkins 1982: 117).

Is Dawkins right? Chapter 2 of *The Extended Phenotype* (1982) is an explicit defence of the meaningfulness of talk of 'genes for' indefinitely complex morphological and behavioural traits. In this, we believe,

---

[6] There is a valuable discussion of Gould's claims in Sober 1984: 227 ff.

Dawkins is faithful to the practice of classical geneticists. Consider the vast number of loci in *Drosophila melanogaster* which are labelled for eye-colour traits—white, eosin, vermilion, raspberry, and so forth. Nobody who subscribes to this practice of labelling believes that a pair of appropriately chosen stretches of DNA, cultured in splendid isolation, would produce a detached eye of the pertinent colour. Rather, the intent is to indicate the effect that certain changes at a locus would make against the background of the rest of the genome.

Dawkins's project here is important not just in conforming to traditions of nomenclature. Remember: Dawkins needs to show that we can sensibly speak of alleles having (environment-sensitive) effects, effects in virtue of which they are selected for or selected against. If we can talk of a gene for X, where X is a selectively important phenotypic characteristic, we can sensibly talk of the effect of an allele on its expected copying frequency, even if the effects are always indirect, via the characteristics of some vehicle.

What follows is a rather technical reconstruction of the relevant notion. The precision is needed to allow for the extreme environmental sensitivity of allelic causation: But the intuitive idea is simple: we can speak of genes for X if substitutions on a chromosome would lead, in the relevant environments, to a difference in the X-ishness of the phenotype.

Consider a species S and an arbitrary locus L in the genome of members of S. We want to give sense to the locution 'L is a locus affecting P' and derivatively to the phrase 'G is a gene for P^*' (where, typically, P will be a determinable and P^* a determinate form of P). Start by taking an *environment* for a locus to be an aggregate of DNA segments that would complement L to form the genome of a member of S together with a set of extra-organismic factors (those aspects of the world external to the organism which we would normally count as part of the organism's environment). Let a set of variants for L be any collection of DNA segments, none of which is debarred, on physico-chemical grounds, from occupying L. (This is obviously a very weak constraint, intended only to rule out those segments which are too long or which have peculiar physico-chemical properties.) Now, we say that L is a locus affecting P in S relative to an environment E and a set of variants V just in case there are segments s, s^*, and s^{**} in V such that the substitution of s^{**} for s^* in an organism having s and s^* at L would cause a difference in the form of P, against the background of E. In other words, given the environment E, organisms who are ss^* at L differ in the form of P from organisms who are ss^{**} at L, and the cause of the difference is the presence of s^* rather than s^{**}. (A minor clarification: while $s*$ and $s**$ are distinct, we do not assume that they are both different from s.)

L is a locus affecting P in S just in case L is a locus affecting P in S relative to any standard environment and a feasible set of variants. Intuitively, the geneticist's practice of labelling loci focuses on the 'typical' character of the complementary part of the genome in the species, the 'usual' extra-organismic environment, and the variant DNA segments which have arisen in the past by mutation or which 'are likely to arise' by mutation. Can these vague ideas about standard conditions be made more precise? We think so. Consider first the genomic part of the environment. There will be numerous alternative combinations of genes at the loci other than L present in the species S. Given most of these gene combinations, we expect modifications at L to produce modifications in the form of P. But there are likely to be some exceptions, cases in which the presence of a rare allele at another locus or a rare combination of alleles produces a phenotypic effect that dominates any effect on P. We can either dismiss the exceptional cases as non-standard because they are infrequent, or we can give a more refined analysis, proposing that each of the non-standard cases involves either (a) a rare allele at a locus L' or (b) a rare combination of alleles at loci L', L'' ... such that locus (a) or those loci jointly (b) affect some phenotypic trait Q that dominates P in the sense that there are modifications of Q which prevent the expression of any modifications of P. As a concrete example, consider the fact that there are modifications at some loci in *Drosophila* which produce embryos that fail to develop heads; given such modifications elsewhere in the genome, alleles affecting eye colour do not produce their standard effects!

We can approach standard extra-genomic environments in the same way. If L affects the form of P in organisms with a typical gene complement, except for those organisms which encounter certain rare combinations of external factors, then we may count those combinations as non-standard simply because of their infrequency. Alternatively, we may allow rare combinations of external factors to count provided that they do not produce some gross interference with the organism's development, and we can render the last notion more precise by taking non-standard environments to be those in which the population mean fitness of organisms in S would be reduced by some arbitrarily chosen factor (say, $1/2$).

Finally, the feasible variants are those which actually occur at L in members of S, together with those which have occurred at L in past members of S and those which are easily attainable from segments that actually occur at L in members of S by means of insertion, deletion, substitution, or transposition. Here the criteria for ease of attainment are given by the details of molecular biology. If an allele is prevalent at L in S, then modifications at sites where the molecular structure favours insertions, deletions, substitutions, or transpositions (so-called hot spots) should count as easily attainable even if some of these modifications do not actually occur.

Obviously, these concepts of 'standard conditions' could be articulated in more detail, and we believe that it is possible to generate a variety of explications, agreeing on the core of central cases but adjusting the boundaries of the concepts in different ways. If we now assess the labelling practices of geneticists, we expect to find that virtually all of their claims about loci affecting a phenotypic trait are sanctioned by all of the explications. Thus, the challenge that there is no way to honour the facts of epigenesis while speaking of loci that affect certain traits would be turned back.

Once we have come this far, it is easy to take the final step. An allele A at a locus L in a species S is for the trait $P*$ (assumed to be a determinate form of the determinable characteristic P) relative to a local allele B and an environment E just in case (a) L affects the form of P in S, (b) E is a standard environment, and (c) in E organisms that are AB have phenotype $P*$. The relativization to a local allele is necessary, of course, because, when we focus on a target allele rather than a locus, we have to extend the notion of the environment—as we saw in the last section, corresponding alleles are potentially important parts of one another's environments. If we say that A is for $P*$ (period), we are claiming that A is for $P*$ relative to standard environments and common local alleles, or that A is for $P*$ relative to standard environments and itself.

Now, let us return to Dawkins and to the apparently *outré* claim that we can talk about genes for reading. Reading is an extraordinarily complex behaviour pattern, and surely no adaptation. Further, many genes must be present, and the extra-organismic environment must be right for a human being to be able to acquire the ability to read. Dyslexia might result from the substitution of an unusual mutant allele at one of the loci, however. Given our account, it will be correct to say that the mutant allele is a gene for dyslexia, and also that the more typical alleles at the locus are alleles for reading. Moreover, if the locus also affects some other (determinable) trait—say, the capacity to factor numbers into primes—then it may turn out that the mutant allele is also an allele for rapid factorization skill, and that the typical allele is an allele for factorization disability. To say that A is an allele for $P*$ does not preclude saying that A is an allele for $Q*$; nor does it commit us to supposing that the phenotypic properties in question are either both skills or both disabilities. Finally, because substitutions at many loci may produce (possibly different types of) dyslexia, there may be many genes for dyslexia and many genes for reading. Our reconstruction of the geneticists' idiom, the idiom which Dawkins wants to use, is innocent of any Mendelian theses about one–one mappings between genes and phenotypic traits.

Visibility

So we can defend Dawkins's thesis that alleles have properties that influence their chances of leaving copies in later generations by suggesting that, in concert with their environments (including their genetic environments), those alleles cause the presence of certain properties in vehicles (such as organisms), and that the properties of the vehicles are causally relevant to the spreading of copies of the alleles. But our answer to question (i) leads naturally to concerns about question (ii). Granting that an allele is for a phenotypic trait $P*$ and that the presence of $P*$ rather than alternative forms of the determinable trait P enhances the chances that an organism will survive and reproduce and thus transmit copies of the underlying allele, is it not $P*$ and its competition which are directly involved in the selection process? What selection 'sees' are the phenotypic properties. When this vague, but suggestive, line of thought has been made precise, we think that there is an adequate Dawkinsian reply to it.

The idea that selection acts directly on phenotypes, expressed in metaphorical terms by Gould (1980a) (and earlier by Ernst Mayr (1963: 184)), has been explored in an interesting essay by Robert Brandon (1984). Brandon proposes that phenotypic traits screen off genotypic traits (in the sense of Wesley Salmon[7]):

$$\Pr\left(O_n / G\&P\right) = \Pr\left(O_n/P\right) \neq \Pr\left(O_n/G\right)$$

where $\Pr(O_n/G\&P)$ is the probability that an organism will produce n offspring given that it has both a phenotypic trait and the usual genetic basis for that trait, $\Pr(O_n/P)$ is the probability that an organism will produce n offspring given that it has the phenotypic trait, and $\Pr(O_n/G)$ is the probability that it will produce n offspring given that it has the usual genetic basis. So fitness seems to vary more directly with the phenotype and less directly with the underlying genotype.

Why is this? The root idea is that the successful phenotype may occur in

---

[7] Brandon refers to Salmon 1971. It is now widely agreed that statistical relevance misses some distinctions which are important in explicating causal relevance. See e.g. Cartwright 1979, Sober 1984: ch. 8, and Salmon 1984.

the presence of the wrong allele as a result of judicious tampering, and, conversely, the typical effect of a 'good' allele may be subverted. If we treat moth larvae with appropriate injections, we can produce pseudo-melanics that have the allele which normally gives rise to the speckled form, and we can produce moths, foiled melanics, that carry the allele for melanin in which the developmental pathway to the emergence of black wings is blocked. The pseudo-melanics will enjoy enhanced reproductive success in polluted woods, and the foiled melanics will be at a disadvantage. Recognizing this type of possibility, Brandon concludes that selection acts at the level of the phenotype.[8]

Once again, there is no dispute about the facts. But our earlier discussion of epigenesis should reveal how genic selectionists will want to tell a different story. The interfering conditions that affect the phenotype of the vehicle are understood as parts of the allelic environment. In effect, Brandon, Gould, and Mayr contend that, in a polluted wood, there is selection for being dark-coloured rather than for the allelic property of directing the production of melanin, because it would be possible to have the reproductive advantage associated with the phenotype without having the allele (and, conversely, it would be possible to lack the advantage while possessing the allele). Champions of the gene's-eye view will maintain that tampering with the phenotype reverses the typical effect of an allele by changing the environment. For these cases involve modification of the allelic environment, and give rise to new selection processes in which allelic properties currently in favour prove detrimental. The fact that selection goes differently in the two environments is no more relevant than the fact that selection for melanic colouration may go differently in Cheshire and in Dorset.

If we do not relativize to a fixed environment, then Brandon's claims about screening off will not generally be true.[9] We suppose that Brandon

---

[8] Unless the treatments are repeated in each generation, the presence of the genetic basis for melanic colouration will be correlated with an increased frequency of grand-offspring, or of great-grand-offspring, or of descendants in some further generation. Thus, analogues of Brandon's probabilistic relations will hold only if the progeny of foiled melanics are treated so as to become foiled melanics, and the progeny of pseudo-melanics are treated so as to become pseudo-melanics. This point reinforces the claims about the relativization to the environment that we make below. Brandon has suggested to us in correspondence that now his preferred strategy for tackling issues of the units of selection would be to formulate a principle for identifying genuine environments.

[9] Intuitively, this will be because Brandon's identities depend on there being no correlation between O_n and G in any environment, except through the property P. Thus, ironically, the screening-off relations only obtain under the assumptions of simple bean-bag genetics! Sober seems to appreciate this point in a cryptic footnote (1984: 229–30).

To see how it applies in detail, imagine that we have more than one environment, and that the reproductive advantages of melanic colouration differ in the different environments. Specifi-

intends to relativize to a fixed environment. But now he has effectively begged the question against the genic selectionist by deploying the orthodox conception of environment. Genic selectionists will also want to relativize to the environment, but they should resist the orthodox conception of it. On their view, the probability relations derived by Brandon involve an illicit averaging over environments (see n. 9). Instead, genic selectionists should propose that the probability of an allele's leaving n copies of itself should be understood relative to the total allelic environment, and that the specification of the total environment ensures that there is no screening off of allelic properties by phenotypic properties. The probability of producing n copies of the allele for melanin in a total allelic environment is invariant under conditionalization on phenotype.

Here, too, the moral of our story is that Dawkinspeak must be undertaken consistently. Mixing orthodox concepts of the environment with ideas about genic selection is a recipe for trouble, but we have tried to show how the genic approach can be thoroughly articulated so as to meet major objections. But what is the point of doing so? We shall close with a survey of some advantages and potential drawbacks.

III. GENES AND GENERALITY

Relatively little fossicking is needed to uncover an extended defence of the view that gene selectionism offers a more general and unified picture of

cally, suppose that E_1 contains m_1 organisms that have P (melanic colouration) and G (the normal genetic basis of melanic colouration), that E_2 contains m_2 organisms that have P and G, and that the probabilities $\Pr(O_n/G\&P\&E_1)$ and $\Pr(O_n/G\&P\&E_2)$ are different. Then, if we do not relativize to environments, we shall compute $\Pr(O_n/G\&P)$ as a weighted average of the probabilities relative to the two environments.

$$
\begin{aligned}
\Pr\left(O_n/G\&P\right) &= \Pr\left(E_1/G\&P\right)\cdot\Pr\left(O_n/G\&P\&E_1\right) \\
&+ \Pr\left(E_2/G\&P\right)\cdot\Pr\left(O_n/G\&P\&E_2\right) \\
&= m_1/\left(m_1 + m_2\right)\cdot\Pr\left(O_n/G\&P\&E_1\right) \\
&+ m_2/\left(m_1 + m_2\right)\cdot\Pr\left(O_n/G\&P\&E_2\right)
\end{aligned}
$$

Now, suppose that tampering occurs in E_2 so that there are m_3 pseudo-melanics in E_2. We can write $\Pr(O_n/P)$ as a weighted average of the probabilities relative to the two environments.

$$
\Pr\left(O_n/P\right) = \Pr\left(E_1/P\right)\cdot\Pr\left(O_n/P\&E_1\right) + \Pr\left(E_2/P\right)\cdot\Pr\left(O_n/P\&E_2\right)
$$

By the argument that Brandon uses to motivate his claims about screening off, we can take $\Pr(O_n/G\&P\&E_1) = \Pr(O_n/P\&E_1)$ for $i = 1, 2$. However, $\Pr(E_1/P) = m_1/(m_1 + m_2 + m_3)$ and $\Pr(E_2/P) = (m_2 + m_3) / (m_1 + m_2 + m_3)$, so that $\Pr(E_1/P) \neq \Pr(E_1/G\&P)$. Thus, $\Pr(O_n/G\&P) \neq \Pr(O_n/P)$, and the claim about screening off fails.

selective processes than can be had from its alternatives. Phenomena anomalous for the orthodox story of evolution by individual selection fall naturally into place from Dawkins's viewpoint. He offers a revision of the 'central theorem' of Darwinism. Instead of expecting individuals to act in their best interests, we should expect an animal's behaviour 'to maximize the survival of genes 'for' that behaviour, whether or not those genes happen to be in the body of that particular animal performing it' (Dawkins 1982: 223).

The cases that Dawkins uses to illustrate the superiority of his own approach are a somewhat motley collection. They seem to fall into two general categories. First are outlaw and quasi-outlaw examples. Here there is competition among genes which cannot be translated into talk of vehicle fitness, because the competition is among co-builders of a single vehicle. The second group comprises 'extended phenotype' cases: instances in which a gene (or combination of genes) has selectively relevant phenotypic consequences which are not traits of the vehicle that it has helped build. Again, the replication potential of the gene cannot be translated into talk of the adaptedness of its vehicle.

We shall begin with outlaws and quasi-outlaws. From the perspective of the orthodox story of individual selection, 'replicators at different loci within the same body can be expected to "cooperate"'. The allele surviving at any given locus tends to be the one best (subject to all the constraints) for the whole genome. By and large, this is a reasonable assumption. Whereas individual outlaw organisms are perfectly possible in groups, and subvert the chances for groups to act as vehicles, outlaw genes seem problematic. Replication of any gene in the genome requires the organism to survive and reproduce, so genes share a substantial common interest. This is true of asexual reproduction, and, granting the fairness of meiosis, of sexual reproduction too.

But there is the rub. Outlaw genes are genes which subvert meiosis to give them a better than even chance of making it to the gamete, typically by sabotaging their corresponding allele (Dawkins 1982: 136). Such genes are *segregation distorters* or *meiotic drive* genes. Usually, they are enemies not only of their alleles but of other parts of the genome, because they reduce the individual fitness of the organism they inhabit. Segregation distorters thrive, when they do, because they exercise their phenotypic power to beat the meiotic lottery. Selection for such genes cannot be

---

Notice that if environments are lumped in this way, then it will only be under fortuitous circumstances that the tampering makes the probabilistic relations come out as Brandon claims. Pseudo-melanics would have to be added in both environments, so that the weights remain exactly the same.

selection for traits that make organisms more likely to survive and reproduce. They provide uncontroversial cases of selective processes in which
the individualistic story cannot be told.

There are also related examples. Altruistic genes can be outlaw-like,
discriminating against their genome mates in favour of the inhabitants of
other vehicles, vehicles that contain copies of themselves. Start with a
hypothetical case, the so-called green beard effect. Consider a gene Q with
two phenotypic effects. Q causes its vehicle to grow a green beard and to
behave altruistically toward green-bearded conspecifics. Q's replication
prospects thus improve, but the particular vehicle that Q helped build does
not have its prospects for survival and reproduction enhanced. Is Q an
outlaw not just with respect to the vehicle, but with respect to the vehicle-
builders? Will there be selection for alleles that suppress Q's effect? How
the selection process goes will depend on the probability that Q's co-
builders are beneficiaries as well. If Q is reliably associated with other gene
kinds, those kinds will reap a net benefit from Q's outlawry.

So altruistic genes are sometimes outlaws. Whether coalitions of other
genes act to suppress them depends on the degree to which they benefit
only themselves. Let us now move from a hypothetical example to the
parade case.

Classical fitness, an organism's propensity to leave descendants in the
next generation, seems a relatively straightforward notion. Once it was
recognized that Darwinian processes do not necessarily favour organisms
with high classical fitness, because classical fitness ignores indirect effects
of costs and benefits to relatives, a variety of alternative measures entered
the literature. The simplest of these would be to add to the classical fitness
of an organism contributions from the classical fitness of relatives (weighted in each case by the coefficient of relatedness). Although accounting of
this sort is prevalent, Dawkins (rightly) regards it as just wrong, for it
involves double bookkeeping, and, in consequence, there is no guarantee
that populations will move to local maxima of the defined quantity. This
measure and measures akin to it, however, are prompted by Hamilton's
rigorous development of the theory of inclusive fitness (in which it is
shown that populations will tend toward local maxima of inclusive
fitness).[10] In the misunderstanding and misformulation of Hamilton's ideas, Dawkins sees an important moral.

Hamilton, he suggests, appreciated the gene selectionist insight
that natural selection will favour 'organs and behavior that cause the

---

[10] For Hamilton's original demonstration, see his 1971. For a brief presentation of Hamilton's
ideas, see Kitcher 1985: 77–87; and for penetrating diagnoses of misunderstandings, see Grafen
1982 and Michod 1984.

individual's genes to be passed on, whether or not the individual is an ancestor' (Dawkins 1982: 185). But Hamilton's own complex (and much misunderstood) notion of inclusive fitness was, for all its theoretical importance, a dodge, a 'brilliant last-ditch rescue attempt to save the individual organism as the level at which we think about natural selection' (ibid. 187). More concretely, Dawkins is urging two claims: first, that the uses of the concept of inclusive fitness in practice are difficult, so that scientists often make mistakes; second, that such uses are conceptually misleading. The first point is defended by identifying examples from the literature in which good researchers have made errors, errors which become obvious once we adopt the gene-selectionist perspective. Moreover, even when the inclusive fitness calculations make the right predictions, they often seem to mystify the selective process involved (thus buttressing Dawkins's second thesis). Even those who are not convinced of the virtues of gene selectionism should admit that it is very hard to see the reproductive output of an organism's relatives as a property of that organism.

Let us now turn to the other family of examples, the 'extended phenotype' cases. Dawkins gives three sorts of 'extended' phenotypic effects: effects of genes—indeed key weapons in the competitive struggle to replicate—which are not traits of the vehicle the genes inhabit. The examples are of artefacts, of parasitic effects on host bodies and behaviours, and of 'manipulation' (the subversion of an organism's normal patterns of behaviour by the genes of another organism via the manipulated organism's nervous system).

Among many vivid, even haunting, examples of parasitic behaviour, Dawkins describes cases in which parasites synthesize special hormones with the consequence that their hosts take on phenotypic traits that decrease their own prospects for reproduction but enhance those of the parasites (see, for a striking instance, Dawkins 1982: 215). There are equally forceful cases of manipulation: cuckoo fledglings subverting their host's parental programme, parasitic queens taking over a hive and having its members work for her. Dawkins suggests that the traits in question should be viewed as adaptations—properties for which selection has occurred—even though they cannot be seen as adaptations of the individuals whose reproductive success they promote, for those individuals do not possess the relevant traits. Instead, we are to think in terms of selectively advantageous characteristics of alleles which orchestrate the behaviour of several different vehicles, some of which do not include them.

At this point there is an obvious objection. Can we not understand the selective processes that are at work by focusing not on the traits that are external to the vehicle that carries the genes, but on the behaviour that the vehicle performs which brings those traits about? Consider a spider's web. Dawkins wants to talk of a gene for a web. A web, of course, is not a characteristic of a spider. Apparently, however, we could talk of a gene for web building. Web building is a trait of spiders, and, if we choose to redescribe the phenomena in these terms, the extended phenotype is brought closer to home. We now have a trait of the vehicle in which the genes reside, and we can tell an orthodox story about natural selection for this trait.

It would be tempting to reply to this objection by stressing that the selective force acts through the artefact. The causal chain from the gene to the web is complex and indirect; the behaviour is only a part of it. Only one element of the chain is distinguished, the endpoint, the web itself, and that is because, independently of what has gone on earlier, provided that the web is in place, the enhancement of the replication chances of the under-lying allele will ensue. But this reply is exactly parallel to the Mayr–Gould–Brandon argument discussed in the last section, and it should be rejected for exactly parallel reasons.

The correct response, we believe, is to take Dawkins at his word when he insists on the possibility of a number of different ways of looking at the same selective processes. Dawkins's two main treatments of natural selection (1976 and 1982) offer distinct versions of the thesis of genic selection-ism. In the earlier discussion (and occasionally in the later one) the thesis is that, for any selection process, there is a uniquely correct representation of that process, a representation which captures the causal structure of the process, and this representation attributes causal efficacy to genic proper-ties. In his (1982), especially in chapters 1 and 13, Dawkins proposes a weaker version of the thesis, to the effect that there are often alternative, equally adequate representations of selection processes, and that, for any selection process, there is a maximally adequate representation which attributes causal efficacy to genic properties. We shall call the strong (early) version *monist genic selectionism* and the weak (later) version *pluralist genic selectionism*. We believe that the monist version is faulty, but that the pluralist thesis is defensible.

In presenting the 'extended phenotype' cases, Dawkins is offering an alternative representation of processes that individualists can redescribe in their own preferred terms by adopting the strategy illustrated in our dis-cussion of spider webs. Instead of talking of genes for webs and their selective advantages, it is possible to discuss the case in terms of the benefits that accrue to spiders who have a disposition to engage in web building. There is no privileged way to segment the causal chain and isolate the (really) real causal story. As we noted two paragraphs back, the analogue of the Mayr–Gould–Brandon argument for the priority of those properties which are most directly connected with survival and reproduction—here the webs themselves—is fallacious. Equally, it is fallacious to insist that the causal story must be told by focusing on traits of individuals which contribute to the reproductive success of those individuals. We are left with the general thesis of pluralism: there are alternative, maximally adequate representations of the causal structure of the selection process. Add to this Dawkins's claim that one can always find a way to achieve a representation in terms of the causal efficacy of genic properties, and we have pluralist genic selectionism.

Pluralism of the kind we espouse has affinities with some traditional views in the philosophy of science. Specifically, our approach is instrumentalist, not of course in denying the existence of entities like genes, but in opposing the idea that natural selection is a force that acts on some determinate target, such as the genotype or the phenotype. Monists err, we believe, in claiming that selection processes must be described in a particular way, and their error involves them in positing entities, 'targets of selection', that do not exist.

Another way to understand our pluralism is to connect it with conventionalist approaches to space-time theories. Just as conventionalists have insisted that there are alternative accounts of the phenomena which meet all our methodological desiderata, so too we maintain that selection processes can usually be treated, equally adequately, from more than one point of view. The virtue of the genic point of view, on the pluralist account, is not that it alone gets the causal structure right, but that it is always available.

What is the rival position? Well, it cannot be the thesis that the only adequate representations are those in terms of individual traits which promote the reproductive success of their bearers, because there are instances in which no such representation is available (outlaws) and instances in which the representation is (at best) heuristically misleading (quasi-outlaws, altruism). The sensible rival position is that there is a hierarchy of selection processes: some cases are aptly represented in terms of genic selection, some in terms of individual selection, some in terms of group selection, and some (maybe) in terms of species selection. Hierarchical monism claims that, for any selection process, there is a unique level of the hierarchy such that only representations that depict selection as acting at that level are maximally adequate. (Intuitively, representations that see selection as acting at other levels get the causal structure wrong). Hierarchical monism differs from pluralist genic selectionism in an interesting way: whereas the pluralist insists that, for any process, there are many adequate representations, one of which will always be a genic representation, the hierarchical monist maintains that for each process there is just one kind of adequate representation, but that processes are diverse in the kinds of representation they demand.[11]

Just as the simple orthodoxy of individualism is ambushed by outlaws and their kin, so too, hierarchical monism is entangled in spider webs. In the 'extended phenotype' cases, Dawkins shows that there are genic representations of selection processes which can be no more adequately illuminated from alternative perspectives. Since we believe that there is no compelling reason to deny the legitimacy of the individualist redescription in terms of web-building behaviour (or dispositions to such behaviour), we conclude that Dawkins should be taken at face value: just as we can adopt different perspectives on a Necker cube, so too, we can look at the workings of selection in different ways (Dawkins 1982: ch. 1).

In previous sections, we have tried to show how genic representations are available in cases that have previously been viewed as troublesome. To complete the defence of genic selectionism, we would need to extend our survey of problematic examples. But the general strategy should be evident. Faced with processes that others see in terms of group selection or species selection, genic selectionists will first try to achieve an individualist representation, and then apply the ideas we have developed from Dawkins to make the translation to genic terms.

Pluralist genic selectionists recommend that practising biologists take advantage of the full range of strategies for representing the workings of selection. The chief merit of Dawkinspeak is its generality. Whereas the individualist perspective may sometimes break down, the gene's-eye view is apparently always available. Moreover, as illustrated by the treatment of inclusive fitness, adopting it may sometimes help us to avoid errors and confusions. Thinking of selection in terms of the devices, sometimes highly indirect, through which genes lever themselves into future generations may also suggest new approaches to familiar problems.

But are there drawbacks? Yes. The principal purpose of the early sections of this essay was to extend some of the ideas of genic selectionism to respond to concerns that are deep and important. Without an adequate rethinking of the concepts of population and of environment, genic representations will fail to capture processes that involve genic interactions or

---

[11] In defending pluralism, we are very close to the views expressed by Maynard Smith (1987). Indeed, we would like to think that Maynard Smith 1987 and the present essay complement one another in a number of respects. In particular, as Maynard Smith explicitly notes, 'recommending a plurality of models of the same process' contrasts with the view (defended by Gould and by Sober) of 'emphasizing a plurality of processes'. Gould's views are clearly expressed in his 1980b, and Sober's ideas are presented in his 1984: ch. 9.

epigenetic constraints. Genic selectionism can easily slide into naive adaptationism as one comes to credit the individual alleles with powers that enable them to operate independently of one another. The move from the 'genes for *P*' locution to the claim that selection can fashion *P* independently of other traits of the organism is perennially tempting.[12] But, in our version, genic representations must be constructed in full recognition of the possibilities for constraints in gene–environment co-evolution. The dangers of genic selectionism, illustrated in some of Dawkins's own writings, are that the commitment to the complexity of the allelic environment is forgotten in practice. In defending the genic approach against important objections, we have been trying to make this commitment explicit, and thus to exhibit both the potential and the demands of correct Dawkin-speak. The return of the gene should not mean the exile of the organism.[13,14]

REFERENCES

Brandon, R. (1984), 'The Levels of Selection', in R. Brandon and R. Burian (eds.), *Genes, Organisms, Populations* (Cambridge, Mass.: MIT Press), 133–41; reproduced as Ch. 9).
Cartwright, N. (1979), 'Causal Laws and Effective Strategies', *Noûs*, 13: 419–37.
Dawkins, R. (1976), *The Selfish Gene* (New York: Oxford University Press).
——(1982), *The Extended Phenotype* (San Francisco: Freeman).
Gould, S. J. (1980a), 'Caring Groups and Selfish Genes', in *The Panda's Thumb* (New York: Norton), 85–92.
——(1980b), 'Is a New and General Theory of Evolution Emerging?', *Paleobiology*, 6: 119–30.
——and Lewontin, R. C. (1979), 'The Spandrels of San Marco and the Panglossian Paradigm: A Critique of the Adaptationist Programme', *Proceedings of the Royal Society, London*, B205: 581–98; repr. in E. Sober (ed.), *Conceptual Problems in Evolutionary Biology* (Cambridge, Mass.: MIT Press, 1984), 252–70.
Grafen, A. (1982), 'How Not to Measure Inclusive Fitness', *Nature*, 298: 425–6.
Hamilton, W. D. (1971), 'The Genetical Evolution of Social Behavior I', in G. C. Williams (ed.), *Group Selection* (Chicago: Aldine), 23–43.

[12] At least one of us believes that the claims of the present essay are perfectly compatible with the critique of adaptationism developed in Gould and Lewontin 1979. For discussion of the difficulties with adaptationism, see Kitcher 1985, 1987.

[13] As, we believe, Dawkins himself appreciates. See the last chapter of his (1982), especially his reaction to the claim that 'Richard Dawkins has rediscovered the organism' (p. 251).

[14] We are equally responsible for this paper, which was written when we discovered that we were writing it independently. We would like to thank those who have offered helpful suggestions to one or both of us, particularly Patrick Bateson, Robert Brandon, Peter Godfrey-Smith, David Hull, Richard Lewontin, Lisa Lloyd, Philip Pettit, David Scheel, and Elliott Sober.

Kettlewell, H. B. D. (1973), *The Evolution of Melanism* (New York: Oxford University Press).

Kitcher, P. (1985), *Vaulting Ambition: Sociobiology and the Quest for Human Nature* (Cambridge, Mass.: MIT Press).

—— (1987), 'Why Not the Best?', in J. Dupré (ed.), *The Latest on the Best: Essays on Optimality and Evolution* (Cambridge, Mass.: MIT Press), 77–102.

Maynard Smith, J. (1982), *Evolution and the Theory of Games* (New York: Cambridge University Press).

—— (1987), 'How to Model Evolution', in J. Dupré (ed.), *The Latest on the Best: Essays on Optimality and Evolution* (Cambridge, Mass.: MIT Press), 119–31.

Mayr, E. (1963), *Animal Species and Evolution* (Cambridge, Mass.: Harvard University Press).

Michod, R. (1984), 'The Theory of Kin Selection', in R. Brandon and R. Burian (eds.), *Genes, Organisms, Populations* (Cambridge, Mass.: MIT Press), 203–37.

Salmon, W. (1971), *Statistical Explanation and Statistical Relevance* (Pittsburgh: Pittsburgh University Press).

—— (1984), *Scientific Explanation and the Causal Structure of the World* (Princeton: Princeton University Press).

Sober, E. (1984), *The Nature of Selection* (Cambridge, Mass.: MIT Press).

—— and Lewontin, R. C. (1982), 'Artifact, Cause and Genic Selection', *Philosophy of Science*, 49: 157–80.

9

THE LEVELS OF SELECTION: A HIERARCHY OF INTERACTORS

ROBERT N. BRANDON

Biologists have long recognized that the biosphere is hierarchically arranged. And at least since 1970 we have recognized that the abstract theory of evolution by natural selection can be applied to a number of elements within the biological hierarchy (Lewontin 1970). But what is it for selection to occur at a given level of biological organization? What is a 'unit of selection'? Is there one privileged level at which selection always, or almost always, occurs? In this chapter I shall try to clarify and partially answer these questions.

GENOTYPES AND PHENOTYPES

As Mayr (1978) has emphasized, evolution by natural selection is a two-step process. According to the received neo-Darwinian view, one step involves the selective discrimination of phenotypes. For instance, suppose there is directional selection for increased height in a population. That means that taller organisms tend to have greater reproductive success than shorter organisms. The reasons for this difference depend on the particular selective environment in which the organisms live. In one population it may be that taller plants receive more sunlight and so have more energy available for seed production; in another, taller animals may be more resistant to predation. Whatever the reason, natural selection requires that there be phenotypic variation (in this case, variation in height). Selection can be thought of as an interaction between phenotype and environment that results in differential reproduction.

But natural selection in the above sense (what quantitative geneticists call 'phenotypic selection') is not sufficient to produce evolutionary

First published in H. Plotkin (ed.), *The Role of Behavior in Evolution* (Cambridge, Mass.: MIT Press, 1988), 51–71. Reprinted by permission.

change. In the case of directional selection for increased height, selection may change the phenotypic distribution in the parental generation (it will do so if selection is by differential mortality); but whether or not that results in evolutionary changes (i.e. changes in the next generation) depends on the heritability of height. That is, it depends on whether or not taller-than-average parents tend to produce taller-than-average offspring and shorter-than-average parents tend to produce shorter-than-average offspring. This is the second step in the two-step process. Of course, height is not directly transmitted from parent to offspring; rather, genes are.[1] Thus, offspring of taller-than-average parents will tend to have genotypes different from those of offspring of shorter-than-average parents. In the process of ontogeny, these genotypic differences manifest themselves as phenotypic differences. And so the phenotypic distribution of the offspring generation has been altered; evolution by natural selection has occurred.

Thus, evolution by natural selection requires both phenotypic variation and the underlying genetic variation. In one step, phenotypes interact with their environment in a way that causes differential reproduction. This leads to the next step, the differential replication of genes. Through ontogeny, this new genotypic distribution leads to a new phenotypic distribution, and the process starts anew.

REPLICATORS AND INTERACTORS

The above description of evolution by natural selection seems perfectly adequate for cases of selection occurring at the level of organismic phenotypes. But during the last twenty-five years there has been increasing interest in the idea that selection may occur at other levels of biological organization. This interest was sparked by V. C. Wynne-Edwards's book *Animal Dispersion in Relation to Social Behaviour* (1962). Wynne-Edwards argued that a major biological phenomenon, the regulation of population size and density, evolves by group selection. In reaction to this thesis, Williams (1966) and Dawkins (1976), argued that selection occurs primarily at the level of genes. The recent flurry of theoretical investigations into kin and group selection has produced some explicitly hierarchical models.[2] It is not obvious how we should apply the genotype–

---

[1] Nuclear genes are not the only means of transmitting traits from parent to offspring. Among other means, cytoplasmic DNA and culture are prominent.

[2] See Brandon and Burian 1984 for a collection of some of the more important papers on questions concerning the levels of selection. The papers of Hamilton (1975), Wimsatt (1980, 1981), and Arnold and Fristrup (1982) offer hierarchical models of selection.

phenotype distinction to describe cases of gametic selection, group selection, or species selection. Thus, Hull (1980, 1981) and Dawkins (1982*a*, *b*) have introduced a distinction between replicators and interactors which is best seen as a generalization of the traditional genotype–phenotype distinction.

Dawkins defines a replicator as 'anything in the universe of which copies are made' (1982*b*: 83). Genes are paradigm examples of replicators, but this definition does not preclude other things' being replicators. For instance, in asexual organisms the entire genome would be a replicator, and in cultural evolution ideas—or what Dawkins (1976) calls memes—may be replicators.

The qualities that make for good replicators are longevity, fecundity, and fidelity (Dawkins 1978). Here longevity means longevity in the form of copies. It is highly unlikely that any particular DNA molecule will live longer than the organism in which it is housed. What is of evolutionary importance is that it produce copies of itself so that it is potentially immortal in the form of copies. Of course, everything else being equal, the more copies a replicator produces (fecundity) and the more accurately it produces them (fidelity), the greater its longevity and evolutionary success.

In explicating Dawkins's notion of replicators, Hull stresses the importance of directness of replication. Although, according to Dawkins, organisms are not replicators, they may be said to produce copies of themselves. This replication process may not be as accurate as that of DNA replication, but none the less there is a commonality of structure produced through descent from parent to offspring. However, there is an important difference in the directness of replication between these two processes. The height of a parent is not directly transmitted to its offspring. As discussed above, that transmission proceeds indirectly through genic transmission and ontogeny. In contrast, genes replicate themselves less circuitously. Both germ-line replication (meiosis) and soma-line replication (mitosis) are physically quite direct. The importance of this is made explicit in Hull's definition of replicators as 'entities which pass on their structure directly in replication' (1981: 33).

As discussed above, evolution by natural selection is a two-step process. One step involves the direct replication of structure. The other involves some interaction with the environment so that replication is differential. The entities functioning in the latter step have traditionally been called 'phenotypes'. But if we want to allow that biological entities other than organisms can interact with their environment in ways that lead to differential replication, then we need to generalize the notion of phenotype. To this end, Hull (1980, 1981) suggests the term 'interactor', which he defines as 'an entity that directly interacts as a cohesive whole with its environment in such a way that replication is differential' (1980: 318).

Although Hull and Dawkins are largely in agreement concerning the replicator–interactor distinction, there are two differences worth noting. The first is purely terminological. Dawkins has not adopted Hull's term 'interactor'; instead he uses the term 'vehicle'. According to Dawkins, a vehicle is 'any relatively discrete entity, such as an individual organism, which houses replicators . . . and which can be regarded as a machine programmed to preserve and propagate the replicators that ride inside it' (1982a: 295). I prefer Hull's term and his definition of it, and so I will use it here.

The second difference is more substantive. Dawkins holds that any change in replicator structure is passed on in the process of replication (1982a: 85; 1982b: 51). Thus, given the truth of Weismannism (the doctrine that there is a one-way causal influence from germ line to body), replicators are supposedly different from most interactors (e.g. organisms). But DNA is capable of self-repair, and so not all changes in DNA structure are passed on in the process of replication. Thus, the property of transmitting changes in structure in the process of replication does not sharply demarcate replicators from interactors. What seems to be important is that replication be direct and accurate. But both directness and accuracy are terms of degree, and if we allow some play in both, then under certain circumstances an organism could be a replicator (which would not preclude its being an interactor as well). For example, Hull (1981: 34) argues that a paramecium dividing into two can be considered a replicator, since its structure is transmitted in a relatively direct and accurate manner.

An important point to note is that the definitions of interactors and replicators are given in functional terms: that is, in terms of the roles these entities play in the process of evolution by natural selection. Nothing in the definitions precludes one and the same entity from being both an interactor and a replicator. For instance, it is likely that the self-replicating entities involved in the earliest evolution of life on this planet were both interactors and replicators (see Eigen et al. 1981). Likewise, in cases of meiotic drive, parts of chromosomes, or perhaps entire chromosomes, can be considered interactors as well as replicators. Hull has suggested, as was mentioned above, that in some cases organisms can be considered replicators as well as interactors. (See Hull 1988 for a more detailed discussion of these issues.)

LEVELS OF SELECTION

Having developed the notions of replicator and interactor, which are generalizations of the notions of genotype and phenotype, we may ask whether the process of evolution by natural selection occurs at other levels of biological organization. This seemingly simple question, however, is ambiguous.[3] Are we asking whether there are replicators other than single genes, or are we asking whether there are interactors other than organismic phenotypes? Both are interesting questions, but many who have addressed the units of selection question have failed to see that there are two separate questions, and thus have confused the two. I shall discuss this in the next section; in this section I shall concentrate on the latter question: that is, the question concerning interactors.

What is it about standard cases of organismic selection that makes them cases of organismic selection?[4] Put another way, what features of standard cases of organismic selection make organisms the interactors? Put still another way, what justifies our claim that in such cases 'natural selection favours (or discriminates against) phenotypes, not genes or genotypes' (Mayr 1963: 184)? Consider again our example of directional selection for increased height. Recall that taller organisms have a higher fitness on average than shorter organisms. Thus there is a positive association between height and fitness. But there is genetic variation in height, so there is also a positive association between certain genes and/or genotypes and fitness. So why not say that natural selection favours phenotypes and genes (or genotypes) equally? Where is the asymmetry between phenotype and genotype?

The asymmetry lies here: reproductive success is determined by phenotype irrespective of genotype. Intuitively, selection 'sees' a 4-foot-tall plant as a 4-foot-tall plant, not as a 4-foot-tall plant with genotype g. This idea can be made precise by using the probabilistically defined notion of *screening off* (Salmon 1971). The basic idea behind the notion of screening off is this: if A renders B statistically irrelevant with respect to outcome E, but not vice versa, then A is a better causal explainer of E than is B. In symbols, A screens off B from E if and only if

---

[3] I believe that Hull (1981), Dawkins (1982b), and I (Brandon 1982) arrived at this conclusion independently.

[4] In this context I prefer the term 'organismic selection' to the more common 'individual selection' because, as Hull has pointed out, interactors at other levels (e.g. groups) must be individuals.

$$P(E, A \cdot B) = P(E, A) = P(E, B).$$

(Read $P(E, A \cdot B)$ as the probability of E given A and B.) If A screens off B from E, then, in the presence of A, B is statistically irrelevant to E; that is,

$$P(E, A) = P(E, A \cdot B).$$

But this relation between A and B is not symmetric. Given B, A is still statistically relevant to E; that is,

$$P(E, B) = P(E, A \cdot B).$$

Thus, where A and B are causally relevant to E, it follows that A's effect on the probability of E acts irrespective of the presence of B, but the same cannot be said of B. The effect B has on the probability of E depends on the presence or absence of A. For our purposes the important point is that proximate causes screen off remote causes from their effects.

Let us return to our case of directional selection for increased height. In that case there is differential reproduction of interactors (organisms) and replicators (genes). But it is obvious that the means through which genes replicate differentially in this case is the differential reproduction of organisms. (In other words, in this case there would be no differential replication of genes without the differential reproduction of organisms.) So the fact to be explained is that taller organisms tend to leave more offspring than shorter organisms. Using the notion of screening off, we can see that this is best explained in terms of differences in height rather than in terms of differences in genotype.

What we need to show is that for any level of reproductive success n, phenotype p, and genotype g,

$$P(n, p \cdot g) = P(n, p) = P(n, g).$$

Gedanken experiments should suffice to show the correctness of both the equality and the inequality. Basically the idea is that manipulating the phenotype without changing the genotype can effect reproductive success. (Castration is the most obvious example.) On the other hand, tampering with the genotype without changing any aspect of the phenotype cannot affect reproductive success. Admittedly, the latter claim is not straight-forwardly empirical. One could tamper with germ-line DNA, say by irradiation, and negatively affect reproductive success without obviously affecting the phenotype. But I would argue that in every such case one could find some aspect of an interactor that had been affected. For example, in many cases of the irradiation of a male, sperm morphology and behaviour are changed. The claim is that a change in the informational content of the genome alone will not make for a change in reproductive success.[5] Thus, the fact that phenotype screens off genotype from reproductive success shows that there is an asymmetry between phenotype and genotype, and that in cases of organismic selection reproductive success is best explained in terms of properties of the phenotype. What is true of this relation between phenotype and genotype obviously holds for the relation between phenotype and gene.

One might worry that our conclusion is a product of our choosing to look at differential reproduction of interactors (organisms) rather than replicators.[6] In our case, taller organisms out-reproduce shorter organisms, and it should be clear that this is the mechanism by which some genes out-reproduce others. But let us change our focus. Let n stand for the realized fitness of a given germ-line gene, let p stand for the phenotype of the organism in which it is housed, and let g stand for some property of the gene (its selection coefficient or whatever else one might think is relevant). Still the phenotype of the organism screens off the genic property from its own reproductive success; that is,

$$P(n, p \cdot g) = P(n, p) = P(n, g).$$

Thus, in our case a particular gene's reproductive success is best explained in terms of the height of the organism in which it is housed.

I have argued that in standard cases of organismic selection the mechanism of selection, the differential reproduction of organisms, is best explained in terms of differences in organismic phenotypes, because phenotypes screen off both genotypes and genes from the reproductive

---

[5] As we saw above, the notions of interactor and replicator are not mutually exclusive; one and the same entity can be both interactor and replicator. Similarly, the notions of genotype and phenotype are not mutually exclusive. The genotype of an organism is a part of its phenotype. Thus, my claim commits me to the position that any change in genotype that does lead to a change in reproductive success must also be a change in the organism's phenotype. This should not be seen as counter-intuitive so long as one realizes that genes (lengths of DNA) have a physical structure.

[6] Sober (1984: 229–30) has raised this objection. He writes: 'Brandon chose an organism's reproductive success. But suppose we choose change in gene frequencies. Then the screening-off relation is inverted. Gene frequencies and genic selection coefficients determine change in gene frequencies, if the population is infinitely large, and confer a probability distribution on future gene frequencies, if drift is taken into account.' This objection is based on a simple equivocation. In the first instance we are concerned with relations among objective probabilities in the real world. That is the sense in which height, not genotype, determines reproductive success. Sober is concerned with the relation between coefficients and variable values in a mathematical model. Mathematical determination in a model does not translate so simply to nature.

success of organisms. Thus, in such cases the interaction between interactor and environment that leads to differential reproduction occurs at the level of the organismic phenotype.

We can now return to the question with which we began this section: Do such interactions occur at other levels of biological organization; are there other levels of interactors besides that of the organismic phenotype? This is ultimately an empirical question. I do not intend to answer it definitively; rather, I shall try to offer the conceptual tools necessary for answering it. But before I offer a general definition of levels of selection, let us consider selection at the group level.

Intuitively, group selection is natural selection acting at the level of biological groups. And natural selection is the differential reproduction of biological entities that is *due to* the differential adaptedness of those entities to a common environment. I have defended this definition elsewhere (Brandon 1978, 1981*b*), but two points are worth re-emphasizing. First, the definition is explicitly causal; thus, it does not include all cases of differential reproduction. For instance, it does not apply to cases where, by chance, a less-well-adapted organism has greater reproductive success than a better-adapted one (who, let us say, was struck by lightning). Second, it applies only to those cases where differences in reproductive success are due to differences in adaptedness to a *common* environment (for further discussion of this point see Damuth 1985, Antonovics *et al.* 1988, and Brandon 1990). This is implicit in the above discussion where I moved from saying that the organism's phenotype best explains its level of reproductive success to saying that *differences* in phenotypes best explain *differences* in reproductive success. This move is valid only if we restrict our attention to organisms, or more generally interactors, within a common environment. This point can be illustrated by a simple example. Suppose we plant two seeds, one in good soil and the other in mildly toxic soil. The first will probably survive longer and produce more seeds than the second; that is, the first will be 'fitter' than the second. But to explain this difference, we must refer to differences in their environments, not to differences in their phenotypes.

In biology we can distinguish at least three different notions of environment. The first I call the 'external environment'. The external environment consists of all elements, both abiotic and biotic, that are external to the evolving population of interest. This is what ecologists typically refer to when they speak of *the* environment. The second notion I call the 'ecological environment'. The ecological environment reflects those features of the external environment that affect the demographic performance of the organisms of interest. It is measured by using the organisms of interest as measuring instruments. Finally, the 'selective environment' reflects those elements of the external environment or of population structure that affect the differential contribution of different types to subsequent generations (i.e. affect differential fitness). Again, the selective environment is measured by using the organisms as measuring instruments, and changes in selective environments are indicated empirically by a genotype–environment (e.g. spatial position) interaction in fitness. In other words, a common selective environment is a population or an area of space or time within which the relative fitness of the varying types remain constant. The selective environment is what is most directly relevant to the theory of natural selection; natural selection occurs within common selective environments. A given selective environment may be quite heterogeneous in terms of its abiotic or its biotic components.[7]

As we have seen above, organismic selection is the differential reproduction of organisms that is due to the differential adaptedness of those interactors to a common environment. Group selection, then, is the differential reproduction of biological groups that is due to the differential adaptedness of those groups to a common environment. Thus, a necessary condition for the occurrence of group selection is that there be differential reproduction (propagation) among groups. But this necessary condition is not sufficient. In order for the differential reproduction of groups to be group selection (i.e. selection at the group level), there must be some group property (the group phenotype) that screens off all other properties from group reproductive success.[8]

It is by no means necessary that such a property exist. For instance, suppose that group productivity or group fitness depends simply on the number of organisms within the group at the end of a certain time period.[9] Suppose further that the adaptedness values of these organisms do not depend in any way on the group's composition. In that case the group

[7] The distinction among external, ecological, and selective environments is introduced in Antonovics et al. 1988. For further discussion see that work or Brandon 1990.

[8] It is not completely clear what sorts of things should count as group properties. Obvious examples include the sorts of things that could not be properties of individual organisms—for instance, the relative frequency of certain alleles within the group, or the phenotypic distribution within the group. Other properties that might be selectively relevant are less obviously group properties. For instance, we may or may not want to count the ability to avoid predation as a group property. Whether or not that is a group property depends on whether the group's ability to avoid predation is something 'over and above' the ability of each individual to avoid predation—that is, on whether there is some group effect on the individuals' abilities to avoid predation.

[9] This need not be the case, but it is assumed in most models. For a review of these models see Wade 1978, Uyenoyama and Feldman 1980, Wilson 1983b, or the introduction to part III of Brandon and Burian 1984. Indeed, in one experimental treatment Wade (1977) selected for groups with the lowest numbers of organisms.

'phenotype' (the distribution of individual phenotypes within the group) would not screen off all non-group properties from group reproductive success. In particular, it would not screen off the aggregate of the individuals' phenotypes; that is, the following relation would not hold:

$$P\left(n,\ G\cdot\left[p_1\cdot p_2\cdots p_k\right]\right)\ =\ P\left(n,\ p_1\cdot p_2\cdots p_k\right),$$

where n is the number of propagule groups, G is the group phenotype, and p_i is the phenotype of the ith member of the group. The equality would hold, but the inequality would not, since the phenotype of each individual within the group would determine that individual's adaptedness, and the adaptedness values of each member of the group would determine the adaptedness value of the group.

In summary, group selection occurs if and only if (1) there is differential reproduction of groups and (2) the group phenotype screens off all other properties (of entities at any level) from group reproductive success. One way to restate (2) is this: differential group reproduction is best explained in terms of differences in group-level properties (differences in group adaptedness to a common selective environment). Still another way to restate (2)—a way that would have seemed question-begging prior to what has been said concerning screening off—is this: the causal process of interaction occurs at the level of the group phenotype.[10]

What has been said about group selection is easily generalizable into the following definition: selection occurs at a given level (within a common selective environment) if and only if (1) there is differential reproduction among the entities at that level, and (2) the 'phenotypes' of the entities at that level screen off properties of entities at every other level from reproductive values at the given level.

A HIERARCHY OF INTERACTORS

What sorts of biological entities fall under the above definition? Organisms certainly do; for ample documentation, see Endler 1986. What about entities at lower levels of biological organization? Eigen *et al.* (1981) have

---

[10] It goes beyond the scope of this chapter to apply this approach to one of the major conceptual problems concerning group selection: viz. whether group selection requires the non-random formation of groups. For arguments that non-random group formation is required, see Hamilton 1975, Maynard Smith 1982, and Nunney 1985. For opposing arguments see Wilson 1983*b* and Wade 1984. For an application of the approach of the present chapter to the problem, see Brandon 1990.

presented a plausible scenario concerning the origin of life. In this scenario, lengths of RNA interact with proteins in a 'primordial soup', and by this selection process the genetic code develops. Thus, in this scenario, there is selection at the level of lengths of RNA. Clearly these bits of RNA qualify as replicators; they replicate their structure directly and accurately. But they are interactors as well. It is their physical structure, their 'phenotype', that determines their adaptedness to given conditions. For instance, in one experimental treatment RNAs were selected under conditions of high concentrations of ribonuclease, an enzyme that cleaves RNA into pieces. In this treatment, variants developed that were resistant to cleavage. 'Apparently the variant that is resistant to this degradation folds in a way that protects the sites at which cleavage would take place' (Eigen *et al.* 1981: 97).

Doolittle and Sapienza (1980) and Orgel and Crick (1980) have argued that intra-genomic selection results in the spread of 'selfish genes': that is, genes that increase their representation in the genome not through their effects on the phenotype of the organisms in which they are housed but rather through their superior replication efficiency within the genome. Such genes may or may not be transcribed, but in general one expects them to have a negative impact on the fitness of organisms because of the energetic costs of excess DNA. Doolittle and Sapienza (1980) describe the selection process by which 'selfish genes' spread as 'non-phenotypic selection'. In the terminology of the present chapter, what they mean is that the level of this selection process is not the organismic phenotype. But 'selfish genes' are interactors. They interact within the cellular environment in a way that leads to differential replication. It is their 'phenotype' (i.e. their physical structure) that matters, not the phenotype of the organism in which they are housed. Similar remarks apply to chromosomes or parts of chromosomes in cases of meiotic drive (Crow 1979).

I have already discussed the possibility of selection at the level of groups. Wade (1977) has created group selection in a laboratory setting. Group selection in nature is more controversial; see Wilson 1983*a* for an illuminating discussion of this controversy and Wilson 1983*b* for a plausible case of group selection in nature. I have not attempted here to answer the empirical question of how prevalent group selection is in nature; rather, I have tried to shed light on the conceptual question of what should count as group selection. For present purposes, however, the important point is that when there is selection at the level of groups, these groups are interactors.

The groups that are relevant to discussions of group selection are relatively small and relatively short-lived. Can selection occur at higher levels of organization: for example, at the species level? There have been many recent discussions of species selection (Stanley 1975, 1979, Gould and Eldredge 1977, Eldredge and Cracraft 1980), but most of these have not clearly distinguished between mere differential replication of species and true species selection. In the useful terminology of Vrba and Gould (1986), they have not distinguished between sorting and selection. Clearly there is sorting at the species level: that is, there is differential replication of species. But is there species selection? Damuth (1985) argues that, in general, a species is not the right sort of entity to participate in a selection process. In cases of organismic selection, selection occurs among organisms inhabiting a common selective environment. Notice that the population consisting of these organisms (the organisms inhabiting a common selective environment) is not necessarily a population united by gene flow (a deme). According to most proponents of species selection (e.g. Stanley (1975)), species selection occurs among species within a clade. But a clade is a genealogical unit rather than an ecological unit, and it is implausible that, in general, the constituents of a clade share a common selective environment.[11] Likewise, different local populations of a species will oftentimes not share a common selective environment. Thus, species in clades are not analogous in the relevant way to organisms within a population of organisms inhabiting a common selective environment. In the first case we have a unit united by gene flow (a species) within a larger genealogically characterized unit (a clade). In the second case we have an interactor (an organism) within a population of interactors united by common selective forces. To have an explanatory hierarchical theory of selection, we need a hierarchy of the right sort of units. Units of selection need not, and usually do not, correspond to units of a genealogical nexus. Damuth argues that local populations of a species within an ecological community are the sort of thing that could be a unit of higher-level selection, and that the community (not the clade) would be the unit within which selection would occur. Again, it should be clear that these higher-level entities (which Damuth calls 'avatars') are interactors.

Nothing in Damuth's argument precludes species selection; his argument is that species *qua* units of gene exchange are not necessarily the sort of entities that participate in the selection process (i.e. interactors). But in some cases there may be differences in species-level properties that do

---

[11] If there is a hierarchy of interactors, then there is a corresponding hierarchy of selective environments. It is virtually inconceivable that all the organisms in a clade would share a common organismic selective environment. But the question that is relevant here is whether species in a clade share a common species-level selective environment. Damuth's point is that there is no reason to expect that they do in general. This point holds even if it is not wildly implausible that in some instances species in a clade do share a common (species-level) selective environment.

lead to differences in speciation and extinction rates. For instance, among marine gastropods, species whose larvae are plankto-trophic have greater larval dispersal than those whose larvae are not, and so have greater gene flow. This leads to lower levels of speciation and extinction (see Jablonski 1986 and references therein). This may represent a genuine case of species selection. Indeed, Jablonski (1986) presents persuasive evidence that during the end-Cretaceous, mass extinction-selection occurred at even higher levels of organization. Clades with more extensive geographic ranges showed higher survivorship than clades with smaller geographic ranges. That this was an emergent property of clades was indicated by the fact that individual species' ranges had no effect on clade survivorship.

(I have argued that selection requires common selective environments. Selection at higher levels, e.g. species and clades, requires much more spatio-temporally extensive selective environments. If mass extinctions are indeed caused by catastrophic events, such as the impact of large meteors, then this may have the effect of homogenizing vast parts of the Earth with respect to certain selective pressures. If this is so, one would expect increased higher-level selection during such periods.)

I have offered a definition of levels of selection, and in this section we have discussed various levels at which selection may occur. Selection may occur among bare lengths of RNA within a 'primordial soup', among lengths of DNA within cells, among chromosomes within cells, among organisms within (selectively homogeneous) populations, among groups of organisms within local populations, among local populations within communities, among species within groups of competing species (which may or may not correspond to clades), and among clades. I have argued in each case that when there is selection at a given level, the entities at that level are interactors. This should not be surprising, since my definition of levels of selection is designed to pick out levels of interaction. Thus, we have a hierarchy of interactors ranging from bare lengths of RNA through organisms to clades. Let me re-emphasize that some of these interactors may also be replicators. The point is that when selection occurs at a given level, the entities at that level must be interactors.

The hierarchy presented here apparently differs from that presented by Lewontin (1970), Hamilton (1975), Wimsatt (1980, 1981), Arnold and Fristrup (1982), and Wade (1984),[12] all of whom agree that in cases of organismic selection the 'unit' of selection is some genetic unit—a gene, an

---

[12] I say 'apparently' for the following reason. If you assume that every copy of a particular genotype has the same phenotype, then in standard cases of organismic selection there will be some genetic unit such that the variance in fitness at that level will be context-independent. But that assumption is not likely to be true for any real population. When the assumption fails,

entire chromosome, or even the entire genome, depending on the amount of epistasis and gene linkage. For illustration I shall present Wimsatt's definition of units of selection, but I believe that what is said about it applies to all the aforementioned works. Wimsatt (1981: 144) defines a unit of selection as 'any entity for which there is heritable context-independent variance in fitness among entities at that level which does not appear as heritable context-independent variance in fitness (and thus, for which the variance in fitness is context-dependent) at any lower level of organization'. Wimsatt, following Mayr (1963) and Lewontin (1974), argued against those (e.g. Williams (1966)) who claimed that in standard cases of selection—cases we would classify as organismic selection—genes are the units of selection. Wimsatt's argument is based on certain general facts about genetic systems. Most important among them is the fact that, in general, genes interact in the way they affect the phenotype. In particular, a given gene's effect on fitness depends on its genetic context. Thus, the variance in fitness at the level of genes is, in general, not context-independent. Wimsatt concludes, again in agreement with Mayr and Lewontin, that the unit of selection in standard cases is a much larger genetic unit: the entire genome. But at other levels of selection Wimsatt's definition coincides with mine. Thus, when I would conclude that there is group-level selection, Wimsatt would say that groups are the units of selection, and when my approach implies that there is selection at the level of lengths of DNA, Wimsatt would say that these lengths of DNA are units of selection. Recall that such 'selfish DNA' acts as an interactor. In fact, Wimsatt's analysis agrees with mine when and only when his units are interactors.

The hierarchy that results from approaches such as Wimsatt's is incoherent. It includes replicators *qua* replicators and interactors *qua* interactors. Interactors and replicators play different roles in the process of evolution by natural selection. In order to resolve the 'units of selection controversy', we need to ask coherent questions. One coherent question is this: At what levels do the interactions between biological entity and environment that lead to differential replication occur? That is, what are the levels of interactors? This is the question my analysis is designed to answer. Using this analysis, I have argued for the plausibility of a hierarchy of interactors ranging from bare lengths of RNA to entire clades. I should point out that this hierarchy may not be exhaustive; there may be other levels of selection than those we have considered.

copies of a given genotype could be partitioned by phenotype in a way that would be relevant to fitness. Thus, the variance in fitness at the genotype level would not be context-independent. If all this is correct, I have no argument with Wimsatt's analysis; my argument would be against his application of that analysis.

A HIERARCHY OF REPLICATORS?

There is a second question one might ask. It has to do with levels of replicators. Is there an interesting hierarchy of replicators corresponding to the hierarchy of interactors? Let us examine the selection scenarios discussed above to see what the replicators are in each case. In the case of the model of Eigen *et al.* (1981), bare lengths of RNA interacted within a 'primordial soup'. Differences in their physical structure resulted in differences in survivorship and rates of replication. Thus, they are interactors. But it is obvious that they are replicators as well—indeed, they are paradigm cases of entities that replicate their structure directly and accurately. In the case of 'selfish genes' (*sensu* Doolittle and Sapienza 1980 and Orgel and Crick 1980), lengths of germ-line DNA interact within a cellular environment in a way that leads to some such lengths dramatically increasing their representation within the genome. Again, these lengths of DNA are clearly interactors; it is equally clear that they are replicators as well. Cases of meiotic drive are similar. There, lengths of chromosomes or whole chromosomes are both interactors and replicators.

The next step in the hierarchy of interactors I have discussed is the level of organismic selection. Here organisms are interactors, but what are the replicators? The answer depends on the mode of reproduction. In sexual reproduction, the genomes of organisms are broken up by segregation and recombination. Thus, only parts of the genome reproduce their structure directly and accurately. What do we call these parts of genomes? Williams defines a gene as 'that which segregates and recombines with appreciable frequency' (1966: 24). (The strength of selection determines what counts as an appreciable frequency.) So the replicators in cases of organismic selection with sexual reproduction are genes (*sensu* Williams). In cases of asexual reproduction, the entire genome is passed on directly from parent to offspring. In such cases we could say that the entire genome was the replicator. Notice that in asexual organisms the genome is a 'gene' in Williams's sense. One could argue, as has Hull (1981), that in asexual reproduction the organism itself is a replicator. It replicates its structure in a fairly direct and accurate manner. In this case, the difference between whole organisms and their genomes is one of degree, and the vagueness of the notion of replicator allows us to say that either is a replicator.

The replicators in cases of group selection also depend on the nature of the reproductive process. At the risk of oversimplification, we can distinguish two basic types of group selection: *intrademic* and *interdemic*. In cases of intrademic group selection, groups are formed during part of the organismic life cycle (e.g. the larval stage) and fitness-affecting interactions occur within these groups. Then the group members disperse into a common mating pool. The process of group selection occurs by differential group dispersal caused by differences in group structure (usually represented in formal models by different relative frequencies of alleles or genotypes). But group structure is not passed on directly to the next generation of groups. Rather, individuals from all the different groups unite in a common mating pool and reproduce sexually. New groups are formed in the next organismic generation at the appropriate stage in the life cycle. The replicators here are simply the replicators in normal sexual reproduction: namely, genes (*sensu* Williams). In cases of interdemic group selection, groups are more or less reproductively isolated. Organismic selection occurs within groups, and group selection occurs between groups by processes of differential group extinction and propagation. Here the replicators are the groups themselves (which is not to say that genes are not also concurrently replicators with respect to the process of organismic reproduction). Group reproduction is a splitting process more similar to ameboid or bacterial reproduction than to sexual reproduction in higher plants and animals.[13]

Cases of what Damuth calls 'avatar selection' (selection among local populations of species within an ecological community) and other selection processes at even higher levels (species selection, clade selection?) are similar to interdemic group selection in that the group itself (avatar, species, clade) is the replicator, because the reproductive process is a splitting process of these higher-level entities. Thus we have the dual hierarchy of interactors and replicators shown in Table 9.1.

I want to make three points concerning this dual hierarchy. The first is that the hierarchy in the 'interactor' column of the table is a fairly neat hierarchy of inclusion, whereas the 'replicator' hierarchy is less neat.[14] The hierarchy of replicators could be made to look neater if we were to adopt Williams's abstract notion of gene (mentioned above). In that sense of gene, all the replicators up to the case of interdemic group selection are genes. But that neatness is illusory if we think of the hierarchy as one of *physical* inclusion. The second point I want to make is that the replicator hierarchy is derivative from the interactor hierarchy in the sense that we need to determine the level of interaction in order to determine the level

---

[13] For further discussion of the distinction between interdemic and intrademic group selection see Wade 1978 or the introduction to part III of Brandon and Burian 1984.

[14] The 'interactor' column is a hierarchy of inclusion if we ignore the first entry (i.e., the case of selection among RNAs within a primordial group). However, since that process is not concurrent with the others, there is a reason to ignore it when our concern is with a hierarchy of concurrent or possibly concurrent selection processes.

TABLE 9.1 *Dual hierarchy of interactors and replicators*

Selection scenario	Interactor	Replicator
Eigen's model for origins of life	Lengths of RNA	Lengths of RNA
'Selfish genes' (Orgel and Crick; Doolittle and Sapienza)	Lengths of DNA	Lengths of DNA
Meiotic drive	Chromosome (or a part thereof)	Chromosome (or a part thereof)
Organismic selection		
asexual reproduction	Organism	Genome (or organism?)
sexual reproduction	Organism	Genes
Intrademic group	Group	Genes
Interdemic group	Group	Group
Avatar selection	Avatar	Avatar
Species selection	Species	Species
Clade selection	Clade	Clade

of replication, but not vice versa. For instance, if we know that group selection is occurring, then we can determine the appropriate replicators, depending on the group reproductive process (intrademic versus interdemic selection). Because of this, the first point is of little import. Given a hierarchy of interactors, we simply let the replicators fall where they may.[15] My final point is that single hierarchies (such as that presented by Wimsatt) that mix interactors and replicators serve to answer neither the question about interactors nor that about replicators. Their incoherence clouds the real issues.

BEHAVIOUR AND SELECTION

A pervasive metaphor in evolutionary biology is that natural selection is like the process of fitting a key to a lock. The physical structure of the lock

[15] One might compare this dual hierarchy with those presented in Eldredge and Salthe 1985, Eldredge 1985, and Salthe 1985. There is one major difference: the hierarchy I have presented is relative to a specific process (viz. selection). Theirs is not. Thus, Eldredge and Salthe take a broader view of interactors than I. According to Hull's definition, which I have adopted, interactors imply selection. But there are many forms of interaction (mass–energy interchange) with the environment that do not necessarily lead to selection. Perhaps, then, my dual hierarchy is a special case of theirs.

is fixed, and the key must be shaped to fit it. Likewise, the features of the external environment are fixed, and natural selection shapes organisms to fit it (Lewontin 1978, 1983). This metaphor founders on the assumption that it is the external environment that determines selection pressures. But, as mentioned above, it is the selective environment, rather than the external environment, that is directly related to natural selection. (For detailed arguments to this effect see Antonovics *et al.* 1988 and Brandon 1990.) Recall that a selective environment is an area or a population within which the relative fitnesses of the competing types within the evolving population are constant. In other words, an area or a population is *selectively homogeneous* if the relative fitnesses do not vary in a significant way within it. Thus, the patterns of selective homogeneity and heterogeneity depend on the organisms in at least two ways. First, differing sensitivities to factors of the external environment will affect the pattern of selective heterogeneity. For instance, some organisms may, while others may not, perceive a given pattern of changes in nitrogen concentrations in the soil as selectively relevant. Thus, the pattern of selective environmental heterogeneity depends on the organisms present and their sensitivity to nitrogen concentrations. Second, by behaviour, organisms can effectively damp out heterogeneity in the physical environment. For instance, egg-laying females within a population of phytophagus insects may choose from many available plant species one particular species on which to lay their eggs. These plants may differ in many ways that would affect the fitness of the insects (e.g. differential nutritional quality, differential protection from predators), but by behavioural choice this potential heterogeneity is damped out. Indeed, this damping of external environmental heterogeneity seems to be one of the major trends of evolution. (The damping need not be behavioural; it can be morphological or physiological as well—consider blubber in sea mammals and warm-bloodedness as examples.)

Not only can behaviour affect the patterns of selection; it can affect the level of selection as well. This is obvious from a consideration of group selection. Group selection requires some grouping of organisms. This grouping may result from external processes acting on passive organisms, but it is more likely to result from active behaviour. Furthermore, some have argued that non-random grouping is a necessary condition for group selection (Hamilton 1975, Maynard Smith 1982, Nunney 1985). Whether or not one accepts this argument, it is widely recognized that non-random group formation is a condition that would increase the evolutionary effectiveness of group selection. Clearly, non-random group formation is most likely to result from specific behaviours among the relevant organisms.

If the selective environment is to be compared to a lock, it must be a malleable lock, one that can be changed to fit the key. By behaviour, organisms can change their selective environment. If these changes are selectively advantageous and if the behaviours are heritable, thcn we can expect the co-evolution of organism and environment. (See Odling-Smee 1988.) This can affect both the patterns of selection at a given level and the level (or levels) at which selection occurs.

CONCLUDING REMARKS

The theory of evolution by natural selection is the only theory we have that can explain the origins and the maintenance of adaptations. If these explanations are to be scientific rather than mere exercises in story telling, then adaptations must be carefully related to the selection processes that produce them (see Brandon 1981a and 1985). For instance, early- to mid-twentieth-century ecology and ethology are notorious for their explanations of organic features in terms of the features' being 'for the good of the species'. Nowadays we understand that benefit to the species is irrelevant to a selection process occurring at the level of organisms. But if there is a hierarchy of interactors, if selection occurs at different levels, then we cannot axiomatically assume that all adaptations are for the good of organisms. That is, we cannot assume that all adaptations are to be explained in terms of their benefits to organisms. Indeed, it is just this assumption that Doolittle and Sapienza (1980) criticize as the 'phenotype paradigm' (by 'phenotype' they mean organismic phenotype). As they point out, it is futile to search for the organismic benefit of the repetitive sequences of DNA that they call 'selfish genes'. It is futile not because the organismic benefit does not exist (in this case it doesn't), but rather, because of the irrelevance of any such benefit to the intracellular selection processes that produce these repetitive sequences. Similarly, the organismic benefit of any product of a higher-level selection process (e.g. group selection) is irrelevant to the explanation of its origin and/or maintenance. Of course, selection processes at different levels may interact (see Arnold and Fristrup 1982 or Vrba and Gould 1986 for discussion). This further complicates our theory of adaptation.

In this essay I have presented a hierarchy of interactors, or rather, a hierarchy of plausible interactors. I have not claimed that the importance of selection at any level other than the organismic has been conclusively demonstrated. Lacking such demonstrations, could one not argue that we should ignore considerations of hierarchical levels in applying the theory of natural selection to the biological world? I can think of only two such arguments, and both are seriously flawed. One is based on the assumption that we can know a priori that the only important level of selection is the organismic. In light of the conceptual and empirical work during the last ten years, that argument cannot be taken seriously (see Brandon and Burian 1984 and Wilson 1983a). The other argument, which is that considerations of parsimony lead us to try to explain all adaptations as products of organismic selection (as suggested in Williams 1966), is based on bad methodology. Once we see that other levels of selection are theoretically possible, we should not adopt a methodology that blinds us to their existence. Ultimately it may be that the only important level of selection is the organismic, and so the major adaptations in nature are organismic adaptations. However, at least for the moment, we need a hierarchical theory of interactors, if only to test the claim that organisms are the only important interactors in evolution.[16]

REFERENCES

Antonovics, J., Ellstrand, N. C., and Brandon, R. N. (1988), 'Genetic Variation and Environmental Variation: Expectations and Experiments', in S. K. Jain and L. D. Gottlieb (eds.), *Plant Evolutionary Biology* (London: Chapman and Hall), 275–303.
Arnold, A. J., and Fristrup, K. (1982), 'The Theory of Evolution by Natural Selection: A Hierarchical Expansion', *Paleobiology*, 8: 113–29.
Brandon, R. N. (1978), 'Adaptation and Evolutionary Theory', *Studies in History and Philosophy of Science*, 9: 181–206.
——(1981a), 'Biological Teleology: Questions and Explanations', *Studies in History and Philosophy of Science*, 12: 91–105.
——(1981b), 'A Structural Description of Evolutionary Theory', in P. Asquith and R. Giere (eds.), *PSA 1980*, ii (East Lansing, Mich.: Philosophy of Science Association), 427–39.
——(1982), 'The Levels of Selection', in P. Asquith and T. Nickles (eds.), *PSA 1982*, i (East Lansing, Mich.: Philosophy of Science Association), 315–22.
——(1985), 'Adaptation Explanations: Are Adaptations for the Good of Replicators or Interactors?', in B. Weber and D. Depew (eds.), *Evolution at a Crossroads: The New Biology and the New Philosophy of Science* (Cambridge, Mass.: MIT Press, A Bradford Book), 81–96.

[16] Vrba (1984) tests the alternative hypotheses of chance, organismic selection, and species selection among some monophylectic taxa of extinct and extant African mammals. Her data support the organismic selection hypothesis, i.e. the hypothesis that the pattern of differential speciation is an *effect* (*sensu* Williams 1966) of organismic selection. My thanks to Richard Burian and Stanley Salthe, who provided helpful comments on an earlier version of this.

Brandon, R. N. (1990), *Adaptation and Environment* (Princeton: Princeton University Press).

—— and Burian, R. M. (1984) (eds.), *Genes, Organisms, Populations: Controversies over the Units of Selection* (Cambridge, Mass.: MIT Press, A Bradford Book).

Crow, J. F. (1979), 'Genes that Violate Mendel's Rules', *Scientific American*, 240/2: 134–46.

Damuth, J. (1985), 'Selection among "Species": A Formulation in Terms of Natural Functional Units', *Evolution*, 39: 1132–46.

Dawkins, R. (1976), *The Selfish Gene* (Oxford: Oxford University Press).

——(1978), 'Replicator Selection and the Extended Phenotype', *Zeitschrift für Tierpsychologie*, 47: 61–76.

——(1982*a*), *The Extended Phenotype* (San Francisco: Freeman).

——(1982*b*), 'Replicators and Vehicles', in King's College Sociobiology Group (eds.), *Current Problems in Sociobiology* (Cambridge: Cambridge University Press), 45–64.

Doolittle, W. F., and Sapienza, C. (1980), 'Selfish Genes, the Phenotype Paradigm and Genome Evolution', *Nature*, 284: 601–3.

Eigen, M., Gardiner, W., Schuster, P., and Winkler-Oswatitsch, R. (1981), 'The Origin of Genetic Information', *Scientific American*, 244/4: 78–94.

Eldredge, N. (1985), *The Unfinished Synthesis* (Oxford: Oxford University Press).

—— and Cracraft, J. (1980), *Phylogenetic Patterns and Evolutionary Process* (New York: Columbia University Press).

—— and Salthe, S. (1985), 'Hierarchy and Evolution', in R. Dawkins and M. Ridley (eds.), *Oxford Surveys of Evolutionary Biology* (Oxford: Oxford University Press), 184–208.

Endler, J. A. (1986), *Natural Selection in the Wild* (Princeton: Princeton University Press).

Gould, S. J., and Eldredge, N. (1977), 'Punctuated Equilibria: The Tempo and Mode of Evolution Reconsidered', *Paleobiology*, 3: 115–51.

Hamilton, W. D. (1975), 'Innate Social Aptitudes of Man: An Approach from Evolutionary Genetics', in R. Fox (ed.), *Biosocial Anthropology* (New York: Wiley), 133–55.

Hull, D. (1980), 'Individuality and Selection', *Annual Review of Ecology and Systematics*, 11: 311–32.

——(1981), 'Units of Evolution: A Metaphysical Essay', in U. L. Jensen and R. Harre (eds.), *The Philosophy of Evolution* (Brighton: Harvester), 23–44.

——(1988), 'Interactors versus Genes', in H. Plotkin (ed.), *The Role of Behavior in Evolution* (Cambridge, Mass.: MIT Press), 19–50.

Jablonski, D. (1986), 'Background and Mass Extinctions: The Alternation of Macroevolutionary Regimes', *Science*, 231: 129–33.

Lewontin, R. C. (1970), 'The Units of Selection', *Annual Review of Ecology and Systematics*, 1: 1–18.

——(1974), *The Genetic Basis of Evolutionary Change* (New York: Columbia University Press).

——(1978), 'Adaptation', *Scientific American*, 239/3: 156–69.

——(1983), 'Gene, Organism and Environment', in D. S. Bendall (ed.), *Evolution from Molecules to Man* (Cambridge: Cambridge University Press).

Maynard Smith, J. (1982), 'The Evolution of Social Behaviour—A Classification of Models', in King's College Sociobiology Group (ed.), *Current Problems in Sociobiology* (Cambridge: Cambridge University Press), 29–44.

Mayr, E. (1963), *Animal Species and Evolution* (Cambridge, Mass.: Harvard University Press).

—— (1978), 'Evolution', *Scientific American*, 239/3: 46–55.

Nunney, L. (1985), 'Group Selection, Altruism and Structured-Dome Models', *American Naturalist*, 126: 262–93.

Odling-Smee, F. J. (1988), 'Phenotypes', in H. Plotkin (ed.), *The Role of Behavior in Evolution* (Cambridge, Mass.: MIT Press), 73–132.

Orgel, L. E., and Crick, F. H. C. (1980), 'Selfish DNA: The Ultimate Parasite', *Nature*, 284: 604–7.

Plotkin, H. (1988) (ed.), *The Role of Behavior in Evolution* (Cambridge, Mass.: MIT Press).

Salmon, W. C. (1971), *Statistical Explanation and Statistical Relevance* (Pittsburgh: University of Pittsburgh Press).

Salthe, S. N. (1985), *Evolving Hierarchical Systems* (New York: Columbia University Press).

Sober, E. (1984), *The Nature of Selection* (Cambridge, Mass.: MIT Press, A Bradford Book).

Stanley, S. M. (1975), 'A Theory of Evolution above the Species Level', *Proceedings of the National Academy of Sciences*, 72: 646–50.

—— (1979), *Macroevolution: Pattern and Process* (San Francisco: Freeman).

Uyenoyama, M., and Feldman, M. W. (1980), 'Theories of Kin and Group Selection: A Population Genetics Perspective', *Theoretical Population Biology*, 19: 87–123.

Vrba, E. S. (1984), 'Evolutionary Pattern and Process in the Sister-Group Alcelaphini-Aepycerotini (Mammalia: Bovidae)', in N. Eldredge and S. M. Stanley (eds.), *Living Fossils* (New York: Springer-Verlag), 62–79.

—— and Gould, S. J. (1986), 'The Hierarchical Expansion of Sorting and Selection: Sorting and Selection Cannot Be Equated', *Paleobiology*, 12: 217–28.

Wade, M. J. (1977), 'An Experimental Study of Group Selection', *Evolution*, 31: 134–53.

—— (1978), 'A Critical Review of the Models of Group Selection', *Quarterly Review of Biology*, 53: 101–14.

—— (1984), 'Soft Selection, Hard Selection, Kin Selection, and Group Selection', *American Naturalist*, 125: 61–73.

Williams, G. C. (1966), *Adaptation and Natural Selection* (Princeton: Princeton University Press).

Wilson, D. S. (1983a), 'The Effect of Population Structure on the Evolution of Mutualism: A Field Test Involving Burying Beetles and their Phoretic Mites', *American Naturalist*, 121: 851–70.

—— (1983b), 'The Group Selection Controversy: History and Current Status', *Annual Review of Ecology and Systematics*, 14: 159–87.

Wimsatt, W. C. (1980), 'Reductionist Research Strategies and their Biases in the Units of Selection Controversy', in T. Nickles (ed.), *Scientific Discovery*, ii (Dordrecht: Reidel), 213–59.

—— (1981), 'The Units of Selection and the Structure of the Multi-Level Genome', in P. Asquith and R. Giere (eds.), *PSA 1980*, ii (East Lansing, Mich.: Philosophy of Science Association), 122–83.

Wynne-Edwards, V. C. (1962), *Animal Dispersion in Relation to Social Behaviour* (Edinburgh: Oliver and Boyd).

10

A CRITICAL REVIEW OF PHILOSOPHICAL WORK ON THE UNITS OF SELECTION PROBLEM

ELLIOTT SOBER AND DAVID SLOAN WILSON

I. INTRODUCTION

Philosophers have produced a large literature aimed at clarifying what a unit of selection is. Rather than launch immediately into the technical details of that literature, we begin with an informal description of what the problem of the units of selection is about. As the positivists used to say, the *explicandum* must be clarified before the adequacy of the *explicans* can be evaluated. Section II explains the issue. Sections III–VII criticize some of the main ideas that have been introduced. Section VIII presents our own take on the problem. Section IX extracts a consequence.

II. BACK TO BASICS

The problem of the units of selection has engaged the attention of evolutionists ever since Darwin. It concerns whether traits evolve because they benefit individual organisms[1] or because they are good for the group in which they occur. More recently, a third alternative has been proposed, which holds that traits evolve because they benefit the genes that code for them (Williams 1966, Dawkins 1976).

The choice that Darwin considered—between the group and the organism as units of selection—was important because of the issue of *evolutionary altruism*. An altruistic trait reduces the fitness of organisms that possess it while benefiting the group in which it occurs. Altruistic traits are bad for

First published in *Philosophy of Science*, 61 (1994): 534–55. Reprinted by permission.

[1] As a terminological convenience, we use 'individual' and 'organism' interchangeably. This does not prejudge the substantive claims that species are individuals (*sensu* Hull 1988) or that groups are sometimes organisms (*sensu* Wilson and Sober 1989, 1994*a*).

the organism but good for the group. If the organism is the exclusive unit of selection, then altruism cannot evolve. However, if the group is sometimes a unit of selection, altruism becomes an evolutionary possibility.

Two consequences of this standard pairing of altruism with the group as unit of selection and selfishness with the individual as unit of selection (Wilson 1990) are worth noting. First, altruism and selfishness are defined by the fitness effects of a behaviour; they have nothing essentially to do with psychological motives. Second, altruism is not the same as helping. Parental care is a type of helping, but if parents who care for their offspring are fitter than parents who do not, then parental care is not an instance of altruism.[2]

In order to fix ideas, it will be useful to apply the contrast between the group and the organism as units of selection to a pair of examples. Our interest here is not in getting the biological details right, but in helping the reader to see the relevant conceptual contrast. Consider, first, why zebras run fast rather than slowly. The answer is that zebras who ran fast were more successful in surviving to reproductive age than were zebras who ran slowly. The trait of running fast evolved because it benefited the organisms who possessed it. Compare this with the barbed stinger of the honey-bee. When a honey-bee stings an intruder to the nest, the bee disembowels itself. The barb did not evolve because it helped bees who had the barb. On the contrary, barbs evolved because they helped the group, and in spite of the fact that they harmed the organisms possessing them. Nests made of individuals with barbed stingers did better than nests made of individuals without barbs.[3]

If the biological details are as stated in these two examples, we should conclude that the individual organism is a unit of selection in the evolution of running speed in zebras, whereas the group is a unit of selection in the evolution of barbed stingers in honey-bees. Generalizing from these two examples, we obtain the following definitions:

The organism was a unit of selection in the evolution of trait T iff one of the factors that influenced T's evolution was that T conferred a benefit on organisms.

---

[2] This is most obvious when one considers species with uniparental reproduction. Also, the present point does not deny the possibility of parent/offspring conflicts of interest (Trivers 1972, Haig 1993).

[3] Readers who think that barbed stingers evolved by kin selection and that kin selection is not a kind of group selection are asked to grant this example for illustrative purposes only. We argue that kin selection is a type of group selection (in which the groups are composed of relatives) in Wilson and Sober (1989, 1994a). This also is the position taken by Seeley (1989); the title of his paper is instructive: 'The Honey Bee Colony as a Superorganism'.

The group was a unit of selection in the evolution of trait T iff one of the factors that influenced T's evolution was that T conferred a benefit on groups.

These two special cases generalize to yield the following formulation:

Objects at level X were units of selection in the evolution of trait T iff one of the factors that influenced T's evolution was that T conferred a benefit on objects at level X.

Although the first two definitions do not describe what it means for the gene to be a unit of selection, the third one does: if a trait evolved because it benefited the gene that coded for it, then the gene was a unit of selection.[4]

We note two consequences of this proposal. First, different traits may evolve for different reasons; for example, the group may be a unit of selection for one trait but not for another. Second, the same trait may evolve for several reasons—several units of selection may be associated with the evolution of a particular trait. Although a monolithic solution to the units of selection problem is possible (e.g. 'the gene is the one and only unit of selection for all traits'), such an approach must be argued for explicitly; it is not dictated merely by the problem's formulation.

We believe that this simple schema is helpful as a point of departure for understanding the units of selection problem. None the less, a number of problems of clarification remain, which we will address in due course.

III. REPLICATORS AND INTERACTORS

Hull (1980, 1981, 1988) has argued that the distinction between *interactor* and *replicator* is central to understanding the debate over the units of selection. Hull's ideas generalize themes explored by Dawkins (1976, 1982). Dawkins distinguished replicators and *vehicles*. Genes are examples of the former and organisms are examples of the latter. Hull substituted the term 'interactor' for Dawkins's 'vehicle', because Hull took Dawkins's terminology to be committed to the selfish gene point of view, according to which the gene, not the organism or the group, is the one and only unit of selection. Hull wanted to formulate a more general framework than Dawkins had enunciated, one in which various positions concerning the units of selection problem could be stated and clarified. Other authors (e.g.

---

[4] It may seem to follow from this that the gene is *always* a unit of selection. This will be discussed in Section VII.

Lloyd (1988), Brandon (1990)) have endorsed Hull's suggestions, and have added more technical proposals of their own.

Hull defines a replicator as 'an entity that passes on its structure directly in replication' and an interactor as 'an entity that directly interacts as a cohesive whole with its environment in such a way that replication is differential' (1980: 318).

The most valuable part of Hull's (and Dawkins's) distinction is that it separates the issue of heredity from the issue of which causal processes underlie differential reproduction. That genes are the units of *heredity* has never been at issue in the units of selection problem. If an altruistic phenotypic trait evolves by a process of group selection, the genes coding for that trait also must evolve. The idea that genes are units of heredity— that they are replicators—is common ground.

Even though the replicator concept is not central to the units of selection problem, it merits philosophical scrutiny. If a replicator is defined as an object that passes on its structure directly in replication, what does 'direct' mean, and what is replication? The process by which genes replicate is intricate. In what sense do parental genes create offspring genes *directly*?[5]

Dawkins holds that sexual organisms are not replicators, but genes are. In what sense do organisms fail to make copies of themselves, whereas genes succeed? Sexual organisms often exhibit less than perfect copying fidelity, although for canalized traits, fidelity is often very high. Human parents have one heart, and their children usually do too. In any event, if organismic reproduction involves imperfect fidelity, why does this mean that organisms are not replicators at all? Why not say, instead, that their replication is imperfect?

Dawkins (1976, 1982) stipulates that replicators obey Weismannian, rather than Lamarckian, principles—they cannot mediate the inheritance of acquired characteristics. This prohibition is illustrated in Figure 10.1. When a mother giraffe lengthens her neck by stretching, this does not induce a mutation in the genes passed along to her offspring that allows them to have long necks without needing to stretch. Phenotypic traits acquired in development do not alter the genes passed along in reproduction.

We have no quarrel with this routine rejection of Lamarckian ideas, although we emphasize that it is an empirical question whether Weismannism is *always* correct. However, we do not see why the concept of replication should be burdened with the stipulation that replicators are

---

[5] The question of how to interpret the idea of *directness* becomes even more pressing when one considers the suggestion (Hull 1988, Dawkins 1982, and Williams 1992) that asexual organisms, populations, and/or species can be replicators.

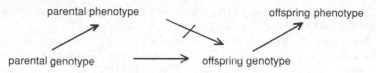

Fig. 10.1. Lamarckian inheritance as precluded according to Weismannism.

Weismannian. If genes occasionally violated Weismannism, would this mean that genes are not replicators?

A second question concerns the relationship of the concept of replicator to the notion of heredity. In biology, heredity is measured by the concept of heritability. When offspring phenotypically resemble their parents, this can be due to their sharing genes or living in similar environments (or both). A phenotypic trait has non-zero (narrow) heritability when parent/offspring resemblance is attributable, at least in part, to shared genes.

As such, heritability is a property of the phenotypic traits of organisms. The same concept applies straightforwardly to groups of organisms. If groups of organisms bud off daughter colonies, daughter colonies may resemble parental populations because of genetic similarities. Groups have heredity in the same sense that individual organisms do. (Maynard Smith 1987 and Ridley 1993 disagree; we pursue the point in Wilson and Sober 1989, 1994b.)

Heritability is essential for natural selection to cause evolution. If running speed in zebras is to evolve by individual selection, offspring organisms must resemble their parents. And if barbed stingers are to evolve in honey-bees by group selection, daughter colonies must resemble their parents. Deciding 'what the replicators are' is not important here, though ensuring that the traits are heritable is of the essence.[6]

If organisms and groups both can possess heritable characters, what is so special about genes? Genes are, by definition, the objects that give the phenotypes of these higher-level objects their (narrow) heritability. None the less, reproduction and parent/offspring resemblance are hardly unique features of genes. Although being a replicator is not the same as being heritable, we see no harm in defining 'the unit of heredity' as the gene.

A separate issue concerning the replicator concept is worth mentioning. Are the pages fed into a copying machine 'replicators'? To be sure, copies are made of them. But do they make copies of themselves? Arguably, the answer is *no*. The pages are *replicated*, but they are not *replicators*. One

---

[6] This is one reason why Darwin was able to develop so many insights about natural selection even though his picture of the mechanism of heredity was completely erroneous.

implication of the term 'replicator' is that replicators control their own destiny. They actively make copies of themselves; they are not passive entities of which copies are made. The idea that genes are replicators may exaggerate the degree of encapsulation that the replication process possesses (Oyama 1985, Lewontin 1992).

We now turn to the interactor concept. In Section II, we introduced running speed in zebras and barbed stingers in honey-bees as working examples of the organism and the group as units of selection. Does the concept of an interactor capture the requisite distinction? Let us see.

These two examples differ in a way that must now be made explicit. In our hypothetical example about running speed, we imagine that the individual organism, but not the group, is a unit of selection. However, in the case of the barbed stinger, we must recognize two units—the trait's evolution is influenced both by the fact that barbed stingers are good for the group and by the fact that a barbless stinger is good for the organism. The group and the organism are units of selection in this instance.

If Hull's proposal is to reflect these ideas, then it must be true, in the first case, that individual zebras, but not zebra herds, directly interact as cohesive wholes with their environment in such a way that replication is differential. In contrast, it must be true, in the second example, that both individual honey-bees and the hives to which they belong directly interact as cohesive wholes with their environment in such a way that replication is differential.

Knowing how to judge an interaction's 'directness' is difficult, however. Presumably, zebras interact directly with the lions that kill them, just as bees interact directly with the bears they sting. But what if one zebra herd goes extinct because all its members are slow, while another survives because all its members are fast? If this is not a case of groups interacting directly as a cohesive whole with their environments, how does this differ from one beehive's going extinct because it does not contain individuals with barbed stingers, while another survives because its members have barbed stingers? For the group to be a unit of selection, more is required than the fact that some groups do better than others. How the idea of 'direct interaction as a cohesive whole' supplies that further ingredient remains unclear.

IV. THE ANALYSIS OF VARIANCE

Wimsatt (1980) proposed the following definition:

A *unit of selection* is any entity for which there is heritable *context-independent* variance in fitness among entities at that level which does not appear as heritable context-dependent variance in fitness (and, thus, for which the variance in fitness is *context-dependent*) at any lower level of organization. (p. 236)

How should the idea of 'context-independence' be understood? Wimsatt explains that the idea of additivity, which has a clear meaning in the statistical method known as the analysis of variance, is a special case of context-independence. Since we do not fully understand the wider meaning of 'context-independence', we focus on additivity.

Consider the fitness relationships that obtain between two loci, each of which has two alleles. Each organism is either *AA*, *Aa*, or *aa* at one locus and *BB*, *Bb*, or *bb* at the other. The fitness of each two-locus genotype may be represented as follows:

		B-locus		
		BB	*Bb*	*bb*
	AA	w_{11}	w_{12}	w_{13}
A-locus	*Aa*	w_{21}	w_{22}	w_{23}
	aa	w_{31}	w_{32}	w_{33}

If w_{i2} is precisely half-way between w_{i1} and w_{i3}, and w_{2j} is precisely half-way between w_{1j} and w_{3j} ($i, j = 1, 2, 3$), then fitness relationships are additive, and Wimsatt's criterion judges the single gene to be the unit of selection.

Wimsatt's criterion entails that the single gene is not the unit of selection in at least two circumstances. First, heterozygotes may fail to be precisely intermediate, though the relationships that obtain within one locus do not depend on what is true at the other. A hypothetical example is provided by the following table of viability fitnesses:

		B-locus		
		BB	*Bb*	*bb*
	AA	0.8	0.7	0.6
A-locus	*Aa*	0.7	0.6	0.5
	aa	0.3	0.2	0.1

Because the *A*-locus exhibits *dominance in fitness* in this case, Wimsatt's criterion entails that the unit of selection is not the single gene, but the single locus genotype.

The second pathway by which the single gene may fail to be the unit of selection, according to Wimsatt's criterion, involves *epistasis in fitness*. This occurs when the fitness relationships among the genotypes at one locus depend on what is true at the other, as is illustrated by the following hypothetical data set:

		B-locus		
		BB	Bb	bb
	AA	0.1	0.2	0.3
A-locus	Aa	0.1	0.3	0.2
	aa	0.2	0.1	0.3

Notice that the fitness ordering of the B-genotypes depends on what is true at the A-locus. Wimsatt's criterion would conclude that the unit of selection in this case is the two-locus genotype, not the single-locus genotype, and not the single gene.[7]

Wimsatt presented his criterion as a criticism of an argument that Williams (1966) advanced and Dawkins (1976) repeated, which aimed to show that the 'meiotically dissociated gene' is the unit of selection. Here is Williams's statement of the argument:

Obviously it is unrealistic to believe that a gene actually exists in its own world with no complications other than abstract selection coefficients and mutation rates. The unity of the genotype and the functional subordination of the individual genes to each other and to their surroundings would seem, at first sight, to invalidate the one-locus model of natural selection. Actually, these considerations do not bear on the basic postulates of the theory. No matter how functionally dependent a gene may be, and no matter how complicated its interactions with other genes and environmental factors, it must always be true that a given gene substitution will have an arithmetic mean effect on fitness in any population. One allele can always be regarded as having a certain selection coefficient relative to another at the same locus at any given point in time. Such coefficients are numbers that can be treated algebraically, and conclusions inferred from one locus can be iterated over all loci. Adaptation can thus be attributed to the effect of selection acting independently at each locus. (Williams 1966: 56–7)

Wimsatt contends that although attending to the frequencies and fitness values of single genes may be useful as a 'bookkeeping' device, this point is irrelevant to whether the single gene is the unit of selection. The appropriate criterion, Wimsatt maintains, is the additivity criterion we have just discussed, which leads to quite different conclusions.

We agree with Wimsatt's criticism of Williams's argument, but disagree with the additivity criterion that Wimsatt proposed. Before explaining the disagreement, we will elaborate on Wimsatt's important critique.

If evolution is defined as change in gene frequency, then evolution by natural selection entails that genes will differ in fitness, *regardless of what the unit of selection is*. If group selection causes an altruistic gene to evolve, that gene will be fitter than the selfish allele it displaces. This means that the units of selection problem is not settled by the mere fact that the different alleles in a population have fitness values that can be 'treated algebraically' (Sober and Lewontin 1982; Sober 1984, 1993; see also Godfrey-Smith and Lewontin 1993 on the irrelevance of a model's dimensionality to the units of selection question).

Let us now apply Wimsatt's additivity criterion to the examples of organismic and group adaptations introduced in Section II. The first point is that the issue of additivity plays no role in explaining why running speed in zebras is an adaptation that evolved for the good of the individual organism. The genes that influence running speed may or may not exhibit dominance or epistasis. These questions are relevant to *how fast* running speed will evolve (as we learn from Fisher's (1958) fundamental theorem). But they simply cut no ice with respect to the problem of whether the organism is the unit of selection in this instance.

We believe that the same conclusion should be drawn when the additivity criterion is applied to the problem of defining what it means for the group to be a unit of selection. To explain why, we need to say a little more about the concepts of altruism and selfishness. Figure 10.2 represents two fundamental facts about these evolutionary concepts. No matter what mix of altruism and selfishness is found in a group, selfish individuals are fitter on average than altruists. Second, increasing the frequency of altruism found in a group raises the fitnesses of altruists and selfish individuals alike. If we define the fitness of a group as the average fitness ($\bar{w}$) of the individuals in the group, then this second point means that groups in which altruism is common are fitter than groups in which altruism is rare.

Figure 10.2 is a standard representation of the fitness relationships of evolutionary altruism and selfishness. When an ensemble of populations, each containing its own mix of altruistic and selfish individuals, satisfies certain further conditions, altruism can increase in frequency by the process of group selection. This suffices for the group to be a unit of selection.

Note that the fitness functions depicted in the figure are *straight lines*. The fitness of the group is here an additive function of the proportion of altruists it contains. However, this additive relation does not prevent the group from being a unit of selection. Of course, it is easy to model the non-additive case. For example, just bend the fitness functions in Figure 10.2 so that groups benefit from additional altruists according to a rule of diminishing returns. We then have an analogue of dominance in fitness, but this makes no difference as to whether the group is a unit of selection.

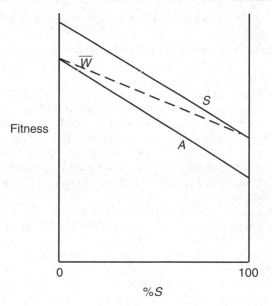

FIG. 10.2. Selfish individuals on average are fitter than altruists, and increasing the frequency of altruism in a group increases the fitness of altruists and selfish individuals alike.

Sober (1984) and Lloyd (1988) agreed with Wimsatt that absence of dominance and epistasis is criterial for the gene to be a unit of selection. Lloyd (1988) and Mayr (1990) also used additivity as a criterion for the group to be a unit of selection, whereas Sober (1984) resisted this conclusion. The schizophrenia implicit in Sober's treatment of the issue was pointed out by D. S. Wilson in conversation, and was independently identified by Walton (1991).

We suggest that additivity is wrong through and through.[8] If running speed in zebras evolved because it benefited individual organisms, this says nothing about the details of how genotypes code for phenotypes. Similarly, group selection does not require that group phenotypes be 'emergent', if emergence entails non-additivity. The group as unit of selection embodies a kind of holism (Sober 1981, Wilson 1988, Wilson and Sober 1994a), but it is a holism that does not demand emergentism.

[8] Richardson (1985), Maynard Smith (1987), Sterelny and Kitcher (1988), and Waters (1991) challenged the plausibility of the additivity criterion as applied to the case of heterozygote superiority (an argument put forward by Sober and Lewontin 1982). Godfrey-Smith (1992) develops further objections to the additivity criterion.

V. THE PATTERN OF VARIATION IN FITNESS

Additivity describes a relationship that can obtain among fitness values. Even if the additivity criterion misdescribes what a unit of selection is, it is worth asking whether some other relationship defined on the observed variation in fitness can be used to characterize what a unit of selection is. We believe that an argument presented by Sober (1984) shows that the answer is *no*. Suppose we investigate a set of populations, each internally homogeneous for height. In the first population, all the organisms are one unit tall. In the second, all are two units tall, and so on. When we measure the fitnesses of these individuals, we find that height is perfectly correlated with fitness. In this case, there is no within-group variance in fitness; all the variation is between groups.

If pattern of variation somehow determined what the unit of selection is, the above information would settle definitively what the unit of selection is in this case. But it does not; two different hypotheses are consistent with the information given. The first proposes that there is individual selection for being tall, in which case the individual is the unit of selection. The second says there is group selection favouring groups with higher average height, which would mean that the group, not the individual, is the unit of selection.

Although pattern of actual variation in fitness does not determine what the unit of selection is in the example of the tall and the short, an experiment that would provide useful evidence is not hard to describe. Suppose we create some heterogeneous groups and then measure the fitnesses of the organisms in them. If tall individuals are equally fit, regardless of the kind of group they inhabit, this is evidence that selection is at the level of individuals. And if tall and short individuals in the same group have the same fitness regardless of their individual phenotypes, this favours the hypothesis of group selection.

The conclusion that Sober (1984) drew about this example—that the actual pattern of variation in fitness does not define what the units of selection are—was challenged by Lloyd (1988) and by Griesmer and Wade (1988). They argued that biologists use background information that allows them to use the observed variation in fitness to infer the units of selection in a given case. To some extent, Sober and his critics were talking past each other. Sober argued that facts about within- and among-group variation, by themselves, do not uniquely determine what the units are. The critics argued that those facts, plus other assumptions, settle the matter. Obviously, these two assertions are compatible.

Even if the actual pattern of variation in fitness does not determine the units of selection, a related criterion might be considered. The new idea is that group selection occurs precisely when the fitness of organisms depends on the kind of group they inhabit. Sober (1984) argued that this criterion is too permissive. In many cases of individual selection, individual fitnesses depend on group composition. For example, suppose that in the evolution of traits A and B, the advantage goes to the common trait; within each population in which the traits are exemplified, individual selection proceeds according to this frequency-dependent rule. Now imagine two populations: in the first A is common, while in the second A is rare. These two populations will evolve in different directions. But this is not an instance of group selection.

Even if this critique is correct, we must recognize that the pattern of variation in fitness exhibited within and among populations is important in several ways. First, pattern of variation may be *evidence* for the existence of different sorts of selection processes. Second, the pattern of variation in fitness helps *predict* how the system will evolve. And finally, pattern of variation *does* play a defining role in a limiting case. Selection at a given level requires variation in fitness at that level. If groups do not vary in fitness, the group cannot be a unit of selection. If the organisms under study do not vary in fitness, the organism cannot be a unit of selection.

Group selection requires more than groups varying in fitness. And it is not enough that they vary in fitness and differ in their rates of extinction and colonization. Rather, what is required is that this pattern of variation obtains *because* of their different traits.

VI. SCREENING-OFF

Brandon (1984, 1990) has argued that the statistical concept of *screening-off* can be used to clarify the concept of a unit of selection.[9] Y is said to screen off X from Z precisely when $P(Z|X\&Y) = P(Z|Y) \neq P(Z|X)$.[10] When this relation obtains, Y and X are related asymmetrically to the task of predicting Z; if you know Y, knowing in addition that X is true would not change your prediction. According to Brandon,

[s]election occurs at a given level (within a common selective environment) if and only if (1) there is differential reproduction among the entities at that level; and (2)

---

[9] Brandon uses 'level of selection' to talk about what we call the units of election problem. He reserves the term 'unit' for another use. We use 'level' and 'unit' interchangeably.

[10] A better formulation of Reichenbach's (1956) idea would treat X, Y, and Z as *variables* that come in *states*. Let '$X = a$' mean that X is in state a. Then Y screens off Z from X precisely when, for all i, j, k, $P(Z = i|Y = j \& X = k) = P(Z = i|Y = j) \neq P(Z = i|X = k)$.

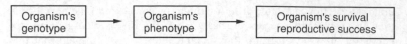

FIG. 10.3. An organism's phenotype is a more proximal cause of an organism's survival and reproduction than its genome.

the 'phenotypes' of the entities at that level screen-off properties of entities at every other level from reproductive values at the given level. (1990: 88)

Brandon applies this criterion in two contexts. First, he argues that it explains why selection standardly acts on an organism's phenotype, not on the genes the organism contains. Second, he argues that it elucidates what group selection is, and how it differs from selection at the level of the individual organism. We consider these in turn.

Mayr (1963: 184) and Gould (1980: 90) emphasize that selection acts 'directly' on the organism's phenotype, and only 'indirectly' on its genes. Gould takes this to undermine the selfish gene point of view—the idea that the gene is the one and only unit of selection. Brandon suggests that the Mayr/Gould point about the directness of selection can be captured via the notion of screening-off, and that the criterion stated above explains why selection typically acts at the level of organismic phenotypes, not at the level of the gene.

Mayr and Gould's causal claim is illustrated in Figure 10.3. Even though this diagram fails to represent the causal role of the environment, the point is that the organism's phenotype is a *more proximal* cause, and its genome is a *more distal* cause of the organism's survival and reproductive success.

In many causal chains, the proximal cause screens off the distal cause from the effect.[11] Is this true in the case at hand? Often it is. If a zebra's fitness is determined by its speed, then fixing a zebra's running speed allows (probabilistic) prediction of its survival and reproductive success; adding information about the genes that endow the zebra with the running speed it has will not alter the prediction.

The simple case of genetic dominance, however, provides an exception.[12] Suppose that individuals with the AA and Aa genotypes are phenotypically indistinguishable, but that both differ from individuals who are aa. Let AA and Aa have the same chance of surviving from egg to adult, and aa individuals have a lower viability. Then individuals who are AA and

[11] This happens often, but not always. When the chain is deterministic, or when it fails to include *all* factors that play a causal role, screening-off can fail. See Sober 1992 for discussion.
[12] We owe this observation to Marsha Ensor, Julie Faulhaber, and Jennifer Hoepner.

Aa have *different* prospects for reproductive success. The reason is that *AA* individuals never produce *aa* offspring, whereas *Aa* individuals sometimes do. Success in reproduction is not measured merely by number of offspring, but by the number of viable, fertile offspring. In this instance, phenotype does not screen off genotype from reproductive success.[13] But, as argued in Section III, the organism can be a unit of selection even when there is dominance.

We turn now to Brandon's application of the screening-off criterion to the task of distinguishing group from individual organism as units of selection. Assuming that the groups in question exhibit differential reproduction, Brandon proposes that the group is a unit of selection precisely when there is 'some group property (the group "phenotype") that screens-off all other properties from group reproductive success' (1990: 87). Modifying Brandon's notation slightly, his idea is that the group is a unit of selection precisely when

$$\text{Exp}(n|G \& P) = \text{Exp}(n|G) \neq \text{Exp}(n|P).$$

Here $\text{Exp}(n|—)$ means that n is the expected number of propagule groups that a group produces (conditional on—),[14] G is the group phenotype, and P is a specification of the phenotypes of the organisms in the group. Brandon remarks that it is not inevitable that this two-part condition be satisfied. He describes a case in which the equality is true, but the inequality is not, the latter because 'the phenotype of each individual within the group would determine that individual's adaptedness, and the adaptedness values of each member of the group would determine the adaptedness value of the group' (ibid.).

We are puzzled as to why the inequality demanded by this criterion should *ever* be true, since the unary and relational properties of individuals evidently *determine* the properties of the group. Consider the case of the honey-bee's barbed stinger. Here the group is a unit of selection; groups

---

[13] Brandon says that 'the notions of genotype and phenotype are not mutually exclusive. The genotype of an organism is part of its phenotype. Thus my claim commits me to the position that any change in genotype that does lead to a change in reproductive success must also be a change in the organism's phenotype' (1990: 84–5). This stipulation would save Brandon's proposal from the problem posed by dominance. However, this proposal endangers Brandon's whole enterprise. If phenotype is to screen off genotype, then phenotype cannot include genotype. Violation of this requirement would mean that some of the conditional probabilities required by the screening-off relation are not defined; if P is to screen off G from something, then all possible combinations of P-states and G-states must have non-zero probability.

[14] Brandon focuses on the number of offspring groups, without taking their census size into account, because he feels that this is essential for the idea of group selection. This formulation will be problematic when parental groups all have the same number of groups as offspring, but differ in the census size of the groups they found.

benefit because they contain individuals who have barbed stingers. Yet, in this instance, the group phenotype is determined by the phenotypes of the individuals in the group.

One possible solution is to restrict what one means by a 'group property' and by a 'property of an organism'. Brandon observes that 'it is not completely clear what should count as group properties' (ibid.), but adds that 'obvious examples include . . . the relative frequency of certain alleles within the group [and] the phenotypic distribution within the group . . .' (ibid.). However, *these* group properties apparently do not screen off, since they are determined by the array of properties that the individuals in the group possess.[15]

VII. ON WHETHER GENIC SELECTIONISM IS BOTH SUBSTANTIVE AND PLAUSIBLE

We have argued on several occasions (e.g. Sober 1984, 1990b, 1993, Wilson and Sober 1989, 1994a) that Dawkins's (1976, 1982) thesis that the gene is the one and only unit of selection is either false or vacuous. Sterelny and Kitcher (1988) have defended Dawkins, arguing that his claim is both non-trivial and plausible. According to them, Dawkins's substantive point is that 'barring complication, the average ability of the genes in the gene pool to leave copies of themselves increases with time' (Sterelny and Kitcher 1988: 340). We believe that this claim is not part of Dawkins's theory, and, in any event, is not biologically plausible.

Dawkins has frequently emphasized that natural selection operating within the confines of a single group will eliminate altruistic characteristics. The same point holds if one talks about *genes*. A gene for altruism will be displaced by a gene for selfishness in this instance.

Whether the issue concerns phenotypes or genes, the relationship between selfishness and altruism is the one depicted in Figure 10.2. The quantity $\bar{w}$ measures the average fitness of the individuals in the population; equivalently, it measures the average fitness of the altruistic and selfish genes in the population. This means that purely within-group selection *reduces* the value of $\bar{w}$. As selfishness displaces altruism, the

---

[15] Brandon connects his screening-off criterion for the units of selection problem with a more general view concerning what constitutes the best explanation of an effect. He suggests that if Y screens off X from Z, then Y is a better explanation of Z than X is. Mitchell (1987) shows that by switching *explananda*, Brandon's criterion can be used to defend the genic point of view. Sober (1992) also develops objections to this proposal. Brandon *et al.* (1994) reply to this criticism.

individuals in the population decline in their average fitness. Indeed, the very same thing is true of genes. Genes become *less fit* as a result of the process of 'subversion from within' that Dawkins has highlighted. This shows, we believe, that Sterelny and Kitcher's positive reconstruction of what is supposed to be non-trivial in Dawkins's theory cannot be sustained.[16] The fitness of genes does *not* increase when genes for selfishness replace genes for altruism (Sober 1990*b*). Ironically, the fitness of genes *can* increase when group selection leads altruism to evolve; but group selection is a process that Dawkins would not touch with a stick.

Sterelny and Kitcher (1988), Kitcher *et al.* (1990), and Waters (1991) have argued that a point in favour of the selfish gene point of view is that *all* selection processes can be represented in terms of single genes and their properties. They note that the same cannot be said of the organism or the group as units of selection. We agree that this difference exists, but we see it as a defect, not a strength. The argument that Kitcher, Sterelny, and Waters present here is a variant of the 'representation argument' advanced by Williams and Dawkins discussed in Section IV. If even group selection can be represented as a kind of genic selection, then genic selectionism is not a substantive alternative to anything. The selfish gene theory is vacuous if it is consistent with any and all types of selection process.

If it is not an automatic truism that the group or the organism is the unit of selection, the same should hold for the gene. An adequate clarification of the units of selection problem should treat these three levels on a conceptual par, not because they are equally correct, but because they should be evaluated by the same standards.[17]

VIII. COMMON FATE

A proper understanding of the units of selection problem must take account of an important symmetry: *just as organisms are parts of groups, so genes are parts of organisms*. The parts of a whole can interact co-operatively, enhancing the fitness of the whole at their own expense.

---

[16] Alternatively, one could interpret Sterelny and Kitcher as saying, not that $\bar{w}$ increases, but that genes of higher than average fitness tend to leave more copies of themselves. This reading turns Dawkins's position into a triviality; it is true even when the group is the unit of selection.

[17] Dawkins treats genes that co-operate with each other and genes that compete with each other as both exemplifying the gene as unit of selection. This contributes to the vacuity of his version of genic selectionism; no matter what a gene does, it is 'acting in its own interest'. No such confusion could arise in connection with the organism/group relationship. An organism that sacrifices its own welfare for the sake of the group differs from an organism that sacrifices group welfare for its own selfish advantage.

Alternatively, the parts can interact competitively, enhancing their own fitness at the expense of the whole in which they reside. In the former case, the parts behave altruistically; in the latter, they behave selfishly.

Dawkins (1976, 1982) has rightly emphasized that evolution does not inevitably produce the highly integrated and well-adapted individual organisms we now observe. If this is to happen, competition among the parts of organisms must be modest. The Mendelian system has largely succeeded in creating this circumstance by making meiosis 'fair'. Each of the genes in an organism standardly has the same chance of making it to the next generation. Within any organism, the genes are identical in fitness.[18] Exceptions to this pattern occur, of course, as in meiotic drive. But these exceptions aside, the genes in an organism have a *common fate* (Sober 1981, Walton 1991, and Wilson and Sober 1994a); this helps explain why organisms were able to evolve into functionally integrated wholes.

When the genes inside an organism sink or swim together, competition can occur *between* organisms, but not *within* them. In such cases, the unit of selection is the organism, not the gene. Meiotic drive, on the other hand, is a genuine case of the gene as unit of selection (as is the dynamics of junk DNA). The gene is sometimes the unit of selection, but very often it is not.[19]

Empirically detectable cases of meiotic drive involve both genic and organismic selection, acting in opposite directions.[20] Within an organism, the driving allele D is fitter than the normal allele N against which it competes. However, organisms with two copies of D do worse than organisms with one or zero. Two causal processes are at work here. In one, genes in the same organism compete against each other; in the other, all the genes in the same organism are in the same boat.

These ideas may be frame-shifted upward one level to provide a perspective on how groups and organisms are related to each other as candidate units of selection. We may begin by asking why groups are often less integrated and adapted than the organisms that are their parts. The answer is that within-group competition is often substantial. Organisms in the same group often compete with each other and have *un*equal chances of surviving and reproducing.

---

[18] Of course, germ-line and somatic copies of the same gene have different probabilities of making it to the next generation. The point is that different germ-line genes in the same organism have the same probability. Buss (1987) discusses how this arrangement evolved.

[19] This proposal is orthogonal to the additivity criterion discussed in Section IV. If the genes inside an organism are bound together by common fate, the gene fails to be a unit of selection, *regardless of whether genotypic fitnesses are additive.*

[20] Without selection against the driving gene, it will go to fixation. If so, the population will contain no heterozygotes, and the scientist will be unable to see that the gene is, in fact, a driving gene.

When group and organismic selection occur together, two types of causal process occur—one within groups, the other between them. Let us consider the example of how altruism (A) and selfishness (S) evolve. Within any group, S individuals are fitter than A individuals. But groups that contain more S individuals do worse than groups that contain fewer. The evolution of the D and N alleles is isomorphic with the evolution of the S and A phenotypes.

Unifying these two examples is the concept of *common fate*. In each case, a process of competition obtains between parts, but a second process binds together the parts in a single whole by common fate. One causal process affects the parts in the same whole differentially, but another lumps together the parts in the same whole and treats them similarly.

We so far have addressed the complicated type of case in which selection occurs at the level of parts *and also* at the level of wholes. However, simpler scenarios in which there is just one unit of selection can be described merely by suppressing variation in fitness at all levels but one. Cases in which group and genic selection do not occur, and organismic selection is the only process at work, are easily described. Similarly, monolithic dynamics can be described for the other candidate levels. In all such cases, identifying the unit(s) of selection involves discovering how parts and wholes vary in fitness *and why they do so*.

Our criterion does not have the consequence that frequency-dependent selection is automatically an instance of group selection, a problem discussed in Section V. It is not enough that groups vary in fitness because of their different internal compositions; for the members of the group to have 'common fate', some property of the group must have the effect of putting them 'in the same boat'.

When wholes compete against wholes and parts also compete against parts, two units of selection exist. The traits that then evolve will often represent compromises between what is good for the whole and what is good for the parts. For example, when selection takes place purely within the confines of a single population, individual selection will lead to a sex ratio in which parents invest equally in the two sexes (Fisher 1958); when the sexes are equally costly, the sex ratio will be 1:1. Alternatively, when selection is purely between groups, they should evolve a heavily female-biased sex ratio in which males are produced only to the extent that they are needed to fertilize the females. The female-biased sex ratios so often observed in nature typically are compromises between these two 'pure' cases. The sex ratio is not a purely individual adaptation, nor did it evolve solely because it benefits the group. It evolved for two, conflicting reasons, and so is imperfect when judged by either monolithic criterion.

A similar compromise solution is evident in observed cases of meiotic drive. The driving gene does not go to fixation (as it would if the gene were the sole unit of selection), nor does the driving gene go extinct (which would happen if the organism were the sole unit of selection).

In Section II, X was said to be a unit of selection in the evolution of trait T precisely when one of the factors that influenced T's evolution was that T benefited Xs. We can now use the concept of 'common fate' to clarify this schematic idea. The crucial notion is *differential* benefit. When a trait is an organismic adaptation (and so the organism was a unit of selection in its evolution), the trait benefited organisms in the sense that organisms who had the trait did better than organisms *in the same group* who did not. Our initial formulation raised the question of whether genes are *always* units of selection, since any trait that evolves seems to 'benefit' the genes that code for it. We can now see that this impression is mistaken. If all the genes in an organism are 'in the same boat', one gene cannot do better than other genes *in the same organism*.

We are not suggesting that the idea of common fate is original with us. Dawkins and Hull recognize that if the units of selection problem concerns what types of adaptations have evolved, then it is an issue about vehicles/interactors, not replicators. We believe that the idea of common fate helps clarify what it takes for genes, organisms, and groups to be vehicles/interactors.

IX. REALISM, PLURALISM, AND CONVENTIONALISM

Some have questioned why it is necessary to choose a single view concerning which units of selection exist in nature. For example, Dawkins (1982) has argued that choice of the gene as the single unit of selection is a matter of convenience, not a matter of fact. Several philosophers (e.g. Cassidy 1978, Sterelny and Kitcher 1988, Kitcher *et al.* 1990, Waters 1991) have elaborated positions of this sort.

We believe that the units of selection problem is factual, not conventional, because different hypotheses about the units of selection typically make contrary predictions about which traits will evolve. For example, pure group selection will lead altruism to evolve, whereas pure individual selection will lead selfishness to evolve instead. Since the mix of characters found in a population is an observable matter, we have here an uncontroversial reason to regard the units of selection problem as non-conventional in character.[21]

As noted in Section VII, Sterelny and Kitcher (1988) and Kitcher *et al.* (1990) argued that the causal processes at work in natural selection can always be described in terms of what happens to genes. We grant this, but the sense of 'causal description' they discuss is not relevant to the empirical problem of deciding what types of adaptation are found in nature. As noted earlier, their argument is simply a version of the representation argument advanced by Williams (1966) and Dawkins (1976) on which we have already commented.

Our 'realist' position with respect to the units of selection problem does not force us to choose between *every* pair of causal descriptions. When zebras evolve a faster running speed, they also evolve a suite of genes that codes for this phenotype. When genotype causes phenotype, both types of trait are causes of survival and reproductive success, and both 'benefit' in the sense that both are made to increase in frequency. There is no need to choose (Mitchell 1987). However, in this instance, the organism, not the gene or the group, is the unit of selection. The genes in a zebra are bound together by a common fate; this is quite consistent with the fact that zebras with one suite of genes run faster than zebras with another.

Although we believe that the units of selection problem is substantive and non-conventional, we recognize that the genic point of view often has heuristic value, apart from whether the gene is in fact a unit of selection. Even when group selection occurs (and so the group is a unit of selection), thinking about evolution from the point of view of a single gene (e.g. ad gene for altruism) can be useful. We are inclined to be pluralists at the level of heuristic approaches and rather more monistic at the level of factual statements about nature.[22]

X. CONCLUDING REMARKS

In this essay, we have tried to canvas some of the themes that have occupied philosophical reflection on the units of selection problem during

---

[21] We take it to be *sufficient* for the dispute between two hypotheses to be non-conventional that they make contrary predictions about observables. We will not discuss here whether this is a necessary condition. We note, however, that competing hypotheses about the units of selection *sometimes* predict the same equilibrium configuration. This happens, for example, when group and individual selection both favour the evolution of a particular trait. However, even in this kind of case, it usually is possible to design a test to discriminate between the competing hypotheses.

[22] The relationship between what happens in nature and which approaches are heuristic exhibits an interesting asymmetry. When group selection is not operating, it is hard to see the

the last ten to fifteen years. However, space limitations prevent us from considering other important issues. For example, Williams (1966) and Dawkins (1976) appealed to a principle of parsimony to justify the single gene as unit of selection, and we have not attended to the role of parsimony considerations in the units of selection problem (but see Sober 1990*a*). And we have not discussed the pragmatic advantages of different frameworks for thinking about selection, or empirical problems of measurement and testing (Lloyd 1986, 1988). We also have had to pass over questions concerning the interpretation of Hamilton's (1964) notion of inclusive fitness and whether kin selection is a type of group selection (Sober 1993, Wilson and Sober 1989, 1994*a*), the status of species selection (Eldredge and Gould 1972, Stanley 1979, Sober 1984, Lloyd 1988, Williams 1992, Lloyd and Gould 1993), and the connection of units of selection issues with the concept of individuality (Hull 1980, Sober 1991, 1993). And we have had to shy away from discussing more general philosophical questions concerning causality, explanation, realism, and conventionalism. We regard all of these problems as interesting and important. It is not for nothing that this problem has excited so much philosophical discussion.[23]

REFERENCES

Brandon, R. (1984), 'The Levels of Selection', repr. in R. Brandon and R. Burian (eds.), *Genes, Organisms, and Populations: Controversies Over the Units of Selection* (Cambridge, Mass.: MIT Press), 133–41; reproduced as Ch. 9.
——(1990), *Adaptation and Environment* (Princeton: Princeton University Press).
——Antonovics, J., Burian, R., Carson, S., Cooper, G., Davies, P., Horvath, C., Mishler, B., Richardson, R., Smith, K., and Thrall, P. (1994), 'Discussion: Sober on Brandon on Screening-Off and the Levels of Selection', *Philosophy of Science*, 61: 475–86.
Buss, L. (1987), *The Evolution of Individuality* (Princeton: Princeton University Press).
Cassidy, J. (1978), 'Philosophical Aspects of the Group Selection Controversy', *Philosophy of Science*, 45: 575–94.
Dawkins, R. (1976), *The Selfish Gene* (New York: Oxford University Press).
——(1982), *The Extended Phenotype: The Gene as the Unit of Selection* (San Francisco: Freeman).

utility of representing processes at the group level. However, when group selection is operating, representing processes at the level of genes can be useful.

[23] We thank the National Science Foundation for financial support (NSF Grant SBE92-12294). We are also grateful to Robert Brandon, Peter Godfrey-Smith, David Hull, Richard Lewontin, and the anonymous referee of *Philosophy of Science* for useful suggestions.

Eldredge, N., and Gould, S. (1972), 'Punctuated Equilibria: An Alternative to Phyletic Gradualism', in T. Schopf (ed.), *Models in Paleobiology* (San Francisco: Freeman), 82–115.

Fisher, R. (1958), *The Genetical Theory of Natural Selection*, 2nd edn. (New York: Dover).

Godfrey-Smith, P. (1992), 'Additivity and the Units of Selection', in D. Hull, M. Forbes, and K. Okruhlik (eds.), *PSA 1992*, i (East Lansing, Mich.: Philosophy of Science Association), 315–28.

—— and Lewontin, R. (1993), 'The Dimensions of Selection', *Philosophy of Science*, 60: 373–95.

Gould, S. (1980), 'Caring Groups and Selfish Genes', in *The Panda's Thumb: More Reflections on Natural History* (New York: Norton), 85–92.

Greisemer, J., and Wade, M. (1988), 'Laboratory Models, Causal Explanations, and Group Selection', *Biology and Philosophy*, 3: 67–96.

Haig, D. (1993), 'Genetic Conflicts in Human Pregnancy', *Quarterly Review of Biology*, 68: 495–532.

Hamilton, W. D. (1964), 'The Genetic Theory of Social Behavior I and II', *Journal of Theoretical Biology*, 7: 1–52.

Hull, D. (1980), 'Individuality and Selection', *Annual Review of Ecology and Systematics*, 11: 311–32.

—— (1981), 'The Units of Evolution—A Metaphysical Essay', in U. Jensen and R. Harré (eds.), *The Philosophy of Evolution* (Brighton: Harvester Press), 23–44.

—— (1988), *Science as a Process: An Evolutionary Account of the Social and Conceptual Development of Science* (Chicago: University of Chicago Press).

Kitcher, P., Sterelny, K., and Waters, W. (1990), 'The Illusory Riches of Sober's Monism', *Journal of Philosophy*, 87: 158–60.

Lewontin, R. C. (1974), *The Genetic Basis of Evolutionary Change* (New York: Columbia University Press).

—— (1992), *Biology as Ideology: The Doctrine of DNA* (New York: Harper Collin Publishers).

Lloyd, E. (1986), 'Evaluation of Evidence in Group Selection Debates', in A. Fine and P. Machamer (eds.), *PSA 1986*, i (East Lansing, Mich.: Philosophy of Science Association), 483–93.

—— (1988), *The Structure and Confirmation of Evolutionary Theory* (New York: Greenwood Press).

—— and Gould, S. (1993), 'Species Selection on Variability', *Proceedings of the National Academy of Science*, 90: 595–9.

Maynard Smith, J. (1987), 'How to Model Evolution', in J. Dupré (ed.), *The Latest on the Best: Essays on Evolution and Optimality* (Cambridge, Mass.: MIT Press), 119–31.

Mayr, E. (1963), *Animal Species and Evolution* (Cambridge, Mass.: Harvard University Press).

—— (1990), 'Myxoma and Group Selection', *Biologisches Zentraleblatt*, 109: 453–7.

Mitchell, S. (1987), 'Competing Units of Selection?: A Case of Symbiosis', *Philosophy of Science*, 54: 351–67.

Oyama, S. (1985), *The Ontogeny of Information: Developmental Systems and Evolution* (New York: Oxford University Press).

Reichenbach, H. (1956), *The Direction of Time*, ed. M. Reichenbach (Berkeley and Los Angeles: University of California Press).

Richardson, R. (1985), 'Biological Reductionism and Genic Selectionism', in J. Fetzer (ed.), *Sociobiology and Epistemology* (Dordrecht: Reidel), 133–60.

Ridley, M. (1993), *Evolution* (Boston: Blackwells Scientific).

Seeley, T. (1989), 'The Honey Bee Colony as a Superorganism', *American Scientist*, 77: 546–53.

Sober, E. (1981), 'Holism, Individualism, and the Units of Selection', in R. Giere and P. Asquith (eds.), *PSA 1980*, ii (East Lansing, Mich.: Philosophy of Science Association), 93–121.

——(1984), *The Nature of Selection: Evolutionary Theory in Philosophical Focus* (Cambridge, Mass.: MIT Press).

——(1990a), 'Let's Razor Ockham's Razor', in D. Knowles (ed.), *Explanation and its Limits* (Cambridge: Cambridge University Press), 73–94.

——(1990b), 'The Poverty of Pluralism', *Journal of Philosophy*, 87: 151–7.

——(1991), 'Organisms, Individuals, and Units of Selection', in A. Tauber (ed.), *Organism and the Origins of Self* (Dordrecht: Kluwer), 273–96.

——(1992), 'Screening-Off and the Units of Selection', *Philosophy of Science*, 59: 142–52.

——(1993), *Philosophy of Biology* (Boulder, Colo.: Westview Press).

——and Lewontin, R. (1982), 'Artifact, Cause, and Genic Selection', *Philosophy of Science*, 47: 157–80.

Stanley, S. (1979), *Macroevolution: Pattern and Process* (San Francisco: Freeman).

Sterelny, K., and Kitcher P. (1988), 'The Return of the Gene', *Journal of Philosophy*, 85: 339–61; reproduced as Ch. 8.

Trivers, R. (1972), 'The Evolution of Reciprocal Altruism', *Quarterly Review of Biology*, 46: 35–57.

Walton, K. (1991), 'The Units of Selection and the Bases of Selection', *Philosophy of Science*, 58: 417–35.

Waters, K. (1991), 'Tempered Realism about the Forces of Selection', *Philosophy of Science* 58: 553–73.

Williams, G. C. (1966), *Adaptation and Natural Selection: A Critique of Some Current Evolutionary Thought* (Princeton: Princeton University Press).

——(1992), *Natural Selection: Domains, Levels, and Challenges* (New York: Oxford University Press).

Wilson, D. S. (1988), 'Holism and Reductionism in Evolutionary Biology', *Oikos*, 53: 269–73.

——(1990), 'Weak Altruism, Strong Group Selection', *Oikos*, 59: 135–40.

——and Sober, E. (1989), 'Reviving the Superorganism', *Journal of Theoretical Biology*, 136: 337–56.

——(1994a), 'Reintroducing Group Selection to the Human Behavioral Sciences', *Behavior and Brain Sciences*, 17: 585–608.

——(1994b), 'Reply to Comments on "Reintroducing Group Selection to the Human Behavioral Sciences"', *Behavior and Brain Sciences*, 17: 639–47.

Wimsatt, W. (1980), 'Reductionistic Research Strategies and their Biases in the Units of Selection Controversy', in T. Nickles (ed.), *Scientific Discovery: Case Studies*, ii (Dordrecht: Reidel), 213–59.

——(1981), 'The Units of Selection and the Structure of the Multi-Level Genome', in R. Giere and P. Asquith (eds.), *PSA 1980*, ii (East Lansing, Mich.: Philosophy of Science Association), 122–83.

PART IV

FUNCTION

INTRODUCTION TO PART IV

DAVID L. HULL

For generations philosophers have engaged in an activity termed 'linguistic analysis': that is, determining in the context of a particular language what a particular term means. A history of attempts by philosophers to clarify the notion of function exhibits all the strengths and weaknesses of this sort of undertaking. The way that the game is played is that an author discusses some traditional examples of function: for example, the function of the heart in vertebrates to pump blood. The author then suggests an analysis of 'function' in terms of some property or mechanism: for example, the frequency with which a system achieves and/or returns to a preferred state in the face of wide but not unlimited changes in the system and/or its environment, the mechanisms that produce this behaviour, such as causal feedback loops, programmes, information, and so on.

Next, counter-examples are presented, some designed to show that the analysis is too broad (i.e. examples that count as functions under this definition, but intuitively are clearly not functions), some that the analysis is too narrow (i.e. examples that exclude particular examples that clearly seem to be functions). Does the analysis fit a child's toy called a 'walking beetle' or 'Old Faithful'? According to our pre-analytic intuitions, the child's toy seems to behave as if it were functionally organized, while 'Old Faithful' does not. If the toy fits our analysis and Old Faithful does not, we are pleased. If not, we have to argue that, contrary to appearances, the toy is not actually a functionally organized system and 'Old Faithful' actually is.

The investigation of these counter-examples necessarily results in attempts to get clearer on the properties used to define 'function'—for example, negative feedback loops. And the game is replayed, this time with negative feedback loops as the concept to be analysed. Typically the first definition suggested is quite short and simple, but as the analysis continues, restrictive clauses are added to restrictive clauses in the effort to neutralize all the various counter-examples that have been raised previously. 'A is B whenever C does not obtain, except in conditions D and E, whenever. . . .'

Such conceptual analyses can help us to see more clearly the range of phenomena that may have been obscure prior to our analysis; but rarely, if ever, do they eventuate in a final, well-articulated, satisfactory analysis. One reason for the indeterminacy of conceptual analysis is that the process gradually does extreme damage to one's intuitions. For instance, I remember when I had intuitions about what counts as a functional system and what does not; but after studying counter-example after counter-example, my intuitions have been so battered that they are no longer of any use whatsoever. Is a compound pendulum at rest an example of a functional system? I have no idea. Others come through this process with their intuitions still intact; how, I don't know.

Another reason why conceptual analyses tend to be so indeterminate is that the subject-matter is frequently amorphous—for example, ordinary understanding. The likelihood that all present-day people, let alone all people throughout time, meant a single thing by a term like 'function' is next to nil. One way to resolve this problem is to narrow one's field of interest: for example, to biology today. But even in present-day biology, 'function' does not have a univocal use. For instance, this term operates differently in evolutionary biology and functional morphology. The move to situate a term in a particular well-structured context is certainly a move in the right direction. At the very least, well-structured contexts give one something to go on in addition to one's intuitions.

Three sorts of phenomena have traditionally been explained in terms of some form of 'teleology': (i) conscious intentional behaviour (she went to town in order to buy a magazine), (ii) the structure of artefacts (the function of the spring is to close the door), and (iii) the organization of living creatures (the function of the heart's beating in vertebrates is to circulate the blood). The task that philosophers have set for themselves is to analyse the preceding sorts of claim to make clear what they do and do not imply. Must each sort be explained in a different way, or can two or more of them be assimilated to the same analysis?

Looking back over the past three decades, two papers on function stand out as being 'seminal' in a literal sense of this term: papers by Larry Wright (1973) and Robert Cummins (1975). Wright presented an aetiological (or causal) analysis of 'function', while Cummins presented a more traditional analysis in terms of certain characteristics of certain complex systems. Both Wright and Cummins originally intended their analyses to apply to all three sorts of phenomena listed above, but subsequent authors have modified these analyses in divergent ways. Some authors have narrowed Wright's analysis to apply only to selection and adaptation in evolutionary biology (Neander 1991; Millikan (1989) came up with this same narrow analysis independently of Wright), while other authors have retained Cummins's original goal, but modified his definition by appending all sorts of provisos to accommodate counter-examples. Wright's analysis and its descendants are short and sweet. Cummins's analysis started off rather complex, and has become increasingly more complex.

Descendants of Wright's analysis (selected effect accounts) are significant in two respects: (a) they are historical, and (b) they are limited to a particular scientific context (evolutionary biology). In attributing a function to a structure in a present-day organism, we are committed to this structure in the past having contributed to the inclusive fitness of the ancestors of this organism and having caused the relevant genes to be selected. Quite obviously, narrowing Wright's account results in the exclusion of many traditional examples of functional claims. In addition, reference to the past, except for the very recent past, means that in most cases, we will never have the relevant data necessary to justify such claims, but at least we know what data would be relevant in deciding which structures count as selected effects and which do not.

One reason why definitions of 'function' in terms of selected effects seem attractive is that the term has been limited to a single, highly articulated theoretical context. As a result, we have more to go on than our intuitions. Millikan (1989) rejects conceptual analysis of the sort found in most philosophical papers for theoretical definitions. The context for her analysis is a particular scientific theory—evolutionary theory. Amundson and Lauder (1994; ch. 11 below) are willing to limit functional claims to biological contexts, but they argue that evolutionary biology is not the *only* legitimate context for functional claims, even in biology. Functional claims are also made in comparative and functional anatomy, and these functional claims are independent of evolutionary biology. Instead, a Cummins-style analysis fits them. Kitcher (1993; Ch. 12 below) goes even further, arguing that even if we limit ourselves to evolutionary biology, we still cannot produce a *single* analysis because of the timing of selection processes. It matters how distant the relevant past turns out to be.

Hoary or not, at this juncture, pluralism raises its head. All the authors in this section claim to be pluralists. Godfrey-Smith (1993; Ch. 13 below), as well as Amundson and Lauder (1994), claim to be pluralists, because they recognize two distinct and legitimate senses of 'function'—Wright functions and Cummins functions. Even though Kitcher (1993) presents an analysis of function in terms of a single criterion (design), he too claims to be a pluralist, because his monistic requirement can be met in two ways—the Wright way and the Cummins way. Obviously, the philosophically correct thing to be is a pluralist, regardless of the position one holds.

REFERENCES

Cummins, R. (1975), 'Functional Analysis', *Journal of Philosophy*, 72: 741–65.
Millikan, R. G. (1989), 'In Defence of Proper Functions', *Philosophy of Science*, 56: 288–302.
Neander, K. (1991), 'Functions as Selected Effects: The Conceptual Analytic Defense', *Philosophy of Science*, 58: 168–84.
Wright, L. (1973), 'Functions', *Philosophical Review*, 82: 139–68.

11

FUNCTION WITHOUT PURPOSE: THE USES OF CAUSAL ROLE FUNCTION IN EVOLUTIONARY BIOLOGY

RON AMUNDSON AND GEORGE V. LAUDER

I. INTRODUCTION

Philosophical analyses of the concept of biological function come in three kinds. One kind defines the function of a given trait of an organism in terms of the history of natural selection which ancestors of the organism have undergone. In this account the function of a trait can be seen as its evolutionary purpose, with purpose being imbued by selective history. A second approach is non-historical, and identifies the function of a trait as certain of its current causal properties. The relevant properties are seen either as those which contribute to the organism's current needs, purposes, and goals (Boorse 1976) or those which have evolutionary significance to the organism's survival and reproduction (Ruse 1971, Bigelow and Pargetter 1987). A third approach has been articulated and defended by Robert Cummins (1975, 1983), mostly in application to psychological theory. Cummins's view is unique in that neither evolutionary nor contemporary purposes or goals play a role in the analysis of function. It has received little support in the philosophy of biology, even from Cummins himself. Nevertheless, we will show that the concept is central to certain ongoing research programmes in biology, and that it is not threatened by the philosophical criticisms usually raised against it. Philosophers' special interests in purposive concepts can lead to the neglect of many crucial but non-purposive concepts in the science of biology.

Karen Neander recently and correctly reported that the selective view of function is 'fast becoming the consensus' (Neander 1991: 168). Larry Wright showed in his canonical (1973) paper on selective function that an intuitively pleasing feature of the view is that citing a trait's function would

First published in *Biology and Philosophy*, 9 (1993–4): 443–69. © Kluwer Academic Publishers. Reprinted with kind permission of Kluwer Academic Publishers.

play a role in explaining how the trait came to exist. Concepts of function similar to Wright's were hinted at by the biologists Francisco Ayala (1970) and G. C. Williams (1966), and later endorsed by the philosophers Robert Brandon (1981, 1990), Elliott Sober (1984), Ruth Millikan (1989), and Karen Neander (1991), among others. (For a good review of the history of philosophical discussions of function see Schaffner 1993: ch. 8.)

The evolutionary, selective account of function is commonly termed the 'aetiological concept', since functions are individuated by a trait's causal history. In the present context the term 'aetiological' may lead to confusion, so we will refer rather to the *selected effect* (SE) account of function. The Cummins style of account will be designated the *causal role* (CR) account (following Neander 1991: 181).

Given the consensus in favour of SE function among philosophers of biology, it is surprisingly difficult to find an unequivocal rejection of Cummins's alternative. This may stem from a recognition that some areas of science (medicine, physiology, and perhaps psychology) require other kinds of function concepts. It does seem generally accepted, however, that SE function is the concept uniquely appropriate *to evolutionary biology*. It is this position which we will attempt to refute.

Ruth Millikan (1989) and Karen Neander (1991) have recently presented arguments in favour of selected effect concepts of biological function. We will pay special attention to these papers for two reasons. First, they express positions on the nature of philosophical analysis which we find valuable, and which we will use in defending CR function. Second, they examine Cummins's account of function in detail. Some of the ideas they develop are shared by other advocates of SE function, and many are novel; all are worthy of analysis. (Unless otherwise stated, all references to Millikan and Neander will be to those papers.)

Millikan examined the source of the criticisms which philosophers had made against the SE theory, and found them to be based in the philosophical practice of conceptual analysis. She declared this practice 'a confused program, a philosophical chimera, a squaring of the circle . . .' among other crackling critiques (p. 290). The search for necessary and sufficient conditions for the common-sense application of terms was not what Millikan was about. Neander similarly rejected conceptual analysis of ordinary language as the goal of the philosophical analysis of function. Indeed, in rereading the debates on function of the 1970s, one is struck by the concern shown by philosophers for consistency with ordinary language. Millikan and Neander replace the old style of ordinary language analysis with somewhat different alternatives. Millikan was interested in a *theoretical definition* of the concept of function, a concept which she labels 'proper function'. Neander instead focused on a conceptual analysis, but not the traditional kind based on ordinary language. Rather, Neander intended to analyse *specialists'* language—in this case the usage of the term 'function' in the language of evolutionary biology. 'What matters is only that biologists implicitly understand 'proper function' to refer to the effects for which traits were selected by natural selection' (p. 176).[1] While each writer intended the analysis of function to be relativized to a theory (rather than to ordinary language), Neander intended the relevant theory to be evolutionary biology, while Millikan located her analysis in the context of her own research project involving the relations among language, thought, and biology (Millikan 1984, 1993).

While the intended status of their resulting analyses differed, Millikan's and Neander's approaches had similar benefits for the SE analysis of function (and also, as we will presently argue, for the CR analysis of function). Both approaches tied function analyses to actual theories, in this way eliminating many ordinary-language-based counter-examples to SE function. Theoretical definitions, such as 'Gold is the element with atomic number 79', need not match ordinary usage, but instead reflect current scientific knowledge about the true nature of the subject-matter. The use of bizarre counterfactuals such as Twin Earth cases and miraculous instantaneous creations of living beings (e.g. lions) were a mainstay of earlier criticisms of SE function. These kinds of cases are irrelevant to evolutionary theory and to the vocabulary of real-world evolutionary scientists. Appeals to pre-Darwinian uses of the term 'function' (e.g. William Harvey said that the function of the heart was to pump blood) are equally irrelevant. After all, Harvey didn't know the atomic number of gold any more than he knew the historical origin of organic design. None the less, gold is (and was) the element with atomic number 79, and (by the SE definition) the heart's blood-pumping function is constituted by its natural selective history for that effect.

We fully approve of these moves. Taking the contents of science more seriously than is philosophically customary is exactly what philosophers of science ought to be doing. We will not question the philosophical adequacy of Millikan's or Neander's approaches, nor their defences of SE theory against its philosophical critics. We will, however, call into question the

---

[1] We take it that Neander intends her analysis to reflect biologists' use of the term 'function', not necessarily their use of the concept defined by Millikan as 'proper function'. Both Millikan (p. 290 n. 1) and Neander (p. 168 n. 1) refer to Neander's widely circulated but unpublished 'Teleology in Biology'. In that paper Neander referred only to the biological concept of 'function' (i.e. not to 'proper function') except when she needed to distinguish between 'a part's proper function and things which it just happens to do fortuitously' (MS, p. 11).

common SE functionalist's belief that evolutionary biology is univocally committed to SE function. We will show that the rejection of ordinary-language conceptual analysis immunizes Cummins-style CR function against some very appealing philosophical critiques—critiques expressed by Millikan and Neander themselves. We will show that a well-articulated causal role concept of function is in current use in biology. It is as immune from Millikan's and Neander's critiques of CR function as their own SE accounts are from ordinary-language opposition.

The field of biology called 'functional anatomy' or 'functional morphology' explicitly rejects the exclusive use of the SE concept of function. To be sure, there are other biological fields in which the SE concept is the common one—ethology is an example. The most moderate conclusion of this semantic observation is only a plea for conceptual pluralism, for the usefulness of different concepts in different areas of research. But further conclusions will be stronger than mere pluralism. We will defend CR function from philosophical refutation. We will show that a detailed knowledge of the selective history (and so the SE function) of specific anatomical traits is much more difficult to achieve than one would expect from the intuitive ease of its application. Finally, we will demonstrate the ineliminability of CR function from certain key research programmes in evolutionary biology.

II. ADAPTATION AND SELECTED EFFECT FUNCTIONS

First, a specification of the selected-effect concept of function:

The function of X is F means
(a) X is there because it does F,
(b) F is a consequence (or result) of X's being there. (Wright 1973: 161; variables renamed for consistency)

Wright intended his analysis to apply equally to intentional and natural selection. When the context is restricted to evolution, and natural selection accounts for the 'because' in (a), something like Neander's definition results:

It is the/a proper function of an item (X) of an organism (O) to do that which items of X's type did to contribute to the inclusive fitness of O's ancestors, and which caused the genotype, of which X is the phenotypic expression, to be selected by natural selection. (Neander 1991: 174; cf. Millikan 1989: 228)

Not surprisingly, there are very closely related concepts within evolutionary biology, particularly the concept of *adaptation*. During the twentieth century there has been some semantic slippage surrounding the term. Describing an organic trait as an adaptation has meant either (1) that it benefits the organism in its present environment (whatever the trait's causal origin), or (2) that it arose via natural selection to perform the action which now benefits the organism. That is, the term 'adaptation' has sometimes, but sometimes not, been given an SE, historical meaning. G. C. Williams gave a trenchant examination to the concept of adaptation, referring to it as 'a special and onerous concept that should be used only where it is really necessary' (1966: 4). In particular, Williams thought it important to distinguish between an adaptation and a fortuitous benefit. These ideas inspired the 'historical concept' of adaptation, according to which the term was restricted to traits which carried selective benefits and which resulted from natural selection for those benefits. Terms such as 'adaptedness' or 'aptness' came to be used to designate current utility, covering both selected adaptations and fortuitous benefits (Gould and Vrba 1982). The onerous term adaptation was reserved for traits which had evolved by natural selection. Robert Brandon recently declared the historical definition of adaptation 'the received view' (1990: 186). Elliott Sober described the concept as follows:

X is an adaptation for task F in population P if and only if X became prevalent in P because there was selection for X, where the selective advantage of X was due to the fact that X helped perform task F. (Sober 1984: 208; variables renamed for consistency)

Sober's task F is precisely what SE theorists would call the function of trait X. Moreover, Williams, Sober, and Brandon, like Millikan and Neander, all refer to a benefit produced by X as the function of X just when that benefit was the cause of selection for X. In other words, for a trait to be an adaptation (historically defined) is *precisely* for that trait to have a function (selected-effect-defined). A trait *is* an adaptation when and only when it *has* a function. The two terms are interchangeable. If a law were passed against the SE concept of function, its use in biology could be fully served by the historical concept of adaptation.

III. FUNCTIONAL ANATOMY AND CAUSAL ROLE FUNCTIONS

The major philosophical competitors to the SE concept of function refer to contemporary causal powers of a trait rather than to the causal origins of that trait. Most of these non-selective analyses also advert to the (contemporary) purposes or goals of a system. The goals are presumed knowable prior to addressing the question of function, so that identifying a trait's function amounts to identifying the causal role played by the trait in the organism's ability to achieve a contemporary goal. Robert Cummins (1975) introduced a novel concept of function in which the specification of a real, objective goal simply dropped out. Since neither current benefits and goals nor evolutionary purposes were relevant, evolutionary history was also irrelevant to the specification of function. Cummins focused on functional analysis, which he took to be a distinctive scientific explanatory strategy. In functional analysis, a scientist intends to explain a *capacity* of a system by appealing to the capacities of the system's component parts. A novel feature of Cummins's analysis is that capacities are not presented as (necessarily) goals or purposes of the system. Scientists choose capacities which they feel are worthy of functional analysis, and then try to devise accounts of how those capacities arise from interactions among (capacities of) the component parts. The functions assigned to each trait (component) are thus relativized both to the overall capacity chosen for analysis and the functional explanation offered by the scientist. Given some functional system s:

X functions as an F in s (or: the function of X in s is to F) relative to an analytical account A of s's capacity to G just in case X is capable of F-ing in s and A appropriately and adequately accounts for s's capacity to G by, in part, appealing to the capacity of X to F in s. (Cummins 1975: 762; variables renamed for consistency)

Cummins's assessments of function do not depend on prior discoveries of the purposes or goals served by the analysed capacities, as do other non-SE theories of function.[2] This creates a problem for Cummins. Prior, extrinsic information about system goals would narrow the list of possible functions to those which *can* contribute to the already-known goal. With no extrinsic criteria to delimit the list of relevant causal properties, Cummins needs some other method of constraining the list of causal powers which are to be identified as functions. Indeed, the problem of constraint gives rise to the most frequent challenge to Cummins's approach; examples will be discussed below. Critics find it easy to devise whimsical 'functional analyses' which trade on the lack of external constraint, and which appear to show

---

[2] A minority of commentators interpret Cummins as surreptitiously introducing goals and purposes by choosing for analysis only traits which are already known to be purposive (Rosenberg 1985: 68, Schaffner 1993: 399 ff.). We interpret Cummins as fully agnostic with regard to purpose, which is why the criticisms being considered are worthy of discussion. Rosenberg appears to be the only philosopher who supports Cummins's account of function for evolutionary biology; he does so partly because of this purposive reading. Whatever Cummins's original intentions, we intend CR function to be both non-historical and non-purposive in its applications.

Cummins's definition of function to be too weak to distinguish between functions and mere effects. To make up for the loss of the external constraint of goal specificity, Cummins offers internal criteria for assessing the scientific significance of a proffered functional analysis. A valuable (as opposed to a trivial) functional analysis is one which adds a great deal to our understanding of the analysed trait. In particular, the scientific significance or value of a given functional analysis is judged to be high when the analysing capacities cited are *simpler* and *different in type* from the analysed capacities. An analysis is also of high value when it reveals a high degree of *complexity of organization* in the system. Functional analyses of very simple systems are judged to be trivial on these criteria. 'As the role of organization becomes less and less significant, the [functional] analytical strategy becomes less and less appropriate, and talk of functions makes less and less sense. This may be philosophically disappointing, but there is no help for it' (ibid. 764). Philosophical disappointment in this messy outcome could be alleviated by requiring an independent specification of goals and purposes prior to any functional analysis. But, as we shall see, such philosophical serenity would carry a high cost for scientific practice.

Cummins's account is of special interest because of its close match to the concepts of function used within functional anatomy. His emphasis on causal capacities of components and the absence of essential reference to overall systemic goals is shared by the anatomists. This is somewhat surprising, since Cummins's chief interest was in functional analysis in psychology. He did assert (without documentation) that biology fit the model, but has written nothing else on biological function (ibid. 760). Other philosophers have recognized non-SE uses of function in biology. Boorse cited physiology and medicine as supporting his goal-oriented causal role analysis (Boorse 1976: 85). Brandon acknowledged the non-historical use in physiology, but disapproved: 'I believe that ahistorical functional ascriptions only invite confusion, and that biologists ought to restrict the concept [to] its evolutionary meaning, but I will not offer further arguments for that here' (Brandon 1990: 187 n. 24). The wisdom of this counsel will be assessed below.

The classic account of the vocabulary of functional anatomy was given by Walter Bock and Gerd von Wahlert (1965). These authors referred to 'the form–function complex' as an alternative to the customary contrast between the two—form *versus* function. This was not merely an attempt at conciliation between advocates of the primacy of form over function and advocates of the converse. Rather, it was a reconceptualization of the task of anatomists, especially evolutionary anatomists. Bock and von Wahlert stated that the form and the function of anatomical traits were *both* at the methodological base, the lowest level, of the functional anatomist's enterprise. The rejection of the contrast between form and function (its replacement with the form–function complex) amounted to a rejection of the SE concept of function itself. In the functional anatomist's vocabulary, form and function were both observable, experimentally measurable attributes of anatomical items (e.g. bones, muscles, ligaments). Neither form nor function was inferred via hypotheses of evolutionary history. The form of an item was its physical shape and constitution. The function of the same item was 'all physical and chemical properties arising from its form . . . providing that [predicates describing the function] do not mention any reference to the environment of the organism' (ibid. 274). This denial of reference to environment eliminates not only the SE concept of evolutionary function, but also the non-historical notion of function as a contribution to contemporary adaptedness or other goal-achieving properties. These implications were intended. Concepts involving biological importance, selective value, and (especially) selective *history* (and therefore Darwinian adaptation) are all at higher and more inferential levels of analysis than that of anatomical function. The intention was not to ignore these higher levels, but to provide an adequate functional-anatomic evidentiary base from which the higher levels can be addressed.

The level of organization above the form–function complex is the character complex. A character complex is a group of features (typically anatomical items themselves seen as form–function complexes) which interact functionally to carry out a common *biological role*. When we reach the biological role, we find ourselves in more familiar Darwinian territory. The biological role of a character complex (or of a single trait) is designated by 'that class of predicates which includes all actions or uses of the faculties (the form–function complex) of the feature by the organism in the course of its life history, provided that these predicates include reference to the environment of the organism' (ibid. 278). At last we find reference to that organism/environment relation which constitutes adaptedness or fitness. The further inference to the SE advocate's concept of evolutionary function involves an additional assertion that the trait's present existence is not fortuitous, but the result of a history of natural selection controlled by the same benefits which the trait now confers in its biological role.

So a chain of inference from anatomical function to evolutionary function involves several steps and additional (i.e. non-anatomical) kinds of data. An evolutionist may not feel the need to start from the anatomical base, of course. Given a simple trait with a known biological role, the evolutionist might feel justified in ignoring anatomical details. But in highly integrated character complexes with long evolutionary histories (e.g. the vertebrate jaw or limb) it is arguably perilous to ignore anatomical function (Wake and Roth 1989).

In one way, Bock and von Wahlert's concept of function is even more radical than Cummins's. Cummins assigns functions only to those capacities of components which are actually invoked in a functional explanation, those which are believed to contribute to the higher-level capacity being analysed. Bock and von Wahlert include *all possible* capacities (causal powers) of the feature, given its current form. Some of these capacities are utilized, and some are not. Both utilized and unutilized capacities are properly called functions. The determination of unutilized functions may require experiments which are ecologically unrealistic, but this is still a part of the functional anatomist's job. Bock and von Wahlert suggest that a functional anatomist might want to experimentally study the functional properties of a muscle at 40 per cent of its rest length, even when it is known that the muscle never contracts more than 10 per cent during the life history of the organism (1965: 274). The relevance of unutilized functions depends on the sort of question being asked. Other anatomists attend primarily to utilized functions. 'The study of function is the study of how structures are used, and functional data are those in which the use of structural features has been directly measured. Functions are the actions of phenotypic components' (Lauder 1990: 318). Bock's special interest in unutilized functions comes from his interest in the phenomenon of pre-adaptation (or exaptation) (Bock 1958, Gould and Vrba 1982). It is often the unutilized functional properties of traits which allow them to be 'co-opted' and put to new uses when the evolutionary opportunity arises.

Apart from the issue of unutilized functions, Cummins's concept of function matches the anatomists'. Functional anatomists typically choose to analyse integrated character complexes which have significant biological roles. An anatomist might choose to analyse the crushing capacity of the jaw of a particular species. Cummins's s is the jaw, and G the capacity to crush things. In the analysis the anatomist might cite the capacity of a particular muscle (component X) to contract, thereby bringing two bones (other components of s) closer together. If the citation of that capacity of X fits together with other citations of component capacities into an 'appropriate and adequate' account of the capacity of the jaw to crush things, then it is proper on Cummins's analysis to say that the function (or a function) of that muscle is to bring those two bones closer together.

We can also apply Cummins's evaluative suggestions to such an analysis. In a valuable functional analysis, the analysing capacities will be simpler and/or different in type from the analysed, and the system's discovered

interesting causal role functions have a history of natural selection. Instant lions would have no such history, but they do not exist in our world. Earthquakes and rainfalls are in our world, but have no such history, and so no complex functional organization. Such imaginative counter-examples might be telling against conceptual analyses of ordinary-language function concepts. But they count neither for nor against CR or SE function theories, so long as those theories are *each* seen as science-based rather than conceptual analyses of ordinary language.

A second and more complex criticism involves the so-called normative role of function ascriptions and the problem of pathological malformations of functional items. Neander considers it the responsibility of a theory of biological function to categorize organic parts such that the categories are able to 'embrace both interspecies and pathological diversity' (p. 181). Millikan endorses at least the latter, and other SE theorists have been concerned with variation and dysfunction as far back as Wright (1973: 146, 151). According to these theorists, only SE function can categorize parts into their proper categories irrespective of variation and malformation. It does so by defining 'function categories'. CR function (like other non-historical theories) cannot define appropriate function categories, and so is unable both to identify diseased or malformed hearts as hearts and to identify the same organ under different forms in different species.

On pathology, Millikan points out that diseased, malformed, and otherwise dysfunctional organs are denominated by the function they would serve if normal. 'The problem is, how did the atypical members of the category that cannot perform its defining function *get* into the same function category as the things that actually can perform the function?' (Millikan 1989: 295; cf. Neander 1991: 180–1). A CR analysis of a deformed heart which cannot pump blood obviously cannot designate its *function* as pumping blood, since it doesn't have that causal capacity. On the other hand, even the organism with the malformed heart has a selective history of ancestors which survived because *their* hearts pumped blood. So the category 'heart' which ranges over both healthy and malformed organs must be defined by SE, not CR, function. On interspecies diversity of form:

The notion of a 'proper function' is the notion of what a part is *supposed* to do. This fact is crucial to one of the most important theoretical roles of the notion in biology, which is that most biological categories are only definable in functional terms. For instance, 'heart' cannot be defined except by reference to the function of hearts because no description purely in terms of morphological criteria could demarcate hearts from non-hearts. (Neander 1991: 180)

The claim that biological categories must be defined by SE functional analyses is a significant challenge to CR functional analysis. If SE function is truly the basis of biological classification, then CR functional analyses must either (1) deal with undefined biological categories, or (2) depend on prior SE functional analyses for a classification of biological traits. We will now argue that SE functionalists are simply mistaken in this claim. SE functions are not the foundation for the classification of basic biological traits. To be sure, CR function does not define basic categories either. The classifications come from a third, non-functional source.

Consider Neander's claim that 'most biological categories are only definable in functional terms'. Hardly a controversial statement, especially in the philosophical literature. Nevertheless, it is utterly false. Perhaps most *philosophically interesting* biological categories are functional (depending on the interests of philosophers). But a glance in any comparative anatomy textbook rapidly convinces the reader (and appals the student) with the ocean of individually classified bones, ligaments, tendons, nerves, etc., etc. We do not mean simply to quibble over a census count of functional versus anatomical terms in biology. Rather, we wish to argue for the importance, often unrecognized by philosophers, of anatomical, morphological, and other non-purposive, but theoretically crucial, concepts in biology. In this case the relevant conceptual apparatus belongs to the field of comparative anatomy.

Many body parts can be referred to either by anatomical or functional characterizations. The human kneecap is a bone referred to as the patella. 'Kneecap' is a (roughly) functional characterization; a kneecap covers what would otherwise be an exposed joint surface between the femur and the tibia. 'Patella' is an anatomical, not a functional, characterization. The patella in other vertebrates need not 'cap' the 'knee' (for example, in species in which it is greatly reduced), and some species might conceivably have their knees capped by bones not homologous to the patella. The category *patella* is not a function category, but an anatomical category. *Kneecap* is a function category. To call a feature a wing is to characterize it (primarily) functionally. To call it a vertebrate forelimb is to characterize it anatomically. The wings of butterflies and birds have common functions but no common anatomy.

The concept of *homology* is central to the practice of evolutionary biology. It is arguably as important as the concept of *adaptation*. Anatomical features which are known (at their naming) to be homologically corresponding features in related species are given common names. A traditional Darwinian definition of homology refers to the common derivation of body parts: 'A feature in two or more taxa is homologous when it is derived from the same (or a corresponding) feature of their common ancestor' (Mayr 1982: 45). This definition has recently come under scrutiny, and a more openly phylogenetic definition (most clearly explicated by Patterson 1982) is often preferred. (See Hall 1984 for discussions of homology.) On this concept, homologous traits are those which characterize natural (monophyletic) clades of species. Thus, the wing of a sparrow is homologous to the wing of an owl, because the character 'wing' (recognized by a particular structural configuration of bones, muscles, and feathers) characterizes a natural evolutionary clade (birds) to which sparrows and owls belong. Wings of sparrows are not homologous to wings of insects, because there is no evidence that a clade consisting of birds + insects constitutes a natural evolutionary unit. This remains true even if 'wing' is characterized functionally, as 'flattened body appendage used in flight'. Whatever the favoured definition of homology, one feature of the concept is crucial: *the relation of homology does not derive from the common function of homologous organs.* Organs which are similar in form not by virtue of phylogeny but because of common biological role (or SE function) are said to be *analogous* rather than homologous. The wings of insects and birds are analogous—they have similar SE functions, and so evolved to have similar gross structure. The forelimbs of humans, dogs, bats, moles, and whales, and each of their component parts—humerus, carpals, phalanges—are homologous. Morphologically, they are the same feature under different forms. Functionally, they are quite distinct.[3]

Comparative anatomy, morphology, and the concept of homology predate evolutionary biology. They provided Darwin with some of the most potent evidence for the fact of descent with modification. (This alone demonstrates the importance of other-than-adaptational factors in evolutionary biology.) So the evolutionary definition of homology mentioned above is a theoretical definition. As with other theoretical definitions, it is subject to sniping from practitioners of conceptual analysis. A philosopher

---

[3] Note that even extremely similar traits may arise by convergent evolution, and that the final test of homology is not similarity but, rather, congruent phylogenetic distribution of the putative homology with other characters providing evidence of monophyly. Thus, the eye of a squid and the eye of vertebrates are very similar in many (but not all) features. The non-homology of squid and vertebrate eyes does not rest on the differences noted between the eyes (virtually all homologous characters have some differences), but rather on the fact that very few other traits support the hypothesis that squids + vertebrates constitute a natural evolutionary lineage. The phylogenetic relationships among species thus provide the basis on which we make decisions about the homology of individual characters. For similar reasons, our statements to the effect that (homologous) traits *characterize* taxa should not be taken to mean that those traits are logically necessary or sufficient conditions for a species' membership in a taxon. Snakes are tetrapods, notwithstanding their leglessness. The phylogenetic distribution of traits other than legs makes it clear that snakes are members of the same monophyletic group as more typically legged tetrapods. See Sober (1993: 178) for a caution against appearances of essentialism in discussions of phylogenetic classification.

could argue (pointlessly) that 'homology' cannot *mean* 'traits which characterize monophyletic clades', since many 1840s biologists knew that birds' wings were homologous to human arms but did not believe in evolution (and so disbelieved that humans and birds shared a clade). SE advocates' usual reply to the William Harvey objection is applicable here. Just as Harvey could see the marks of biological purpose without knowing the origin or true nature of biological purpose, so pre-Darwinian anatomists could see the marks of homology without knowing the cause and true nature of homology itself.

But if anatomical items are not anatomically categorized by function, how are they identified? There are several classical (pre-Darwinian) ways of postulating homologies. Similarity in structure may suggest homology. Second, the 'principle of connectedness' states that items are identical which have identical connections or position within an overall structural pattern. Third, structurally diverse characters may be recognized as homological by their common developmental origin in the embryo. Mammalian inner ear bones and reptile jaw-bones can be seen (if you look *very* carefully) to arise out of common embryological elements. If you look more closely still, the reptilian jaw-bones can be seen to be homologous to portions of the gill arches of fish. The important point is that if anatomical parts had to be identified by their common biological role or SE function, all interesting homologies would be invisible. Darwin would have lost crucial evidence for descent with modification.

The fact that anatomical or morphological terms typically designate homologies shows that they are not functional categories. There is some casual use of anatomical terms by biologists, especially when formal analogies are striking. Arthropods and vertebrates each have 'tibias' and 'thoraxes', but the usage is self-consciously metaphorical between the groups; dictionaries of biology have two separate entries. The anatomical unit is, for example, the *vertebrate tibia*.

There is indeed a set of important biological categories which group organic traits by their common biological roles or SE functions. The most general of these apply to items which have biological roles so broadly significant in the animal world that they are served by analogous structures in widely divergent taxa. Among such concepts are gut (and mouth and anus), gill, gonad, eye, wing, and head (but not skull, an anatomical feature only of vertebrates). Also in the group is that all-time favourite of philosophical commentators on function—the heart. These are presumably what Neander had in mind as typical 'biological categories', and they are reasonably regarded as 'function categories' in Millikan's sense. They are analogical (as opposed to homological) in implication. Narrower function categories occur also (e.g. kneecap and ring finger), but are of limited scientific interest.

The importance of the above function categories comes from the fact that they all apply to features which result from evolutionary convergence—the selective shaping of non-homologous parts to common biological roles. It might be argued that homologous organs or body parts can be categorized by function as well. For example, *kidney* is not listed among the above function terms. Kidneys do all perform similar functions, but properly-so-called (i.e. by scientific biological usage), they exist only as homologues in vertebrates. Analogous organs exist in molluscs, but are only informally called 'kidneys'. 'The excretory organs are a pair of tubular metanephridia, commonly called kidneys in living species' (Barnes 1991: 345). But isn't 'kidney' a function category? Well, kidneys do all perform common functions (in vertebrates). But they are also homologous. This means that we could identify all members of the category 'kidney' by morphological criteria alone (morphological connectedness and developmental origin). So, at least in that sense, 'kidney' is not a function category, or at least not *essentially and necessarily* a function category. Unlike hearts, kidneys can be picked out by anatomical criteria alone. Identifying the function of kidneys amounts to discovering a (universal) functional fact about an anatomically defined category.

Even full-fledged, cross-taxon functional categories like 'heart' can often be given anatomical readings within a taxon. That is, *the vertebrate heart* can be treated as an anatomical category like the kidney. Vertebrate hearts, like kidneys, do have common functions. But they are identifiable within the taxon by their anatomical features alone. For example, mammalian heart muscle (as well as that of many other vertebrates) has a unique structure with individual cardiac muscle cells connected electrically in specialized junctional discs. The histological structure of mammalian cardiac muscle could not be mistaken for any other tissue. Thus, it is incorrect to suggest that hearts that characterize natural evolutionary clades cannot be characterized by anatomical criteria. This situation will obtain just when all of the members of the functional category are homologous within the taxon. Since all *vertebrate* hearts are homologous, they can be identified by anatomical criteria, notwithstanding the name they share with their molluscan analogues. Similarly, tetrapod hearts can be defined by unique anatomical features, as can amniote hearts and mammal hearts. The nested phylogenetic pattern (vertebrates: tetrapods: amniotes: mammals) is thus mirrored in the nested set of anatomical definitions available for vertebrate hearts. This is not surprising, as it is nested sets of similarities that provide evidence of phylogeny. On the other hand, *insect wing* cannot be treated as an anatomical category, for the simple reason that the wings of all insect taxa are probably not homologous.

Again, the point is not to quibble over the word-counts of biological concepts which are function categories and those which are not. The question is this: Do the observations of Millikan, Neander, and other SE advocates on function categories imply that CR functional anatomists will be dependent on SE functionalists in order to characterize their subject-matter? Does the existence of biological function categories mean that a reliance on causal role function will leave functional anatomists unable to identify dysfunctional hearts as hearts, a malformed tibia as a tibia? Is it true, as Neander reports, that 'no description purely in terms of morphological criteria could demarcate hearts from non-hearts'?

These claims, taken as critiques of CR functional anatomy, are almost completely groundless.[4] Morphologists are able to identify anatomical items by anatomical criteria, ignoring SE function, and do so frequently. Are hearts impossible to define by 'morphological criteria alone'? It is hard to know what Neander means by this. Criteria actually *used by morphologists*—for example, connection, micro-structure, and developmental origin—certainly *are* capable of discriminating between hearts and non-hearts within vertebrates. Perhaps by 'morphological criteria' Neander has in mind the gross physical shapes of organs. To be sure, hearts have quite different shapes and different numbers of chambers in different vertebrate species. But no practising morphologist uses gross shape as the 'morphological criterion' for an organ's identity. Even a severely malformed vertebrate heart, completely incapable of pumping blood (or serving any biological role at all), could be identified as a heart by histological examination.

Complaining about the absence of necessary and sufficient gross physical characteristics for a morphological identification of *vertebrate heart* is surely an unwarranted philosophical intrusion on science. Such an argument should only be offered by someone practising the 'confused program, philosophical chimera' of ordinary-language conceptual analysis. Morphologists can get along quite well without providing necessary and sufficient conditions for hearthood which would satisfy conceptual analysts. There is no doubt that the philosophers among us could play the

---

[4] There is one felicitous application of Neander's claim about the inadequacies of morphological criteria to designate hearts. Since the category 'heart' is used across major taxonomic differences, a vertebrate taxonomist unfamiliar with molluscs might well not be able to use *vertebrate* morphological criteria to identify a *molluscan* heart. And, to get only slightly bizarre, it is possible to imagine discovering a new taxon of animals which have organs functionally identifiable as hearts, but which fit the morphological criteria for hearts of no known taxon. We agree with the SE functionalist's point in this rather limited set of cases.

conceptual analyst's game, and dream up a bizarre case in which a miraculously deformed vertebrate's heart happened to have bizarre embryonic origins and histology, and was located under the poor creature's kneecap. The organism, if real, would baffle the anatomists just as the instant lion would baffle Darwin. But post-ordinary-language philosophers do not indulge in that style of philosophy. Anatomy *as it is practised* requires no input from SE functionalists or from biological students of adaptation in order to adequately classify and identify the structures and traits with which it deals.

SE functionalists are not the only philosophers whose emphasis on purposive function is associated with an under-appreciation of anatomical concepts. Daniel Dennett shows the same tendency. Dennett argued for the indeterminacy of (purposive) functional characterizations. He brought up Stephen Jay Gould's famous example of the panda's thumb. Gould (1980) had observed that the body part used as a thumb by the panda was not anatomically a digit at all, but an enlarged radial sesamoid, a bone from the panda's wrist. Dennett's comment: 'The panda's thumb was no more *really* a wrist bone than it is a thumb' (Dennett 1987: 320). The problem with this claim is that while 'thumb' is a functional category, 'radial sesamoid' (or 'wrist bone') is an anatomical one. Even if Dennett were correct about functional indeterminacy, anatomical indeterminacy would require a separate argument, nowhere offered. Dennett's arguments for functional indeterminacy involved the optimality assumptions he claimed were present in all functional ascriptions. Such arguments carry no weight in anatomical contexts. Such an unsupported application of a point about function to an anatomical category reflects the widespread philosophical presumption that biology is almost entirely the study of purposive function. (See Amundson 1988, 1990, on Dennett's defences of adaptationism.)

To be fair, we must acknowledge that Millikan and Neander, like other SE functionalists, were primarily interested in *purposive* concepts of function, not in *all possible* function concepts. And it is true that SE function provides an analysis of purpose which is lacking in CR function. But their interests in purpose can lead SE functionalists to overestimate the value of purposive concepts. It is simply false that anatomists require purposive concepts in order to properly categorize body parts. Anatomical categorizations of biological items already embrace interspecies and pathological diversity without any appeal to purposive function. Anatomical distinctions are not normally based on CR function *either*, to be sure. Functional anatomists *per se* do not categorize body parts. Rather, they study the capacities of anatomical complexes which have already been categorized by comparative anatomists. Causal role functional anatomy proceeds un-encumbered by demands to account either for the categorization or for the causal origins of the systems under analysis.

V. THE ELIMINABILITY OF CAUSAL ROLE FUNCTIONS

In this and the following two sections we will consider whether CR func-tions, as studied in functional anatomy, can be eliminated from evolution-ary biology in favour of SE functions. We will find them ineliminable.

First, let us consider the simplest case. Is it possible that there is a one-to-one correspondence between SE functions and CR functions? Perhaps CR functions just *are* SE functions seen through jaundiced non-historical and non-purposive lenses. To examine this possibility, let us suppose that we could easily identify which character complexes serve their present biological roles in virtue of having been selected to do so. (Not at all a trivial assumption, as will soon be seen.) What would be the relation between the biological role(s) played by a character complex (e.g. a jaw) and the CR functions which characterize the actions of its component parts? Bock and von Wahlert offer an answer: 'Usually . . . the biological roles of the individual features are the same as those of the character complex' (Bock and von Wahlert 1965: 272). Taking the jaw as a character complex which has as one of its biological roles the mastication of food, each component muscle, bone, etc. of the jaw shares in the food mastica-tion biological role.

But if the biological roles, and hence the SE functions, of the compo-nents of a character complex are the same as those of the overall complex itself, the CR functions of the components cannot be the same as their SE functions. All components of a complex have the *same* biological role/SE function, but each plays a *different* causal role within the character com-plex. So on this account SE functions cannot replace CR functions. Per-haps this result is to be expected. Bock and von Wahlert are, after all, functional anatomists. But if advocates of SE function hope to oppose this result, and refute the special significance of CR function, they presumably must argue that the activities of each component of a character complex are individually subject to the SE definition of function.

One consideration which might tempt an SE advocate in this direction is Millikan's observation, mentioned above, that all items in this world with functional complexity have undergone histories of natural selection. (Or, in the case of artefacts, were created by organisms which have such a history.) Notice, however, that the generalization *Functionally complex items have selective histories* does not by itself imply that a positive selective influence was responsible for every causal property of every component of the functional complex. Bock and von Wahlert could accept the generalization, but still distinguish biological role from CR function.

Indeed, there are many reasons to reject the identification of CR functions as merely non-historically viewed SE functions. For example, some functional anatomists wish to examine *unutilized* CR functions; clearly an unutilized function is not one which can be selected for. Further, the identification of CR with SE functions would define pre-adaptations (or exaptations) out of existence. But the question of the existence of currently utilized but unselected-for pre-adaptations (exaptations) or other selectively unshaped causal properties must be decided on the basis of evidence, not by definitional fiat.

We will not further belabour this implausible position; perhaps no SE advocate would take it anyhow. The point of this and the previous section is only that CR functions cannot be definitionally or philosophically eliminated. More interesting questions remain. Why do anatomists *need* to deal with causal role functions? Why can't they get along with purposes and selected effects?

VI. APPLICABILITY OF SELECTED EFFECT FUNCTION TO RESEARCH IN FUNCTIONAL ANATOMY

A major concern of practising functional anatomists is the utility of concepts such as function and biological role. In day-to-day research, how are functions to be identified and compared across species, and how, in practice, are we to identify the biological role of a structure? By specifying that function is that effect for which a trait was selected, SE functionalists have placed anatomists in a difficult position. In order to be able to label a structure with a corresponding function, a functional morphologist must be able to demonstrate, first, that selection acted on that structure in the population in which it arose historically, and second, that selection acted specifically to increase fitness in the ancestral population by enhancing the one specific effect that we are now to label a function of the structure. There are at least three areas in which practical difficulties arise in meeting these conditions.

First, as biologists have long recognized (e.g. Darwin 1859: ch. 6), structures may have more than one function, and these functions may change in evolution. If such change occurs, are we to identify the function of a structure as the effect for which it was first selected? If selection changes to alter the SE function of a structure through time, how are functional morphologists to identify which SE function should be applied to a structure? A recent example that points out some of the difficulties of an SE concept of function in this regard is the analysis of the origin of insect wings performed by Kingsolver and Koehl (1985). Although efforts to estimate the past action of selection (as discussed below) are fraught with difficulty, Kingsolver and Koehl used aerodynamic modelling experiments in an effort to understand the possible function of early insect wings. Do short-winged insect models obtain any aerodynamic benefit from their short wings? In other words, is it likely that selection acted on very small wings to improve aerodynamic efficiency and enhance the utility of the small wings for flight, eventually producing larger-winged insects? If so, then it would be possible to argue that the SE function of insect wings is flight. However, Kingsolver and Koehl (1985: 488) found that short insect wings provided no aerodynamic advantage, and argued that 'there could be no effective selection for increasing wing length in wingless or short-winged insects'. These authors did find, however, that short wings provided a significant advantage for thermo-regulation; short wings specifically aided in increasing body temperature, which is important for increasing muscle contraction kinetics and allowing for rapid movements. Based on these data, then, one might hypothesize that insect wings originated as a result of selection for improved thermo-regulatory ability, and that only subsequently (when wings had reached a certain threshold size) did selection act to improve flight performance.

If we identify the function of insect wings as that effect for which they were *first* selected, then we would say that the function of insect wings is thermo-regulation. It might be argued that in fact, the earliest wing-like structures actually are not proper wings, and that modern insect wings really do have the SE function of flight because at some point there was selection for improved flight performance. But this fails to recognize the size continuum of morphological structures that we call insect wings, the fact that large wings even today are used in thermo-regulation, the structural homology of large and small wings, and the virtual impossibility of identifying the selection threshold in past evolutionary time. If we cannot identify the threshold, we will not know when to change the SE function of wings from thermo-regulation to flight. Examples such as this illustrate the difficulty of assuming that the present-day roles or uses of structures are an accurate guide to inferring past selection and hence SE function.

The SE theory of function does not rule out the existence of changing patterns of selection on a given structure, nor the existence, in principle, of several SE functions for one structure. However, the complexities of this common biological situation for the association of an SE function with a specific structure have not been adequately addressed or appreciated.

Second, there are enormous practical difficulties in determining just what the selected effect of a structure was in the first place. Many structures are ancient, having arisen hundreds of millions of years ago. During this time, environments and selection pressures have changed enormously. How are we to reconstruct the ancient selected effect? The example of insect wings given above represents a best-case scenario in which we are able to make biophysical models and use well-established mathematical theories of fluid flow to estimate the likely action of selection. But many structures (particularly in fossils) are not amenable to such an analysis. Even with modern populations, studies designed to show selection on a given trait are difficult, and are subject to numerous alternative interpretations and confounding effects (Endler 1986, Arnold 1986). Functional morphologists do not have the luxury of simply asserting that the SE function of structure X is F (as philosophers so regularly do with the heart): there must be direct evidence that selection acted on structure X for effect F.

Third, there is considerable difficulty in determining that selection is acting (or acted) on *just* the structure of interest, even in extant taxa. Such difficulties are, for all practical purposes, insurmountable when dealing with fossil taxa or ancient structures. For the SE function of a structure to be identified, it is critical to be able to show that selection acted on that particular structure. However, as has been widely documented (e.g. Falconer 1989, Rose 1982), selection on one trait will cause manifold changes in many other traits through pleiotropic effects of the gene(s) under selection. Thus, selection for increased running endurance in a population of lizards may have the concomitant effect of increasing heart mass, muscle enzyme concentrations, body size, and the number of eggs laid, despite the fact that selection was directed only at endurance.

In fact, many phenotypic features are linked via common developmental and genetic controls, and this pattern of phenotypic interconnection makes isolation of any single trait and its selected effect very difficult (Lauder *et al.* 1993). If biologists had a ready means of locating the specific trait that is (or was) being acted on by selection, then the SE definition of function would be easy to apply. In actuality, due to pleiotropy, one typically sees a response in many traits to any particular selective influence. In laboratory selection experiments, the selected effect is known, and it is relatively easy to separate the selected trait from correlated responses. But in wild populations, one observes changing mean values of numerous traits in response to selection, and it is extremely difficult to separate the individual trait that is responding to selection from those that are exhibiting a correlated response.

It is also important to recognize that in extant species, the selected effect may be easier to identify than the trait acted upon by selection. This might seem counter-intuitive at first, since so many studies of adaptation proceed by first identifying a trait, and only then searching for its selective advantage(s). The difficulty of identifying the trait arises because of the correlation of the many biological traits that influence selected effects or organismal performance, and the hierarchical nature of physiological causation. Consider one powerful method for the study of selection in nature: the analysis of cohorts of individuals in a population and their demographic statistics by following individuals through time (Endler 1986). For example, if one marks individual insects in a population and measures their fitness (e.g. mating success) and their performance on an ecologically relevant variable (say, maximum flight duration), one may well find that the mean flight duration increases in the population through time due to selection against individuals that cannot remain aloft long enough to successfully mate. (Such selection might be demonstrated using the statistical methods proposed by Arnold (1983, Arnold and Wade 1984, Lande and Arnold 1983).) Here we have strong evidence that selection is operating, and an identified selected effect (increased flight duration). But what is the trait X on which selection is acting? Suppose, as we mark the individual insects, we also take a number of measurements of morphology (such as body size, eye diameter, wing length and area). We can now examine these morphological variables to see if we observe changes in these population means that are correlated with changes in flight duration. If we find that only one variable, wing area, shows an increase in mean value that is correlated with the increase in flight performance through time, then we may be willing to conclude that wing area is trait X, the trait for which the SE function is 'increasing flight duration'.

Unfortunately, an example of this type would be truly exceptional. The common result is that *many* variables are usually correlated with changes in performance and fitness. It is almost certain, in fact, that many aspects of muscle physiology, nervous system activity, flight muscle enzyme concentrations and kinetics, and numerous other physiological features would show correlated change in mean values with the increase in flight duration. In addition, body length and mass are likely to show positive correlations, as are wing length, area, and traits that have no obvious functional relevance to flight performance (such as leg length). If we cannot identify the causal relationships among these correlated variables to single out the one that was selected for, we will be unable to assign a trait X to the SE function already identified. We have a SE function, but we do not know which trait to hang it on. The fact that pleiotropic effects are so pervasive in biological systems causes severe problems in applying the definition of SE function.

Two issues relate to the analysis of traits that might be selected for in an example such as the one discussed above. First, we might choose only to measure traits on individuals which a priori physiological and mechanical considerations suggest should bear a functional relationship to the demonstrated performance change. Thus, we might decide not to measure variables such as leg length, since it is difficult to identify a physiological model in which increasing leg length would cause increased flight duration. Choosing variables based on an a priori model will certainly help narrow the universe of possible traits, but the remaining number of physiologically and mechanically relevant traits will still be very large. A second complexity in picking the trait that has been selected for arises from the hierarchical nature of physiological processes. A change in a performance characteristic (such as flight duration) may result from changes at many levels of biological design (Lauder 1991): muscle mass and insertions could change, muscle contraction kinetics could change by changing the proportion of different fibre types, enzyme concentrations within fibre types could be altered, and many features of the nervous system could be transformed. These different types of physiological traits have a hierarchical relationship to each other (in addition to a possible pleiotropic relationship) that represents a causal chain: changes at any one or more of these levels of design could account for a performance change at the organismal level. Yet, each of these features must be a distinct trait X in the SE definition, and we are unlikely in most cases to be able to identify the particular trait, or particular combination of traits, that was selected for. Of course, flight duration itself might well be considered as a trait, subject to selection and the same hierarchical patterns of underlying physiological variation as any other trait. In this case, the very same difficulties would obtain: we would need to be able to document selection *on that trait* (flight duration) in order to apply the SE concept of function.

These considerations show why anatomists are rarely able to identify which of the causal role functions of a given trait are its SE functions—that is, which (if any) are the effects for which the trait was selectively favoured. But, as the next section will show, anatomists cannot afford to abandon CR functions simply because SE function assignments are unavailable. Important research programmes are at stake.

VII. RESEARCH PROGRAMMES IN WHICH CAUSAL ROLE FUNCTION IS CENTRAL

Several aspects of current research in functional and evolutionary morphology make crucial and ineliminable use of the concept of CR function. Anatomists often write on 'the evolution of function' in certain organs or mechanical systems, and may do so with no reference to selection or to the effects of selection (e.g. Goslow *et al.* 1989, Lauder 1991, Liem 1989, Nishikawa *et al.* 1992). Rather, in these papers functional morphologists mean to consider how CR functions have changed through time, in the same manner that morphologists have traditionally examined structures in a comparative and phylogenetic context to reconstruct their evolutionary history. Indeed, a significant contribution of the field of functional anatomy (which has blossomed in the last twenty years by adopting physiological techniques to measure CR functions in different species) has been to treat functions as conceptually similar to structures. For example, Lauder (1982) and others (e.g. Wake 1991, Lauder and Wainwright 1992) have argued that CR functions may be treated just like any other phenotypic trait, and analysed in a historical and phylogenetic context to reveal the evolutionary relationship between structure and function.

So, like SE functionalists, CR functional anatomists and morphologists are interested in history. But unlike SE functionalists, anatomists do not *define* a trait's function by its history. CR function is non-historically defined. The historical interests of evolutionary morphologists are not directed towards the evolutionary mechanism of selection or the analysis of adaptation. The relation between the approaches to history taken by SE functionalists and anatomical functionalists parallels the two major explanatory modes used in the analysis of organismal structure and function. These have been termed the *equilibrium* and the *transformational* approaches (Lauder 1981, Lewontin 1969). Studies of organismal design conducted under the equilibrium view study structure in relationship to environmental and ecological variables. Such analyses are appropriate for investigating current patterns of selection and for interpreting biological design in terms of extant environmental influences. The goal of equilibrium studies is to understand extrinsic influences on form (such as temperature, wind velocity, or competition for resources), and these studies are designed to clarify current patterns of selection and hence adaptation (Bock 1980, Gans 1974). Equilibrium studies tell us little about the history of characters, however (Lewontin 1969), as the very nature of the methodology presumes (at least a momentary) equilibrium between organismal design and environmental stresses.

Many studies in functional morphology, especially in the last ten years, have adopted the transformational approach (Lauder 1981), in which historical (phylogenetic) patterns of change in form are explicitly analysed for the effects of intrinsic design properties. Here, the focus is not on adaptation, selection, or the influence of the environment, but rather on the effect that specific structural configurations might have on directions of evolutionary transformation. For example, a functional morphologist might ask: Does the possession of a segmented body plan in a clade have any consequences for subsequent evolutionary transformation in design? Under a transformational research programme one might examine a number of lineages, each of which has independently acquired a segmented body plan, to determine if subsequent phylogenetic diversification within each lineage shows any common features attributable to the presence of segmentation (regardless of the different environmental or biophysical influences on each of the species). In fact, segmentation, or more generally, the duplication or repetition of parts, appears to be a significant vehicle for the generation of evolutionary diversity in form and function, by allowing independent specialization of structural and functional components (Lauder and Liem 1989). An exemplary transformational study is Emerson's (1988) analysis of frog pectoral girdles, in which she showed that the initial starting configuration of the pectoral girdle in several clades was predictive of subsequent changes in shape. This transformational regularity occurred despite the different environments inhabited by the frog species studied. Transformational analyses by functional morphologists are historical in character: they focus on pathways of phylogenetic transformation in design which result from the arrangement of structures and the causal roles of those structures.

Functional morphologists also view organismal design as a complex interacting system of structures and functions (Liem and Wake 1985, Wake and Roth 1989). Indeed, the notion of 'functional integration', which describes the interconnectedness of structures and their CR functions, is central to discussions of organismal design and its evolution. The extent to which individual components of morphology can be altered independently of other elements without changing the (CR) functioning of the whole is one aspect of this current research (Lauder 1991). Given a structural configuration involving many muscles, bones, nerves, and ligaments, for example, all of which interact to move the jaws in a species, one might ask what effect changing the mass of just one muscle will have on the action (CR function) of the jaws as a whole. Some arrangements of structural components will have limited evolutionary flexibility due to the necessity of performing a given function such as mouth opening: even minor alterations in design may have a deleterious effect on the performance of such a critical function. This implicates CR functions as agents of evolutionary constraint. We could also enquire about possible components in a functionally integrated system that might theoretically be changed while maintaining the function of the whole system. Do predicted, permitted changes correspond to patterns of evolutionary transformation actually seen? The comparison of predicted and actual pathways of transformation is but one part of a larger effort to map a theoretical 'morphospace' of *possible* biological designs. By defining basic design parameters for a given complex morphological system, a multi-dimensional morphospace may be constructed (e.g. Bookstein *et al.* 1985, Raup and Stanley 1971). Comparing this theoretical construct with the extent to which actual biological forms have filled the theoretically possible space allows the identification of fundamental constraints on the evolution of biological design. A frequent finding is that large areas of the theoretically possible morphospace are unoccupied, and explaining this unoccupied space is a key task of functional and evolutionary morphology.

For these reasons, it is difficult to envision how the concept of a CR function so integral to both transformational analysis and functional integration, could be eliminated from the conceptual armamentarium of functional morphologists without also eliminating many key research questions.

VIII. CONCLUSION

Our rejection of some of Millikan's and Neander's conclusions should not disguise our strong agreement with their stance on the relation between the practices of science and philosophy. We heartily agree that conceptual analyses of ordinary language are inappropriately used to critique the concepts of a science. Indeed, most of our defences of CR function against ordinary-language conceptual analysis are versions of the ones used first be Millikan or Neander as they defended SE function against the same opponent. We differ from them not on the proper uses of philosophy, but on the needs and practices of biology.

We are more pluralistic than most philosophical commentators on function. We do not consider the SE concept of function, or its near-synonym the historical concept of adaptation, to be biologically or philosophically illegitimate. Our reservations about the application of purposive concepts in biology are primarily epistemological. As Williams said of adaptation, SE function in biology is 'a special and onerous concept that should be used only where it is really necessary'. Causal role function in anatomy, if less philosophically fertile than selected-effect function, is on much firmer epistemic footing. It also happens to be ineliminably involved in ongoing research programmes. This alone ought to establish its credentials.

Given comparative anatomy to categorize its subject-matter, and ecological or ethological studies of biological role to suggest which character complexes to analyse, functional anatomy is subject to none of the conceptual analyst's critiques of CR function. It is just as immune from philosophical refutation as Millikan's and Neander's science-based theory of SE function. The adequacy of each account is to be assessed not by its ability to fend off the facile imaginations of conceptual analysts, but to deal with real-world scientific issues.

Finally, a recent recommendation from Elliott Sober: 'If function is understood to mean adaptation, then it is clear enough what the concept means. If a scientist or philosopher uses the concept of function in some other way, we should demand that the concept be clarified' (Sober 1993: 86). We submit that Sober's challenge has now been met.[5]

REFERENCES

Amundson, R. (1988), 'Logical Adaptationism', *Behavioral and Brain Sciences*, 11: 505–6.
—— (1990), 'Doctor Dennett and Doctor Pangloss: Perfection and Selection in Psychology and Biology', *Behavioral and Brain Sciences*, 13: 577–84.
Arnold, S. J. (1983), 'Morphology, Performance, and Fitness', *American Zoologist*, 23: 347–61.
—— (1986), 'Laboratory and Field Approaches to the Study of Adaptation', in M. E. Feder and G. V. Lauder (eds.), *Predator–Prey Relationships: Perspectives and Approaches from the Study of Lower Vertebrates* (Chicago: University of Chicago Press).
—— and Wade, M. J. (1984), 'On the Measurement of Natural and Sexual Selection: Theory', *Evolution*, 38: 709–19.
Ayala, Francisco J. (1970), 'Teleological Explanations in Evolutionary Biology', *Philosophy of Science*, 37: 1–15.

[5] We received valuable comments on an earlier version of this essay from Elliott Sober, Ruth Millikan, Robert Brandon, and an anonymous referee. Kenneth Schaffner generously supplied a pre-publication copy of the chapter we cited and helpful observations on various function concepts. The work was supported by NSF grants SBE-9122646 (to Amundson) and IBN91-19502 (to Lauder).

Barnes, R. D. (1991), *Invertebrate Zoology* (Fort Worth: Harcourt Brace Jovanovich).

Bigelow, J., and Pargetter, R. (1987), 'Functions', *Journal of Philosophy*, 84: 181–96.

Bock, W. J. (1958), 'Preadaptation and Multiple Evolutionary Pathways', *Evolution*, 13: 194–211.

—— (1980), 'The Definition and Recognition of Biological Adaptation', *American Zoologist*, 20: 217–27.

—— and von Wahlert, G. (1965), 'Adaptation and the Form–Function Complex', *Evolution*, 19: 269–99.

Bookstein, F., Chernoff, B., Elder, R., Humphries, J., Smith, G., and Strauss, R. (1985), *Morphometrics in Evolutionary Biology* (Philadelphia: Academy of Natural Sciences).

Boorse, C. (1976), 'Wright on Functions', *Philosophical Review*, 85: 70–86.

Brandon, R. N. (1981), 'Biological Teleology: Questions and Explanations', *Studies in History and Philosophy of Science*, 12: 91–105.

—— (1990), *Adaptation and Environment* (Princeton: Princeton University Press).

Cummins, R. (1975), 'Functional Analysis,' *Journal of Philosophy*, 72: 741–65; excerpts repr. in Ned Block (ed.), *Readings in Philosophy of Psychology* (Cambridge, Mass.: MIT Press), 185–90.

—— (1983), *The Nature of Psychological Explanation* (Cambridge, Mass.: MIT Press).

Darwin, C. (1859), *On the Origin of Species* (London: John Murray).

Dennett, D. C. (1987), *The Intentional Stance* (Cambridge, Mass.: MIT Press).

Emerson, S. (1988), 'Testing for Historical Patterns of Change: A Case Study with Frog Pectoral Girdles', *Paleobiology*, 14: 174–86.

Endler, J. (1986), *Natural Selection in the Wild* (Princeton: Princeton University Press).

Falconer, D. S. (1989), *Introduction to Quantitative Genetics*, 3rd edn. (London: Longman).

Gans, C. (1974), *Biomechanics: An Approach to Vertebrate Biology* (Philadelphia: J. B. Lippincott).

Goslow, G. E., Dial, K. P., and Jenkins, F. A. (1989), 'The Avian Shoulder: An Experimental Approach', *American Zoologist*, 29: 287–301.

Gould, S. J. (1980), *The Panda's Thumb* (New York: W. W. Norton).

—— and Vrba, E. S. (1982), 'Exaptation—A Missing Term in the Science of Form', *Paleobiology*, 8: 4–15; reproduced as Ch. 4.

Hall, B. K. (1984) (ed.), *Homology* (San Diego: Academic Press).

Kingsolver, J. G., and Koehl, M. A. R. (1985), 'Aerodynamics, Thermoregulation, and the Evolution of Insect Wings: Differential Scaling and Evolutionary Change', *Evolution*, 39: 488–504.

Lande, R., and Arnold, S. J. (1983), 'The Measurement of Selection on Correlated Characters', *Evolution*, 37: 1210–26.

Lauder, G. V. (1981), 'Form and Function: Structural Analysis in Evolutionary Morphology', *Paleobiology*, 7: 430–42.

—— (1982), 'Historical Biology and the Problem of Design', *Journal of Theoretical Biology*, 97: 57–67.

—— (1990), 'Functional Morphology and Systematics: Studying Functional Patterns in an Historical Context', *Annual Review of Ecology and Systematics*, 21: 317–40.

Lauder, G. V. (1991), 'Biomechanics and Evolution: Integrating Physical and His-
torical Biology in the Study of Complex Systems', in J. M. V. Rayner and R. J.
Wootton (eds.), *Biomechanics in Evolution* (Cambridge: Cambridge University
Press), 1–19.
—— and Liem, K. F. (1989), 'The Role of Historical Factors in the Evolution of
Complex Organismal Functions', in D. B. Wake and G. Roth (eds.), *Complex
Organismal Functions: Integration and Evolution in Vertebrates* (Chichester: John
Wiley and Sons), 63–78.
—— and Wainwright, P. C. (1992), 'Function and History: The Pharyngeal Jaw
Apparatus in Primitive Ray-finned Fishes', in R. Mayden (ed.), *Systematics, His-
torical Ecology, and North American Freshwater Fishes* (Stanford, Calif.: Stanford
University Press), 455–71.
—— Leroi, A., and Rose, M. (1993), 'Adaptations and History', *Trends in Ecology
and Evolution*, 8: 294–7.
Lewontin, R. C. (1969). 'The Bases of Conflict in Biological Explanation', *Journal of
the History of Biology*, 2: 35–45.
Liem, K. F. (1989) 'Respiratory Gas Bladders in Teleosts: Functional Conservatism
and Morphological Diversity', *American Zoologist*, 29: 333–52.
—— and Wake, D. B. (1985), 'Morphology: Current Approaches and Concepts', in
M. Hildebrand, D. M. Bramble, K. F. Liem, and D. B. Wake (eds.), *Functional
Vertebrate Morphology* (Cambridge, Mass.: Harvard University Press).
Mayr, E. (1982), *The Growth of Biological Thought* (Cambrige, Mass.: Harvard
University Press).
Millikan, R. G. (1984), *Language, Thought, and Other Biological Categories* (Cam-
bridge, Mass.: MIT Press).
—— (1989), 'In Defense of Proper Functions', *Philosophy of Science*, 56: 288–302.
—— (1993), *White Queen Psychology and Other Essays for Alice* (Cambridge,
Mass.: MIT Press).
Neander, K. (1991), 'Functions as Selected Effects: The Conceptual Analyst's De-
fense', *Philosophy of Science*, 58: 168–84.
—— (MS), 'Teleology in Biology' (Woollongong University, Australia).
Nishikawa, K., Anderson, C. W., Deban, S. M., and O'Reilly, J. (1992), 'The Evolu-
tion of Neural Circuits Controlling Feeding Behavior in Frogs', *Brain, Behavior,
and Evolution*, 40: 125–40.
Patterson, C. (1982), 'Morphological Characters and Homology', in K. A. Joysey
and A. E. Friday (eds.), *Problems of Phylogenetic Reconstruction* (London: Aca-
demic Press), 21–74.
Raup, D. M., and Stanley, S. M. (1971), *Principles of Paleontology* (San Francisco:
W. H. Freeman and Co.).
Rose, M. R. (1982), 'Antagonistic Pleiotropy, Dominance, and Genetic Variation',
Heredity, 48: 63–78.
Rosenberg, A. (1985), *The Structure of Biological Science* (Cambridge: Cambridge
University Press).
Ruse, M. (1971), 'Function Statements in Biology', *Philosophy of Science*, 38: 87–95.
Schaffner, K. (1993), *Discovery and Explanation in Biology and Medicine* (Chicago:
University of Chicago Press).
Sober, E. (1984), *The Nature of Selection* (Cambridge, Mass.: MIT Press).
—— (1993), *Philosophy of Biology* (Boulder, Colo.: Westview Press).
Wake, D. B., and Roth, R. (1989), *Complex Organismal Functions: Integration and
Evolution in Vertebrates* (Chichester: John Wiley and Sons).

Wake, M. H. (1991), 'Morphology, the Study of Form and Function, in Modern Evolutionary Biology', *Oxford Surveys in Evolutionary Biology*, 8: 289–346.

Williams, G. C. (1966), *Adaptation and Natural Selection* (Princeton: Princeton University Press).

Wright, L. (1973), 'Functions', *Philosophical Review*, 82: 139–68.

12

FUNCTION AND DESIGN

PHILIP KITCHER

I

The organic world is full of functions, and biologists' descriptions of that world abound in functional talk. Organs, traits, and behavioural strategies all have functions.[1] Thus the function of the *bicoid* protein is to establish anterior–posterior polarity in the drosophila embryo; the function of the length of jack-rabbits' ears is to assist in thermo-regulation in desert environments; and the function of a male baboon's picking up a juvenile in the presence of a strange male may be to appease the stranger, or to protect the juvenile, or to impress surrounding females. Ascriptions of function have worried many philosophers. Do they presuppose some kind of supernatural purposiveness that ought to be rejected? Do they fulfil any explanatory role? Despite a long, and increasingly sophisticated, literature addressing these questions, I believe that we still lack a clear and complete account of function ascriptions. My aim in what follows is to take some further steps towards dissolving the mysteries that surround functional discourse.

I shall start with the idea that there is some unity of conception that spans attributions of functions across the history of biology and across contemporary ascriptions in biological and non-biological contexts. This unity is founded on the notion that the function of an entity *S* is *what S is designed to do*. The fundamental connection between function and design is readily seen in our everyday references to the functions of parts of artefacts: the function of the little lever in the mousetrap is to release the metal bar when the end of the lever is depressed (when the mouse takes

First published in *Midwest Studies in Philosophy*, xviii, ed. Peter A. French, Theodore E. Uehling, Jun. and Howard K. Weltstein. © 1993 by the University of Notre Dame Press, Notre Dame, Indiana. Reprinted by permission of the publisher.

[1] I shall sometimes identify the bearers of functions simply as 'entities', sometimes, for stylistic variety, talk of traits, structures, organs, behaviours as having functions. I hope it will be obvious throughout that my usage is inclusive.

the cheese), for that is what the lever is designed to do (it was put there to do just that). I believe that we can also recognize it in pre-Darwinian perspectives on the organic world, specifically in the ways in which the organization of living things is taken to reflect the intentions of the Creator: Harvey's claim that the function of the heart is to pump the blood can be understood as proposing that the wise and beneficent designer foresaw the need for a circulation of blood, and assigned to the heart the job of pumping.

Now examples like these are precisely those that either provoke suspicion of functional talk or else prompt us to think that the concept of function has been altered in the course of the history of science. Even though we may retain the idea of the 'job' that an entity is supposed to perform in contexts where we can sensibly speak of systems fashioned and/ or used with definite intentions—paradigmatically machines and other artefacts—it appears that the link between function and design must be broken in ascribing functions to parts, traits, and behaviours of organisms. But this conclusion is, I think, mistaken. On the view I shall propose, the central common feature of usages of function—across the history of enquiry, and across contexts involving both organic and inorganic entities— is that the function of S is what S is designed to do; design is not always to be understood in terms of background intentions, however; one of Darwin's important discoveries is that we can think of design without a designer.[2]

Contemporary attributions of function recognize two sources of design: one in the intentions of agents and one in the action of natural selection. The latter is the source of functions throughout *most* of the organic realm—there are occasional exceptions, as in cases in which the function of a recombinant DNA plasmid is to produce the substance that the designing molecular biologist intended. But, as I shall now suggest, the links to intentions and to selection can be more or less direct.

II

Imagine that you are making a machine. You intend that the machine should do something, and that is the machine's function. Recognizing that the machine will only be able to perform as intended if some small part

---

[2] This aspect of Darwin's accomplishment is forcefully elaborated by Richard Dawkins (1987). Although I have reservations about Dawkins's penchant for seeing adaptation almost everywhere in nature, I believe that he is quite correct to stress Darwin's idea of design without a designer.

does a particular job, you design a part that is able to do the job. Doing the job is the function of the part. Here, as with the function of the whole machine, there is a direct link between function and intention: the function of X is what X is designed to do, and the design stems from an explicit intention that X do just that.

It is possible that you do not know everything about the conditions of operation of your machine. Unbeknownst to you, there is a connection that has to be made between two parts if the whole machine is to do its intended job. Luckily, as you were working, you dropped a small screw into the incomplete machine, and it lodged between the two pieces, setting up the required connection. I claim that the screw has a function, the function of making the connection. But its having that function cannot be grounded in your explicit intention that it do that, for you have no intentions with respect to the screw. Rather, the link between function and intention is much less direct. The machine has a function grounded in your explicit intention, and its fulfilling that function poses various demands on the parts of which it is composed. You recognize some of these demands, and explicitly design parts that can satisfy them. But in other cases, as with the luckily placed screw, you do not see that a demand of a particular type has to be met. Nevertheless, whatever satisfies that demand has the function of so doing. The function here is grounded in the contribution that is made towards the performance of the whole machine and in the link between the performance and the explicit intentions of the designer.

Pre-Darwinians may have tacitly relied on a similar distinction in ascribing functions to traits and organs. Perhaps the Creator foresaw all the details of the grand design, and explicitly intended that all the minutest parts should do particular things. Or perhaps the design was achieved through secondary causes: organisms were equipped with abilities to respond to their needs, and the particular lines along which their responses would develop were not explicitly identified in advance. So the Creator intended that jack-rabbits should have the ability to thrive in desert environments, and explicitly intended that they should have certain kinds of structures. However, it may be that there was no explicit intention about the length of jack-rabbits' ears. Yet, because the length of the ears contributes to the maintenance of roughly constant body temperature, and because this is a necessary condition of the organism's flourishing (which is an explicitly intended effect), the length of the ears has the function of helping in thermo-regulation.

Understanding this distinction enables us to see how earlier physiologists could identify functions without engaging in theological speculation.[3] Operating on the presupposition that organisms were de-signed to thrive in the environments in which they are found, physiologists could ask after the necessary conditions for organisms of the pertinent types to survive and multiply. When they found such necessary conditions, they could recognize the structures, traits, and behaviours of the organisms that contributed to satisfaction of such conditions as having precisely such functions—without assuming that the Creator explicitly intended that those structures, traits, and behaviours perform just those tasks.

I have introduced this distinction in the context of machine design and of pre-Darwinian biology because it is more easily grasped in such contexts. I shall now try to show how a similar distinction can be drawn when natural selection is conceived as the source of design, and how this distinction enables us to resolve important questions about functional ascriptions.

III

We can consider natural selection from either of two perspectives. The first, the organism-centred perspective, is familiar. Holding the principal traits of members of a group of organisms fixed, we investigate the ways in which, in a particular environment or class of environments, variation with respect to a focal trait, or cluster of focal traits, would affect reproductive success. Equally, we can adopt an environment-centred perspective on selection. Holding the principal features of the environment fixed, we can ask what selective pressures are imposed on members of a group of organisms. In posing such questions we suppose that some of the general properties of the organisms do not vary, and consider the obstacles that must be overcome if organisms with those general properties are to survive and reproduce in environments of the type that interests us.

So, for example, we might consider the selection pressures on mammals whose digestive systems are capable of processing vegetation but not meat (or carrion) in an environment in which the accessible plants have tough cellulose outer layers. Holding fixed the very general properties of the animals that determine their need to take in food and the more particular features of their digestive systems, we recognize that they will not be able to survive to maturity (and hence will not be able to reproduce) unless they

---

[3] The fact that the intentions of the Creator are in the remote background in much pre-Darwinian physiological work is one of the two factors that allow for continuity between pre-Darwinian physiology and the physiology of today. As I shall argue later, appeals to selection as a source of design are kept in the remote background in contemporary physiological discussions.

have some means of breaking down the cellulose layers of the plants in their environments. Thus the environments impose selection pressure to develop some means of breaking down cellulose. Organisms might respond to that pressure in various ways: by harbouring bacteria that can break down cellulose or by having molars that are capable of grinding tough plant material. If our mammals do not have an appropriate colony of intestinal bacteria, but do have broad molars that break down cellulose, we may recognize the molars as their particular response to the selection pressure, and ascribe them the function of processing the available plants in a way that suits the operation of their digestive systems. At a more fine-grained level, we may hold fixed features of the dentition, and identify properties of particular teeth as having functions in terms of their contributions to the breakdown of cellulose.

This illustration can serve as the prototype of a style of functional analysis that is prominent in physiology and in general zoological and botanical studies. One starts from the most general evolutionary pressures, stemming from the competition to reproduce and concomitant needs to survive to sexual maturity, to produce gametes, to identify and attract mates, and so forth. In the context of general features of the organisms in question and of the environments they inhabit, we can specify selection pressures more narrowly, recognizing needs to process certain types of food, to evade certain kinds of predators, to produce particular types of signals, and so forth. We now appreciate that certain types of complex structures, traits, and behaviours enable the organisms to satisfy these more specific needs. *Their* functions are specified by noting the selection pressures to which they respond. The functions of their constituents are understood in terms of the contributions made to the functioning of the whole. Here, I suggest, we have a mixture of evolutionary and mechanistic analysis. There is a link to selection through the environment-centred perspective from which we generate the selection pressures that determine the functions of complex entities, and there is a mechanistic analysis of these complex entities that displays the ways in which the constituent parts contribute to total performance.

I claim that understanding the environment-centred perspective on selection enables us to draw an analogous distinction to that introduced in Section II, and thus to map the diversity of ways in which biologists understand functions. However, before offering an extended defence of this claim, two important points deserve to be made.

First, the environment-centred perspective has obvious affinities with the idea that organisms face selective 'problems', posed by the environment, an idea that Richard Lewontin has recently criticized (Lewontin

1982 and Lewontin and Leoins 1985). According to Lewontin, there is a 'dialectical relationship' between organism and environment that renders senseless the notion of an environment prior to and independent of the organism to which 'problems' are posed. Lewontin's critique rests on the correct idea that there is no specifying which parts of the universe are constituents of an organism's environment, without taking into account properties of the organism. In identifying the environment-centred perspective, I have explicitly responded to this point, by proposing that the selection pressures on organisms arise only when we have held fixed important features of those organisms, features that specify limits on those parts of nature with which they causally interact. Quite evidently, if we were to hold fixed properties that could easily be modified through mutation (or in development), we would obtain an inadequate picture of the organism's environment and, consequently, of the selection pressures to which it is subject. If, however, we start from those characteristics of an organism that would require large genetic changes to modify—as when we hold fixed the inability of rabbits to fight foxes—then our picture of the environment takes into account the evolutionary possibilities for the organism and offers a realistic view of the selection pressures imposed.

Second, as we shall see in more detail below, recognizing a trait, structure, or behaviour of an organism as responding to a selection pressure imposed by the environment (in the context of other features of the organism that are viewed as inaccessible to modification without severe loss of fitness), we do not necessarily commit ourselves to claiming that the entity in question originated by selection or that it is maintained by selection. For it may be that genetic variation in the population allows for alternatives that would be selectively advantageous, but are fortuitously absent. Thus the entity is a response to a genuine demand imposed on the organism by the environment, even though selection cannot be invoked to explain why it, rather than the alternative, is present. In effect, it is the analogue of the luckily placed screw, answering to a real need, but not itself the product of design. I shall be exploring the consequences of this point below.

IV

The simplest way of developing a post-Darwinian account of function is to insist on a direct link between the design of biological entities and the operation of natural selection. The function of X is what X is designed to do, and what X is designed to do is that for which X was selected. Since the publication of a seminal article by Larry Wright (1973), aetiological accounts of function have become extremely popular.[4] Wright claimed that the function of an entity is what explains why that entity is there. This simple account proved vulnerable to counter-examples: if a scientist conducting an experiment becomes unconscious because gas escapes from a leaky valve, then the presence of the gas in the room is explained by the fact that the scientist is unconscious (for otherwise she would have turned off the supply), but the function of the gas is not to asphyxiate scientists.[5] Such objections can be avoided by restricting the form of explanations to explanations in terms of selection, so that identifying the function of X as that for which X was selected enables us to preserve Wright's idea that functions play a role in explaining the presence of their bearers without admitting those forms of non-selective explanation that generate counter-examples.[6] However, this move forfeits one of the virtues of Wright's analysis: to wit, its recognition of a common feature in attributions of functions to artefacts and to organic entities.

There are other issues that aetiological analyses of functional ascriptions must confront, issues that arise from the character of evolutionary explanations. First is the question of the *time* at which the envisaged selection regime is supposed to act. Second, we must consider the *alternatives* to the entity whose presence is to be explained, and the extent of the role that selection played in the singling out of that entity.[7] If these issues are neglected—as they frequently are—the consequence will be either to engage in highly ambiguous attributions of function or else to fail to recognize the demands placed on functional ascription.

Selection for a particular property may be responsible for the original presence of an entity in an organism or for the maintenance of that entity.[8] In many instances, selection for P explains the initial presence of a trait *and* the subsequent maintenance of that trait: the initial benefit that led to the trait's increase with respect to its rivals also accounts for its superiority over alternatives that arose after the original process of fixation. But, as a host of well-known examples reveals, this is by no means always the case. To cite one of the most celebrated instances, feathers were apparently

---

[4] For further elaboration, see Millikan 1984, Neander 1991, and Godfrey-Smith 1993.

[5] This example stems from Boorse 1976.

[6] This way of evading the trouble is due to Millikan 1984.

[7] These issues are broached by Godfrey-Smith (1993). He and I are in broad agreement about questions of timing and diverge in our approaches to the second cluster of questions.

[8] Here, and in the ensuing discussion, I permit myself an obvious shorthand. In speaking of the origination of an entity in an organism, I do not, of course, mean to refer to the mutational and developmental history that lies behind the emergence of the entity in an individual organism but in the process that culminates in the initial fixation of that entity in members of the population. I hope that this abbreviatory style will not cause confusions.

originally selected in early birds (or their dinosaur ancestors) for their role in thermo-regulation; after the development of appropriate musculature (and other adaptations for flight), the primary selective significance of feathers became one of making a causal contribution to efficient flying.

Faced with examples in which the properties for which selection initially occurs are different from those for which there is selection in maintaining a trait, behaviour, or structure, the aetiological analysis must decide which of the following conditions is to govern functional attributions:

1. The function of X is Y only if the initial presence of X is to be explained through selection for Y.
2. The function of X is Y only if the maintenance of X is to be explained through selection for Y.
3. The function of X is Y only if both the initial presence of X and the maintenance of X are to be explained through selection for Y.

But deciding among these three conditions is only the beginning of the enterprise of disambiguating the aetiological analysis of function. Just as the properties important in initiating selection may not be those that figure in maintaining selection, it is possible that an entity may be *maintained* by selection for different properties at different times. Hence, both (2) and (3) require us to specify the appropriate period at which the maintenance of X is to be considered. I believe that there are two plausible candidates with respect to (2)—namely, the present and the recent past—and that the most well-motivated version of (3) requires that the character of the selective regime is constant across all times. Thus we obtain:

2a. The function of X is Y only if selection of Y has been responsible for maintaining X in the recent past.
2b. The function of X is Y only if selection for Y is currently responsible for maintaining X.
3. The function of X is Y only if selection of Y was responsible for the initial presence of X and for maintaining X at all subsequent times up to and including the present.

A consequence of adopting (1)—which effectively takes functions to be *original* functions—is that two of Tinbergen's (1968) famous four why-questions are conflated: there is now no distinction between the 'why' of evolutionary origins and the 'why' of functional attribution. In those biological discussions in which an aetiological conception of function is most apparent (ecology, and especially behavioural ecology), Tinbergen's distinction seems to play an important role. Thus I doubt that an aetiological analysis based on (1) reflects much that is significant in biological practice.

Aetiological analyses clearly based on (3) can sometimes be found in the writings of those who are critical of unrigorous employment of the notion of function. So, for example, Stephen Jay Gould's and Elisabeth Vrba's (1982) contrast between functions and 'exaptations' seems to me to thrive on the idea that specification of functions must rest on the presupposition that selection has been operating in the same way in originating and maintaining traits (and, indeed, that traits maintained by selection were originally fashioned by selection). Because there is frequently no available evidence for this presupposition, adoption of aetiological conception based on (3) can easily fuel scepticism about ascriptions of function.

I suspect that some biologists do tacitly adopt an aetiological conception of function founded on (3), and that their practice of ascribing functions is subject to Gould's strictures. Others plainly do not. Thus, Ernst Mayr (1976), explicitly recognizes the possibility of change of function over evolutionary time, suggesting that he acknowledges *two* notions of function, one ('original function') founded upon (1) and another ('present function') based on some version of (2). For biologists who draw such distinctions, Gould's criticisms will seem to claim novelty for a point that is already widely appreciated. (Of course, one of the most prominent features of the debates about adaptationism is the opposition between those who believe that the criticisms tiresomely remind the evolutionary community of what is already well known and those who contend that what is professed under attack is ignored in biological practice.)[9]

The most prevalent concept of function among contemporary ecologists is, I believe, an aetiological concept founded on some version of (2). Claims about functions are founded on measurements or calculations of fitness, and the measurements and calculations are made on *present* populations. Faced with the question 'Do you believe that the properties for which selection is now occurring are those that originally figure in the fixation of the trait (structure, behaviour)?', sophisticated ecologists would often plead agnosticism. Their concern is with what is currently occurring, and they are happy to confess that things may have been different in a remote past that is beyond their ability to observe and analyse in the requisite detail. Hence the concept of function they employ is founded on the link between functions and contemporary processes of selection that maintain the entities in question, a link recorded in (2).

But which version of (2) should they endorse? Here, I believe, philosophical analyses reveal unresolved ambiguities in biological practice. An

---

[9] The point that biologists often ignore in practice the strictures on adaptationist claims that they recognize in theory is very clearly expressed in Gould and Lewontin 1979.

account of functions that effectively endorses (2b) has been proposed by John Bigelow and Robert Pargetter (1987) (who, idiosyncratically it seems to me, attempt to distance themselves from Wright and other aetiological theorists). My own prior discussions of functional ascriptions presuppose a concept based on (2a), and this notion of function has been thoroughly articulated by Peter Godfrey-Smith (1993).[10] On what basis can we decide among these accounts?

As Godfrey-Smith rightly notes, a 'recent history' notion of function, committed to (2a), gives functional ascriptions an explanatory role. Identifying the function of an entity outlines an explanation of why the entity is now present by indicating the selection pressures that have maintained it in the recent past. Arguing that philosophers ought to identify a concept that does some explanatory work, he concludes that (2a) represents the right choice. But this seems to me to be too quick. The conception of function defended by Bigelow and Pargetter, founded on (2b), is perhaps most evident in those biological discussions in which the recognition that a trait is functional supports a prediction about its future presence in the population. Yet the 'forward-looking' conception also allows ascriptions of function to serve as explanations of why the trait will continue to be present. There is still an explanatory project, but the *explanandum* has been shifted from current presence to future presence.

Biological practice seems to me to be too various for definitive resolution of these differences. Sometimes attributions of function outline explanations of current presence, sometimes offer predictions about the course of selection in the immediate future, sometimes sketch explanations of the presence of traits in succeeding generations. Moreover, since it is often reasonable to think that the environmental and genetic conditions are sufficiently constant to ensure that the operation of selection in the recent past was the same as the selection seen in the present, it will be justifiable to combine the main features of the 'recent past' and 'forward-looking' accounts to found a notion of function on a combination of (2a) and (2b):

2c. The function of X is Y only if selection of Y is responsible for maintaining X both in the recent past and in the present.

In situations in which there is reason to think that the action of selection has been constant across the relatively short time periods under consideration, use of a notion of function founded on (2c) will allow functional attributions to play a role in all the explanatory and predictive projects I have considered.

[10] For my own commitments to a similar view see Kitcher 1988, 1990.

If biological practice overlooks potential ambiguities with respect to the timing of the selection processes that underlie attributions of function, it is even more silent on issues about the competition involved in such processes. What are the alternatives to the biological entity whose presence is due to selection? And to what extent is selection the *complete* explanation of the presence of that entity?

Ecologists working on pheromones in insects or on territory size in birds can sometimes specify rather exactly the set of alternatives they consider. Holding fixed certain features of the organisms they study, features that would, they suppose, only be modifiable by enormous genetic changes that render rivals effectively inaccessible, they can impose necessary conditions that define a set of rival possibilities: pheromones must have such-and-such diffusion properties, territories must be able to supply such-and-such an amount of food, and so forth. In light of these constraints, they may be able to construct a mathematical model showing that the entity actually found in the population is optimal (or, more realistically, 'sufficiently close' to the optimum).[11] A different strategy is to consider alternatives that arise by mutation in populations that can be observed, and to measure the pertinent fitness values. Either of these approaches will support claims about selection processes that have occurred/are occurring in the recent past or the present. In both instances there may be legitimate concern that unconsidered alternatives might have figured in historically more remote selection processes, either because the organisms were not always subject to the constraints built into the mathematical model or because the genetic context in which mutations are now considered is quite different from the genetic contexts experienced by organisms earlier in their evolutionary histories. So far this simply underscores our previous conclusions about the greater plausibility of analyses based on some version of (2).

But now let us ask how exactly selection is supposed to winnow the alternatives. Suppose we ascribe a function to an entity X, basing that function on a selection process with alternatives $X_1, \ldots, X_{11}$. Must it be the case that organisms with X have higher fitness than organisms with any of the X_i? On a strict aetiological analysis of functional discourse, this question should be answered affirmatively: where selection is the *complete* foundation of the design that underlies X's function, X is favoured by selection over *all* its rivals. Thus, on the strongest version of an aetiological conception, functional ascriptions should be based either on recognition that X has greater fitness than all the alternatives arising by mutation in current populations, or on an analysis that shows X to be strictly optimal.

[11] See e.g. the discussion of Geoffrey Parker's ingenious and sophisticated work on copulation time in male dungflies in Kitcher 1985: ch. 5.

I believe that some biologists—particularly in ecology and behavioural ecology—make functional claims in this strong sense and attempt to back them up with careful and ingenious observations and calculations.[12] None the less, there is surely room for a less demanding account of biological function.

Consider two possibilities. First, our optimality analysis shows that, while X is reasonably close to the optimum, it is theoretically sub-optimal. We do not know enough about the genetics and developmental biology of the organisms under study to know whether mutations providing a genetic basis for superior rivals could arise in the population. Under these circumstances, one cannot claim that the presence of X is entirely due to the operation of selection. It may be that X is present because theoretically possible mutants have not (recently) arisen, and selection, acting on a limited set of alternatives, has fixed X. Second, we may be able to identify actual rivals to X that are indeed superior in fitness but that have fortuitously been eliminated from the population. During the period that concerns us (present or recent past), organisms bearing some entity X_i have arisen, and these have had greater fitness than organisms bearing X. By chance, however, such organisms have perished. Here, we can go further than simply recognizing an inability to support the strong claim about optimality—we recognize that X is definitely sub-optimal, and that its presence is not the result of selection alone.

Nevertheless, many biologists would surely be uninterested in these possibilities or actualities, regarding X as having the function associated with the selective process, even if it were possibly, even definitely, sub-optimal. There are various ways of weakening the requirement that X's fitness be greater than those of alternatives. We might demand the X be fitter than *most* alternatives, that it be fitter than the *most frequently occurring* alternatives, and so forth. It requires only a little imagination to devise scenarios in which an entity is inferior in fitness to most of its rivals and/or to its most frequently occurring rivals, even though it may still be ascribed the function associated with the selection process.

Imagine that there is a species of moth that is protected from predatory birds through a camouflaging wing pattern that renders it hard to perceive when it rests on a common environmental background. We observe the population and discover a number of rival wing colourations, none of which ever occurs in substantial numbers. Less than half of these alternatives are absolutely disastrous, and organisms with them are vulnerable to predation, and quickly eliminated. Investigating the others, we find, to our

---

[12] See the examples given in section VI below.

surprise, that they prove slightly superior to the prevalent form, in affording improved camouflage, without any deleterious side-effects. However, as the result of various events that we can identify—disruptions of habitat, increased concentrations of predators in areas in which there is a high frequency of the mutants—these alternatives are eliminated as the result of chance. None the less, although it is somewhat inferior to most of its rivals, the common wing pattern still has the function of protecting the moth from predation.

I think that it is obvious what we should say about this and kindred scenarios. The impulse to recognize X as having a function can stem from recognition that X is a response to an identifiable selection pressure, *whether or not the presence of X is completely explicable in terms of selection*. Thus, instead of trying to weaken the conditions on aetiological conceptions of function, I suggest that we can accommodate cases that prove troublesome by drawing on the distinctions of Sections II and III. I shall now try to show how this leads to a rich account of functional ascriptions that will cover practice in physiology as well as in those areas in which the aetiological conception finds its most natural home.

V

Entities have functions when they are designed to do something, and their function is what they are designed to do. Design can stem from the intentions of a cognitive agent or from the operation of selection (and, perhaps, recognizing how unintuitive the notion of design without a designer would have seemed before 1859, from other sources that we cannot yet specify). The link between function and the source of design may be direct, as in instances of agents explicitly intending that an entity perform a particular task, or when the entity is present because of selection for a particular property (that is, its presence is completely explained in terms of selection for that property). Or the link may be indirect, as when an agent intends that a complex system perform some task and a component entity makes a necessary causal contribution to the performance, or when organisms experience selection pressure that demands some complex response of them, and one of their parts, traits, or behaviours makes a needed causal contribution to that response. As noted in the previous section, there are also ambiguities about the time period throughout which the selection process is operative. It would be easy to tell a parallel story about agents and their intentions.

I have noted that the strong aetiological conception—that based on a direct link between function and the underlying source of design (in this case, selection)—is very demanding. While some ecologists undoubtedly aim to find functions in the strong sense, much functional discourse within ecology, as well as in other parts of biology is more relaxed. Imagine practising biologists accompanied by a philosophical Jiminy Cricket, constantly chirping doubts about whether selection is *entirely* responsible for the presence of entities to which functions are ascribed. Many biologists would ignore the irritating cavils, contending that the attribution of function is unaffected by the possibilities suggested by philosophical conscience. It is enough, they would insist, that genuine demands on the organism have been identified and that the entities to which they attribute functions make causal contributions to the satisfaction of those demands. What is wrong with the relaxed attitude?

Functional attributions in the strong sense have clear explanatory work to do. They indicate the lines along which we should account for the presence of the entities to which functions are ascribed. To say that the function of X is F is to propose that a complete explanation of the presence of X (at the appropriate time) should be sought in terms of selection for F. Once we relax the demands on functional ascriptions, the role of selection is no longer clear; indeed, a biologist may explicitly allow that selection has not been responsible for maintaining X (or, at least, not completely responsible). But there is a different type of explanatory project to which the more lenient attributions contribute. They help us to understand the causal role that entities play in contributing to complex effects.

Here we encounter a central theme of the main philosophical rival to the aetiological conception, lucidly articulated in an influential article by Robert Cummins (1975). For Cummins, functional analysis is about the identification of constituent causal contributions in complex processes. This style of activity is prominent in physiological studies, where the apparent aim is to decompose a complex 'organic function' and to recognize how it is discharged. I claim that Cummins has captured an important part of the notion of biological function, but that his ideas need to be integrated with those of the aetiological approach, not set up in opposition to it.

When we attribute functions to entities that make a causal contribution to complex processes, there is, I suggest, always a source of design in the background. The constituents of a machine have functions because the machine, as a whole, is explicitly intended to do something. Similarly with organisms. Here selection lurks in the background as the ultimate source of design, generating a hierarchy of ever more specific selection pressures, and the structures, traits, and behaviours of organisms have functions in virtue of their making a causal contribution to responses to those pressures.

Without recognizing the background role of the sources of design, an account of the Cummins variety becomes too liberal. Any complex system can be subjected to functional analysis. Thus we can identify the 'function' that a particular arrangement of rocks makes in contributing to the widening of a river delta some miles downstream, or the 'functions' of mutant DNA sequences in the formation of tumours—but there are no genuine functions here, and no functional analysis. The causal analysis of delta formation does not link up in any way with a source of design; the account of the causes of tumours reveals *dysfunctions*, not functions.

Recognizing the liberality of Cummins-style analyses, proponents of the aetiological conception drag evolutionary considerations into the foreground. In doing so they make *all* projects of attributing functions focus on the explanation of the presence of the bearers of those functions. However, important though the theory of evolution by natural selection undoubtedly is to biology, there are other biological enterprises, some even continuous with those that occupied pre-Darwinians, which can be carried out in ignorance of the details of selective regimes. Thus the conscience-ridden biologists who offer more relaxed attributions of function can quite legitimately protest that the niceties of selection processes are not their primary concerns: without knowing what alternatives there were to the particular valves that help the heart to pump blood, they can recognize both that there is a general selection pressure on vertebrates to pump blood and that particular valves make identifiable contributions to the pumping. Selection, they might say, is the background source of design here, but it need not be dragged into the foreground to raise questions that are irrelevant to the project they set for themselves (understanding the mechanism through which successful pumping is achieved).

I believe that the account I have offered thus restores some unity to the concept of function through the recognition that each functional attribution rests on some presupposition about design and a pertinent source of design. But it allows for a number of distinct conceptions of function to be developed, based on sources of design (intention versus selection), time relation between source of design and the present, and directness of connection between source of design and the entity to which functions are ascribed. This pluralism enables us to capture the insights of the two main rival philosophical conceptions of function, and to do justice to the diversity of biological projects.

Does it go too far? In their original form, actiological accounts were vulnerable to counter-example, and the resolution invoked selection *ad hoc*. Am I committed to supposing that the leaky valve that asphyxiates the scientist has the function of so doing? No. For there is no explaining the presence of the valve in terms of selection for ability to asphyxiate scientists; nor is there any selection pressure on a larger system to whose response the action of the valve makes a causal contribution. Even though the account I have offered is more inclusive than traditional aetiological conceptions, it does not seem to fall victim to the traditional counter-examples.

VI

I have tried to motivate my account of function and design by alluding to some quickly sketched examples. This strategy helps to elaborate the approach, but invites concerns to the effect that a more thorough investigation of biological practice would disclose less ambiguity than I have claimed. To alleviate such concerns, I now want to look at some cases of functional attribution in a little more detail.

I shall start with two examples that are explicitly concerned with evolutionary issues. The first concerns a 'functional analysis of the egg sac' in golden silk spiders (Christenson and Wenzl 1980). The orb-weaving spider *Nephila clavipes* lays its eggs under the leaf canopy, covers them with silk, and weaves a loop of silk around twig and branch which holds the sac in place. The authors of the study investigate the functions of components of the egg-laying behaviour. I shall concentrate on the spinning of the loop.

Christenson and Wenzl (1980: 1114) write:

> The functions of the silk loop around the attachment branch were assessed by examining clutches that fell to the ground. We found 19 of the 59 egg sacs that fell due to naturally occurring twig breakage; 84.2% (16) failed to produce spiderlings, 13 because of ground moisture and subsequent rotting, and 3 because of predation. . . . The remaining three sacs had fallen a few weeks prior to the normal time of spring emergence; the spiderlings appeared to disperse and inhabit individual orbs.[13] In contrast to those that fell, sacs that remained in the tree were dry and appeared relatively safe from predation. Only 4.5% (15 of 353) showed unambiguous signs of predation, that is, some damage to the silk such as a tear or a bore hole.

[13] I should note here that spiderlings typically overwinter in the egg sac, so that the period of a few weeks represents a fall only a *short* time before the usual time of emergence. Thus the successful instances are those in which the normal course of development is only slightly perturbed.

I interpret this passage as demonstrating a marked fitness difference between spiders who perform the looping operation that attaches the egg sac to twig and branch and those who fail to do so. Christenson and Wenzl are tacitly comparing the normal behaviour of *N. clavipes* with mutants whose ability to weave an attachment loop was somehow impaired. Their emphasis on evolutionary considerations is evident not only in their detailed measurements of survivorships, but also in the framing of their analysis and in their final discussion. The authors begin by noting that '[f]unctional analyses of behaviours are often speculative due to the difficulty of demonstrating that the behaviour contributes to the individual's reproductive success, and what the relevant selective agents might be' (ibid. 1110). They conclude by contending that 'Female *Nephila* maximize their reproductive efforts, in part, through the construction of an elaborate egg sac' (ibid. 1115). This study is thus naturally interpreted as deploying the strong aetiological conception of function, linking function directly with selection and proposing that the entities bearing functions are optimal.

Similarly, a study of the function of roaring in red deer by T. Clutton-Brock and S. Albon (1979) explicitly connects the attribution of function to claims about selection. The authors begin by examining a traditional proposal:

A common functional explanation is that displays serve to intimidate the opponent. . . . This argument has the weakness that selection should favour individuals which are not intimidated unnecessarily and which adjust their behaviour only to the probability of winning and the costs and benefits of fighting. (p. 145)

Here it seems that a necessary condition on the truth of an ascription of function is that there should not be possible mutants that would be favoured by selection. The same strong conception of function is apparent later in the discussion, when Clutton-Brock and Albon consider the hypothesis that roaring serves as an advertisement, enabling stages to assess others' fighting ability. Although their careful observations indicate that stags rarely defeat those by whom they have been out-roared, they recognize that their data leave open other possibilities for the relation between roaring and fighting ability. They suggest that fighting and roaring may both draw on the same groups of muscles, so that roaring serves as an 'honest advertisement' to other stags. But they note that this depends on assuming that 'selection could not produce a mutant which was able to roar more frequently without increasing its strength or stamina in fights' (ibid. 165). I interpret the caution expressed in their discussion to be grounded in recognition of the stringent conditions that must be met in showing that a form of behaviour maximizes reproductive success, and thus their reliance on the strong version of the aetiological conception.

I now turn to two physiological studies in which the connection to evolution is far less evident. Here, there are neither detailed measurements of the fitnesses (or proxies such as survivorships) of rival types of organism (as in the study of golden silk spiders) or connections with mathematical models of a selection process (as in the investigation of the roaring of stags). Instead, the authors undertake a mechanistic analysis of the workings of a biological system. Consider the following discussion of digestion in insects:

> Food in the midgut is enclosed in the peritrophic membrane, which is secreted by cells at the anterior end of the midgut in some insects or formed by the midgut epithelium in most. It is secreted continuously or in response to a distended midgut, as in biting flies. It is likely that the peritrophic membrane has several functions, although the evidence is not conclusive. It may protect the midgut epithelium from abrasion by food or from attack by microorganism or it may be involved in ionic interactions within the lumen. It has a curious function in some coleopterous larvae, where, in various ways, it is used to make the cocoon. (McFarlane 1985: 64)

The interesting point about this passage is that it could easily be accepted by a biologist ignorant of or hostile to evolutionary theory. So long as one has a sense of the overall life of an insect and of the conditions that must be satisfied for the insect to thrive, one can view the peritrophic membrane as making a causal contribution to the organism's flourishing. Of course, Darwinians will view these conditions as grounded in selection pressures to which insects must respond, but physiology can keep this Darwinian perspective very much in the background. It is enough to recognize that insects must have a digestive system capable of processing food items, that the passage of food through the system must not abrade the cells lining the gut, and so forth. I suggest that this, like so many other physiological discussions, presupposes a background picture of the selection pressures on the organisms under study, and analyses the causal mechanisms that work to meet those pressures, without attending to the fitness of alternatives that would have to be considered to underwrite a claim about the operation of selection.

Finally, I turn to a developmental study of sexual differentiation in *Drosophila* (Kaulenas 1992: sec. 2.3). The problem is to understand simultaneously how an embryo with two X chromosomes becomes a female, how an embryo with one X chromosome becomes a male, and how the organism compensates for the extra chromosomal material found in females. The author summarizes a complex causal story, as follows:

> The primary controlling agent in sex determination and dosage compensation is the ratio between the X chromosomes to sets of autosomes (the X : A ratio). This ratio is 'read' by the products of a number of genes; some of which function as numerator

elements, while others as denominator elements. Two of the numerator genes have been identified [*sisterless a (sis a)* and *sisterless b (sis b)*] and others probably exist. The denominator elements are less clearly defined. The end result of this 'reading' is probably the production of DNA-binding proteins, which, with the cooperation of the *daughterless (da)* gene product (and possibly other components) activate the *Sex lethal (sxl)* gene. This gene is the key element in regulating female differentiation. One early function is autoregulation, which sets the gene in the functional mode. Once functional, it controls the proper expression of the *doublesex (dsx)* gene. The function of *dsx* in female somatic cell differentiation is to suppress male differentiation genes. *Dsx* needs the action of the *intersex (ix)* gene for this function. Female differentiation genes are not repressed, and female development ensues. (ibid. 17)

Here is a causal story about how female flies come to express the appropriate proteins in their somatic cells. The elements of the story concern the ways in which particular bits of DNA code for proteins that either activate the right genes or block transcription of the wrong ones. In the background is a general picture of how selection acts on sexually reproducing organisms, a picture that recognizes the selectively disadvantageous effects of failing to suppress one set of genes (those associated with the distinctive reactions that occur in male somatic cells) and of failing to activate the genes in another set (those whose action is responsible for the distinctive reactions of female somatic cells). The functions of the specific genes identified by Kaulenas are understood in terms of the causal contributions they make in a complex process. There is no attempt to canvass the genetic variation in *Drosophila* populations or to argue that the specific alleles mentioned are somehow fitter than their rivals. The discussion takes for granted a particular type of selection pressure—thus adopting the environment-centred perspective on evolution—and considers only the causal interactions that result in a response to that selection pressure. The causal analysis is vividly presented in a diagram (reproduced in Fig. 12.1), which shows the kinship between the type of mechanistic approach adopted in this study and the analysis of complex systems designed by human beings. Selection furnishes a context in which the overall design is considered, and, within that context, the physiologist tries to understand how the system works.

I offer these four examples as paradigmatic of two very different types of biological practice offering ascriptions of function. I hope that it is evident how introducing the strong aetiological conception within the last two would distort the character of the achievement, rendering it vulnerable to sceptical worries about the operation of selection that are in fact quite irrelevant. By the same token, it is impossible to appreciate the line of argument offered in the explicitly evolutionary studies without recognizing

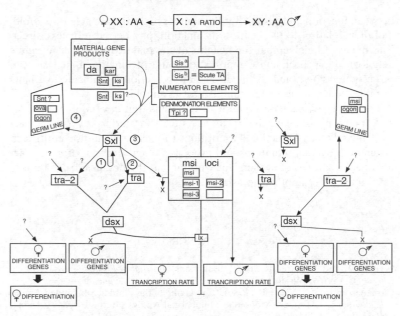

Fig. 12.1. Diagram illustrating the interrelationships of the genes involved in the control of sexual differentiation and dosage compensation in *Drosophila*. From Kaulenas 1992: 18.

the stringent requirements that the strong aetiological conception imposes. There are undoubtedly many instances in which the notion of function intended is far less clear. I believe that keeping our attention focused on paradigms will be valuable in the work of disambiguation.

VII

Philosophical discussions of function have tended to pit different analyses and different intuitions against one another without noting the pluralism inherent in biological practice.[14] On the account I have offered here, there is indeed a unity in the concept of function, expressed in the connection

[14] As I have argued elsewhere, biological practice is pluralistic in its employment of concepts of gene and species and in its identification of units of selection. See Kitcher 1982, 1983, and Sterelny and Kitcher 1988.

between function and design, but the sources of design are at least twofold, and their relation to the bearers of function may be more or less direct. This means, I believe, that the insights of the main competitors, Wright's aetiological approach and Cummins's account of functional analysis, can be accommodated (and, as the discussion in Section IV indicates, variants of the aetiological approach can also be given their due).

The result is a general account of functions that covers both artefacts and organisms. I believe that it can also be elaborated to cover the apparently mixed case of functional ascriptions to social and cultural entities, in which both explicit intentions and processes of cultural selection may act together as sources of design. But working out the details of such impure cases must await another occasion.[15]

REFERENCES

Bigelow, J., and Pargetter, R. (1987), 'Functions', *Journal of Philosophy*, 84: 181–96.
Boorse, C. (1976), 'Wright on Functions', *Philosophical Review*, 85: 70–86.
Christenson, T., and Wenzl, P. (1980), 'Egg-laying of the Golden Silk Spider, *Nephila clavipes* L. (Araneae, Araneidae): Functional Analysis of the Egg Sac', *Animal Behaviour*, 28: 1110–18.
Clutton-Brock, T., and Albon, S. (1979), 'The Roaring of Red Deer and the Evolution of Honest Advertisement', *Behaviour*, 69: 145–68.
Cummins, R. (1975), 'Functional Analysis', *Journal of Philosophy*, 72: 741–65.
Dawkins, R. (1986), *The Blind Watchmaker* (New York: Norton).
Godfrey-Smith, P. (1993), 'Functions: Consensus without Unity', *Pacific Philosophical Quarterly*, 74: 196–208; reproduced as Ch. 13.
Gould, S. J., and Lewontin, R. C. (1979), 'The Spandrels of San Marco and the Panglossian Paradigm: A Critique of the Adaptationist Programme', *Proceedings of the Royal Society of London*, B205: 581–98.
——and Vrba, E. S. (1982), 'Exaptation—A Missing Concept in the Science of Form', *Paleobiology*, 8: 4–15; reproduced as Ch. 4.
Kaulenas, M. (1992), *Insect Accessory Reproductive Structures: Function, Structure, and Development* (New York; Springer).
Kitcher, P. (1982), 'Genes', *British Journal for the Philosophy of Science*, 33: 337–59.
——(1984), 'Species', *Philosophy of Science*, 51: 308–33.
——(1985), *Vaulting Ambition: Sociobiology and the Quest for Human Nature* (Cambridge, Mass.: MIT Press).

[15] I am extremely grateful to the Office of Graduate Studies and Research at the University of California–San Diego for research support, and to Bruce Glymour for research assistance. My thinking about functional attributions in biology has been greatly aided by numerous conversations with Peter Godfrey-Smith. Despite important residual differences, I have been much influenced by Godfrey-Smith's careful elaboration and resourceful defence of an aetiological view of functions.

—— (1988), 'Why Not the Best?', in J. Dupré (ed.), *The Latest on the Best: Essays on Optimality and Evolution* (Cambridge, Mass.: MIT Press), 77–102.

—— (1990), 'Developmental Decomposition and the Future of Human Behavioral Ecology', *Philosophy of Science*, 57: 96–117.

Lewontin, R. C. (1982), 'Organism and Environment', in H. C. Plotkin (ed.), *Learning, Development, and Culture* (Chichester: John Wiley), 151–70.

—— and Levins, R. (1985), *The Dialectical Biologist* (Cambridge, Mass.: Harvard University Press).

McFarland, J. (1985), 'Nutrition and Digestive Organs', in M. Blum (ed.), *Fundamentals of Insect Physiology* (Chichester: John Wiley), 59–90.

Mayr, E. (1976), 'The Emergence of Evolutionary Novelties', in *Evolution and the Diversity of Life* (Cambridge, Mass.: Belknap Press of Harvard University Press), 88–113.

Millikan, R. G. (1984), *Language, Thought, and Other Biological Categories* (Cambridge, Mass.: MIT Press).

Neander, K. (1991), 'Functions as Selected Effects: The Conceptual Analyst's Defense', *Philosophy of Science*, 58: 168–84.

Sterelny, K., and Kitcher, P. (1988), 'The Return of the Gene', *Journal of Philosophy*, 85: 335–58; reproduced as Ch. 8.

Tinbergen, N. (1968), 'On War and Peace in Animals and Man', *Science*, 160: 1411–18.

Wright, L. (1973), 'Functions', *Philosophical Review*, 82: 139–68.

13

FUNCTIONS: CONSENSUS WITHOUT UNITY

PETER GODFREY-SMITH

I. TWENTY YEARS

The year 1993 marked the twentieth anniversary of the publication of
Larry Wright's article 'Functions' (1973), an article which decisively reori-
ented the functions debate.

Wright's article did not answer all the questions philosophers have
asked about functions, but it did answer some of them, and it showed the
way forward to answering more. Much of the literature since 1973 has,
in effect, engaged in the refinement of Wright's original idea. Many
writers do not think of themselves as doing this; indeed, several have
actively resisted this interpretation.[1] None the less, since 1973 there
has been a convergence towards a view of functions which has Wright's
idea at its core.[2] I think of this trend as an example of real progress in
philosophy.

In this essay I will sketch what I see as the view towards which the
literature is converging. One feature of the theory which should reasonably be regarded as controversial is a bifurcation within it. On my view,
functions as analysed by Wright and functions as analysed by Robert
Cummins are both real, and important, and distinct. Philip Kitcher (1993)
has argued that the concept of *design* can unify these two conceptions of
function. I will resist this move towards unification. Although some will
find a bifurcation unattractive, unity is not always a good thing.

First published in *Pacific Philosophical Quarterly*, 74 (1993): 196–208.

[1] Bigelow and Pargetter 1987 and Millikan 1989*b* are examples.
[2] Works contributing to the consensus which are not discussed elsewhere in this essay include
Neander 1991, Brandon 1990, Mitchell 1989, Sober 1984, and Griffiths 1992.

II. WRIGHT'S TWO ADVANCES

Wright said:

The function of X is Z means
(a) X is there because it does Z,
(b) Z is a consequence (or result) of X's being there (1976: 81).

The most striking thing about this formula is its simplicity. Through the 1960s philosophers became accustomed to long and intricate definitions of functions—at least six lines long and four variables deep. And whether or not the biological phenomenon known as Cope's Rule is generally true in nature, it is admirably illustrated by most philosophical lineages: definitions of a given concept get physically bigger through time, not smaller. Yet Wright's definition of function was shorter than its predecessors. This poses a small puzzle in the history of philosophy: why, given all that had gone before, was it possible to defend a two-line theory at that point?

Earlier analyses of functions were driven in large part by general assumptions made about explanation. For writers such as Hempel, a functional explanation has to explain the presence of the functionally characterized entity, and the explanation has to conform to something like the D-N model. The D-N model, or deductive-nomological model of explanation, which was dominant through the 1960s, understands explanations as inferences. An *explanans* is a set of premises, including a law of some kind, which confers either deductive certainty (or, for I-S explanations, high probability) on the *explanandum* (Hempel and Oppenheim 1948). So if functional explanation was to be genuine, citing the function of the heart, for instance, had to *imply* the existence of hearts, given some other premises about the containing system. But although the heart's function is pumping blood, and people need blood to be circulated, it is not possible to infer the existence of hearts from their blood-pumping ability, and no reasonable amount of fine-tuning of the envisaged argument will make this so.

Wright dismisses this conception of explanation, both in general and as applied to functions (this is most clear in his 1976 discussion). Although it is not presented in this way, Wright's conception of explanation is related to ideas developed in detail by writers such as Wesley Salmon (1984) and Bas van Fraassen (1980). An explanation cites factors which rule out or make less probable certain alternative events to the *explanandum*. Which ones are to be ruled out depends on the context in which the explanation is offered. It is not necessary to rule out *all* alternatives. Wright's analysis of functions is made against a background of a liberal conception of explanation. Once we have this conception of explanation, it is clear that sometimes we can explain the presence or persistence of entities by citing certain of their effects or dispositions. Whenever this is possible, these effects are those entities' functions. So one of the advances in Wright's analysis is an instance or an application of general progress made around that time with respect to explanation.

The other step forward in Wright's analysis does not have to do with philosophical currents outside the functions industry. Wright's analysis is driven in large part by constant attention to what he calls the 'function/ accident distinction'. When attending to this distinction, one insists that there is a definite difference between a function and a fortuitous benefit. Something can have beneficial effects, or make a useful contribution to a containing system, but these are not functions unless the thing in question is there *because* of these effects. Otherwise they are accidental, fortuitous benefits. This is the manœuvre in Wright which disposes of the whole range of analyses of functions based upon contributions to goals. A mere contribution to a goal is not a function unless it is not fortuitous, unless this contribution explains why the thing is there. But this requirement of explanatory salience is apparently now bearing the whole weight of the concept of function, and goals drop out of the picture. In some respects Wright here does for philosophy what G. C. Williams did for biologists in his *Adaptation and Natural Selection* (1966): motivate vigilant attention to the difference between fortuitous benefit and genuine adaptation.

III. A CONSENSUS VIEW

The simplicity of Wright's analysis was also intended to reflect the ease of application characteristic of the concept of function. But subsequent discussion has indicated that the schema he proposed was left too simple, or made excessive demands on contextual factors. I will run through a sequence of objections and modifications, which are designed specifically to improve the theory's analysis of functional discourse in biology.

Lineages

Here is a counter-example modified from some used by Boorse (1976). Consider a small rock holding up a larger rock in a fast-moving stream. If the small rock did not support the larger rock, it would be washed away. Holding up the big rock is the thing the small rock does; that explains why it is there. So on Wright's original analysis this is the function of the small rock.

In Ruth Millikan's (1984) analysis of biological functions, this type of problem is immediately avoided, by restricting the entities to which functions are ascribed to those which exist within lineages defined by relations of reproduction or replication. Very roughly, the function of something is whatever past tokens of that reproductively defined type did that explains the existence of present tokens. In Millikan's account it is also explicitly required that the explanation make reference to a selection process.

These modifications deal effectively with many otherwise troubling counter-examples, such as Boorse's. However, it is important to note a less attractive consequence of explicitly building them into the analysis. Once we make an appeal to lineages defined by reproduction we begin to lose the *generality* of Wright's view. For example, if we do not build these restrictions into the analysis, then the concept of function used in Dretske's recent work on meaning and explanation (1988) can be understood as Wright's concept in a different setting. Dretske says that an inner state C can have the function of indicating an external condition F if C has been recruited as cause of some motion M because it indicates F. This is easily understood as an instance of Wright's basic formula: the thing C does that explains why it is where it is, why it has been recruited, is indicating F. C *qua cause of M* is there because it indicates F. So indicating F is C's function (see also Godfrey-Smith 1992). Thus Dretske's work, too, can be seen as part of the unacknowledged consensus.

However, we can only understand Dretske's concept as an instance of Wright's general view if we do not build into the analysis an explicit appeal to reproduction or replication. Dretske is most interested in 'recruitment' of inner indicators that results from individual learning, and in these cases it is very hard to see C as a member of a reproductively defined lineage whose earlier members indicated F and were recruited for this reason. Dretske's view fits the basic Wright formula, not the Millikan-style modifications. This is not to say that the modified concept of function has no applications in philosophy of mind—these applications are much of Millikan's motivation for developing the concept. But the use of biological or 'teleonomic' concepts of function in philosophy of mind is made more complex when Wright's formula is augmented in this way. On the other hand, if we do not build in these additional requirements, we have a harder time with counter-examples such as Boorse's. In this discussion I will assume that an explicit appeal to selection processes and reproductively defined lineages is appropriate.

Past and Present

Another set of problems derive from the facts of biological usage. In biological discussion it is common to make an explicit distinction between 'evolutionary' and 'functional' explanations for a trait. Tinbergen is often cited for this, especially by behavioural biologists.[3] But on Wright's analysis of functions this distinction should not exist. Horan (1989) appeals to this fact about biological usage to motivate a selection-based account of functions which is forward-looking rather than backward-looking. The best forward-looking theory I know is the propensity theory of Bigelow and Pargetter (1987): functions do not derive from a past history of selection, but from present propensities to succeed under selection.

My view is that looking forward is a mistake; it is better to look backward in a slightly different way. Functions can be seen as effects of a trait which have led to its maintenance during *recent* episodes of natural selection. The distinction between 'functional' and 'evolutionary' explanations can be cast as a distinction between the explanation for the original establishment of the trait, and the explanation, which may be different, for its recent maintenance (Godfrey-Smith 1994). Thus we can make sense of biological usage while retaining the idea that in giving a function we are, *ipso facto*, giving an explanation for why the functionally characterized thing exists now.

Cummins Functions

Once a modified version of Wright's theory is in place, the explanatory role of many function statements in fields like behavioural ecology is clear. But there remain entire realms of functional discourse, in fields such as biochemistry, developmental biology, and much of the neurosciences, which are hard to fit into this mould, as functional claims in these fields often appear to make no reference to evolution or selection. These are areas in which the attractive account of functions has always been that of Robert Cummins (1975). On Cummins's analysis, functions are not effects which explain why something is there, but effects which contribute to the explanation of more complex capacities and dispositions of a containing system.

Although it is not always appreciated, the distinction between function and *malfunction* can be made within Cummins's framework, as well as within Wright's. If a token of a component of a system is not able to do whatever it is that other tokens do, that plays a distinguished role in the

---

[3] Tinbergen (1963) acknowledges Julian Huxley. Mayr 1961 is another early source.

explanation of the capacities of the broader system, then that token component is malfunctional. The concept of malfunction is context-dependent on Cummins's view, just as the concept of function in general is.

My view of this issue derives from Millikan (1989a). We should accept both senses of function, and keep them strictly distinct. All attempts to make one concept of function work equally for behavioural ecology and physiology are misguided. On this view, 'Wright functions' and 'Cummins functions' are both effects which are distinguished by their explanatory importance. The difference is in the type of explanation. So if it is claimed, for instance, that the function of the myelin sheaths round some brain cells is to make possible the efficient conduction of signals over long distances, it may not be obvious which explanatory project is involved. This may be intended as an explanation of why the myelin is there, or it could be part of an explanation of how the brain manages to perform certain complex tasks. Sometimes the same assignment of functions will be made from both perspectives, but this does not mean the questions are the same.

I conjecture that it has often been the suspicion that there must be underlying unity between function ascriptions in diverse fields that has led to people holding back from accepting that Wright found the key to understanding the most philosophically troublesome concept of function. I realize that many people will find a fused or unified concept of function more attractive; they will prefer an account on which it is at least clearer why diverse biological discourses use the same word, 'function'. I will spend the rest of this essay criticizing this longing for unity.

IV. FALSE UNITY

A view of functions which has many ideas in common with the view I am defending, but which holds out for more unity, is defended skillfully by Philip Kitcher in 'Function and Design' (1993).

On Kitcher's view, different modes of functional characterization are unified by the concept of 'design', where human intention and natural selection are equally sources of design. Kitcher claims that all biological attributions of function take place in a context characterized by design. But design can be relevant to attributions of function in more and less direct ways. One way is the way analysed by Wright: we can explain the presence of some component of a system in terms of what it does, in terms of a selective history. This is a 'direct' case.

There are also explanations which appeal to design more indirectly. We can consider an organic system which is, overall, the product of design, and then examine how its workings relate to 'demands' made by the environment. If some part of a system is a 'response to an identifiable selection pressure' (p. 270 this volume), then it has a function, whether we believe that component is itself the product of selection or not. The origins of the component, the reason why that particular part is there, do not enter into it.

The explanatory project in which such functions are used is similar to that of Cummins. The aim is to understand how the component plays a role in the system's dealing with its environment. So these functions are a subset of Cummins functions as originally understood—in many ways a core subset. According to Kitcher, whenever Cummins-style functional analysis is really done, there is a 'source of design in the background' (p. 271 this volume). In a science such as physiology, 'Selection furnishes a context in which the overall design is considered, and, within that context, the physiologist tries to understand how the system works' (p. 276 this volume).

I agree that many aspects of biological usage in areas at some remove from evolution are accurately described by this analysis. This seems to me to be about as good as a unified theory of functional discourse in biology can be. But I do not think it is right. Let us focus more closely on cases where design plays an 'indirect' role, in particular on the crucial cases where a part of a system makes a contribution to the system's dealing with its environment without being itself the product of selection.

There are, roughly speaking, two sources for traits of organisms which fall into this second category. The sources are *chance* and *constraint*. Kitcher's two explicit examples of traits which are part of a response to an environment's demands, but which do not have a Wright-style selective history, both involve chance. In one example, similar to examples discussed in debates over Wright's analysis, a screw falls into a machine and by chance makes an essential connection between two parts. The designer of the machine did not realize this connection was necessary, so without the luckily falling screw, the machine would not work. Kitcher says the screw has the function of making that connection.

Kitcher also discusses a biological case, in which a moth has a wing pattern that provides some camouflage from predatory birds, but which is inferior to other patterns. We know the superior patterns are genetically possible alternatives, because they are seen in low frequency in some areas. But, we discover, the superior rival patterns have never taken over the population because of a range of unlucky breaks. The better mutants have tended to arise in areas where predation is especially heavy, and so on.[4] Kitcher says that even when we find out that the camouflage pattern does not have a pure Wright-style history in this way, it is still natural to say that the pattern has the function of camouflaging the moths from predatory birds. This is still a contribution the pattern makes to the organism's response to environmental demands.[5]

Now let us look at how Kitcher's proposal handles a case from another family of unselected organic properties, properties due to constraint.

In Richard Levins's classic (1968) discussion of evolution in changing environments, he claims that the following pattern is common in invertebrates: high temperatures speed up development (as long as the temperatures are not so high that they simply break the system), and the final result is a smaller adult body size. This seems to be a physiologically inevitable consequence of at least many invertebrate metabolic systems. This fact has some interesting consequences. Consider the situation of some different types of fruit-fly. Suppose, first, as is reasonable, that the adaptive significance of size in fruit-flies has much to do with avoiding desiccation, the loss of moisture. When it is dry, you need to be somewhat bigger than normal to avoid drying out. Then it is possible for the basic facts about temperature and metabolism to either work for the fly or against it, depending on the structure of the environment.

First, suppose that the hot areas in the fly's habitat also tend to be the humid ones, and the cool ones are the dry ones. Then physiological inevitability works in the fly's favour. Whichever way the metabolism is fine-tuned, it will always be the case that when it is dry, the fly will wind up larger than it will when it is humid, and this is just what it needs. The fly gets a certain kind of developmental plasticity for free; there is a pre-established harmony between its metabolic properties and the environment.

On the other hand, if the hot areas are also the dry ones, and the cool areas are humid, then the basic facts of metabolism work *against* the fly. When it is hot and dry, and the fly needs to be larger, it will wind up small. Levins discusses two actual species of fly, which exemplify these

---

[4] See Brandon 1990 for detailed discussion of cases where environmental diversity contributes to the outcome of selection in this type of way. Some versions of this situation are cases of Simpson's paradox (Cartwright 1979).

[5] On my version of Wright's view, and probably on Wright's, this pattern does have a Wright function in any case, as long as some significant (contextually determined) range of alternatives *were* beaten out via selection. This is a consequence of the general liberalism about explanation which goes with Wright's view. This move does not trivialize the theory; there has to actually be a range of alternatives beaten out, and whether a given range is a 'significant' range is determined by the general standards applicable for causal explanation.

alternatives. Flies from a Middle East population enjoy the pre-established harmony, as there humidity is correlated with heat. Around Puerto Rico, though, the dry areas are the warmer areas, and the flies must deal with a natural antipathy between the facts of development and environment.[6]

For now let us focus on the situation of the lucky flies, the flies whose metabolism makes them big when it is good to be big and smaller otherwise. On Kitcher's view of functions, as far as I can see, the physiological facts about enzymes and reaction rates that bring about this relationship have the *function* of adjusting the flies' size to spatial variation in their environment. These physiological properties are properties of a system which is the product of 'design', and these properties are part of the way the fly deals with variable aspects of the environment. There is an identifiable environmental demand here, a selection pressure which cannot be evaded or side-stepped without large changes to the fly's basic architecture. The fly's biochemical properties provide a 'response' to this pressure, in that they are properties that produce phenotypic plasticity in the flies which enables them to deal with this environmental demand. However, these metabolic properties are entirely inevitable, given the general structure of the fly's physiology. They are the product of architectural constraint, and the fact that they work for the fly's benefit is simply a stroke of luck. Elsewhere they make the fly's life even harder.

So, I claim, a theory of biological functions which has anything to do with concepts of 'design', a theory which is not explicitly as liberal as Cummins's, should not recognize a case such as this as functional. The basic biochemical properties which cause the flies to change adult size with temperature do not have the function of altering the flies' size to deal with the problem of moisture loss. I am not saying this simply because these biochemical properties of the fly are not *always* useful. That is the case with many truly functional properties. I claim this is not a functional property because it is physiologically inevitable; it is the product of constraint.

It might be objected that so far I have just emitted some Wright-style intuitions, the intuition that an effect is not a function if it does not explain why the thing is there. We brace for what Bigelow and Pargetter (1987) called 'the dull thud of conflicting intuitions'. So I will try to justify these claims with some more theoretical considerations.

On the view I am presenting, the functions literature is heading towards a view in which the analysis of functional discourse is bifurcated, and

---

[6] The differences between flies in these two situations show up in interesting ways when flies from warm and cool areas are raised at a single temperature in the laboratory (Levins 1968: 65–9).

Wright functions and Cummins functions are both recognized. The recognition of this disunity is itself progressive. The concept of function was bequeathed to post-Darwinian science from an earlier conceptual scheme. The original concept of function probably did have a close connection to the concept of design, and was (for all I know) a fairly unified concept.

But the categories we recognize now should be determined, of course, by our own world-view. The analyses of Wright and Cummins locate functional attribution within two distinct explanatory modes which are legitimate parts of our contemporary world-view. Natural and artificial selection exist, and the attributes of various things can be explained in terms of selective histories. Complex, organized systems also exist, and have global capacities which may be explained in terms of the capacities of component parts. These are two legitimate explanatory modes within the sciences. Crucially for us, these are two *different* explanatory modes within science. There is not some single explanatory project, distinct from others, which encompasses these two modes. They are two different kinds of understanding we can have of a system. This is why I view Kitcher's proposal as offering a false unity, a unity which should be resisted in the interests of maintaining an accurate understanding of different explanatory strategies in the sciences.

I would like to approach this point from several different directions. Kitcher claims that every time Cummins-style functional characterization is (seriously) done, there is 'a source of design in the background' (p. 271 this volume). My point is that even if this is true, this should not be respected by a philosophical analysis of functions. It should not be respected because there is nothing scientifically special about contributions to capacities, *qua* contributions to capacities, in systems which are the product of design—as opposed to contributions to capacities in systems which are not the product of design. This is not to say that there are not *some* differences between capacities of components of systems that are the product of design, and capacities of components of systems that are not. Components of systems which are the product of design are often themselves the products of design—products of selection, at least. That is to say, the components of these systems often have *Wright* functions; they are there because of the effects and capacities they have. But this is an additional fact, over and above the mere fact that the component is within a system which is the product of selection. Part of the point of Wright's analysis is to stress the fact that there is a real difference between being a part of a certain kind of system and making a useful contribution to its working, on the one hand, and being in that system *because* of this useful contribution, on the other.

To put the point yet another way: Kitcher discusses the example of a contribution made by a chance arrangement of rocks to the structure of a river delta downstream. He says that on Cummins's original analysis these rocks can have the function of widening the delta, given the right specification of the system and so on. Kitcher says this is an inappropriate consequence for Cummins's view to have, and this problem is solved by restricting Cummins-style functional analysis to systems which are the products of design. My point is that even if this is intuitive, and even if it re-unifies the concept of function, it should be resisted by the philosopher of science. A contribution to a system has the same real status, *qua* contribution made to a system, whether the system is a river and its surrounds or the intricacies of human vision. The difference between the two systems is that the components of the visual system have Wright functions as well.

Let us also return briefly to Levins's lucky flies. The facts of biochemistry have a Cummins function in these flies. They make a contribution to the capacities and dispositions of the fly when confronted with a variable environment. However, they have this Cummins function when the flies are in an environment where the biochemical facts work for them and also where the biochemical facts work against them. Whether the fly is lucky or unlucky makes no difference; the biochemistry has effects on the system either way. On Kitcher's view the only case in which these effects are functions is the case in which the effects are beneficial, and help the organism meet the 'demands' of the environment. The problem here is not that this marks a distinction without a difference—in one case the biochemical facts are good and in the other they are bad; that is a real difference. The point is that attention to this difference, in this context, distorts our understanding of these systems. Kitcher's view assimilates the properties of the biochemistry of the lucky flies to those properties of the fly which have genuine Wright functions. But the lucky flies exhibit *bogus* design in this case; theirs is in no real sense a 'response' to the environment. Thus the important distinction between selected effects and fortuitous benefits is blurred.

Once Cummins functions are recognized and understood within familiar cases, which concern systems which are complex and highly adapted, such as the nervous system, a question arises concerning the links between these cases and more peripheral ones. Fairly peripheral cases include some seen in community ecology, where the function of a predator may be to regulate the numbers of some other species. Here we have already left the domain in which systems' components have Wright functions as well, on standard conceptions of evolution. Then there are extremely peripheral cases, such as the rock and the river delta. My proposal, which I think is in line with

Cummins's original attitude (1975: 764), is that once Cummins functions have been recognized and the explanatory mode which utilizes them has been understood, they should be allowed to roam freely, even into the farthest periphery.

Kitcher discusses a case where Cummins functions can be attributed, and which is not peripheral or 'stretched' by Cummins's own criteria, but in which some may want to resist functional attribution of any kind. This is the case of the contribution made by some particular mutant DNA sequence in the development of a tumour. Because the DNA sequence goes wrong in some particular way, the cancer as a whole has certain properties. It is not, Kitcher says, the function of these aspects of the mutation to produce certain characteristics in the cancer. On the view I have presented, we have to say that this is a case where components of the system have both Wright functions and Cummins functions, and some of the Cummins functions—those determined by our explanatory interest in the cancer—are opposed to the Wright functions. The Wright functions of this stretch of DNA have to do (we suppose) with regulating cell division in a particular way, which keeps the number of cells of this type at a certain level. When the mutation produces a tumour, and this tumour becomes the subject of a certain sort of investigation, the Cummins function of this bit of DNA, relative to that investigation, is a Wright *mal*function. On Kitcher's view, the only functions here are those stemming from the design properties of the system. In no sense are the causally salient effects of the cancer-causing mutation regarded as functions, even if they are part of a complex system which we want to understand. I recognize the intuitive appeal in Kitcher's view here, and this must be weighed against the arguments I have presented for the disunified view. The most important of these arguments, again, concern the need to recognize the real difference between the two modes of scientific understanding in which Wright functions and Cummins functions play a role.

Lastly, it might be asked: on my view, what reason is there to use the word 'function' for both Wright and Cummins functions? What do the concepts have in common that justifies this usage? My reply is: there is no strong reason for using the same word. Both types of function are 'explanatorily important properties of components of systems', but this is a very broad category. I doubt if linguistic reform is possible here, as both types of functional ascription are deeply embedded in biological usage. At least let philosophers do the right thing when we analyse functional characterization: let no philosopher join what science has put asunder.[7]

---

[7] My thinking on these matters has been greatly influenced by discussions with Philip Kitcher and Richard Francis. A version of this essay was presented at the Pacific APA, 1993, and discussion at that meeting was also very helpful.

REFERENCES

Bigelow, J., and Pargetter, R. (1987), 'Functions', *Journal of Philosophy*, 84: 181–96.
Boorse, C. (1976), 'Wright on Functions', *Philosophical Review*, 85: 70–86.
Brandon, R. (1990), *Adaptation and Environment* (Princeton: Princeton University Press).
Cartwright, N. (1979), 'Simpson's Paradox', *Nous*, 13: 419–37.
Cummins, R. (1975), 'Functional Analysis', *Journal of Philosophy*, 72: 741–65.
Dretske, F. (1988), *Explaining Behavior* (Cambridge, Mass.: MIT Press).
Godfrey-Smith, P. (1992), 'Adaptation and Indication', *Synthese*, 92: 283–312.
—— (1994), 'A Modern History Theory of Functions', *Nous*, 28: 344–62.
Griffiths, P. (1992), 'Adaptive Explanation and the Concept of a Vestige', in P. Griffiths (ed.), *Trees of Life: Essays in Philosophy of Biology* (Dordrecht: Kluwer), 111–31.
Hempel, C. G., and Oppenheim, P. (1948), 'Studies in the Logic of Explanation', *Philosophy of Science*, 15: 135–75.
Horan, B. (1989), 'Functional Explanations in Sociobiology', *Biology and Philosophy*, 4: 131–58.
Kitcher, P. S. (1993), 'Function and Design', *Midwest Studies in Philosophy*, 18: 379–97; reproduced as Ch. 12.
Levins, R. (1968), *Evolution in Changing Environments* (Princeton: Princeton University Press).
Mayr, E. (1961), 'Cause and Effect in Biology', *Science*, 134: 1501–6.
Millikan, R. G. (1984), *Language, Thought, and Other Biological Categories* (Cambridge, Mass.: MIT Press).
—— (1989a), 'An Ambiguity in the Notion "Function"', *Biology and Philosophy*, 4: 172–6.
—— (1989b), 'In Defence of Proper Functions', *Philosophy of Science*, 56: 288–302.
Mitchell, S. (1989), 'The Causal Background of Functional Explanation', *International Studies in the Philosophy of Science*, 3: 213–29.
Neander, K. (1991), 'The Teleological Notion of "Function"', *Australasian Journal of Philosophy*, 69: 454–68.
Salmon, W. (1984), *Scientific Explanation and the Causal Structure of the World* (Princeton: Princeton University Press).
Sober, E. (1984), *The Nature of Selection* (Cambridge, Mass.: MIT Press).
Tinbergen, N. (1963), 'On the Aims and Methods of Ethology', *Zeitschrift für Tierpsychologie*, 20: 410–33.
Van Fraassen, B. (1980), *The Scientific Image* (Oxford: Clarendon Press).
Williams, G. C. (1966), *Adaptation and Natural Selection* (Princeton: Princeton University Press).
Wright, L. (1973), 'Functions', *Philosophical Review*, 82: 139–68.
—— (1976), *Teleological Explanations* (Berkeley: University of California Press).

PART V

SPECIES

INTRODUCTION TO PART V

DAVID L. HULL

The species concept must serve two goals in biology—as a fundamental unit in evolution and a fundamental unit of classification. In biological evolution species are the things that evolve, that split successively through time to form the phylogenetic tree. Species are also one level in the taxonomic hierarchy. Organisms are grouped into species, species into genera, genera into families, and so on. Considerable unification can be brought about in biology if the basic units of classification can be made to coincide with the basic units of evolution, assuming something like 'basic units' exist. One assumption that pervades discussions of the species problem is that a single level of organization exists across all organisms, that can be properly termed the 'species level'. The goal is to discover this level and the mechanisms that produce it.

As is usually the case when a single entity must serve two or more functions, tensions arise. Evolutionary biologists are interested in finding out how the evolutionary process works. They investigate such problems as how much gene flow among various populations of the same species is necessary to keep all these populations integrated into the same species, how frequent interspecific hybridization actually is, and the effects of geographic isolation on species. Systematists are interested in recognizing species, diagnosing them, and ordering them in a taxonomic hierarchy. Stability is an important desideratum for them, because the groupings that they produce are to be used by *all* biologists, not just evolutionary biologists. De Queiroz and Donoghue (Ch. 15) show the difficulties that arise when a criterion used traditionally for higher taxa (monophyly) is extended to the species category.

From Darwin on, various systematists have resisted the identification of the basic units of evolution with the basic units of classification. One school can be called 'idealists', for want of a better term. They view species in the same atemporal way as physicists view the physical elements. They want to order living species according to something like the periodic table. For idealists species are as atemporal as gold. All atoms with the atomic number 79 are atoms of gold, regardless of place, time, origin, or condition. Biological species also seem to exhibit certain timeless relations. The goal is to find the key that can individuate and order species the way that atomic number orders the physical elements (for a present-day example of this school, see Webster and Goodwin 1996).

Evolutionary biologists want to understand the evolutionary process. If species turn out to be very cryptic and variable, integrated by a variety of mechanisms, then so be it. Systematists feel a stronger need for their species concept to be applicable than do evolutionary biologists. They must produce classifications that biologists—all biologists—can use in their work, even if this order is an over-simplification from an evolutionary perspective. For a species concept to be applicable, species must be recognizable in nature. Thus, systematists by necessity are interested in epistemological issues: for example, how we are to recognize a species as a species.

Thirty years or so ago, a school of systematists (numerical pheneticists) arose that put recognition first. Species are just one level of a hierarchy ordered according to degree of similarity. Species are groups of organisms exhibiting X amount of similarity, genera are more inclusive groups of organisms exhibiting Y amount of similarity, and so on. For them, phylogenetic descent is irrelevant. Sessile organisms with sessile propagules living in Patagonia and Norway would belong to the same species if they exhibit the same degree of morphological similarity. Their isolation from each other would pose no problems whatsoever. As strong as the desire is to make systematics as operational as possible, most systematists are willing to take on the more ambitious task of discerning species as the things that evolve.

A third issue that has arisen in the context of controversies over species is their ontological status. What sorts of things are species? A distinction that has characterized Western thought since its beginnings in ancient Greece is between individuals and classes, and the classic example of this distinction is between an individual organism (such as Gargantua) and the species to which it belongs (*Gorilla gorilla*). Gargantua was born at a particular place and time, grew to adulthood, and eventually died. His name simply denoted him. Just because he was named 'Gargantua', he need not be an especially large gorilla. Species and all higher taxa are quite a different sort of thing. They are classes whose names can be defined by a list of characters. Classes can be defined in spatio-temporal terms so that they are located at particular times and places—for example, all the gold bars in Fort Knox. But classes are important because they can be general.

The classes that are important in science are those that are spatio-temporally unrestricted, so that they can function in laws of nature. Discovering new physical elements has always been considered an important event in physics. The physical elements are part of the warp and woof of nature. In the golden years of systematics, discovering new species was also treated with some respect, especially if these species were gigantic, exotic, or closely related to *Homo sapiens*. However, after Darwin, systematists were forced to acknowledge that the number of species, past and present, is huge. Millions upon millions of species have evolved and gone extinct. Millions more will do the same in the future. If statements of the traits that characterize each of these millions upon millions of species are considered laws of nature, this notion becomes trivialized beyond redemption. No one is going to be awarded a Nobel Prize for discovering yet another species of fruit-fly.

Throughout the history of systematics, even before the advent of the concept of evolution, biologists have thought that species are in some sense 'special'. They do not represent just a certain level of similarity. Perhaps higher taxa are not 'real', but species are. First Ghiselin (1974), then Hull (1976), explained this feeling that species are different from higher taxa in terms of their being more closely akin to individuals than to classes. Like all individuals, they have a beginning and ending in time, a certain location (commonly termed the 'range' of the species), and exhibit various degrees and sorts of integration. Species are the entities over which biological laws operate.

As this distinction was exposed to intense scrutiny, systematists and evolutionary biologists have concluded that the sharp distinction between individuals and classes is too simple. Species are clearly not classes, unless 'class' is defined so broadly that everything from a bare particular to Richard Nixon becomes a class (or a set). Species are more like individuals, but individuals come in a variety of kinds. One of the topics in the recent literature on the species problem concerns the sorts of integration and cohesiveness that characterize biological species. Does a single level of integration and cohesiveness exist across all organisms, from single-celled to multi-cellular organisms, from sexual to asexual organisms, from plants and protists to animals?

As with the topics treated in earlier parts of this anthology (e.g. Parts III and IV), the issue of pluralism has entered into discussions of the species problem. Each side attempts to define the other as holding extremely radical views. Monists complain that pluralists have provided no principled reasons to limit the luxuriant growth of alternative perspectives and explanations. Species are anything that anyone chooses to make them, from the eternal and immutable species of Aristotelian philosophy to the operational taxonomic units of the numerical pheneticists or even the divinely created species of the Creationists! Pluralists, in their turn, complain that monists think that nature can be divided up in one and only one way. More than this, each monistic kind must be definable by a single criterion. For example, Kitcher (1993; reproduced above as Chapter 12) considers his unitary analysis of 'function' in terms of design 'pluralist', because two sorts of design can be found in nature.

Actually, both pluralism and monism are needed in science—the generation of new hypotheses and their critical winnowing. Sometimes scientists in a particular area get locked so firmly into a single way of viewing the world that they find it impossible to generate new solutions to unsolved problems. In such circumstances, calls for pluralism are warranted. However, sometimes scientists in a particular area are inundated with alternative explanatory schemas. Every phenomenon can be explained in myriad ways. In such circumstances a strong dose of monism can't hurt. Such a heuristic interpretation of the continuing dispute between monists and pluralists, however, is not likely to satisfy more metaphysically inclined philosophers (for general discussions, see Dupré 1993, Rosenberg 1994, and Ereshefsky 1992, reproduced here as Ch. 16).

In this connection, Mishler and Brandon (1987; reproduced here as Ch. 14) distinguish between grouping and ranking. Certain criteria are used to decide which organisms go together in the same species. For example, numerical pheneticists provide a list of characteristics for each group of organisms. Any organisms that have enough of these characteristics (are similar enough to each other) belong to the same species. In addition, numerical pheneticists define ranks by decreasing degrees of overall similarity—how similar is similar enough? A high degree of similarity is required for species, a lesser degree for genera, and so on. Mishler and Brandon, to the contrary, argue for a species definition that is monistic with respect to its grouping criterion (monophyly) and pluralistic with respect to its ranking criteria. Current debates over the species category turn on disagreements about which characteristics are to be used for grouping and which for ranking.

REFERENCES

Dupré, J. (1993), *The Disunity of Things: Metaphysical Foundations of the Disunity of Science* (Cambridge, Mass.: Harvard University Press).

by Mishler and Donoghue 1982) that species taxa as currently delimited often do not meet this criterion of individuality (even though they may meet one or both of the two criteria listed above).

Cohesion

We have designated 'cohesion' to refer to situations where an entity behaves as a whole with respect to some process. In such a situation, the presence or activity of one part of an entity need not directly affect another, yet all parts of the entity respond uniformly to some specific process (although details of the actual response in different parts of the entity may be different because of the operation of other processes). Examples of this type of causal interaction include the failure of a corporation due to a stock-market crash, developmental canalization in biological systems, and processes of density-independent natural selection. Clearly, species taxa as currently delimited may show cohesion as defined in this way, or integration, or both, or neither.

Problems with Application of Individuality to Species

It should be clear from the above examples that, despite its philosophical appeal, the 'species as individual' concept developed by Ghiselin and Hull cannot be applied in its simplistic form to most species taxa as currently delimited; nor, we would argue, could taxonomic practice be revamped so as to make it generally applicable (see Mishler and Donoghue 1982 for further arguments and examples). The major reasons for this inapplicability are two: the plethora of causal processes acting on biological entities and the lack of correspondence between either these processes or patterns resulting from them.

As pointed out by Van Valen (1982) and Holsinger (1984) among others, a great number of processes impinge on organisms and groups of organisms. A non-exhaustive list would include breeding relationships, competition, geological change, developmental canalization, symbioses, and predation. Entities can simultaneously behave as individuals with respect to different processes, at different levels of inclusiveness (Holsinger 1984). Furthermore, groups of organisms defined by aspects of individuality with respect to one process are often not congruent with groups defined with respect to a second process (Mishler and Donoghue 1982).

Mary Williams's recent attempt (1985) to link her concept of 'Darwinian subclan' with Ghiselin and Hull's formulation of species as individuals fails for both of these reasons. Her whole argument rests on the assumption that all biological species are in the domain of a legitimate interpretation of 'Darwinian subclan', or in other words, that species are Darwinian subclans. However, this amounts to the assumption that species are cohesive units with respect to (at least some) selective forces—that is, that organisms within a species are all acted upon by those same forces. This flies in the face of much of what is known about selection. For example, a species ranging over a geographical cline would hardly qualify as a Darwinian subclan. For a more theoretical example, consider the intrademic models of kin and group selection (Wilson 1980). Here the population units that are cohesive with respect to selection are generally much smaller than the local population, much less the entire species. It is possible, even likely, that species will be Darwinian subclans for some period of their existence (especially at their origin), but this does not help Williams's argument. She needs this to be generally true. However, current knowledge of evolutionary processes does not back her up.

The upshot is that species taxa often are not integrated or cohesive because of particular selective regimes. Other processes causing integration and/or cohesion of species taxa include gene flow and developmental canalization (Van Valen 1982, Mishler 1985). As mentioned above, species taxa as currently recognized may not be integrated or cohesive in any sense (although, as will be discussed below, this situation might be changed by revision of taxonomic practice). Furthermore, there is no reason to believe that reproductive processes and selective processes pick out the same units in nature (Mishler and Donoghue 1982, Holsinger 1984)—a correspondence necessary to relate Williams's Darwinian subclans to Mayr's biological species concept.

To summarize this section, it is useful to consider the nature of various examples of biological entities with differing degrees and aspects of individuality, to drive home the point that application of the simple dichotomy between individuals and classes has obscured important distinctions. Are there important biological groupings that are spatio-temporally localized but neither integrated nor cohesive? Yes, monophyletic higher taxa, called 'historical entities' by Wiley (1980), and Darwinian clans, as formalized by Williams (1970), would usually fit such a description. Mayr (1987) suggests that species often represent an intermediate kind of entity (which he terms a 'population') that have spatio-temporal localization but weak integration and cohesion. Thus the distinction made above can admit to differing degrees of integration or cohesion, ranging from strong (in a paradigmatic individual organism) to weak or absent.

Are there important biological groupings that are integrated and/or cohesive but not spatio-temporally localized? Yes, groups defined by their participation in processes, such as plant communities, pollinator guilds, trophic levels, mixed-species feeding flocks, or C_4 photosynthesizers, may be highly integrated, cohesive, or both, and yet lack any temporal boundaries. Further examples are given by polyphyletic or paraphyletic taxonomic groupings. Such groups may be cohesive because of ecological factors or shared developmental programmes, but lack a unique beginning (in the case of polyphyletic groups) or a unique end (in the case of both kinds of groups). Integration and cohesion do seem to require some form of spatio-temporal connectedness, but, as our examples illustrate, this does not imply temporal boundaries. Does it strictly imply spatial boundaries? We think it does; in any case we cannot think of any plausible examples of integrated and/or cohesive entities lacking spatial boundaries.

THE PHYLOGENETIC SPECIES CONCEPT

The search for a satisfactory concept of species is complicated by the need to simultaneously reconcile recent advances in evolutionary theory with recent advances in systematic theory, with empirical requirements of objectivity and testability, and with constraints imposed by the formal Linnaean nomenclatorial system. Before discussing one recently proposed solution, there is a need to introduce and clarify two important subjects: pluralism and the distinction between grouping and ranking.

Pluralism

As a number of authors have pointed out, controversies in evolutionary biology over causal agents generally do not involve claims that all but one favoured agent are impossible. Rather, a number of causal agents are acknowledged to be possible, and controversy centres around which agent is the 'most important' (Gould and Lewontin 1979, Beatty 1985).

The result of this situation in evolutionary biology has been a number of calls for 'pluralism', meaning generally to keep an open mind about which particular causal agent is to be invoked as an organizing principle in any particular case. The case of species concepts has heard similar calls (Mishler and Donoghue 1982, Kitcher 1984a, b).

However, in the case of species, two very different sorts of 'pluralism' have been advocated; thus confusion has resulted. Both sorts of pluralism are based on the fact that many different (and non-overlapping) groups of organisms are functioning in important biological processes (see discussion by Holsinger (1984, 1987)). Both sorts of pluralism deny that a universal species concept exists. However, they differ in their application to particular biological cases. Kitcher's (1984*a*, *b*) brand of pluralism implies that there are many possible and permissible species classifications for a given situation (say the *Drosophila melanogaster* complex), depending on the needs and interests of particular systematists. In contrast, Mishler and Donoghue's (1982) brand of pluralism implies that a single, optimal, general-purpose classification exists for each particular situation, but that the criteria applied in each situation may well be different. This latter meaning of pluralism, we would argue, is close to the use of the term by Gould and Lewontin (1979). Furthermore, we would also argue that its use results in perfectly reasonable and rigorous scientific solutions to particular problems. The only caveat is that problems (such as difficult species complexes) that seem at least superficially similar may require different criteria for solution.

Ghiselin (1987) has unfortunately confused these two uses of 'pluralism' and tarred them both with a broad brush. Also unfortunately, he has engaged in *ad hominem* attacks (by suggesting that pluralists are lazy, incompetent, dishonest, and generally not engaged in science at all) and fallacious arguments. Despite his unsupported assertion that the biological species definition is 'fully applicable to plants', numerous botanists (and others) have published careful empirical and theoretical analyses of the difficulties with applying the biological species concept (see Mishler and Donoghue 1982 for references). Problems having to do with lack of correspondence between patterns resulting from different causal processes, and the gradual nature of breeding discontinuities in plants, cannot be waved aside casually.

To further distinguish between the two meanings of 'pluralism' and to clarify the proper usage of the term with respect to biological theories, it is necessary to examine connections with the concept of parsimony. It is natural and correct for scientists to have a bias towards monism, because of the fundamental scientific tenet of economy in hypotheses. Hull's (1987) arguments for consistency in using cessation of gene flow as a uniform definition of the species category carry a lot of weight (see also arguments by Sober (1984)). The burden of proof rests squarely on someone who argues that the current domain of explanation of a monistic theoretical concept must be broken into smaller domains, each with its own explanatory concept. Note that this sort of pluralism (which is the sort advocated by Gould and Lewontin (1979) and Mishler and Donoghue (1982)) is 'pluralistic' only during the transition as a prevailing monistic concept is broken up. Once controversy settles and the transition is complete, you are left with a greater number of explanatory concepts, each quite monistic within its proper domain. Parsimony considerations weigh in balance against the need to provide proper explanations for biological diversity. As scientists, we strongly attempt to minimize the number of theoretical concepts (to one if possible) allowed to delimit (for example) basic taxonomic units. Yet we should grudgingly grant status to additional concepts if the need for them is proved in particular cases.

This use of pluralism is clearly not the use advocated by Kitcher (1984a, b). He implies a sort of 'permanent pluralism', where an indefinitely large number of theoretical concepts (limited only by interests of particular biologists) remain acceptable within a single domain. We share the scepticism of Sober (1984), Hull (1987), and Ghiselin (1987) towards this meaning of pluralism. Its use with respect to species concepts would seem to rob systematics of any objective way of choosing between conflicting classifications or of any use of species as units of comparison. Therefore, in what follows, we use 'pluralism' in the sense of Mishler and Donoghue (1982).

Grouping versus Ranking

All species concepts must have two components: one to provide criteria for placing organisms together into a taxon ('grouping') and another to decide the cut-off point at which the taxon is designated a species ('ranking'). This distinction (as detailed by Mishler and Donoghue (1982), Donoghue (1985), and Mishler (1985)) has often not been recognized (but see a similar distinction made by Mayr 1982: 254). Taking the biological species concept as an example, its grouping component is 'organisms that interbreed'. But since such groups are found at many levels of inclusiveness, especially if 'potentially interbreeding' is added to the grouping criterion, a ranking component is needed which usually is something like 'the largest grouping in which effective interbreeding occurs in nature'.[2]

Since both components are implicit in any adequate species concept, confusion is likely to result if the distinction between them is ignored. Thus Hull's (1987) argument that using patterns of gene flow to define species will result in 'a consistently genealogical perspective' is unsound. It depends on whether reproductive criteria are used for grouping or for

[2] As pointed out by Hull (personal communication), when the distinction between grouping and ranking has previously been made, it was often blurred. This may often be because researchers use variations on the same theme for both grouping and ranking: e.g. patterns of morphological similarity or of gene exchange. As will be apparent below, we advocate distinctly different criteria for grouping than for ranking.

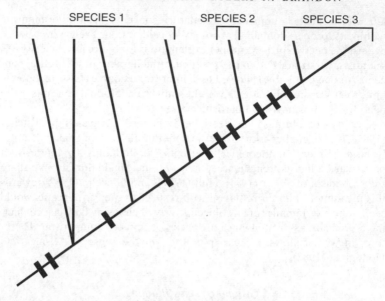

FIG. 14.1. A hypothetical cladogram showing three named species. Synapomorphies are shown as cross-bars; autapomorphies are not shown. Species 1 is paraphyletic.

ranking. Both Rosen (1979) and Donoghue (1985), among others, have nicely shown that the use of reproductive criteria in grouping can easily result in non-monophyletic taxa, in contrast to the genealogical units Hull (along with us) hopes for. The 'recognition concept of species' (Paterson 1985), wherein species are defined by the possession of a common fertilization system, suffers from a similar problem, in that non-monophyletic taxa often result (see Fig. 14.1, where species 1 may well be definable by reproductive criteria, but is not monophyletic).

Further objections to various prevailing species concepts have been given by Mishler and Donoghue (1982), Donoghue (1985), and Mishler (1985). These authors made the following points. (1) None of the dozens of species concepts held currently by various authors can provide grouping criteria able to produce truly genealogical species classifications (including, curiously enough, species concepts advocated by cladists, a group dedicated to genealogical classification). (2) In order to reflect the diversity of causal agents directing evolutionary differentiation in different lineages, no universal ranking criterion can be found.

An Alternative Concept of Species

An alternative perspective on species as genealogical, theoretically significant taxa has been developed by Mishler and Donoghue (1982), Donoghue (1985), and Mishler (1985), and called the 'phylogenetic species concept' (not to be confused with the concept proposed by Cracraft (1983), with the same name). This concept explicitly recognizes a grouping and a ranking component, is monistic with respect to grouping, yet pluralistic (in the sense advocated above) with respect to ranking, and produces species taxa with at least some aspects of individuality.

The grouping criterion advocated by Mishler and Donoghue is monophyly in the cladistic sense. Further discussion of the meaning of 'monophyly' is needed (see below), because the term is not normally applied to species in a substantive way by cladists. For now it suffices to say that 'monophyly' here is taken to refer to a grouping that had a single origin and contains (as far as can be empirically determined) all descendants of that origin.

Monophyletic groupings as roughly defined above exist at all levels of inclusiveness; thus a ranking criterion for species is needed as the basal systematic taxon (i.e. the least inclusive monophyletic group recognized in a particular classification). It is here that Mishler and Donoghue have advocated a pluralistic adjustment in the number of ranking criteria allowable for consideration in particular cases. They argued that the currently favoured monistic ranking concept of absolute reproductive isolation is not the most appropriate for all groups of organisms. The ranking concept to be used in each case should be based on the causal agent judged to be most important in producing and maintaining distinct lineages in the group in question. The presence of breeding barriers might be used, but so might selective constraints or the action of strong developmental canalization (Mishler 1985). In the great majority of cases, little to nothing is actually known about any of these biological aspects. In such cases grouping (estimation of monophyletic groups) will proceed solely by study of patterns of synapomorphy (i.e. shared, derived characters), and a practical ranking concept must be used until something becomes known about biology. This preliminary and pragmatic ranking concept will usually be the size of morphological gaps (i.e. number of synapomorphies along any particular internode of a cladogram) in most cases, a concept in accord with current taxonomic practice.

The phylogenetic species concept (PSC) of Mishler and Donoghue can be summarized as follows:

A species is the least inclusive taxon recognized in a classification, into which organisms are grouped because of evidence of monophyly (usually, but not restricted to, the presence of synapomorphies), that is ranked as a species because it is the smallest 'important' lineage deemed worthy of formal recognition, where 'important' refers to the action of those processes that are dominant in producing and maintaining lineages in a particular case.

Relating the PSC back to the earlier discussion of individuality, it is clear that species so defined (as with monophyletic taxa at all levels) will at least meet the restricted spatio-temporal criterion of individuality. They may or may not be integrated or cohesive. However, these criteria may often prove useful in ranking decisions. Since the strength of integrative or cohesive bonds tends to gradually weaken as more and more inclusive groups of organisms are taken (see e.g. the discussion in Mayr 1987), it may be possible in many cases to objectively fix the species level as the *most* inclusive monophyletic group that is integrated or cohesive with respect to 'important' processes. Again, 'important' has a context-dependent meaning, and will often not refer to reproductive criteria. It may often be difficult to apply this standard, especially if macro-evolutionary processes occur (even rarely) involving groups at high taxonomic levels (Gould 1980, Jablonski 1986). If so, integrated and/or cohesive groups may occur at *much* more inclusive levels than anyone would wish to name as basal taxonomic units.

The problem of (at least partial) non-comparability of species taxa in different groups of organisms is a real one (Sober 1984, Hull 1987, Ghiselin 1987). However, as pointed out by Mishler and Donoghue (1982), this has always been the case, despite the fact that many users of species taxa—ecologists, philosophers, palaeobiologists, biogeographers, for example—remain blissfully unaware. This difficult situation has not come about because (as suggested by Ghiselin (1987)) systematists working with organisms other than birds are incompetent, but rather reflects a fact of nature. The pluralistic ranking concept of the PSC was proposed to allow different biological situations to be explicitly treated. Persons interested in studying some biological process simply cannot avoid the responsibility of learning enough about the systematics of the organisms they are studying to ensure that the entities being compared are truly comparable with respect to that process.

To take one example that has been widely recognized (Mayr 1987), asexual organisms present insurmountable difficulties for the biological species concept. One proposed solution has been to deny that such organisms form species (Bernstein *et al.* 1985, Eldredge 1985, Hull 1987, Ghiselin 1987). This *reductio ad absurdum* of the biological species concept demonstrates how a monistic ranking (and grouping) concept based on inter-breeding criteria can obscure actual patterns of diversification. One of us happens to work on a genus of mosses (*Tortula*, see Mishler 1985 for references), in which frequently sexual, rarely sexual, and entirely asexual lineages occur. The interesting thing is that the asexual lineages form species that seem comparable in all important ways with species recognized in the mostly asexual lineages and even in the sexual lineages.[3] It just happens in this case that potential interbreeding or lack thereof seems of little or no importance in the origination and maintenance of diversity. The application of the PSC here is able to reflect an underlying unity that the biological species concept could not.

Indeed, there seems to be a fundamental confusion at the heart of the biological species concept and its insistence that only sexual organisms can form species. *Potential* interbreeding and the lack thereof (i.e. breeding barriers) can be observed in nature, and so can be used as a ranking criterion for species. But why should it be so used, or rather, why should it be the only ranking criterion used? We suspect that part of the rationale stems from a confusion over the roles of potential interbreeding and actual interbreeding.

Actual interbreeding is a process. It results in lineages (but not always lineages important enough to be named species—for example, short-lived hybrid populations). The process of (actually) interbreeding also inevitably leads to a certain amount of integration. In sexual species it undoubtedly is one of the important processes holding the species together. But *potential* interbreeding is not a process, and therefore has no effect on the integration or cohesion of species. The dispersed parts of a sexual species are not bound together by this non-process; they may be bound together by sharing common environments or common developmental programmes, but they cannot be bound together by 'potential interbreeding'.

In general, the potential to interbreed is based on organisms sharing common environments and common developmental programmes. The processes that result in groups of organisms sharing such features and in discontinuities between such groups are multifarious, and are not restricted to sexual organisms. Organisms share common developmental programmes because they share a common ancestor. Reproduction is a relevant process here, but not necessarily sexual reproduction.

---

[3] A similar result has been arrived at by Holman (personal communication), based on comparisons between bdelloid rotifers (which are exclusively parthenogenic) and monogonont rotifers (which occasionally reproduce sexually). Using numbers of synonymous species names as an index of taxonomic distinctness of species, he has shown that bdelloid species are apparently more consistently recognized by taxonomists than are monogonont species.

It is our argument that the PSC is superior to the biological species concept (or to the evolutionary species concept of Simpson (1961) and Wiley (1978), which is similar in these ways to the biological species concept) in two fundamental ways. First, monophyly as a grouping criterion is superior to ability to interbreed, because it will lead to a consistently genealogical classification. Second, the pluralistic ranking concept of the PSC is superior to the monistic insistence on breeding barriers of the biological species concept because it can more adequately reflect evolutionary causes of importance in different groups.

Other cladistic species concepts, such as the 'phylogenetic species concept' of Cracraft (1983), which is very similar to the species concept of Nelson and Platnick (1981), are also inferior to the PSC of Mishler and Donoghue, but for somewhat different reasons. The grouping concept used by the former authors (i.e. a cluster of organisms defined by a unique combination of primitive and derived characters) does not rule out the possibility of paraphyletic species, unlike the PSC (see next section). Furthermore, the concepts of Cracraft and Nelson and Platnick (in addition to the concept of Rosen (1979), that *does* use presence of synapomorphies as a grouping criterion) are incomplete, in that they lack a ranking criterion. It is not sufficient to say that a species is the smallest diagnosable cluster (Cracraft 1983) or even monophyletic group, because such groups occur at all levels, even *within* organisms (e.g. cell lineages). Some judgement of the significance of discontinuities is needed.

Monophyly

One final area in need of clarification is the concept of monophyly. Traditionally, the cladistic definition of monophyly (which we favour) has not been applied to the species level. Henning (1966) did not do so because he was committed to a biological species concept, and thought that there was a clean break at the species level, with reticulating genealogical relationships predominating below and diverging genealogical relationships predominating above. Later cladists (e.g. Wiley 1981) have followed Hennig and defined a monophyletic taxon as one that originated in a single species and that contains all descendants of that species. Species are taken to be monophyletic a priori; therefore it is argued that they need not possess synapomorphies or really be monophyletic in the sense of higher taxa (e.g. Wiley 1981). One major reason for this is the supposed problem on 'ancestral' species.

It is our view that, properly clarified, there are no insurmountable problems with applying the concept of monophyly explicitly to species (as the basal systematic taxon). Furthermore, this application *must* be carried out in order to have a consistently genealogical classification.

Monophyly should be redefined in such a way as to apply to species:

A monophyletic taxon is a group that contains all and only descendants of a common ancestor, originating in a single event.

'Ancestor' here refers, not to an ancestral species, but to a single individual. By 'individual' here, we do not necessarily mean a single organism, but rather an entity (less inclusive than the species level) with spatio-temporal localization and with either cohesion or integration or both (as defined above). In particular cases this ancestral individual could be a single organism, a kin group, or a local population. We would argue that it would never be a whole species, because we share the widespread view that new species come about only via splitting, not by any amount of anagenetic change.

The originating 'event' of a monophyletic group referred to in the definition above could be due to the spatio-temporally restricted action of a number of different causes. These could include, in different cases, the origin of an evolutionary novelty which causes a new monophyletic group to be subject to a different selective regime than the rest of the 'parent' species or which causes a disruption of the normal developmental canalization of the 'parent' species. These could also include acquisition of an isolating mechanism or even the origin of a new species by hybridization between parts of two 'parent' species. This diversity of causes for evolutionary divergence reinforces the need for a pluralistic ranking concept.

Some examples of the application of this concept should clarify the definition. It is thought at the present time that a common mode of speciation is via peripheral isolation. In such a case, the peripherally isolated part of the species, if spatio-temporally localized (say, on the same island at the same time) and either cohesive, integrated, or both (say, by inter-breeding and sharing a common niche), would qualify as a monophyletic group under our definition. This would be true even if several rather unrelated members of the original species were the founders of the peripheral population, as long as the above conditions obtain. On the other hand, if two similar but *non*-spatio-temporally connected peripheral populations (say, on two different islands) have been established by members (even closely related ones) of the original species, these two populations would have to be considered as two separate monophyletic groups. They are two separate monophyletic groups, because they originated in two different events. Hybrid speciation provides similar examples. If two original species produce a hybrid population in one place (say, a single valley) at one time (say, in a single breeding season), and if this hybrid population behaves as an integrated and/or cohesive entity, then it is a perfectly good monophyletic group under our definition. However, if similar hybrids are produced elsewhere in the ranges of the two original species, or if hybrids are produced in the *same* locality but discontinuously in time (i.e. if the first hybrid population goes extinct *before* the new hybrids are produced), then the separate hybrid populations would have to be considered as separate monophyletic groups, and could not be taken together and named as a new species. Note that this conclusion is directly opposite that of Kitcher (1984*b*: 314–15). The implications of our concept of monophyly for the *original* species in the above examples will be discussed below.

This concept of monophyly is, of course, only a grouping criterion. It does not imply that any particular peripheral isolate or hybrid population *must* be recognized as a species. It only specifies the genealogical conditions under which such groups *can* be recognized if the ranking criterion applied in a particular case supports recognition at the species level. The grouping and ranking criteria can thus be seen to interact in producing a species classification. Note that a corollary of the PSC is that not all organisms will belong to a formal Linnaean species, since some monophyletic groups (e.g. hybrid populations that arise, but then quickly go extinct) will not be judged to be 'important' monophyletic groups. The hybrid organisms in such a case would not formally belong to either original species.

The definition of monophyly given above solves the problem perceived by Hennig (1966), Wiley (1981), and Cracraft (1983) with 'ancestral species'. No such things exist. Only parts of an original species give rise to new ones, as in the above examples. If a currently recognized species is found to be paraphyletic, because parts of it can be demonstrated to be more closely related to another species (Fig. 14.1; see also discussions and diagrams of such a situation in Bremer and Wanntorp 1979, Avise 1986), then the paraphyletic species should be broken up into smaller monophyletic species.

Note that if species 1 (Fig. 14.1) is actually integrated by gene flow, then over time its cladistic structure should approach that of species 1 in Figure 14.2. Moreover, over an even longer time in such a truly integrated species, patterns of character distribution should even out such that no autapomorphies remain to distinguish lineages within the species, and species 1 would be represented in a cladogram by a single line (albeit still without any synapomorphies to distinguish it as a species). In systematic studies, a situation is frequently encountered (Fig. 14.2) in which a number of unre-

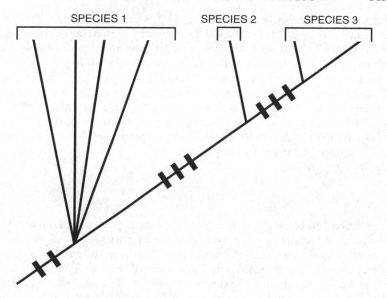

Fig. 14.2. A hypothetical cladogram showing three named species. Synapomorphies are shown as cross-bars; autapomorphies are not shown. Species 1 is metaphyletic.

solved lineages exist, one or more of which are deemed worthy of recognition as separate species, and the rest of which have traditionally been considered a species taken together. This type of situation has been confused with paraphyly. However, it is actually a case of a taxon (e.g. species 1 in Fig. 14.2) with an uncertain status between paraphyly and monophyly. With further study, synapomorphic characters may be found uniting some part of species 1 with the lineage of species 2 and 3 (as in Fig. 14.1). If that becomes the case, species 1 truly is paraphyletic and must be broken up. On the other hand, further study may demonstrate synapomorphies uniting all of the lineages in species 1, thus making it an unproblematic phylogenetic species.

It has been cogently argued by Donoghue (1985) that a group such as species 1 in Figure 14.2 could acceptably be named a species in a tentative and pragmatic way, pending further study designed to resolve the relationships, as long as a special convention was followed to indicate the uncertain status of the species (Donoghue suggests marking the binomial name of all such species with an asterisk). This solution is practical, because it avoids unnecessary naming of highly localized species (if, for example, all recognizable lineages in species 1, Fig. 14.2, were formally named). It is also probably unavoidable, since if speciation by peripheral isolation occurs frequently, such situations may often be in principle unresolvable, as discussed above. Donoghue (1985) suggested calling this type of species a 'metaspecies', to clearly distinguish it from a known monophyletic species. Following the prefix he suggested, we suggest the need for a new term, 'metaphyly', to refer to the status of groups that are not known to be either paraphyletic or monophyletic. Although beyond the scope of the present essay, this term would clarify similar situations with respect to higher taxa, and may thus prove more widely useful.

CONCLUSION

The 'species problem' as discussed here involves a search for a definition of the basal systematic unit that will be at once practical, provide optimal general-purpose classifications, and reflect the best current knowledge about evolutionary processes. We have claimed that the PSC will fulfil these criteria. However, we certainly have not claimed that *all* important biological entities can be recognized using the PSC.

As pointed out clearly by Holsinger (1984), a multitude of interesting biological entities, often non-overlapping, are behaving as (at least partial) individuals with respect to a multitude of interesting processes in any particular group of organisms. While we do need to settle on criteria for recognizing formal taxa for our Linnaean taxonomic system (including species), we are of course in no way prohibited from informally naming and studying other entities of interest that do not fit the formal taxonomic system—that is, as long as different types of entities are explicitly distinguished from each other.[4]

REFERENCES

Avise, J. C. (1986), 'Mitochondrial DNA and the Evolutionary Genetics of Higher Animals', *Philosophical Transactions of the Royal Society, London*, B312: 325–42.

[4] We dedicate this essay to Ernst Mayr, even though he probably disagrees with much of its contents. At different times and in different ways, we both were profoundly affected by our interactions with him during our graduate careers at Harvard. We thank him for his advice, insights, and patience. We also thank David Hull and Marjorie Grene for comments that helped to clarify certain aspects of the paper. Eric Holman kindly allowed us to cite his unpublished data on rotifers.

Beatty, J. (1985), 'Pluralism and Panselectionism', in P. D. Asquith and P. Kitcher (eds.), *PSA 1984*, ii (East Lansing, Mich.: Philosophy of Science Association), 113–28.

Bernier, R. (1984), 'The Species as Individual: Facing Essentialism', *Systematic Zoology*, 33: 460–9.

Bernstein, H., Byerly, H. C., Hopf, F. A., and Michod, R. E. (1985), 'Sex and the Emergence of Species', *Journal of Theoretical Biology*, 117: 665–90.

Bremer, K., and Wanntorp, H.-E. (1979), 'Geographical Populations or Biological Species in Phylogeny Reconstruction?', *Systematic Zoology*, 28: 220–4.

Cracraft, J. (1983), 'Species Concepts and Speciation Analysis', *Current Ornithology*, 1: 159–87.

Donoghue, M. J. (1985), 'A Critique of the Biological Species Concept and Recommendations for a Phylogenetic Alternative', *Bryologist*, 88: 172–81.

Eldredge, N. (1985), *Unfinished Synthesis: Biological Hierarchies and Modern Evolutionary Thought* (New York: Oxford University Press).

Ghiselin, M. J. (1974), 'A Radical Solution to the Species Problem', *Systematic Zoology*, 23: 536–44.

——(1987), 'Species Concepts, Individuality, and Objectivity', *Biology and Philosophy*, 2: 127–43.

Gould, S. J. (1980), 'Is a New and General Theory of Evolution Emerging?', *Paleobiology*, 6: 119–30.

——and Lewontin, R. C. (1979), 'The Spandrels of San Marco and the Panglossian Paradigm: A Critique of the Adaptationist Programme', *Proceedings of the Royal Society, London*, B205: 581–98.

Haffer, J. (1986), 'Superspecies and Species Limits in Vertebrates', *Zeitschrift für Zoologische Systematics und Evolutionsforschung*, 24: 169–90.

Hennig, W. (1966), *Phylogenetic Systematics* (Urbana, Ill.: University of Illinois Press).

Holsinger, K. E. (1984), 'The Nature of Biological Species', *Philosophy of Science*, 51: 293–307.

——(1987), 'Discussion: Pluralism and Species Concepts, or When Must We Agree With Each Other?', *Philosophy of Science*, 54: 480–5.

Hull, D. L. (1976), 'Are Species Really Individuals?', *Systematic Zoology*, 25: 174–91.

——(1978), 'A Matter of Individuality', *Philosophy of Science*, 45: 335–60.

——(1984), 'Can Kripke Alone Save Essentialism? A Reply to Kitts', *Systematic Zoology*, 33: 110–12.

——(1987), 'Genealogical Actors in Ecological Roles', *Biology and Philosophy*, 2: 168–84.

Jablonski, D. (1986), 'Background and Mass Extinctions: The Alternation of Macroevolutionary Regimes', *Science*, 231: 129–33.

Kitcher, P. (1984a), 'Against the Monism of the Moment', *Philosophy of Science*, 51: 616–30.

——(1984b), 'Species', *Philosophy of Science*, 51: 308–33.

Kitts, D. B. (1983), 'Can Baptism Alone Save a Species?', *Systematic Zoology*, 32: 27–33.

——(1984), 'The Names of Species: A Reply to Hull', *Systematic Zoology*, 33: 112–15.

Mayr, E. (1982), *The Growth of Biological Thought* (Cambridge, Mass.: Harvard University Press).

Mayr, E. (1987), 'The Ontological Status of Species', *Biology and Philosophy*, 2: 145–66.

Mishler, B. D. (1985), 'The Morphological, Developmental, and Phylogenetic Basis of Species Concepts in Bryophytes', *Bryologist*, 88: 207–14.

——and Donoghue, M. J. (1982), 'Species Concepts: A Case For Pluralism', *Systematic Zoology*, 31: 491–503.

Nelson, G., and Platnick, N. I. (1981), *Systematics and Biogeography: Cladistics and Vicariance* (New York: Columbia University Press).

Paterson, H. E. H. (1985), 'The Recognition Concept of Species', in E. S. Vrba (ed.), *Species and Speciation*, Transvaal Museum Monograph, 4 (Pretoria: Transvaal Museum,), pp. 21–9.

Rieppel, O. (1986), 'Species are Individuals: A Review and Critique of the Argument', *Evolutionary Biology*, 20: 283–317.

Rosen, D. E. (1979), 'Fishes from the Uplands and Intermontane Basins of Guatemala: Revisionary Studies and Comparative Geography', *Bulletin of the American Museum of Natural History*, 162: 267–376.

Ruse, M. (1987), 'Species: Natural Kinds, Individuals, or What?', *British Journal of the Philosophy of Science*, 38: 225–42.

Simpson, G. G. (1961), *Principles of Animal Taxonomy* (New York: Columbia University Press).

Sober, E. (1984), 'Discussion: Sets, Species, and Evolution: Comments on Philip Kitcher's "Species"', *Philosophy of Science*, 51: 334–41.

Van Valen, L. M. (1982), 'Integration of Species: Stasis and Biogeography', *Evolutionary Theory*, 6: 99–112.

Wiley, E. O. (1978), 'The Evolutionary Species Concept Reconsidered', *Systematic Zoology*, 27: 17–26.

——(1980), 'Is the Evolutionary Species Fiction?—A Consideration of Classes, Individuals, and Historical Entities', *Systematic Zoology*, 29: 76–80.

——(1981), *Phylogenetics: The Theory and Practice of Phylogenetic Systematics* (New York: John Wiley).

Williams, M. B. (1970), 'Deducing the Consequences of Evolution: A Mathematical Model', *Journal of Theoretical Biology*, 29: 343–85.

——(1985), 'Species are Individuals: Theoretical Foundations for the Claim', *Philosophy of Science*, 52: 578–90.

Wilson, D. S. (1980), *The Natural Selection of Populations and Communities* (Menlo Park, Calif.: Benjamin Cummings).

15

PHYLOGENETIC SYSTEMATICS AND THE SPECIES PROBLEM

KEVIN DE QUEIROZ AND MICHAEL J. DONOGHUE

[T]he task of 'ordering' (and what means the same thing, of systematics) lies in considering the unit as a member of an ordered whole. It is a fact . . . that no unit exists as a member of only one whole.

Therefore it is possible to arrange animated natural things in numerous different systems, depending on which of these different relationships has been investigated. The differences among all these systems are determined by the particular relationships of which they are a concrete expression. All these different systems are, fundamentally, equally justified so long as they are a proper expression of the membership position that an object of nature possesses within the framework of the totality, for the dimension that was chosen as the basis for the particular system.

The different systems . . . are not unrelated to one another. The relations between them . . . can themselves be made the subject of scientific systematic investigation. On the other hand, it is not basically a scientific task to combine several systems so created, because one and the same object cannot be presented and understood at the same time in its position as a member of different totalities.

Hennig, *Phylogenetic Systematics*

INTRODUCTION

Darwin established the fact of evolution—the process of descent with modification—and its product, phylogeny. Although he predicted that taxonomies would become, 'as far as they can be so made, genealogies' (Darwin 1859: 486), the widespread acceptance of evolution did not lead to a major re-evaluation of the goals, principles, and methods of taxonomy. Instead, existing taxonomies simply were reinterpreted in evolutionary terms. That is, the reality of previously recognized taxa was taken for

First published in *Cladistics*, 4 (1988): 317–38. Reprinted by permission. The authors share equal responsibility; order of authorship is arbitrary.

granted, and evolutionary concepts and mechanisms were formulated to account for their existence (Stevens 1984, de Queiroz 1988).

During the 'modern synthesis' several authors, Mayr and Simpson in particular, explored the link between taxonomy and evolutionary theory. Their widely accepted conclusion was that species are fundamentally different from taxa at both higher and lower categorical levels. Species, unlike other taxa, are not only an outcome of evolution; they actually function in a direct way in the evolutionary process: as gene pools in the case of Mayr, and as lineages extending through time in the case of Simpson. Species were seen to exist as wholes—that is, to be real things—whereas other taxa were viewed as subjective and arbitrary (Mayr 1963: 600–1, 1969*b*: 91–2; Simpson 1961: 188–91).

From the perspective of developing evolutionary systematics, perhaps the most significant aspect of the views of Mayr and Simpson was that existing species taxa were not taken as given. Although these concepts may have been formulated initially as theories to explain the existence of groups having common morphologies or ecologies, they quickly became prescriptions about how the species category should be defined, and as such they necessitated a re-evaluation of the status of existing taxa (Donoghue 1985). Because the species category was defined in such a way that its members would be participants in the evolutionary process, the basal taxonomic unit became a fundamental evolutionary unit (e.g. Simpson 1961; Hull 1965, 1976, Mayr 1969*a*, 1982).

This outlook contrasts sharply with an alternative view in which species concepts are treated as theories meant to explain the existence of already recognized taxa (e.g. Mishler and Donoghue 1982: 494), a perspective that has hindered the development of systematics. By accepting the reality of previously recognized taxa, concepts associated with important biological processes are relegated to the role of after-the-fact explanations for the existence of these taxa, instead of functioning as central tenets from which real entities and the methods for their discovery are deduced (cf. de Queiroz 1988).

Hennig (1966) did for the development of evolutionary systematics above the 'species level' what Mayr and Simpson had done with regard to 'species'. That is, he changed the role of evolution as it relates to 'higher' taxa, from an after-the-fact interpretation of the order already manifest in taxonomy to a central tenet from which he deduced what entities exist as its natural outcome (de Queiroz 1985). According to Hennig, the products of evolution above the 'species level' are groups composed of ancestral species and all of their descendants—complete systems of common ancestry—clades—monophyletic groups. Inasmuch as monophyletic groups are a natural outcome of the process of evolutionary descent, they are real and exist as wholes outside of the minds of taxonomists.

Hennig's concept of monophyly was seen by some later authors to have implications not only for taxa at 'higher' categorical levels but also for those at the 'species' level. In particular, Rosen (1978, 1979) and Bremer and Wanntorp (1979) argued that reproductive compatibility might be lost in a mosaic pattern among the populations descended from a common ancestor in such a way that the ability to interbreed, as a retained ancestral trait, would be uninformative about recency of common ancestry. Consequently, if organisms or populations were assigned to species taxa on the basis of this ability, then some species would be paraphyletic. This conclusion has led some authors to argue against species concepts based on interbreeding and to develop species concepts based on monophyly (Mishler and Donoghue 1982; Cracraft 1983, 1987; Ackery and Vane-Wright 1984; Donoghue 1985; Mishler and Brandon 1987; McKitrick and Zink 1988). They argue that there is not (or at least there should not be) a basic difference between species and other taxa; some monophyletic groups are simply more inclusive than others.

In short, a tension has developed around species concepts that involves ideas central to evolutionary biology in general and phylogenetic systematics in particular (cf. Løvtrup 1987: 172–3). Here we explore some manifestations of this tension and their significance for phylogenetic systematics, especially as they bear on a choice among alternative species concepts. Nevertheless, we advocate neither a new species concept nor any existing one. Instead, we develop a way of *looking at* the species problem that builds upon the conceptualization of systematics expressed in the epigraph. Central to this view is a consideration of different kinds of entities that exist in nature and their relationships to one another.

MONOPHYLY

Tension between the significance of interbreeding and common descent is evident in discussions of the kinds of entities to which the concept of monophyly properly applies. Some arguments simply define the conflict out of existence. Platnick (1977), Willmann (1983) and Ax (1987), for example, considered it inappropriate to enquire whether species are monophyletic, paraphyletic, or polyphyletic, claiming that these terms apply only to groups of species. This position unnecessarily restricts the concept of monophyly, and overlooks the fact that species themselves are 'groups' (groups of organisms). Regardless of precedents set by previous authors, there is no biological reason not to view monophyly, paraphyly, and polyphyly as general concepts wherein the units of common ancestry are unspecified. Thus, these terms can be applied not only to groups of species, but also to groups of any entities that reproduce and thus form ancestor–descendant lineages. Under this view it is legitimate to ask whether a particular organism is or is not a monophyletic group of cells, whether a particular population is a monophyletic group of organisms, or whether a particular species taxon is a monophyletic group of populations—as legitimate as it is to enquire whether a particular 'higher' taxon is or is not a monophyletic group of species.

Wiley (1977, 1979) attempted to resolve the conflict between interbreeding and monophyly in another way. He claimed that species 'are a priori monophyletic by their very nature' (Wiley 1979: 214). In effect, his proposition is that because species have 'a real existence in nature', therefore they are monophyletic. But this implies that there is only one kind of existence. If 'species' and monophyletic groups exist in different ways, then 'species' can exist without being monophyletic.

Other authors allow that it is legitimate to enquire whether species are monophyletic, but, unlike Wiley, they conclude that some species—namely, ancestral ones—are paraphyletic. Brothers (1985) coupled this idea with the notion that asexual organisms form evolutionary species (sensu Simpson 1961, Wiley 1978), and concluded that paraphyletic higher taxa are meaningful evolutionary groups. This follows from his assertion that the relationship between asexual species and their component organisms is analogous to that between higher taxa (including paraphyletic ones) and their component species.

Brothers's argument hinges on the false premiss that paraphyletic sexual and asexual 'species' exist in the same way. Paraphyletic asexual 'species', however, are not unified by interbreeding, as are sexual 'species'; instead, they are defined solely by phenetic similarities and gaps (Brothers 1985: 36). In fact, the only connection between sexual and asexual 'species' in Brothers's argument is that both are supposedly accommodated under the evolutionary species concept. The evolutionary species concept (Simpson 1961), however, refers to 'a single lineage of ancestral descendant populations' (Wiley 1978: 18); and to equate the kinds of lineages formed by sexual and asexual organisms under the term 'evolutionary species' is to confuse two different uses of 'population'. Only the unjustified acceptance of phenetically delimited, paraphyletic collections of asexual organisms as 'real evolutionary species' supports Brothers's contention that paraphyletic higher taxa are acceptable evolutionary groups (see Donoghue 1987 for additional discussion).

The arguments of Wiley and of Brothers are similar in one important respect—both tacitly assume that different kinds of entities exist in the same way: monophyletic groups and 'species' in the case of Wiley, sexual and asexual 'species' in the case of Brothers. Others—for example, Eldredge and Cracraft (1980)— have argued that there is a fundamental difference between sexual species and monophyletic higher taxa. They allow that some species—namely, ancestors—are not monophyletic, but they consider this to be acceptable because species exist in a different way: namely, as individuals. For Eldredge and Cracraft (1980: 90), monophyletic groups exist, but are not necessarily individuals, whereas species exist *because* they are individuals.

INDIVIDUALITY

The concept of individuality has figured prominently in many recent discussions of species concepts, including several of those discussed above. That organisms are not the only kind of biological 'individuals' follows from accepting that living matter is organized into wholes that are themselves parts of more inclusive wholes. Although Ghiselin (1966, 1974, 1981, 1985) and Hull (1976, 1977, 1978) deserve credit for popularizing and developing the idea that species are appropriately viewed as individuals in the philosophical sense, very similar ideas were set forth independently by Hennig (1966) and Griffiths (1974), whose discussions of the individuality of biological taxa stem from the writings of even earlier authors (i.e. Woodger 1952, Gregg 1954).

The concept of individuality is commonly illustrated by contrasting individuals with classes and describing characteristics of each (Ghiselin 1974; Hull 1976, 1977, 1980, 1981). Classes have members; individuals have parts. Classes are spatio-temporally unrestricted; individuals are localized in space and time. The names of classes are usually defined 'intensionally' (i.e. by listing the attributes that are necessary and sufficient for membership); the names of individuals are proper names, and can only be defined 'ostensively' (i.e. by showing the object to which the name is given). The members of a class are similar, in that they share at least the attributes that define the class name; the parts of an individual need not be, and frequently are not, similar. Beyond this general characterization, however, there are more and less restricted concepts of individuality. Thus, according to Hull (1978) and Wiley (1981), individuals must not only be spatio-temporally localized but also must be continuous and cohesive. These last two terms require special attention, as they bear directly on the existence of different kinds of entities that have organisms as parts.

Continuity

There are at least two different forms of continuity: current and historical. Wiley (1981) made current continuity an explicit component of his concept of individuality. He did not, however, distinguish between current continuity and cohesion, for he considered both to result from the same process (at least in sexual species): namely, reproductive ties among organisms. In contrast, Ghiselin (1974) explicitly rejected current continuity as a necessary component of individuality, arguing that the United States of America is an individual nation, despite the physical discontinuity between Alaska and the remainder of the continental United States. The truth of this example notwithstanding, at least some kinds of individuals (e.g. multicellular organisms) result from direct physical connections among their parts, and in these cases continuity is inescapable. It appears, then, that whether current continuity is a necessary component of individuality depends upon the nature of the phenomenon conferring individuality.

Historical continuity has been identified as the unbroken chain of descent from a common ancestor (e.g. Ghiselin 1980, Wiley 1981). While this applies to some kinds of individuals (e.g. monophyletic groups), it does not seem to be a necessary component of individuality. An organism, for example, does not cease to be an individual when it receives an organ transplant; nor does a population of interbreeding organisms cease to be an individual when it receives immigrants. As in the case of current continuity, it seems that whether historical continuity is necessary for individuality depends on the nature of the phenomenon conferring individuality.

Cohesion

The presence or absence of cohesion has been considered an important difference between 'species' and monophyletic higher taxa (e.g. Hennig 1966, Wiley 1981, Ghiselin 1985). Unfortunately, ambiguities still plague this critical issue, some of which are clarified by considering the meaning of 'cohesion' and the biological phenomena that might confer it.

'Cohesion' is commonly used to mean 'sticking together' (e.g. *Webster's New International Dictionary*, 2nd edn.); thus, cohesion is a property that might confer individuality by uniting parts to form a whole. The cells that make up a multi-cellular organism are physically stuck together, but at the 'species level' cohesion is less obvious. According to Wiley (1981), cohesion among the parts of a species composed of sexually reproducing organisms is maintained by reproductive ties (see also Brooks and Wiley 1986: 48–9). In contrast with the biological species concept (e.g. Mayr 1942), however, only actual interbreeding matters in this context (cf. Hull 1965). If cohesion is conferred by interbreeding, then the potential to interbreed allows only the potential to cohere. That interbreeding is widely considered to be the process conferring 'species level' cohesion is evident from the commonly stated view that asexual organisms do not form 'species' (e.g. Bernstein et al. 1985: 328). As Hull (1980) put it, 'strictly asexual organisms form no higher-level entities; organism lineages are the highest level lineages produced'.

Other than sexual reproduction, no biological process has been identified that might confer cohesion at the 'species level'. Although interactions other than interbreeding seem to confer cohesion on groups of organisms that make up colonies or symbiotic partnerships, these entities are never called 'species'. Several other phenomena have been suggested as 'species level' agents of cohesion, but such proposals confuse cohesion with constraint or inertia. Wiley (1981), for example, considered that stasis maintains cohesion among the parts of 'species' composed of either sexual or asexual organisms (see also Mishler 1985, Mishler and Brandon 1987). Stasis may result from either extrinsic or intrinsic *constraints*, such as stabilizing selection or the resilience of developmental systems. Although such phenomena may cause organisms to remain similar, this is not the same as 'sticking together'. When discussing biological individuals having organisms as their parts, cohesion must refer to interactions among those organisms. Shared genetic or developmental programmes, or common mate recognition systems (Paterson 1978, 1985), or any other properties that organisms might have in common, no matter how biologically significant, are not interactions among those organisms.[1] Although some of these properties may allow cohesive interactions to occur among organisms, they do not, by themselves, constitute cohesion.

Although cohesion has often been associated with individuality, it is not required by every version of the concept. Thus, according to Ghiselin (1974), an individual is simply 'a particular thing'. This is compatible with the view taken by Hennig (1966), Patterson (1978), Ghiselin (1969, 1980,

---

[1] Vrba (1985) suggested a 'fundamental compatibility between the "individual" and "recognition concepts" of species', at least in part because both 'draw on the reproductive activities among organisms'. Nevertheless, under Paterson's recognition concept, species are classes defined by similarities and differences in characters that make fertilization *possible*; actual reproduction, or even actual mating, are not required (see Donoghue 1987).

1985), Griffiths (1974), and Hull (1976) that monophyletic higher taxa are individuals, despite the fact that they do not exhibit cohesion among their parts, each being made up of independently evolving lineages (Wiley 1980, 1981). Wiley (1980, 1981) stressed this basic distinction by coining the term 'historical group' for monophyletic higher taxa (which are historically continuous), and restricting 'individual' to cohesive entities such as 'species'. Ghiselin (1985) accepted this distinction, but preferred a classification in which 'individual' includes both non-cohesive historical groups and cohesive units (which he called 'integrated wholes').

We conclude that several different kinds of entities have been called individuals. Consequently, the individuality 'revolution' (Ghiselin 1987) may be partially responsible for obscuring significant distinctions between them. The view has developed that individuals are things with a real existence in nature; for this reason, if something is said to be an individual, it seems to gain significance. Simply asserting that something is an individual, however, does little to clarify the nature of its existence. Inasmuch as one kind of individual may be significant for one theory but not for another, it is necessary to go beyond individuality and answer the question 'individual what?'. In the next section our aim is to focus attention away from individuality *per se*, and instead to explore those phenomena that confer existence on certain entities that have been identified as individuals.

SYSTEMS

The nature of existence of wholes is clarified by adopting the perspective of systematics formulated by Griffiths (1974) and has been discussed recently be de Queiroz (1988). These authors distinguished between classification, the ordering of entities into classes, and systematization, the ordering of entities into systems. Classification and systematization differ fundamentally, in that classes are groups whose members belong to those groups because they share some attribute(s), whereas systems are wholes that derive their existence from some natural process through which their parts are related (de Queiroz 1988). Ghiselin (1974) pointed out that the term 'individual' can designate systems at various levels of integration, which suggests that the different kinds of entities previously identified as individuals might be viewed as kinds of wholes deriving their existence from different underlying natural processes. This perspective facilitates discrimination among different kinds of individuals by focusing directly upon the natural processes responsible for their existence.

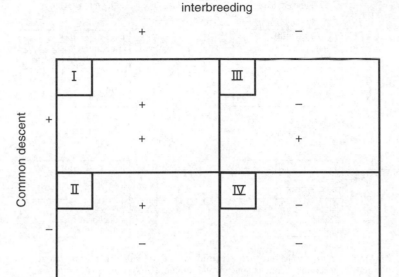

FIG. 15.1. Possible relations between cohesive wholes resulting from the process of interbreeding and monophyletic groups resulting from the process of common descent. The presence of each process is symbolized by +, and absence by −.

We have identified two processes through which organisms are related: interbreeding and common descent. Exploration of the systems resulting from these different processes and the relations between them is facilitated by constructing a table with interbreeding (resulting in one kind of cohesive whole) along one axis and common descent (resulting in monophyletic groups) along the other (Fig. 15.1). Entities in the upper left-hand box of this table (labelled I) are both cohesive and monophyletic; entities in the lower left-hand box (II) are cohesive but not monophyletic; and so forth. The word 'individual' has been applied by one author or another to entities in each of the first three boxes. Any entities in box IV are either systems deriving their existence from some natural process other than interbreeding or common descent, or they are recognized simply because their members share certain traits, and therefore are classes. In the latter case, regardless of the importance of the traits upon which such groups are based, they do not qualify as systems. Unless their parts are related through some natural process, such classes do not exist as wholes. Included

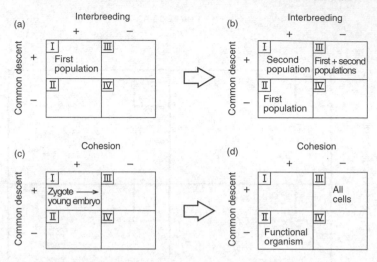

Fig. 15.2. Entities followed through time as categorized in Fig. 15.1. (a) A young population descended from a single gravid female. (b) System (a) after the establishment of a new population. (c) A young embryo descended from a single zygote. (d) System (c) after some cells have died and been sloughed off.

here are paraphyletic and polyphyletic higher taxa, which must be viewed either as aggregations or collections of less inclusive wholes, or as parts (incomplete systems) of more inclusive wholes.

Because two different processes are being considered, a group of entities that forms a system resulting from one of the processes may or may not also form a system resulting from the other (cf. Holsinger 1984). This point is easily visualized by using the table in Figure 15.1 to follow groups of entities through time (Fig. 15.2). Suppose, for example, that we begin with a gravid female of some kind of sexually reproducing organism. She and her offspring establish a population within which there is steady interbreeding between component organisms. For the sake of simplicity, let us further suppose that no deaths occur. For a time this population resides in box I—it is both a monophyletic group of organisms and a group that is cohesive as a result of interbreeding (Fig. 15.2a). Now imagine that at some later time another gravid female leaves this population and successfully establishes a new population that is geographically separated from the first, so that the two populations are reproductively isolated by distance. At this point (Fig. 15.2b) the first population shifts into box II—it is no longer a monophyletic group of organisms; however, it remains a cohesive entity. The first and second populations, taken as a unit, now occupy box III, because together they constitute a monophyletic but non-cohesive group. Finally, the newly established population begins its existence in box I.

The point of this exercise is that there may be switches between boxes or states of existence, and one is free to focus attention on entities belonging to any of the classes in the table of possibilities. Thus, we might choose to focus on interbreeding systems, on those resulting from common descent, or on both. There is no right or wrong in this; one is not better than another, or generally more significant. The entities in the upper row of boxes and those in the left-hand column all exist, but they exist in different ways: that is, they exist as the outcome of different processes. Furthermore, in box I, wholes deriving their existence from one of the processes correspond precisely with (have the same parts as) wholes deriving their existence from the other process.

It is worth noting that the framework developed above is a general one, which is to say that other forms of cohesion and common descent may occur at different levels of organization. For example, instead of following groups of organisms, one might focus on groups of cells (Fig. 15.2c, d). Following the first few mitotic divisions, the group of cells making up an embryo is integrated into a cohesive whole by physical and chemical interactions; these also form a monophyletic group of cells descended from the zygote. This group of cells therefore exists in box I (Fig. 15.2c). At a later time during development (Fig. 15.2d), some cells die and are sloughed off the embryo (or perhaps the embryo is split into two cohesive wholes, as in the case of identical twins). After this point we might choose to follow the fate of the functioning organism, which remains a cohesive whole, but is no longer a monophyletic group of cells. Alternatively, we might focus on the set of *all* cells descended from the zygote, even though these are no longer all integrated in one functioning body.[2] Traditionally, attention has been focused on the cohesive organism, but there may be some purposes for which it is necessary to keep track of the monophyletic group of cells—for example, in studying the frequency of somatic mutations.

The foregoing analysis emphasizes that the tension surrounding species concepts results from there being different kinds of real biological entities.

[2] A monophyletic group consists of an ancestor and all of its descendants. Thus, the dead and sloughed-off cells are part of the monophyletic group of cells descended from the zygote, although they are no longer part of the functioning organism. Similarly, monophyletic groups of organisms include dead organisms, even though these are no longer parts of interbreeding populations, and monophyletic 'higher taxa' include extinct and unknown subgroups.

Some of these entities exist as an outcome of a process conferring cohesion, while others exist as an outcome of descent from a common ancestor. And sometimes an entity that exists as the consequence of one of these processes happens to correspond exactly with one that exists as a consequence of the other. Before we can explore how these conclusions bear on the species problem, it is first necessary to examine some assumptions and limits of phylogenetic systematics.

PHYLOGENETIC SYSTEMATICS

Adopting the view that systematics is the discovery of entities that derive their existence from some underlying natural process implies that phylogenetic systematics is that kind of systematics in which the process of interest is evolutionary descent (de Queiroz 1988). The methods of phylogenetic systematics are based on the premiss that there exists an evolutionary tree and, therefore, a nested hierarchical pattern of relationships. This implies that it is inappropriate to apply cladistic methods to entities that are expected *not* to be related in a nested hierarchical pattern: that is, entities related in some other pattern, such as a reticulum of intersecting sets. In other words, there are identifiable limits to the sensible application of phylogenetic methods, boundaries beyond which it is fruitless to proceed.

The exact nature of these limits depends on the properties of the entities under investigation. In the case of sexually reproducing organisms, a limit is set by the level at which continually branching (diverging) relations give way to predominantly reticulate relations resulting from interbreeding. It is inappropriate to enquire about phylogenetic relationships among actually interbreeding organisms, because here the pattern of relationships is not a nested hierarchy (cf. Hennig 1966: 18–19). Phylogenetic methods break down in this case, because an assumption underlying the principle that shared derived characters provide evidence of phylogenetic relationship (i.e. of monophyly) is violated. Thus, in the case of sexual dimorphism, grouping by shared derived characters may lead to the false conclusion that the males (for example) within a population of interbreeding organisms form a monophyletic group. The problem in this case is that sex-linked traits of the males are being interpreted as synapomorphies at the wrong level, a fact that would become evident upon examining the distribution of these traits among parents and their offspring.

Populations themselves, by contrast with their component organisms, may show a branching pattern of relationship to one another. Indeed, using populations as terminal taxa will potentially yield the finest possible resolution of phylogenetic relationships among sexually reproducing organisms. Populations, therefore, have a special role as 'basal units' in the phylogenetic systematics of organisms.[3] This role is entirely independent of whether these units are monophyletic, but instead is an outcome of the process of interbreeding.

In the case of organisms that reproduce only asexually, the limits of phylogenetic analysis are different. Here, in contrast to the reticulate relationships that result from sexual reproduction, the pattern of common ancestry among asexual organisms forms a nested hierarchy. Whether asexual organisms are monophyletic or paraphyletic groups of cells, relationships among them are amenable to phylogenetic analysis, because these organisms are cohesive wholes that form diverging lineages.[4]

Hennig (1966: 29–32) delimited the scope of phylogenetic systematics in distinguishing parts of 'the total structure of hologenetic relationships'. His figure 6 (our Fig. 15.3) shows semaphoronts linked into semaphoront groups (individual organisms) through ontogenetic relationships, and organisms linked through 'tokogenetic relationships' into species. Phylogenetic relationships were limited by Hennig to those above the level of interbreeding groups—to relationships among 'species'. Most of Hennig's discussion assumed a sexual mode of reproduction. Regarding cases of asexual reproduction, he noted that the differences between ontogenetic, tokogenetic, and phylogenetic relationships are blurred. Nevertheless, he concluded that even in asexual groups 'it is possible to delimit in the fabric of hologenetic relationships an area that lies between the more or less

---

[3] Throughout this essay we mean by 'population' units within which interbreeding between organisms of different sub-units is sufficient such that the relationships among these sub-units are reticulate, while relationships among the units themselves are predominantly diverging. Consequently, the units that qualify as populations depend upon the time-scale under consideration. Over short time periods, there may be a diverging pattern of relationships among demes, and this is potentially recoverable through cladistic analysis. This pattern of relationships, however, may be obliterated over longer periods of time if there is sufficient gene flow among demes so that they function together as a single population.

[4] The cohesion responsible for the existence of individual organisms, whether sexual or asexual, does not involve reticulate patterns of descent among their component parts. Therefore, phylogenetic analysis can be extended down to the level of cells, or even to parts of cells (e.g. organelles, chromosomes, 'genes'). This is done, for example, in analysing the propagation of somatic mutations within and between the meristems of plants (Whitham and Slobodchikoff 1981, Klekowski et al. 1986; see Buss 1983, 1987, for examples in animals). However, parts that reproduce can begin diverging prior to the divergence of more inclusive wholes. This occurrence, in conjunction with differential sorting of variant parts within higher-level lineages, can result in non-correspondence between phylogenetic relationships among entities at different levels (Kawata 1987). For example, mitochondria, which form lineages perpetuated through maternal germ cells, can exhibit different patterns of phylogenetic relationships from the populations of organisms in which they reside (Neigel and Avise 1986).

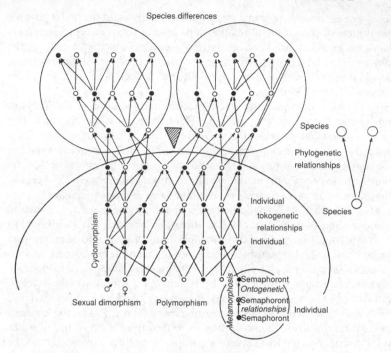

Fig. 15.3. The total structure of hologenetic relationships and the differences in form associated with its individual parts. After Hennig 1966: 31, fig. 66. Reprinted with the permission of University of Illinois Press.

unequivocally phylogenetic relationships on the one hand and the ontogenetic relationships on the other', and that 'this area naturally corresponds to the species category of organisms with bisexual reproduction' (Hennig 1966: 44).

Hennig's discussion of hologenetic relationships in sexual organisms is insightful, as is his recognition that the difference between reproduction and development is not always entirely clear in the case of asexual organisms (cf. Janzen 1977 and the 'genet'/'ramet' terminology of botanists, e.g. Harper 1977). Nevertheless, we disagree with his views on the status of asexual 'species' and the limits of their phylogenetic relationships. In asexual organisms tokogenetic relationships have a fundamentally different structure than they do in sexual forms, each organism being the direct descendant of one, rather than two parents. In such cases there are no systems deriving their existence from interbreeding as there are in sexually reproducing organisms. Consequently, in obligately asexual groups, phylogenetic relationships correspond precisely with tokogenetic relationships, both being relationships among individual organisms (i.e. life cycles *sensu* Bonner 1974).

SPECIES

If we endeavour to practise systematics in the sense of Griffiths, then species names (or the names of any systematic taxa) should refer to the individual members of one of the classes of entities that exist as the outcome of some natural process. But this still leaves open different possibilities, because distinct classes of entities relevant to phylogenetic systematics derive their existence from both interbreeding and common descent.

We will illustrate these possibilities with a hypothetical situation. Suppose that we have identified all of the separate populations within a particular monophyletic group and that the phylogenetic relationships among these populations have been assessed using cladistic methods (Fig. 15.4). In actuality, the relationships might be more completely resolved than those shown in Figure 15.4, but for the sake of the following discussion we

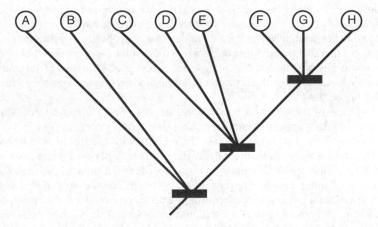

Fig. 15.4. A cladogram of eight populations (A–H); interbreeding occurs within each population, but not among populations. Although certain monophyletic groups of populations exist, the populations themselves are not necessarily monophyletic.

will assume that the organisms in some of the populations are not differentiated from one another, and therefore some relationships cannot be resolved. Indeed, we expect that cladograms of populations will often be less than fully resolved (Arnold 1981, Donoghue 1985; also see discussion below of direct ancestry under 'Species Concepts Based on Monophyly'). This case provides a framework for considering several possibilities for the application of the term 'species'. We will use it to illustrate the consequences of adopting each of several alternative species concepts. It is not our intent, however, to advocate one of these concepts over the others. Instead, we accept the validity of each one and explore its implications for phylogenetic systematics and taxonomic conventions.

Species Concepts Based on Interbreeding

One possibility, which might be considered even without any knowledge of cladistic relationships, would be to apply species names to each of the eight separate populations (A–H, Fig. 15.4). This alternative focuses on the systems that exist as a result of interbreeding at the present time, without considering what might happen to them in the future or their phylogenetic relationships to one another. In effect, this is a narrow version of the biological species concept.

Equating species with actually interbreeding groups of organisms would be useful to many biologists, since these entities are presumed to play a special role in the evolutionary process (e.g. Futuyma 1986). Furthermore, the entities recognized as species under this concept are significant from the perspective of phylogenetic systematics, since, as we argued above, populations are the least inclusive units appropriate for use as terminal taxa when analysing phylogenetic relationships among sexually reproducing organisms.

In view of the fact that populations are not always monophyletic, this concept might appear to entail a double standard concerning the criterion of monophyly. This is not the case. In keeping with the tradition in which species are seen as fundamentally different from other taxa, the names of species simply would designate an entirely different kind of entity than the names of other taxa in the phylogenetic system (de Queiroz 1988). The 'higher' taxa, as systems of common ancestry, would be members of the category 'monophyletic group', but members of the species category, as interbreeding systems, might not be monophyletic. In short, there would be two different classes of systems formally recognized as taxa. That groups of actually interbreeding organisms are not always monophyletic is not, by itself, a reason to avoid designating such groups as species; evolutionary descent is not the only process through which organisms are related, nor is monophyly the only form of existence.

Perhaps the main difficulties with this species concept are practical ones. It is often very difficult to determine the limits of actual interbreeding, especially since the degree of gene flow varies in space and time, and there need be no correspondence between interbreeding and morphological or ecological divergence (Mishler and Donoghue 1982, Donoghue 1985). Beyond this methodological problem, adoption of this concept would probably lead to conflicts with traditional species taxa. If species names were applied to all separate populations, there would be many more species than are currently recognized. Furthermore, organisms that reproduce exclusively by asexual means could not be considered to be parts of species.

There is a well-known alternative to applying species names to actually interbreeding groups of organisms: namely, to have species names represent potentially interbreeding groups of organisms—the broad (and standard) version of the biological species concept. This alternative is conceptually related to the first, and because there is presumably a continuum of reproductive interactions—from frequent to rare to none at all—these two concepts grade into one another.

In order to explore this alternative, suppose that in addition to the information represented in Figure 15.4 we also know the potential of organisms in each of the eight populations to interbreed with one another and produce fertile offspring. In particular, suppose that members of populations A–E can successfully interbreed (even though they are not actually interbreeding), and that members of populations F–H also can interbreed among themselves, but that interbreeding is not possible between organisms from the two different groups of populations (Fig. 15.5). If the species category is defined on the basis of the potential to interbreed, then species names would be given to these two groups of populations (A–E and F–H).

Delimiting species on the basis of the potential to interbreed is appealing, in that it attempts to capture the idea that species exist through evolutionary time rather than being manifestations of current gene flow. Moreover, loss of the potential to interbreed guarantees that the entities are functioning as separate evolutionary units. In these respects, the potentially interbreeding species concept is similar to the evolutionary species concepts of Simpson (1961) and Wiley (1978), which emphasize the existence of species through time by viewing them in terms of their fates as lineages. One might argue, for example, that populations among which there is potential but currently no actual interbreeding might come back in

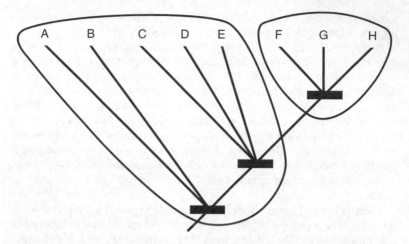

FIG. 15.5. A cladogram of separate populations (as in Fig. 15.4) showing potentially interbreeding groups. Organisms within the enclosed groups of populations (A–E and F–H) can potentially interbreed; interbreeding is not possible between organisms from the different groups.

contact in the near future, at which time there would be sufficient gene flow that the populations would fuse, and any differentiation between them would disappear. In other words, given enough time, these populations would be in contact often enough that they would function together as a single unit in evolution.

Despite this appeal, defining the species category in terms of potential interbreeding also has theoretical drawbacks. Units recognized strictly on this basis need not be, and perhaps often will not be, cohesive in the short run or even in the long run. Species based on potential interbreeding may be simply collections or classes, the members of which are functioning and will always function as separate units in the evolutionary process. Consequently, the processes responsible for 'speciation' (i.e. irreversible reproductive closure) under this concept are not necessarily the same as those responsible for the origin of separate evolutionary units. Furthermore, as noted earlier, potentially interbreeding groups defined solely by the retained ability to interbreed might be paraphyletic; in other words, they might not be systems of common ancestry any more than interbreeding systems. Such demonstrably paraphyletic groups (e.g. populations A–E in Fig. 15.5) obscure information on common ancestry, which in turn hinders the study of historical biogeography and character evolution. It is not clear how the recognition of such units, which are neither cohesive nor monophyletic, and which are delimited on the basis of what might or might not occur in the future, can be used in testing theorics about evolutionary processes (W. Maddison in Vlijm 1986).

Potential interbreeding as a criterion for circumscribing species has practical advantages over the first alternative, because it avoids the technically difficult task of assessing which organisms are actually interbreeding with one another. Furthermore, in contrast to giving species names to populations, it probably would not greatly increase the number of species now recognized, and might even substantially reduce the number in some groups. Nevertheless, defining the species category in terms of potential interbreeding is plagued by its own practical difficulties, particularly when it is viewed as an attempt to identify separately evolving lineages. It is, after all, difficult to determine which organisms will and will not be able to interbreed successfully on the basis of morphological, behavioural, or ecological similarities and differences, and the results of laboratory experiments cannot always be extrapolated to natural circumstances. But even if these problems could be solved, it still would be difficult, if not impossible, to predict future developments such as the duration of persistence of potential interbreeding or changes in geographic ranges that might bring populations into contact. Information about such developments must be available if separately evolving lineages are to be identified accurately, and to the extent that the future cannot be predicted, lineage concepts of species can only be applied retrospectively.[5]

Species Concepts Based on Monophyly

A second set of possibilities focus on evolutionary descent. Here species taxa are some subset of those groups thought to be monophyletic, whether or not they are cohesive. Thus, species would be systems of the same sort as 'higher' taxa in the phylogenetic system, and the species category would designate one rank in a hierarchy, all the ranks of which would be applied to monophyletic taxa. The process of delimiting such species might proceed as before, with the identification of appropriate basal units (populations in the case of sexually reproducing organisms) and the assessment of phylogenetic relationships among them (Fig. 15.4). Under the requirement

---

[5] Throughout this essay, 'lineage' refers to a single ancestor–descendant sequence, and is not to be equated with 'monophyletic group'. Monophyletic groups are often composed of multiple lineages (de Queiroz 1988).

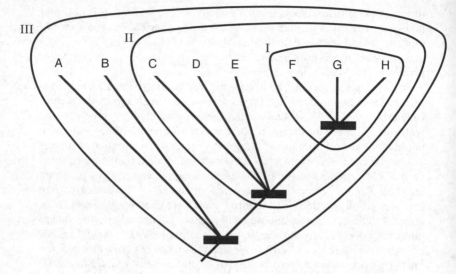

FIG. 15.6. A cladogram of separate populations (as in Fig. 15.4) showing three monophyletic groups of populations (I–III).

that all taxa, including species, be monophyletic groups, the groups labelled I, II, and III in Figure 15.6 would qualify. But which one(s) of these monophyletic groups ought to be assigned to the species category?

One possibility is to recognize as species all and only the smallest (least inclusive) monophyletic groups—either individual populations or groups of populations. In our example, the clade labelled I would therefore be recognized as a species, but clades II and III could not be species for at least two reasons. First, they are not the smallest monophyletic groups, and second, recognizing one or both of them (as well as clade I) as species would result in species nested within one another, which would take away the meaning of categorical ranks altogether. Thus, if clade I is a species, then clades II and III must be 'higher' taxa, in which case the lowest-ranking monophyletic taxon to which any of the populations A–E could be assigned would be a 'higher' taxon. In short, it will be possible to assign all organisms/interbreeding populations to one or more monophyletic taxa, but it will not be possible to assign all such entities to monophyletic taxa of species rank.

This conclusion is not simply a function of having chosen at the outset to recognize only the smallest monophyletic groups as species; the same result obtains even when more inclusive monophyletic groups are recognized as species. For example, we might choose to recognize clade II as a species, but then it would not be possible to assign populations A and B to a monophyletic taxon of the species category. Neither does the problem result from incomplete information about phylogeny, for some population(s) may be ancestral to others, and hence paraphyletic. Although identification of ancestral populations is generally a difficult task, such populations presumably exist. Even if their status as direct ancestors cannot be demonstrated, they are likely to appear in cladograms as parts of unresolved polytomies or as single branches without diagnostic apomorphies.

Although not assigning all organisms or populations to taxa of species rank violates a long-standing convention, this alone is insufficient grounds for rejecting a definition of the species category based on monophyly. If the goal of systematics is to depict relationships accurately, then any traditions that interfere with this goal should be abandoned.

There is, however, a way of emphasizing monophyly in the definition of the species category while also providing for the assignment of the vast majority of organisms to species taxa. This is achieved by introducing a convention that allows the recognition as species of single basal entities, or groups of basal entities, whose monophyletic status is uncertain (Donoghue 1985). For example, the relationships of populations C, D, and E in Figure 15.4 are unclear. Together they may form a monophyletic group, or this group may be a paraphyletic assemblage—characters support neither hypothesis (Fig. 15.7). Following Donoghue (1985) and Gauthier et al. (1988), potential paraphyly should be distinguished from demonstrated paraphyly, in which there is evidence that some populations are more closely related to populations placed in another taxon. Demonstrably, paraphyletic groups would not be recognized as taxa under this convention. Nevertheless, the following kinds of groups might be recognized temporarily on the grounds that they may be monophyletic: (1) populations lacking autapomorphies, or (2) groups of populations that are not differentiated from one another and that lack the diagnostic apomorphies of any clade(s) nested within the least inclusive monophyletic group to which they belong (e.g. the group of populations [C, D, E] in Fig. 15.6). If this proposal were adopted, it might be desirable to give such groups of uncertain status a special designation (Donoghue (1985) proposed 'metaspecies', and suggested that their names be marked with an asterisk). These measures are intended to ensure that such taxa would be treated cautiously until their relationships are better understood. The 'meta-species' convention allows an unambiguous reflection of phylogenetic

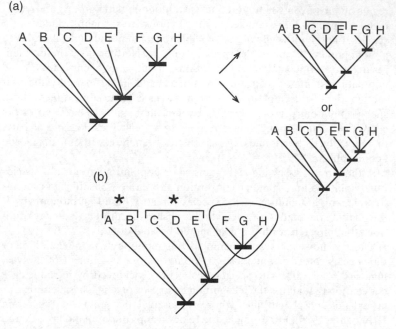

FIG. 15.7. Cladogram of populations (as in Fig. 15.4). (a) Possible resolutions of the unresolved group of populations C–E. In the upper resolution, C–E is found to be monophyletic; in the lower resolution, C–E is found to be 'positively' paraphyletic. (b) Application of the 'metaspecies' convention, where asterisks symbolize metaspecies A–B and C–E; populations F–H form a 'positively' monophyletic species.

relationships, in so far as these are known, while also allowing unresolved groups of organisms to be assigned to taxa of species rank.

Nevertheless, if the criterion of monophyly is to be applied consistently to basal entities as well as to groups of such entities, then it still would not be possible to recognize as species (or monophyletic taxa of any rank) populations that are known to be ancestral to others (perhaps through direct observation of an immigrant establishing a new population). Such populations are paraphyletic (although not necessarily as evidenced by characters), and therefore would not be covered by the metaspecies convention, because they contradict the fundamental idea of a species concept based on monophyly. As pointed out by Hennig (1966: 71), an ancestral species, before it gives rise to any descendants, is equivalent (as a monophyletic group) to the group composed of itself and all the species descended from it considered at a later time. Thus, the ancestral population of a monophyletic group recognized as a genus is part of that genus but not of any less inclusive monophyletic taxon. Nevertheless, given the difficulty of identifying ancestors (Wiley 1981), leaving these unassigned to taxa of species rank is unlikely to cause great practical problems.

The case in which 'species' refers to the least inclusive monophyletic groups has several other practical difficulties. One problem is that it is not an easy task to construct cladograms using populations (or organisms) as terminal units (cf. Arnold 1981), and this degree of resolution is far from being achieved in most groups. Another problem is that there would probably be more species recognized if this approach were adopted than we are presently accustomed to; for example, many groups that are presently accorded subspecific ranks would qualify as species (Cracraft 1983). In asexual organisms the situation would be even more extreme.

A more general objection to defining the species category as one level in the hierarchy of monophyletic groups arises from considering the usefulness of categorical ranks. Linnaean taxonomy requires two distinct activities: grouping, the discovery/identification of groups, and ranking, the assignment of a Linnaean categorical rank to each one of these groups (Donoghue 1985).[6] In systematics (*sensu* Griffiths), the activity of grouping corresponds with systematization itself, but the significance of ranking is less clear. Although monophyletic taxa exhibit a nested, hierarchical pattern of relationships, which is exactly the same kind of pattern used in Linnaean taxonomies, categorical ranks add no information about monophyly that is not already contained in a cladogram or an indented taxonomy (Eldredge and Cracraft 1980; Gauthier et al. 1988).

There is also the problem that the very existence of categorical ranks encourages spurious comparisons between entities assigned the same rank but that are not otherwise comparable (Gauthier et al. 1988). One possible solution to this problem is to have ranks reflect the absolute ages of groups (Hennig 1966), but this proposition has not been accepted by most systematists. Another possibility suggested by Hennig (1969) and Griffiths (1974, 1976), among others, is that the categorical ranks of Linnaean taxonomy be abandoned. If ranking serves no purpose other than perpetuating

---

[6] The term 'grouping' is unfortunate, in that it seemingly implies that the 'groups' are formed by some human activity, rather than existing independently and being discovered by humans. Furthermore, the term 'group' itself, as in 'monophyletic group' and 'historical group', is unfortunate, in that it seems to imply that such entities are collections rather than more inclusive wholes. At best, such usage reflects viewing the whole not from its own level but from the level of the less inclusive entities (parts) of which it is composed.

tradition, the difficulties associated with it provide a compelling reason for considering the possibility of abandoning ranks.

It is ironic that the possibility of eliminating ranks, which arose from considering 'species', like other taxa, to be monophyletic groups, reopens the alternative of a species concept based on interbreeding. If 'species' simply denotes one hierarchical rank within the category 'monophyletic group', and if ranks are unnecessary, then why should some monophyletic groups be called 'species'? In other words, while the abandonment of categorial ranks in no way hinders the representation of monophyletic groups, it also frees the term 'species' to represent some other, entirely different category. And, if freed in this way, why not use the term to designate some kind of interbreeding group?

A Disjunctive Species Concept

An alternative to species concepts based on interbreeding, as well as those based on common descent, is to base the concept on both of these processes—a disjunctive definition of the species category (cf. Hull 1965, Løvtrup 1987). Under such a concept, species *either* would be populations (whether monophyletic or not), *or* they would be monophyletic (but not para- or polyphyletic) groups of populations. This alternative allows all populations/organisms to be assigned to species-level taxa. It has other implications as well. First, to the extent that only populations are recognized as species in sexually reproducing groups, it converges on the approach of recognizing every separate population as a species. Second, some 'species' would be different kinds of entities than other 'species', as well as taxa at other levels in the phylogenetic system. From the viewpoint of cladistic analysis, these considerations may seem unimportant, because both populations and monophyletic groups are appropriate terminal taxa. In any case, this problem could be remedied by introducing a new set of conventions (along the lines of 'metaspecies') to distinguish these different kinds of entities from one another. However, in view of the confusion that might be generated by mixing the processes of interbreeding and common descent, a disjunctive definition may create more problems than it solves, especially if its only benefit is maintaining the tradition of assigning all populations/organisms to taxa of the species category.

CONCLUSIONS

Our analysis implies that neither populations nor monophyletic groups are generally more real or significant than the other; instead, their relative significance varies with the particular theoretical context. We therefore agree with the tenet that 'there is no unique relation which is privileged in that the species taxa it generates will answer to the needs of all biologists and will be applicable to all organisms' (Kitcher 1984: 309). Nevertheless, we reject the brand of pluralism that applies different criteria or even different combinations of criteria on a case-by-case (group-by-group) basis in an attempt to achieve a single, optimal, general-purpose taxonomy. Attempting to reflect a combination of processes, so as to provide species taxa significant in all contexts, will only result in confusion over what species taxa represent, and how they might be used.

From the viewpoint of phylogenetic systematics, each of the species concepts we have considered designates units that can be used as terminal taxa, and each one also has consequences. Disjunctive species concepts, because they mix different classes of systems, result in species taxa that are not comparable. Such concepts are at odds with the unambiguous representation of different kinds of systems. Species concepts based on interbreeding entail the absence of species in organisms that reproduce only asexually. Within this category of species concepts, potentially interbreeding groups of organisms may be neither monophyletic nor cohesive; that is, they may not represent unitary evolutionary entities, and they may exhibit cladistic relationships among their included populations. Species concepts based on actual interbreeding may result in recognizing as separate 'species' entities that over longer time periods function together as a single evolutionary unit. Finally, if all species are to be monophyletic, then some organisms are not parts of species, although in contrast with species concepts based on interbreeding, these organisms are not asexuals but members of ancestral populations.

In considering these consequences, a given reader may see some as insurmountable difficulties and others as simple facts of life. However, which consequences are viewed as problems and which ones as facts will differ, depending on one's point of view. This is the species problem. Given this state of affairs, we can imagine several possible fates for the term 'species'. One possibility is that it may become restricted to one of the classes of real biological entities, such as those resulting from interbreeding or those resulting from common descent. Which of these concepts is favoured depends not only on the theoretical context but also on whether 'species' is viewed as the name of a class of real biological entities or as the name of a rank in a hierarchy within a class of real biological entities. Alternatively, 'species' may continue to be used as a general term referring to an assemblage of several classes sharing nothing more than having been conflated historically. Realistically, the use of the term 'species' will be determined as much by historical and sociological factors as by logic and biological considerations. In any case, the entities deriving their existence from different natural process are all valid objects of investigation. Acknowledging this fact and exploring the relations among the different kinds of entities is central to both biology and systematics.[7]

REFERENCES

Ackery, P. R., and Vane-Wright, R. I. (1984), *Milkweed Butterflies: Their Cladistics and Biology* (Ithaca, NY: Cornell University Press).

Arnold, E. N. (1981), 'Estimating Phylogenies at Low Taxonomic Levels', *Zeitschrift für Zoologische Systematik und Evolutionsforschung*, 19: 1–35.

Ax, P. (1987), *The Phylogenetic System. The Systematization of Living Organisms on the Basis of their Phylogenesis* (New York: John Wiley).

Bernstein, H., Byerly, H. C., Hopf, F. A., and Michod, R. E. (1985), 'Sex and the Emergence of Species', *Journal of Theoretical Biology*, 117: 665–90.

Bonner, J. T. (1974), *On Development. The Biology of Form* (Cambridge, Mass.: Harvard University Press).

Bremer, K., and Wanntorp, H.-E. (1979), 'Geographic Populations or Biological Species in Phylogeny Reconstruction?', *Systematic Zoology*, 28: 220–4.

Brooks, D. R., and Wiley, E. O. (1986), *Evolution as Entropy* (Chicago: University of Chicago Press).

Brothers, D. J. (1985), 'Species Concepts, Speciation, and Higher Taxa', in E. S. Vrba (ed.), *Species and Speciation*, Transvaal Museum Monograph, 4 (Pretoria: Transvaal Museum), 35–42.

Buss, L. W. (1983), 'Evolution, Development and the Units of Selection', *Proceedings of the National Academy of Sciences*, 80: 1387–91.

——(1987), *The Evolution of Individuality* (Princeton: Princeton University Press).

Cracraft, J. (1983), 'Species Concepts and Speciation Analysis', in R. Johnston (ed.), *Current Ornithology* (New York: Plenum Press), 159–87.

——(1987), 'Species Concepts and the Ontology of Evolution', *Biology and Philosophy*, 2: 329–46.

Darwin, C. (1859), *On the Origin of Species by Means of Natural Selection* (London: John Murray).

de Queiroz, K. (1985), 'The Ontogenetic Method for Determining Character Polarity and its Relevance to Phylogenetic Systematics', *Systematic Zoology*, 34: 280–99.

——(1988), 'Systematics and the Darwinian Revolution', *Philosophy of Science*, 55: 238–59.

[7] We thank D. Cannatella, P. Cantino, J. Carpenter, A. de Queiroz, J. Gauthier, D. Hull, A. Larson, W. Maddison, D. Wake, and E. Wiley for helpful comments and discussion. Alan de Queiroz provided the idea of representing the different processes in a table. This essay is a highly modified offshoot of an unpublished manuscript by K. de Queiroz, M. Donoghue, and A. de Queiroz that was cited by Donoghue (1985). A comparison of Donoghue 1985 with the present essay will reveal some differences in outlook; we leave it to the reader to determine which of the views expressed in these papers are compatible.

Donoghue, M. J. (1985), 'A Critique of the Biological Species Concept and Recommendations for a Phylogenetic Alternative', *Bryologist*, 88: 172–81.

——(1987), 'South African Perspectives on Species: An Evaluation of the Recognition Concept', *Cladistics*, 3: 265–74.

Eldredge, N., and Cracraft, J. (1980), *Phylogenetic Patterns and the Evolutionary Process* (New York: Columbia University Press).

Futuyma, D. J. (1986), *Evolutionary Biology*, 2nd edn. (Sunderland, Mass.: Sinauer).

Gauthier, J. A., Estes, R., and de Queiroz, K. (1988), 'A Phylogenetic Analysis of *Lepidosauromorpha*', in R. Estes and G. Pregill (eds.), *Phylogenetic Relationships of the Lizard Families* (Palo Alto, Calif.: Stanford University Press), 15–98.

Ghiselin, M. T. (1966), 'On Psychologism in the Logic of Taxonomic Controversies', *Systematic Zoology*, 15: 207–15.

——(1969), *The Triumph of the Darwinian Method* (Berkeley: University of California Press).

——(1974), 'A Radical Solution to the Species Problem', *Systematic Zoology*, 23: 536–44.

——(1980), 'Natural Kinds and Literary Accomplishments', *Michigan Quarterly Review*, 19: 73–88.

——(1981), 'Categories, Life, and Thinking', *Behavioral and Brain Sciences*, 4: 269–313.

——(1985), 'Narrow Approaches to Phylogeny: A Review of Nine Books of Cladism', *Oxford Surveys of Evolutionary Biology*, 1: 209–22.

——(1987), 'Hierarchies and their Components', *Paleobiology*, 13: 108–11.

Gregg, J. R. (1954), *The Language of Taxonomy: An Application of Symbolic Logic to the Study of Classificatory Systems* (New York: Columbia University Press).

Griffiths, G. C. D. (1974), 'On the Foundations of Biological Systematics', *Acta Biotheoretica*, 23: 85–131.

——(1976), 'The Future of Linnaean Nomenclature', *Systematic Zoology*, 25: 168–73.

Harper, J. L. (1977), *Population Biology of Plants* (New York: Academic Press).

Hennig, W. (1966), *Phylogenetic Systematics* (Urbana, Ill.: University of Illinois Press).

——(1969), *Die Stammesgeschichte der Insekten* (Frankfurt am Main: Senchenberg Naturforschung Gesellschaft).

Holsinger, K. E. (1984), 'The Nature of Biological Species', *Philosophy of Science*, 51: 293–307.

Hull, D. L. (1965), 'The Effect of Essentialism on Taxonomy—Two Thousand Years of Stasis (II)', *British Journal of the Philosophy Science*, 61: 1–18.

——(1976), 'Are Species really Individuals?', *Systematic Zoology*, 25: 174–91.

——(1977), 'The Ontological Status of Species as Evolutionary Units', in R. Butts and J. Hintikka (eds.), *Foundational problems in the special sciences* (Dordrecht: Reidel), 91–102.

——(1978), 'A Matter of Individuality', *Philosophy of Science*, 45: 335–60.

——(1980), 'Individuality and Selection', *Annual Review of Ecology and Systematics*, 11: 311–32.

——(1981), 'Units of Evolution: A Metaphysical Essay', in V. J. Jensen and R. Harré (eds.), *The Philosophy of Evolution* (Brighton: Harvester), 23–44.

Janzen, D. H. (1977), 'What are Dandelions and Aphids?', *American Naturalist*, 111: 586–9.

Kawata, M. (1987), 'Units and Passages: A View for Evolutionary Biology and Ecology', *Biology and Philosophy*, 2: 415–34.

Kitcher, P. (1984), 'Species', *Philosophy of Science*, 51: 308–33.

Klekowski, E. J., Mohr, H., and Kazarinova-Fukshansky, N. (1986), 'Mutation, Apical Meristems and Developmental Selection in Plants', in J. P. Gustafson, G. L. Stebbins, and F. J. Ayala (eds.), *Genetics, Development, and Evolution* (New York: Plenum Press), 79–113.

Løvtrup, S. (1987), 'On Species and Other Taxa', *Cladistics*, 3: 157–77.

McKitrick, M. C., and Zink, R. M. (1988), 'Species Concepts in Ornithology', *Condor*, 90: 1–14.

Mayr, E. (1942), *Systematics and the Origin of Species* (New York: Columbia University Press).

——(1963), *Animal Species and Evolution* (Cambridge, Mass.: Harvard University Press).

——(1969a), 'The Biological Meaning of Species', *Biological Journal of the Linnean Society*, 1: 311–20.

——(1969b), *Principles of Systematic Zoology* (New York: McGraw-Hill).

——(1982), *The Growth of Biological Thought* (Cambridge, Mass.: Harvard University Press).

Mishler, B. D. (1985), 'The Morphological, Developmental and Phylogenetic Basis of Species Concepts in Bryophytes', *Bryologist*, 88: 207–14.

——and Brandon, R. N. (1987), 'Individuality, Pluralism, and the Phylogenetic Species Concept', *Biology and Philosophy*, 2: 397–414; reproduced as Ch. 14.

——and Donoghue, M. J. (1982), 'Species Concepts: A Case for Pluralism', *Systematic Zoology*, 31: 491–503.

Neigel, J. E., and Avise, J. C. (1986), 'Phylogenetic Relationships of Mitochondrial DNA under Various Demographic Models of Speciation', in S. Karlin and E. Nevo (eds.), *Evolutionary Processes and Theory* (New York: Academic Press), 515–34.

Paterson, H. E. H. (1978), 'More Evidence against Speciation by Reinforcement', *South African Journal of Science*, 74: 369–71.

——(1985), 'The Recognition Concept of Species', in E. S. Vrba (ed.), *Species and Speciation*, Transvaal Museum Monograph, 4 (Pretoria: Transvaal Museum), 21–9.

Patterson, C. (1978), 'Verifiability in Systematics', *Systematic Zoology*, 27: 218–22.

Platnick, N. I. (1977), 'Monotypy and the Origin of Higher Taxa: A Reply to E. O. Wiley', *Systematic Zoology*, 26: 355–7.

Rosen, D. E. (1978), 'Vicariant Patterns and Historical Explanations in Biogeography', *Systematic Zoology*, 27: 159–88.

——(1979), 'Fishes from the Uplands and Intermontane Basins of Guatemala: Revisionary Studies and Comparative Geography', *Bulletin of the American Museum of Natural History*, 162: 267–376.

Simpson, G. G. (1961), *Principles of Animal Taxonomy* (New York: Columbia University Press).

Stevens, P. F. (1984), 'Metaphors and Typology in the Development of Botanical Systematics 1690–1960, or the Art of Putting New Wine in Old Bottles', *Taxon*, 33: 169–211.

Vlijm, L. (1986), 'Ethospecies. Behavioral Patterns as an interspecific barrier', *Actes X Congrès International de Aracnologie*, 2: 41–5.

Vrba, E. S. (1985), 'Introductory Comments on Species and Speciation', in E. S.

Vrba (ed.), *Species and Speciation*, Transvaal Museum Monograph, 4 (Pretoria: Transvaal Museum), pp. ix–xviii.

Whitham, T. G., and Slobodchikoff, G. N. (1981), 'Evolution by Individuals, Plant–Herbivore Interactions, and Mosaics of Genetic Variability: The Adaptive Significance of Somatic Mutations in Plants', *Oecologia (Berl.)*, 49: 287–9.

Wiley, E. O. (1977), 'Are Monotypic Genera Paraphyletic?—A Response to Norman Platnick', *Systematic Zoology*, 26: 352–5.

——(1978), 'The Evolutionary Species Concept Reconsidered', *Systematic Zoology*, 27: 17–26.

——(1979), 'Ancestors, Species, and Cladograms—Remarks on the Symposium', in J. Cracraft and N. Eldredge (eds.), *Phylogenetic Analysis and Paleontology* (New York: Columbia University Press), 211–25.

——(1980), 'Is the Evolutionary Species Fiction?—A Consideration of Classes, Individuals and Historical Entities', *Systematic Zoology*, 29: 76–80.

——(1981), *Phylogenetics. The Theory and Practice of Phylogenetic Systematics* (New York: John Wiley).

Willmann, R. (1983), 'Biospecies und Phylogenetische Systematik', *Zeitschrift für Zoologische Systematik und Evolutionsforschung*, 21: 241–9.

Woodger, J. H. (1952), 'From Biology to Mathematics', *British Journal of the Philosophy of Science*, 3: 1–21.

16

ELIMINATIVE PLURALISM

MARC ERESHEFSKY

I. INTRODUCTION

The species category plays two intimately connected roles in biology. The first occurs in biological systematics. Systematists attempt to provide a taxonomy of life using, for the most part, the Linnaean framework. Species taxa are the basal units in that taxonomy; higher taxa (such as genera, families, and classes) are composed of species taxa, and form more inclusive units. The second major role of the species category occurs in evolutionary biology. While systematists attempt to provide a taxonomy of the organic world's diversity, evolutionists attempt to explain why that diversity exists. An essential part of that explanation is that species taxa are 'the evolutionary units' of the organic world—groups of organisms that evolve as units due to their exposure to common evolutionary forces (Mayr 1970, Dobzhansky 1970).

Given the fundamental role of the species category, a proper definition of that category would seem crucial for systematics and evolutionary biology. Unfortunately, biologists widely disagree on how to define the species category. A recent anthology on species (Ereshefsky 1992) contains no fewer than eight prominent definitions, and these eight are just a small sample of the dozens of definitions found in the current biological literature. Of course, disagreement over the nature of species is nothing new. Since and before Linnaeus, biologists have disagreed on the nature of species. For example, in a letter to botanist Joseph Hooker, Darwin writes:

It is really laughable to see that different ideas are prominent in various naturalists' minds, when they speak of 'species'; in some, resemblance is everything and descent of little weight—in some resemblance seems to go for nothing, and Creation the reigning idea—in some, descent is the key,—in some sterility an unfailing test, with others it is not worth a farthing. (Darwin 1887: ii. 88)

First published in *Philosophy of Science*, 59 (1992): 671–90. Reprinted by permission.

Biologists and philosophers have taken one of two approaches to the diversity of species definitions found in the biological literature. Some consider the species problem an unfinished debate, in which the proper definition needs to be weeded from the improper ones (see e.g. Hull 1987, Ghiselin 1987, and Mayr 1987). Others hold that there is no common and distinctive attribute of all species taxa; thus the species category is heterogeneous (see Ruse 1969, 1987; Dupré 1981; Mishler and Donoghue 1982; Kitcher 1984a, b, 1987; Mishler and Brandon 1987). The first group of authors advocate species monism, the second promote species pluralism.

In this essay, I take up the cause of species pluralism. Though others have already advocated species pluralism, their versions are defective. Thus I offer an alternative brand of pluralism. In particular, my aim here is fourfold: first, to provide a comprehensive argument for species pluralism; second, to answer various monist objections to pluralism; third, to offer an alternative form of species pluralism, which I call 'eliminative pluralism'; and fourth, to show that eliminative pluralism is an improvement over previous forms of pluralism.

II. THE CASE FOR SPECIES PLURALISM

Biologists offer various definitions of the species category. Many of those definitions fall within three general approaches to species.[1] The first approach, the interbreeding approach, is best known through Mayr's biological species concept. According to Mayr (in his most widely accepted version of that concept), 'species are groups of interbreeding natural populations that are reproductively isolated from other such groups' (1970: 12). In other words, a species is the most extensive group of organisms that interbreed and produce fertile offspring. Furthermore, the members of a species are separated from all other organisms by 'isolating mechanisms' (ibid. 55 ff.). These mechanisms either prevent interbreeding between interspecific organisms or prevent the production of fertile offspring if such interbreeding does occur. A more recent species concept that falls within the interbreeding approach is Paterson's (1985) mate recognition concept. Paterson utilizes the interbreeding half of the biological species concept,

---

[1] These approaches do not contain all of the definitions currently proposed by biologists (for a more complete survey see Ereshefsky 1992). My case for pluralism turns on there being more than one viable approach to species. Thus I have limited my survey to those approaches that I take to be the most viable. The addition of further viable approaches only strengthens my argument.

but drops any reference to reproductive isolation. According to Paterson, species are interbreeding groups whose members contain similar mate recognition systems: namely, behavioural and morphological characteristics that allow organisms to recognize conspecific mates (e.g. the chemical signals of wasps, the light signals of fireflies, even the stigmas of orchids). Despite their differences, both Mayr's and Paterson's species concepts capture the heart of the interbreeding approach: species are groups of biparental organisms that share common fertilization systems. (For a Mayrian view of the difference between Mayr's and Paterson's concepts, see Mayr 1988.) Other proponents of the interbreeding approach to species include Dobzhansky (1970), Carson (1975), Ghiselin (1974), and Eldredge (1985).

According to proponents of the interbreeding approach, species are stable taxonomic units, because the members of a species exchange genetic material through sexual reproduction. Ehrlich and Raven (1969), Van Valen (1976), and Andersson (1990), however, disagree. They argue that the stability of a species is primarily due to environmental forces rather than interbreeding. Thus these authors promote an ecological approach to species. For example, according to Van Valen, 'A species is a lineage . . . which occupies an adaptive zone minimally different from that of any other lineage in its range and evolves separately from all lineages outside its range' (1976: 235). In other words, each species occupies its own distinctive adaptive zone, or niche, and the distinct set of selection forces in each zone is responsible for the maintenance of species as separate taxonomic units. It is worth emphasizing that, according to the ecological approach, species must be lineages and not merely groups of organisms that occupy the same adaptive zone.

The interbreeding and the ecological approaches to species stem from work in evolutionary biology, whereas the third approach to species, the phylogenetic approach, flows out of biological systematics. According to Mishler and Donoghue (1982), Cracraft (1983), and Mishler and Brandon (1987), organisms should be classified according to propinquity of descent. In particular, each taxonomic group, whether it be a species, a genus, and so on, should contain all and only the descendants of a common ancestor, Such taxonomic groups are called 'monophyletic taxa'. The notion of monophyly, however, is not enough to provide a phylogenetic definition of the species category. Monophyletic taxa occur up and down the evolutionary continuum; according to the above authors, all taxa, species, genera, orders, even all life on this planet (assuming a common origin), form monophyletic taxa. Thus Cracraft, Mishler, Donoghue, and Brandon offer various ranking criteria for distinguishing which monophyletic taxa are species taxa. For example, according to Mishler and Donoghue, 'species ranking criteria could include group size, gap size, geological age, ecological and geographical criteria, degree of intersterility, tradition and possibly others' (1982: 499). In other words, some monophyletic taxa are ranked as species because their organisms interbreed or share common ecological and developmental factors. Other monophyletic taxa are ranked as species on the basis of morphological gaps between their organisms and those of other taxa. Cracraft, on the other hand, defines a species taxon as 'the smallest diagnosable cluster of individual organisms within which there is a parental pattern of ancestry and descent' (1983: 170). According to Cracraft, diagnosable clusters of organisms are, for the most part, 'defined by uniquely derived characteristics' (ibid.). (Cracraft allows that some diagnosable clusters may be defined by 'unique combinations of primitive and derived characteristics' (ibid.).) Though proponents of the phylogenetic approach offer different ranking criteria, they tend to agree that species are basal monophyletic taxa.

These three general approaches to species (the interbreeding, ecological, and phylogenetic) are diverse, but they all assume that species are lineages. By 'lineage' I mean either a single descendant–ancestor sequence of organisms or a group of such sequences that share a common origin. In philosophical jargon, these approaches assume that species are spatio-temporally continuous or historical entities. (See Section IV below on why this assumption is essential for any post-Darwinian definition of the species category.) Beyond the assumption that species are lineages, the three approaches provide different pictures of species and the organic world. In particular, they provide incompatible taxonomies of the organic world. Consider a small-scale example.

Suppose we want to determine the correct taxonomy of the insects that live on the side of a mountain. The insects consist of three populations, A, B, and C (Fig. 16.1a). Each population forms a single basal monophyletic taxon (in other words, each is 'the smallest diagnosable cluster of individual organisms within which there is a parental pattern of ancestry and descent' (Cracraft 1983: 170)). The organisms in B and C share a common ecological niche, while the organisms in A occupy their own niche. Concerning breeding behaviour, the organisms in A and B can successfully interbreed and produce fertile offspring. But the organisms in C reproduce asexually; their females reproduce via parthenogenesis, thus their eggs do not require fertilization. So what is the correct taxonomy of the insects in this plot? According to the phylogenetic approach, it is a taxonomy consisting of three species; A, B, and C (Fig. 16.1b). According to the ecological approach, it is the taxonomy consisting of two species: A and B + C

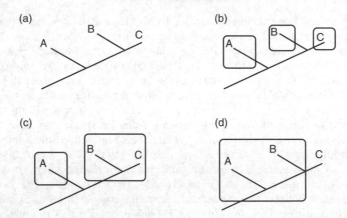

FIG. 16.1. A phylogenetic tree with three populations, A, B, and C. (a) The phylogenetic tree. (b) The phylogenetic tree with three phylogenetic species, A, B, and C. (c) The phylogenetic tree with two ecological species, A, and $B + C$. (d) The phylogenetic tree with one interbreeding species, $A + B$.

(Fig. 16.1c). According to the interbreeding approach, it is the taxonomy consisting of a single species: $A + B$ (Fig. 16.1d). (Because the organisms in C reproduce asexually, that population forms no species according to the interbreeding approach.) Hence these general approaches to species provide three different taxonomies of the insects in the plot.[2]

Consider this problem on a much larger scale: namely, that of trying to provide a taxonomy of all the organisms on this planet. Because biologists disagree on the correct approach to the species category, they provide different taxonomies of the organic world. Moreover, these taxonomies are *incompatible*, in that they often classify the same organisms into different lineages (see references in n. 2). Such incompatibility occurs in two ways (and can be illustrated with Fig. 16.1). First, an organism may belong to two lineages, where one lineage is properly contained in another; for example, a member of the phylogenetic species A is also a member of the interbreeding species $A + B$. Second, an organism may belong to two lineages that are disjoint; for example, an organism in population B be-

---

[2] Though this example is hypothetical, it is based on empirical studies showing that interbreeding, ecological, and monophyletic lineages often do not correspond in nature. For the discontinuity between interbreeding and ecological lineages, see Templeton 1989 and Grant 1981. For the discontinuity between interbreeding and monophyletic lineages, see de Queiroz and Donoghue 1988 and Frost and Hillis 1990: 96–7. For the discontinuity between ecological and monophyletic lineages, see Mayr 1982: 230 and Ridley 1986: 32 ff.

longs to both the ecological species $B + C$ and the interbreeding species $A + B$. The type of pluralism I am highlighting here should be distinguished from a more moderate form suggested in the literature. According to Mishler and Donoghue (1982) and Mishler and Brandon (1987) (see Section IV below), there are a number of legitimate species approaches, but different approaches apply to different organisms, and no more than one approach is applicable to an organism. The result is a unique taxonomy of the organic world. I am presenting a more radical picture of the organic world. Different species approaches often classify the same organisms into different lineages. Consequently, there are a number of incompatible taxonomies of that world.

What does a monist make of all this? A monist would insist that only one correct approach to species exists, and consequently only one correct taxonomy of the organic world exists. I disagree on both counts. In what follows I argue for a plurality of equally legitimate, though incompatible, taxonomies of the organic world.

First, I should point out that the argument for pluralism offered here is ontological, not epistemological. Species pluralism, according to current evolutionary theory, is a real feature of the world, and not merely a feature of our lack of information about that world. Others (e.g. Cartwright (1983) and Levins (1968)) provide epistemologically motivated arguments for pluralism. According to these arguments, the world is exceedingly complex, and we have limited cognitive abilities; thus we posit a plurality of simplified and inaccurate models and taxonomies. These arguments nevertheless allow that there may be a single correct taxonomy of the world, and perhaps in time we will acquire enough information to discover that taxonomy.

An ontological argument for species pluralism, however, can be found in contemporary evolutionary theory. Specifically, evolutionary theory provides the following picture of the organic world. All of the organisms on this planet belong to a single genealogical tree. The forces of evolution segment that tree into a number of different types of lineages, often causing the same organisms to belong to more than one type of lineage. The evolutionary forces at work here include interbreeding, selection, genetic homeostasis, common descent, and developmental canalization (see Templeton 1989 for a discussion of these forces). The resultant lineages include lineages that form interbreeding units, lineages that form ecological units, and lineages that form monophyletic taxa. (Interbreeding units are the result of interbreeding (Mayr 1970); ecological units are the result of environmental selection (Van Valen 1976); and basal monophyletic taxa owe their existence to common descent (de Queiroz and

Donoghue 1988).) So the forces of evolution segment the tree of life into a plurality of incompatible taxonomies: one taxonomy consisting of interbreeding units, another consisting of ecological units, and a third consisting of monophyletic taxa. Of course, this picture of evolution could be wrong; perhaps some of the above-mentioned forces do not exist, or those forces lack the ability to produce stable taxonomic entities. These are, after all, empirical matters. But given what current evolutionary theory tells us, the forces of evolution segment the tree of life into different and incompatible taxonomies. (Holsinger (1984) presents a similar picture of evolution.)

Proponents of monism may allow different types of basal lineages in the world, but they would contend that one type of lineage is more important for understanding the course of evolution; thus only that type of lineage should be designated by the term 'species'. For example, Eldredge (1985: 200–1) and Ghiselin (1989: 74–5) argue that lineages with sexual organisms are much more important in the course of evolution. As a result, they suggest that only interbreeding units should be called 'species'.

This suggestion and others like it should be rejected. If we are to understand how evolution has occurred on this planet, we must study the various types of theoretically important lineages in the world. No particular type of lineage is prior in that study. Consider Eldredge and Ghiselin's suggestion that sexual organisms are the most important in the course of evolution on this planet. As Eldredge (1985: 200–1) and Ghiselin (1989: 74) point out, the occurrence of recombination in sexual organisms provides sexual species with greater genetic flexibility than asexual species. Consequently, sexual species frequently out-compete asexual ones. I agree that the important differences between sexual and asexual species cannot be ignored. However, the competitive ability of sexual organisms (in certain circumstances) should not blind us to the fact that most organisms in the history of this planet have been asexual.[3] Nor should it cause us to ignore the existence of stable taxonomic lineages consisting of asexual organisms (see the proceedings of a recent symposium on asexual organisms (Mishler and Budd 1990)). A taxonomy containing only interbreeding units provides an inadequate framework for studying life's diversity. The same applies to a taxonomy consisting of only monophyletic taxa or just ecological units. A proper systematic study of life requires each of these taxonomies. Consider the sorts of theoretically important information each taxonomy offers. A taxonomy of monophyletic taxa provides a framework for examining ge-

---

[3] According to Hull, '[F]or the first three-quarters of life on Earth, the primary, possibly sole form of reproduction was asexual' (1988: 429). Furthermore, if one looks at Earth's present biota, sexuality 'turns out to be rare on every measure suggested by evolutionary biologists— number of organisms, biomass, amount of energy transduced, and so on' (ibid.).

nealogy. A taxonomy of interbreeding units offers a framework for examining the effect of sex on evolution. And a taxonomy of ecological units provides a structure for observing the effect of environmental selection forces. A systematic study that considers just one of these taxonomies provides an overly coarse-grained picture of evolution.

Thus far I have argued that the tree of life consists of three types of basal lineages, and that these lineages give rise to distinct taxonomies of the organic world. One might accept my representation of current evolutionary thinking, yet be hopeful that a fourth parameter common to all three types of base lineages will be discovered. Such a parameter would define a fourth type of base lineage to which the other types of lineages could be reduced, resulting in a single correct taxonomy of the organic world. I cannot foreclose the possibility of such an empirical discovery, but a closer look at current biological thinking offers reasons for doubting the existence of such a parameter.

Suppose, for example, one were to suggest that once biologists have performed enough genetic analyses (e.g. like the current Human Genome Project), they will find that overall genetic similarity is a parameter common to interbreeding, ecological, and monophyletic lineages. This suggestion, however, is problematic. If different species concepts classify the same group of organisms such that one lineage is fully contained in another, then it is impossible that both lineages consist of organisms with the most overall genetic similarity. Recall the example illustrated in Figure 16.1. A is a monophyletic taxon, and $A + B$ forms an interbreeding unit. Consequently, the organisms in A and $A + B$ cannot both have the most overall genetic similarity. For if the organisms in A have the most overall genetic similarity, then the organisms in $A + B$ must have less overall genetic similarity. Given that some basal lineages are contained in others, then not all basal lineages consist of organisms with the most overall genetic similarity. Nature also gets in the way in the attempt to align interbreeding and monophyletic lineages with lineages containing the most overall genetic similarity. In some situations, lineages with the most overall genetic similarity do not form monophyletic taxa (Frost and Hillis 1990: 96, Futuyma 1985: 311). In other situations, lineages with the most overall genetic similarity do not form interbreeding units (Futuyma 1985: 220, Mayr 1970: 321, and Frost and Hillis 1990: 95).[4]

Alternatively, a reductionist might complain that I have addressed this

[4] Thus Mayr, a proponent of the interbreeding approach, writes: 'Species difference [cannot] be expressed in terms of the genetic bits of information, the nucleotide pairs of the DNA. That would be quite as absurd as trying to express the differences between the Bible and Dante's *Divina Commedia* in terms of the difference in the frequency of the letters of the alphabet used in the two books. The meaningful level of integration is well above that of the base code of information, the nucleotide pairs' (1970: 321–2).

problem from the wrong direction. I should be looking for some common genetic factor in the three types of lineages in question, rather than overall genetic similarity. For, as Futuyma points out, 'species owe their existence to specific characters governed by specific genes' (1985: 223). However, this approach does not provide a common basis to reduce the three types of lineages either. The genes that Futuyma (a proponent of the interbreeding approach) thinks define species are those that affect sexual behaviour. Yet in some instances, an alteration in the genetic content of an organism can affect its ecological adaptiveness but not its sexual behaviour. For example, being heterozygous rather than homozygous for haemoglobin S in a malaria zone affects one's ecological adaptiveness, but not one's choice of mates (see Futuyma 1985: 75–6). Furthermore, in instances where genes controlling morphological distinctiveness and reproductive behaviour are separable (Mayr 1970: 322), mutations affecting the former but not the latter can cause the existence of new monophyletic taxa that are not distinct interbreeding units. And, in instances where genetic material governing morphology and ecological behaviour is separable (Futuyma 1985: 238), mutations affecting the former but not the latter can cause the existence of new monophyletic taxa that are not distinct ecological units. The upshot is that the genetic factors governing the distinctive features of interbreeding, ecological, and monophyletic lineages are separable. Thus the reduction of these types of lineages to their underlying genetic bases results in three separate genetic taxonomies. In other words, the plurality of types of lineages at the macroscopic level is just transferred to a plurality of types of genetic factors at the microscopic level. The attempt to find a common genetic factor that unifies the three types of lineages in question fails.

The results of this section can be summarized as follows. The forces of evolution produce at least three different types of basal lineages (interbreeding, ecological, and monophyletic) that cross classify the organic world. Each of these lineages is equally important in the evolution of life on this planet. Moreover, according to current biological thinking, there is no fourth parameter to which these types of lineages can be reduced. Consequently, the tree of life on this planet is segmented into a plurality of incompatible, but equally legitimate, taxonomies.

III. CRITICISMS OF PLURALISM

Some philosophers (e.g. Hull (1987, 1988, 1989) and Sober (1984)) and biologists (e.g. Hennig (1966), Ghiselin (1969, 1987), and Cracraft (1983,

1987)) take a dim view of species pluralism. In this section I consider and answer their objections to pluralism. In doing so, I further explicate eliminative pluralism, and lay a general foundation for a pluralistic approach to taxonomy.

The Communication Objection

Systematists often point out that 'the necessity of classifications has long been recognized . . . for the very communication of general ideas' (Ghiselin 1969: 79; also see Mayr 1969: 89, Eldredge and Cracraft 1980: 165 ff.). With this presumption of the goal of systematics, Ghiselin provides the following argument against pluralism:

Whatever standard one does take for ranking taxonomic groups, it should be clear that systematists work at cross purposes when they do not agree on any such criteria. If a common standard were recognized, the system would be more informative by far, and the goal of natural classification would be better served. (1969: 85)

Similarly, Hull writes that 'terming a hodgepodge of different units "species" serves no useful purpose. . . . If pluralism entails confusion and ambiguity, I am forced to join with Fodor's . . . Granny in her crusade to stamp out creeping pluralism' (1987: 181; also see Hull 1989: 313). This objection to pluralism can be codified in the following argument. Species pluralism entails that the term 'species' is ambiguous. If the term 'species' is ambiguous, then confusion will set in when biologists discuss the nature of species, for biologists will mean different things by 'species'. Such confusion should be avoided. Thus species pluralism should be avoided.

I agree with Kitcher's (1984b: 326–7) response to this argument. Species pluralism does not 'unlock the doors of Babel', and plunge biological discourse into confusion. Different species concepts often pick out different taxa in a single biological situation. To guard against confusion, biologists merely need to be explicit concerning the concept being used when referring to a group of taxa as 'species'. Indeed, in professional journals biologists usually are explicit concerning what they mean by 'species'.

I would like to offer a stronger response to the communication objection. Instead of referring to basal lineages as 'species', biologists should categorize those lineages by the criteria used to segment them: interbreeding units, monophyletic units, and ecological units. The term 'species' is superfluous beyond the reference to a segmentation criterion; and when the term is used alone, it leads to confusion. The term 'species' has outlived its usefulness, and should be replaced by terms that more accurately describe the different types of lineages that biologists refer to as 'species'. For example, Grant (1981) suggests using the terms 'biospecies' and 'ecospecies' for the lineages picked out by the interbreeding and ecological approaches. Add to these the term 'phylospecies' for the lineages highlighted by the phylogenetic approach. Replacing 'species' with such terms better serves the goal of communication placed on systematics. Accordingly, I call the form of species pluralism advocated here 'eliminative pluralism': eliminate the term 'species', and replace it with a plurality of more accurate terms.

The Inconsistency Objection

Some authors (e.g. Hennig (1966), Hull (1987), and Cracraft (1983, 1987)) are unhappy with the pluralistic result that there are incompatible taxonomies of the organic world. Hennig, for example, writes that 'if systematics is to be a science it must bow to the self-evident requirement that objects to which the same label is given must be comparable in some way' (1966: 154). In a similar vein, Hull objects to Kitcher's version of species pluralism, because it does not provide 'a consistent treatment of the evolutionary process' (1987: 180). However, I would counter that the existence of incompatible taxonomies of the organic world does not provide an inconsistent view of evolution. Eliminative pluralism assumes that there is one genealogical tree of life, but that the tree is segmented by different evolutionary forces into different lineages (often with the same organisms belonging to more than one type of lineage). As a result, the tree of life is segmented into different taxonomies: one taxonomy consisting of interbreeding units, another consisting of ecological units, and a third consisting of monophyletic taxa. The resultant taxonomies are taxonomies of *different aspects* of the tree of life. Hence they are not inconsistent with one another. Moreover, each of these taxonomies is internally consistent: one taxonomy consists only of interbreeding units, another consists only of basal monophyletic taxa, and so on. So eliminative pluralism provides a fully consistent treatment of evolution.

The No Criteria Objection

Ghiselin (1987: 135–6) and Hull (1987: 180, 1989: 313, personal communication) believe that pluralism is an overly liberal approach to science. In particular, they contend that pluralists provide no criteria for discerning legitimate from illegitimate taxonomies. What, for example, discriminates between taxonomies based on current evolutionary theory from those based on idealistic morphology, or even Creationism? As Hull sees it, pluralism places no checks on science. In rougher terms, Ghiselin views pluralism as an instance of lazy thinking that results in an attitude of 'anything goes' (see Ghiselin 1987: 135–6). To answer this objection, I offer candidate criteria that a pluralist can use for determining whether a taxonomy is legitimate. These criteria are similar to those standardly used in determining whether a theory is scientific (see e.g. Laudan 1984). Before I present the criteria, I need to introduce some terminology and state explicitly some common presumptions of scientific classification.

A taxonomy is produced by what I call a 'taxonomic approach'. Such an approach constructs a taxonomy by a set of principles. Those principles come in two forms. 'Sorting principles' sort the constituents of a theory into basic units. 'Motivating principles' justify the use of sorting principles. Consider the biological species concept. It constitutes a taxonomic approach for producing a taxonomy of the organic world. Its sorting principles roughly assert: Sort organisms that can interbreed and produce fertile offspring into a single species, sort organisms that reproduce sexually but cannot interbreed into different species, and sort organisms that reproduce asexually into no species. The motivating principle of the biological species concept assumes that the process of interbreeding causes stability within lineages of organisms that interbreed. (By 'stability' I mean that the organisms of lineage evolve as a unit or share a common stasis.) In brief, the motivating principle of the biological species concept sets out the causal factor responsible for the existence of the lineages in question.

The general idea behind motivating principles can be described as follows. A taxonomy (biological or otherwise) consists of entities that are the nodes of causal processes. Those entities are either the result of a common type of causal process, or they are objects that have a similar causally efficacious property. So motivating principles either cite the causal processes that give rise to lineages or the similar causally efficacious nature of those lineages. The three taxonomic approaches considered here (the interbreeding, ecological, and phylogenetic) contain motivating principles that cite the causal process primarily responsible for the type of lineages in question. The interbreeding approach cites the process of interbreeding, the ecological approach highlights environmental selection pressures, and the phylogenetic approach focuses on the process of descent from common ancestry.

Notice that this conception of motivating principles does not violate the common observation that no universal generalizations exist whose predicates are the names of species taxa (see e.g. Hull 1976, 1978). Nothing in the conception of motivating principles forces a taxonomic approach to assign some qualitative property to all the members of a species taxon.

However, the notion of motivating principles does suggest that there may be universal generalizations whose predicates are the names of *types* of basal taxonomic units. That is, there may be universal generalizations containing such predicates as 'biospecies', 'phylospecies', or 'ecospecies'. So though no laws exist about particular species taxa, there may very well be laws about types of species taxa.

I now turn to the criteria that a taxonomic approach must satisfy to be considered legitimate. Ideally, such a list of criteria would provide individually necessary and jointly sufficient criteria. The following list, however, is merely a first stab; further criteria may be needed to properly complete the list.

First, the motivating principles of a taxonomic approach should be empirically testable. Put simply, such principles should have an empirical basis. For example, in determining the legitimacy of the interbreeding approach, biologists should be able to determine empirically if interbreeding is an important causal factor in the stability of lineages of organisms that interbreed.

Second, the sorting principles of a taxonomic approach should produce a single internally consistent taxonomy. In other words, a taxonomic approach should be unambiguous. Ambiguity can occur in two ways: the base units of a taxonomic approach can be ambiguous, and a taxonomic approach can produce an ambiguous classification—that is, it can produce more than one taxonomy with no way of discriminating which is the correct one (see Ridley 1986: 6–7). Ambiguity of the first type can give rise to ambiguity of the second type: if a taxonomic approach allows a heterogeneous class of base units, then that approach will produce inconsistent taxonomies. The criterion of internal consistency is designed to avoid both kinds of ambiguity. All taxa designated as 'species' (or more precisely, 'phylospecies', 'biospecies', and so on) *within* a taxonomic approach should be comparable along the appropriate parameters. If that requirement is met, then a taxonomic approach provides a single consistent taxonomy of the organic world.

Third, the motivating principles of a taxonomic approach should be consistent with well-established hypotheses in other scientific disciplines. For example, a taxonomic approach in biology should not violate any well-established laws in biochemistry or geology. This criterion opens a nest of standard problems in the philosophy of science. The question of what is a well-established hypothesis is none other than the problem of confirmation. The question of what constitutes a scientific discipline brings up the demarcation problem of discerning science from non-science. I mention these problems in passing to indicate that questions concerning the legitimacy of a taxonomic approach are intricately tied to central questions in the philosophy of science.

Fourth, the motivating principles of a taxonomic approach should be consistent with and derivable from the tenets of the theory for which the taxonomy is produced. In particular, a taxonomic approach in biological systematics should be derivable from well-established tenets in evolutionary theory. For example, in the case of the interbreeding approach, the motivating principle that interbreeding can cause stability in lineages should be an extension of what evolutionary theory tells us about the stability of lineages in general.

Though this list of criteria may be incomplete as it stands, it nevertheless does a good job at ruling out paradigm-illegitimate taxonomic approaches. Consider three such approaches. A Creationist taxonomic approach contains motivating principles that are neither empirical nor consistent with the tenets of evolutionary theory or well-established tenets in other disciplines (e.g. carbon dating in geology). Thus a Creationist approach is illegitimate because it violates criteria 1, 2, and 3. Taxonomic approaches based on idealistic morphology (e.g. those advocated by Goethe and Richard Owen; see Mayr 1982: 457–8) rely on typological thinking. Typological thinking, however, is incompatible with current evolutionary biology, and has been replaced with population thinking (see Sober 1980). Consequently, approaches based on idealistic morphology are illegitimate, because they violate criterion 4. Phenetic taxonomic approaches (e.g. Sneath and Sokal 1973) produce a number of inconsistent taxonomies of the organic world (see Hull 1970 and Ridley 1986). Hence such approaches are illegitimate, because they violate criterion 2.

On the other hand, I contend that the three approaches to species discussed in this essay—the interbreeding, ecological, and phylogenetic— do satisfy the four criteria for legitimate taxonomic approaches. Of course, proponents of a particular taxonomic approach argue that the other approaches are defective (see e.g. Cracraft 1983, Ghiselin 1987, and Ridley 1990). For the most part, these arguments are based on the false premiss that a single taxonomic approach is supposed to provide a universal definition for all basal lineages. (That premiss was cast in serious doubt by the arguments of Section II above.[5]) I would like to make one final point concerning the above criteria. By adopting such criteria, a worker is not

---

[5] Some proponents of the phylogenetic approach (e.g. Ridley (1986, 1990)) argue that the ecological and interbreeding approaches are ambiguous, and thus violate criterion 2. Perhaps some versions of those approaches are ambiguous, but some are not. Templeton's (1989) notions of genetic and demographic exchangeability, for example, provide unambiguous definitions of basal interbreeding and ecological units.

committed in an a priori fashion to the existence of a number of legitimate taxonomies, for it is possible that only one taxonomic approach satisfies the above criteria. The empirical world ultimately decides whether pluralism within a particular discipline is appropriate.

In this section, I have completed two tasks. First, I have answered several objections to taxonomic pluralism. Second, and more importantly, I have further explicated eliminative pluralism, and laid a general foundation for a taxonomic pluralism.

IV. SPECIES PLURALISM AND SPECIES OF PLURALISM

As mentioned earlier, the idea of species pluralism is not new. A number of philosophers and biologists (Ruse 1969, 1987; Dupré 1981; Mishler and Donoghue 1982; Kitcher 1984a, b, 1987; Mishler and Brandon 1987) have already advocated species pluralism. In this section, I sketch some problems facing earlier forms of species pluralism. I end the section by indicating how eliminative pluralism avoids those problems.

Ruse (1969, 1987) offers the most conservative form of species pluralism. Ruse acknowledges that different species concepts provide different criteria for sorting organisms. But Ruse claims: 'There are different ways of breaking organisms into groups, and they *coincide!* The genetic species is the morphological species is the reproductively isolated species is the group with common ancestors' (1987: 238). In other words, Ruse believes that the various species concepts offered by biologists pick out the same set of taxa. Ruse's motivation for establishing a coincidence among species concepts is his belief that such a coincidence would indicate the naturalness (or reality) of species taxa. Following Whewell and Hempel (see Ruse 1987 for references), Ruse takes consilience to be a mark of reality: that is, an indication that a classification is natural rather than artificial. So, according to Ruse, if various species concepts citing different biological properties pick out the same taxa, we have good reason to believe that those taxa are real.

The consilience of various species concepts is the ideal that some evolutionists hoped for (e.g. Mayr 1969: 28). However, as illustrated in this essay, nature has stymied that ideal. Groups of organisms that have the most overall genetic similarity often are not groups of interbreeding organisms (Futuyma 1985: 220, Mayr 1970: 321, and Frost and Hillis 1990: 95). Many monophyletic taxa are not interbreeding units (Mishler and Donoghue 1982, Frost and Hillis 1990). Many groups of organisms that form ecological units are not interbreeding units (Templeton 1989, Grant

1981). This lack of consilience is not limited to a few borderline cases. Consider the case of asexual organisms. Most organisms in the history of this planet have been asexual (see n. 3). As a result, a major discrepancy divides species concepts that recognize both sexual and asexual taxa (e.g. ecological and phylogenetic concepts) from concepts that recognize only sexual taxa (e.g. interbreeding concepts). Given this lack of consilience, a form of species pluralism that requires consilience should be rejected.

One other item from Ruse's version of pluralism is worth mentioning. Ruse assumes that the alleged coincidence of species concepts indicates the naturalness of species taxa. However, a major point of this essay is that the naturalness of some objects does not lie at the intersection of various scientific concepts. In particular, the lack of consilience among various species concepts does not show that the taxa they pick out are not real. The taxa are real; they just do not fall under a single category (the species category).

Mishler and Donoghue (1982) and Mishler and Brandon (1987) offer a more liberal form of species pluralism. Unlike Ruse, they recognize that different species concepts often pick out different groups of organisms. Recall their phylogenetic species concept, which requires that all species taxa form monophyletic taxa. But Mishler and Donoghue recognize that 'because different factors may be "most important" in the evolution of different groups, a universal criterion for delimiting fundamental, cohesive evolutionary units does not exist' (1982: 495). Thus, some species taxa owe their existence to reproductive factors; other species taxa are the result of ecological forces; still others are due to homeostatic inertia. Their phylogenetic species concept is monistic in that all species taxa are monophyletic, but it is pluralistic in that different types of processes cause lineages to be species. The result is a single taxonomy of the organic world consisting of different types of basal monophyletic lineages.

Mishler, Donoghue, and Brandon's form of pluralism presents two problems. First, it requires that all species taxa form monophyletic taxa. Consequently, any taxa that are not monophyletic, despite their forming good interbreeding or ecological units, should not be formally recognized. As mentioned previously, however, instances of non-monophyletic basal taxa that form interbreeding or ecological units abound (see e.g. Tajima 1983, Neigel and Avise 1986, de Queiroz and Donoghue 1988, and Frost and Hillis 1990). Mishler, Donoghue, and Brandon's form of pluralism is inadequate, because it ignores non-monophyletic basal taxa that satisfy classic population-genetic parameters for specieshood (namely, gene flow and exposure to common selection regimes).

The second problem with their form of species pluralism is its commitment to a *single* taxonomy of the organic world. As noted in Section II, different species approaches often cross-classify the same group of organisms. As a result, different species approaches produce incompatible taxonomies of the organic world. This incompatibility is not limited to the discrepancy between a taxonomy containing only monophyletic taxa (as in the case of Mishler, Donoghue, and Brandon) and a taxonomy containing both monophyletic and non-monophyletic taxa. Even within a strictly monophyletic taxonomy, there are monophyletic interbreeding and ecological units that are not coextensive (see Templeton 1989). Because Mishler, Donoghue, and Brandon's pluralism does not allow for the existence of incompatible but empirically significant taxonomies, their pluralism does not go far enough.

Another form of species pluralism is found in Dupré (1981). Dupré describes his version of pluralism as 'promiscuous realism'. 'The realism derives from the fact that there are many sameness relations that serve to distinguish classes of organisms in ways that are relevant to various concerns; the promiscuity derives from the fact that none of these relations is privileged' (1981: 82). What are some of those sameness relations? One sameness relation is the phenetic measurement of overall similarity (ibid. 82–3, also 89–90). Other sameness relations consist of more limited ranges of properties: for example, the properties of 'texture or flavor' that gourmets use to classify organisms (ibid. 83). Then there are the familiar properties of interbreeding behaviour (ibid. 85–7) and phylogenetic relations (ibid. 87–9). Dupré's form of pluralism is certainly more liberal than the forms advocated by Mishler, Donoghue, and Brandon and by Ruse, but Dupré's pluralism is too promiscuous. Taxonomies based on cooking lore are taken on a par with those based on contemporary evolutionary biology. Dupré's pluralism is just the sort of pluralism that Ghiselin and Hull worry about in their 'no criterion objection' (see Section III): it legitimizes taxonomies that are in no way based on scientific reasoning. Dupré's pluralism needs to be supplemented by criteria for judging the adequacy of sameness relations; otherwise it condones any taxonomic approach.

Recently Philip Kitcher has become a prominent advocate of species pluralism (see 1984*a*, *b*, 1987, and 1989). Kitcher organizes the species concepts he accepts as legitimate into two types: historical and structural (1984*b*: 321 ff.). Historical species concepts require that species are genealogical entities. The interbreeding, ecological, and phylogenetic approaches are historical concepts: each requires that species taxa form historically (spatio-temporally) continuous entities. Structural species concepts, on the other hand, do not require that species taxa form historically continuous entities. Instead, structural concepts require that the organisms of a species have important functional similarities; Kitcher suggests genetic, chromosomal, or developmental similarities.

Those familiar with the biological literature may wonder why Kitcher accepts structural species concepts as legitimate, for all currently proposed species concepts fall under the historical heading—they all require that species form historically continuous entities (even Kitcher's (1984b; 325) own taxonomy of species concepts illustrates this). Be that as it may, Kitcher wants to stress that biological practice could, and should, allow the legitimacy of historical *and* non-historical (structural) species concepts. To make the legitimacy of non-historical species concepts intuitive, Kitcher cites a hypothetical case of lizard lineages (1984b: 314–15). The lineages are spatio-temporally disconnected from one another; nevertheless their organisms are very similar along morphological, behavioural, ecological, and genetic parameters. Kitcher writes that 'to hypothesize "sibling species" in this case (and in like cases) seems to me not only to multiply species beyond necessity but also to obfuscate all the biological similarities that matter' (ibid. 315). Thus Kitcher suggests that we allow the existence of spatio-temporally disconnected species taxa and accept the legitimacy of non-historical (structural) species concepts. (Ruse (1987: 235–6) provides a similar argument.)

Kitcher's argument for the legitimacy of non-historical species concepts overlooks the theoretical reason that biologists reject such concepts. Hull (1976, 1978, 1987) presents this reason in the following argument (also see Sober 1984). Since the inception of evolutionary theory, species taxa have been considered evolutionary units: that is, groups of organisms capable of evolving. The evolution of such groups requires that the organisms of a species taxon be connected by heredity relations. Heredity relations, whether they be genetic or not, require that the generations of a taxon be historically connected; otherwise information will not be transmitted. The upshot is that if species taxa, or any taxa, are to evolve, they must form historically connected entities. By allowing non-historical species concepts, Kitcher's pluralism falls outside the domain of evolutionary biology and should be rejected.[6]

---

[6] Kitcher (1989) has recently offered a further argument for accepting non-historical species concepts. As Kitcher points out, asserting that species are historical entities does not sufficiently specify the nature of species; for each organism, population, and all of life on this planet is a historical entity. To understand fully the nature of species, we need sufficient conditions that distinguish species from other historical entities. Kitcher then shows that the conditions for segmenting the tree of life into species are vague and problematic. As a result, Kitcher concludes that we should allow the legitimacy of non-historical species concepts (1989: 204). But it is important to note that our lack of fully adequate conditions for segmenting the tree of life into

In summary, Dupré's and Kitcher's forms of pluralism are too liberal, while the forms advocated by Mishler, Donoghue, and Brandon and by Ruse are not liberal enough. Eliminative pluralism charts a middle course between these forms of pluralism. It acknowledges that the forces of evolution create different types of basal taxa. It also recognizes that these different types of taxa give rise to taxonomies that cross-classify the organic world. Eliminative pluralism, however, is prudent enough to place constraints on pluralism: only taxonomic approaches that satisfy the criteria suggested in Section III are allowed into the store of legitimate taxonomic approaches. Moreover, eliminative pluralism avoids ambiguity by designating different types of taxa with different terms, and it preserves consistency by requiring that taxonomic approaches be internally consistent. Some may view eliminative pluralism as just a complicated form of monism. If that is the case, then the arguments of this essay have been successful.[7]

REFERENCES

Andersson, L. (1990), 'The Driving Force: Species Concepts and Ecology', *Taxon*, 39: 375–82.
Carson, L. (1975), 'The Species as a Field for Genetic Recombination', in E. Mayr (ed.), *The Species Problem* (Washington: American Association for the Advancement of Science), 23–38.
Cartwright, N. (1983), *How the Laws of Physics Lie* (Oxford: Oxford University Press).
Cracraft, J. (1983), 'Species Concepts and Speciation Analysis', in R. Johnston (ed.), *Current Ornithology* (New York: Plenum Press), 159–87.
—— (1987), 'Species Concepts and the Ontology of Evolution', *Biology and Philosophy*, 2: 329–46.
Darwin, F. (1987) (ed.), *The Life and Letters of Charles Darwin, Including an Autobiographical Chapter*, 3rd edn. (London: John Murray).
de Queiroz, K., and Donoghue, M. (1988), 'Phylogenetic Systematics and the Species Problem', *Cladistics*, 4: 317–38; reproduced as Ch. 15.
Dobzhansky, T. (1970), *Genetics and the Evolutionary Process* (New York: Columbia University Press).

species in no way nullifies the requirement that species must be historical entities. (In more general terms, a condition's insufficiency does not imply that it is unnecessary.) Species, whether they be basal interbreeding, ecological, or phylogenetic taxa, are historical entities. We are just uncertain on how to draw the boundaries of such taxa. Indeed, such boundaries may be naturally vague (see Ereshefsky 1991).

[7] I thank David Hull, Brent Mishler, and Elliott Sober for their detailed comments on an earlier draft of this essay. Financial support was provided by the Department of Philosophy at Washington University, St Louis, in the form of a Mellon Post-Doctorate Fellowship.

Dupré, J. (1981), 'Natural Kinds and Biological Taxa', *Philosophical Review*, 90: 66–90.

Ehrlich, P., and Raven, P. (1969), 'Differentiation of Populations', *Science*, 165: 1228–32.

Eldredge, N. (1985), *Unfinished Synthesis: Biological Hierarchies and Modern Evolutionary Thought* (New York: Oxford University Press).

——and Cracraft, J. (1980), *Phylogenetic Patterns and the Evolutionary Process: Method and Theory in Comparative Biology* (New York: Columbia University Press).

Ereshefsky, M. (1991), 'Species, Higher Taxa, and the Units of Evolution', *Philosophy of Science*, 58: 84–101.

——(1992), (ed.), *The Units of Evolution: Essays on the Nature of Species* (Cambridge, Mass.: MIT Press).

Frost, D., and Hillis, D. (1990), 'Species in Concept and Practice: Herpetological Applications', *Herpetologica*, 46: 87–104.

Futuyma, D. (1985), *Evolutionary Biology*, 2nd edn. (Sunderland, Mass.: Sinauer Associates).

Ghiselin, M. (1969), *The Triumph of the Darwinian Method* (Berkeley and Los Angeles: University of California Press).

——(1974), 'A Radical Solution to the Species Problem', *Systematic Zoology*, 23: 127–43.

——(1987), 'Species Concepts, Individuality, and Objectivity', *Biology and Philosophy*, 2: 127–43.

——(1989), 'Sex and the Individuality of Species: A Reply to Mishler and Brandon', *Biology and Philosophy*, 4: 73–6.

Grant, V. (1981), *Plant Speciation*, 2nd edn. (New York: Columbia University Press).

Hennig, W. (1966), *Phylogenetic Systematics* (Urbana, Ill.: University of Illinois Press).

Holsinger, K. (1984), 'The Nature of Biological Species', *Philosophy of Science*, 51: 293–307.

Hull, D. (1970), 'Contemporary Systematic Philosophies', *Annual Review of Ecology and Systematics*, 1: 19–54.

——(1976), 'Are Species Really Individuals?', *Systematic Zoology*, 25: 174–91.

——(1978), 'A Matter of Individuality', *Philosophy of Science*, 45: 335–60.

——(1987), 'Genealogical Actors in Ecological Roles', *Biology and Philosophy*, 2: 168–83.

——(1988), *Science as a Process: An Evolutionary Account of the Social and Conceptual Development of Science* (Chicago: University of Chicago Press).

——(1989), 'A Function for Actual Examples in Philosophy of Science', in M. Ruse (ed.), *What the Philosophy of Biology Is: Essays Dedicated to David Hull* (Dordrecht: Kluwer), 309–21.

Kitcher, P. (1984a), 'Against the Monism of the Moment: A Reply to Elliott Sober', *Philosophy of Science*, 51: 616–30.

——(1984b), 'Species', *Philosophy of Science*, 51: 308–33.

——(1987), 'Ghostly Whispers: Mayr, Ghiselin and the "Philosophers" on the Ontological Status of Species', *Biology and Philosophy*, 2: 184–92.

——(1989), 'Some Puzzles about Species', in M. Ruse (ed.), *What the Philosophy of Biology Is: Essays Dedicated to David Hull* (Dordrecht: Kluwer), 183–208.

Laudan, L. (1984), *Science and Values: An Essay on the Aims of Science and their*

Role in Scientific Debate (Berkeley and Los Angeles: University of California Press).

Levins, R. (1968), *Evolution in Changing Environments: Some Theoretical Explorations* (Princeton: Princeton University Press).

Mayr, E. (1969), *Principles of Systematic Zoology* (New York: McGraw-Hill).

——(1970), *Populations, Species, and Evolution: An Abridgment of Animal Species and Evolution* (Cambridge, Mass.: Harvard University Press).

——(1982), *The Growth of Biological Thought: Diversity, Evolution, and Inheritance* (Cambridge, Mass.: Harvard University Press).

——(1987), 'The Ontological Status of Species: Scientific Progress and Philosophical Terminology', *Biology and Philosophy*, 2: 145–66.

——(1988), 'The Why and How of Species', *Biology and Philosophy*, 3: 431–42.

Mishler, B., and Brandon, R. (1987), 'Individuality, Pluralism, and the Phylogenetic Species Concept', *Biology and Philosophy*, 2: 397–414; reproduced as Ch. 14.

——and Budd, A. (1990), 'Species and Evolution in Clonal Organisms—Introduction', *Systematic Botany*, 15: 70–85.

——and Donoghue, M. (1982), 'Species Concepts: A Case for Pluralism', *Systematic Zoology*, 31: 491–503.

Neigel, J., and Avise, J. (1986), 'Phylogenetic Relationships of Mitochondrial DNA under Various Demographic Models of Speciation', in S. Karlin and E. Nevo (eds.), *Evolutionary Processes and Theory* (Orlando, Fla.: Academic Press), 515–34.

Paterson, H. (1985), 'The Recognition Concept of Species', in E. Vrba (ed.), *Species and Speciation* (Pretoria: Transvaal Museum), 21–9.

Ridley, M. (1986), *Evolution and Classification: The Reformation of Cladism* (London: Longman).

——(1990), 'Comments on Wilkinson's Commentary', *Biology and Philosophy*, 4: 447–50.

Ruse, M. (1969), 'Definitions of Species in Biology', *British Journal for the Philosophy of Science*, 20: 97–119.

——(1987), 'Biological Species: Natural Kinds, Individuals, or What?', *British Journal for the Philosophy of Science*, 38: 225–42.

Sneath, P., and Sokal, R. (1973), *Numerical Taxonomy: The Principles and Practice of Numerical Classification* (San Francisco: Freeman).

Sober, E. (1980), 'Evolution, Population Thinking and Evolution', *Philosophy of Science*, 47: 350–83.

——(1984), 'Sets, Species, and Evolution: Comments on Philip Kitcher's "Species"', *Philosophy of Science*, 51: 334–41.

Tajima, F. (1983), 'Evolutionary Relationships of DNA Sequences in Finite Populations', *Genetics*, 105: 437–60.

Templeton, A. (1989), 'The Meaning of Species and Speciation: A Genetic Perspective', in D. Otte and J. Endler (eds.), *Speciation and its Consequences* (Sunderland, Mass.: Sinauer Associates), 3–27.

Van Valen, L. (1976), 'Ecological Species, Multispecies, and Oaks', *Taxon*, 25: 233–9.

PART VI

HUMAN NATURE

.

INTRODUCTION TO PART VI

MICHAEL RUSE

'What is man that thou art mindful of him?', asked the psalmist. 'A being made in the image of God', came back the reply. But what is the image of God, we all ask. Does God have sexuality? And if so, is God male, as the language of the Bible suggests? Or is God female, akin to Mother Earth, as some feminists respond? Perhaps it is better to stay away from the body altogether. Human beings are God-like inasmuch as they have immortal souls; and whatever else this means, it means that we humans have a rational faculty—something distinguishing us from the brutes and plants and other living things. This would have been a happy answer for the Greek philosophers, who put much emphasis on our ability to think, arguing that in it lies the essence, the true, real, distinguishing feature, of humankind.

But what of babes in arms? Or the old man or woman suffering from Alzheimer's disease? Are we to say that they are not human? The decent person recoils from such a conclusion. Are we all less than human when we are under anaesthetic? When we are asleep? The great English philosopher John Locke (1990, bk. III, pt. vi, p. 12), in the seventeenth century, drew back from such a 'realist' or 'essentialist' definition of human nature, opting rather for 'nominalism', where words mean only that to which they apply. Any difference between men and changelings 'is only known to us, by their agreement, or disagreement with the complex *idea* that the word *Man* stands for'. But this suggestion has difficulties of its own. We are all sure that Roy Rogers is a man. Why are we equally sure that Trigger is a horse? If it is all a matter of words, why should not Trigger be a man also?

That this problem of 'universals' in the human context is more than just a matter of esoteric philosophical interest is shown at once by festering sores which afflict society today. Is abortion wrong? Killing human beings is wrong. We all agree to that. But is the foetus a human being? There's the rub! Women are human beings, surely. But are they equal in their humanity to men? If so, then why will the Catholic Church not sanction women priests? Homosexuals are human like the rest of us. But in their actions and desires they are not like the rest of us. Should their behaviour therefore be censured as inappropriate— inappropriate to humans, that is?

Many have thought that biology can surely put these and like issues to rest. The purpose of this part of the anthology is to show that matters are not quite this simple. Biology is surely pertinent. I imagine that most of us, on learning that 99 per cent of our DNA is shared by the higher apes, have thought again about our relationship with the animals—or at least felt the need to mount an argument as to why the genetic similarity should not be taken as decisive. Likewise, a major reason why no serious person today could be a Nazi is that we know that Jews and Gypsies are simply not that different from anyone else. There was probably as much genetic difference between Hitler and Goebbels as between Hitler and a rabbi in a Polish ghetto. But biology is certainly not everything. If it were, the abortion debate would be past history.

I have said, *a* major reason why no serious person today could be a Nazi is biological. But should I have allowed that biology is pertinent at all? Surely, the only reason for not being a Nazi is that being a Nazi is morally wrong? The facts of biology are irrelevant. If there were all the difference in the world between Hitler and the rabbi, it would still have been wrong for the former to have sent the latter to Auschwitz. John Maynard Smith, one of the world's most distinguished evolutionists, who tells us in his chapter (17) how his whole view of life was formed by growing up during the Nazi era, wrestles with issues like this. Ultimately, he is uncomfortable with a hard-line position either way. Science has to be important. But what he calls 'myth'—one's moral framework—is no less important. The trick is finding the right blend of the two.

David Hull, a philosopher, is very wary of attempts to found morality on some disinterested objective universal biologically grounded human nature, and he uses biology to make his point! He shows (Ch. 18) just how much variability there is in a species, the human species in particular, and he argues that any attempt to find a shared essence of what it is to be human is bound to fail. Does this mean that attempts to link biology and morality are doomed to failure? Not necessarily. Citing the sociobiologists Michael Ruse and Edward O. Wilson, perhaps morality is a contingent part of our being, and perhaps there are different moralities for different people. What we cannot conclude is that there is one standard for all, and that everyone who fails to fit—the handicapped, women, homosexuals—is thereby abnormal and in the wrong.

The distinguished feminist historian of science Evelyn Fox Keller gives us a brief history of the way in which research developed on the whole question of gender and science (Ch. 19). If there is a moral to be drawn from her story, it is that any attempt to come to fair decisions about differences between males and females will be very difficult, given the ways in which traditional science built biases right into its theorizing. However, she does leave one with the feeling that satisfactory answers may be more possible now than they were just a few years ago. However, the psychologist Susan Oyama (Ch. 20) argues that, paradoxically, one might be misled by the very people intent on liberating our thinking. In discussing claims about the innate nature of women, she shows that it is not just insensitive male chauvinists who think that there might be something special and 'feminine' about women, but that this is indeed an assumption of certain contemporary feminist writers. Oyama concludes by warning against simplistic attempts to put human nature at the feet of either the genes (nature) or the environment (nurture). The true answer, one which will apply to both men and women, requires a subtle combination of both.

Finally, in a piece written especially for this volume, the philosopher Edward Stein (Ch. 21) looks at the current debate between sexologists, who tend to think that sexual orientation is a function of biology and hence an 'essential' aspect of human nature, and gay activists, who are much influenced by the writings of the late Michel Foucault, and who argue that such orientation is a 'construction', something made by society rather than found in reality. Stein refers to some of the most recent scientific work, specifically that of the neurobiologist Simon LeVay, who claims to find difference between the brains of male heterosexuals and male homosexuals. It is Stein's contention that work like this can never be definitive, because it has biases away from constructivism and towards essentialism built right into its methodology. He concludes with a plea for a more pluralistic approach, feeling that in the quest to understand human nature, essentialists have things to learn from constructivists.

REFERENCE

Locke, John (1990), *Drafts of the 'Essay Concerning Human Nature' and Other Philosophical Writing*, ed. P. Nidditch and G. A. Rogous (Oxford: Oxford University Press).

17

SCIENCE AND MYTH

JOHN MAYNARD SMITH

At twenty, I thought 'we should do without myths and confine ourselves to science...but it really won't do'.

Maynard Smith

Recently, after giving a radio talk on Charles Darwin, I received through
the post a pamphlet by Don Smith entitled 'Why Are There Gays At All?
Why Hasn't Evolution Eliminated Gayness Millions of Years Ago?' The
pamphlet points to a genuine concern: the prevalence of homosexual
behaviour in our species is not understood, and is certainly not something
that would be predicted from Darwinian theory. Smith wrote the pamphlet
because he believes the persecution of gays has been strengthened and
justified by the existence of a theory of evolution that asserts gays are unfit
because they do not reproduce. He also believes that gays can be protected
from future persecution only if it can be shown that they have played an
essential and creative role in evolution. His argument is that, in evolution,
novelty arises when individuals adopt mating habits different from those
typical of their species: it is summed up on a button that reads: 'Sexual
deviation is the mainspring of evolution'.

I do not find this argument particularly persuasive, but that is not the
point I want to make. I think Smith would have been better advised to have
written: 'If people despise gays because gayness does not contribute to
biological fitness, they are wrong to do so. It would be as sensible to
persecute mathematicians because an ability to solve differential equations
does not contribute to fitness. A scientific theory—Darwinism or any
other—has nothing to say about the value of a human being.'

The point I am making is that Smith is demanding of evolutionary
biology that it be a myth: that is, a story with a moral message. He is not
alone in this. Elaine Morgan's book *The Descent of Woman* is an account
of the origin of *Homo sapiens* that is intended to give mythical support to

First published in *Natural History*, 93/11 (1984): 10–24. © 1984, American Museum of Natural
History. Reprinted with permission from *Natural History* (1984).

the women's movement by emphasizing the role of the female sex and, in particular, the mother–child bond. She claims, with reason, that many other accounts of human evolution have, perhaps unconsciously, placed undue emphasis on the role of males. Earlier, George Bernard Shaw wrote *Back to Methuselah* (1922) avowedly as an evolutionary myth, because he found in Darwinism a justification of selfishness and brutality, and because he wished instead to support the Lamarckian theory of the inheritance of acquired characters, which he saw as justifying free will and individual endeavour.

We should not be surprised by Don Smith, Elaine Morgan, and Bernard Shaw. In all societies, people have constructed myths about the origins of the universe and of humans. The function of these myths is to define our place in nature, and thus to give us a sense of purpose and value. Since Darwinism is, among other things, an account of human origins, is it any wonder that it is expected to carry a moral message?

The people and the objects that figure in a myth stand not only for themselves, but also as symbols of other things. To some extent, myths and their symbolic components develop simply because human beings find it difficult to accept any input as meaningless. Shown an ink blot, we see witches, bats, and dragons. This refusal to accept input as mere noise lies at the root of divination by tarot cards, tea leaves, the livers or shoulders of animals, or the sticks of the *I Ching*. It may also account for the strangely late development of a mathematical theory of probability or of any scientific theory with a chance element. As anthropologist Dan Sperber has written: 'Symbolic thought is capable, precisely, of transforming noise into information.'

Another—and in the present context, more important—function of myths is to give moral and evaluative guidance. Some myth making is quite conscious. In *Back to Methuselah*, for example, Shaw deliberately invented a story that would have the moral effect he desired. More usually, however, I suspect that a myth-maker conceives a story that moves him or her in a particular way—at its lowest, it reinforces prejudices, and at its highest, to borrow Aristotle's words, it evokes feelings of pity and fear. People repeat myths because they hope to persuade others to behave in certain ways.

This raises the question of why we use myths rather than simple statements of instruction. Why do we talk of King Alfred and the cakes, for example, instead of saying that people in important positions should be modest? Perhaps a story whose meaning has to be puzzled out or guessed carries more conviction than a mere instruction. What we imagine is more important than what we are told.

Sometimes, I find it hard to discover how far people distinguish stories intended to give moral guidance from those meant simply to supply technical help. Confusion seems particularly likely to crop up when rituals are involved. For example, if, before going into battle, a man sharpens his spear and undergoes ritual purification (or, for that matter, cleans his rifle and goes to mass), he may regard the two procedures as equally efficacious. Indeed, they may well be so, one in preparing the spear and the other himself. If we regard the former as more practical, we do so only because we understand metallurgy better than psychology.

Despite the difficulty, most people do try to distinguish procedures and technical instructions that alter the external world from procedures and stories intended to alter our own state of consciousness or persuade us that certain things are right. Indeed, we take some trouble when educating our children to give hints about which category of information is being transmitted. For example, a surprisingly large proportion of the stories read aloud to children, particularly those with a moral message, are about talking animals or even talking steam-engines. It is as if we wanted to be sure that the stories are not taken literally.

While such efforts may be successful in many spheres of human endeavour, the examples of Don Smith and Bernard Shaw show how hard it is for many people to separate science, and especially evolution, from myth. One reaction to this difficulty is to assert that there is no difference, that evolution theory has no more claim to objective truth than Genesis. Many scientists would be enraged by such an assertion, but rage is no substitute for argument. In the last century, it was widely held that the scientific method, conceived of as establishing theories by induction from observation, led to certain knowledge. Darwin and Einstein have robbed us of that certainty—or have liberated us from that prison. If, as Darwin showed, there is not a fixed and finite number of things in the universe, each with a knowable essence, then induction is logically impossible. Einstein, in turn, showed that what scientists had been most confident of—classical mechanics—was at worst false and at best a special case of a more general theory. After that twin blow, certain knowledge is something we can expect only at our funerals.

But it is one thing to admit that scientific knowledge cannot be certain, and another to claim that there is no difference between science and myth. Karl Popper, perhaps the most influential contemporary philosopher of science, has told us that it was the impact of Einstein, and in particular the wish to distinguish Einstein's theory from those of Freud, Adler, and Marx, that led him to propose falsifiability as the criterion for separating science from pseudo-science. If a theory is scientific, he suggested, observations can be conceived of which, if they were accepted, would show the theory to be false. In contrast, he suggests that no conceivable pattern of human behaviour could falsify Freudian theory.

Popper's views have been attacked, primarily on the grounds that there are no such things as theory-free observations. Every observation is subject to interpretation, conscious and unconscious. Consequently, there can never be certain grounds for rejecting a scientific theory, and hence the distinction between science and pseudo-science disappears.

This criticism seems to me largely to miss the point. If Popper were claiming that scientific knowledge were certain, then the impossibility of certain falsification would indeed be damaging. But he makes no such claim. He insists on two things. First, a scientific theory must assert that certain kinds of events cannot happen, so that the theory is falsified if these events are subsequently observed, and second, there is inevitably a logical asymmetry in any attempt to test a theory, so that a theory can be falsified but cannot be proved true by the acceptance of observation.

There is, however, a tide of ideas that would deny the distinction. The emotional force behind this tide derives, in part, from an entirely proper disgust at some of the consequences of technology in the modern world and, in part, from an equally proper wish to treat the ideas of other peoples as of equal value to our own. What is common to these two reactions is the conviction, which I share, that scientific theories are not the only kind of ideas that we need. A frequently drawn corollary of this conviction, which I do not share, is that scientific ideas are not distinguishable from other ideas.

One source of the belief that science and myth can be lumped together lay, surprisingly, in Marxism, a philosophy that has led to two very different interpretations of science. One interpretation was pioneered by the British physicist J. D. Bernal. For Bernal, the crucial thing about science was that it made socialism possible by providing the techniques needed to satisfy people's wants. He saw science as being distorted under capitalism—for example, by being pressed into the service of military research; nevertheless, he does not appear to have thought that capitalism would prevent science from making progress toward an understanding of nature. In this, his views coincided with those of Marx himself, who largely excluded science from the set of ideas—for example, about religion, philosophy, and law—that he saw as reflecting the class interests of those who held them.

Thus Bernal regarded science as the greatest hope for the future, and would have rejected any suggestion that science is indistinguishable from myth. However, another thread within Marxism has led to a different end.

In 1931, the Russian B. Hessen argued that not only had Newton been influenced by the technical problems of his day (for example, gunnery and navigation), but also that the form his theory took reflected contemporary society. Such a view is perhaps more easily understood and more obviously true when applied to Darwin, whose theory did recognize in the natural world the processes of competition predominant in the society of his day. Indeed, both Darwin and Wallace stated that they borrowed their essential concept from the economist Malthus. If, then, this second thread of Marxist thinking argues, major scientific theories merely project on to nature features of contemporary society, they have more in common with myths than most scientists would readily accept.

Here, it seems to me, a crucial distinction must be made between the psychological sources of a theory and the testing of it. If Darwin's ideas, or Newton's, were accepted because they were socially appealing, then indeed science and myth would be indistinguishable. But I do not think that they were. They were accepted because of their explanatory power and ability to withstand experimental test. Of course, new ideas in science sometimes come from analogies with society, just as, in one scientific discipline, such as biology, they arise by analogy with others, such as physics and engineering. But what matters for the progress of science is not where the ideas come from, but how they are treated.

Society influences the development of science through both the problems that seem worth solving and the resources available for their solution. I have little doubt that society also influences scientists, both as individuals and groups, by making some ideas seem worth pursuing and others implausible or unpromising. For example, my own caution about applying to humans ideas drawn from a study of animal societies—a caution that contrasts with the enthusiasm of such scientists as E. O. Wilson and Richard Alexander—probably arose because I grew up under the shadow of Hitler and the Nazi theories of racial superiority and biological determinism, and not because of anything internal to biology or sociology.

There is, however, a caricature inherent in the externalist view of science that I reject emphatically. This is the idea that we can evaluate a scientific theory by reference to the society in which it was born, or to the moral and political conclusions that might be drawn from it. Once accept that view and science is dead, as genetics died in Russia in 1948, when Stalin supported Lysenko's Lamarckian views against the Mendelians. Stalin took his position partly in the hope of quick returns in agricultural productivity, and partly because Lysenko's belief in the inheritance of acquired characters seemed to accord better with Marxism than did the orthodox—and, as it happens, more nearly correct—Mendelian doctrine that hereditary characteristics are transmitted from parent to offspring by genes, and that the genetic message is independent of changes induced in the body of the parent during its lifetime.

Today, the belief that there are no objective criteria whereby one can choose between rival theories (and hence, by implication, that one can allow one's prejudices full rein) derives largely, I think, from the work of Thomas Kuhn, although the conclusion is far from the one that he himself would wish to draw. Kuhn sees science as divided into 'normal' and 'revolutionary' periods. In a period of normal science, members of a scientific community agree about what assumptions can be made, what problems are worth solving, what will count as a solution, and what experimental methods should be used. Most important, they share a 'paradigm', or set of exemplary solutions to problems, that can be used as a standard. Revolutions occur when, usually as a result of long-continued failure to solve certain problems within the accepted frame, a fundamentally new set of assumptions and procedures replaces the old.

All this seems to bear some resemblance to reality; it also bears some resemblance to Popper's remark that 'there is much less accumulation of knowledge in science than there is revolutionary changing of scientific theories'. The main difficulty lies in Kuhn's account of how one paradigm replaces another. He speaks of a 'paradigm debate' in which the proponents 'fail to make complete contact with each other's viewpoints', and in which they 'see the world differently'. Again, there is much in what he says. During my lifetime in science, I have engaged in too many arguments, in which I and my opponent have talked right past one another, not to recognize this.

The fallacy is to suppose that because two scientists are unable to understand each other fully, there is no rational way, given time, of settling the issue between them. With the passage of time, choosing between two theories or two methods of approach becomes easier. Eventually, one or the other approach is more successful in overcoming its difficulties, or, as in the case of the particle and wave interpretations of light, a third theory is developed and subsumes them both. The trouble is that scientists must often commit themselves before the evidence is in. In Darwin's words, one must have 'a theory by which to work'. This is what gives an air of irrationality to the procedure, and has led some people to suppose that the choice between scientific theories is arbitrary.

Consider an example. After the rediscovery of Mendel's laws in 1900, a debate broke out between the Mendelians and the earlier school of biometricians, headed by Karl Pearson. Pearson refused to accept the new theory, at least in part because it was incompatible with his previously held philosophical view that the business of science is merely to describe the world and not to imagine hypothetical entities such as genes. As the historian Bernard Norton has recently pointed out, Pearson understood that Mendelian inheritance could account for the phenomena of continuous variability, which he had been studying, but still rejected it on philosophical grounds—an interesting illustration of how philosophical preconceptions can be a poor guide to scientific practice. Yet, despite all this, no one today would doubt the utility either of Pearson's statistical methods or of Mendelian genetics.

If theories were genuinely incommensurable, and rational choice between them impossible, progress in science would not be expected. Kuhn himself accepts the reality of scientific progress, but only in the sense that the explanatory power of scientific theories has increased: he doubts whether it is sensible to say that science draws closer to the truth about what is 'really there'. I will return to this point in a moment.

Before leaving Kuhn, I want to suggest that, despite his insights, his insistence on a distinction between normal and revolutionary science, and on the incommensurability of paradigms, has been exaggerated. The major scientific revolution during my working life has been the rise of molecular biology, which has all the characteristics of a new 'disciplinary matrix' in Kuhn's sense—new scientists, new problems, new experimental methods, new journals, new textbooks, and new culture heroes. But where was the incommensurability? I myself was raised in the older discipline of classical genetics, and have never mastered the experimental methods of the new. Yet my almost immediate reaction to the Watson–Crick paper was that a mystery within my field had been cleared up. Those of us trained in classical genetics sometimes had difficulty in learning the new techniques, but there were few conceptual difficulties and no paradigm debate.

Perhaps the birth of molecular genetics was not a Kuhnian revolution. As it happens, I suspect that before we make much progress in developmental biology, a bigger conceptual revolution may be needed than the transition from classical to molecular genetics. All the same, if molecular biology could be born without the full panoply of a paradigm debate, where does that leave the concepts of normal and revolutionary science?

The history of genetics also forces us to look again at Kuhn's suggestion that progress in science is progress in explaining, but not progress in knowing what the world is really like. To me, the change from the concept of the gene as a Mendelian factor to the gene as a piece of a chromosome, and thence to the gene as a molecule of DNA, docs look like progress in knowing what the world is like. But perhaps that is a question I should leave to the philosophers.

I would not have spent so much time discussing the difference between scientific theories and myths if the difference between them were obvious. Indeed, they have much in common. Both are constructs of the human mind, and both are intended to have a significance wider than the direct assertions they contain. Popper suggested falsifiability as the criterion distinguishing them, and I think he was right. However, we can often also distinguish them by their function: the function of a scientific theory is to account for experience—often, it is true, the rather esoteric experience emerging from deliberate experiment; the function of a myth is to provide a source and justification for values. What, then, should be the relation between them?

Three views are tenable. The first, sometimes expressed as a demand for 'normative science', is that the same mental constructs should serve both as myths and as scientific theories. If I am right, this widely held view underlies the criticism of Darwinism from gays, from the women's movement, from socialists, and so on. It explains the preference expressed by some churchmen for 'Big Bang' as opposed to 'steady state' theories of cosmology. Although well-intentioned, it seems to me pernicious in its effects. Applied to evolution theory, it means either that we must embrace Darwinism and draw from it the conclusion that gays are unnatural and social services wicked, or that we must embrace Lamarckism whether or not the genetic evidence supports it. Normative science will be bad morality or bad science, and most probably both.

The second view is that we should do without myths and confine ourselves to science. This is the view I held at the age of 20, but it really won't do. If, as I now believe, scientific theories say nothing about what is right, but only about what is possible, we need some other source of values, and that source has to be myth, in the broadest sense of the term.

The third view, and I think the only sensible one, is that we need both myths and scientific theories, but that we must be as clear as we can be about which is which. In essence, this was the view urged by the French molecular biologist Jacques Monod in *Chance and Necessity* (1972). Oddly, Monod was almost universally derided by his critics for arguing that one can derive values from science, when in fact he argued the precise opposite. His case was that there is no place in science for teleological, or value-laden, hypotheses. Yet, to do science, one must first be committed to some values—not least, to the value of seeking the truth. Since this value cannot be derived from science, it must be seen as a prior moral commitment, needed before science is possible. So far from values being derived from science, Monod saw science as depending on values.

Although I disagree with some aspects of his book, I agree with Monod on two basic points. First, values do not derive from science, but are necessary for the practice of science. Second, we should distinguish as clearly as we can between science and myth. We should make this distinction, not because we could then discard the myths and retain only science, but because the roles they play are different. Scientific theories tell us what is possible; myths tell us what is desirable. Both are needed to guide proper action.

18

ON HUMAN NATURE

DAVID L. HULL

Generations of philosophers have argued that all human beings are essentially the same—that is, they share the same nature—and that this essential similarity is extremely important. Periodically, philosophers have proposed to base the essential sameness of human beings on biology. In this essay I argue that if 'biology' is taken to refer to the technical pronouncements of professional biologists—in particular, evolutionary biologists—it is simply not true that all organisms that belong to *Homo sapiens* as a biological species are essentially the same. If 'characters' is taken to refer to evolutionary homologies, then periodically a biological species might be characterized by one or more characters which are both universally distributed among, and limited to, the organisms belonging to that species, but such states of affairs are temporary, contingent, and relatively rare. In most cases, any character universally distributed among the organisms belonging to a particular species is also possessed by organisms belonging to other species; and, conversely, any character that happens to be limited to the organisms belonging to a particular species is unlikely to be possessed by all of them.

The natural move at this juncture is to argue that the properties which characterize biological species at least 'cluster'. Organisms belong to a particular biological species because they possess enough of the relevant properties or enough of the more important relevant properties. Such unimodal clusters do exist, and might well count as 'statistical natures', but in most cases the distributions that characterize biological species are multi-modal, depending on the properties studied. No matter how desperately one wants to construe biological species as natural kinds characterizable by some sort of 'essences' or 'natures', such multi-modal distributions simply will not do. To complicate matters further, these clusters of properties, whether uni- or multi-modal, change through time. A character state

First published in *PSA 1986* (East Lansing, Mich.: Philosophy of Science Association, 1986), ii. 3–13. Reprinted by permission.

(or allele) which is rare may become common, and one that is nearly universal may become entirely eliminated. In short, species evolve, and to the extent that they evolve through natural selection, both genetic and phenotypic variation are essential. Which particular variations a species exhibits is a function of both the fundamental regularities which characterize selection processes and numerous historical contingencies. However, variation as such is hardly an accidental characteristic of biological species. Without it, evolution would soon grind to a halt. *Which* variations characterize a particular species is to a large extent accidental; *that* variation characterizes species as such is not.

The preceding characterization depends on the existence of a criterion for individuating species, in addition to character covariation. If species are taken to be the things which evolve, then they can, and must, be characterized in terms of ancestor–descendant relations, and in sexual species these relations depend on mating. The organisms that comprise sexual species form complex networks of mating and reproduction. Any organism that is part of such a network belongs to that species, even if the characters it exhibits are atypical or in some sense aberrant. Conversely, an organism that happens to exhibit precisely the same characters as an organism belonging to a particular species might not itself belong to that species. Genealogy and character covariation are not perfectly coincident, and when they differ, genealogy takes precedence. The priority of genealogy to character covariation is not negated by the fact that species periodically split or bud off additional species. To the extent that speciation is 'punctuational', such periods will be short, and involve only a relatively few organisms; but inherent in species as genealogical entities is the existence of periods during which particular organisms do not belong unequivocally to one species or another. *Homo sapiens* currently is not undergoing one of these periods. The genealogical boundaries of our species are extremely sharp. The comparable boundaries in character space are a good deal fuzzier. As a result, those who view character covariation as fundamental and want our species to be clearly distinguishable from other species accordingly are forced to resort to embarrassing conceptual contortions to include retardates, dyslexics, and the like in our species, while keeping bees and computers out.

The preceding observations about species in general and *Homo sapiens* in particular frequently elicit considerable consternation. Biological species cannot possibly have the characteristics that biologists claim they do. There *must* be characteristics which all and only people exhibit, or at least *potentially* exhibit, or which all *normal* people exhibit—at least potentially. I continue to remain dismayed at the vehemence with which these views are expressed in the absence of any explicitly formulated biological foundations for these notions. In this essay I argue that biological species, including our own, do have the character claimed by evolutionary biologists, and that attempts to argue away this state of affairs by reference to 'potentiality' and 'normality' have little if any foundation in biology. Perhaps numerous ordinary conceptions exist in which an organism that lacks the genetic information necessary to produce a particular enzyme nevertheless possesses this enzyme potentially. I am equally sure that there are conceptions of normality according to which worker bees are abnormal. But these ordinary conceptions have no foundation in biology as a technical discipline. To make matters even worse, I do not see why the existence of human universals is all that important. Perhaps all and only people have opposable thumbs, use tools, live in true societies, or what have you. I think that such attributions are either false or vacuous, but even if they were true and significant, the distributions of these particular characters is largely a matter of evolutionary happenstance. I, for one, would be extremely uneasy to base something as important as human rights on such temporary contingencies. Given the character of the evolutionary process, it is extremely unlikely that all human beings are essentially the same, but even if we are, I fail to see why it matters. I fail to see, for example, why we must all be essentially the same to have rights.

To repeat, in my discussion of human nature, I am taking 'human' to refer to a particular biological species. This term has numerous other meanings, which have little or nothing to do with DNA, meiosis, and what have you. Nothing that I say should be taken to imply anything about ordinary usage, common-sense conceptions, or what 'we' are inclined to say or not to say. In particular, I am not talking about 'persons'. The context of this essay is biology as a scientific discipline. Within biology itself several different species concepts can also be found. I am concerned only with those doctrines which claim to be based on the nature of *Homo sapiens* as a biological species. Those authors who are not interested in what biologists have to say about biological species or who are content with conceptual pluralism for the sake of conceptual pluralism will find nothing of interest here.

I. UNIVERSALITY AND VARIABILITY

All concepts are to some extent malleable, and data can always be massaged, but in some areas both activities are more narrowly constrained than in others. For example, it is much harder to argue for genetic than for cultural universals, because the identity of alleles is easier to establish than the identity of cultural practices. However, if biological species are characterized by a particular sort of genetic variability, then one might be justified in exposing claims that cultural traits are immune to a similar variability to closer scrutiny. I certainly do not mean to imply by the preceding statement that I think that cultural variability is in any sense caused by genetic variability. Rather, the reason for introducing the topic of genetic variability is that geneticists have been forced to acknowledge it in the face of considerable resistance, the same sort of resistance that confronts comparable claims about cultural variability. If there are any cultural universals, one of them is surely a persistent distaste for variability. But if genetic variability characterizes species, even though everyone is absolutely certain that it does not, then possibly a similar variability characterizes cultures, even though the parallel conviction about cultures is, if anything, stronger.

For example, Kaplan and Manners remark that a 'number of anthropologists have even attempted to compile lists of universal cultural characteristics. Presumably such cultural universals reflect in some sense the uniform psychological nature of man. But the search for cultural universals has invariably yielded generalizations of a very broad, and sometimes not particularly illuminating nature—such as, all cultures prefer health to illness; or, all cultures make some institutional provision for feeding their members; or, all cultures have devices for maintaining internal order' (1972: 151). Massive evidence can be presented to refute the claim that all human beings have essentially the same blood type. A parallel response to the claim that all cultures prefer health to illness is more difficult, because of the plasticity of such terms as 'health' and 'illness'. My argument is analogical. Both population geneticists and anthropologists have been strongly predisposed to discount variability. Genetics is sufficiently well developed that geneticists have been forced to acknowledge how variable both genes and traits are, both within species and between them. The social sciences are not so well developed. Hence, it is easier for them to hold fast to their metaphysical preferences.

One reason for anthropologists searching so assiduously for cultural universals is the mistaken belief that some connection exists between universality and innateness. For example, in a paper on the *human* nature of human nature, Eisenberg states that 'one trait common to man everywhere is language; in the sense that only the human species displays it, the capacity to acquire language must be genetic' (1972: 126). In the space of a very few words, Eisenberg elides from language being common to man everywhere (universality), to the capacity to acquire language being the result of successive interactions between its genes, current phenotypic make-up, and successive environments. The reaction norm for a particular genotype is all possible phenotypes that would result, given all possible sequences of environments in which the organism might survive. Needless to say, biologists know very little about the reaction norm for most species, our own included. To estimate reaction norms, biologists must have access to numerous genetically identical zygotes, and be able to raise these zygotes in a variety of environments. When they do, the results are endlessly fascinating. Some reaction norms are very narrow—that is in any environment in which the organism can develop, it exhibits a particular trait, and only that trait. Sometimes reaction norms turn out to be extremely broad. A particular trait can be exhibited in a wide variety of states, depending on the environments to which the organism is exposed. Sometimes a reaction norm starts off broad, but rapidly become quite narrow. Some reaction norms are continuous; others disjunctive. Sometimes most organisms occupy the centre of the reaction norm; sometimes they are clustered at either extreme, and so on. Everything that could happen, in some organism or other, does happen.

In spite of all the preceding, the conviction is sure to remain that in most cases there must be some normal developmental pathway through which most organisms develop or would develop if presented with the appropriate environment, or something. But inherent in the notion of a reaction norm are alternative pathways. Because environments are so variable in both the short and the long term, developmental plasticity is absolutely necessary if organisms are to survive to reproduce. Any organism that can fulfil a need in only one way in only a narrowly circumscribed environment is not likely to survive for long. Although there are a few cases in which particular species can fulfil one or two functions in only highly specialized ways, both these species and their specialized functions are relatively rare.

But, one might complain, there *must* be some significant sense of 'normal development'. There is a fairly clear sense of 'normal development', but it is not very significant. As far as I can see, all it denotes is that development pathway with which the speaker is familiar in recent, locally prevalent environments. We find it very difficult to acknowledge that a particular environment which has been common in the recent past may be quite new and 'aberrant', given the duration of the species under investigation. Throughout most of its existence, a species may have persisted in very low numbers and only recently boomed to produce high population density, and high population density might well switch increasing numbers of organisms to quite different developmental pathways. During this transition period, we are likely to look back on the old pathway as 'normal' and decry the new pathway as 'abnormal'; but as we get used to the new alternative, just the opposite intuition is likely to prevail. Although the nuclear family is a relatively new social innovation, and is rapidly disappearing, to most of us it seems 'normal'. Any deviation from it is sure to produce humanoids at best.

From the evolutionary perspective, all alleles which we now possess were once more than just rare: they were unique. Evolution is the process by which rare alleles become common, possibly universal, and universally distributed alleles become totally eliminated. If a particular allele must be universally distributed among the organisms belonging to a particular species (or at least widespread) in order to be part of its 'nature', then natures are very temporary, variable things. From the human perspective, evolutionary change might seem quite slow. For example, blue eyes have existed in the human species from the earliest recorded times, yet less than 1 per cent of the people who belong to the human species have blue eyes. Because people with blue eyes can see no better than people with brown eyes, one plausible explanation for the increase of blue eyes in the human population is sexual selection. It might well take thousands of generations for a mutation to replace what was once termed the 'wild type' and become the new 'wild type'. Early on one allele will surely be considered natural, while later on its replacement will be held with equal certainty to be natural. Human memory is short. From the evolutionary perspective, claims about 'normal' genes tend to be sheer prejudice arising from limited experience.

If by 'human nature' all one means is a trait which happens to be prevalent and important for the moment, then human nature surely exists. Each species exhibits adaptations, and these adaptations are important for its continued existence. One of our most important adaptations is our ability to play the knowledge game. It is important that enough of us play this game well enough, because our species is not very good at anything else. But this adaptation may not have characterized us throughout our existence, and may not continue to characterize us in the future. Biologically, we will remain the same species, the same lineage, even though we lose our 'essence'. It should also be kept in mind that some non-humans play the knowledge game better than some humans. If those organisms that are smarter than some people are to be excluded from our species, while those people who are not all that capable are kept in, something must be more basic than mental ability in the individuation of our species. Once again, I am discussing *Homo sapiens* as a biological species, not personhood. Although in a higher and more sophisticated sense of 'human being' retardates are not human beings, from the crude and pedestrian biological perspective, they are unproblematically human.

The central notion of normality relative to human nature, however, seems to be functional. When people dismiss variation in connection with human nature, they usually resort to functional notions of normality and abnormality. Perhaps someone has produced a minimally adequate analysis of 'normal function', but I have yet to see it. As the huge literature on the subject clearly attests, it is difficult enough to give an adequate analysis of 'function', let alone 'normal function'. In general, structures and functions do not map neatly on to each other, nor can they be made to do so. A single structure commonly performs more than one function, and, conversely, a single function can be fulfilled by more than one structure. If one individuates structures in terms of functions and functions in terms of structures, then the complex mapping of structures and functions can be reduced, possibly eliminated, but only at considerable cost. For example, no matter how one subdivides the human urogenital system, there is no way to work it out so that a particular structure is used for excretion and another structure is used for reproduction. No amount of gerrymandering succeeds without extreme artificiality. Nor has anyone been able to redefine functional limit so that excretion and reproduction turn out to count as a single function.

Like it or not, a single structure can perform more than one function, and one and the same function can be performed by more than one structure. Nor is this an accidental feature of organisms. In evolution, organisms must make do with what they've got. An organ evolved to perform one function might be commandeered to perform another. For example, what is the normal function of the hand? We can do many things with our hands. We can drive cars, play the violin, type on electronic computers, scratch itches, masturbate, and strangle one another. Some of these actions may seem normal, others not; but there is no correlation between common-sense notions of normal functions and the functions which hands were able to fulfil throughout our existence. Any notion of 'the function of the hand' which is sufficiently general to capture all the things that we can do with our hands is likely to be all but vacuous, and surely will make no cut between normal and abnormal uses. About all a biologist can say about the function of the human hand is that anything that we can do with it is 'normal'. A more restricted sense of normality must be imported from common sense, society, deeply held intuitions, or systems of morals. Some might argue that this fact merely indicates the poverty of the biological perspective. If so, so be it, but this is the topic of my essay.

A few additional examples might help us to see the huge gap that exists between biological senses of 'function' and the various senses of this term as it is used in other contexts. A major topic in the biological literature is the function of sexual reproduction. What is the function of sex? The common-sense answer is reproduction, but this is not the answer given by biologists. Biologically, first and foremost, the primary function of sex is to increase genetic heterogeneity. 'But that is not what I mean! When I say that the biological function of sex is reproduction, I do not mean "biological" in the sense that biologists use this term but in some other, more basic sense.' Is being sexually neuter functionally normal? Well, it is certainly normal among honey-bees. Most honey-bees are neuter females. Many species, especially social species, exhibit reproductive strategies that involve some organisms becoming non-reproductives. What counts in biological evolution is inclusive fitness. It is both possible and quite common for organisms to increase their inclusive fitness by not reproducing themselves. 'But I am talking about human beings, not honey-bees.' From the perspective of common-sense biology, human non-reproductives such as old maids and priests may be biological abnormal, but from the perspective of professional biology, they need not be.

Finally, having blue eyes is abnormal in about every sense one cares to mention. Blue-eyed people are very rare. The inability to produce brown pigment is the result of a defective gene. The alleles which code for the structure of the enzyme which completes the synthesis of the brown pigment found on the surface of the human iris produce an enzyme which cannot perform this function. As far as we know, the enzyme produced performs no other function either. However, as far as sight is concerned, blue eyes are perfectly functional, and as far as sexual selection is concerned, downright advantageous. What common sense has to say on these topics, I do not know. My own common-sense estimates about what 'we' mean when 'we' make judgements on such topics depart so drastically from what analytic philosophers publish on these topics that I hesitate to venture an opinion lest I mark myself as being linguistically abnormal.

III. CONCLUSION

Because I have argued so persistently for so long that particular species lack anything that might be termed an 'essence', I have got the reputation of being totally opposed to essentialism. To the contrary, I am rather old-fashioned on this topic (see Dupré 1986 for a more contemporary view). In fact, I think that natural kinds do exist, and that they exhibit characters which are severally necessary and jointly sufficient for membership. More than this, I think that it is extremely important for our understanding of the natural world that such kinds exist. All I want to argue is that natural kinds of this sort are very rare, extremely difficult to discover, and that biological species as evolving lineages do not belong in this category. Just because one thinks that species are not natural kinds, it does not follow that one is committed to the view that there are no natural kinds at all. One misplaced example does not totally invalidate a general thesis.

In fact, I think that the species category might very well be a natural kind, and that part of its essence is variability. If variability is essential to species, then it follows that the human species should be variable, both genetically and phenotypically, and it is. That *Homo sapiens* exhibits considerable variability is not an accidental feature of our species. Which particular variations we exhibit is largely a function of evolutionary happenstance; the presence of variability itself is not. Nor does it help to switch from traditional essences to statistically characterized essences. If the history of phenetic taxonomy has shown anything, it is that organisms can be subdivided into species as Operational Taxonomic Units in indefinitely many ways if all one looks at is character covariation. Compared to many species, our species is relatively isolated in character space. Perhaps a unimodal distribution of characters might be found which succeeds in placing all human beings in a single species and in keeping all non-humans out. If so, this too would be an evolutionary happenstance, and might well change in time.

But why is it so important for the human species to have a nature? One likely answer is to provide a foundation for ethics and morals. If one wants to found ethics on human nature, and human nature is to be at least consistent with current biological knowledge, then it follows that the resulting ethical system will be composed largely of contingent claims. The only authors of whom I am aware who acknowledge this state of affairs, and are still willing to embrace the consequences that flow from it, are Michael Ruse and E. O. Wilson. Ruse and Wilson propose to base ethics on the epigenetic rules of mental development in human beings. They acknowledge that these rules are the 'idiosyncratic products of the genetic history of the species and as such were shaped by particular regimes of natural selection. . . . It follows that the ethical code of one species cannot be translated into that of another. No abstract moral principles exist outside the particular nature of individual species.' (1986: 186).

19

GENDER AND SCIENCE: ORIGIN, HISTORY, AND POLITICS

EVELYN FOX KELLER

Historians of science may be tempted to treat gender and science as a sub-speciality within the history of science—indeed, to judge from new course titles, bibliographies, session headings in history of science conferences, one of the fastest growing sub-specialities in the field. But to make sense of this sub-speciality, and especially to understand the extraordinary interest the term (and the subject) have attracted in recent years, both in this country and abroad, it is necessary to begin by locating it first, not within the history of science, but in relation to contemporary feminist theory—in other words, in the context out of which it in fact emerged.

A feminist historian, Mary Poovey, introduced the term 'border cases' to denote historical phenomena that, by virtue of their location on the 'border between two defining alternatives', constitute privileged sites for examining the ideological work of gender. 'Border cases', she wrote, 'mark the limits of ideological certainty' (Poovey 1989: 12). Peter Galison (1997) a historian of science, introduced the kindred notion of 'trading zones' to call attention to the extensive traffic across borders (in his case, between experiment, instruments, and theory), and hence to the impossibility of clear demarcations. Both notions might be invoked to highlight the cultural and historical specificity of disciplinary perspectives in general, and, at the same time, to problematize the very concept of disciplinary boundaries. The issue of gender and science, I suggest, is a border case *par excellence*. It sits not on one border, but on multiple borders—indeed, on the borders between feminist theory and all the scientific and meta-scientific disciplines. It is also a trading zone, a domain of cross-talk, exchange, and struggle. By its very existence, it calls into question the borders of all these disciplines.

Given my personal involvement with the history of both the term and the subject, I have found it difficult to review that history as a disinterested

First published in *OSIRIS*, 10 (1995): 27–38. Reprinted by permission.

observer. What follows, therefore, is a frankly first-person perspective on the history and problems of gender and science.

I. GENDER AND SCIENCE

As far as I can tell, the phrase 'gender and science' first made its appearance in an article I published under that title in 1978, not, as it happened, in a history of science journal, but in a psychoanalytic journal. By that title I sought to rouse readers out of a certain habitual complacency, explaining my concerns as follows:

The historically pervasive association between masculine and objective, more specifically between masculine and scientific, is a topic that academic critics resist taking seriously. Why is that? Is it not odd that an association so familiar and so deeply entrenched is a topic only for informal discourse, literary allusion, and popular criticism? How is it that formal criticism in the philosophy and sociology of science has failed to see here a topic requiring analysis? . . .
    The survival of [such] mythlike beliefs in our thinking about science . . . ought, it would seem, to invite our curiosity and demand investigation. Unexamined myths . . . have a subterranean potency; they affect our thinking in ways we are not aware of, and to the extent that we lack awareness, our capacity to resist their influence is undermined. The presence of the mythical in science seems particularly inappropriate. What is it doing there? From where does it come? And how does it influence our conceptions of science, of objectivity, or, for that matter, of gender? (Keller 1978: 410)

Although I was technically the author of this article, I did not then, and I do not now, regard the ideas behind it as originating with me. In the introduction to the book that the article eventually became, I tried to emphasize the ways in which these questions derived from the logic of a collective endeavour we had just begun to call 'feminist theory', and in one of the early drafts I even tried to spell out the sense in which they were not mine at all, but rather, 'ours'. Feminists have recently become as suspicious of the first-person plural as they were earlier of the impersonal pronoun, so I need to try to be quite clear about my notion of a collective 'we'. Who, and what, did I have in mind? Certainly not historians of science—I am not sure I even knew any at that time. Rather, what I had in mind was a very local collectivity of women academics who had actively participated in what has been called 'the women's movement' (recall that 1975 was the designated 'International Year of Women'), who called themselves 'feminists', who were involved in 'consciousness-raising' groups, and who had begun to deploy their heightened consciousness in radical theoretical critiques of the disciplines (and the worlds) from which they had come.

By the mid-1970s, works in 'feminist theory', the name we gave to this collective endeavour, began to appear in anthropology and sociology (Rosaldo and Lamphere 1974, Orther 1974, Reiter 1975, Rubin 1975, Smith 1974), history (Kelly-Godol 1976), literature (Millett 1970, Showalter 1977, Gilbert and Gubar 1979), and psychoanalysis (Chodorow 1974, Miller 1976)—though not yet in any discipline relating to the natural sciences.

The first step was to appropriate the term 'gender' to underscore and elaborate Simone de Beauvoir's dictum that 'one is not born a woman'. In a classic and self-conscious deployment of naming as a form of political action, they (we) redefined 'gender', in contradistinction to sex, to demarcate the social and political, hence variable, meanings of 'masculinity' and 'femininity' from the biological or presumably fixed categories of 'male' and 'female'. The function of this redefinition was to redirect attention away *from* the meaning of sexual difference and *to* the question of how such meanings are deployed. To quote Donna Haraway, 'Gender is a concept developed in order to contest the naturalization of sexual difference' (1991: 131). Very quickly feminists began to see, and as quickly to exploit, the analytic power of this distinction for exploring the force of gender and gender norms, not only in the making of men and women, but also as silent organizers of the cognitive and discursive maps of the social and natural worlds that we, as humans, simultaneously inhabit and construct—even of those worlds that women rarely enter.

Even, that is, the world of the natural sciences. It was just a matter of time before feminists who were involved in these conversations and reading these papers, and who knew something about the natural sciences, would take on, as they say, the 'hard' case. I may have been the first to use the term 'gender and science', but I was hardly alone in recognizing the kind of questions now brought into view, which, once in view, demanded analysis. By the late 1970s, a generation of feminists from a range of different disciplines, who brought with them varying senses of a collective 'we', were taking due note of the traditional naming of the scientific mind as 'masculine' and the collateral naming of nature as 'feminine', and accordingly calling for an examination of the meaning and consequences of these historical connotations.[1] We hoped by that route both to undermine these traditional dichotomies and to pave the way for a restoration or relegitimation of just those values that had been excluded from science by virtue of being labelled 'feminine'. All of us were variously fuelled, and to various degrees, by what Donna Haraway calls 'paranoid fantasies and

---

[1] I might mention Elizabeth Fee, Donna Haraway, Sandra Harding, Hilde Hein, Leigh Star, Carolyn Merchant, and Helen Longino.

academic resentments' (1991: 183), by what I might call 'utopian aspirations', as well as by more straightforward recognitions of intellectual (and soon, even academic) opportunities. For some a suspicion that 'objectivity' might be a code word for 'domination' went hand in hand with the fantasy that we had hold of a lever with which we could not only liberate women, but also turn our disciplines upside down—perhaps even change the world. (Some of us were humbler: I, for one, merely thought of changing science.) In other words, it was a pretty heady time.

The first book-length response came from Carolyn Merchant (1980) writing as a historian of science, as a Marxist, as a feminist, and as a committed environmentalist. In *The Death of Nature* she focused squarely on the significance of the metaphor of nature as woman—for science, for capitalism, for women, and for nature—in the displacement of organicist by mechanist world-views. By arguing that this displacement implied, at least to the users of that language, a symbolic act of violence both against nature-as-woman and against woman-as-nature, her work played a major role in mobilizing, at least briefly, a coalition between feminists and environmentalists.

By the time my own book on gender and science came out, I too had discovered history, and I sought to tie such shifts in metaphors of nature (and simultaneously of mind and knowledge) to changing conceptions of individuality, selfhood, and masculinity—changes that were themselves neither epiphenomenal nor causal, but deeply enmeshed in the social, economic, and political changes of the time. Especially, I attempted to argue for the confluence of new definitions of masculinity and new conceptions of what constituted a 'proper' and epistemologically 'productive' relation between 'mind' and 'nature'. I sought, in sum, to locate the popular equation between 'masculine' and 'objective' in a particular historical transition. But to understand the relationship between 'objectivity' and domination, I returned to the psychodynamics of individual development, adding to my earlier foray an analysis of the relations between love, knowledge, and power (Keller 1985).

By the early 1980s, quite a bit of feminist literature about science had already been written, out of widely varying conceptions of feminism. Besides the work growing directly out of feminist theory, another literature, on the history of women in science, was growing with equal rapidity directly out of the history of science—most notably, Margaret Rossiter's (1982) book on women scientists in America (my own book on Barbara McClintock (Keller 1983) might also belong here). There was also a body of work, mostly by biologists, devoted to a critical examination of scientific constructions of 'woman'. Perhaps the earliest of these was the edited volume *Genes and Gender*, published by Ruth Hubbard and Marian Lowe in 1979. Ruth Bleier's *Science and Gender* appeared in 1981, and Anne Fausto-Sterling's *Myths of Gender* in 1985. By grouping these different literatures together, one could now begin to offer courses under the rubric 'gender and science'—provided, that is, one expanded the meaning of the term 'gender' to include women and sex. Such an expansion-cum-coalition was strategic; it seemed reasonable enough at the time, especially since other problems were more pressing. One such problem was academic shelter, as it were. Where would one put such a course, composed of so many different disciplinary and intellectual agendas? And as became increasingly clear over the 1980s, even of different political agendas? I was fortunate; I had access to a science and technology studies programme that was hospitable. Others tried women's studies: eventually, some even tried history of science programmes. And indeed, it was with the entry of this catch-all label of gender and science into the history of science—as titles for courses, bibliographies, and conference proceedings—that I originally attempted to begin this paper. In other words, at precisely the point where the use of the rubric to cover a loose coalition of assorted works on women, sex, gender, and science threatened to break down altogether. Let me explain.

II. WOMEN, SEX, AND GENDER

Even at the very beginning, the slippage between women and gender had been a source of discomfort for feminists. For one thing, and most trivially, the equation of *women* with *gender* is a logical error—in fact, as Donna Haraway points out, exactly the same kind of error involved in equating *race* with people of colour (1991: 243). Whatever the term is taken to mean, when we use it to apply to people, strictly speaking, we mean it to apply to at least most, if not all, people. In fact, my own interest in gender and science focused neither on women nor on 'femininity', but on men and conceptions of 'masculinity'. What, then, invites the elision, and why would feminists permit such an elision? One answer to the first question is immediately made clear by the analogy with race: women are culturally and historically marked by their sex or gender in a way that men are not, much as people of colour are marked by their race. But the principal reasons that feminists at least initially tolerated, and to some extent even supported, the slippage between women and gender were twofold: first, the primary concern with which most of us had begun was with the force of gender on women's lives; second, even when our concerns moved outward, as they conspicuously did in feminist theory, that slippage was endured out of simple expedience. More recently, however, at least in most parts of the academy, it has widely come to be seen necessary to mark explicitly the distinction between women and gender: witness the renaming of a number of feminist research programmes as 'programme for the study of women and gender'. I want to suggest that this distinction has now become especially necessary in the study of gender and science.

A decade ago a loose coalition of works on women, sex, gender, and science could be held together by the common denominator of feminism— by a commitment to the betterment of women's lives. Differences in other commitments—differences, say, in disciplinary, theoretical, or political perspectives—might well pale by comparison—as long, that is, as that common denominator remained primary, as long as the women whose lives we sought to improve seemed to have common needs, and as long as we continued to see ourselves as a subversive force operating from the margins and interstices of academic life. In the intervening years, however, in part because of the very successes of contemporary feminism, all these differences became both more visible and more pressing as we, and our concerns, began to move into established academic and disciplinary niches. Today it has become conspicuously evident that not all women have the same interests or needs; so, too, it has become evident that not all scholars who call themselves 'feminist' have common or even reconcilable theoretical and disciplinary agendas.

These remarks surely pertain to feminist scholarship in general, but I would like to try to spell out their particular relevance for the present status of the catch-all 'gender and science' in the US academy, and especially in history of science. First, in so far as our interest in the history of women in science was initially motivated by a protest against a history of exclusion and by a political quest for equity, the dramatic changes that have occurred in the participation of at least some women in science over these fifteen years need to be noted. I do not mean to suggest that women have achieved equity in the sciences, but rather that, where fifteen years ago gender appeared as the principal axis of exclusion, today a glance at the racial and ethnic profile of the increased numbers of women who have entered scientific professions over these years suggests that it no longer does. Furthermore, from the perspective of those who have broken through the gender barrier and have now forged alliances within the scientific enterprise, it is not at all obvious how a continued focus on gender, especially given its emphasis on criticism, might be in their interest. These, I think, are some of the issues that Londa Schiebinger was referring to when she stressed the importance of context in what she somewhat elliptically calls 'arguments for and against gender differences' (1989: 273); they are also the issues contributing to the increasing difficulty experienced by so many historians of science who attempt to teach the disparate subjects of women, sex, and gender in science together under one rubric.[2]

But if the primacy of gender as an occupational barrier in the sciences has receded, and its utility as a critical wedge been blunted by occupational success, recognition of both its cultural and analytic importance has in other areas only increased—attesting, once again, to the successes of contemporary feminism. It is, alas, true that feminist theory has not proved itself powerful enough to change the world, as I once hoped it might. It has, however, radically, and I think irrevocably, changed the landscape of a number of academic disciplines. Some more so than others. Notably not the natural sciences, nor—at least, not yet—the philosophy of science; hardly at all the social studies of science, and only some areas of the history of science.

For feminist theory to realize more fully its analytic promise in the history of science, obviously historians need to start reading its literature. But also, I suggest, we need a new taxonomy: 'gender and science' needs to be disaggregated into its component parts. Schematically, these might be described as those studies examining the history of (1) women in science, (2) scientific constructions of sexual difference, and (3) the uses of scientific constructions of subjects and objects that lie both beneath and beyond the human skin (or skeleton). Each of these subjects has by now accumulated a rich literature in its own right, and requires its own reconfiguration into new kinds of 'trading zones'. Here I will focus on the third of these, 'gender *in* science'—trading not between assorted studies of women, sex, and gender in science, but between historical studies of gender, language, and culture in the production of science. For such studies of gender *in* science, I want to make a particular plea for a consolidated, two-way effort toward integrating a number of analytic perspectives that are currently (mis)perceived as disjoint.

It will be evident that in proposing a disaggregation of what has come to be known as gender and science, and focusing only on gender in science, I have made my task considerably easier, for I have now bracketed most of the literature usually considered under this term. My aim, however, is not to privilege my own or any other particular political or intellectual agenda; rather, it is to provide room for all the different concerns these sub-

---

[2] In the session on syllabi for courses in gender and science at the 1990 History of Science Society meeting, such difficulty was widely attested.

categories separately raise. I want especially to side-step the question of whether one is for or against gender differences, and to allow for the possibility of being either both or neither. The principal point is that I take the role of gender ideology to be but one aspect of the constitutive role of language, culture, and ideology in the construction of science, and hence, though the roots of such analyses have been, and must continue to lie, in feminist theory, I take their place in the history of science proper to be just one part of that more general enquiry. I suggest that work in this area has not only raised novel kinds of questions for historians, but also offers some novel models of, and sites for, historiographic analysis that might even be of use to historians of science who are not women, who may not even be gendered, and, possibly, who do not necessarily think of themselves as feminists.

III. GENDER IN SCIENCE

Let me now illustrate this claim with a few examples of the kinds of questions feminists have raised about the implications of a gendered vocabulary in scientific discourse, proceeding from those that have relatively straightforward implications for the reading of scientific texts to those with rather more indirect implications. In all of these examples metaphors of gender can be seen to work, as social images in science invariably do, in two directions: they import social expectations into our representations of nature, and by so doing they simultaneously serve to reify (or naturalize) cultural beliefs and practices. Although the dynamics of these two processes are almost surely inextricable, many feminists focus on the latter, emphasizing their effects (usually negative) on women; here my focus will be on the former, on their influence on the course of scientific research.

I will start where concerns about women, sex, and gender are most likely to intersect, in analyses of past and current work in the biology of reproduction and development. Many of these analyses reduce to a common basic form, the identification of synecdochal (or part for whole) errors of the following sort: (1) the world of human bodies is divided into two kinds, male and female (i.e. by sex); (2) additional, extra-physical properties are culturally attributed to those bodies—active or passive, independent or dependent, primary or secondary (read gender); and (3) the same properties that have been ascribed to the whole are then attributed to the sub-categories of, or processes associated with, these bodies. Often, though not necessarily, these analyses are undertaken from the vantage-point of present, presumably superior, knowledge.

Undoubtedly the most conspicuous examples of such synecdoches are found in the history of theories of generation. Nancy Tuana (1989) has sought to augment the existing literature on reproductive theories from Aristotle to the preformationists by focusing on the imposition of prevailing views of women (i.e. as passive, weak, and generally inferior) on to their roles in reproduction. Thomas Laqueur's more recent and more probing analysis, *The Making of Sex* (1990), adds substantially to such an effort. And some authors have undertaken corresponding analyses of contemporary discussions of fertilization. Scott Gilbert and his students, for example, have traced the language of courtship rituals in standard treatments of fertilization in twentieth-century textbooks (Biology and Gender Study Group 1989).

Emily Martin (1991) has continued this effort by tracking the 'importation of cultural ideas about passive females and heroic males into the "personalities" of gametes' (p. 500) in the most recent technical literature. This is how the argument goes. Conventionally, the sperm cell has been depicted as 'active', 'forceful', and 'self-propelled', qualities that enable it to 'burrow through the egg coat' and 'penetrate' the egg, to which it 'delivers' its genes and 'activate[s] the developmental program'. By contrast, the egg cell 'is transported', 'swept', or merely 'drifts' along the fallopian tube until it is 'assaulted', 'penetrated', and fertilized by the sperm (pp. 489–90). The technical details that elaborate this picture have, until the last few years, been remarkably consistent: they provide chemical and mechanical accounts for the motility of the sperm, their adhesion to the cell membrane, and their ability to effect membrane fusion. The activity of the egg, assumed non-existent, requires no mechanism. Only recently has this picture shifted, and with that shift, so too has shifted our technical understanding of the molecular dynamics of fertilization. One early and self-conscious marking of this shift by two researchers in the field, Gerald and Helen Schatten, appeared in 1983:

The classic account, current for centuries, has emphasized the sperm's performance and relegated to the egg the supporting role of Sleeping Beauty.... The egg is central to this drama, to be sure, but it is as passive a character as the Grimm brothers' princess. Now, it is becoming clear that the egg is not merely a large yolk-filled sphere into which the sperm burrows to endow new life. Rather, recent research suggests the almost heretical view that sperm and egg are mutually active partners. (p. 29)

And indeed, the most current research on the subject routinely emphasizes the activity of the egg cell in producing the proteins or molecules necessary for adhesion and penetration. At least nominal equity (and who, in 1994, could ask for anything more?) seems even to have reached the most recent editions of *The Molecular Biology of the Cell*, where 'fertilization' is defined as the process by which egg and sperm 'find each other and fuse' (Alberts *et al.* 1994: 868).

For historians of science, this recapitulation may raise more questions than it supplies answers, but one question is critical: What *is* the relation between the shift in metaphor in these accounts, the development of new technical procedures for representing the mechanisms of fertilization, and the concurrent embrace of at least nominal gender equity in the culture at large? If nothing else, tracking the metaphors of gender in this literature has provided us with an ideal site in which, with more extensive analysis, we can better appreciate and perhaps even sort out the complex lines of influence and interactions between cultural norms, metaphor, and technical development. Such a task does not require us to embrace the most recent version as correct, either scientifically or politically; witness the work of Frederick Churchill (1979) and John Farley (1982) on nineteenth-century debates about sexual reproduction. It is merely necessary to register the extent to which gender ideologies are implicated in the construction of (at least some) scientific stories.

Similar reviews could be provided of biological accounts of sex determination: Anne Fausto-Sterling, for example, has explored the language of presence and absence in recent discussions of the sex gene (1989). Or of the relationship between cytoplasm and nucleus over the past hundred years, in which, far from coincidentally, the cytoplasm has at least tacitly been routinely figured as female, and the nucleus as male. Jan Sapp (1987) has given us an excellent account of the history of cytoplasmic inheritance, but he neglected to note the significant marks of gender in this history, some of which were noted by Scott Gilbert and his students (Biology and Gender Study Group 1989). Indeed, the history of depictions of cytoplasm and nucleus is remarkably parallel to that for the egg and sperm: like the reformulations of the gametes' roles, recent support of cytoplasmic inheritance and of an important role for cytoplasmic determinants in development coincides with the rise in the 1980s of an ideology of gender equality. There are even some rumblings indicating a similar theoretical shift in the most recent discussion of sex genes. I group these examples together because of the similarity of their structure and the simplicity of their morals.

Other examples, in which readings of gender are less closely tied to readings of biological sex, are correspondingly less straightforward in both their structure and their implications. One general area of the history of science that has attracted particular interest among feminist scientists is that of the relation between genetics and developmental biology, and here an undercurrent of resistance to genetic determinism and a corresponding championship of the organizing models of developmental biology can clearly be seen. The arguments of Lynda Birke (1984), Ruth Bleier (Biology and Gender Study Group 1989), Scott Gilbert's study group (see Birke and Silverton 1984), and Ruth Hubbard (see Hubbard and Lowe 1979) especially come to mind.[3] Similar (or related) preferences can be seen in analyses of brain and behaviour science, as carried out by Helen Longino and Ruth Doell (1983);[4] of mechanism and organicism, examined by Merchant; and even in certain analyses of models in physics, explored by Stephen Kellert (1993). The common denominator in these discussions might be described as a preference for interactionist, contextual, or global models over linear, causal, or 'master molecule' theories.[5] The question is: What does gender have to do either with these concerns or with these preferences?

One link to gender of particular relevance to discussions of genetics and development can be traced in the tacit coding, already suggested, of the cytoplasm (and, more generally, of the body) as female and the corresponding coding of the nucleus or gene as male. Just as in the story about egg and sperm, such codings carry with them traces of social relations between male and female, inviting the suspicion that even when merely implicit, they have been silently working to support correspondingly hierarchical structures of control in biological debates. To date, only fragmentary evidence has been brought to bear on this suspicion, but it provides sufficient support, I think, to demonstrate the need to attend to gender markings in future, more detailed, investigations of these subjects.[6]

But feminist concerns about 'master molecule' theories do not necessarily depend on the allocation of gender labels to the constituent elements of debate. For some critics the concern is more explicitly political, based on the fear that such hierarchical structures in biology are themselves rationales for existing social hierarchies; or that they reflect values supporting not only social but also individual constraint. Such concerns led Helen Longino (1990) to develop her sophisticated philosophical analysis of the ways in which political and social values inevitably enter into theory choice. Other scholars have employed (as I have) yet a different kind of

---

[3] See also Keller 1983.
[4] See also Longino 1990.
[5] The term 'master molecule' was originally invoked by David Nanney (1957) in protest over the conception of genes as 'dictatorial elements in the cellular economy'. Only later was it appropriated by feminists for a larger critique. In the late 1980s it was reappropriated by the NSF in celebration of the successes of DNA.
[6] For further discussion see Keller 1993.

argument, seeing in these hierarchical models the expression of a mind-set predicated on control and domination, unconsciously projecting its own sense of self and other on to representations of processes operating in the natural world. Today I find this argument by projection to be unduly limited—above all, by its failure to take into account the particular kinds of material consequences that models or metaphors of domination have and, accordingly, the particular kinds of material ambitions such models support. As I have argued elsewhere (Keller 1992), master molecule theories are not only psychologically satisfying; they are also remarkably productive—productive, that is, in relation to particular kinds of aims.

IV. STILL TO BE DONE

By focusing on mind-sets and metaphors, feminists have made visible the possibility of alternative mind-sets and metaphors—by itself no mean feat for our thinking about science. But now we need a further analysis of the ways in which metaphors work to bridge the gulf between representing and intervening, of how they help to organize and define research trajectories. Such studies in language and science might be said to constitute one of the most interesting new frontiers in the history of science, and our understanding of the conceptual dynamics of gendered metaphors has been well assisted by the sophistication which Gillian Beer (1990), Ludmilla Jordanova (1989), and Nancy Stepan (1986) have brought to their own work at this frontier.

More generally, and of utmost importance for historians of science, is the need for deep and thoroughgoing contextualization of all the early work that was undertaken with such broad strokes. Feminist scholars may have been among the first in modern times to raise the meaning of objectivity as a central issue for investigation, but it has taken more mainstream historians of science to turn this question into a significant historiographic pursuit, as in the extremely interesting work on the history of objectivity by Lorraine Daston (1992),[7] Peter Dear (1992), and Theodore Porter (1992a, b). Now, however, it is time to integrate gender issues into this admirably careful and context-sensitive historiographic work, and perhaps even to acknowledge how the inclusion of gender can transform the very questions we ask.

Still, 'context' is a big word, and it points in many different directions, a fact which the generic notion of historical specificity, even with the

---

[7] See also Daston and Galison 1992.

addition of gender, does not always capture. It may fail to do justice to the need to attend to specificities of local disciplinary or even sub-disciplinary interests, or to the specificities of local social (national, ethnic, or racial) interests. Donna Haraway offers us some useful ways of talking here that, for many, resonate with current political and intellectual priorities simultaneously: she defines feminist objectivity as *situated knowledges*, and stresses the need for 'partial perspectives'. And indeed, it is with Haraway's radical, though controversial, deployment of the method of partial perspectives that I will end this all too cursory review.

In *Primate Visions* Haraway attempts, in contrast to a subject-rooted approach, a subject-free unpacking of how 'love, power, and science [are] entwined in the constructions of nature in the late twentieth century' (1989: 1) and she pursues this aim through an insistent scrutiny of the politics of narratives. As Gregg Mitman writes, 'Her craft is the art of storytelling; her model for the construction of scientific knowledge is "contested narrative fields." Science, and in this instance primatology, is a story about nature, a tale circumscribed by its narrator, but one constantly evolving as new storytellers enter' (Mitman 1991: 164). Haraway vigorously eschews the idea that gender can be understood independently of the politics of race and class, and her subject is an ideal one for making this case. As she makes abundantly clear, constructions of 'nature' in twentieth-century primatology, like constructions of gender, are profoundly implicated in twentieth-century politics of race and colonialism. The fact that the (mostly white) women who entered the field in the 1970s and 1980s took the lead in restructuring traditional narratives does not, for Haraway, provide support for the idea of 'a feminist science' (and with this I agree), but rather demonstrates once again the dependence of scientific narratives on their authors' historical 'positioning in particular cognitive and political structures of science, race, and gender' (1989: 303). Haraway's very method precludes telling or even hoping for one coherent story, and many readers may be left feeling a bit too destabilized. But her reach toward a post-modern historiography has not only provided new models for working with 'gender' in science; it also suggests new models for any politically oriented analysis of science.

Undoubtedly, my choice of examples is idiosyncratic, and I have omitted much work that has had an enormous impact on other disciplines.[8] But the

___

[8] One subject especially close to my own heart that I have not discussed is the impact that feminist analyses of the personal and subjective dimensions of science have had on other disciplines; nor have I discussed the potential value of these analyses for the history of science—were it not for the intellectual and political climate that currently prevails. Interestingly, it was the gender coding of these dimensions as 'feminine' that led us to examine their exclusion from investigation in the first place; might, I wonder, the same gender coding still be operative?

moral of my account will, I hope, be clear. For the future of gender in science in the particular context of the history of science, I want to add now, complementing Haraway's emphasis on fractures and partiality, a plea for affiliation and integration. Fifteen years ago it took an organized effort on the part of feminists to rouse historians' attention to the marks and significance of gender. Working from the strengths of their political consciousness of gender and their irreverence for received boundaries operating both within and between disciplines, feminist theorists have brought home powerful lessons about the cognitive and institutional politics of gender that even historians of science can no longer ignore. But to move the analysis of gender in science forward in this discipline, I suggest that these strengths need now to be integrated with the strengths of more conventional historiographic scholarship. There needs to be a lot more traffic and two-way exchange in this trading zone if it is to do the work it is capable of doing.

REFERENCES

Alberts, B., *et al.* (1994), *Molecular Biology of the Cell* (New York and London: Garland Press).
Beer, G. (1990), 'Translation or Transformation? The Relations of Literature and Science', *Notes and Records of the Royal Society of London*, 44: 81–99.
Biology and Gender Study Group (1989), 'The Importance of Feminist Critique for Contemporary Cell Biology', in Tuana (1989), 172–87.
Birke, L., and Silverton, J. (1984) (eds.), *More than the Parts: Biology and Politics* (London: Pluto Press).
Bleier, R. (1984), *Science and Gender: A Critique of Biology and its Themes on Women* (New York: Pergamon).
Chodorow, N. (1974), 'Family Structure and Feminine Personality', in Rosaldo and Lamphere (1974), 43–66.
Churchill, F. (1979), 'Sex and the Single Organism: Biological Theories of Sexuality in Mid-Nineteenth Century', *Studies in the History of Biology*, 3: 139–77.
Daston, L. (1992), 'Objectivity and the Escape from Perspective', *Social Studies of Science*, 22: 597–618.
——and Galison, P. (1992), 'The Image of Objectivity', *Representations*, 40: 81–128.
Dear, P. (1992), 'From Truth to Disinterestedness in the Seventeenth Century', *Social Studies of Science*, 22: 619–31.
Farley, J. (1982), *Gametes and Spores: Ideas about Sexual Reproduction, 1750–1914* (Baltimore: Johns Hopkins University Press).
Fausto-Sterling, A. (1985), *Myths of Gender: Biological Theories about Men and Women* (New York: Basic Books).
——(1989), 'Life in the XY Corral', *Women's Studies International Forum*, 12/3: 319–31.

Galison, P. (1997), 'The Trading Zone: Coordinating Action and Belief', in *Image and Logic: The Material Culture of Modern Physics* (Chicago: University of Chicago Press), ch. 9.

Gilbert, S. M., and Gubar, S. (1979), *The Madwoman in the Attic: The Woman Writer and the Nineteenth-Century Imagination* (New Haven: Yale University Press).

Haraway, D. (1989), *Primate Visions: Gender, Race, and Nature in the World of Modern Science* (New York: Routledge).

——(1991), *Simians, Cyborgs, and Women: The Reinvention of Nature* (New York: Routledge).

Hubbard, R., and Lowe, M. (1979) (eds.), *Genes and Gender* (Staten Island, NY: Gordian Press).

Jordanova, L. (1989), *Sexual Visions: Images of Gender in Science and Medicine between the Eighteenth and Nineteenth Centuries* (Madison: University of Wisconsin Press).

Keller, E. F. (1978), 'Gender and Science', *Psychoanalysis and Contemporary Thought*, 1/3: 409–33.

——(1983), *A Feeling for the Organism: The Life and Work of Barbara McClintock* (San Francisco: Freeman).

——(1985), *Reflections on Gender and Science* (New Haven: Yale University Press).

——(1992), 'Critical Silences in Scientific Discourse', in *Secrets of Life, Secrets of Death: Essays on Language, Gender, and Science* (New York: Routledge), 73–92.

——(1993), 'Rethinking the Meaning of Genetic Determinism', Tanner Lecture, University of Utah.

Kellert, S. (1993), *In the Wake of Chaos: Unpredictable Order in Dynamical Systems* (Chicago: University of Chicago Press).

Kelly-Godol, J. (1976), 'The Social Relations of the Sexes: Methodological Implications of Women's History', *Signs*, 1/4: 809–23.

Laqueur, T. (1990), *Making Sex: Body and Gender from the Greeks to Freud* (Cambridge, Mass.: Harvard University Press).

Longino, H. (1990), *Science as Social Knowledge: Values and Objectivity in Scientific Inquiry* (Princeton: Princeton University Press).

——and Doell, R. (1983), 'Body, Bias, and Behavior', *Signs*, 9/2: 206–27.

Martin, E. (1991), 'The Egg and the Sperm: How Science has Constructed a Romance Based on Stereotypical Male–Female Roles', *Signs*, 16: 485–501.

Merchant, C. (1980), *The Death of Nature: Women, Ecology, and the Scientific Revolution* (New York: Harper & Row).

Miller, J. B. (1976), *Toward a New Psychology of Women* (Boston: Beacon Press).

Millett, K. (1970), *Sexual Politics* (New York: Doubleday).

Mitman, G. (1991), review of D. Haraway, *Primate Visions*, *Isis*, 82: 163–5.

Nanney, D. (1957), 'The Role of Cytoplasm in Development', in W. D. McElroy and H. B. Glass (eds.), *The Chemical Basis of Heredity* (Baltimore: Johns Hopkins University Press).

Ortner, S. B. (1974), 'Is Female to Male as Nature is to Culture?', in Rosaldo and Lamphere (1974), 67–87.

Poovey, M. (1989), *Uneven Development: The Ideological Work of Gender in Mid-Victorian England* (Chicago: University of Chicago Press).

Porter, T. M. (1992a), 'Objectivity as Standardization: The Rhetoric of Impersonality in Measurement, Statistics, and Cost–Benefit Analysis', *Annals of Scholarship*, 9: 19–60.

—— (1992*b*), 'Quantification and the Accounting Ideal in Science', *Social Studies of Science*, 22: 633–52.

Reiter, R. R. (1975) (ed.), *Toward an Anthropology of Women* (New York: Monthly Review).

Rosaldo, M. Z., and Lamphere, L. (1974) (eds.), *Woman, Culture, and Society* (Stanford, Calif.: Stanford University Press).

Rossiter, M. (1982), *Women Scientists in America: Struggles and Strategies to 1940* (Baltimore: Johns Hopkins University Press).

Rubin, G. (1975), 'The Traffic in Women: Notes on the Political Economy of Sex', in Reiter (1975), 157–210.

Sapp, J. (1987), *Beyond the Gene: Cytoplasmic Inheritance and the Struggle for Authority in Genetics* (New York: Oxford University Press).

Schatten, G., and Schatten, H. (1983), 'The Energetic Egg', *The Sciences*, 23/5: 28–34.

Schiebinger, L. (1989), *The Mind Has No Sex? Women in the Origins of Modern Science* (Cambridge, Mass.: Harvard University Press).

Showalter, E. (1977), *A Literature of their Own: British Women Novelists from Bronte to Lessing* (Princeton: Princeton University Press).

Smith, D. (1974), 'Women's Perspective as a Radical Critique of Sociology', *Sociological Inquiry*, 44: 7–13.

Stepan, N. (1986), 'Race and Gender: The Role of Analogy in Science', *Isis*, 77: 261–77.

Tuana, N. (1989) (ed.), *Feminism and Science* (Bloomington, Ind.: Indiana University Press).

20

ESSENTIALISM, WOMEN, AND WAR: PROTESTING TOO MUCH, PROTESTING TOO LITTLE

SUSAN OYAMA

Recently some biological theorists and feminists have converged on 'essentialist' accounts of war that are strangely similar in certain ways. At first glance it seems to be an unlikely development, given the frequency with which we have seen anti-feminists and feminists line up on opposite sides of the nature–nurture rift. At second glance, though, perhaps the association is not so surprising after all. We seem to be in the midst of a pendulum swing 'back to nature' and away from environmentalism. This movement, in turn, is probably part of a more general trend in the United States toward conservatism and a certain brand of romanticism, though the issue is a good deal more complex than one might think. Apart from the current emphasis on so-called traditional values, though, the convergence on 'biological' views reflects some very common and pervasive beliefs about genes and environment, biology and learning, that are as evident in environmentalist approaches as they are in biological ones.

By 'essentialist', I mean an assumption that human beings have an underlying universal nature, one that is more fundamental than any variations that may exist among us, and that is in some sense always present— perhaps as 'genetic propensity'—even if it is not discernible. People frequently define this pre-existing nature in biological terms, and they believe it will tend to express itself even though it might be somewhat modified by learning, and thus might be partially obscured by a sort of cultural veneer. (For a good discussion of this theme in feminism, see Alison Jaggar 1983: ch. 5. Janet Sayers (1982: 148) criticizes essentialist feminism, and Anne Fausto-Sterling (1985: 195) refers to human sociobiology as a 'theory of essences'; Ruth Bleier (1984: ch. 1) criticizes both. Of these authors, Jaggar is perhaps most successful in transcending the

First published in E. Hunter (ed.), *Genes and Gender: A Challenge to Genetic Explanations* (San Francisco: W. H. Freeman, 1991), 64–76. Reprinted by permission.

biology–culture opposition, but all are aware of the mischief it has caused for scientists and non-scientists alike.)

When I say that environmentalist and biological approaches share many assumptions about nature and nurture, I mean that they have often argued about which and how many traits were genetic and which were learned; but, in doing so, they have accepted the premiss that genes and learning were properly treated as alternative explanations for human characteristics and actions. They also tended to agree that the possibility of change was somehow illuminated by their disputes.

In a metaphor that is revealing in more than one way, sociobiologist David Barash compares the relationship of nature and nurture to two people wrestling. As they tumble about, their limbs entwine so that it is hard to tell which is which. However entangled they may become, the combatants do not merge; they are separate persons in competition, and our imperfect powers of observation do not change that fact (1981: 12).

Though they routinely declare that the nature–nurture dichotomy is meaningless and that the effects of biology and culture cannot be clearly distinguished, scholars of all stripes (including not only sociobiologists like Barash but many of sociobiology's critics as well—see discussion and references in Oyama 1981, 1982) continue to treat them as separate sources of living form and behaviour: some things are (mostly) programmed by our genes, others are (mostly) programmed by our environments. We will return to this conceptual problem later.

Let's look first, however, at some examples of essentialist accounts of women and war. The first several examples come from scholars who have offered us their biological views of human behaviour and society, while the last two come from a recent collection of feminist writings.

THE ARGUMENT

Lionel Tiger and Robin Fox say that war 'is not a human action but a male action; war is not a human problem but a male problem'. If nuclear weapons could be curbed for a year and women could be put into 'all the menial and mighty military posts in the world', these authors declare, there would be no war. They immediately concede that this proposition is but a fantasy, and a totally unrealistic one at that, because the human 'biogrammar' (a term they use more or less the way others use 'genetic programme') ensures that such a thing could never happen. Men, they say, have evolved as hunters who band into groups and turn their aggressiveness out against common enemies or prey (1971: 212–13). Political structures in modern societies are formed on this primeval hunting model, and, naturally, men dominate these structures as well. Tiger speculates that women, who do not bond and co-operate as effectively as men, could be given positions in government by special mandate. He feels, however, that it might be quite unwise to expect that even this effort could effect much change (1970: 270–2); presumably, attempts to subvert biologically natural tendencies are not likely to succeed.

Barash accepts the idea that males bond and exclude women from political power (1981: 187–9), though he emphasizes the grounding of male aggression in the competition for reproductive opportunities and reproductively relevant resources (ibid. 174, e.g.). He argues that women are only allowed political power if they are in some sense 'desexed', by age and/or physical unattractiveness. Otherwise, men refuse to recognize a woman's authority, even when she manages to gain admittance to the male 'club' (ibid. 189–90).

Another writer on biological topics, Melvin Konner, also suggests that, because they are less aggressive than men, women should be placed in authority, in order to 'buffer' or 'dampen' violent conflict between nations (1982: pp. xviii, 126, 420). He seems to reject hunting bands as the evolutionary explanation for human aggression, though he does cite Tiger's work, and allows that 'something happens when men get together in groups; it is not well understood, but it is natural, and it is altogether not very nice' (ibid. 203–6). Like Barash, Konner is more impressed with the notion that male aggression is explained by competition for the reproductive resources provided by females: eggs and parental care (ibid. ch. 12).

The suggestion that women might be more peaceful than men in positions of power is thus immediately and quite effectively undone by the theorists' other assumptions about natural differences between women and men, and the social consequences of these differences. It is a rather neat irony—that the qualities that might save the world are kept out of the public sphere by the very biological order that produces them. (Konner does not say that women must always be excluded from power, though he says that it is pointless to use violent female rulers of the past as models for the future, because they 'have invariably been embedded in and bound by an almost totally masculine power structure, and have gotten where they were by being unrepresentative of their gender' (1982: 126). He does not say how to implement his suggestion that 'average' women be allowed to control the world's arsenals.)

The argument that women are inevitably excluded from political life also appears, of course, in past and present anti-feminist writings on the necessity of patriarchy (see Sayers 1982 for discussion, and Goldberg 1973 for a relatively recent example). Partly because biological arguments have often been associated with reactionary politics, it is now common for theorists to declare their liberal values, deny that biological treatments are necessarily either deterministic or conservative, and emphasize that biological explanation is not the same as moral approval. Then the theorists typically call on us to know our natures in order to transcend them. Barash, for example, asks whether we can use our understanding to overrule the biological 'whisperings from within' (1981: 198). At the same time, they often warn against trying to challenge the boundaries and constraints our genes set for us. Charles Lumsden and E. O. Wilson warn that trying to escape these constraints risks the 'very essence of humanness'. They advise us to learn what the limits are, and to set our goals within them (1981: 359–60), while Barash declares that denying natural sex differences is 'likely to generate discontent' (1981: 116; for a critique of the language of constraints and limits, see Oyama 1985).

The relationship between politics and science is a complex one, and, though I think it can be argued that at any particular time some scientific approaches tend to be associated with and/or imply certain attitudes toward the moral and political worlds, there is no direct link between reactionary values and an interest in, for example, sociobiological analysis. Biologists become quite as annoyed at having their politics misrepresented as anyone else, and to assume that someone who emphasizes biological bases of human behaviour is automatically a 'crypto-Nazi' is to engage in just the sort of reductionist thinking I am criticizing in this paper. A crucial link between one's scientific and political views is one's conception of will and possibility, and this notion is rarely made explicit. Just because the relationships are so complicated, however, it becomes very important to make assumptions explicit whenever possible, for it is these hidden assumptions that structure the arguments and invite the conclusions.

Examples of feminist essentialism are found in Pam McAllister's collection *Reweaving the Web of Life* (1982). In 'The Prevalence of the Natural Law within Women', Connie Salamone describes women's roles in protecting both the species' young and the natural law that governs the world. This role, if it is not subverted by male values, endows females with a special affinity to other animals, and tends to give rise to concern over animal rights and vegetarianism. Salamone contrasts the female 'aesthetic of untampered biological law' with 'the artificial aesthetic of male science' (1982: 365–6). In 'Patriarchy: A State of War', Barbara Zanotti invokes Mary Daly's concept of women's *biophilia* (love for life), and describes the history of patriarchy as the history of war. She asserts that, in making war,

patriarchy attacks not the opposing military force but women, who represent life. Soldiers, she suggests, are encouraged to identify military aggression with sexual aggression, so that 'the language of war is the language of
gynocide' (1982: 17).

Here I have addressed some versions of the argument: that women are
inherently less aggressive than men, war is caused by male aggression,
and women are thus somehow more capable than men of bringing about
peace—or, at least, if they were in power, women would be less destructive
than men. It is a skeletal argument, of course, abstracted from very different sorts of writings.

The transition from individual to international conflict is not necessarily
direct; for Tiger and Fox (1971), Barash (1981), and Zanotti (1982), for
example, war is specifically the aggression of male-bonded men in groups,
and Tiger (1970: 219) distinguishes between individual and group violence.
Furthermore the connection between aggression and peacemaking is not
clear. Especially in the work of the male scientists cited above, it is *lack* of
aggressiveness, rather than any positive quality, that is emphasized. Even
in a world in which women are traditionally defined by their deficits, I'm
not sure that peacemaking and peacekeeping should be seen as merely
passive (to invoke another loaded dichotomy) results of low levels of
aggression. Recent radical feminists are more likely to point out positive
female qualities of nurturance, sensitivity to connections, and peaceful-
ness. (I follow Jaggar (1983: ch. 5) and Sayers (1982) in using the term
'radical' here. For the purposes of this essay, it entails a tendency to speak
of essential feminine qualities in a positive, even celebratory way, rather
than insisting on women's basic similarities to men.)

LUMPING: HOW TO IGNORE IMPORTANT DISTINCTIONS

The arguments all require that aggression be somehow unitary. They
depend on, and encourage, certain kinds of illegitimate 'lumping'. One sort
of lumping is definitional; that is, all sorts of behaviour, feelings, intentions,
and effects of actions are grouped together as aggressive. Tiger's definition, for example, is so broad that it embraces all 'effective action which is
part of a process of mastery of the environment'; violence is one outcome
of such aggressive activity, but not the only one (1970: 203). That women
do not bond in aggressive groups implies, then, that they are less capable
of 'effective action' in the service of 'mastery of the environment'—a
sweeping statement indeed, and one that bodes ill for any political action
on the part of women.

Another approach that allows us to treat aggression as a uniform quantity is cross-species lumping. Very different phenomena often are equated, so that, for example, mounting or fighting in rodents, territoriality in fish or birds, and hunting, murder, or political competition in humans are all 'aggressive' behaviours. Then there is the lumping of levels of analysis. The activity of nations and institutions is reductively collapsed to the level of individuals, or even of hormones or genes. Finally, there is developmental lumping. Activity levels in new-born babies, rough-and-tumble play in young children, fighting or delinquency in teenagers, and decisions of national leaders in wars are viewed as somehow developmentally continuous, or 'the same'. Sex differences in these behaviours are then seen as manifestations of basic sex differences in aggressiveness (see Money and Ehrhardt 1972 for a flawed, but highly influential, treatment of some sex differences; see also critiques by Bleier (1984), Fausto-Sterling (1985), and this and other volumes in the Genes and Gender series; also see Klama 1988 for treatment of more general issues in aggression studies).

Much of this lumping depends on very common modern-day versions of preformationism and essentialism. Today we think of preformationism as an archaic relic of outmoded thought, and we snicker about the absurd idea that there could be little people curled up in sperm or egg cells. But replacing curled-up people with curled-up blueprints or programmes for people is not so different. That is, whether we speak of aggression in the genes or coded instructions for aggression in genes, we haven't made much conceptual progress. What is central to preformationist thought is not the literal presence of fully formed creatures in germ cells but, rather, a way of thinking about development—development as revelation of preformed nature or essence, as expression of pre-existing programme or plan, rather than as contingent series of constructive interactions, transformations, and emergences. It is a view that makes real development irrelevant, since the basic 'information', or form, is there from the beginning, a legacy from our evolutionary ancestors (see Oyama 1985 for fuller treatment of these issues).

Nor is the basic reasoning much changed by the less deterministic-sounding language of biological predispositions, propensities, or limits to flexibility; the assumptions underlying these apparently more moderate formulations are not substantially different from the more dichotomous ones that people ridicule these days. Similarly, saying that, of course, nature combines with, or interacts with, nurture shows continued reliance on a biological nature defined by genes before development begins and moderated or deflected by an external, environmental nurture. Even though no one claims our natures to be absolutely uniform and immutable, the somewhat softer language of genetic predispositions and tendencies shares the logical weaknesses of strict determinism, even if it seems to give us more possibilities for change. One problem is that, in this more moderate sort of account, the genes still define the boundaries within which action is possible, and they still constitute the ultimate source of control. (Barash likens us to horses being ridden by genetic riders who give us considerable freedom, but who remain in firm command (1981: 200).) A puzzle for those who hold this view is how to conceptualize a 'we' that is pitted against our genes in a struggle for control over 'our' behaviour. Another puzzle, for feminists who embrace the argument for inherent male aggression and dominance, is how to mobilize for change in a world populated by inherently aggressive and dominant males.

WHAT IS THE POINT OF THIS CRITIQUE?

It is important to be clear about what I am *not* doing when I criticize the nature–nurture opposition or the lumping of different definitions, species, levels of analysis, and developmental phenomena that often accompany it. I am not saying that aggression, however it is defined, is unimportant. I am not denying that nations are composed of individuals, and that individuals are composed of cells, chemicals, and so on. I am not denying that understanding these parts might help us to understand the wholes of which our world is composed. I am not, therefore, rejecting research on individuals, hormones, neurons, and genes, including those of other species. I am not making an environmentalist argument—that biology is irrelevant, that genes don't count, and that everything about our behaviours can be changed (these three are not the same argument, and one of our problems is that we tend to lump them at the same time that we lump biological arguments). I am not even denying certain constancies or similarities among individuals within and across societies, though often we are rather cavalier with our methods of demonstrating these likenesses.

What I *am* saying is that analysis should be conducted in the interests of the eventual synthesis of a complex, multi-levelled reality (just as temporary lumping—of diverse essentialist treatments of aggression, for example—can serve the elaboration of a more complex argument). The levels I have in mind here are not like onion skins that can be stripped away to reveal a more basic reality. After all. when you take away enough of an onion's layers, there's nothing left to reveal. Rather, they are levels of analysis whose interrelationships must never be assumed, but discovered.

We will never understand the role of genes and hormones in individual lives or of individuals in society unless we move beyond traditional oppositions. We will never gain insight into the possibilities of different developmental pathways if we assume them to be fixed, on the basis of an inappropriate argument. This is the point at which the environmental and biological determinists, as well as the more moderate 'in-betweenists', are unwitting allies: they usually agree on what it would *mean* for something to be biological or cultural, even as they argue about relative contributions of genes and learning.

If we want to use scientific analysis to answer questions, we must know what questions we are asking, or we'll never know what evidence could help us answer them. And if we want to fight the good fight, we must know what the enemy is, or we will waste precious time and energy that may not be ours to waste. (Note: I have said *what* the enemy is, not *who*, because in this case I am concerned with ways of thinking, not people, that make our task harder.)

QUESTIONS, CONCERNS, AND ANSWERS

We reveal a great deal of confusion when we ask if something is biological. We might be asking about the chemical processes associated with some behaviour, for instance—this is a matter of the level of analysis, and such questions can be asked about any behaviour, learned or unlearned, common or uncommon, fixed or labile. We might be asking about development: Does a given behaviour, for example, seem to be learned? Is it present at birth? (These questions are not the same thing, since learning can be pre-natal.) We might be asking about evolutionary history, which in turn, resolves into several kinds of questions: Is the behaviour present in phylogenetic relatives? When did it appear in our own evolutionary line, and why? We might be asking what role, if any, a character now plays in enhancing survival and reproduction. We might be asking whether variation in a character is heritable in a given population (whether differences in the character are correlated with genetic differences). This last question has to do with population genetics, which is useful if one wants to know about the possibility of artificial or natural selection in the population.

These are very different questions, for which different evidence is relevant, and they do not exhaust the catalogue of biological queries. None has any automatic bearing on any other, and lumping them together as genetic or biological simply creates confusion and faulty inferences. Often, however, a person asking whether some trait is biological is not interested in these particular questions at all, but has something else in mind. She or he is concerned about the inevitability of a trait, or its unchangeability in the individuals evincing it, or its goodness, justifiability, or naturalness, or perhaps the consequences of trying to change or prevent it (will it come bursting out as soon as we drop our guard? Will intervention do more harm than good?). Scientists frequently share these concerns and the confusions that link them to biology.

None of the scientific questions listed earlier—about evolution, developmental timing, or process or level of analysis—is relevant to these underlying concerns. Our misguided but deeply embedded beliefs about genes and biology, however, cloud the issue. *Genetic* and *biological*, in fact, are often effective synonyms for *inevitability*, *unchangeability*, and *normality*. The common concept of genetic control and guidance of development implies that fate, or at least the range of potential fates, is set before birth. But persons must develop, and development is the result of a whole system; there is no clear way to see the role of genes as more basic, formative, directive, controlling, or limiting than other aspects of the system, and the role of any particular factor depends on its interrelations among the others.

Concerns about inevitability are really about *possible* developmental pathways, not about past or present ones. (Even wondering whether a present state of affairs is immutable implies wanting to know *what would happen if...*, and wondering whether it was inevitable implies wanting to know *what would have happened if....*) When Barash speculates that male parenting in humans, as in other mammals, is 'not nearly as innate as modern sexual egalitarians' think, he seems to be saying something about the probability of reaching certain personal and political goals, and he certainly seems to believe that 'innateness' (a concept he never defines satisfactorily) has something to do with the difficulties that he thinks 'sexual egalitarians' will encounter (1981: 88).

Because any pathway is the function of an entire developmental system, which includes much more than genes, its qualities are not predictable from genes alone. We would have to understand development extremely well to know whether the necessary conditions for constancy *or* change in patterns of aggressive behaviour would be present in any particular alternative world. To say anything intelligible about possible relationships among nations, we would need to know a great deal about issues that are not in any simple way related to particular sorts of individual aggressiveness.

Inevitability is not predictable from observations at the morphological or biochemical levels of analysis. It is not predictable from the role of learning in the development of a behaviour or from its time of appearance. It is not predictable from phylogenetic history, a pattern of heritability in some population, prevalence in certain environments, or even universality. That is, none of these traditional scientific biological questions is relevant to the concerns that most often motivate the questions. To ask biology to address concerns about desirability, furthermore, is to ask science to do our moral work for us.

We must decide what kind of world we want, and why. We won't necessarily succeed in bringing it about, but we shouldn't be deterred prematurely from trying because of biological evidence of whatever variety, either because we believe the biological, in any of its senses, is fixed, or because we believe it is dangerous to tamper with what we think of as natural. Similarly, we shouldn't be complacent about natural features we might value (virtues that are thought basically feminine in this world won't necessarily persist in the one that's coming). There is a tendency to view the biological as static, but it is, in fact, historical at all levels. When I say 'history', I am referring to contingency, interaction, possibility, and change. (The habit of asking whether some feature of our world is the result of biology or history is thus deeply mistaken.) When we ask about biology, though, our concerns tend to be mythological, not historical. Here I do not mean myth as wrong, or 'bad', science (though it might be), but as a way of thinking that reveals ultimate truth, eternal necessity, and legitimacy.

Lionel Tiger, chronicler of male bonding and aggression, refers to *Lord of the Flies*, the widely read story of a group of English schoolboys. Marooned on an island, the lads rapidly degenerate into a horde of savage little creatures. Apparently, the author of the book, William Golding, has said he wanted to construct a myth, a tale that would give the key to the whole of life and experience (cited in Tiger 1970: 207). The feminist Zanotti, too, accepts the centrality of male bonding to individual and social life in her claim that, in making war, men are eternally attacking and destroying women (1982: 17). Both theorists invoke unchanging essence to explain gender, relations between men and women, and, thus, the world. But it is a static world in which ancient tragedies are played out again and again according to primal necessity, not a historical world in which necessity and nature arise by process and then give way to other necessities and natures. Nature, then, should not be seen as one term in the traditional nature–nurture, genes–environment, biology–culture pair. It is not a *cause* of development, but rather an emerging *product* of development.

PLAYING THE GAME

It should be clear by now how I feel about several common strategies for dealing with biological arguments. When someone says, 'It's biological,' we reply, 'No, it's not, it's cultural,' when instead we should be asking why the cultural and the biological are treated as alternatives in the first place, and just what we (and they) really mean by either explanation. I call this the 'Protesting Too Much Syndrome', because we are often afraid that the trait in question *is* biological in one or more of the mistaken senses described above. Or someone says that we are innately inferior, and we counter, 'No, we're not, *you* are,' rather than rejecting the assumption of essential nature that allows *any* pronouncements of this sort. This second strategy I have dubbed the 'Protesting Too Little Syndrome'. It entails agreeing that differences *are* biological, but reversing the evaluative polarity. Male nature is bad; female nature is good. While it offers the momentary satisfaction of turning the tables, it is based on all the mistaken ideas about nature and nurture that, I think, get us all into so much difficulty—too great a price to pay for Mother Nature's favour. The solution is not to protest precisely the correct amount, or to find the degree of biological determination that is 'just right', like some Goldilocks trying to find comfort in a house that is not her own. Rather, it is to protest a whole lot *about the very rules of the discourse.* We mustn't allow the argument to be defined for us. Instead, we must be reflective enough to rethink it.

I am not saying that we ought to throw out everything and start from the ground up. We couldn't do it if we wanted to. But when there are ample grounds for doubting the validity of a conceptual framework or a set of issues, as is the case with the nature–nurture complex, we do ourselves no favour by blindly accepting the terms of the game. Some of our gravest problems come, after all, from letting others set terms for us. The burden of clarification certainly does not rest entirely with women, but, if we shirk our part, how can we do justice to the struggle?

Instead of pitting one mythical account against another, instead of searching for a morally or emotionally resonant evolutionary past to explain the present, and then projecting it into the future, we must focus on real historical processes whose courses are not foreseeable on the basis of any account of nature as manifested in hunter-gatherers, baboons or chimps, hormones, brain centres, or DNA strands. I speak here of individual developmental history, as well as historical change on the societal level, for it is within these processes that nature and possibility are defined.

ARE WOMEN LESS AGGRESSIVE AND, HENCE, LESS WARLIKE?

I could say much about aggression and about women, and maybe even a little about war, but in this essay I haven't said much about any of them. Perhaps the reason is that the essentialist theories I have been discussing don't say much about these topics either. Instead, I have focused on the ways we think about these topics. War is about politics, diplomacy, economics, and historical continuity and change in relations among people, not about brain centres, testosterone, or rough-and-tumble play. It is like a fight between individuals only by analogy, just as certain encounters between groups of ants is war only by analogy. Perhaps it is significant that when sociobiologist David Barash co-authors a book on preventing nuclear war (Barash and Lipton 1982), it contains nothing about different capacities and contributions of males and females, but instead gives lists of very pragmatic suggestions for effective action. Obviously, I would never claim that women have no role in national and international politics, but neither can I make sense of the notion that we ought to be somehow inserted into public life because of some mythic direct line to life, peace, and love. Men are not a plague, and women are not a cure.

REFERENCES

Barash, D. P. (1981), *The Whisperings Within* (New York: Harper and Row).
——and Lipton, J. E. (1982), *Stop Nuclear War!* (New York: Grove Press).
Bleier, R. (1984), *Science and Gender: A Critique of Biology and its Theories on Women* (New York: Pergamon Press).
Fausto-Sterling, A. (1985), *Myths of Gender: Biological Theories about Women and Men* (New York: Basic Books).
Goldberg, S. (1973), *The Inevitability of Patriarchy* (New York: Morrow).
Jaggar, A. M. (1983), *Feminist Politics and Human Nature* (repr. Totowa, NJ: Rowman and Allanheld).
Klama, J. (1988), *Aggression: The Myth of the Beast Within* (New York: Wiley).
Konner, M. (1982), *The Tangled Wing: Biological Constraints on the Human Spirit* (New York: Holt, Rinehart and Winston).
Lumsden, C. J., and Wilson, E. O. (1981), *Genes, Mind, and Culture* (Cambridge, Mass.: Harvard University Press).
McAllister, P. (1982) (ed.), *Reweaving the Web of Life* (Philadelphia: New Society Publishers).
Money, J., and Ehrhardt, A. A. (1972), *Man and Woman, Boy and Girl* (Baltimore: Johns Hopkins University Press).
Oyama, S. (1981), 'What Does the Phenocopy Copy?', *Psychological Reports*, 48: 571–81.

Oyama, S. (1982), 'A Reformulation of the Concept of Maturation', in P. P. G. Bateson and P. H. Klopfer (eds.), *Perspectives in Ethology* (New York: Plenum), 101–31.
——(1985), *The Ontogeny of Information: Developmental Systems and Evolution* (Cambridge: Cambridge University Press).
Salamone, C. (1982), 'The Prevalence of the Natural Law within Women: Women and Animal Rights', in McAllister (1982), 364–75.
Sayers, J. (1982), *Biological Politics* (London: Tavistock).
Tiger, L. (1970), *Men in Groups* (New York: Vintage Books).
——and Fox, R. (1971), *The Imperial Animal* (New York: Holt, Rinehart and Winston).
Zanotti, B. (1982), 'Patriarchy: A State of War', in McAllister (1982), 16–19.

21

ESSENTIALISM AND CONSTRUCTIONISM ABOUT SEXUAL ORIENTATION

EDWARD STEIN

I. INTRODUCTION

In the past several years, scientific research into human sexual orientation—in particular, research that claims male homosexuality is innate—has garnered a great deal of attention among scientists and the general public. Although research of this sort has been ongoing for over a hundred years (Bullough 1994), its current popularity stems from Simon LeVay's neuro-anatomical research (LeVay 1992) and two genetic studies (Bailey and Pillard 1991, Hamer et al. 1993). Although some have criticized this scientific approach (see e.g. Byne and Parsons 1993 and Fausto-Sterling 1992), most scientists doing work on sexual orientation claim there is strong support for the view that sexual orientations are innate, very difficult to change, and as coming in two or three flavours (heterosexual, homosexual, and perhaps bisexual).

At the same time as this scientific paradigm for thinking about sexual orientation has been emerging, lesbian and gay studies, a new interdisciplinary field, has developed within the humanities and some of the social sciences (see Abelove et al. 1993). Within lesbian and gay studies, the reigning—but not uncontested (see e.g. Boswell 1982–3 and Mohr 1992)—paradigm for thinking about sexual orientation is *constructionism*, a view with roots in the work of Michel Foucault (1978) and in various approaches to sociology (e.g. McIntosh 1968). Constructionism about sexual orientation emphasizes the historical and cultural contingencies of sexual orientation and sexuality (Stein 1992*b*). Constructionism is in conflict with the reigning scientific approach to sexual orientation, because such scientific research assumes *essentialism*, the view that our contemporary categories of sexual orientation can be applied to people in any culture and at any point in history. What should be made of this conflict between the two

Written for this anthology and published by permission of the author.

models for thinking about human sexual orientation used by these two important approaches to human sexuality?

In this essay, I explicate the debate between essentialism and constructionism about sexual orientation. This debate has been conflated with questions concerning the causes of sexual orientation and questions about how the sexual desires of people in other cultures ought to be understood. These questions, while related to essentialism and constructionism, are distinct from it. In the next section, I develop precise characterizations of essentialism and constructionism. Following that, I explain the relationship between this debate and the question of whether our categories of sexual orientation can be applied to historical cultures (Section III) and between this debate and the question of whether a person's sexual orientation is the result of genetic or environmental factors (Section IV). Having differentiated these questions, I discuss in detail Simon LeVay's neuroanatomical research, to show how scientific research does not provide evidence for essentialism, but rather assumes it, and thereby ignores a whole range of plausible theories concerning human sexual desires (Section V).

II. NATURAL HUMAN KINDS

Underlying the conflict between essentialism and constructionism is a question concerning *natural kinds*. A natural kind is a grouping of entities that plays a central role in the correct scientific laws and explanations. A group is a natural kind in virtue of the properties its members share or the functions they play independent of how we conceive of them. For example, if current chemical, physical, and physiological theories are correct, then gold, electrons, haemoglobin, and hearts are natural kinds, while chairs, teddy bears, and diet soft drinks are not.

Some groups that we think are natural kinds probably will turn out not to be. As a historical example, consider *phlogiston*. Until the late 1700s, scientists thought that phlogiston was an element found in high concentrations in substances that burned while exposed to air. Air that absorbed a lot of phlogiston would not support further combustion, and was not suitable for living creatures to breathe, because air was supposed to remove phlogiston from the body through respiration. We now know that there is no such substance as phlogiston. Although phlogiston was thought to be a natural kind, the scientific laws in which phlogiston played an explanatory role have turned out to be false. There is no single substance that has the properties attributed to phlogiston. The category 'phlogiston' fails to pick out any actual stuff. I call such groupings 'empty kinds'.[1]

Some natural kinds apply to people. I call these groupings of people that play a central role in scientific explanations and laws 'natural human kinds' (for a related discussion, see Hacking 1986). Examples of natural human kinds include groupings of people by blood types—for example, people with blood type AB—and groupings of people by genetic structures—for example, people with XY sex chromosomes. People with blood type AB constitute a natural human kind, because having blood type AB plays a role in laws about what sorts of people a person can donate blood to and receive blood from; a person with blood type AB can, for example, receive transfusions of red blood cells from a person of any blood type. Just as there are many groups of things that are *not* natural kinds, so there are many groups of people that are not natural human kinds; I call them 'social' human kinds. Examples include registered members of the Democratic Party, matriculated students at an Ivy League college, and convicted felons. These groupings may play a social role, but they do not play an explanatory role in scientific explanation. Similarly, just as there are empty natural kinds, so there are empty human kinds: that is, groupings of people supposed to be natural human kinds, but which fail to match the way nature actually groups people. As an example, consider the concept of a hysteric woman. The idea that there are women who suffer from a particular disease called 'hysteria' dates back to early Egyptian culture, and continued to the middle of the twentieth century. The symptoms of this disease included uncontrollable outbursts, as well as symptoms of any number of bodily diseases, though most typically spasms, paralysis, swelling, blindness, and deafness. Originally thought to be caused by a 'wandering' womb, hysteria came to be seen as a specific neurotic illness primarily affecting women. Many people were diagnosed as having hysteria when, in fact, they had some other unidentified illness or were depressed, tired, or nervous. Today, no one believes that hysteria is an actual medical condition. The hysteric is, in a sense, a human kind, but, as there are no hysterics, the hysteric is an *empty human kind*. Other examples of empty human kinds include witches and warlocks, namely women and men, respectively,

---

[1] This makes things seem simpler than they actually are. It is in practice hard to distinguish between an empty kind and a natural kind that has been mischaracterized. Consider the following example: for centuries, people thought that the earth was the centre of the universe and that the heavenly bodies—including the planets, the sun, and the stars—revolved around the earth. After Copernicus, we think that the sun is the centre of the solar system, and that the earth and planets revolve around it. In the shift from the Ptolemaic world-view to the Copernican world-view, it is not clear whether the group of 'heavenly bodies' was determined to be an empty kind, or whether we learned that it is a natural kind but that it includes the earth.

with supernatural powers. These are empty human kinds, I would argue, because people with such supernatural powers simply do not exist.

It is relatively straightforward to use these notions to define constructionism about sexual orientation and its opposite, essentialism.[2] Essentialism about sexual orientation is the view that the categories of sexual orientation refer to natural human kinds, while constructionism about sexual orientation is the view that the categories of sexual orientation do not refer to natural human kinds. Some people think that there are no natural kinds. From this it follows that there are no natural *human* kinds, and, thus, that our categories of sexual orientation do not refer to natural human kinds. While the position that there are no natural human kinds is an interesting general philosophical position (known as 'anti-realism', as well as by other names), it is not an interesting position with regard to sexual orientation. Constructionism about sexual orientation at least implicitly involves the claim that the categories of sexual orientation do not refer to natural kinds, but that other categories, like being a proton or being a person with blood type B, do. If there are no natural kinds, then there is nothing special about sexual orientation; sexual orientations would not be natural kinds, but neither would having a Y chromosome or being a proton. The point here is that constructionism is an interesting position *about sexual orientation* only if there are some natural kinds. Given this, constructionism about sexual orientation is the view that, although there are some natural kinds, the categories of sexual orientation do not refer to natural human kinds, but instead refer to empty human kinds or social human kinds. Having clarified what essentialism and constructionism claim, I turn to a discussion of how these views are frequently misunderstood.

III. A HISTORICAL EXAMPLE

Consider Aristophanes' speech in the *Symposium*. In this speech, Aristophanes offers a myth about the origins of love according to which the human race was once made up of three sexes: 'male and female, ... [and] a third which partook of the nature of both, [called] ... "hermaphrodite". ... a being which was half male and half fe-

---

[2] Some philosophers use the term 'essentialism' for the view that a grouping is a natural kind if and only if all and only members of that kind share an intrinsic property (or a set of intrinsic properties). Essentialism in the sense I mean here does not necessarily entail this sense of the term. Following Boyd (1991), I hold that some natural kinds are defined by a cluster of relevant causal conditions, none of which is necessary or sufficient.

male' (Plato 1935: 189d–e). The human race, besides having an additional sex, looked rather different. Each human was 'globular in shape, with . . . four arms and legs, and two faces, both on the same cylindrical neck, and one head, with one face on one side and one on the other, and four ears, and two lots of privates' (ibid. 189c–190). Because these humans were too powerful, Zeus split each human in half down the middle leaving 'each half with a desperate yearning for the other' (ibid. 191). The three globular types of humans gave rise, once they had been divided, to different types of people, defined by the type of person (other half) he or she was longing for:

The man who is a slice of the hermaphrodite sex . . . will naturally be attracted by women . . . and the women who run after men are of similar descent. . . . But the woman who is a slice of the original female is attracted by women rather than by men . . . while men who are slices of the male are followers of the male. (ibid. 191d–e)

A natural interpretation of this myth is that Aristophanes is talking about male heterosexuals, female heterosexuals, lesbians, and gay men, respectively, and that he is arguing that sexual orientation is an important, inborn, defining feature of a person (Boswell 1982–3). This reading of the myth depends on the appropriateness of applying our categories of sexual orientation to people living in Athens centuries ago.

There is, however, an alternative interpretation of Aristophanes' myth. Some scholars have argued that Aristophanes' speech is not about sexual orientation at all (Halperin 1990). This alternative reading draws support from historical evidence that the Greeks thought about people's sexual interests in a quite different way than we do. In Attic Greece, a person's social status—that is, whether the person was a citizen or a non-citizen, a slave or a free person, an adult or a child, a woman or a man—was important to how the culture viewed his or her sexual interests. In terms of law and social custom, a citizen was allowed to penetrate but not be penetrated by non-citizens (that is, slaves, children, women, and foreigners), and was not allowed to penetrate or to be penetrated by other citizens (Kaplan 1997). This historical evidence is supposed to indicate that Aristophanes and his contemporaries could not have used anything like our categories of sexual orientation, because these categories simply did not exist in their vocabulary. According to this view, it is anachronistic to interpret Aristophanes as talking about lesbians, heterosexuals, and gay men; it would be like claiming that someone in Attic Greece was talking about telephones or computers.

Supporters of the view that Aristophanes was talking about homosexuals in roughly the sense of the term as it is used by most people in contemporary North America (the first interpretation) might admit that, due to the social institutions of Attic Greece, homosexual desire there took a very different form than it takes in our society (in particular, it took an 'age-asymmetric' form, whereby adult males were primarily attracted to teen-aged boys). There are, on this view, homosexuals and heterosexuals in both contemporary times and in Attic Greece, but these types of people express their sexual desires in different ways in different eras. Aristophanes can, on this view, still be interpreted as talking about heterosexuals, lesbians, and gay men, even though he is talking about social structures surrounding sexual behaviours that seem foreign to us. The general point is that just because there are different social structures surrounding some human phenomenon in the past and the present does not necessarily mean that the two phenomena are different. Consider pregnancy. In earlier times, people thought about pregnancy in ways very different from how we do today. For example, many people in Attic Greece believed that males were solely responsible for the material production of offspring; they thought that each sperm contained a small human being inside of it. Females, on this view, were just incubators that enabled sperm to develop into human beings. On this view, the male provided the seed, while the female provided fertile soil in which the seed could grow. The fact that the Greeks believed this does not mean that females in Attic Greece failed to contribute genetic material to their offspring in the form of eggs. Even though the Greeks had a dramatically different understanding of pregnancy than we do today, and even though pregnancy had a quite different meaning in their culture than in ours, pregnancy then and now is the same biological phenomenon. I can say with confidence that a female in Attic Greece became pregnant when an egg fertilized by a sperm became attached to the wall of her uterus. Returning to sexual orientation, someone who favoured the first interpretation would reply to the second interpretation of Aristophanes' speech in a way that parallels my points about pregnancy: just because homosexuality took a different form and was understood differently does not mean that our categories of sexual orientation fail to apply to people in Attic Greece. Our contemporary medical and biological categories apply to them; why should our sexual categories fail to do so?

A friend of the second interpretation might reply to this line of argument by admitting that some of our categories (for example, the category of pregnant women) are appropriate to apply to people from Attic Greece. She could argue, however, that sexual orientation is not like these concepts. The difference between us and the Greeks is not just that we have different concepts concerning sexual orientation; we also live our lives in different ways. We and the Attic Greeks have different *forms of desire*; gender is among the most important features of a person that determine whether we might be attracted to him or her; for the Greeks, age, social class, and citizenship status were of greater or equal importance.

The question of how Aristophanes' speech in the *Symposium* should be interpreted (as well as similar interpretive questions) is often connected to the debate between essentialists and constructionists. Essentialists think that our categories of sexual orientation are applicable to people in every culture, because the categories refer to natural human kinds. The first interpretation of Aristophanes' speech—according to which he is talking about heterosexuals and homosexuals—thus fits well with essentialism about sexual orientation. By contrast, constructionists say that it does not make sense to apply the categories of sexual orientation to other cultures, such as Attic Greece, because our categories of sexual orientation do not refer to natural human kinds. Constructionists admit that there were people in Attic Greece who had sex with people of the same sex, and perhaps even that there were people who were sexually attracted *primarily* to people of the same sex, but they deny that this entails that there were homosexuals (in our sense of the term) in Attic Greece.[3] To apply our sexual-orientation terms to another culture, we need to have evidence that people in that culture had sexual orientations in roughly our sense of the term. In fact, say constructionists, no one in any culture before the mid-1800s had sexual orientations.

What, then, is the relationship between the essentialist–constructionist debate and the interpretive question about Aristophanes' speech in the *Symposium*? If the first reading of Aristophanes' speech is right, and Aristophanes is talking about lesbians, gay men, and heterosexuals, this shows that we share our categories for thinking about sexual orientation with the people of Attic Greece. This alone does not show that essentialism about sexual orientation is true. A person who believed that witch was a natural human kind would not establish the truth of this belief by showing that people in another culture—say, the colonists of Massachusetts in the 1600s—shared his or her categories. Contemporary believers in witches could use the same category that the colonists did, but this would not prove that witch is a natural human kind. Although showing that the Greeks had the same categories that we do does not establish the truth of essentialism, it does count as inductive evidence for this thesis. If almost

---

[3] I am here assuming that sexual orientations have more to do with a person's underlying desires and dispositions than with his or her actual sexual behaviour. People have sexual orientations even before they have engaged in any sexual behaviour and many people engage in sexual behaviour that does not 'fit' with their sexual orientation.

every culture divides people up into the same categories, then this suggests (but does not prove) that these categories capture some truth about human nature.

However, even if we were sure that Aristophanes is not talking about lesbians, gay men, and heterosexuals, this would not establish the truth of constructionism about sexual orientation. No one in Attic Greece talked about blood types, but, then and now, the categories of blood type still pick out natural human kinds. That most cultures lack a concept does not establish the truth of constructionism with respect to this concept. A culture might not have the concept, but the people in that culture might still fit it. If not, we could eliminate epilepsy, for example, just by eliminating the concept (Stein 1992b).

Just as the appropriateness of the first interpretation does not establish the truth of essentialism, and the appropriateness of the second interpretation does not establish the truth of constructionism, the truth of essentialism does not entail the truth of the first interpretation, and the truth of constructionism does not entail the appropriateness of the second interpretation. First, suppose that essentialism is true, and that sexual orientations are natural human kinds. It is perfectly consistent with this that the Greeks lacked these categories, and hence, that Aristophanes could not have been talking about homosexuals and heterosexuals in the *Symposium*. A parallel point can be made about pregnancy. It is perfectly consistent with the fact that females in Attic Greece contributed genetic material to their offspring through eggs that the Greeks lacked a concept of a human egg. Second, suppose that constructionism is true, and that our categories of sexual orientation do not refer to natural human kinds. It is consistent with this that the Greeks might still have the same categories of sexual orientation that we do. Two cultures can have the same category, even if the category does not refer to a natural kind.

The debate between essentialism and constructionism has often been equated with the debate over how various cultures categorized people in terms of their sexual desires. It is easy to see why this has happened. Our access to the sexual orientation of people in cultures that no longer exist (like Attic Greece) is dramatically limited—we cannot observe their sexual behaviour, and we cannot ask them about their sexual desires. Our access is limited to their expressions of their desires in the form of their writings, art, laws, and the like. This limitation on our access to other cultures should not, however, cause us to confuse what is at issue between essentialism and constructionism. Even if the Greeks lacked our categories, our categories still might refer to natural kinds; even if the Greeks had our categories, our categories still might not refer to natural kinds. There are, however, some links between the interpretive question and the ontological one. First, if we find that people in many other cultures have the same categories as we do, this provides some suggestive evidence for essentialism. Second, keeping in mind the constructionist thesis helps us appreciate cultural differences. This is a useful antidote to simply seeing all cultures through the lens of our categories of sexual desire; in Section IV, I will show that this is important for science as well as for history.

IV. NATURE VERSUS NURTURE

The debate between essentialists and constructionists is also often understood as reducing to whether sexual orientation is the result of a person's genetic make-up or his or her environment. As typically discussed, this question of nature versus nurture is based on a false dichotomy. No human trait is strictly the result of genes or strictly the result of environmental factors; all human traits are partly the result of both. There are genetic factors that effect even the most seemingly environmental traits, like, for example, what subject a person will study at university. On the other hand, environmental and developmental factors contribute to the development of even the most seemingly genetic traits—for example, eye colour. Although every human trait is affected by both genetic and environmental factors, there does seem to be some variance in degree—that my eyes are hazel and that my blood type is B are more tightly constrained by genetic factors than that in college I studied philosophy and that I associate the word 'tree' with tall, leafy plants found in forests. I have genes that make it almost certain that my eyes will be hazel, but I do not have genes that make it almost certain that I will study philosophy—I do not even have genes that make it almost certain that *if* I live in a culture where there are colleges that offer philosophy degrees, I will study philosophy. The nature–nurture debate about sexual orientation, properly understood, concerns where sexual orientation fits on the continuum between eye colour, on the one hand, and field of study, on the other.

Many people assume that essentialism entails that sexual orientation is strongly constrained by genes, and constructionism entails that sexual orientation is primarily shaped by the environment. Neither is the case. In general, it is possible for a category to refer to a natural human kind without it being the case that a person fits that category in virtue of her genes. For example, being a person with tuberculosis is a natural human kind, as is being a person who is immune to polio, but whether or not one is a member of either group is *not* genetically determined. Whether one is, has to do with the presence of a particular rod-shaped bacterium or certain antibodies in the bloodstream, respectively. Neither condition is determined genetically, but both are associated with natural human kinds. With respect to sexual orientation, if a simplified version of the Freudian theory of the origins of male sexual orientation were true, and a man has the sexual orientation he does in virtue of his relationship with his parents—namely, if he has a resolved or unresolved Oedipal complex—then essentialism about sexual orientation would be true. Once a boy has settled into a particular Oedipal status, he has a naturalistically determinate sexual orientation: his brain instantiates a particular psychological state that makes him a heterosexual or homosexual. In virtue of this psychological state, certain scientific laws apply to him. If this theory were right, then sexual orientations would be natural human kinds, but they would not be primarily genetic. This shows that the truth of essentialism does not entail the truth of nativism; not all natural human kinds have a genetic basis. The same example will suffice to show that if sexual orientation is shaped primarily by environmental factors, the truth of constructionism does not necessarily follow. If this simple Freudian theory were true, then sexual orientation would not be innate, but constructionism would be false (because sexual orientation, in virtue of its psychological basis, would be a natural human kind). The simple connections commonly thought to hold between the nature of the categories of sexual orientation and the cause of sexual orientation do not in fact hold: essentialism does not entail nativism, and environmentalism does not entail constructionism.

The commonly held view that there is a connection between the essentialism–constructionism debate and the nature–nurture debate about sexual orientation is not, however, completely wrong. If sexual orientation is innate, then constructionism about sexual orientation is false, and essentialism is true. If a person's sexual orientation is primarily determined by genetic factors, then there are natural human kinds associated with sexual orientations in virtue of the genes responsible for sexual orientation. If this were the case, constructionism would be false. This does not, however, constitute an objection to constructionism; it just makes clear that constructionism—and essentialism—are *empirical* theses—in particular, they are empirical theses related to whether sexual orientation is innate. If nativism is true, then constructionism is false, and essentialism is true. The contrapositive of this claim—the truth of constructionism entails that nativism about sexual orientation is false—follows directly from the same line of argument: if sexual orientation is not a natural human kind, then it cannot be innate.

V. THE EXAMPLE OF LeVAY'S NEURO-ANATOMICAL RESEARCH

In the previous section, I concluded that the issue between essentialism and constructionism is empirical; if, for example, sexual orientation is innate, then essentialism is true. It might seem to follow from this that current scientific research on sexual orientation, which claims that sexual orientation is innate, supports essentialism. In fact, rather than proving it, current scientific research *assumes* essentialism. As an example of this, I consider LeVay's widely cited neuro-anatomical study.

LeVay (1992) reported a neurological difference between heterosexual and gay men. LeVay examined the interstitial nuclei of the anterior hypothalamus (INAH) of the brains of forty-one people. Nineteen were inferred to be homosexual or bisexual men (based on accounts of their sexual activity in their medical records), who died of complications due to AIDS; six were men of undetermined sexual orientation who also died of AIDS (LeVay presumed they were heterosexual); ten were men also of undetermined sexual orientation who died of other causes (and were also presumed to be homosexual); and six were women, all presumed to be heterosexual, one of whom died of AIDS, and five of whom died from other causes. LeVay reported that one of the four sections of the INAH—INAH3—was significantly smaller in homosexual and bisexual men than in the men of undetermined sexual orientation. On this basis, he concluded that there is a correlation between hypothalami and sexual orientation, and further, that this correlation suggests there is something biological about sexual orientation. For various reasons concerning the details of its experimental design, this study should be viewed with great scepticism (see Byne and Parsons 1993, Byne 1996, Stein, in press). However, for the purposes of this essay, I want to focus on two general features of LeVay's project, features that it shares with most research on sexual orientation in the reigning scientific paradigm: namely, its assumption that homosexuality is a form of sex inversion and its simplistic picture of what a sexual orientation is. Examining these features of LeVay's project underscores that essentialism is an assumption, not a conclusion, of his study. Doing so also points to the role which constructionism can play in ferreting out the cultural biases implicit in scientific research on sexual orientation.

LeVay's study is premised on seeing sexual orientation as a trait with two forms: a male form that causes sexual attraction to women (shared by heterosexual men and lesbians) and a female form that causes sexual attraction to men (shared by heterosexual women and gay men). I call this assumption that homosexuality results from a sex-reversed brain or physiology the 'inversion assumption'. This assumption is evident, for example, in the equation of same-sex sexual activity in men with effeminacy. Research premissed on the inversion assumption typically proceeds by first trying to identify sex differences, and then seeing if any alleged sex difference is reversed in homosexuals.

There are alternatives to the inversion assumption. Perhaps, from the physiological point of view, gay men and lesbians should be grouped together, and heterosexual men and women should be grouped together. This would be the case if heterosexuals had a brain structure or physiology that disposed them to be sexually attracted to people of the opposite sex, while lesbians and gay men had a brain that disposed them to be sexually attracted to people of the same sex. Or, more plausibly, there might be no interesting generalizable differences in brains that correlate with these categories of sexual orientation. This would be the case if sexual orientation is not simply dimorphic but rather takes many different forms. The conscious and unconscious motivations associated with sexual attraction could differ even among individuals of the same sex and sexual orientation. A myriad of experiences (and subjective interpretations of them) could interact to lead different individuals to the same relative degree of sexual attraction to men, women, or both. Because sexual attraction to women, for example, could be driven by various different psychological factors, there is no reason to expect that all individuals attracted to women should share any particular physiology that distinguishes them from individuals attracted to men. This does not imply that a person's sexual orientation is not represented in her brain. Sexual orientation must of course be represented in the brain, but so must all of my other desires, proclivities, and preferences, including whether I prefer Brahms or Beethoven and whether I enjoy tennis (preferences that are probably not innate—that is, preferences that are closer to the origins of one's field of study than one's eye colour).

When reviewing evidence like LeVay's that is based on the inversion assumption, it is important to remember both that there are alternatives to this assumption, and that it is as yet scientifically unsupported. This is not to deny that the inversion assumption is intuitively plausible to many people in our culture. This is not because the assumption is intrinsically more plausible than either of the two rivals I mentioned; rather, it is because of our cultural bias towards thinking that gay men are feminine and lesbians are masculine. This cultural view is hardly universal. For example, in some cultures, sexual activity between males was associated with warrior status (see e.g. Watanabe and Iwata 1989), while in others receptive samesex sexual activity was universally practised by males and believed to be essential to their virility (Herdt 1984). The fact that the inversion assumption would not have been plausible in some cultures does not show that it is false. Together, however, the lack of scientific evidence to support the inversion assumption and the fact that it is not culturally universal suggest that it may be based as much on our cultural biases as anything else.

Turning to the second general feature of LeVay's study that I want to consider, note that one of the reasons why LeVay looked to the hypothalamus stems from an analogy with rodents: hormonal exposure in the early development of rodents exerts organizational influences on their hypothalamus, which determines the balance between male-typical and female-typical patterns of mating behaviours displayed in adulthood. There are two problems with this analogy. First, it is unclear whether humans and rodents have relevantly similar hypothalami (Byne 1996). Second, it is highly problematic to extrapolate from behaviour of rodents to psychological phenomena in humans. According to some researchers and many popular accounts, a male rat that is castrated at birth and subsequently shows a sexually receptive posture called 'lordosis' when mounted by another male is homosexual. This receptive posture, however, is basically a reflex. A neonatally castrated male will assume the same posture if a handler strokes its back. Further, while the male who displays lordosis when mounted by another male and the female who mounts another female are counted as homosexual, the male that mounts another male is considered heterosexual, as is the female that displays lordosis when mounted by another female. In this laboratory paradigm, sexual orientation is defined in terms of specific behaviours and postures. By contrast, in the human case, both the male who penetrates and the male who is penetrated (as well as the male who engages in or desires to engage in various acts that do not involve penetration at all) are counted as gay. Sexual orientation in humans is defined not by what 'position' one takes in sexual intercourse, but by one's pattern of erotic responsiveness and the sex of one's preferred sex partner. Some researchers acknowledge the problem of equating behaviours of rodents with their sexual desires, and employ a variety of strategies to reveal the sexual *preferences* of animal (for example, an animal is given the choice of pressing a bar that will allow access to a male conspecific or pressing a bar that will allow access to a female conspecific; if the animal presses the bar to gain access to the animal of the same sex, this establishes that the animal is homosexual). Even these studies may, however, have little to do with human sexual orientation. In rodents, in order for a genetic male rodent to behave as a female rodent typically does with respect to partner preference (as well as sexual position), he must be exposed to extreme hormonal abnormalities that are unlikely to occur outside the neuro-endocrine laboratory. First, he must either be castrated as a neonate, depriving him of androgens, or particular androgen-responsive regions of his brain must be destroyed. In addition, in order to activate the display of female-typical behaviours and preferences, he must also be injected with estrogens in adulthood (Byne 1996). Because gay men and lesbians have hormonal profiles that are indistinguishable from those of their heterosexual counterparts (Meyer-Bahlburg 1984), it is difficult to see how this situation has any bearing on human sexual orientation.

LeVay's study begins with a particular picture of sexual orientation. He assumes that homosexuality results from a more general sex inversion, and, in virtue of the analogy to rodents, that human sexual orientation can be usefully analysed in terms of the sexual positions one takes and the preferences for the company of males or females one exhibits. Neither of these assumptions is adequately justified, although both fit with a commonly accepted picture of sexual orientation (see LeVay 1996; also see Stein, in press). Constructionism challenges us to look beyond our cultural assumptions about sexual orientation to the possibility that our categories of sexual orientation are not natural human kinds. This may be a difficult thing to do, because we see ourselves and others through the categories of sexual orientation; these categories may thus seem natural when, in fact, they are not. Like the inversion assumption, essentialism is an assumption, not a conclusion, of current scientific research on sexual orientation.

VI. CONCLUSION

It is easy to picture what it would be like for essentialism to be true, but it is trickier to imagine how constructionism could be true. If the categories of sexual orientation are not natural kinds, how could we explain the fact that some people are attracted primarily to people of the same sex, while others are not? Presumably, we would do so in the same way we would explain other differences among people that do not seem to be primarily explicable in terms of their genetic make-up, like, for example, differences in musical tastes, dietary preferences (Halperin 1990), and sleep positions (Stein 1992a). We do not think, for example, that categories associated with whether a person typically sleeps on her back or her stomach play a role in any significant scientific laws, yet we can imagine giving an explanation of how such an 'orientation' develops. Constructionism about sexual orientation does not preclude an explanation of how people's sexual tastes develop; it just precludes that such explanations will involve natural kinds associated with the categories of sexual orientation. Scientists exploring sexual orientation need to take special care not to introduce unjustified cultural assumptions about the organization and nature of human sexual desires into their research. Constructionism (and the field of lesbian and gay studies from which it emerges) can make an important contribution to such research, because it highlights these assumptions.[4]

REFERENCES

Abelove, H., Barale, M., and Halperin, D. (1993) (eds.), *The Lesbian and Gay Studies Reader* (New York: Routledge).
Bailey, J. M., and Pillard, R. (1991), 'A Genetic Study of Male Sexual Orientation', *Archives of General Psychiatry*, 48: 1089–96.
Boswell, J. (1982–3), 'Revolutions, Universals and Sexual Categories', *Salmagundi*, 58–9: 89–113.
Boyd, R. (1991), 'Realism, Anti-Foundationalism and the Enthusiasm for Natural Kinds', *Philosophical Studies*, 61: 127–48.
Bullough, V. (1994), *Science in the Bedroom: The History of Sex Research* (New York: Basic Books).
Byne, W. (1996), 'Biology and Sexual Orientation: Implications of Endocrinological and Neuroanatomical Research', in R. Cabaj and T. Stein (eds.), *Comprehensive Textbook of Homosexuality* (Washington: American Psychiatric Press), 129–46.
——and Parsons, B. (1993), 'Sexual Orientation: The Biological Theories Reappraised', *Archives of General Psychiatry*, 50: 228–39.
——and Stein, E. (1997), 'Ethical Implications of Scientific Research on the Causes of Sexual Orientation', *Health Care Analysis*, 5: 136–48.
Fausto-Sterling, A. (1992), *Myths of Gender: Biological Theories about Women and Men*, rev. edn. (New York: Basic Books).
Foucault, M. (1978), *The History of Sexuality*, Vol. 1: *An Introduction*, trans. R. Hurley (New York: Random House).
Hacking, I. (1986), 'Making Up People', in T. Heller, M. Sosna, and D. Wellbery (eds.), *Reconstructing Individualism: Autonomy, Individuality, and the Self in Western Thought* (Stanford, Calif.: Stanford University Press), 222–36; repr. in Stein, (1992), 69–88.
——(1991), 'A Tradition of Natural Kinds', *Philosophical Studies*, 61: 109–26.
Halperin, D. (1990), *One Hundred Years of Homosexuality and Other Essays in Greek Love* (New York: Routledge).
Hamer, D., *et al.* (1993), 'A Linkage between DNA Markers on the X Chromosome and Male Sexual Orientation', *Science*, 261: 321–7.

---

[4] For further discussion of these issues, see Stein (forthcoming). This essay benefited from the comments of David Hull and a collaboration with Bill Byne on a related project (Byne and Stein 1997).

Herdt, G. (1984), 'Semen Transactions in Sambia Culture', in G. Herdt (ed.), *Ritualized Homosexuality in Melanesia* (Berkeley: University of California Press), 167–210.

Kaplan, M. (1997), *Sexual Justice: Democratic Citizenship and the Politics of Desire* (New York: Routledge).

LeVay, S. (1992), 'A Difference in the Hypothalamic Structure between Heterosexual and Homosexual Men', *Science*, 253: 1034–7.

——(1996), *Queer Science: The Use and Abuse of Research into Homosexuality* (Cambridge: MIT Press).

McIntosh, M. (1968), 'The Homosexual Role', *Social Problems*, 16: 182–92; repr. in Stein (1992), 25–42.

Meyer-Bahlburg, H. (1984), 'Psychoendocrine Research on Sexual Orientation: Current Status and Future Options', *Progress in Brain Research*, 71: 375–97.

Mohr, R. (1992), 'The Thing of It Is: Some Problems with Models for the Social Construction of Homosexuality', in R. Mohr (ed.), *Gay Ideas* (Boston: Beacon), 221–42.

Plato (1935), *Symposium*, trans. M. Joyce (New York: Everyman's Library).

Stein, E. (1992a), 'The Essentials of Constructionism and the Construction of Essentialism', in Stein (1992b), 325–54.

——(1992b) (ed.), *Forms of Desire: Sexual Orientation and the Social Constructionist Controversy* (New York: Routledge).

——(Forthcoming), review of LeVay (1997), *Journal of Homosexuality*.

——(Forthcoming), *Sexual Desires: Science, Theory and Ethics* (New York: Oxford University Press).

Watanabe, T., and Iwata, J. (1989), *The Love of the Samurai: A Thousand Years of Japanese Homosexuality* (London: Gay Men's Press).

PART VII

ALTRUISM

INTRODUCTION TO PART VII

MICHAEL RUSE

Darwinians have always realized that, when it comes to success in the struggle for existence, behaviour is as important as structure. The fruit-fly with a flawed mating dance is no less handicapped than the fruit-fly with missing legs. What is often really important is *social* behaviour. In many cases, this is no great cause for comment. When a mother puts in effort for the benefit of her offspring, clearly she is aiding her own reproductive success. When a pack of wolves hunt together, likewise they are helping themselves. A full-grown moose is too big and too strong for a single wolf, but when wolves work together, the prey is vulnerable, and all of the hunters benefit.

But often the social behaviour is puzzling, if not outrightly paradoxical. Why would a young bird return to the parental nest, helping in the rearing of its siblings, rather than striking out on its own? Why would an organism act as a sentinel—often at great cost to itself—when it might easily look immediately to its own well-being? Why, most puzzling of all, do the female workers in the hymenoptera (ants, bees, and wasps), devote their whole lives to the good of the nest and to their mother's offspring, entirely forgoing their own reproduction? And why is it that only the females are in this situation, and not the males?

The obvious general answer to questions such as these is that although the individual may not benefit as such, the group (usually the species) does gain by the individual's sacrifice or 'altruism'. Yet, from Charles Darwin on, Darwinian evolutionists have realized that there are major problems with such a quick and easy answer. Most problematic is the fact that altruistic behaviour lays itself open to exploitation, with its apparent rapid elimination through cheating. If one has two organisms, one putting effort into the welfare of others and the other devoting its labours exclusively to its own end (although it may accept the aid of others), then, all other things being equal, the altruism is doomed to rapid evolutionary extinction. Selection is a short-term process, in that it does not think of the long-range gains. Even though the group may benefit in the long run from the altruist's behaviour, selection will never let genes for self-sacrifice get established. (See also Part III on 'Units of Selection'.)

Darwin himself worried a great deal about this problem, particularly in the cases of the hymenoptera and of humans. As Alexander Rosenberg shows in his introduction to the topic (Ch. 22), at some level Darwin felt obliged to resort to a group-selective solution—that somehow selection can overcome the problems of exploitation. However, he was always wriggling, trying to find mitigating or explanatory circumstances. In the case of the hymenoptera, Darwin argued that since nests are made up of related individuals, in some wise it is legitimate to treat the group as a super-individual, all of whose members are working towards the same end. A similar explanation was invoked for humans—selection occurring between bands of relatives—although at the same time Darwin speculated on a kind of proto-version of what has come to be called 'reciprocal altruism'. Help given demands and evokes help returned. You scratch my back, and I will scratch yours—and if you do not return the favour, expect no further help from me or my relatives and friends!

As Rosenberg explains, the real breakthrough came only in the 1960s, when the English (then) graduate student William Hamilton saw that the hymenopteran situation is a function of the peculiar mating system of ants, bees, and wasps. Because only females have fathers (males are virgin-born), sisters are more closely related to each other (3/4) than mothers and daughters are to each other (1/2). Hence, the self-sacrificing behaviour of sterile workers is in a sense nothing of the kind. They are 'selfishly' promoting their own biological ends by raising sisters rather than daughters. (Males have no such skewed relationships, and so no biological reason to help others, as do the workers.) With this breakthrough, the whole Darwinian sub-discipline of the study of animal social behaviour ('sociobiology') came into being. As also did a host of philosophical questions about when, and in what sense, one can properly speak of 'altruism', and when, and in what sense, one can properly speak of 'selfishness', and whether concepts which apply in the animal world can ever, and in what circumstances, be applied to humans.

It is to some of these philosophical problems that the essays of Elliott Sober (Ch. 23) and David Sloan Wilson (Ch. 24) are directed. Sober's main point is as important as it is straightforward: one must be very careful about moves from biological or evolutionary notions of altruism to (what he calls) 'vernacular altruism', the disinterested giving of human beings to others. Most importantly, human altruism requires a psychological factor, an intention to do something for someone. Biological altruism does not. But there are other differences. Evolutionary altruism is directed towards reproductive benefit. There is no reason whatsoever why human or vernacular altruism should have anything at all to do with reproduction. Indeed, it may point in the other direction. Also, evolutionary altruism is comparative (as things always are when one thinks in selective terms). I cannot be an evolutionary altruist in isolation. It is a concept which occurs only inasmuch as one does something better than something or someone else (who is thus, by definition, 'selfish'). In the human case, vernacular altruism is absolute. It does not matter what you do or think. My actions and intentions are to be judged in their own right. One important consequence of Sober's discussion is that, even if one thinks that selection is always for individual benefit, it does not follow that vernacular altruism could never evolve under the control of selection. It is a fallacy to think that because of our biology we are all necessarily born selfish, in the human sense.

Wilson continues this discussion, stressing the extent to which metaphors (altruism, selfishness) are introduced into biological discussions. His point is not that metaphors are wrong—Darwin himself appreciated how much he relied on metaphor (natural selection, struggle for existence, tree of life, and many more)—but that one must be careful to recognize when one is using metaphors, and not think that they are literal descriptions of matters of fact. At the end of his discussion, Wilson asks a number of interesting questions, showing that more work will be needed both biologically and philosophically. Why are motives so important? Could there be a biological pay-off? Why do we have psychological (or Sober's vernacular) altruism at all? Is there a biological benefit? Or is it simply a limitation of the human brain? Could it not be that psychological altruism is a compromise or making the best job in limited circumstances? Perhaps in theory there are better ways for humans to interact socially, but psychological altruism is the only tool available to natural selection. These and related questions show that social behaviour will be as much a topic for evolutionists of the future as it has been for those of the past.

22

ALTRUISM: THEORETICAL CONTEXTS

ALEXANDER ROSENBERG

In *Sociobiology: The New Synthesis* E. O. Wilson (1975: 578) defines altruism as 'self-destructive behavior performed for the benefit of others'. More specifically, sociobiology treats behaviour as altruistic whenever the behaviour increases the reproductive fitness of another at the expense of one's own reproductive fitness. At the outset of his touchstone of contemporary behavioural biology, Wilson identifies altruism as the 'central theoretical problem of sociobiology', and asks, 'how can altruism, which by definition reduces personal fitness, possibly evolve by natural selection?' (1975: 3). The problem is apparently one of explaining how the actual is possible. Altruism, like co-operation in general, is an obvious feature of human and infra-human behaviour. Indeed, sociality requires it. And yet, if altruism reduces fitness, in the evolutionary long run it should have been expunged, not enhanced. So we face a choice between exempting human and some infra-human behaviour from the constraint of natural selection and finding a way of rendering it consistent with Darwin's theory. This is Wilson's problem.

The biological problem of altruism is vexed by a prior terminological controversy. Altruism, as commonly understood, is, by definition, action that advantages another by design. It is an 'aetiological' concept, which carries a definitional commitment to a motive—an intentional cause. But this motive is missing in Wilson's definition. And few sociobiologists suppose that the occurrences of actions and their intentional causes are explainable by natural selection. Nevertheless, sociobiological altruism might be relevant to motivated altruism. It may explain a genus—altruism-motivated or unmotivated—of which motivated altruism is a species.

But even if what the sociobiologist means by altruism has nothing to do with motivated altruism, Wilson's stipulative definition still describes an

First published in E. F. Keller and E. Lloyd (eds.), *Keywords in Evolutionary Biology* (Cambridge, Mass.: Harvard University Press, 1992), 19–28. Reprinted by permission.

important phenomenon with which evolutionary theory must come to terms. For other-regarding behaviour—no matter its cause—seems endemic to mammalian life, and essential for human society. We need an explanation of how it can be consistent with the theory of natural selection.

In addition to differences about motivation, we should bear in mind another significant difference between biological altruism and motivated altruism. What ordinary language means by altruism is sacrifice of self-interest—that is, the interests of the individual. Biological altruism involves not what we would ordinarily recognize as the interests of the individual but, rather, the interests of its offspring, immediate and distant. For motivated altruism's disservice to the interests of the self, biological altruism substitutes disservice to the future survival of the line of descent, the biological lineage of which the self is a member; the coin in which biological altruism is measured is evolutionary fitness—differential rates of reproduction. Sometimes the interests of the self and of its descendants coincide. But not always. For a sociobiological account of altruism to shed much light on motivated altruism among humans, it must link the interests of a lineage with that of an individual.

The problem of altruism, and more generally of co-operation among humans, is one Charles Darwin broached in *The Descent of Man* (1871). Broadly, his answer to the question of how other-regarding behaviour is possible over the long haul of natural selection appealed to the group as the unit of selection. Darwin linked reproductive fitness to motivated altruism via the emotion of sympathy. Identifying the immediate cause of other-regarding behaviour as the feeling of sympathy, Darwin wrote:

In however complex a manner this feeling may have originated, as it is one of high importance to all those animals which aid and defend each other, it will have been increased, through natural selection; for those communities, which include the greatest number of the most sympathetic members, will flourish best and rear the greatest number of off-spring. (p. 82)

. . . an instinctive impulse, if it be in any way more beneficial to a species than some other or opposed instinct, would be rendered the more potent of the two through natural selection; for the individuals which had it most strongly developed would survive in large numbers. (p. 84)

Thus altruism persists among individuals, despite its cost in individual fitness, because it enhances fitness of the group. This account of individual co-operation by appeal to group selection hinged on an important explanatory success of *On the Origin of Species* (1859). There Darwin noted that 'many instincts of very difficult explanation could be opposed to the theory of natural selection,—cases in which we cannot see how an instinct could possibly have originated . . . I will not here enter on these several cases, but will confine myself to one special difficulty, which at first appears to me insuperable, and actually fatal to my whole theory. I allude to the neuters or sterile females in insect communities . . . from being sterile, they cannot propagate their kind' (p. 257). Neuter castes are relevant to the problem of altruism, because they represent the extreme case of sacrificing reproductive fitness for the advantage of others. Darwin argued that 'some insects and other articulate animals in a state of nature occasionally become sterile; and if such insects had been social, and it had been profitable to the community that a number should have been annually born capable of work, but incapable of procreation, I can see no very great difficulty in this being effected by natural selection' (p. 258). Sterility is the sacrifice of all of the organism's reproductive interests to the benefit of the community and its fertile members. It is the height of biological altruism, although in ordinary terms it hardly counts as self-destructive behaviour. Darwin's explanation of the sterile castes reveals that in the case of the social insects the individual is not the organism, but the colony or hive that it belongs to. This supra-organismic individual is not itself biologically altruistic. The colony's fitness-maximizing strategy, however, involves the production of sterile castes that aid non-sterile ones.

Among organisms much below man in cognitive powers, there is of course no role for sympathy as the cause of self-sacrifice. And Darwin did not know anything about how genetic information can control such behaviour. But at the level of organization at which sympathy can emerge, Darwin also hinted at another factor that would encourage the natural selection of sympathetic feelings and the action that stems from them: kinship. In *The Descent of Man* he wrote:

It is evident in the first place, that with mankind the instinctive impulses have different degrees of strength; a young and timid mother urged by the maternal instinct will, without a moment's hesitation, run the greatest danger for her infant, but not for a mere fellow-creature. (1871: 87)

[The social virtues] are practiced almost exclusively in relation to the men of the same tribe; and their opposite are not regarded as crimes in relation to the men of other tribes. No tribe could hold together if murder, robbery, treachery, etc. were common; consequently such crimes within the limits of the same tribe are . . . branded with everlasting infamy . . . ; but excite no such sentiment beyond these limits. (1871: 93)

Subsequent discussion of the problem of altruism has focused on these two tendencies in Darwin's thought: the role of the group as the unit of selection and the influence of kinship on behaviour. Group-selectionist accounts of altruism and other individual fitness-reducing traits fell out of favour among biological theorists after the rediscovery of Gregor Mendel at the turn of the century. R. A. Fisher, J. B. S. Haldane, and Sewall Wright all recognized the biological possibility of group selection for traits that are individually maladaptive, but held that the process requires conditions unlikely to be found in nature. Accordingly, they held group selection to be unsatisfactory as an explanation for altruism or other co-operative instincts. Even the emergence of sterile castes in the social insects need not require group selection: as Darwin himself suggested, it could be the result of an optimal individual reproductive strategy, one that involves producing a mix of fertile offspring and sterile offspring that aid the survival and reproduction of the fertile ones.

For a long time the role of altruism as a cause of group selection was more widely accepted outside the circle of mathematical population geneticists. Among advocates of the efficacy of group selection in the evolution of co-operation were Peter Kropotkin (1902), A. M. Carr-Sanders (1922), W. C. Allee (1955), Konrad Lorenz (1966), and V. C. Wynne-Edwards, whose book *Animal Dispersion in Relation to Social Behaviour* (1962) reopened widespread discussion of group selection as a Darwinian explanation of altruism in particular and of co-operation in general. Wynne-Edwards was struck by the fact that populations appear to regulate their sizes to levels well below the environment's carrying capacity, and that when populations approach carrying capacity, members lower their offspring numbers. Wynne-Edwards's explanation involved the concept of altruistic self-restraint on the part of individual organisms who reproduce at rates below those optimal for their individual fitness, but optimal for the survival of the group of which they are members. Individual fitness maximization involves consumption of resources at the expense of other members of the group. Altruistic self-restraint ensures the survival of the whole group. It should be noted that this explanation for population growth limits had already been undermined in favour of an individual fitness-maximizing strategy in David Lack's study of clutch size among finches (1954). Lack noted that having smaller clutch sizes could be a more adaptive strategy than having larger ones, because this enables the parent to divide investment in offspring into small numbers of larger bundles, thus increasing the likelihood that individual offspring will survive to adulthood. The availability of explanations for apparent altruism that appeal to its advantages for individual selection has motivated much of the discussion since the work of Lack and Wynne-Edwards.

After G. C. Williams's attack on the notion of group selection, even the benefits of altruism for selection at the level of the individual became controversial. Williams (1966) held that adaptational significance should not be accorded to a trait at any level above that necessary to explain the persistence of the trait in question. Following this principle, he argued that the locus of selection is always the individual gene—so that if altruism is to be given an evolutionary explanation, it must be shown how it can be adaptive for the individual allele. And, as J. Maynard Smith showed at about the same time (1964), selfish individuals or genes could subvert altruistic groups, no matter how these altruistic groups emerged, by individual or group selection. Start with a group of altruists, regardless of how they may have emerged, and introduce an egoist to the group. Over time the egoist will traduce the co-operators, and because the egoist's strategy advantages it at the expense of the others, over the long haul the egoist and its egoistic offspring will reproduce at greater rates, and eventually swamp the altruists in the group. Altruism is an 'invadable' strategy. Given variation and selection, strategies that can be invaded will be.

By the late 1960s Darwin's other idea, that altruism might be selected for because of kinship between the altruist and the recipient of altruism, came to dominate discussions of how the emergence of co-operation might be explained. The idea is due principally to W. D. Hamilton (1964), though it can be found in G. C. Williams and D. C. Williams (1957). Hamilton argued that nature will select for the strategy that leaves the largest number of copies of the gene that codes for it. In the case of sexually reproducing organisms, this will usually be direct offspring—sons and daughters. But it need not be. A strategy that sacrifices a son or daughter to save two siblings, three nephews or nieces, or a parent will ensure the survival of just as many or more copies of the individual's genes as will a strategy of saving a son or daughter. Accordingly, nature will select for 'inclusive fitness', which combines the organism's fitness with the fitness contribution (adjusted by some coefficient of relatedness) of each of its kin—the other organisms with whom it shares copies of the same genes. If nature engages in 'kin selection' by selecting for 'inclusive fitness', then altruism may emerge as an adaptive strategy for an individual that is part of a group of kin. Not only will feeding, protecting, teaching, and otherwise devoting resources to direct offspring be an adaptationally optimal strategy for individuals, but so will devoting resources to kin in proportion to their consanguinity. Thus we may explain Darwin's observation in *The Descent of Man* that the social virtues are always practised by men of the same tribe.

There are two problems with this neat picture. First, how do individual kin-altruists identify other animals as close enough kin, and what do they do in cases of uncertainty? This problem is especially difficult for males, because the uncertainty of paternity is always greater than the uncertainty of maternity, and may sometimes be substantial. (For a discussion of this problem as it touches sociobiology, see Alexander 1979.) Second, why does an altruist provide resources at its own expense to *unrelated* organisms? This latter is the important question for sociobiology. As human societies have evolved, they have been characterized more and more by co-operation, and less and less by kinship. Either human society slips the leash of selection for the fittest strategies, or else we need to find an explanation for the emergence of altruism other than its inclusive fitness.

It is easy to hypothesize that altruism emerges as a reciprocal strategy, in which individuals are co-operative in order to secure co-operation from others on a later occasion. The term 'reciprocal altruism' was introduced in an account of the emergence of non-kin altruism by Robert Trivers (1971). Altruistic actions offered in the expectation of reciprocation are, however, vulnerable to disappointment; indeed, it appears that the optimal strategy for a fitness maximizer is to accept altruistic benefits from others, but to decline to make them. Accordingly, altruism in the expectation of reciprocation will be displaced in the long run by egoism.

Trivers and others had noticed that among animals, including humans, the problem of reciprocal altruism reflects what game theorists and others have called a 'prisoner's dilemma'. Suppose two partners in crime are apprehended, and are then separately interrogated by the police. Each is offered the following 'deal': if neither confesses, both will be imprisoned for two years; if one confesses and the other stands mute, the confessing prisoner will spend one year in gaol, and the non-confessing prisoner will be sentenced to ten years; if both confess, each will be imprisoned for five years. What should each partner do? Confession is the 'dominant' strategy—the strategy that gives the best pay-off no matter what the other partner does. Not to confess means that one can be played for a sucker, and by confessing, one at least stands a chance to free-ride on the other partner's refusal to confess. The situation is a dilemma, however, because the dominant strategy for each partner leads to an outcome less desirable than another attainable one. Both prefer the two-year sentences each will receive if both stand mute to the five-year sentences each will receive if both confess. But any reason to think that standing mute is a good strategy for oneself is a reason to think that the other partner will also decide to stand mute, and this is an inducement to confess and get off with only one year in prison. The point can be illustrated graphically (see Fig. 22.1).

The preference rankings for the two partners are as follows:

Partner 1: (a) > (b) > (c) > (d)
Partner 2: (d) > (b) > (c) > (a)

Notice that although both rank outcome (b) second, the dominant strategy for each results in the attainment of outcome (c), which both rank third. Whence the dilemma: for either partner the only reason to act so as to

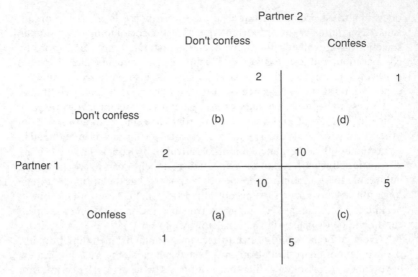

Fig. 22.1. An example of a game-theoretical prisoner's dilemma.

attain outcome (b) is just another reason to confess so as to stand a chance to attain the most preferred alternative, and therefore increases the likelihood of outcome (c). The pay-off to free-riding and the cost of being taken for a sucker both exclude altruism as an available strategy to the rational agent maximizing utility. To the extent that nature selects for maximizing fitness, it will discourage altruism and encourage free-riding whenever the pay-offs in nature are like those in the prisoner's dilemma. When are the pay-offs in nature like this? Not in gaol time or money or utilities, of course, but in numbers of offspring the individual leaves as a result of its choice and its opponent's choice. These pay-offs arise whenever opportunities to co-operate have a structure that results in the two preference rankings given above. The actual pay-offs can be any amount of any commodity, as long as the two players have the stipulated order of preferences. This will happen whenever there is a chance of being taken for a sucker combined with an opportunity to free-ride: possible cases include signalling the presence of predators or silently fleeing instead, sharing food or hogging it, respecting unprotected territory or encroaching on it, and raising offspring or procreating without staying around to raise the offspring. Opportunities for altruism by one agent are opportunities for free-riding by another, and it appears that the costs and benefits make free-riding an adaptational strategy for individuals and altruism a maladaptive one. If occasions for social interaction are like prisoner's dilemmas, then there is a great impediment to the emergence of altruism or co-operation. The problem is worse than Darwin imagined.

It remained for Hamilton, together with Robert Axelrod, to suggest a way in which reciprocal altruism might evolve even when the structure of altruistic opportunities reflects the pay-offs of the prisoner's dilemma. The first thing to note is that opportunities to co-operate or not come repeatedly in nature. The prisoner's dilemma is a situation animals face over and over again with a relatively small number of other animals. Only when we reach complex industrialized human societies does the frequency with which any two people can expect to play a prisoner's dilemma game a second time fall to anything close to zero (and, of course, if altruism is ever an adaptive strategy, it surely must have become one before the onset of contemporary industrial society). It is in the search for an optimal strategy for repeated or 'iterated' prisoner's dilemma situations that Hamilton and Axelrod (1981) found a potential solution to Wilson's problem of how altruism might have emerged.

Employing computer simulations of Axelrod's (1984), they showed that, under certain circumstances, in an iterated prisoner's dilemma, the optimal strategy for the individual is one called 'tit-for-tat': co-operate in game one, and then in each subsequent round do what the other player did in the previous round. Thus, if the other player tries to free-ride in round one, the best response is to refuse to co-operate in round two. If in round two the other party changes strategies and co-operates, in round three one should return to co-operating as well.

Axelrod's computer simulation pitted players using a variety of strategies for playing the prisoner's dilemma against each other. If, among a number of players using different strategies, the ones with the lowest pay-offs are eliminated, say, after every five turns, then in the end, after enough turns, all remaining players will be using tit-for-tat. In the long run no strategy generates a higher pay-off than tit-for-tat. Tit-for-tat is an effective strategy, in part because it is clear—opponents can easily tell what strategy a player is using; it is nice—it begins by co-operating; and it is forgiving—it retaliates only once for each attempt to free-ride on it. If playing tit-for-tat in an iterated prisoner's dilemma is a significant individually adaptive strategy, and can be transmitted from generation to generation (genetically or otherwise), then, in the long run, reciprocal altruism can be established even among animals that have neither kinship ties nor even common membership in the same biological species. Tit-for-tat is altruism in the expectation of reciprocation, with the threat of retaliation just in case co-operation is not forthcoming. Note that when a group of players play tit-for-tat among themselves, reciprocating regularly, they and their strategy are not invadable by players using an always free-ride–never co-operate strategy. Players who do not co-operate will do better on the first round with each of the tit-for-tat-ers, but will do worse on each subsequent round, and in the long run will be eliminated. In Maynard Smith's (1982) terms, tit-for-tat is an evolutionarily stable strategy: if it gets enough of a foothold in a group, it will expand until it is the dominant strategy, and once it is established, it cannot be overwhelmed by another strategy.

It is important to bear in mind that tit-for-tat is an optimal strategy for maximizing the individuals's pay-off (evolutionary or otherwise) only under certain conditions. Among these conditions are some that are easily satisfied in evolutionary contexts and others that are much harder to be sure of. Most important is the requirement that the number of games any two players play with each other cannot be known to either. If two players play an iterated prisoner's dilemma a certain number of times, and the number is common knowledge to both, then though they may agree to co-operate on each round, on the last round each will certainly find it irrational to co-operate, for the last round is like a single-case prisoner's dilemma game. But this means that the next-to-last round now becomes the last chance to free-ride on the other party's co-operation, and the last chance to be taken for a sucker. Therefore, in this penultimate round, neither party will co-operate—and so on, for each game working back to the very first one. Common knowledge about the exact number of iterations unravels the optimization of the tit-for-tat strategy, and destroys any chance of co-operation. Even knowledge of the probability that some future game will be the last game can lead to a breakdown, if the pay-off to free-riding on that game is worth the risk that further games will be played after all. Another important condition reflects the trade-off between immediate gains and long-term gains. For tit-for-tat to be the best strategy, the long-term pay-off to co-operation must be greater than the short-term gain from free-riding. Tit-for-tat is a strategy that sacrifices current-round opportunities to free-ride for future benefits from reciprocation by the other player. As the economist would put it, the discounted value of future pay-offs to co-operation must be greater than the value of present pay-offs to free-riding. Still a third condition for tit-for-tat to get started among large numbers of players is that players be able to recognize each other and recall previous games in the iteration. This requirement is of course unnecessary when the game is played repeatedly with only one other partner. And if tit-for-tat becomes fixed among a small number of players, it will not be invadable as the number of players grows, even though the ability to identify players and remember previous rounds does not increase. Moreover, there is no requirement that the tit-for-tat strategy be the conscious result of calculation. Unlike motivated altruism, tit-for-tat governs behaviour, not its causes.

How might tit-for-tat strategies actually emerge and spread in nature, especially among organisms of limited cognitive power? The best the sociobiologist can do by way of answering this question is to point to the power of nature to provide variations in behaviour and to point to environmental exigencies as being sharp enough to select the most optimal behavioural routines no matter how fortuitously they may emerge. If there are iterated prisoner's dilemmas in nature, and if the adaptational advantages of tit-for-tat strategies are powerful enough, nature will find them, for it has world enough and time.

It is of course unclear whether apparently co-operative behaviour evidenced by animals within and across species actually constitutes reciprocal altruism, and still less whether it is strategic behaviour in accordance with a tit-for-tat rule. But the tit-for-tat solution to iterated prisoner's dilemmas does solve Wilson's problem: it shows how evolutionary altruism is possible in at least some possible environments ruled by natural selection for individual fitness. The work of converting this abstract possibility into an explanation of actual co-operative behaviour, let alone motivated altruism, has yet to be completed. The story will have to link the adaptational advantages of reciprocal altruism for fitness maximization to the actual institutions of co-operation and the widespread occurrence of motivated altruism. It may establish these links via Darwin's mechanism: sympathy and other social motives persist because the behaviour they cause contributes to the fitness of the individual and its lineage. Or sociobiological theory may find less obvious connections between the self that common sense tells us altruism sacrifices and the line of descent that biological altruism may benefit.

REFERENCES

Alexander, R. D. (1979), *Darwinism and Human Affairs* (Seattle: University of Washington Press).
Allee, W. C. (1955), *Cooperation among Animals, with Human Implications* (New York: Shuman).
Axelrod, R. (1984), *The Evolution of Cooperation* (New York: Basic Books).
Carr-Sanders, A. M. (1922), *The Population Problem: A Study in Human Evolution* (Oxford: Clarendon Press).

458 ALEXANDER ROSENBERG

Darwin, C. (1859), *On the Origin of Species* (repr., Baltimore: Penguin Books, 1968).

—— (1871), The *Descent of Man and Selection in Relation to Sex* (London: John Murray).

Hamilton, W. D. (1964), 'The Genetical Evolution of Social Behavior, I and II', *Journal of Theoretical Biology*, 7: 1–52.

—— and Axelrod, R. (1981), 'The Evolution of Cooperation', *Science*, 211: 1390–6.

Kropotkin, P. (1902), *Mutual Aid: A Factor in Evolution* (London: Heinemann).

Lack, D. (1954), *The Natural Regulation of Animal Numbers* (Oxford: Oxford University Press).

Lorenz, K. (1966), *On Aggression* (London: Methuen).

Maynard Smith, J. (1964), 'Group Selection and Kin Selection', *Nature*, 201: 1145–7.

—— (1982), *Evolution and the Theory of Games* (Cambridge: Cambridge University Press).

Trivers, R. L. (1971), 'The Evolution of Reciprocal Altruism', *Quarterly Review of Biology*, 46: 35–57.

Williams, G. C. (1966), *Adaptation and Natural Selection* (Princeton: Princeton University Press).

—— and Williams, D. C. (1957), 'Natural Selection of Individually Harmful Social Adaptations among Sibs with Special Reference to the Social Insects', *Evolution*, 11: 32–9.

Wilson, E. O. (1975), *Sociobiology: The New Synthesis* (Cambridge, Mass.: Harvard University Press).

Wynne-Edwards, V. C. (1962), *Animal Dispersion in Relation to Social Behaviour* (Edinburgh: Oliver and Boyd).

23

WHAT IS EVOLUTIONARY ALTRUISM?

ELLIOTT SOBER

In this essay I want to clarify what biologists are talking about when they talk about the evolution of altruism. I'll begin by saying something about the common-sense concept. This familiar idea I'll call 'vernacular altruism'. One point of doing this is to make it devastatingly obvious that the common-sense concept is very different from the concept as it's used in evolutionary theory. After that preliminary, I'll describe some features of the evolutionary concept. Then I'll conclude by briefly considering what explanatory relation might obtain between vernacular altruism and evolutionary altruism.

Although the points I'll make are rather elementary ones, their interest is not restricted to those who have never heard of the evolutionary problem. The reason for this is that there is some amount of confusion about evolutionary altruism among evolutionary biologists themselves. Sociobiologists sometimes confuse vernacular and evolutionary altruism, as when they argue that people cannot really be altruists in the vernacular sense, on the grounds that evolutionary altruism cannot be a reality.[1] It also is common for biologists to think that Trivers's (1971) concept of reciprocal altruism describes a form of evolutionary altruism. My view is that Trivers's concept does not describe a form of evolutionary altruism at all. The idea that 'reciprocal altruism isn't altruism' may sound like a contradiction, but it is an idea I will defend in what follows. And lastly, there is a paradox that is absolutely central to the evolutionary concept, one which has not been widely appreciated.

Another reason for reviewing some of these ideas is that they directly parallel an idea that social scientists have thought about a great deal. Although vernacular and evolutionary altruism are quite separate matters, their similarities are very much in evidence when we consider what stu-

First published in *Canadian Journal of Philosophy*, suppl. vol. 14 (1988): 75–99. Reprinted by permission.

[1] See Kitcher 1985: ch. 11 for discussion of this error.

dents of game theory call the tragedy of the commons (or the prisoners' dilemma). So, besides separating biology from the social sciences in one sense, I want to bring them together in another.

I. VERNACULAR ALTRUISM

The first and most obvious difference between the vernacular and the evolutionary concept of altruism is this: to be a vernacular altruist, you have to have a mind. But biologists can discuss the question of evolutionary altruism for any organism you please, whether it has a mind or not.

The reason I say that a mind is essential for the common-sense concept is that vernacular altruism has to do with motives. Doing someone a good turn is not definitive of this sort of altruism. If I aim at harming you, but by mistake do you some good, that does not make me an altruist. Likewise, if I aim at helping you, but my plans get messed up, I nevertheless may be an altruist. So altruism, whatever else it is, has to do with the motive of benefiting others.

The second simple feature of the common-sense concept that we should note is that the aimed-for benefits do not have to be reproductive benefits. If I know that you love to play the piano, I may give you a volume of Beethoven sonatas out of the goodness of my heart. I am an altruist here, but the good I have done you does not enhance your evolutionary fitness. In fact, it may be true that time at the piano is time away from reproduction; so in love are you with the piano, that you would rather play the piano than make babies. If so, my gift diminishes your prospects for reproductive success. But I may have been a vernacular altruist none the less.

The third component of this familiar concept is a little less obvious. If I give you the volume of sonatas out of the goodness of my heart, I may thereby count as an altruist. Now suppose that unbeknownst to me someone else gives you *two* volumes of sonatas. This donor has given away more than I have. We might want to say that he behaved *more* altruistically than I did. Notice that this is a comparative judgement. My present point, though, is that this comparative claim does not show that I am not an altruist.

Vernacular altruism is an 'absolute' concept, not a comparative one. An altruist is someone who acts from certain sorts of motives. If follows that whether *I* am an altruist does not conceptually depend on what *you* do or on what your motives are. Altruism is an intrinsic property. It's more like the concept of being a millionaire than it is like the concept of being rich.

I have noted three properties of our common-sense concept. It is essentially psychological. It does not essentially involve reproduction. And it is not essentially comparative. This last point, recall, does not mean that we never say that some people are more altruistic than others. Rather, the idea is that in calling people altruists, we are making a comment on their motives, not comparing their motives with those of others.

I have been working so far with the idea that altruists are people who act on the basis of their desire to help others. However, a moment's thought shows that this is not sufficient, even if it is necessary. I may give you some money because I want you to have it. But if my want is itself a consequence of some selfish desire, then I will not be an altruist. For example, we do not describe ordinary buying and selling as displays of altruism. Yet notice that in voluntary exchange, each party wants the other to have the goods or the cash. If we interrupt an exchange of this sort and ask, 'Do you really want the other person to have this thing?', each party would sincerely answer 'Yes'. But altruism is not involved, because each has this want only because it is a means to the selfish end of getting the cash or the goods.

What, then, is the extra ingredient? An altruist, it would seem, must not just have an other-directed desire, but must have this desire in a non-instrumental way. The good of the other must be an end, not just a means to some selfish satisfaction. But here we seem to run up against a banal truism: people want to have their desires satisfied. The altruist wants to help others. The selfish individual wants to keep the cookies for himself. But both, in so far as they engage in rational deliberation, select actions that maximize their chances of getting the most of what they want. Does this mean that vernacular altruism is really an illusion—that the distinction we wish to draw between genuine other-directedness and genuine selfishness dissolves?

This question I will not try to answer here.[2] However, I will note two constraints that an adequate explanation of the difference between vernacular altruism and selfishness must obey. First, the distinction must not run afoul of the truism that people act so as to satisfy the desires they have. That people act on the basis of their own desires is a fact about the *subject* of desires. But this truism about the subject of desires is quite separate from the question of what the *contents* of desires are. Whether I am an altruist concerns *what* I want; the issue is not decided by the obvious fact that it is *I* who does the wanting. The second constraint that an adequate account must respect is that selfish actions can sometimes include motives that involve the welfare of others. This is the point illustrated by the

[2] I have attempted to do so, however, in Sober 1989.

example of buying and selling. We cannot conclude that people are never altruistic because they always act so as to satisfy their own desires; but neither can we conclude that people are sometimes altruistic just because their preferences include benefiting others.

II. DARWINIAN SELECTION

I now want to review some simple facts about Darwinian selection, ones that will allow the issue of evolutionary altruism to emerge clearly. I said in the previous section that vernacular altruism is essentially psychological, not essentially reproductive, and not essentially comparative. Evolutionary altruism is just the opposite (see Table 23.1). The first two contrasts may be sufficiently obvious. Evolutionary altruism can occur in organisms that don't have minds; and evolutionary altruism involves the donation of reproductive benefits. Evolutionary altruism has to do with the reproductive consequences of behaviour, not with the proximate mechanism (psychological or otherwise) that guides that behaviour.[3] This is why the concept of evolutionary altruism can apply to creatures with minds as well as to those without.

The third contrast may be a little less transparent. But before it can be clarified, we must review some fundamental facts about how Darwinian selection works.

Let us imagine that there are two kinds of organisms in a single population. We imagine that the two characteristics are heritable. All this means is that parents tend to resemble their offspring. This may be because parents transmit genes to their offspring; or it may be because parents teach their children to be like them. The mechanism of inheritance does not matter; only the fact of heritability is essential.[4]

We imagine, further, that the two traits have different consequences for survival and reproduction. That is to say, we imagine that one of the traits is fitter than the other. We now have the pre-conditions for a process of Darwinian selection—heritable variation in the fitnesses of organisms.

So as to make this somewhat abstract formulation more concrete, let us imagine that we are talking about a herd of deer. The two traits are fast and slow. Sexual reproduction complicates our simple picture of what

---

[3] See Sober 1985 for discussion of the difference between what Ernst Mayr has called 'proximal' and 'ultimate' explanations of biological traits.

[4] Here I use 'heritability' in a sense that is broader than that customary in population genetics. The genetical concept is intended to isolate the correlation of parents and progeny attributable to genetic transmission. See Falconer 1981 for discussion.

TABLE 23.1

	Vernacular altruism	Evolutionary altruism
Essentially psychological	YES	NO
Essentially reproductive	NO	YES
Essentially comparative	NO	YES

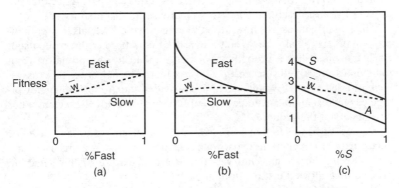

FIG. 23.1. A comparison of the fitnesses of a herd of deer with respect to how fast they can run.

heritability means here—if offspring are to resemble their parents, what should happen when one parent is fast and the other is slow? To avoid this complication, let us imagine that the organisms reproduce by asexual cloning. Running speed is unerringly transmitted by the simple rule of like reproducing like.

How are we to compare the fitnesses of the two traits? I am interested in how the two traits allow organisms to avoid being caught and eaten by predators. There are several different fitness relationships we can consider.

First, let's imagine that your chance of being caught is simply determined by whether you are fast or slow. That is, we are imagining that your vulnerability to predators is not affected by whether you live in a fast or a slow herd, or whether the speed you happen to have is common or rare. In this case the fitness relationship of the two traits is frequency-independent, as shown in Figure 23.1a.

What will happen in a population of slow individuals, if a fast mutant (or migrant) is introduced? The newcomer will be fitter than the other individuals, and so will be more reproductively successful. In consequence, the fast trait will increase in frequency. In the next generation, it will still be true that fast individuals are on average fitter than slow ones, so the trait will increase in frequency once again. This will continue until fast goes to 100 per cent representation in the population.

At the beginning of the process, all the deer were slow; at the end, all are fast. Given our assumption about how the predators behave, the individuals in the population are better off at the end than the individuals were at the beginning. The average fitness of the organisms in the population (called '$\bar{w}$') is represented in Figure 23.1a by a dotted line. Notice that the process I've just described leads to an increase in this quantity.

This quantity measures how fit, on average, the individuals in a population are. But it can also be taken to measure the welfare of the group itself. Each individual has a probability of being killed by the predator; if all individuals are killed, the group goes extinct. The selection process we have just described, it would seem, has provided the group with an advantage. By increasing the average level of fitness of individuals, selection has also benefited the group.

We now need to see that increases in $\bar{w}$ and group advantages are not necessary consequences in Darwinian selection. We can see this by asking the following question: What was the essential feature of this selection process that allowed fast to supplant slow?

The answer (assuming heritability as we have done all along) is simply that *fast is fitter than slow*. This comparative fact suffices. But a few changes in the graph shown in Figure 23.1a will allow us to see that fast can replace slow without $\bar{w}$ ending up higher at the end of the process than it was at the beginning.

Let us suppose, to modify our example, that fast individuals are always better off than slow ones, but that the advantage importantly depends on the rarity of fast individuals. Predators prefer to chase down slow*er* individuals. It isn't that they are too slow to catch the fast ones; it's that they are too lazy to bother, when slower prey present themselves. When fastness is rare, fast individuals do enormously better than slow ones. But when fastness is very common, the advantage is slight. And when fastness has gone to 100 per cent, predators catch them as readily as they caught the slow ones when the slow ones were the only things around to eat; the predators just have to run a little faster to do this, but this is something easily within their grasp. This fitness relationship is shown in Figure 23.1b. Notice that the fitnesses are frequency-dependent, and that $\bar{w}$ is no higher at the end of the process than it was at the beginning.

In both Figure 23.1a and 23.1b, fast is fitter than slow. This comparative fact is enough to ensure in both cases that fast replaces slow. The figures differ, however, in the question of what happens to average fitness. In

Figure 23.1a, it goes up; in Figure 23.1b, it rises momentarily, only to fall back to where it began.

A third example will illustrate this point in an even more extreme way. Let us consider two traits S and A, whose fitnesses arc depicted in Figure 23.1c.[5] What will happen when an S individual is dropped into a population of A individuals? Since S is fitter than A, S will increase in frequency. In the next generation, the same fitness relationship obtains, so S continues to increase. The process will take S all the way to 100 per cent. But notice that $\bar{w}$ steadily declines. The organisms at the end of the process are less fit than the organisms in the beginning. It is important to grasp the bleakness of the process depicted in Figure 23.1c. Natural selection can lead a population right to extinction. The fitter replace the less fit, and the whole process plummets downhill. If Figure 23.1a portrays an optimistic vision of selection the improver, Figure 23.1c provides a pessimistic picture of selection the destroyer.

The three figures have in common the thing that is fundamental to Darwinian selection—*comparative* fitness determines the population's trajectory. This leaves totally unspecified what happens to *absolute* fitness along the way; it is with respect to this quantity ($\bar{w}$) that the three graphs differ.

Figure 23.1c depicts the essentials of the concepts of evolutionary selfishness (S) and altruism (A). We can interpret this graph as showing that there are two causal factors that affect an individual's fitness. First, it is better to be selfish than to be altruistic. Second, it is better to live among altruists than among selfish individuals. Altruists thus provide a group advantage—they benefit those with whom they live, even though altruists would be better off being selfish.

So it is nice to have altruists around. But the fact of the matter is that Darwinian selection predicts that there should be no such thing. Selfish spitefulness will triumph: a trait that makes things worse for everyone will spread to fixation, as long as it makes things worse for non-bearers of the trait than it does for bearers of the trait. Imagine, for example, a trait in a plant population that causes its bearer to leach a toxic chemical into the soil. As long as the poison hurts non-bearers of the trait more than it hurts bearers of it, the trait will spread. The mirror image is that a trait that boosts everyone's reproductive prospects cannot evolve, if it benefits non-bearers more than it benefits bearers. Imagine a trait that causes the plants that have it to leach an insecticide into the soil. If non-bearers of the trait are benefited more—either because the chemical makes them more

---

[5] Ignore the numbers labelling the *y*-axis in Fig. 23.1c for now.

immune, or because non-leachers do not incur the energetic cost of providing the chemical—the trait cannot evolve by Darwinian selection.

The definition of altruism I have given is essentially comparative. An altruistic trait is one that is related to the alternative trait (which we call 'selfish') by the fitness function shown in Figure 23.1c. Within a group, selfish individuals do better than altruists, but everybody benefits in a group by having lots of altruists around.

In this respect, evolutionary altruism differs from the vernacular variety. Consider a trait that leads individuals who have it to give away one unit of benefit to each of the individuals with whom they live. Is this trait an instance of evolutionary altruism? No answer can be given until the alternative traits are specified. If the other individuals in the population give away no benefits at all, then the single-unit donors are altruists. If, on the other hand, the other individuals give away two units of benefit, then the single-unit donor is selfish.

An immediate consequence of this example is that we should not equate altruism with donation. In a population of single-unit donors and double-unit donors, both traits involve donation, but only one of them is altruistic. In a sense, every altruist is a donor, but not every donor is an altruist.

This helps show why Trivers's (1971) idea of reciprocal altruism really does not involve evolutionary altruism at all. Let's imagine a population of beavers who co-operate to build a dam. The dam is very important to the beaver way of life, but what is to prevent cheating beavers from enjoying the benefits of the dam without helping to build it? As stated so far, the answer is *nothing*. If the population consists of two types of individuals— one helps build and the other does not—and both can enjoy the benefits of the dam once it exists, we have an example of altruism and selfishness. Darwinian selection should eliminate the builders, perhaps to the detriment of builders and non-builders alike.

But suppose the traits present in the population are different. Let us imagine that the builders are able to prevent the non-builders from enjoying the benefits of the dam. Builders assassinate cheaters, we might imagine. The game is now different, because the players are different. In this case, the builders will be fitter than the non-builders, so Darwinian selection will maintain the building behaviour.

In this example, the builders co-operate. Non-builders, we are imagining, do not. But the builders are not evolutionary altruists; and the non-builders are not evolutionarily selfish.

The vengeful builders are reciprocal altruists, in Trivers's sense. They do things that benefit others, but punish individuals who do not reciprocate. The point to focus on is that within the single beaver population, vengeful builders are fitter on average than the individuals who do not build. Vengeful building is just a variety of Darwinian selfishness. Given the choice between being a vengeful builder and an atomistic non-builder, an individual would quite selfishly prefer to be a builder. This is why reciprocal altruism is not altruism.

I want to emphasize that I have no interest in quibbling over words here. My reason for saying that reciprocal altruism is not altruism is motivated by a desire to clearly distinguish different kinds of causal processes. Individual selection can produce reciprocal altruists, but it cannot produce altruism in the sense defined in Figure 23.1c. We should recognize this fact about individual selection, not obscure it by lumping together two quite different kinds of characters. In saying this, I think I am following Trivers's (1971) own observation that his model is intended 'to take the altruism out of altruism'.

Notice that applying the contrast between altruism and selfishness to a natural population can be quite difficult. When you go out in the woods and see all the beavers in a group co-operating to build a dam, you have no idea whether the trait in question should be called altruistic. You first have to ask yourself what the other traits were against which the one you observe was competing. This may take some imagination, because you have to envisage what variation was found in the ancestral population for natural selection to act upon. Unfortunately, selection frequently destroys the kind of evidence that is needed to reconstruct its history; selection requires variation to proceed, but typically it destroys the pre-conditions for its own existence.

III. THE TRAGEDY OF THE COMMONS

The Darwinian treatment of evolutionary altruism subverts the idea that natural selection must improve fitness. It is interesting to note that precisely the same phenomenon can arise in a very different domain. Rather than think of the natural selection of organisms, let us consider rational agents who deliberate about actions with a clear view of the consequences of what they do. When agents are fully informed and rationally deliberate, shouldn't they end up better off than they would be if they were irrational? The tragedy of the commons (also known as the prisoners' dilemma) in game theory provides a negative answer to this question, for reasons isomorphic with the Darwinian analysis of evolutionary altruism.

Let's imagine that you are deciding whether to put an emission control device on your car. We suppose that this is not a matter of law, but of

TABLE 23.2 *A consumer's dilemma*

	States of the world	
Acts	Everybody else buys one	Nobody else buys one
You buy	3	1
You don't buy	4	2

individual choice. The cost to you is modest—$20. But what are the bene-
fits? That depends on what other people do. If no one buys the device, it
won't be worthwhile for you to buy one. Though the atmosphere would
improve infinitesimally, the gain is so trivial that you'd rather save the $20.
On the other hand, if everybody else buys the device, the atmosphere will
be very good. But here again, the improvement in the atmosphere that
would be added if you also bought the device would be trivial. Again,
you'd rather save the $20.

Your preferences, with 4 indicating best and 1 indicating worst, are
shown in Table 23.2. The rational act in this game is to not buy the device.
That action 'dominates' the alternative; whatever everybody else does,
you're better off not buying (4 > 3 and 2 > 1).

But here is the rub: everybody else has the same preferences, so each
other agent rationally decides not to buy the device. What is the result?
The group ends up with no one getting the device, which means that
everybody receives two units of value. Notice that everybody is now worse
off than they would have been if they had all decided to buy; in that case,
the pay-off for each would have been three units.

This problem has the following paradoxical property. The rational action for each individual to choose is known in advance to make all the
players worse off than they would have been if they had all chosen the
irrational action.

In the consumers dilemma table, I represented only two extreme states
of the world—everybody else buys a device and nobody else buys a device.
But there are intermediate frequency ranges—90 per cent buys, 80 per
cent buys, and so on. The full game is not specified by a two-by-two table,
but by a two-by-infinite table, so to speak. However, there is a simpler
representation: merely use the fitness function for selfishness and altruism.
Buying a device is altruistic; not buying is selfish. The pay-offs from the
table are inscribed as entries on the graph shown in Figure 23.1c. The

problem of evolutionary altruism is an instance of the general game-theoretic problem. Instead of rational deliberation, we have natural selection. And instead of preferences concerning dollar outlay and pollution, we have benefits computed in the currency of survival and reproductive success.

The fact that the problem posed by this decision problem has a rather depressing solution is not necessarily cause for despair. It is not carved in stone that human beings must play the game I have just described. For example, it is an assumption of this game that actions and states of the world are independent. Your buying an emission control device is independent of whether anybody else does. But suppose we pass a law that says that everybody has to do the same thing. Then we have a new game, with the only possible outcomes being the ones on the main diagonal of the previous table. The result is that we all choose to buy the device, which is a much cheerier prospect than the one obtained initially.[6]

There is an important truth behind the misleading idea that there are various ways of 'solving' the prisoners' dilemma problem. In the game as initially described, there is exactly one rational solution, which leads to a deleterious universal selfishness. The rational kernel, though, is that it is within the power of rational agents to restructure the games they play. The important thing to remember is that the solution to a game is contingent on the assumptions that went into defining the problem. If the assumptions can be changed, so too may the solution. What is inevitable within the framework of one game may not be within the framework of another.

Although human beings can consciously restructure the games they play, organisms in general do not have this ability when it comes to the problems posed by natural selection. Still, there is nothing absolute about the negative verdicts we have reached so far about evolutionary altruism. I have said that evolutionary altruism cannot evolve, if the game being played is Darwinian selection. But there has been a tradition of thinking in biology—one which has waxed and waned in the course of the development of evolutionary theory from Darwin to the present—that says that altruism is a reality, which means that Darwinian selection is not the game that organisms always play. We now need to examine this non-Darwinian idea. For our grasp of the concept of evolutionary altruism will be incomplete unless we see clearly how it is connected to the idea of group selection.

[6] Another reformulation of the problem is provided by the iterated prisoners' dilemma, which is explored in Axelrod 1984.

IV. SIMPSON'S PARADOX

It is a basic rule about natural selection, both in the simple format we have considered so far and in the context of the more complicated models we will consider now, that a trait must have a higher fitness if it is to increase in frequency. This is as true for altruism as it is for speed in the deer example. But we have already seen that within any group, altruism is less fit than selfishness. This is a matter of definition. How, then, can altruism evolve by natural selection?

To see that this is possible, one must grasp a paradox. Let us now consider not one group, but an ensemble of many groups. Within each group, altruists do worse on average than selfish individuals. But this fact does not guarantee that altruism is less fit when you average over the ensemble of groups. What is true within each group need not be true overall.

This is a concept that is very hard to grasp; we are so used to thinking that what happens in the part must translate directly into what happens in the whole. How can an organism get bigger if each of its parts gets smaller? That, I grant, does sound impossible. Suppose I told you that in every state of the USA, Democrats were declining in frequency and Republicans increasing. Would it follow that Democrats are becoming rarer in the US taken as a whole? The knee-jerk reaction here is to say that what happens in the part must happen in the whole. We now must see that this need not be so.

Let's start with some very simple examples of how this decoupling of part and whole can occur. Imagine an audience in which men are on average taller than women. Is it possible to divide this audience into two groups, so that within each group, women are taller than men? Here's an example of how this can happen (Table 23.3). There are 100 females and 100 males in total. The female average is 5.5 units of height; the male average is 8.5. We then split the total population of 200 individuals into two groups. The first contains ten females and ninety males; the ten women are 10 units tall and the men are 9. The second group contains ninety females and ten males, with average heights of 5 and 4, respectively. The heights of

TABLE 23.3

Group 1	Group 2	Global average
10(F): 10	90(F): 5	100(F): 5.5
90(M): 9	10(M): 4	100(M): 8.5

TABLE 23.4

	Department 1	Department 2	
Applicants	10(M)	90(M)	= 100(M) total
	90(F)	10(F)	= 100(F) total
Rejection rate	90%	10%	
Number rejected	9(M)	9(M)	= 18(M) total
	81(F)	1(F)	= 82(F) total

the women and men within each group are given. Notice that males are taller on average, though women are taller within each group.

Another example of this phenomenon I owe to Nancy Cartwright (1979). She reports that the University of California at Berkeley was once investigated for discriminating against women in admission to graduate school. The reason for the suspicion was that women were turned down far more frequently than men. However, when departments were investigated one at a time, it emerged that the rejection rates of women and the rejection rates of men within each department were the same. Women were not turned down more often than men in biology, in philosophy, in physics, or in any other department. But in the whole university of which these departments are parts, they were.

Let's construct a hypothetical example to see how this is possible (Table 23.4). We imagine that 100 men and 100 women apply to the two departments. Notice that in each department, a woman has the same chance of admission as a man. Yet women are turned down more often overall, because they disproportionately apply to a department with a very high rejection rate.

The phenomenon I have been discussing is sometimes called 'Simpson's paradox', in tribute to a statistician who wrote about it in the 1950s (Simpson 1951). However, the phenomenon has been noticed by statisticians for a long time.[7]

Let us review the two examples. In the first one concerning height, women were taller on average than men within each group, but men are taller than women overall. In the second, each academic department rejects women no more often than it rejects men, yet women are rejected more often overall. In both cases, we make two comparisons. First, we compare male and female averages within each group. Then, we compare

---

[7] Skyrms (1980: 107) cites Edgeworth, Pearson, Bravais, and Yule as having noted the phenomenon.

the overall male average with the overall female average. The inequality within groups need not be maintained when we average across groups.

I hope these examples give you a feel for the pattern involved in Simpson's paradox. Now let's ask a separate question: What causes Simpson's paradox to arise? What allows inequalities within groups to reverse when we take the overall averages in these examples? The answer is *correlation*. In the first case, tall women tend to be found in the taller group. In the second, women tend to apply to departments with high rejection rates. If the male average and the female average were the same across groups, Simpson's paradox would disappear.

We now can show why Simpson's paradox is at the heart of the idea that group selection can allow altruism to evolve. Let's imagine that we have not one group, but an ensemble of them. In each there is some mixture or other of selfish individuals and altruists. We need to consider two questions: first, are selfish individuals fitter than altruists within each group? Second, are selfish individuals fitter than altruists, when we average over the ensemble of groups?

The answer to the first question is *yes*, given the fitness functions shown in Figure 23.1c. No matter what the frequency of altruism is in a group, altruists do less well than selfish individuals in the same group. But how are we to answer the second question? How are we to calculate and compare the overall fitnesses of altruists and selfish individuals?

Just to illustrate how Simpson's paradox applies here, let's imagine that our ensemble consists of two groups made up of 100 individuals each. The first is 1 per cent selfish; the second is 99 per cent selfish. In Table 23.5, I've written the within-group fitnesses and the overall fitnesses (rounded off, for simplicity) given by Figure 23.1c. Altruism is less fit within each group, but more fit when one averages over the ensemble of groups.

I mentioned before that models of natural selection of the sort we are considering imply that fitter traits increase in frequency. The present example is no exception. I stipulated that the two-population ensemble begins with 50 per cent altruists and 50 per cent selfish individuals. What will happen to the frequencies of the traits in the next generation? Within each group, altruism will decline in frequency because it is less fit. But across the

TABLE 23.5

Group 1	Group 2	Global average
1(S): 4	99(S): 2	100(S): 2
99(A): 3	1(A): 1	100(A): 3

ensemble of groups, altruism will increase in frequency, because it is on average fitter. So if we census the two-population ensemble after one generation has passed, altruism will have increased in frequency.

What will happen if we follow the system over many generations? If the two groups remain intact—that is, if there is no extinction or splitting of groups to found colonies—then the two groups will grow larger and larger (assuming that the fitness values shown in Figure 23.1c represent reproduction above replacement level). Within each group, altruism will decline. So sooner or later, altruism must disappear from the two-population ensemble. The increase in frequency in the first generation was momentary; starting with 50 per cent altruists and 50 per cent selfish individuals who are distributed into groups in the way described, altruism will initially increase. But sooner or later, the pattern that Dawkins (1976) once called 'subversion from within' must take its toll.

So we still have not seen how altruism can evolve and be maintained. But we are on the right track. One condition is before us: altruism must be fitter overall than selfishness, if it is to increase in frequency. How can this be achieved? As in the other examples of Simpson's paradox, the key idea is correlation. What is essential is that like live with like. Altruists must associate with each other more frequently than would be expected if association were at random. This could be achieved by having relatives live together; or it could happen if similar individuals preferred each other's company, regardless of whether they are relatives.

But like living with like is not enough. Even if the two groups just described are subgroups, subversion from within will drive altruism to extinction as the kin reproduce. What is essential is that the groups fragment and found colonies.[8]

To see how this might happen, let's imagine that a group goes extinct if there are more than 50 per cent selfish individuals in it. Imagine further that when a population reaches a certain census size, it fragments into many small subgroups, which then start growing. Notice that colonies are always founded by individuals from the same parent population. And founded colonies may not have exactly the same frequency of altruism as their parents. Imagine that a parent population reaches the fission size of 1,000 and then splits into fifty colonies of twenty individuals each. The parent population, we may suppose, is 75 per cent altruistic. What will the

---

[8] Although many biologists believe that Hamilton's (1964) concept of inclusive fitness allows self-sacrifice among relatives to be treated as a form of individual selection, I believe that this is a mistake. A single kin group that holds together for many generations will experience subversion from within just as much as a group of unrelated individuals. Hamilton (1975) recognizes this very point: an inclusive fitness treatment is *not* an argument against group selection. See Wilson and Sober (1989) for further discussion.

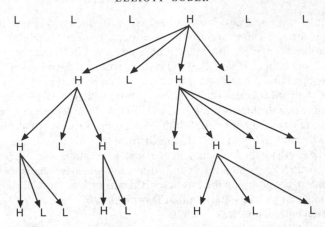

FIG. 23.2. The sort of fragmentation of population that can give rise to group selection.

fifty offspring colonies be like? Probably, some will be less than 75 per cent altruistic, whereas others will contain more than 75 per cent altruists.

One more ingredient is needed, if altruism is to evolve and be maintained by group selection. It is the factor of timing. Suppose that selfishness is sufficiently fitter than altruism that if a group holds together for fifty generations, selfishness will go to 100 per cent in it, no matter what the initial frequency was in the group. In this case, altruism will disappear if a parent population fragments and founds colonies less often than once every fifty generations. The fragmentation will come too late; by then, altruism will have disappeared. So groups must found colonies sufficiently often, how often being determined by how fast selfishness is displacing altruism within each group.

Figure 23.2 shows an example of group selection of the sort just described, whereby altruism can evolve and be maintained. Groups found colonies at a good clip, and groups with low frequencies of altruists go extinct. If the numbers are right, one will find that in every generation of this process, altruism is represented.

V. EVOLUTION AND THE GENEALOGY OF MORALS

What connection is there between vernacular altruism and evolutionary altruism? I noted early on that it is possible for an individual to be an evolutionary altruist without being a vernacular one. Traits that are group-beneficial but individually deleterious, like the example of the plant that leaches an insecticide, need not be psychological. I also pointed out that an individual can be a vernacular altruist without being an evolutionary one; this is what I am when I give you the piano sonatas out of the goodness of my heart, thereby distracting you from the business of reproduction.

Besides these simple distinctions, however, there is the question of what connection human morality has with natural selection. If systematically altruistic behaviour (not just the occasional transfer of a volume of music) is a reality, what does this imply about our evolutionary past?

A strict Darwinian, in the current sense of that term, will deny the existence of evolutionary altruism. The reason is that the trait implies the existence of a selection process that the Darwinian rejects. But even the strictest of Darwinian may sometimes lapse from the Darwinian straight and narrow. This is what Darwin himself did when he considered the evolutionary consequences of vernacular altruism. In *The Descent of Man*, Darwin formulated the issue in terms of his characteristic calculus of individual advantage:

It is extremely doubtful whether the offspring of the more sympathetic and benevolent parents, or of those which were the most faithful to their comrades, would be reared in greater number than the children of selfish and treacherous parents of the same tribe. He who was ready to sacrifice his life, as many a savage has been, rather than betray his comrades, would often leave no offspring to inherit his noble nature. The bravest men, who were always willing to come to the front in war, and who freely risked their lives for others would on average perish in larger numbers than other men. (Darwin 1871: 163)

But rather than concluding that vernacular altruism does not exist, Darwin argued that what is bad for the individual may be good for the group:

It must not be forgotten that although a high standard of morality gives but a slight or no advantage to each individual man and his children over the other men of the same tribe, yet that an advancement of well-endowed men will certainly give an immense advantage to one tribe over another. (ibid. 166)

Darwin's assumption here seems to be that vernacular altruism was under the direct control of natural selection. The trait is present now because, historically, there was selection for it. Darwin went the route of group selection because he did not doubt the trait's reality; some of his latter-day followers, on the other hand, have accepted the assumption, but have concluded that vernacular altruism cannot exist, on the ground that individual selection is the name of the game.[9]

[9] For example, Dawkins (1976: 3) asserts that human beings are 'born selfish' and Barash (1979: 135, 167) says that 'real, honest-to-God altruism simply doesn't occur in nature', and that 'evolutionary biology is quite clear that "What's in it for me?" is an ancient refrain for all life, and there is no reason to exclude *Homo sapiens*'.

However, there is another possibility that needs to be considered, which rejects the idea that a trait, if it exists now, must have been under direct selective control. It is the idea of evolutionary spin-off. Human beings now have the ability to do trigonometry; yet no one supposes that there must have been selection for that ability in our ancestral past. Rather, it is far more plausible to think that there was selection for some other suite of mental characteristics. Perhaps there was selection for increased intelligence and language use. Once these traits evolved, and human beings subsequently found themselves in environments rather unlike the ancestral ones, various spin-off properties became visible.[10]

This is the scenario that Peter Singer (1981) explores in his book *The Expanding Circle*. Perhaps the ability to reason abstractly evolved because of its individual advantageousness. But once in place, this intelligence led human beings to see that rational considerations oblige them to take the interests of others as seriously as they take their own. If something like this is right, then vernacular altruism may find its pedigree not in evolutionary altruism, but in the sophisticated thoughts and feelings that a mind produced by individual selection was first able to formulate.

I will not evaluate the plausibility of this spin-off explanation of vernacular altruism. My point here is a conceptual one. Even if we suppose that group selection never happened—that selection is always selection for traits that are individually advantageous—it does not follow that vernacular altruism could not have evolved. It is one thing to hold that all selection is individual selection, quite another to maintain that all characters are under direct selective control.

VI. CONCLUDING REMARKS

Evolutionary altruism is a kind of trait. In our plant example, it involves leaching an insecticide into the soil; in a species of crow, it might involve issuing warning cries. Traits are altruistic, so altruism is a trait of a trait. The evolutionary problem is to see whether the physiological, behavioural, and morphological traits found in nature are examples of evolutionary altruism.

The trait in question must not be confused with the trait of vernacular altruism. Although it is possible to propose a causal connection between evolutionary altruism and vernacular altruism, as Darwin did in one direc-

---

[10] The difference between direct selective control and spin-off is explained by Sober (1984) in terms of the distinction between 'selection of' and 'selection for'. Gould and Lewontin (1979) use the term 'spandrel' to mark the concept of evolutionary spin-off.

tion, and some contemporary sociobiologists have done in the other, this is not inevitable. Evolutionary altruism does not imply vernacular altruism; nor does vernacular altruism imply evolutionary altruism. Group selection can lead to the evolution and maintenance of evolutionary altruism. Darwinian selection cannot. Although altruists are by definition less fit than selfish individuals within the same group, this does not settle the question of their comparative fitnesses when we average over the ensemble of groups. To see why this is so, one must grasp the meaning of Simpson's paradox. Once this is achieved, one can understand how altruism can evolve, given the right assumptions about like living with like and appropriate rates of extinction and colonization.

All this is not to say one word about whether evolutionary altruism is found in nature. My concern here has been to say what altruism is, not whether it exists. I have mentioned that evolutionary opinion has swung back and forth on this question. At the moment, Darwinism is the dominant mode of thought; although group selection and altruism are not treated with total scorn by all biologists, it is a small minority of biologists that takes the idea seriously.

The reasons for this opinion bear examining. Sometimes opposition to group selection is based on spurious arguments. It is sometimes suggested that altruism cannot evolve simply because, by definition, altruists are less fit than selfish individuals within each group. An understanding of Simpson's paradox should make us immune to the attractions of this *non sequitur*. However, even if Darwinism is right in rejecting group selection, it is important that it do so for the right reasons. There are substantive questions here about natural selection that need to be resolved; removing confused arguments may help biologists see these questions for what they are.

REFERENCES

Axelrod, R. (1984), *The Evolution of Cooperation* (New York: Basic Books).
Barash, D. (1979), *The Whisperings Within* (Harmondsworth: Penguin).
Cartwright, N. (1979), 'Causal Laws and Effective Strategies', *Noûs*, 13: 419–37.
Darwin, C. (1871), *The Descent of Man, and Selection in Relation to Sex* (London: J. Murray; repr. Princeton: Princeton University Press, 1981).
Dawkins, R. (1976), *The Selfish Gene* (Oxford: Oxford University Press).
Falconer, D. (1981), *Introduction to Quantitative Genetics* (London: Longman).
Gould, S., and Lewontin, R. (1979), 'The Spandrels of San Marco and the Panglossian Paradigm: A Critique of the Adaptationist Programme', *Proceedings of the*

Royal Society of London, B205: 581–98; repr. in E. Sober (ed.), *Conceptual Issues in Evolutionary Biology* (Cambridge, Mass.: MIT Press, 1984), 252–70.

Hamilton, W. (1964), 'The Genetic Evolution of Social Behavior', *Journal of Theoretical Biology*, 7: 1–52.

——(1975), 'Innate Social Aptitudes of Man: An Approach from Evolutionary Genetics', in R. Fox (ed.), *Biosocial Anthropology* (London: Malaby Press), 37–67.

Kitcher, P. (1985), *Vaulting Ambition: Sociobiology and the Quest for Human Nature* (Cambridge, Mass.: MIT Press).

Simpson, E. H. (1951), 'The Interpretation of Interaction in Contingency Tables', *Journal of the Royal Statistical Society*, B13: 238–41.

Singer, P. (1981), *The Expanding Circle* (New York: Farrar, Straus, and Giroux).

Skyrms, B. (1980), *Causal Necessity* (New Haven: Yale University Press).

Sober, E. (1984), *The Nature of Selection* (Cambridge, Mass.: Bradford/MIT Press).

——(1985), 'Methodological Behaviorism, Evolution, and Game Theory', in J. Fetzer (ed.), *Sociobiology and Epistemology* (Dordrecht: D. Reidel), 181–200.

——(1989), 'What Is Psychological Egoism?', *Behaviorism*, 17: 89–102.

Trivers, R. (1971), 'The Evolution of Reciprocal Altruism', *Quarterly Review of Biology*, 46: 35–57.

Wilson, D., and Sober, E. (1989), 'Reviving the Superorganism', *Journal of Theoretical Biology*, 136: 337–56.

24

ON THE RELATIONSHIP BETWEEN EVOLUTIONARY AND PSYCHOLOGICAL DEFINITIONS OF ALTRUISM AND SELFISHNESS

DAVID SLOAN WILSON

The word 'altruism' and associated words such as 'selfishness', 'spite', and 'co-operation' are familiar to everyone as descriptions of human conduct. The same words are used routinely by evolutionists to describe the behaviour of non-human species. At first glance this practice seems unobjectionable. Charges of anthropomorphism aside, if we see a baboon endangering its life to protect another baboon from a leopard, or if we see a baboon doing everything it can to put another baboon between itself and a leopard, it seems reasonable to use the same language that we would use if the baboons had been humans.

At second glance, however, the relationship between the evolutionary and everyday meanings becomes more complex. In particular, evolutionary definitions are supposed to be based solely on fitness effects. If a behaviour increases the fitness of a recipient at the expense of the actor's fitness, it is termed 'altruistic', regardless of what the actor felt or thought as it performed the behaviour. By contrast, everyday meanings depend largely on the motives of the actor. If we see someone benefit others at the economic, social, or material expense of himself, we may still regard him as selfish if we think that he derives pleasure from his action or if he regards his action as part of a broader scheme in which he, the actor, stands to gain. Definitions that are based on the actor's motives will hereafter be called 'psychological' definitions.[1]

Since motives figure so largely in everyday meanings of altruism and

First published in *Biology and Philosophy*, 7 (1992): 61–8. © 1992 Kluwer Academic Publishers. Reprinted with kind permission from Kluwer Academic Publishers.

[1] Bertram (1982), Kitcher (1985), and Sober (1988) make a similar distinction between evolutionary and psychological definitions, but their development of the theme is somewhat different from mine, as discussed below.

associated words, it is remarkable that the words can remain intuitive when stripped of their motivational content. It is equally remarkable that, in borrowing the words from common language, evolutionists have changed the defining criterion from motives to effects largely without comment. In this essay I make two points, one elementary and the other more subtle. First, behaviours that are altruistic in the evolutionary sense can be psychologically either selfish or altruistic. No simple relationship exists between the forces of natural selection that cause behaviours to evolve and the proximate mechanisms that elicit the behaviours. Second, evolutionary definitions remain intuitive by invoking a cryptic form of motivation, based not on what the actor thinks or feels but on what the evolutionist must account for while calculating gene frequency change. Reliance on motives as a metaphor may explain why several conflicting definitions of altruism exist among evolutionists, each with their own intuitive appeal.

I. FROM FITNESS EFFECT TO MOTIVES OF THE ACTOR

As mentioned above and described in detail below, evolutionists frequently disagree on whether a particular behaviour is altruistic or on whether altruism in general is rare or common in nature. Nevertheless, all evolutionists agree that behaviours defined as altruistic based on fitness effects can, in principle, evolve. Thus, the fact that a behaviour is *evolutionarily successful* does not imply that it is necessarily *selfish*. Let us therefore consider a behaviour, defined as altruistic based on fitness effects, that is also evolutionarily successful. Any behaviour that evolves must include a proximate mechanism that actually causes the organism to display the behaviour. Psychological definitions of altruism and selfishness are based on these proximate mechanisms. Our question is: Are behaviours that are altruistic in the evolutionary sense necessarily caused by proximate mechanisms that are altruistic in the psychological sense?

The answer is 'No'. The simplest case to consider is pleasure, which can be broadly regarded as a physiological/psychological mechanism, designed in part by natural selection, that causes organisms to behave in ways that are evolutionarily successful. But 'evolutionarily successful' is not the same as 'selfish'! If increasing the fitness of others at the expense of one's own fitness is evolutionarily successful in the long run, it can be proximately motivated by the same feelings of pleasure that accompany successful selfishness. From the evolutionary standpoint, defining selfishness as 'the pursuit of personal pleasure' would be as meaningless as defining selfishness as 'everything that evolves' (see also Sober 1993).

A less obvious, and therefore more interesting, case involves cognition, which can be broadly regarded as a system of mental operations, designed in part by natural selection, that causes organisms to behave in ways that are evolutionarily successful. Consider an imaginary experiment in which a human subject is asked to choose between two behaviours, A and B. If he elects A, he must donate one dollar, and five dollars will be given to a member chosen at random from his group. If he elects B, no money will be taken from him. In both cases he stands to receive five dollars from other A-types in his group. We assume that the subject is psychologically selfish, and quickly decides to be a B-type. Now we inform the subject that two groups exist, one with 20 per cent A-types and one with 80 per cent A-types. If he elects to be an A-type, he will be placed in groups 1 and 2 with probability 0.2 and 0.8 respectively, and vice versa if he elects to be a B-type. Our subject now embarks upon the following mental calculation: 'In group 1, I can expect to receive $0.8(5) = 4$ dollars from other A-types, while in group 2 I can expect to receive only $0.2(5) = 1$ dollar. It is true that as an A-type I will lose a dollar, but I also have an 80 per cent chance of being in group 1, yielding an expected gain of $0.8(4) + 0.2(1) - 1 = 2.4$ dollars. As a B-type I will keep my dollar, but I also will probably end up in group 2, yielding an expected gain of $0.2(4) + 0.8(1) = 1.6$ dollars.' Our subject now elects to be an A-type, despite the fact that he is psychologically selfish and did not include the welfare of others in his mental calculation.

Readers familiar with the evolutionary literature will recognize that, if we substitute the word 'offspring' for 'dollars', the two-group version of the experiment is equivalent to a structured population model for the evolution of altruism in which A-types increase the fitness of others at the expense of themselves, but nevertheless evolve because of a clustering process that causes altruists to interact primarily with other altruists (Wilson 1983, 1989).[2] Nevertheless, it is possible for a utility-maximizing organism to elect to be an A-type without directly considering the welfare of others by combining both the effect of its own behaviour and the population structure in a single measure of fitness averaged across all

---

[2] Another major difference between evolutionary and psychological definitions is that the currency of altruism in evolutionary models is always fitness, whereas in common language it can be any desirable commodity (Sober 1988). As Sober (1988) points out, however, evolutionary and economic models based on prisoner's dilemma and tragedy of the commons scenarios are otherwise parallel.

contexts. In effect, the self becomes a *representative* A-type or a *representative* B-type, existing in the capacity of both actor and other, which makes the inclusion of others as a separate category redundant. In this fashion, a thinking organism whose mental operations are psychologically selfish can, in principle, adopt any evolutionarily successful behaviour.[3]

The fact that evolutionarily successful behaviours are not necessarily selfish, and that proximate mechanisms are designed to elicit evolutionarily successful behaviours regardless of whether they are selfish or altruistic, destroys any hope for a simple relationship between definitions based on fitness effects and definitions based on motives. Not only can altruistic behaviours (in the evolutionary sense) be selfishly motivated (in the psychological sense), but the reverse is also true; individuals that care truly for others can be selfish in the evolutionary sense (Sober 1993). These observations are elementary, but they are not sufficiently appreciated by evolutionists or philosophers interested in the concept of altruism.

II. MOTIVES AS A METAPHOR IN EVOLUTIONARY DEFINITIONS

Given the problems outlined above, it is a wonder that evolutionary discussions of altruism and associated words appear as natural as they do. Consider the following passage from Nunney (1985: 226):

Suppose that you are offered two financial options. Under the selfish option you receive 10 dollars and keep it all. Under the benevolent option you receive 10 million dollars but 6 million must be given to a neighbor. Given that neighborhoods are random samples of a large population, the choice is clearly the benevolent option, a choice based purely on individual greed and not on the general benefit to the neighborhood. Replacing money by fitness, it can be seen that benevolence spreads by individual selection because a net gain of 4 million units of fitness is superior to a net gain of 10 units of fitness. The 6-million gain of the neighbor is irrelevant.

In the first half of this passage, a person chooses to receive 4 million dollars rather than ten dollars, despite the fact that a neighbour, randomly chosen from a large population, will receive even more than himself. This is

[3] Kitcher (1985) and Sober (1988) also discuss the tenuous connection between evolutionary and psychological altruism. Both assert that psychological altruism can exist even if evolutionary altruism never evolves, either because the proximate mechanisms that motivate evolutionarily selfish behaviour are psychologically altruistic, or because there is more to human behaviour than fitness maximization. My main point, that evolutionarily altruistic behaviour can be psychologically selfish, is complementary to theirs.

because the person is motivated entirely by self-interest, and ignores his effects on others (positive or negative) in his decision. The second half of the passage concerns an evolutionary model in which A-types have an effect d on themselves and an effect r on members of their group. If groups are formed at random from a large population, then the probability of the recipients being A will equal the frequency of A in the large population. Any effect on recipients, positive or negative, on average will not alter the global frequency of types. Thus, only effects on self (represented by d) produce evolutionary change, and effects on others (represented by r) appear irrelevant. If groups are not formed at random, then the probability of the recipients being A can exceed the frequency of A in the large population, and positive effects on others ($r > 0$) can be selected despite negative effects on self ($d < 0$). Such behaviours are altruistic according to Nunney.

Notice that the proximate mechanisms that motivate the A-types are never an issue. The similarity between the first and second half of the passage is between a *human actor* on the one hand, who cares only about himself, and an *evolutionary process* on the other, in which only absolute effects on self are relevant to gene frequency change. Thus, despite the fact that Nunney's definitions of altruism and selfishness are based on fitness effects, they owe their intuitive appeal to a metaphor based on motives. When groups are formed at random, the products of natural selection are *like* the decisions of a psychologically selfish individual that cares only about its own absolute fitness. When groups are formed non-randomly, the products of natural selection are *like* the decisions of a psychologically altruistic individual that values the welfare of others.

As with all metaphors, this one has the advantage of endowing the unfamiliar (a model) with properties of the familiar (human decision making). Unfortunately, at least two other methods exist to calculate gene frequency change that are equally amenable to the same metaphor.

(a) Hamilton (1964) invented a new measure called 'inclusive fitness', that includes the effects of a gene not only on itself, but on all copies that are identical by descent, which by definition are present only in genetic relatives. Thus, an individual can increase 'its' inclusive fitness by aiding relatives, even at the expense of its personal fitness. Hamilton's measure is widely regarded as the utility that well-adapted organisms strive to maximize. But by replacing the utility of personal fitness with the utility of inclusive fitness, the boundary of selfishness can be pushed outward to include aid-giving to relatives. The products of natural selection are *like*

C. The Evolution of Psychological Altruism

When are organisms with feelings of pleasure based on empathy and with cognitive processes in which others are perceived as valuable in their own right more fit than organisms for whom others are psychologically merely tools for manipulation? As we have seen, the answer to this question is not as simple as determining when benefiting others at the expense of self is evolutionarily successful. The evolutionary literature is therefore curiously silent when it comes to explaining the biological roots of psychological altruism. Several possibilities suggest themselves. Perhaps psychological altruists are so popular as associates that they succeed despite frequent exploitation. Perhaps clustering mechanisms that segregate altruists from non-altruists are so consistent that the ability to calculate when to be altruistic (in terms of effects) is unnecessary. Or perhaps the 'bounded rationality' of the human brain (Simon 1983) makes simple empathy an efficient rule of thumb, compared to Machiavellian thought (Ruse 1986).[7]

These and other questions become interesting when we realize that the evolution of altruism, defined in terms of fitness effects, does not by itself determine the proximate mechanisms that elicit the altruistic behaviours.[8]

REFERENCES

Bertram, B. C. R. (1982), 'Problems with Altruism', in P. Bateson (ed.), Current Problems in Sociobiology (Cambridge: Cambridge University Press), 251–68.
Hamilton, W. D. (1964), 'The Genetical Evolution of Social Behavior, I and II', Journal of Theoretical Biology, 7: 1–52.
Kitcher, P. (1985), Vaulting Ambition (Cambridge, Mass.: MIT Press).
Michod, R. (1982), 'The Theory of Kin Selection', Annual Review of Ecology and Systematics, 13: 23–55.
Nunney, L. (1985), 'Group Selection, Altruism and Structured-Deme Models', American Naturalist, 126: 212–30.
Ruse, M. (1986), Taking Darwin Seriously (Oxford: Basil Blackwell).
Simon, H. (1983), Reason and Human Affairs (Stanford, Calif.: Stanford University Press).

---

[7] This issue is dealt with in the following two subsequent publications: Wilson, D. S., and Sober, E. (1994), 'Reintroducing Group Selection to the Human Behavioral Sciences', Behavioral and Brain Sciences, 17: 585–654; Sober, E., and Wilson, D. S. (in press), Unto Others: The Evolution of Altruism, (Cambridge, Mass.: Harvard University Press).

[8] I thank Lee Dugatkin, Elliott Sober, and George Williams for helpful discussions. Supported by grant no. DE-FG02-89ER60884, awarded by the Ecological Research Division, Office of Health and Environmental Research, US Department of Energy.

Sober, E. (1988), 'What Is Evolutionary Altruism?', in B. Linsky and M. Matthen (eds.), *New Essays on Philosophy and Biology*, Canadian Journal of Philosophy supplementary volume 14, 75–99; reproduced as Ch. 23.

—— (1989), 'What Is Psychological Egoism?', *Behaviorism*, 17: 89–102.

—— (1993), 'Evolutionary Altruism, Psychological Egoism and Morality: Disentangling the Phenotypes', in M. Nitecki (ed.), *Evolutionary Ethics* (Albany, NY: State University of New York Press), 199–216.

Wilson, D. S. (1980), *The Natural Selection of Populations and Communities* (Menlo Park, Calif.: Benjamin Cummins).

—— (1983), 'The Group Selection Controversy: History and Current Status', *Annual Review of Ecology and Systematics*, 14: 159–89.

—— (1989), 'Levels of Selection: An Alternative to Individualism in Biology and the Human Sciences', *Social Networks*, 11: 257–72.

—— (1990), 'Weak Altruism, Strong Group Selection', *Oikos*, 59: 135–40.

—— and Dugatkin, L. A. (1991), 'Altruism', in E. F. Keller and L. Lloyd (eds.), *Keywords in Evolutionary Biology* (Cambridge, Mass.: Harvard University Press), 29–33.

—— and Sober, E. (1989), 'Reviving the Superorganism', *Journal of Theoretical Biology*, 136: 337–56.

PART VIII

THE HUMAN GENOME PROJECT

INTRODUCTION TO PART VIII

MICHAEL RUSE

In 1953, James Watson and Francis Crick discovered that the deoxyribonucleic acid (DNA) molecule is a double helix, and that each strand is made from copies of a number of smaller molecules ('bases'), strung along a backbone. The order of these copies contains the genetic information which is passed on from generation to generation, and which is the blueprint for the building of each individual organism. Very shortly, the 'code' (the exact way in which the sub-molecules carry their information) was cracked, and much work was done (and many Nobel Prizes won!) as 'molecular biologists' started working out the ways in which the information of the DNA molecule is translated into the building processes of the cells of the body.

Then, in the 1970s, powerful new ('recombinant DNA') techniques were developed, enabling biologists to read the ordering of the bases of individual organisms, and thus was launched the dream of uncovering the entire genetic information to be found in the DNA (the 'genome') of organisms of some particular species. Since we are human beings, it was natural that interest was focused on finding the full story of our own genetics. And so was born the 'Human Genome Project', which hopes to read the full ordering of the bases of human DNA. Although the initial intention was to concentrate exclusively on *Homo sapiens*, soon the project was extended to cover other organisms: the mouse and the fruit-fly, as well as microorganisms (yeast and the bacterium to be found in the human gut, *E. coli*). Molecular biologists working on other organisms complained (and not simply from self-interest) that unless one has other models against which to compare humans, much of the information directly on humans will be of limited value. To what extent, for instance, is our biology unique? And to what extent do we share things with other organisms (who might therefore be used to test hypotheses about humans)?

The Human Genome Project is biology on a mega-scale and, as you might expect, has attracted much attention, not all of it favourable. Proponents argue that unlimited benefits lie at the end of the rainbow. When once we have the human genome fully read, we will be able to perform marvels of medicine, spotting deleterious genes (that is, genes that cause unwanted physical effects), and removing or repairing them before they take full effect. Critics are nothing like so certain, and argue that the benefits are small, or illusory, the dangers are real and great, and the project exists primarily to support biologists in moving forward in the work that they find interesting. Combined with this, critics find that inter-scientific rivalry is a major motive force: there is a large measure of satisfaction among biologists today that it is not only physicists who have large and expensive projects, drawing on taxpayers' funds.

Marga Vicedo's contribution (Ch. 25) lays out some of the basic facts about the Human Genome Project (HGP), arguing that from a philosophical perspective there are two basic issues: one ethical, concerning the extent to which (if successful) the HGP is truly going to lead to benefits and the extent to which it might be misused morally; the other epistemological, concerning the question of 'reduction', especially inasmuch as this centres on the desirability of trying to explain things in science by reference to ever-smaller entities. It is her contention that these are complex questions with no ready answers, a conclusion which is backed firmly by our four other contributors.

Philip Kitcher (Ch. 26) is essentially favourable to the HGP. He sees genuine medical benefits, and not only from the work being done on humans. In this sense, he is supportive of the overall project across different organisms. However, he also sees problems lurking ahead, particularly in the moral realm, as people come to have more control over the nature of the children to which they give birth. Is this going to be eugenics by another name (or by no name at all)? This is precisely the fear of Diane Paul (Ch. 27), who sorts out some of the more extreme (and less likely) effects of the HGP—we are hardly on the path towards a revitalized National Socialism—from some of the less extreme (and more likely) effects—for instance, the pressure that there will be on parents to have children whom society deems genetically acceptable. She fears that many of these issues are simply being ignored, and will be settled (in unsatisfactory ways) by default. The ways in which we have already dealt with complex biological defects gives her no cause for optimism.

Elisabeth Lloyd (Ch. 28) is another who has questions about the optimistic forecasts made by HGP enthusiasts. She too worries that the social side to medicine is being dismissed or belittled. She points out an unfortunate slippage in discussions between the notion of 'normal' meaning average or typical (as in 'This is the normal gene') and 'normal' meaning desirable (as in 'This unfortunate does not have the normal gene'). What starts as a matter of scientific fact has a nasty habit of turning into a medical judgement.

Finally, we have Alexander Rosenberg (Ch. 29), who takes us right into the epistemological realm. He is not keen on the HGP at all. He thinks that it is scientifically flawed—very much more limited than its supporters allow—and that it exists simply to bolster the egos and prospects of molecular biologists. Through a stunning analogy with the information one might expect from a stack of telephone directories, he explains precisely why he thinks the project is misconceived and constricted, and he shows why it is that he entertains little hope for its future. Although his arguments are epistemological—that is, about the nature of the science itself— there is a strong moral undercurrent to his discussion, appalled as he is that we are all being forced to pay for science which he considers essentially useless and most definitely misdescribed. You may not agree with everything that Rosenberg says, but he does present a powerful critique which needs answering.

25

THE HUMAN GENOME PROJECT: TOWARDS AN ANALYSIS OF THE EMPIRICAL, ETHICAL, AND CONCEPTUAL ISSUES INVOLVED

MARGA VICEDO

I. HISTORICAL BACKGROUND AND ORIGINS OF THE HUMAN GENOME PROJECT

The major goal of the Human Genome Project (HGP) is to locate all the genes of the human genome and establish the base sequences of all its DNA. The project to map and sequence the different genes of an organism is not new. The novelty of this task is the creation of an interdisciplinary and multinational enterprise to obtain all the data about the chemical constitution of our genome and to establish a common language to unify our knowledge. When criticisms are raised about the HGP, therefore, they should be directed at the way in which the programme is to be developed: namely, at how, when, where, by whom, etc. It is not a question of *whether* the project should be carried out, but of *how* it should be done. Before offering a detailed evaluation of the HGP, it is important to consider briefly its aims and origins.

Since the establishment of the chromosome theory of Mendelian inheritance around 1920, geneticists have aimed to map the genomes of different organisms—that is, to identify the location of specific genes in the set of chromosomes composing the genome of a given organism. Given that little was known about chromosomes at that time, the methods used to establish the specific location of a gene were much more indirect and inaccurate than the ones employed today. The usual method in the study of heredity in the first decades of this century was the experimental crossing of different varieties of animals and plants. By experimenting with organisms of large progeny, geneticists could follow the destiny of specific

First published in *Biology and Philosophy*, 7 (1992): 255–78. © 1992, Kluwer Academic Publishers. Reprinted with kind permission from Kluwer Academic Publishers.

hereditary traits through many different generations. When geneticists discovered the existence of genetic linkage—that is, the tendency of some genes to be inherited together—they concluded that those genes had to be located near each other in the same chromosome. Studying the frequency of their expression in the phenotype of an organism, T. H. Morgan and his group at Columbia University established maps of the position of the genes in the different chromosomes of drosophila. The discovery of the existence of sex-linked traits also helped to locate certain genes in the sex chromosomes.

These methods were also used to map human chromosomes. To study inheritance in humans, scientists use family pedigrees. When dealing with large families, the study of genealogies can show the patterns of inheritance of certain traits. By following the expression of a trait in individuals of different generations, it is sometimes possible to locate the gene responsible for that trait in a given chromosome. The genes responsible for haemophilia and colour-blindness, for example, have been located in the X chromosome by this method. The obvious limitation of this method is that it can only be used to locate the genes carried by the sex chromosomes.

When chromosome analysis techniques improved, it became possible to locate several genes through the analysis of the bands which showed up in chromosomes treated with certain chemicals. By comparing the bands in the chromosomes of different individuals, geneticists were sometimes able to correlate chromosomal abnormalities with certain hereditary illnesses. However, given the length of the human DNA and the fact that very small changes can cause biochemical alterations, those methods are not completely reliable. Until the development of the somatic hybridization techniques and, much more importantly, recombinant DNA techniques, it was not possible to locate genes with precision. In the 1970s, the amazing advances obtained by using the recombinant DNA techniques developed by S. Cohen and H. Boyer created an atmosphere of optimism regarding the viability of large research projects in genetics. The most ambitious of those has undoubtedly been the initiative to map and sequence the human genome.

The human genome is composed of twenty-three pairs of chromosomes. Each chromosome consists of two strands of DNA. DNA is made up of four different nucleotide bases: adenine, thymine, cytosine, and guanine. The two strands of DNA that form each chromosome are tied together by specific base pairing: adenine always pairs with thymine, and cytosine with guanine. The human haploid genome, which does not include heterozygosis and variation among individuals, is constituted of 3 billion base pairs, and it is thought to have between 50,000 and 100,000 genes. The length of the different genes can range from 10,000 to 2 million base pairs. Those genes might be coded in only some 5 per cent of the total DNA, however. The rest of the human DNA is so-called junk DNA, for which no function is yet known.

The project of mapping and sequencing the human genome had been 'in the air' for several years. In 1984, at a meeting organized by the US Department of Energy (DOE), several researchers pondered the benefits of mapping the human genome. One year later, the chancellor of the University of Santa Cruz, Robert Sinsheimer, organized a meeting to deal with that project. In 1986, the Nobel Laureate Renato Dulbecco published an editorial arguing that significant advances in cancer research could be accomplished if scientists knew the exact sequence of bases of human genes. Finally, that same year, several researchers met at the Cold Spring Harbor laboratories in New York and set up the US project as it now exists. In the United States, the national programme to carry out this endeavour is called the Human Genome Initiative (HGI).

The various tasks involved in the project could not be carried out by only one research centre. It was, therefore, necessary to find a national agency to co-ordinate the many laboratories which would be involved in the HGP. The DOE had already expressed an interest. This agency had been researching the effects of low-level exposure to radiation on genetic material, and in 1983 the DOE created the Genbank, the biggest chromosome bank in the USA, located at Los Alamos. In 1985, the DOE founded the National Gene Library Project, a chromosome library, in its laboratories at Los Alamos and Lawrence Livermore. However, the National Institutes of Health (NIH) were also interested in the HGP. Some researchers saw them as the obvious place to carry out the project, since they were the major US facilities for conducting biological research with medical implications.

Finally, in 1988, the two agencies—the DOE and NIH—agreed to combine their activities. Presently, the National Science Foundation and the US Department of Agriculture also collaborate in some aspects. In 1989, the NIH established the division in charge of the HGP as an independent unit—the National Center for Human Genome Research (NCHGR) headed by James Watson. As the director of the Cold Spring Harbor Laboratory, Watson, who was awarded the Nobel Prize in 1953, together with F. Crick and M. Wilkins, for their discovery of the structure of DNA, has consistently promoted research on molecular genetics, and specifically on the human genome.

The need to develop sophisticated technology, the magnitude of the project, and the world-wide interest in the results made this initiative an international project requiring the co-operation of many laboratories. Scientists and administrators in the US recognized the need to include other countries in the HGP, but were not willing to wait for their commitment. They decided to go ahead and integrate different approaches and people along the way. The Human Genome Organization (HUGO), and international organization based in Switzerland, was established in 1988. Its main task is to co-ordinate the work being carried out by all the groups involved, and regulate access to the data and the distribution of material among the laboratories. HUGO is expected to establish divisions in the USA, Canada, Japan, and several European countries. HUGO will also support research on the ethical and social issues related to the HGP, thereby encouraging historical, sociological, and philosophical studies, and organizing meetings where biologists and non-biologists can discuss the aims of the project and the strategies to reach them.

So far, the countries working on the HGP are Great Britain, USSR, Italy, Japan, USA, France, Canada, and Australia. The USA is the only country where the project has been set up as a national objective; all the other countries are involved to a much more limited extent. The mapping and sequencing of the human genome is also included in UNESCO's Major Programme Area II for the years 1990–5, with a Committee for Scientific Co-ordination led by Santiago Grisolía. This programme aims to promote international co-operation, and avoid the manipulation of the project by the wealthier countries. The co-operation of all the centres involved, under the supervision of UNESCO and HUGO, will be necessary for the development of the programme. This task will not be easy, since many countries recognize the need for co-operation, but some of them clearly desire to maintain their world leadership in biochemistry. HUGO, moreover, is not yet an organization with real powers (see Watson 1990). (See Cantor 1990, McKusick 1989, Watson 1990, and Grisolía 1990 for detailed historical presentations of the HGP.)

II. TOWARDS AN EVALUATION OF THE HUMAN GENOME PROJECT

The issues that should be addressed to evaluate the HGP can be classified into four different categories:

1. the feasibility of the project: economical and organizational issues;
2. the ethical questions regarding the development of the project and the potential uses of the information obtained;

3. the project's scientific value: empirical issues;
4. conceptual issues relevant to understanding the nature and scope of the project.

While these categories are useful for analysis, they are not in fact completely distinct. Questions in section 3 deal with the so-called internal factors, that is, those concerned with the scientific value of the project. In contrast, questions raised in category 2 fall under what is known as external evaluation: that is, the value of the project for the society in which it is developed. However, many issues in (1) affect both internal and external aspects of the programme. The financial aspects of the enterprise, for example, affect both the viability of the project itself and the impact it will have on society. Nevertheless, the several criticisms of the HGP can be classified into these categories. Each can fruitfully be analysed separately for reasons of clarity.

A. The Feasibility of the Project

The major criticism of the feasibility of the project is directed at its enormous cost. In 1977 the estimated cost of sequencing was $1/base. The aim was to reduce the cost to $0.01/base. If this was achieved, the final cost of the HGP would be around $60 million (Smith and Hood 1987: 936). According to the US Department of Health and Human Services and the US Department of Energy (1990), the current cost of DNA sequencing is about $2 per base pair. This is estimated for laboratories which perform sequencing routinely and for sequences whose accuracy has been confirmed. According to these agencies, this cost must be reduced below 50 cents a base pair for DNA sequencing to be cost-effective. Although I agree that the initiative should be set up to reduce the costs as much as possible, I do not think that the expense is a conclusive reason to halt it. The HGP is expensive by comparison with othe projects in the biological area, but so are many research programmes in other areas (high-energy physics, space research, and military research, for example)—some of them with apparently less scientific interest and higher social risks than the HGP.

Another criticism related to the total expense is that the HGP will take money away from other interesting research programmes in biology. This is an understandable fear given that research with a strong social impact and medical implications (like cancer or AIDS research) tends to receive more money than other projects. Given that research budgets are fixed, to give money to a project excludes the possibility of spending it on another enterprise. But, as R. Lewontin has pointed out (personal communication), the opponents of the HGP are opening a Pandora's box when they appeal to this argument. The weakness of this argument is that all scientific research uses money which could be used instead for social expenditures. Even opponents of the HGP would have a difficult time defending their use of money against the demands for pressing social problems, like homelessness. The process of allocating money to science and other social goals is a complex and important issue, and it would be unfair to single out the HGP to blame for relative distribution of resources.

In assessing the project's feasibility, I believe that the co-ordination of the different tasks and the co-operation among all the research groups is much more problematical. Some regulatory guide-lines could be established to secure the smooth functioning of the project, but the scientists concerned hold different views on this issue. J. Watson, for example, thinks that the groups will develop rules to co-ordinate their efforts as the investigations proceed. Other researchers, like Walter Gilbert (Harvard), think that clear rules should provide all participating members access to the results. Others suggest that the need for groups to communicate to obtain mutual benefits will force them to co-operate (Roberts 1990b). Elke Jordan believes that the HGP's goals will be unattainable unless it is 'built on teamwork, networking and collaboration'. In his opinion, 'This makes sharing and co-operation an ethical imperative' (Jordan 1990).

Co-operation is fundamental to the project, in order to stay within reasonable financial and time limits and to avoid the fragmentation of results. Necessity may indeed force researchers to work co-operatively, leaving aside personal and national interests. However, given the intense competition for scientific resources and prestige, it might be too optimistic to reject any regulation. Competition, power relations, and nationalistic struggles for scientific hegemony, commercial interests, etc. are all part of the scientific scene. In the interest of the HGP it would be convenient to prevent these factors from undermining the project's goals. Consequently, it seems advisable to set certain guide-lines. I am not referring to the establishment of an ethical code for scientists, but to the need to encourage the scientists involved in the HGP to perform non-competitive science.

Co-operation will also be essential for the basic task of establishing a common language. As in any other research project, the results depend on the inputs. If, as is currently the case, the data entered in the databanks are fragmentary and codified in different languages, and researchers do not supply the information necessary to interpret their meaning, the HGP may be building a library similar to the one described by the Argentinian writer

J. L. Borges: a universal library containing all the books in the world but where nobody knows how they are catalogued or how to translate the languages they are written in. It should be kept in mind that different techniques and languages are used in the HGP, and there is not always an unambiguous way of translating data obtained by different methodologies. Leslie Roberts (1989: 1438) described the situation as follows:

True, the broad outlines are clear—a physical map shows the actual distance, ideally measured in nucleotide bases, between landmarks distributed along the chromosomes. Genes can then be located within those landmarks. But researchers constructing pieces of this map have yet to agree on what the landmarks should be. And without a common set of landmarks, mapping the chromosomes is a bit like building a road through a mountain: if tunnelers at both sides don't use the standard benchmarks that mark elevation from sea level, they're likely to end up with shafts that don't meet.

In my opinion, the most difficult task for the HGP at the present time is to find a common language to construct a meaningful dictionary (see Olson *et al.* 1989, Roberts 1989, and Stephens *et al.* 1990).

Olson *et al.* (1989) presented a promising method to obtain a common format for maps: the use of STS, or sequence-tagged site, which is a DNA sequence that is unique and can be used as an unambiguous marker. But, according to the US Department of Health and Human Services and the US Department of Energy joint report (1990: 14): 'The STS proposal is still under discussion in the scientific community and few, if any, mapping projects have started to use the STS system.' An additional problem will be the cost of using this methodology. Herein, in my opinion, lies one of most important problems for the elaboration of meaningful maps of the human genome.

B. Relevant Ethical Questions

The discussion about the HGP is too often couched in global terms of support or rejection of technological advances. Vickers (1990), for example, claims that 'The first issue seems to be whether it is ethical to deny access to knowledge simply because that knowledge *might* be misapplied, but where it is clear that the potential benefits are enormous'. This argument, however, could be easily turned around: one should not accept the pursuit of a research programme simply because the resulting knowledge *might* bring some benefits, when it is clear that the potential misuses are enormous. This is not a fruitful way to pose the question, because both the potential benefits and the potential risks of the HGP are enormous. The challenge will be to find a way to assess them rationally and make decisions accordingly.

We should distinguish several issues here. When discussing the ethical and social implications of the HGP, people often raise the general problems related to biotechnological research. But biotechnology, genetic engineering, and even human genetic engineering, all include a wide range of techniques and programmes. The analysis of their implications in a global way leads only to polarized positions defended either from a non-critical acceptance of any technological advance or from an equally non-critical rejection of the new technologies. The debate about the social implications of the new technologies in genetics should not be set up in terms of global acceptance or rejection. The artificial fabrication of proteins, human genome mapping, the manipulation of germinal cells, etc. should not be grouped together in any simple way. The HGP itself, moreover, does not aim to manipulate and modify genetic material. It therefore does not belong to the realm of biotechnology, although the maps and sequences will be used as data to develop techniques to manipulate the genetic material. Thus, the ethical problems raised by the HGP are related to biotechnology issues only as far as the data obtained will be used to develop specific practices of genetic diagnosis and therapy.

These practices are already being implemented, and geneticists make advances in their development independently of the HGP. The data from the HGP may be useful to them, but the evaluation of their utility and consequences cannot, and should not, be tied to the HGP for two main reasons: first, scientists need not wait for the whole sequencing of the human genome to perform genetic diagnosis and therapy; they only need to know the location of the gene responsible for the phenotypic trait they are interested in. Second, the HGP can be rejected or accepted independently of the implementation of genetic diagnosis and therapy. One could support the latter and be against the HGP; conversely, one could support the HGP and be more cautious about the implementation of genetic diagnosis and therapy.

We have to admit, nevertheless, that it is impossible to separate the HGP completely from the use that people will make of the data it gathers. On the one hand, there is not a sharp distinction between basic and applied research that allows us to say that the HGP is only basic research. On the other hand, even if such a distinction were possible, it would not be applicable in this case. A lot of data for the HGP will come from research done previously to implement genetic diagnosis and therapy. Moreover, most scientists involved in the HGP promote it by pointing out the future benefits in biomedical applications. The transfer of the technology to medical applications is also strongly encouraged by the agencies supporting the HGP.

Among the ethical and social implications of the HGP, those which will arise from the use of the data and the techniques developed in applied areas therefore must also be considered. In summary, when the HGP's social implications are discussed, attention should be paid to the application of its results, but the discussion about the potential harms and benefits of genetic diagnosis and therapy should not wait for the development of the HGP. It is beyond the scope of this essay to deal with the specific issues related to genetic therapy and diagnosis. The interested reader is referred to Council for Responsible Genetics (1990a), Lappé (1987), Nelkin and Tancredi (1989), and Suzuki and Knudtson (1989).

The very development of the project also raises several problematic issues from a moral viewpoint. For example, just as the human anatomic map is a property of humankind, it is to be hoped that the genetic map will be too. Its realization must be an integrated task without discrimination against the less developed countries and with the data made freely available to everybody. At the beginning, this was not clear to all the participants. W. Gilbert (Harvard University) founded his own company, Genome Corporation, to patent the data on the genome and sell them (see Palca 1987). If this were allowed, it is clear that pharmaceutical and chemical corporations would soon dominate this practice. The marketing of genetic data could even lead to the illegal exchange of useful information, as happens with the illegal exchange of organs for transplant. Legal action is being taken to avoid this situation. Although the issues are not clear among the participants, because some emphasize the commercial importance of the project (Cantor 1990: 51), while others understate it (McKusick 1989: 912), action is being taken to prevent data on the human genome from being patented.

There will also be problems in establishing limits to restrict access to the data, to determine who should use them and how. The Nobel Laureate J. Dausset has defended the HGP on the basis that it will allow scientists to predict when an individual will develop an illness, or at least will have a predisposition to do so, and predictive medicine is a first step towards preventive medicine. By the same reasoning, we could also start developing a preventive ethics by examining the predictable problems that will arise. Many difficult situations arise because people wait until a morally problematic situation develops, and then quick decisions must be taken.

We can take the case of Ulysses as a model here. Ulysses did not pass between Scylla and Charybdis because he was able to resist the sirens' songs, thus showing a moral strength not possessed by the other sailors. Rather, he was successful because he admitted his weakness and adopted measures to resist temptation. He decided to tie himself to the mast to avoid taking a decision that he would regret later on. This action is recognized as moral prudence. To adopt measures to avoid leaving difficult decisions to luck or future circumstances is both rational and morally recommendable. And to suggest that researchers predict potential risks is not to be against the HGP. It is simply to argue for a preventive ethics which will allow maximum benefit from the project.

Perhaps the first basic task from a moral point of view is for the scientists to inform society about the development of the initiative and its implications. Lack of information always raises suspicion, and leads to misunderstandings. Thus, the first responsibility for geneticists is to publicize in lay language the aims and means of the HGP. Their second responsibility is to assess realistically the value of the project, and to avoid making empty promises. Since it is important for the public to know what the scientific expectations are, let us then turn to this area.

C. The Scientific Value of the HGP: Empirical Issues

The Human Genome Project involves several tasks. First, it is necessary to develop the technology for the mapping of genes among the total DNA and for the sequencing of DNA. The sequencing has been postponed until scientists have completed the genetic and physical maps. As a short-term goal, geneticists aim to obtain genetic maps of the different chromosomes. These maps are usually elaborated by using the recombination frequency of the different genes. The greater the distance between the two genes in a chromosome, the higher the recombination frequency. Geneticists now use DNA markers, such as restriction fragment-length polymorphisms, but the methodology to elaborate a genetic map is basically the same: whether scientists use genes or DNA markers, genetic maps are constructed by analysing the frequency with which two 'markers' are inherited together. Genetic maps show only relative distances between the genes, however; they do not offer the specific location of the genes. To identify specific locations, geneticists must construct physical maps in which the distances between genes correspond to the actual physical distances in the chromosomes. Once they have the physical maps, the following step is to sequence the DNA.

There are several techniques to sequence through enzymatic and chemical methods. Using those techniques, geneticists had sequenced 12 million DNA bases by 1987, the total DNA in the US Genbank, a databank which grows by 30 per cent every year. In 1987, the total human DNA sequenced was about 0.03 per cent of the human genome (Smith and Hood 1987).

Sequencing methods are repetitive, slow, and expensive, and must be used by trained researchers. For these reasons, there is widespread agreement to delay human DNA sequencing until the techniques are less expensive and faster. Several approaches allow for investigation by easy-to-automate methods. In Japan, researchers are working on the automation of existing techniques for the work to be performed by non-specialists. The truth, however, is that there are as yet no fast and reliable automated methods for sequencing.

The HGP also requires research in other areas of genetics besides human genetics. The study of the genomes of simpler organisms like bacteria, yeast, and worms is necessary, to use them as models in the analysis and interpretation of the data on the human genome. The HGP involves research in other disciplines too. It is necessary to organize the data in databases, for it is essential that those data are codified in such a way that access to them is easily obtained. Therefore, researchers will have to develop the information technology necessary to deal efficiently with the data. The development of information technology will also be necessary to analyse the different sequences, to compare sequences, and to compare data on the human genome with that from other genomes, all essential steps to locate the position of the genes and discover their function. When the physical maps are ready, and the sequencing technology has been developed, the systematic sequencing of chromosomes will start. Huge amounts of data will accumulate while researchers try to best store and organize them in a manageable way. When all the data are organized meaningfully, scientists will finally begin to study how that chemical information contributes to the development of human organisms.

To evaluate the promise of the HGP, it is first necessary to clarify what information the mapping and sequencing of the human genome will provide. This project will indicate the sequence of bases of all the chromosomes of our genome: that is, it will produce a massive amount of chemical information about how our genetic information is coded. If it is carried out accurately, it will lead to a genetic dictionary. It is important to point out, however, that scientists will then need to interpret these biochemical data. In many respects, meaningful questions can be asked only after the mapping and sequencing is done.

The human genome may be compared to a literary text. The HGP would be like the syntactic analysis of that text. It will facilitate a study of the signs used in its construction, which words are used, and where specific syntactic constructions are employed. Like a syntactic analysis, the HGP may even suggest the function of different elements in the global text/ genome. To know the meaning or the words in a text, however, we have to then develop a semantic analysis. In human genetics, the integration of biochemical data with embryology and developmental biology will be necessary to interpret the genes' role in the formation of an organism. Later on, the analysis of specific genomes in interrelation with specific environments will permit a pragmatic analysis like that involved in studying a text: the significance of specific constructions in specific situations. Thus, when the HGP is finished, we will certainly *not* know how the human genome functions; much less will we know how 'to build an organism', as is sometimes argued (e.g. Smith and Hood 1987: 939). This, of course, does not mean that the HGP is not important.

To avoid any risk of self-delusion, we should nevertheless be realistic about the knowledge that the HGP will offer. I mentioned earlier the great quantity of redundant DNA which exists in our genome. Thus, scientists will have to locate the genes among all the DNA which is functionally silent. Even once the genes are located, scientists will still know little about how they operate. In the genome of any organism there are many interrelations. We do not know how genes switch 'on' and 'off', or how they regulate their functions. Therefore, mapping a genome and knowing its base sequence does not amount to knowing how to build that organism. Much more will have to be learned in embryology and developmental biology before scientists can decipher all the mysteries of a genome.

It is hoped that the information obtained through the HGP will help to locate the genes responsible for genetic disorders. More than 3,000 human illnesses are caused by genes, but most of these genes have not been identified or located yet. It should be pointed out, however, that locating a gene responsible for a disorder does not amount to knowing how to cope with the disorder. Knowing the sequence of DNA bases of the human genome will also help to understand human polymorphisms. Their analysis promises the discovery of important clues to our immunological system and our evolutionary history. Kowledge of the structure of human DNA will help to discover some facts about our evolutionary history, since knowledge of the genetic variations between individuals in a group and between groups might lead to a better understanding of evolutionary processes (see Moya 1991 on the significance of the HGP for evolutionary genetics). It is also possible that a better knowledge of the human genome will help in cancer research, as Dulbecco has emphasized (Dulbecco 1986, Dulbecco and Chiaberge 1988).

Since genes are only a small part of the total DNA in the human genome, some criticisms point to the fact that much time and money will be employed sequencing DNA which does not codify any proteins. This is the so-called junk DNA. In the human genome 80–95 per cent might be junk DNA. Although it is true that this DNA does not code for proteins, we cannot assume it is worthless until we have discovered its role in cellular mechanisms. It might well play an important role in cellular maintenance or regulation, and it could have crucial evolutionary functions. I think that the existence of junk DNA should be a reason for, and not against, the human genome project. In the first place, it was thanks to the recombinant DNA techniques that scientists discovered junk DNA. The HGP might help us understand what it does and why it is there. Secondly, the function of a gene depends also on its genetic milieu—that is, the DNA surrounding it. Therefore, it may be important to identify all the DNA, even if our main interest lies only in functional genes. Thirdly, there is not yet a method to differentiate functional DNA from junk DNA. Thus, it might be easier and cheaper to sequence all DNA systematically.

Another issue is whether scientists should establish an order of priorities in the analysis of chromosomes. Some argue that geneticists should first locate and sequence those genes known to be responsible for specific genetic disorders. I think that an order of priorities will naturally emerge. As a matter of fact, the more 'attractive' pieces of DNA for researchers (for several different reasons) are those where specific genes are located. The HGP is unlikely to be merely a systematic analysis, from chromosome 1 to chromosome 23. The HGP will more probably consist of the grouping of all the data known about genes and chromosomes obtained by different research teams. As I understand it, the HGP is really a call to assemble all the fragmentary knowledge about the human genome, the sooner to have a genetic map, much as we have the Vesalius anatomic map: that is, a map of an abstract genome, where the data will be useful later to study specific genes in the genome of specific individuals. It is important to point out that the final result will be a composite of a number of sequences from different individuals. Thus, it is an open question whether there will be a consensus about what the 'human genome' is.

In Section I, I mentioned the problem of finding a common language to store the data of the HGP in a manageable and meaningful way. This is important not only to avoid fragmentation and duplication of effort, but to define which problems the HGP will be able to tackle. This point has already been made by Stephens et al. (1990: 243): 'With the increase in data generated by mapping efforts, our concept of genome organization will change, with ramifications for a common co-ordinate system, for nomenclature, and for database management. These issues deserve careful consideration, since the way we organize and record data will limit the questions we can pose.' It will be important to establish a plan according to which scientists can set up different goals and ways to accomplish them. If the HGP aims first at gathering a large amount of data, only then to try to organize and interpret them, it runs the risk of finding one day that much of the data is meaningless for many important questions. As a leading geneticist of our time, F. Jacob, has remarked: 'In the relation between theory and experience, it is always the former one to initiate the dialogue. It is the theory what determines the shape of the question, that is, the limits of the answer' (Jacob 1970: 16). For the HGP to accomplish important goals, it is necessary that scientists progressively reflect on the answers they are aiming at to be able to search for the data relevant to their questions.

Several scientists have claimed that the HGP will offer the clue to the understanding of 'human nature', as if the sequence of nucleotides of a genome determined an organism. Some critics have reacted against this simplistic view of living systems and, specifically, of human beings. In the next section, I will analyse some of these criticisms.

III. CONCEPTUAL ISSUES: REDUCTIONISM, DETERMINISM, AND MOLECULAR GENETICS

A. Reductionism and Molecular Genetics

The reductionism implicit (and sometimes explicit) in the HGP has been criticized in several recent works: Council for Responsible Genetics 1990*b*, Nelkin and Tancredi 1989, Lappé 1987, Tauber and Sarkar 1992, Suzuki and Knudtson 1989, and Teitelman 1989. Philosophers have discussed at great length reductionism and its relation to the unity of science, the search for a systematic and simple picture of the world, the hierarchy of the diverse special sciences, and the role of physical explanations, etc. Reductionism has also been a long-standing concern among biologists. Ever since the disputes between materialists and vitalists and the discussions about the role of teleological explanations in biology, to the debates about sociobiology, reductionism has been present in many biological disputes (see Ruse 1989 for a detailed discussion of the alleged reductionistic character of sociobiology). Thus, to do full justice to this topic would require a book-length analysis. Here, I will briefly discuss the different senses of reductionism, and suggest how they may bear on the Human Genome Project. For two important reasons, the discussion about the HGP includes a general analysis of molecular genetics. First, the HGP is a research programme in molecular genetics; hence, if molecular genetics is found to be reductionist, any of its projects must also be so. Second, some authors have criticized the HGP because they take it to be the culmination of the reductionist approach characteristic of molecular biology (e.g. Keller 1990, Tauber and Sarkar 1992).

Following Ayala (1974), I will distinguish between ontological, methodological, and epistemological reductionism. Contrary to the doctrine of vitalism, ontological reductionism maintains that all living systems are composed of nothing but physico-chemical entities and processes. Methodological reductionism claims that the most promising approach in science is to search for the basic constituents of any process and its explanation in physico-chemical terms. Lastly, epistemological reductionism suggests that the laws and theories of one science can be derived from a lower-level, more fundamental science. Usually, the suggestion is that physics will be able to account for all scientific domains in the long run.

Let's start with *ontological reductionism*. The issue of whether living systems are composed of nothing but physico-chemical entities, or whether they are endowed with some *élan vital* that non-living systems are lacking, was a substantive question at some point in the history of science. Today, however, virtually all scientists accept ontological reductionism as one of the basic tenets of their endeavours. Thus, in light of Ayala's definition, to criticize a scientific programme because it is reductionistic from an ontological viewpoint—that is, because it admits no more things in the world than those accepted by current scientific wisdom—amounts to nothing less than an attack on scientific orthodoxy in all fields.

Now, let's turn to *methodological reductionism* and molecular genetics. By using the strategies and research techniques of molecular genetics, the HGP would automatically be reductionistic if the methodological approach of molecular genetics is shown to be so. Indeed, molecular genetics is not only considered to be reductionistic in this sense (Mohr 1989, Tauber and Sarkar 1992, etc.), but it is often presented as the best example of the success of such an approach to nature (Oppenheim and Putnam 1958, Ruse 1989). To analyse the validity of these claims, we first need to clarify what is meant by methodological reductionism. In his 1974 essay Ayala suggested that the following questions contain the core issues: 'In the study of life phenomena, should we always seek explanations by investigating the underlying processes at lower levels of complexity, and ultimately at the level of atoms and molecules? Or must we seek understanding from the study of higher as well as lower levels or organization?' (p. viii). Given the strong sense given by Ayala to the formulation of methodological reductionism when asking whether the strategy of looking at the lower levels of complexity is *always* the best one, I think that few, if any people, will hold this position.

It is important to note here that a strategy that aims to account for a process at the lowest possible level does not necessarily exclude analysis at higher levels. The question 'Is it a good strategy of research to look for lower-level accounts of phenomena?' is a methodological one. However, the issue about whether the account found at the lower level would be sufficient to explain the whole process under study pertains to the domain of epistemological reductionism, and it will be analysed later.

Let's explore the question of whether it is a wise scientific approach to search for the smallest components of a system and their physico-chemical underpinnings. Scientists (Medawar 1974) and philosophers (Oppenheim and Putnam 1958, Popper 1974, Ruse 1989) have defended reductionism as an analytical tool for research. In Popper's view, reductionism is defensible not as a philosophy, but as a method that led to staggering successes even when the sought-for reduction was not accomplished. His point is that even failed attempts at reduction often lead to the discovery of new and deeper problems. Therefore, according to Popper, scientists should adopt the 'position of a critical reductionist who continues with attempted reductions even if he despairs of any ultimate success' (1974: 270).

Along similar lines, the fruitfulness of the reductionistic methodology was defended by Oppenheim and Putnam (1958). In their view, 'the assumption that unitary science can be attained through cumulative microreduction recommends itself as a working hypothesis' (p. 8). Among the various values that Putnam and Oppenheim see in this working hypothesis (see p. 16) is the fact that it is fruitful 'in the sense of stimulating many different kinds of scientific research. By way of contrast, belief in the irreducibility of various phenomena has yet to yield a single accepted scientific theory' (ibid.). According to the authors mentioned above, in a nutshell, a reductionistic methodology—that is, the belief in the possibility of reduction and the methodology consequently adopted to attempt the reduction—is more fruitful heuristically than an anti-reductionist methodology.

The pragmatic justification of methodological reductionism has good support from several cases in the history of science (the development of molecular genetics being a clear one). Methodological issues, however, should not be approached at such a high level of generality and abstraction. Several philosophers have recently started to point out that methods are local and content-specific (Miller 1987, Sober 1988). The question of whether one should adopt a reductionist approach will depend entirely on the subject-matter and the aims of a specific study. The case of molecular genetics has from the beginning entailed a search for the physico-chemical components or basis of genetic phenomena. Almost by definition, molecular genetics aims at a lower-level account of genetics and, therefore, must adopt reductionistic strategies. A further query would be to ask how fruitful is the strategy of looking for the lowest level of complexity in genetic research. But the issue then would shift from whether molecular genetics is reductionistic to whether scientists should do molecular studies of genetic phenomena in the first place. The amazing and quick successes of molecular biology convincingly indicate that they should.

To conclude this section, the relation between methodological reductionism and molecular biology is clear. Yes, molecular biology is reductionistic from a methodological point of view. The adoption of this stance, moreover, has proved to be extremely successful. Finally, the justification for reductionism in molecular genetics is not only pragmatic, since the very core of the enterprise makes it reductionistic in the sense being discussed here. Molecular genetics is set up to discover the physico-chemical basis of genetic phenomena based on the assumption that their study will help to improve our understanding of those phenomena. The success of the discipline suggests the fruitfulness and validity of the methodology adopted for that task. Thus, molecular genetics is 'reductionistic' in this sense, but it is hardly a sound criticism to censure a discipline for doing what it is supposed to be doing and for being successful at it.

Consider now the more complex issue of *epistemological reductionism*. Epistemological reductionism is sometimes stated as a global thesis about the reducibility of higher levels to lower levels of analysis, and sometimes as a more local thesis about the reducibility of a particular scientific theory to the laws and postulates of another one. The question of whether the laws of a science reduce to the laws of another can be understood as whether one level of analysis is reducible to the other, but it is also sometimes cast in terms of inter-theoretical reduction as an analysis of the dynamics of science. In the latter sense it merely aims at understanding the internal dynamics of theory change.

I do not think that the issue can be meaningfully posed in global terms. Whether one level of analysis can, and will, be reducible to a lower one will depend on which specific levels are involved. One can think that chemistry could be reduced to physics, but not that sociology will be reduced to physics. Consider the specific case of molecular genetics, which almost by definition attempts to describe the physical bases of the structure and organization of living systems. We must decide whether the aim of this enterprise is to offer a physico-chemical description of the structure and dynamics of the phenomenon under consideration, or to explain the phenomenon by reference *only* to its physico-chemical basis. The latter aim would presuppose epistemological reductionism, and any research programme that adopted this strict methodology would be impoverished not by the specific use of the techniques and projects selected, but by the shortcomings—if any—of epistemological reductionism. So, the questions arc: Is or has molecular biology been reductionistic in this latter sense? And what are the limitations of epistemological reductionism?

In a study of the historical development of genetics since the ground-breaking discovery of the molecular structure of DNA by Watson and Crick in 1953, Schaffner (1977) has argued that reductionism was periph eral to the development of molecular genetics. Others, including some of the main scientists involved, have argued that reductionism was a central aim of this new approach to heredity (e.g. Mohr 1989). The question is whether the techniques, concepts, and strategies of molecular biology were employed to explain and reduce the biological system under study by giving a complete account of their structural components only in physical terms. Has this been the central approach of molecular genetics? I do not think so. I support Olby's view that 'the chief feature of molecular biology has been its interdisciplinarity' (1990: 511). Olby believes, nevertheless, that despite the interaction between physics, chemistry, and biology, 'it would be a perverse historian or philosopher who refused to accept that the description of concepts like gene, mutation, specificity and develop-ment in molecular terms represents a form of reduction' (p. 512).

But this last point, in my opinion, is a different issue. There is no doubt that molecular genetics has offered a description of the physico-chemical structure of many biological phenomena. It is another question, however, whether molecular genetics has shown, or even has tried to show, that the physico-chemical characterization of a system is sufficient to account for the behaviour and properties of that system and, therefore, that higher-level accounts of the phenomenon are irrelevant or redundant. It is one thing to argue that the physico-chemical description of a given system is important, or even necessary, to understand its nature and functioning, but it is a completely different matter to argue that this description is *sufficient* to understand that process or system. The first argument only supports the search for the lower characterization of a process, which ultimately leads to its description in physico-chemical terms. The latter argument asserts that once we know the lower level, explanations at higher levels would be irrelevant, and we could proceed to explain the phenomena in a different direction: from the bottom up. But I want to emphasize that one could maintain that going from a macro-level to a micro-level is a possible, interesting, or even necessary step to understand a process, without there-by being committed to the view that the micro-level analysis would be sufficient to understand the system under consideration. In many cases, one might suspect that at the lower level there are multiple physical mechanisms that may underlie a given process or state at the higher level. If so, then we would say that the higher level was supervenient on the lower level.

Not everybody agrees, however, that molecular genetics is not historically rooted in a reductionistic tradition. Mohr, for example, states that 'the program of molecular biology has been strictly reductionistic from the very beginning until the present day. The goal has always been the unity of biological sciences: to explain the major elements of the classical biological disciplines in terms belonging to the most fundamental biological science, molecular biology' (1989: 141).

But I do not think that the search for physico-chemical explanations in biology necessarily rests on a unification of science through the reduction of the different biological sciences to molecular biology (see Oppenheim and Putnam 1958 for a defence of the thesis that the unity of science can only be accomplished through micro-reductions). Along with Fodor (1974: 107), I believe that 'the classical construal of the unity of science has really misconstrued the *goal* of scientific reduction. The point of reduction is *not* primarily to find some natural kind predicate of physics co-extensive with each natural kind predicate of a reduced science. It is, rather, to explicate the physical mechanisms whereby events conform to the laws of the special sciences.' And, as Fodor further argues: 'there is no logical or epistemological reason why success in the second of these projects should require success in the first, . . . the two are likely to come apart *in fact* wherever the physical mechanisms whereby events conform to a law of the special sciences are heterogeneous'.

Reduction in the sense here defined by Fodor is precisely what molecular genetics has been successfully doing so far. Take, for example, the case of the discovery of DNA. The DNA model explicates the physico-chemical mechanism underlying the law of segregation. The whole story of its discovery shows how scientists aimed to find a chemical structure which could account for the biological properties of genes (see Schaffner 1977 for details).

Now, an additional issue is whether molecular geneticists are currently making claims of epistemological imperialism by arguing that the natural kinds of molecular genetics will perfectly map on to the natural kinds of higher biological levels, and thus be sufficient to substitute for them. As the Fodor quotation points out, even when a lower level of analysis can offer a physical description of every phenomenon at a higher level, this does not show that a reduction of their natural kinds has been accomplished, because the natural world could be such that the natural kinds of each level do not perfectly match. Thus, even if molecular genetics gave a description of every genetic phenomenon, it could still be the case that the natural kinds of genetics would not be coextensive with physico-chemical natural kinds. Besides, to fully understand the causal pathways between the genotype and the phenotype, scientists will need to complement the molecular account with evolutionary explanations, as Beatty (1980) has argued.

Two additional points deserve mention. First, even when a micro-description of a higher-level phenomenon is achieved, this does not make the accounts given at that higher level unnecessary. In fact, as Fodor (1974) has shown in the case of psychology, and Sober (1984) has argued regarding the concept of fitness in evolutionary biology, most generalizations made at the higher levels do not require an account of the physical properties that in each case exemplify the validity of the general concept or law at the higher level. One of the examples Sober (1984: 49) presents is the constant-viability model of selection. This model gives an account of human populations subject to malaria and anaemia. The model predicts a balanced polymorphism in those populations where the heterozygote is fitter than either homozygote. By using the concept of fitness, this model can be applied to any population, independently of the physical reasons that make the heterozygote fitter in each specific case. Thus, not only does the description of the lower level fail to make the higher-level description irrelevant, but, for some purposes, the higher-level generalizations make the particular details of the lower level irrelevant.

The second point is that these are empirical questions. Whether this type of reduction is possible in general, and specifically in the case of molecular genetics, depends on how the world is put together (see Fodor 1974 and Kitcher 1984 for positions claiming that anti-reductionism is made true by the way the world is constructed, not by any shortcomings of our cognitive system). The question of whether the natural kinds of two different levels are coextensive has to be answered, then, by scientific research. (Note that this fact makes the adoption of a reductionist approach useful at a certain stage of research, because an attempt to reduce is needed to find out whether this is possible or not.)

Presently, molecular genetics cannot do such a reduction, and is therefore not reductionistic in this sense. But somebody may believe that molecular genetics will reduce the higher biological levels in genetics in the future. Do molecular biologists believe that reduction in the strong sense is a goal that *can* be accomplished and a task that molecular biology *should* aim to carry out? The answer depends on the beliefs of particular scientists. If one wants to analyse this question with regards to a global reference to molecular biology, one faces the problem of selecting certain individuals and justifying why they are taken as representative of this discipline. Thus, there must be adequate grounds for selecting the beliefs of a few people as representative of what the discipline is doing and should do in the future.

But let us assume that we can agree to analyse the beliefs of a certain group of people. In the case of the HGP, one faces the further difficulty of deciding whether to take their words at face value. Very often, at the initial stage of a research programme, particular beliefs are confused with personal hopes or with propagandistic claims. With this point in mind, can we accuse the HGP of being reductionistic because some of its practitioners are making naïve and optimistic claims about what it will accomplish? Such pronouncements are likely to be only window-dressing devices for 'selling' the project to the public and funding agencies. In fact, among those people making these claims are some of the best biologists of our time, and it would be surprising if they did not know the limits of the current knowledge in their specialities.

Critics sometimes pick up on some remarks which are 'reductionistic' only when taken out of context. Tauber and Sarkar (1992), for example, criticize the Human Genome Project, and molecular biology in general, on the basis of some of Gilbert's pronouncements about the project. They point out that the protein structure problem and the need to compare sequences from different organisms make the data of the HGP of very limited use for physiological genetics and evolutionary genetics respectively. Gilbert (1987), however, acknowledges these very same two areas as frontier problems, although he also thinks that the HGP will help to solve them. Gilbert, in fact, draws a pretty realistic analogy for understanding the HGP:

The direct analogy to biology's human genome project is the development of the large synchrotrons in the field of physics. These large accelerators . . . represent research tools, built by specialists, by which experimental physics can be done. The human genome project will create such research tools: the physical map and the sequence. Understanding the sequence, and understanding the genes it contains, is a goal of human biology. The sequence itself is only an aid to that end. (p. 34)

Often when biologists discuss the details, they reveal a good grasp of the limitations of the results they will get, and of the 'instrumental' value of the project.

Yet, I do agree with the criticism that some scientists' claims could give rise to simplistic and deterministic policies regarding medical and genetic issues. Scientists have a moral responsibility to present their research programmes in a realistic way, although they may not always fulfil this responsibility. But this point is different from saying that this project or a whole discipline is 'reductionistic', because reductionism concerns our beliefs about the world and the appropriate way of studying it.

In conclusion, although molecular geneticists aim to give a physico-chemical account of genetic phenomena, this does not make it reductionistic in the sense of trying to give a complete and sufficient account. Whether molecular genetics will ever be reductionistic in this stronger sense remains an open issue. One can analyse the beliefs of individual scientists about the future development of the discipline; however, these do not affect what the research programme of molecular genetics is presently able to do.

In the case of the HGP, it is obvious that the map and sequence of the human genes will not afford any reduction between the biological functions of genes and their physico-chemical description. The HGP should not properly be called reductionistic. The pejorative characterizations of the HGP as reductionistic suggest that the simplistic and misleading claims made about the project by some of its participants will promote a simplistic conception of human biology. It is sometimes argued that the HGP will offer us 'The sequence of the human genome, containing the instructions for building human organisms' (Smith and Hood 1987: 939), 'the recipe to construct human beings' given that 'nucleotides make us what we are' (Zinder 1990), or the keys to understand 'human nature' (Watson 1990). It is possible that these assertions are made only to attract social interest and the attention of agencies to fund the project. They are, nevertheless, unsupported by our current knowledge of biology. This point concerns not the epistemological claims about a discipline, but the ethical position of scientists as experts when addressing the public. While this latter isssue is important, it should not be confused with the epistemological problem of reductionism.

B. Determinism

Some authors have also criticized the HGP's potential for encouraging biological determinism: that is, the thesis that what an organism is and does is the result of its biology. Lappé (1987: 8), for example, points out that 'in the popular mind genes are widely perceived as emissaries of biological destiny', and warns that 'a bank of genetic information will contain what may be perceived as immutable facts about a person. Except for technical errors, a genetic profile is likely to be a permanent fixture of one's biological legacy' (ibid. 7). The Council for Responsible Genetics (1990b) report has also warned against the understanding of organisms as 'readouts' of their genes.

Current biological theories, however, do not support biological determinism, because organisms are the result of complex interactions between their biology and the environment in which they develop. Besides, as I suggested above, the HGP will deal only with the syntax of the human genome. To understand the role of the genes in the organism's final phenotype will require much more research in embryology, genetics, and development and evolutionary biology. At the present stage of development of the HGP, the opinions of molecular geneticists about 'human nature' should not be taken as the last word on the role of biology in human life. Critics of the HGP may fear, nevertheless, the possibility that by studying the human genome, lay people will perceive a strong emphasis on the role of genes in an individual's phenotype. While this may happen, I do not think that this problem should be avoided by stopping the HGP or any other project in biology.

Stich (1978) argued that in order to defend the curtailment of DNA research because it could produce knowledge which might be misused, with disastrous consequences, we would need to support the following moral principle: 'If a line of research can lead to the discovery of knowledge which might be disastrously misused, then that line of research should be curtailed' (p. 203). But, as he rightly pointed out, to accept that principle would entail the abandonment of almost all current scientific research, because all research potentially leads to dangerous knowledge. By analogy, to ban the HGP because the knowledge it may yield might be misunderstood would suggest the banning of many research projects because often their results will be misunderstood by non-specialists. The solution is not to stop research, but to try to educate non-specialists about the meaning of the results of biological research. In addition, scientists should be aware of their responsibility to present their results accurately, and of the dangers of talking 'as scientists' about issues which are beyond the scope of their expertise.

From a scientific viewpoint, the HGP probably will provide a genetic archive, a huge library whose organization we do not know. The chromosomes of the genome are highly complicated, both functionally and structurally. The genes located in each chromosome are regulated by the action of other genes and by the interrelations with other cellular components. Cells, in their turn, make up tissues, and these form organisms. Thus, an organism is not reducible to its genes, not even at the biological level. An individual, moreover, is not reducible to its biology, and an organism's biology is not reducible to nucleotide sequences.

A famous geneticist has written that the HGP will allow us to spell 'human' (Zinder 1990), but the truth is that to know the nucleotide bases of a genome will not even tell us how to spell 'being', much less 'human being'. The genetic code does not contain the clue to decipher what makes us human, because human nature—whatever that might be—is not reducible to biology, much less to biochemistry, much less even to a sequence of nucleotides. As Suzuki and Knudtson (1990: 338) point out, although geneticists succeeded in sequencing the genome of the OX174 virus almost a decade ago, they are still far from understanding viruses. The full complexity of living human beings will likewise not be understood by analysing their genomes. As a matter of fact, such empty promises may, in the long run, only damage the public perception of the HGP.

IV. CONCLUDING REMARKS

Given the complexity of the human genome, the aim of the HGP, to map and sequence the human genome is undoubtedly an ambitious scientific project. This project is also challenging from an organizational point of view, because to carry it out will require the co-operation of several disciplines, laboratories, and countries. Finally, many social and moral concerns arise regarding the development and use of the project. The HGP is clearly a challenge on several fronts.

I have argued here that it is important to keep different issues separated. Many ethical questions have been raised about the HGP. I have pointed only to the problems that are posed by the development of the project, not to those arising from its application. At the present stage of the HGP the main ethical issue is for scientists to make a realistic assessment of the project's expected accomplishments. This is the basis on which to assess the economic and organizational problems, and on which to analyse the problems that may arise in the use of the information collected. Any decision about whether it will be worth spending money and resources, and any analysis of how to weigh the potential benefits of the project against its possible harms, has to rely on an adequate analysis of the scientific value of the project. As we have seen, in that analysis empirical and conceptual issues are involved.

The HGP has also been criticized for its reductionism. The philosophical analysis of notions such as reductionism and determinism can contribute to the clarification of important concerns. I have tried to show here that global characterizations of the HGP as 'reductionistic' are too simplistic. Detailed discussion of the several meanings of the term are needed to assess the value of reduction, instead of taking this feature of a programme as a pejorative characterization. I argued that the HGP cannot be characterized as reductionistic in any of the ways I have analysed here. The unrealistic promises made by some scientists promote a simplistic view of human biology. This, however, should be separated from the epistemological and methodological issues concerning the value of reductionism in molecular genetics.[1]

REFERENCES

Ayala, F. J. (1974), Introduction to F. J. Ayala and T. Dobzhansky (eds.), *Studies in the Philosophy of Biology* (London: Macmillan), pp. vii–xiv.
——(1987), 'Two Frontiers of Human Biology: What the Sequence Won't Tell Us', *Issues in Science and Technology*, 3: 51–6.
Baltimore, D. (1987), 'Genome Sequencing: A Small-Science Approach', *Issues in Science and Technology*, 3: 48–50.
Beatty, J. (1980), 'The Insights and Oversights of Molecular Genetics: The Place of the Evolutionary Perspective', *PSA 1980* (East Lansing, Mich.: Philosophy of Science Association), i. 341–55.
Cantor, C. R. (1990), 'Orchestrating the Human Genome Project', *Science*, 248: 49–51.
Consell Valencià de Cultura (1990), *El Proyecto del Genoma Humano* (Valencia: Monografies del Consell Valencià de Cultura).
Council for Responsible Genetics (1990*a*), 'Position Paper on Genetic Discrimination' (Boston: Committee for Responsible Genetics).
——(1990*b*), 'Position Paper on Human Genome Initiative', (Boston: Committee for Responsible Genetics).
Crow, J. F. (1991), 'Alternative Theories of Molecular Evolution' (MS).
Culliton, B. J. (1990), 'Mapping *Terra Incognita* (*Humani Corporis*)' *Science*, 250: 210–12.
Dulbecco, R. (1986), 'A Turning Point in Cancer Research: Sequencing the Human Genome', *Science*, 231: 1055–6.
——and Chiaberge, R. (1988), *Ingeneri della Vita. Medicine e morale nell'era del DNA* (Sperling & Kupfer Editori S.p.A). I have used the Spanish translation: *Ingenieros de la Vida. Medicina y ética en la era del ADN* (Madrid: Piramide, 1989).
Fodor, J. A. (1974), 'Special Sciences (Or: The Disunity of Science as a Working Hypothesis)', *Synthesis*, 28: 97–115.
Gilbert, Walter (1987), 'Genome Sequencing: Creating a New Biology for the Twenty-First Century', *Issues in Science and Technology*, 3: 26–35.

[1] I wish to thank J. Cornick, J. Crow, S. Grisolía, J. Ilerbaig, R. Lewontin, D. Nelkin, M. Ruse, E. Sober, M. Solovey, and an anonymous referee for insightful discussions and comments on an earlier draft of this essay. Mark Solovey also helped me to put this into readable English. All remaining flaws are, of course, mine.

Goosens, W. K. (1978), 'Reduction by Molecular Genetics', *Philosophy of Science*, 45: 73–95.

Grene, M. (1974), 'Reducibility: Another Side Issue?', in *The Understanding of Nature: Essays in the Philosophy of Biology* (Dordrecht: D. Reidel), 53–73.

Grisolía, S. (1989), 'Mapping the Human Genome', *Hastings Center Report*, 19: 18–19.

—— (1990), 'Presente y Futuro del Genoma Humano', *Política Científica*, 24: 52–8.

—— (1991), 'UNESCO Program for the Human Genome Project', *Genomics*, 9: 404–5.

Hood, L., and Smith, L. (1987), 'Genome Sequencing: How to Proceed', *Issues in Science and Technology*, 3: 36–46.

Hull, D. (1973), *Philosophy of Biological Science* (Englewood Cliffs, NJ: Prentice-Hall).

Jacob, F. (1970), *La Logique du vivant: une histoire de l'héridité* (Paris: Gallimard). I have used the Spanish translation: *La Lógica de lo Viviente* (Barcelona: Salvat, 1986).

Jordan, E. (1990), 'Impact of the HGP on Biological Science: Sharing of and Access to Data', paper presented at the II Workshop on International Co-operation for the Human Genome Project: Ethics, 12–14 Nov. 1990, Valencia, Spain.

Keller, E. F. (1990), 'Physics and the Emergence of Molecular Biology: A History of Cognitive and Political Synergy', *Journal of the History of Biology*, 23: 389–409.

Kitcher, P. (1984), '1953 and All That: A Tale of Two Sciences', *Philosophical Review*, 93: 335–73.

Lappé, M. (1987), 'The Limits of Genetic Inquiry', *Hastings Center Report*, 17: 5–10.

Lewin, R. (1987), 'National Academy Looks at Human Genome Project, Sees Progress', *Science*, 235: 747–8.

Lewontin, R. C., Rose, S., and Kamin, L. J. (1984), *Not in Our Genes: Biology, Ideology, and Human Nature* (New York: Pantheon).

McKusick, V. A. (1989), 'Mapping and Sequencing the Human Genome', *New England Journal of Medicine*, 320: 910–15.

Marshall, E. (1990), 'Data Sharing: A Declining Ethic?', *Science*, 248: 952–7.

Medawar, P. (1974), 'A Geometric Model of Reduction and Emergence', in F. J. Ayala and T. Dobzhansky (eds.), *Studies in the Philosophy of Biology* (London: Macmillan), 57–63.

Miller, R. (1987), *Fact and Method: Explanation, Confirmation and Reality in the Natural and the Social Sciences* (Princeton: Princeton University Press).

Mohr, H. (1989), 'Is the Program of Molecular Biology Reductionistic?', in P. Hoyningen-Huene and F. M. Wuketits (eds.), *Reductionism and Systems Theory in the Life Sciences* (Dordrecht: Kluwer), 137–59.

Moya, A. (1991), 'Dos Aproximaciones al Proyecto Genoma', *Arbor*, 544: 175–80.

Nelkin, D., and Tancredi, L. (1989), *Dangerous Diagnostics: The Social Power of Biological Information* (New York: Basic Books).

Olby, R. (1990), 'The Molecular Revolution in Biology', in R. C. Olby *et al.* (eds.), *Companion to the History of Modern Science* (New York: Routledge), 503–20.

Olson, M., Hood, L., Cantor, C., and Bolstein, D. (1989), 'A Common Language for Physical Mapping of the Human Genome', *Science*, 245: 1434.

Oppenheim, P., and Putnam, H. (1958), 'Unity of Science as a Working Hypothesis', in H. Feigl, M. Scriven, and G. Maxwell (eds.), *Minnesota Studies in the Philosophy of Science* (Minneapolis: University of Minnesota Press), ii. 3–36.

Palca, J. (1987), 'Human Genome Sequencing Plan Wins Unanimous Approval in U.S.', *Nature*, 326: 429.
——(1989), 'Conflict over Conflict of Interest', *Science*, 245: 1440.
Popper, K. R. (1974), 'Scientific Reduction and the Essential Incompleteness of All Science', in F. J. Ayala and T. Dobzhansky (eds.), *Studies in the Philosophy of Biology* (London: Macmillan), 259–84.
Quintanilla, M. A. (1989), 'La Biotecnologia reclama normas deontológicas de validez universal', *Tendencias Científicas y Sociales*, 14: 7.
Roberts, L. (1989), 'New Game Plan for Genome Mapping', *Science*, 245: 1438–40.
——(1990a), 'Genome Project: An Experiment in Sharing', *Science*, 248: 953.
——(1990b), 'The Great Clone Giveaway', *Science*, 248: 956.
Ruse, M. (1976), 'Reduction in Genetics', in R. S. Cohen *et al.* (eds.), *PSA 1974* (Dordrecht: Reidel), 653–70.
——(1989), 'Sociobiology and Reductionism', in P. Hoyningen-Huene and F. M. Wuketits (eds.), *Reductionism and Systems Theory in the Life Sciences* (Dordrecht: Kluwer), 45–83.
Sarkar, S., and Tauber, A. I. (1991), 'Fallacious Claims for the Human Genome Project', *Nature*, 353: 691.
Schaffner, K. F. (1967), 'Approaches to Reduction', *Philosophy of Science*, 34: 137–47.
——(1977), 'Reduction, Reductionism, Values, and Progress in the Biomedical Sciences', in R. G. Colodny (ed.), *Logic, Laws, & Life. Some Philosophical Complications* (Pittsburgh: Pittsburgh University Press), 143–71.
Schlessinger, D. (1990), 'Current State and Prospects of Physical Mapping of Complex Genomes', presented at the II Workshop on International Co-operation for the Human Genome Project: Ethics, 12–14 Nov. 1990, Valencia, Spain.
Smith, L., and Hood, L. (1987), 'Mapping and Sequencing the Human Genome: How To Proceed', *Bio/Technology*, 5: 933–9.
Sober, E. (1984), *The Nature of Selection* (Cambridge, Mass.: MIT Press).
——(1988), *Reconstructing the Past, Parsimony, Evolution, and Inference* (Cambridge, Mass.: MIT Press).
Stephens, J. C., Cavanaugh, M. L., Gradie, M. I., Mador, M. L., and Kidd, K. K. (1990), 'Mapping the Human Genome: Current Status', *Science*, 250: 237–44.
Stich, S. P. (1978), 'The Recombinant DNA Debate', *Philosophy and Public Affairs*, 7: 187–205.
Suzuki, D., and Knudtson, P. (1989), *Genetics. The Clash Between the New Genetics and Human Values* (Cambridge, Mass.: Harvard University Press).
Tauber, A., and Sarkar, S. (1992), 'The Human Genome Project: Has Blind Reductionism Gone Too Far?', *Perspectives in Biology and Medicine*, 35: 220–35.
Teitelman, R. (1989), *Gene Dreams. Wall Street, Academia, and the Rise of Biotechnology* (New York: Basic Books).
Thorpe, W. H. (1974), 'Reductionism in Biology', in F. J. Ayala and T. Dobzhansky (eds.), *Studies in the Philosophy of Biology* (London: Macmillan), 109–38.
US Department of Health and Human Services & US Department of Energy (1990), *Understanding Our Genetic Inheritance. The U.S. Human Genome Project: The First Five Years. FY 1991–1995* (Washington: USDHHS and USDOE).
Vicedo, M. (1991), Review of D. Nelkin and L. Tancredi, *Dangerous Diagnostics: The Social Power of Biological Information*, S. H. Snyder, *Brainstorming: The*

Science and Politics of Opiate Research, and R. Teitelman, *Gene Dreams: Wall Street, Academia, and the Rise of Biotechnology*, *Isis*, 82: 161–73.

——(1991) 'The Human Genome Project: Predictive Medicine and Preventive Ethics', *Arbor*, 544: 181–207.

Vickers, T. (1990), 'A View from Britain', presented at the II Workshop on International Co-operation for the Human Genome Project: Ethics, 12–14 Nov. 1990, Valencia, Spain.

Watson, J. D. (1990), 'The Human Genome Project, Past, Present, and Future', *Science*, 248: 44–9.

Wimsatt, W. C. (1976), 'Reductive Explanation: A Functional Account', *Boston Studies in the Philosophy of Science*, 32: 671–710.

Workshop on International Co-operation for the Human Genome Project (1988), 'Valencia Declaration on the Human Genome Project', in Consell Valencià de Cultura 1990: 147.

Workshop on International Co-operation for the Human Genome Project: Ethics (1990), 'Valencia Declaration on Ethics and the Human Genome Project', in Consell Valencià de Cultura 1990.

Zinder, N. D. (1990), 'The Genome Initiative: How To Spell Human', *Scientific American*, 263: 96.

26

WHO'S AFRAID OF THE HUMAN GENOME PROJECT?

PHILIP KITCHER

I. INTRODUCTION

The Human Genome Project (henceforth HGP) arouses strong feelings. Enthusiasts view it as the culmination of the extraordinary progress of molecular biology during the past half-century and as the beginnings of a revolutionized biology and medicine (Gilbert 1992, Caskey 1992). Detractors charge that it is wasteful, misguided, and pregnant with possibilities for social harm (Lewontin 1992, Hubbard and Wald 1993). In a short essay it is plainly impossible to explore all the sources of disagreement. My aim will be to consider the debate about the purely scientific merit of the HGP, to sketch the nature of the short-term issues which the project raises, and, finally, to identify what I take to be the deepest and most difficult questions.

II. THE SCIENTIFIC PAY-OFF

The official goals of the HGP are to map and sequence the human genome and the genomes of other organisms (so-called model organisms). Much of the criticism of the project focuses on the alleged uselessness of the full human sequence (Tauber and Sarkar 1992, Lewontin 1992, Rosenberg 1995). Little controversy surrounds the construction of detailed genetic maps in human beings and other organisms—and for good reason. Weekly, sometimes daily, discoveries of genes implicated in various human diseases and disorders testify to the power that new maps have brought. Outsiders frequently fail to recognize the success already exhibited by the strategy of positional cloning. Even though biomedical researchers may

First published in *PSA 1994* (East Lansing, Mich.: Philosophy of Science Association, 1995), ii. 313–21. Reprinted by permission.

initially be entirely ignorant about the physiological processes that go awry in a particular disease, knowing how that disease is transmitted in a sufficiently large sample of families, they can sometimes isolate the locus that is responsible. The strategy is to find genetic markers (paradigmatically RFLPs) associated with the transmission of the disease, confine the locus to a particular chromosomal region (typically of the order of a megabase or two), pick out candidate genes, and, ultimately, clone and sequence the desired gene. Knowledge of the gene may then yield enough understanding of the protein to provide insight into the causal basis of the disease (as, for example, discovery of the gene whose mutations are implicated in cystic fibrosis suggested that the problem might have to do with transportation across cell membranes). Genetic maps, which assign markers to relative positions on chromosomes, are crucial to the initial stages of positional cloning; physical maps, which provide a collection of clones whose relative positions are known, are vital to the enterprise of hunting out the genes. Even with these aids, finding genes implicated in single-locus traits is not easy: it takes perseverance, imagination, and a bit of luck to find the candidate genes and to settle on the right one. But without the maps, the entire enterprise would be impossible. (Perhaps improved maps will help positional cloning move to the next stage, the analysis of polygenic traits; for a lucid analysis of the problems currently bedevilling this area, see Lander and Schork 1994.)

Currently, much of the work carried out under the auspices of the HGP is directed at producing better maps. The other major enterprises of the moment are those of trying to find ways of improving sequencing technology, and sequencing parts of the genomes of some model organisms (notably the yeast *Saccharomyces cerevisiae* and the nematode worm *Caenorhabditis elegans*). Contrary to what some people appear to believe, nobody is currently involved in trying to produce large chunks of contiguous human sequence. It is even quite possible that nobody will ever try to sequence an entire human chromosome. To understand the rationale for current practice, we need only consider the present difficulties of large-scale sequencing and the ways in which sequence information about human beings might be put to useful ends.

At present, standard gel electrophoretic methods can generate about 500 bases of sequence per lane, and large-scale genome centres with the appropriate machines to read the sequence run gels with 36 lanes. Typically, they manage one gel per day. So, you might think that it would be possible to produce, per centre per day, 18,000 bases' worth of sequence. However, to obtain 'finished sequence' (i.e. sequence that can be held, with reasonable confidence, to be error-free), it is crucial to sequence the same DNA more than once—those involved often talk of 'sevenfold redundancy'. So the very best that a genome centre is currently likely to be able to yield is between 2,000 and 3,000 bases of finished sequence a day. In conversations with directors of centres, I have usually been given the estimate of a megabase of finished sequence per year (projected for 1994–5, not yet achieved when the conversations took place in 1993–4); technicians who prepared the DNA and tended the machines were more conservative, foreseeing 600 or 700 kilobases of finished sequence per year. Even assuming the more optimistic rate, and supposing, charitably, that there will be thirty centres world-wide committed to the task, it would take about a century to produce the full 3 billion bases of the human genome sequence.

The genomes of some model organisms are much smaller. *S. cerevisiae* has a genome of around 15 megabases, *C. elegans* of around 100 megabases, *Drosophila* about 150 megabases. (The mouse, like other mammals, has about 3 billion bases). Even without significant improvements, we can expect the full sequence of the yeast genome to be available by the end of the decade; with predictable developments, the nematode genome should be sequenced by 2005, and the fruit-fly genome may also be manageable. (Currently, the longest stretch of contiguous sequencing is 2.2 megabases from the nematode (Wilson *et al.* 1994); two yeast chromosomes have already been sequenced (Dujon *et al.* 1994, Johnston *et al.* 1994). But to attack anything as massive as the human genome, significant refinements of current sequencing technology are required.

Some people believe that the improvements will come, that the current bottle-necks in automated sequencing (the preparation of the DNA and the analysis of the data) will be freed up. Indeed, genome researchers hope that the stimulus of the HGP will make large-scale sequencing a simple routine, so that a full human sequence could be achieved in one or two years at the very end of the official project, and so that the biologists of the next century would be able to commission sequence as easily as they now order glassware or strains of flies. However, even if the full human sequence never becomes manageable, it would be wrong to declare the HGP a failure. To appreciate this point, we need to turn to the issue of what exactly sequence data would do for us.

Complete sequencing is hailed as the most direct method of finding all the genes, where 'gene' is understood as picking out those parts of DNA that are transcribed. Genes are sparsely scattered in the genomes of complicated organisms (they occur at greater density in yeast, and in the soil amoeba *Dictyostelium discoideum*). Perhaps 5 per cent of human DNA consists of coding regions. The most appropriate strategy for discovering the coding 'islands' in the 'ocean' of non-coding sequence, is to make use of what we already know about the sequences of genes (some human, the vast majority non-human). As the projects of sequencing yeast and the nematode have already shown, hitherto unsuspected genes can be discovered by searching for open reading frames (stretches of DNA that proceed for a long while without a stop codon), and investigating to see whether mRNAs corresponding to parts of these reading frames can be found. (In eukaryotes, mRNAs will only contain the sequences corresponding to exons.) As a database of gene sequences is found, one can then run computer programs to search for homologous sequences. *Human genes are likely to be extracted from human sequence, by looking for parts of our DNA that are similar to the sequences of known genes.*

This strategy of gene hunting is biologically faithful. We share 40 per cent of our genes with yeast (in our evolutionary history it wasn't necessary to reinvent the ways of carrying out the kinds of processes common to all cells), and it is probable that a very high percentage of our coding regions are very like coding regions in flies: the director of the genome centre devoted to mapping and sequencing *Drosophila*, Gerald Rubin, has estimated, in conversation, that 99 per cent of our coding regions are akin to some coding region in the *Drosophila* genome. In dealing with other organisms, we can carry out experimental manipulations that will reveal whether mutation at a putative locus exerts any phenotypic effect (although so-called knock-out experiments may require refined abilities to pick out small phenotypic changes). Rounding up lots of genes in other species, we will have a basis for searching for homologues in stretches of human DNA sequence. So, *precisely because the HGP is not restricted to our own species*, we have a significant chance of parlaying human sequence data into the identification of human genes.

Can we go any further? Of course, given the genetic code, once you know the nucleotide sequence of a gene (and of its division into exons and introns), you can derive the amino acid sequence of the protein it encodes. However, as critics of the HGP point out, there is currently no known way of moving from a linear sequence of amino acids to the three-dimensional structure of the protein: the protein-folding problem is presently unsolved. Plainly, a general solution (or even a bundle of partial solutions) to the protein-folding problem would be welcome. Lacking such resources, those who make use of future sequence data will have to emulate the strategy for finding human genes in their investigations of human proteins. Again, we have a database of proteins whose three-dimensional configurations are known—indeed, we often have knowledge not just of protein structure but also about the functional domains of the proteins and of the roles they play in intra-cellular interactions. Given a new sequence of amino acids, the appropriate technique is to seek out similar sequences among proteins whose structures (and possibly functions) are already known. Just this kind of research has already brought significant advances in our knowledge about the effects of genes implicated in disease: after cloning and sequencing a new gene, running the computer search exposes the protein it encodes as a cell-surface membrane protein or a DNA repair enzyme. To the extent that organisms play variations on a few simple themes, we can expect the sequence data to yield conclusions about human genes, about the structures and functions of human proteins, and *sometimes* about the kinds of interactions that underlie a particular physiological process.

So human sequence data would not necessarily be the jumble of gibberish that detractors of the HGP often predict. Critics are quite right to point out that the route from a sequence of As, Cs, Gs, and Ts to biological understanding is by no means direct (Tauber and Sarkar 1992, Lewontin 1992, Rosenberg 1995), but err in supposing that there is no route at all. There are already promising signs that all sorts of genes common to a wide range of organisms can be identified, and that the roles of corresponding proteins can be pinned down—through detailed investigation of model organisms. The real excitement of the HGP, at the moment, and possibly throughout its whole history, consists in the development of maps, with their consequences for positional cloning, and in the sequencing of the small genomes of some model organisms. Quite probably, the political allies of the HGP would be appalled at the thought that the major triumph of the project may be the detailed understanding of the genomes of *S. cerevisiae*, *C. elegans*, and *D. melanogaster*; but the biological importance of these sequencing projects should not be underestimated. Knowledge of the yeast sequence will answer many basic questions about gene regulation in eukaryotes, as well as providing insight into the intra-cellular processes common to all organisms. Knowledge of the nematode sequence, coupled to the fate map for the nematode already constructed, will teach us just how different genes are switched on in the development of an organism with a 'closed' developmental programme. Knowledge of the fruit-fly sequence, combined with the enormous understanding of early development in *Drosophila* achieved by Christiane Nüsslein-Vollhard and her coworkers (see Lawrence 1992 for an overview), will instruct us in the patterns of gene activation in development in an organism with a more open ontogeny. We can expect to be able to apply some of the new understanding of physiology and development to our own species.

I suggested earlier that the full human sequence might be difficult to obtain. Is the struggle to achieve it really necessary? Well, as noted earlier, improved sequencing technology would be a great boon to the biology of the twenty-first century, and, if we are out to get all the human genes, or as many as possible, then the most direct approach seems to be to sequence the entire human genome. However, if all else fails, there is an indirect strategy. Molecular biologists know tricks for extracting mRNAs from cells; rounding up human mRNAs, one can then use the enzyme reverse transcriptase to generate the corresponding cDNAs, using these to probe the genome to find coding regions. The *éminence noire* of the HGP, Craig Venter, is currently pursuing this approach, tagging pieces of human coding sequence, and inspiring much official ire with his efforts to patent them. Very probably, Venter will find a large number of human genes; it is very improbable that he will discover all of them, or even as many as could be garnered by grinding out the full sequence and running the appropriate computer programs. (If you start by extracting mRNAs from cells, you are unlikely to pick out all the mRNAs unless you have access to all cell types at all stages of development.) However, Venter's approach is a valuable fall-back in case sequencing technology remains stuck near its present rates. A concerted programme of hunting mRNAs, combined with the crucial information about the sequences of model organisms, could deliver the preponderance of the insights we might expect from the more problematic route of full human sequencing.

The moral of my story should be obvious. What is vital to the HGP is the enterprise of mapping and the sequencing of the genomes of the manageable model organisms. Truth in advertising would suggest that the HGP be renamed: calling it the 'Genomes Project' would be more accurate. However, since the ghost of Senator Proxmire probably still haunts the halls in which Congressional funds are allotted, it is unlikely that representatives will hear testimonials to the importance of the nematode genome. We are probably stuck with a misleading name for a valuable enterprise.

III. SHORT-TERM TROUBLES

The great public appeal of the HGP stems from its expected medical payoffs. However, even in the cases of the diseases that have been most successfully explored with the new tools of molecular biology, much more is promised than is currently delivered. As the example of Huntington's disease makes apparent (and as the older example of sickle-cell anaemia shows even more vividly), significant knowledge about the molecular basis of a disease is compatible with profound ignorance about how to cope with it. In the short term, the medical applications of positional cloning and other work informed by the HGP will largely consist in the provision of tests. Genetic tests are becoming available on a broad scale. Some can be used to diagnose or to disambiguate the various forms of a disease. Others give us the power to make predictions, sometimes determinate, typically probabilistic, about the future phenotype of a person, about the characteristics of the person who would develop from a foetus, about the expected distribution of traits among offspring. The advantages of improved diagnostic tests tend to be underrated, but there are many diseases with ambiguous symptoms that can be more effectively treated if known therapies are introduced early: the extension of the median life span for cystic fibrosis patients stems, in part, from better diagnostic techniques. Similarly, patients can sometimes be spared useless or harmful treatments, or steered to the appropriate treatments, if the particular basis of their case of the disease is understood: dietary therapies are worthless for those people whose high cholesterol levels can be traced to a malfunctioning receptor protein in the liver.

The predictive tests have generated most discussion, both because enthusiasts have seen them as a great step forward in preventative medicine and because critics have recognized the possibilities for trouble. Francis Collins, current director of the HGP at the NIH, sometimes advertises the medical benefits by describing a future in which infants are routinely tested at birth and given a 'genetic report card'. The idea, of course, is that by knowing the kinds of diseases and disabilities that are likely to strike, we can take evasive action. That is a wonderful idea—to the extent that there is anything to be done to lower the chances, or to the extent that people can make better lives for themselves by knowing what will befall them.

The trouble is that, in the majority of instances, there is no *medical* benefit that comes from foreknowledge. If you find out that you have a cancer-predisposing allele, then your doctor may advise you to avoid fatty foods and pollutants, and to take exercise. That is good advice for anyone, and the only function of the genetic test is, perhaps, to stiffen your resolve. Recent celebrations of the identification of an allele implicated in early-onset breast and ovarian cancer (*BRCA1*) have tended to suggest that this will bring important benefits to all women, not just members of families 'at risk'. This, however, is a mistake. Mammograms are able to detect small tumours in the breasts of older women, only rarely in women in their twenties and thirties. If a young woman has the type of breast tissue that allows for mammogram reading, then, given the high incidence of breast cancer, she would be well-advised to have regular mammograms whether or not she carries a mutant *BRCA1* allele. The test may be worthwhile for women in afflicted families, but, given the limitations of current strategies of detection, it is unlikely to benefit the rest of the female population. By contrast, where there are effective monitoring techniques, as with colon cancer, recognition of increased risk will be valuable. Presently, such cases are a small minority.

Many who consider the likely spate of genetic tests draw inspiration from the practice of PKU testing. Here, at first sight, we seem to have an almost perfect example. The test enables doctors to divide the population into two subgroups, one of which would be harmed unless given a low-phenylalanine, high-tyrosine diet, the other of which would be equally damaged by the special diet. However, as Diane Paul has shown (Paul 1995), the history of PKU testing reveals significant problems. The unpleasantness of the diet, its cost, the initial lack of knowledge about how long it should be continued, the absence of any systematic programme of following those originally diagnosed—all these factors make the medical solution far less effective than it might have been. Yet I don't think that the example confronts philosophers with any profound ethical problems. The morals of the history ought to be clear, and we should work to ensure that problems with medical solutions in principle do not fail us in practice because of a lack of social support. The real difficulties here are ones of practical politics rather than moral philosophy: it's abundantly obvious that people diagnosed with PKU need to be provided with far more than the initial evaluation; families may require economic assistance to buy the special food, continued counselling, and so forth; in other words, we can see what should be done, even though making sure that it is done is quite another matter.

In general, for reasons I detail elsewhere (Kitcher 1996), full-scale genetic testing is likely to be expensive, if it is done properly. The information already available about the effectiveness of genetic counselling is sobering, and recent studies have focused on the particular problems encountered by minority groups (Rapp 1994). Facing an immediate future in which the power to test will greatly outstrip the power to treat, it is important to resist the commercial pressures to introduce new tests without any attention to the benefits they might bring or the impact they might have on the lives of those tested. Without some independent body that will advise doctors on the exact nature of the information that a new test can yield, and of the extent to which test information can improve patients' lives, there is real danger that people will be swept into tests that they do not understand and given information that will prove harmful to them. Without serious attention to the problem of improving existing forms of genetic counselling, benefits that might be realized will not be achieved.

(These points are made forcefully in Nelkin and Tancredi 1989 and Holtzman 1989.)

The practice of widespread genetic testing can easily prove harmful in other ways. As genetic information about individuals is collected, various groups within society will clamour for that information. Insurers will want to know the genetic risks of those who apply for insurance; employers may demand the 'genetic report card' as a condition of hiring. It is sometimes suggested that these problems should be solved by insisting on the privacy of genetic information, but this seems to me the wrong way to look at the situation. We feel it important that certain types of information about ourselves should be kept private, solely for the sake of privacy: we share with our intimates certain feelings or experiences that should not be broadcast: ' 'twere profanation of our joyes / to tell the laietie our love,' as Donne puts it. But genetic information isn't like this. There is nothing intrinsically valuable in keeping secret the fact that I carry a particular sequence of As, Cs, Gs, and Ts at a certain chromosomal region. If I need to keep it private, it is because others might use the information to take advantage of me. The appropriate solution is thus not to engage in some (probably hopeless) attempt to dam up the flow of information, but to limit the kinds of things that can be done with genetic information.

In the large, it is fairly clear how to set the limits. There are straightforward arguments (Daniels 1994, Kitcher 1996) for denying health and disability insurance companies the right to use genetic information either in deciding eligibility or in setting premiums. General health risks are only occasionally relevant to a person's qualifications for a job, and the dangers that arise in specific work-place situations are best tackled by cleaning up the working environment (where that is feasible) or creating job opportunities for those with unlucky genotypes (see Draper 1991, Nelkin and Tancredi 1989, Kitcher 1996). Again, the issue is not one of balancing interests in a morally tricky situation, but rather of finding ways to implement the deliverances of relatively elementary lines of ethical argument.

Many of those most enthusiastic about the HGP recognize the need to address these short-term problems. They would rightly remind us that it would be myopic to be so overwhelmed by the gravity of the problems that we simply retreat from applying molecular genetics altogether. Not only do biomedical advances enable people to avoid bringing into the world children who would inevitably suffer, but they will also point to new forms of therapy for major diseases. Critics (e.g. Lewontin 1992) retort that the promise is overblown. However, even though the future of molecular medicine is unpredictable, there are encouraging signs in newly developing treatments for cystic fibrosis, and in the first steps of gene replacement therapy (see Kitcher 1996 for an assessment of a rapidly changing situation).

From the very beginning of the HGP, funds were set aside to examine the ineptly named 'ELSI issues' ('ELSI' is an acronym for 'ethical, social, and legal implications'). Many scientists involved in the project are inclined to say, in unguarded moments, that ELSI is an expensive piece of public relations, and many of them believe that their own perspective on the issues is as good as that of the publicly funded 'experts'. That attitude is buttressed by the apparent lack of progress in coping with the short-term problems faced by the HGP: many genome scientists have attended conference after conference at which the same speakers introduce the same problems as topics that must be discussed during the next years. Ironically, the trouble isn't that the problems are hard to solve, but that the obvious general lines of solution—which scientists can see as clearly as professional medical ethicists, lawyers, or sociologists—call for policies that are unlikely to be popular with an electorate whose principal concern seems to be to cut the role of government. Add to this the fact that there is no ELSI body with any power to recommend legislative action, and the frustrating character of the current situation is patent. Scholars funded by ELSI can explore difficult questions about rare situations in which obtaining genetic information involves a conflict of rights (monozygotic twins at risk for Huntington's, exactly one of whom wants to be tested), or they can carry out ever more sophisticated studies of the current difficulties with genetic counselling. None of this will go anywhere until there is not only a vehicle for political action, but also a means of convincing the electorate that, without some rather obvious forms of governmental regulation, molecular medicine is likely to wreck a significant number of their children's lives.

IV. THE DEEPER ISSUES

There are indeed deeper philosophical questions posed by the HGP, and exposing these questions enables us to understand some of the fiercer opposition to the project. One very obvious consequence of the HGP is that it will greatly enhance the power of prospective parents to know in advance the probabilities that the child who would result from a particular conception would have specific characteristics. On a quite unprecedented scale, our descendants may engage in the enterprise of choosing people.

Sometimes the scenarios are fantasies that outstrip any foreseeable reality. Exercises in imitating Frankenstein are surely remote: the difficulties of gene replacement therapy, wearyingly familiar to anyone who has read the literature—problems of sending the right DNA to the right cells, problems in regulating its expression—mean that our actual power to modify genes, now and in the next decades, is roughly equivalent in its precision to the crudest forms of electroconvulsive therapy. Gene replacement has so far been undertaken in desperate circumstances, when niceties of regulation are not an issue. Visions of the future in which prospective parents manipulate the genes of embryos produced by IVF can be discounted.

However, even if they cannot select *for*, our descendants will be able to select *against*. If the foetus has some combination of alleles that causes them concern, then they can choose to terminate the pregnancy. Waiving general worries about the use of abortion, there are surely some instances in which that choice appears warranted: foetuses can already be tested for genetic conditions that inevitably lead to neural degeneration and early death (Tay–Sachs disease) or that invariably cause severe retardation and acute behavioural problems (Lesch–Nyhan syndrome). The trouble is that we can readily conceive a future in which the practice of selection shades into something far less benign. Unlike some women in northern India, it is unlikely that many American couples will use foetal testing to select against (female) sex. Whether they will abstain from using information about propensities for same-sex preference or obesity is another matter.

So a significant worry is that we are started on a course that will lead our successors to some dark venture in eugenics. Unlike some past chapters in the history of eugenics, the future is not likely to see some central body dictate the reproductive decisions of individuals (or of couples). Yet, even if we honour a principle of reproductive freedom, it is important that the practice of foetal selection be conducted in an informed and responsible fashion, that our descendants make their decisions in a contest that allows them to attend to the morally relevant factors, not to be swept along (as, perhaps, women in northern India are swept) by pervasive social prejudices.

Champions of the HGP are optimistic that the appropriate moral climate can be achieved. When their optimism is articulated, it rests, I believe, on a sense that responsible reproductive decisions are made by reflecting on the qualities of the lives that would be brought into being, and on the impact those lives would have on the qualities of other lives. There is no moral calculus that can be mechanically applied to yield an answer, but it is important to the moral status of the decision that the considerations of the qualities of lives be paramount. So they envisage a world in which the overwhelming majority of genetic pre-natal tests are used responsibly, and in which the general tendency of the practice is to terminate pregnancies that would have led to unfortunate lives.

Critics are worried about this optimistic vision for a number of reasons. First, they recognize that the history of eugenics reveals how easy it is for a seemingly benign practice to decay into something morally repugnant. Second, they hold that much popular thinking is gripped by simple forms of genetic determinism, so that there will be a recurrent tendency to treat genes as doom. They expect this to be manifested in credulous acceptance of claims about the determination of behavioural traits. Third, they believe that the very social factors that currently make the short-term issues surrounding the HGP so hard to solve, in particular the reluctance to provide funds for social services, will make pre-natal testing an acceptable, cheap, way of eliminating people who could have lived productive lives, people with disabilities that can be overcome with social support. (Claims of these kinds, and supporting arguments, can be found in Lewontin 1992 and Hubbard and Wald 1993.)

Those with direct experience of the tragedies of severe genetic afflictions are likely to view these responses as the resurgence of Luddism. Are we to forswear an instrument that could do much good because we fear that it might be abused? Yet the critics' point cuts deeper. If the underlying rationale for a practice of pre-natal testing lies in the possibility of improving the quality of human lives, then we must recognize that the prevention of lives that would be damaged by unlucky alleles takes only a small step. There are far more children whose lives are smashed in their early years through social causes than who suffer from the ravages of defective proteins. So the critics confront the HGP with a serious question: why is public money expended on trying to prevent the birth of children with rare combinations of alleles, while social programmes that could benefit many more go wanting?

This question sets the stage for a debate that has only been joined obliquely in the rhetorical furore around the HGP. The root criticism of the project should not be that it is scientifically worthless, or that the short-term problems are insoluble. Rather, the worry is that the ultimate rationale for the project is cast in terms of improvements to the quality of human lives, and it is inconsistent for a society that tolerates large social causes of reduced quality of life to offer any such rationale. The best case against the HGP would demand that it be accompanied by an equally serious commitment to attacking social problems that lower the quality of human lives.

I do not have space here to do more than indicate the main lines of argument in an important, and difficult, controversy. There is a pragmatic response to the critics' argument, that emphasizes the need to do the local good we can (develop the biomedical promise of the HGP) and not hold it hostage to grander social ventures with uncertain prospects (and here there is likely to be a recitation of the failures of past expensive programmes). Pragmatists mix optimism with pessimism. For, on the one hand they are hopeful that future generations can pursue a practice of choosing people that is free and responsible, that social prejudices can be countered, and the darker byways of eugenics avoided. At the same time, they doubt that attempts to remove the social causes of misery will succeed. The critics' position is a contrary mixture of fear and confidence. Concerned that the allegedly 'free and responsible' reproductive decisions will not be sustained, they remain optimistic about the potential of social programmes.

Plainly, these issues involve far more than the particularities of the HGP. They are deep and difficult, turning on fundamental questions in political and moral philosophy, as well as on details from both biology and social science. The HGP serves just as the occasion for re-examining them. So, if I am right, the philosophically important questions about the HGP have gone largely unnoticed in the numerous discussions about the project—which is not, of course to deny that some of the other questions raised here and in the contemporary literature are also significant. My aim has simply been to do what the HGP has so far done best: to wit, produce a clear map (for much more detailed exploration of the terrain, see Kitcher 1996b).

REFERENCES

Caskey, C. T. (1992), 'DNA-Based Medicine: Prevention and Therapy', in D. Kevles and L. Hood (eds.), *The Code of Codes* (Cambridge, Mass.: Harvard University Press), 112–35.
Daniels, N. (1994), 'The Genome Project, Individual Differences, and Just Health Care', in T. Murphy and M. Lappé (eds.), *Justice and the Human Genome Project* (Berkeley: University of California Press), 110–32.
Draper, E. (1991), *Risky Business* (New York: Cambridge University Press).
Dujon, B., *et al.* (1994), 'Complete DNA Sequence of Yeast Chromosome XI', *Nature*, 369: 371–8.
Gilbert, W. (1992), 'A Vision of the Grail', in D. Kevles and L. Hood (eds.), *The Code of Codes* (Cambridge, Mass.: Harvard University Press), 83–97.
Holtzman, N. A. (1989), *Proceed with Caution* (Baltimore: Johns Hopkins University Press).

Hubbard, R., and Wald, E. (1993), *Exploding the Gene Myth* (Boston: Beacon).

Johnston, M., *et al.* (1994), 'Complete Nucleotide Sequence of *Saccharomyces cerevisiae* Chromosome VIII', *Science*, 265: 2077–82.

Kitcher, P. (1996*a*), *Choosing Genes, Changing Lives* (New York: Simon and Schuster).

——(1996*b*), *The Lives to Come* (London: Penguin).

Lander, E. S., and Schork, N. J. (1994), 'Genetic Dissection of Complex Traits', *Science*, 265: 2037–48.

Lawrence, P. (1992), *The Making of a Fly* (Oxford: Blackwell).

Lewontin, R. C. (1992), 'The Dream of the Human Genome', in *Biology as Ideology* (New York: Harper), 59–83.

Nelkin, D., and Tancredi, L. (1989), *Dangerous Diagnostics* (New York: Basic Books).

Paul, D. (1995), 'Toward a Realistic Assessment of PKU Screening', *PSA 1994* (East Lansing, Mich.: Philosophy of Science Association), ii. 322–8.

Rapp, R. (1994), 'Amniocentesis in Socio-cultural Perspective', *Journal of Genetic Counselling* (in press).

Rosenberg, A. (1995), 'Subversive Reflections on the Human Genome Project (East Lansing, Mich.: Philosophy of Science Association), ii. 329–35.

Tauber, A., and Sarkar, S. (1992), 'The Human Genome Project: Has Blind Reductionism Gone Too Far?', *Perspectives in Biology and Medicine*, 35: 220–35.

Wilson, R., *et al.* (1994), '2.2 Mb of Contiguous Nucleotide Sequence from Chromosome III of *C. elegans*', *Nature*, 368: 32–8.

27

IS HUMAN GENETICS DISGUISED EUGENICS?

DIANE B. PAUL

As a historian of modern genetics, I am often asked whether human genetics represents disguised, or incipient, or possibly a new kind of eugenics. Those who pose the questions may not be certain how to define eugenics, but they are almost always convinced that it is a bad thing, one which should be prevented. Indeed, fear of a eugenics revival appears to be a principal anxiety aroused by the Human Genome Project (HGP), in Europe as well as the United States. Acknowledging this concern, project advocates insist that the mistakes of the past will not be repeated. Thus, at the Human Genome I Conference at San Diego in October 1989, James B. Watson, the HGP's first director, told the audience: 'We have to be aware of the really terrible past of eugenics, where incomplete knowledge was used in a very cavalier and rather awful way, both here in the United States and in Germany. We have to reassure people that their own DNA is private and that no one else can get at it' (quoted in Davis 1991: 262).

While almost everyone agrees that eugenics is objectionable, there is no consensus on what it actually *is*.[1] Indeed, one can be opposed to eugenics, and for almost anything. As Sir Isaiah Berlin remarked about the protean uses of 'freedom', its meaning 'is so porous that there is little interpretation that it seems able to resist' (1969: 121). To denounce eugenics is to signal that one is socially concerned, morally sensitive (and if a geneticist, perhaps worthy of public trust). But it does not predict one's stance on any particular reproductive issue.

In 1990, the International Huntington Association and the World Federation of Neurology adopted guidelines, based on the recommendations

---

First published in Robert F. Weir, Susan C. Lawrence, and Evan Fales (eds.), *Genes, Humans, and Self-Knowledge* (Iowa City: University of Iowa Press, 1994), 67–83. 1994 University of Iowa Press. Reprinted by permission of the University of Iowa Press.

[1] More than twenty years ago, L. C. Dunn remarked in his presidential address to the American Society of Human Genetics that eugenics had tended 'to become all things to all people' (1962: 3).

of a joint committee, for the use of predictive genetic tests. The committee considered the refusal to test women who 'do not give complete assurance that they will terminate a pregnancy where there is an increased risk' of Huntington disease to be acceptable policy (p. 37). Was the committee endorsing eugenics? Some would say yes, while most of its members would certainly be appalled by the suggestion.[2] But, as we will see, there is no objective answer—nor can there be—to the question of whether such a policy constitutes eugenics.

At present, arguments about the relationship of eugenics and human genetics rarely converge. One person thinks they have nothing in common, while another considers the former merely an extension of the latter, not necessarily because they disagree on the facts (though of course they sometimes do), but because they employ the term in conflicting ways, often without noticing. After noting that pre-natal diagnosis inevitable involves the systematic selection of foetuses, Abby Lippman charges: 'Though the word "eugenics" is scrupulously avoided in most biomedical reports about prenatal diagnosis, except where it is strongly disclaimed as a motive for intervention, this is disingenuous. Prenatal diagnosis presupposes that certain fetal conditions are intrinsically not bearable' (1991: 24–5). Conversely, most geneticists employ a narrow definition that identifies eugenics with a social *aim* and often coercive *means*. Both broad and narrow definitions serve a political purpose. The former associates genetic medicine with odious practices, and thus arouses our suspicions; the latter dissociates it from these practices, and thus reassures.

Francis Galton, who coined the word 'eugenics', defined it as 'the study of the agencies under social control that may improve or impair the racial qualities of future generations, either physically or mentally' (1883: 44). However, it is less often identified as a science than as a social movement or policy, as in Bertrand Russell's definition: 'the attempt to improve the biological character of a breed by deliberate methods adopted to that end' (1924: 255). After all, its practical applications, especially the sterilization laws adopted in America and Germany, and Nazi racial policies, explain why people worry about the relationship of eugenics to human genetics. Had its advocates confined themselves to study, eugenics would not now be a source of anxiety.

However, eugenics cannot be *defined* in terms of the social policies that account for its sordid reputation. We know from the historiography of the last decade that people in many countries and across a wide political and

---

[2] Thus Philip R. Reilly, one of the seven American members, is the author of *The Surgical Solution* (1991).

social spectrum advocated policies to genetically improve the 'race'. Eugenics has come to be associated with its most infamous practices in the United States and Germany, and hence with racism and political reaction. But there were eugenics movements of quite different character in France, Brazil, and Russia, among other countries.[3] Even in the Anglo-American world and in Wilhelmine and Weimar Germany, eugenics had a more diverse constituency and set of means and aims than one would imagine from the requisite reviews of the subject in discussions of 'social implications of the new genetics'. Some eugenics enthusiasts favoured, and others repudiated, compulsion, while eugenics was invoked variously in support of capitalism and socialism, pacifism and militarism, patriarchy and women's liberation.[4] Its social content has been infinitely plastic.

Notwithstanding their disparate political perspectives, all eugenicists did agree that individual desires should be subordinated to a larger public purpose. Whether of the right, left, or centre, they assumed that reproductive decisions have social consequences, and thus are a matter of valid social concern. This history explains one conventional line demarcating eugenics from something else (often human genetics). Policies are characterized as eugenic if their intent is to further a social or public purpose, such as reducing costs or sparing future generations unnecessary suffering. Expansion of genetic services motivated by concern for the quality of the *population* would be eugenic by this definition, while the same practices motivated by the desire to increase the choices available to individuals would not be.

Unfortunately, this criterion requires a knowledge of motives, which may not be obvious, and are sometimes mixed. Indeed, genetic services are generally justified on one of two very different grounds: that they increase the options available to families and/or that they reduce the burden of genetic disease in the community, thus saving money.[5] Dennis Karjala asserts that 'all cost/benefit reasoning in the reproductive rights area is essentially eugenics' (1992: 160). That is a plausible claim, although it is not easy to determine the real intention(s) behind the expansion of pre-natal testing and other genetic services. If a social purpose is the litmus test of eugenics, we must assess the importance of different aims, which are not always made explicit—and when they are, may disguise the truth. It is surely easier to defend abortion (which is the object of pre-natal testing) in the language of choice than that of cost savings.

[3] For an overview of these lesser-known movements see Adams 1990.
[4] See Kevles 1985 for an overview of Anglo-American eugenics.
[5] For an example of explicit cost–benefit analysis, see Chapple *et al.* 1987. See also Angus Clarke's critique of this approach (1990) and Sarah Bundey's letter in reply (1990).

A definition in terms of *intention* is also at odds with the use of 'eugenic' to describe the *effects* of individual action or social policy. The latter definition is implicit when a practice is characterized as eugenic because 'we are effectively changing the gene frequency by lowering the number of offspring with "defective" genes' (ibid.). If consequences are properly described as eugenic, motive is no longer germane. Individuals do not ordinarily intend to benefit the gene pool by their reproductive choices. But private decisions may, taken collectively, have population effects. If the word 'eugenic' appropriately describes consequences, and not just intentions, it casts a very wide net. It would make perfect sense, given this usage, to call abortion following pre-natal screening 'eugenics'—whatever the motivations for individual decisions or government funding. A definition broad enough to include unintended consequences will necessarily incorporate most medical genetics, or even individual mating decisions (Carlson 1986: 531).

The recent discussion of 'back-door' eugenics implicitly depends on a definition of eugenics in terms of effects. A number of critics have warned of a resurgence of eugenics as the unintended result of individual choices. In their view, the real danger arises not from state policy, but from our increased ability to *select* the kind of children we want. Thus the new eugenics will result from a multitude of voluntary decisions, or even demands for tests and screens, rather than from social policy designed with eugenic aims in view (Duster 1990: p. x).

Most commentators, however, still restrict the term 'eugenics' to policies pursued for a social purpose. They often add an additional criterion: there must be an element of coercion. According to these definitions, the state 'interferes with', or 'controls', or 'imposes' particular reproductive options, as in one definition of eugenics 'as any effort to interfere with individuals' procreative choices in order to attain a societal goal' (Hollzman 1989: 223). Eugenics is often demarcated from genetics by this criterion. Thus a participant at the 1991 International Congress of Human Genetics asserted: 'Eugenics presumes the existence of significant social control over genetic and reproductive freedoms. Genetics does not require any special control over genetics or reproductive freedom' (Ledley 1991).

However, if we apply the label only to programmes involving some form of coercion, we exclude a large number of individuals and policies ordinarily associated with eugenics. Many eugenicists, especially in Britain, stressed the voluntary nature of their proposals.[6] Virtually all 'positive' eugenics (which seeks to increase the incidence of desirable traits, rather

---

[6] On this point, see especially Soloway 1990.

than reducing that of undesirable ones) would be excluded by this definition. Francis Galton would no longer be a eugenicist. Nor would H. J. Muller or William Shockley, notwithstanding their schemes for the (voluntary) insemination of women with the sperm of especially estimable men.

Moreover, there is no value-neutral answer to the question of whether a policy is coercive. Coercion has different meanings in different political traditions. To classical liberals (like John Stuart Mill or Isaiah Berlin) or libertarian conservatives (such as Milton Friedman), a decision is voluntary if there are no formal, legal barriers to choice. Freedom is thus defined negatively, as the absence of restraint. Coercion, on the other hand, 'implies the deliberate interference of other human beings' with actions a person would otherwise take (Berlin 1969: 123). One is coerced only if actively prevented from attaining a goal. To liberals in the tradition of T. H. Green or John Dewey, however, as well as to socialists, coercion is not simply a matter of legal barriers: we are free to choose only when we have the practical ability to agree or refuse to do something. From their standpoint, a *situation*, such as economic need, may also be coercive.[7]

For conservatives, then, the potential parents of a severely disabled child are free to abort the foetus or bring it to term. From a contemporary liberal or socialist standpoint, choice may be lacking, given the medical and other costs of caring for such a child. On this view, parents could be coerced into aborting a foetus by the threatened loss of insurance coverage or lack of social services (Wilfond and Fost 1990: 2781). (This is not to imply that pressure would evaporate with national health insurance. Even in a socialized system, 'if there is no confidence in the willingness of society to care for their child once they are unable to do so, parents may choose to terminate a pregnancy against their own wishes and beliefs' (Clarke 1990: 1146)[8].) Whether parents are 'free' to choose in these situations is a question that will necessarily be answered differently from different political standpoints.

Given competing (and sometimes implicit) definitions of eugenics, which themselves reflect larger (and unacknowledged) differences in political ideology, it is not surprising that arguments about whether an individual action, social policy, or unintended effect is or is not eugenics often fail to engage. The confusion would be reduced if definitions that do obvious violence to history (or contradict common sense) were excluded. A definition of 'eugenicist' that bars Francis Galton is, on the face of it, absurd.

---

[7] Faden and Beauchamp (1986) employ a strict definition.
[8] Clarke is writing of the situation in Great Britain.

It would also help if those who invoked the term at least specified what they meant by it. But the attempt to draw a line that clearly demarcates all policies acknowledged to be eugenic from those that are not will likely prove as fruitless as the analogous attempts to demarcate 'science' from 'non-science'.[9] I would like to suggest another, potentially more productive tack. Let us ask what scenarios people actually fear when they express anxiety about a resurgence of eugenics, and to evaluate which of them (if any) are likely, which (if any) are possible, and which (if any) are improbable.

Concerns about eugenics typically fall into one of three distinct classes. The first is fear of direct government programmes. Those alarmed at the prospect of state intervention often cite the Nazi experience, though they usually expect the analogue to be less brutal. Thus the biologist Salvador Luria (1989) questioned whether the HGP will transform 'the Nazi program to eradicate Jewish or otherwise "inferior" genes by mass murder . . . into a kinder, gentler program to "perfect" human individuals by "correcting" their genomes'. The activist Jeremy Rifkin, supported by some religious and disability rights groups, has called for a moratorium on human gene therapy research until such time as the NIH establishes a 'Human Eugenics Advisory Committee' to evaluate its implications (see Roberts 1989). These critics fear a slippery slope leading to state action. They believe the government might try to design workers less susceptible to environmental insults, or to redesign us in other ways. 'You could see genetic engineering of human beings from the foetal level on up,' suggests Andrew Kimball, who directs Rifkin's Foundation on Emerging Technologies. 'If they found that your child who's in kindergarten was predisposed to shyness, they would alter that child not to be shy. . . . These technologies have an enormously eugenic potential' (quoted in Saletan 1989: 18).

How realistic is the fear of direct government intervention? In respect to coercive gene therapy, not very. Even if it were technically feasible, large-scale gene implantation would be an extraordinarily expensive kind of eugenics. And cost saving has always been the strongest motive for eugenics. In the early decades of the century, visitors at state fairs and expositions were warned of the price of leaving heredity to chance and apprised 'that every fifteen seconds a hundred dollars of your money went for the care of persons with bad heredity' (Kevles 1985: 62). College (and even high school) textbooks carried the message that 'the cost of caring for those who cannot care for themselves because of their bad breeding is very heavy—perhaps two hundred million or more a year' (Castle et al. 1912:

_____

[9] The history of these attempts is traced in Laudan 1988.

309). Economic considerations explain why only one of the thirty state sterilization statutes adopted between 1907 and 1931 extended to the non-institutionalized—and why the rate of sterilization accelerated during the Depression.[10] Cost–benefit considerations today are less crude and, given sensitivities about abortion, sometimes less explicit. But they remain a principal incentive for state provision of genetic services (Modell 1991). And it will be a long time, if ever, before gene therapy or any form of 'positive' eugenics can be promoted as a way to save money.

Cost–benefit analysis could, however, produce a stimulus for 'negative' eugenics. After all, the plus side of the cost–benefit ledger is represented by the number of terminations achieved for a specified condition. The more women who are screened, and affected foetuses aborted, the more efficient the genetic service. Hence cost–benefit arguments for state support of these programmes provide an incentive to expand genetic testing and 'maximise the rate of terminations of pregnancy for "costly" disorders' (Clarke 1990: 1146).

American states already sponsor many pre-natal and new-born screening programmes, including some that are mandatory. For example, every state tests for phenylketonuria (PKU); parental consent is rarely required (Acuff and Faden 1991). Ultrasound screening is also routinely performed (without consent) on virtually every woman who sees a doctor early enough in her pregnancy (Lippman 1991: 21–2). Cost–benefit considerations help explain the vast expansion in the number of women who now undergo pre-natal testing. They provide a powerful inducement to test more women, for more disorders, at an earlier age. As Neil Holtzman and Andrew Rothstein note, 'Avoiding the conception of an infant at risk for a genetic disease—or avoiding the birth of a fetus prenatally diagnosed as having one—will often be less expensive than clinical management' (1992: 457).

Pressures to routinize screening will certainly increase with the development of tests that predict which individuals will bear offspring with genetic disorders. Individuals with the autosomal dominant gene for Huntington disease, polycystic kidney disease, and a rapidly growing list of other late-onset disorders can now be identified before they suffer symptoms and make reproductive decisions. Carriers of a number of autosomal recessive disorders, such as Tay–Sachs disease, sickle cell anaemia, and most recently, cystic fibrosis (CF), can also be identified. CF is the most common severe genetic disease affecting Caucasians, with an incidence of about 1 in 2,500 live births. About 1 in 25 white persons (or more than 8 million

---

[10] The exception was North Carolina. See Reilly 1991.

people) are asymptomatic carriers (Wilfond and Fost 1990: 2777). The ability to screen for common autosomal recessive diseases offers a solution to the problem that has bedevilled eugenicists since the first decade of this century.

Negative eugenics always faced a formidable practical barrier: the fact that most genes responsible for defects are recessive and thus hidden in apparently normal carriers. Policies such as sterilization and segregation only prevent the affected from breeding, and therefore work very slowly. Various schemes have been proposed to overcome this difficulty. In the 1930s, considerable attention was directed to identifying heritable antigens in the blood linked to genes for mental or moral defects, in the (disappointed) hope that these antigens could serve as genetic markers for the traits of interest.

Other schemes have relied on the phenomenon of partial dominance—that is, the fact that nearly all mutant genes in humans have some phenotypic effect. The most ambitious programme of this type was proposed by H. J. Muller in 1949 (Muller 1950).[11] Unlike the early eugenicists, Muller understood that we all carry harmful genes. But he also realized we did so in different degree (to be precise, that the number of individuals carrying slightly harmful genes would form a Poisson series, with eight as its average). Partial dominance allows for the possibility of a potentially efficient means of selection. In principle, genotypes can be 'surveyed', and those individuals falling in one tail of the distribution identified. Muller calculated that the genetic *status quo* could be maintained if the most mutant 3 per cent of the population refrained from reproducing.

When Muller proposed this scheme, the technology of heterozygote detection did not yet exist. But in the same year James Neel untangled the genetics of sickle cell anaemia, and Linus Pauling the biochemistry, work that made possible the first sickle cell screening programmes in 1971. Three years earlier, Pauling had suggested that all young people be tested for heterozygosity of the sickle cell and other deleterious genes, and that a symbol be tattooed on the foreheads of those found to be carriers. 'If this were done,' he remarked, 'two young people carrying the same seriously defective gene in single dose would recognize this situation at first sight, and would refrain from falling in love with one another. It is my opinion that legislation along this line, compulsory testing for defective genes before marriage, and some form of public or semi-public display of this possession, should be adopted' (Pauling 1968: 269).[12]

[11] For a detailed discussion of this article see Paul 1987.
[12] See also his 1963.

Sickle cell screening was strongly promoted in the African-American community, where one in twelve persons is a carrier, and in a few states it was made mandatory. While the demand for these programmes arose in part from within the black community, enthusiasm faded as the result of instances of discrimination against some of those identified as carriers. Screening programmes for thalassaemia and Tay–Sachs disease enjoyed much greater success (in part as the result of lessons learned from the problems with sickle cell screening). But these disorders are concentrated in particular ethnic groups—Americans of African, Mediterranean, and Ashkenazi Jewish descent. With the capacity to identify carriers of genes for common disorders, such as CF, the incentive for broad population screening has vastly increased. Wilfond and Fost note that 'the potential for CF carrier screening programs will create an entrepreneurial opportunity that will dwarf all previous screening programs' (1990: 2777). Indeed, biotechnology companies have begun offering tests, notwithstanding the 1990 and 1992 recommendations of the American Society of Human Genetics against routine screening where there is no family history of CF, and pilot programmes to evaluate population screening are under way.

It is easy to predict that as we can screen more accurately and cheaply for more common disorders, the use of genetic tests will become increasingly routinized. Will individuals be pressured to make particular reproductive decisions as a result? That would qualify as eugenics by most definitions. It would in any case be a very unhappy development. And it is a likely one, though not as the result of *state* intervention. As Karjala notes, 'Given the natural revulsion that most people feel for interference through mandatory testing or, even worse, mandatory abortion, the issues [of "genetic freedom and genetic responsibility"] are likely to be raised obliquely' through the health insurance system, HMO policies, or doctor pressure (1992: 159).

A number of linked developments—the trend toward respect for patient rights in medicine, the rise of the women's and disability rights movements, the adoption of a broad jurisprudence of privacy and reproductive freedom—have converged to produce wide acceptance of the principle of reproductive autonomy.[13] Pauling's comments appear far more startling now than they would have in the 1960s. Of course, there are still individuals, of varying political perspectives, who express varying degrees of doubt about the consequences of allowing autonomy to trump other values.[14] An extreme standpoint is represented by Margery Shaw, past president of

[13] For a discussion of the evidence for this claim see Paul 1992: esp. 676–7.
[14] Modest doubts are expressed by Capron (1990). Maura Ryan raises feminist objections in her 1990.

the American Society of Human Genetics. Shaw replied no to the question of 'whether or not a defective fetus should be allowed to be born' (1984*b*: 1) and expressed optimism that 'parental rights to reproduce will diminish as parental responsibilities to unborn offspring increase' (ibid. 9).[15] But hers is a now unfashionable minority position. And even she would rely on tort liability—not legislation—to bring about the desired result.

In 1927, Justice Oliver Wendell Holmes commented, in his opinion upholding a Virginia sterilization order: 'We have seen more than once that the public welfare may call upon the best citizens for their lives. It would be strange if it could not call upon those who already sap the strength of the State for these lesser sacrifices . . . in order to prevent our being swamped with incompetence. . . . The principle that sustains compulsory vaccination is broad enough to cover cutting the fallopian tubes' (quoted in Cynkar 1981: 1419). Since the 1940s, however, the Court has moved in a very different direction. In 1942, it unanimously overturned a sterilization law in an opinion that termed procreation 'one of the basic civil rights of man'.[16] The court further expanded the scope of privacy and reproductive freedom in *Griswold* v. *Connecticut* (1965), where it struck down a law prohibiting the use of contraceptives; in *Eisenstadt* v. *Baird* (1972), where it held that 'if the right of privacy means anything, it is the right of the *individual*, married or single, to be free from unwarranted governmental intrusion in matters so fundamentally affecting a person as the decision whether to bear or beget a child'; and of course in *Roe* v. *Wade* (1973). Even if *Roe* were reversed, it would be a long way back to Justice Holmes. Moreover, without access to abortion, there would be much less demand for pre-natal diagnosis. State intervention would in any case be fiercely resisted both by feminists, who are committed to the principle of reproductive choice, and their opponents, who oppose abortion, and by the disability rights movement. It would have to defeat powerful, organized, and highly diverse social forces.

But, as noted earlier, pressures can be indirect and involve actors other than the state. Holtzman and Rothstein remark that 'although we may not soon reach the stage of compulsory eugenics legislation, denying health-care coverage because of genotype could exert pressure on at-risk families to avoid having children with disabilities, despite the families' wishes' (1992: 458). In fact, these are the anxieties most commonly voiced when people say they fear 'eugenics'. They worry that tests, screens, and therapies will be introduced and promoted because they have the potential to

---

[15] See also Shaw 1990 and 1984*a*.
[16] *Skinner* v. *Oklahoma*, 316 U.S. 535 (1942).

generate profits for biotechnology companies, savings for employers and life and health insurers, and protection against malpractice suits for physicians—and that they will lack realistic alternatives to the decision to be tested or to abort a foetus identified as 'defective' as a result of policies adopted by these quasi-public or private actors. The problems will intensify as carrier and predictive tests for common disorders (representing large markets) become more reliable. That a test is mandated by an insurer, rather than the state, does not necessarily make its consequences less drastic.

At present, few insurers make use of genetic tests. That situation will likely change as these tests become reliable and cheap, and as more is learned about dispositions to common disorders, such as cancer and hypertension (Billings *et al.* 1992.) In 1989, the American Council of Life Insurance issued a report to its member companies (the largest providers of life and health insurance in the USA) to prepare them for the day when people routinely learn their genetic profiles. It suggested that when individuals have access to knowledge about predispositions to disease, the companies must also. Otherwise, individuals who know that they are at high risk of developing a disease will load up on insurance (Bishop and Waldholz 1990: 297–8).[17] The courts are not likely to bar insurers from acquiring and using information on risks. Thus the US Court of Appeals for the Fifth Circuit recently held that a self-insured employer could limit health benefits for AIDS after an employee was diagnosed with the disease, a decision that 'could apply to people in whom genetic tests indicate a high probability of future disease' (Holtzman and Rothstein 1992: 457).

Decisions may also be strongly affected by new 'standards of care' adopted by professional organizations in response to (exaggerated or real) fear of malpractice suits. In her survey of state legislation and current practice standards, Katherine Acuff notes that 'legislation is by no means the only, or even the predominant, route to ensuring adoption of policies regarding prenatal or newborn screening. . . . [Many] testing procedures have been incorporated into routine medical practice based upon the pronouncements of professional organizations, without the spur of legislative mandate' (1991: 133).

Thus in 1985 the Department of Professional Liability of the American College of Obstetricians and Gynecologists (ACOG) alerted ACOG members to the purported need to inform *all* pregnant patients of the availability of maternal serum alpha-foetoprotein (MSAFP) screening. The physicians were told it was 'imperative that every prenatal patient be

---

[17] For a useful discussion of this issue, see Greely 1992.

advised of the advisability of this test and that your discussion about the test and the patient's decision with respect to the test be documented in the patient's chart' (ACOG 1982). As George Annas and Sherman Elias note, the rationale for the alert was legal, not medical: 'to give the physician "the best possible defense" in a medical malpractice suit premised on the birth of a baby with a neural tube defect' (1985: 1375). Indeed, for a variety of reasons, including high false negative and false positive rates associated with the test and the need for appropriate counselling, ACOG had concluded that routine screening was of dubious value and should not be implemented in the absence of appropriate counselling and follow-up services (ibid. 1374).[18]

The doctrine of 'informed refusal' (codified in the 1980 case of *Truman* v. *Thomas*) has also contributed to the expansion of testing. It is relatively easy to document informed consent. Moreover, there is essentially no (medical) risk attached to the procedure. If a woman is tested, there is thus no potential problem of liability for the physician. But it is relatively hard to document that she gave an informed 'no'.[19]

In the absence of public policy designed to prevent it, reproductive decisions will often be driven by the conjoined interests of powerful non-state entities, such as physicians, lawyers, insurers, and biotechnology firms.[20] These are entities over which the public has limited control—precisely because they are private. In some formulations, this is a more subtle version of the concern about state action. But it may also be its converse. Thus Robert Wright suggests that the real threat is not a government programme to breed better babies. 'The more likely danger', he writes, 'is roughly the opposite; it isn't that the government will get involved in reproductive choice, but that it won't. It is when left to the free market that the fruits of genome research are most assuredly rotten' (1990: 27). In any case, this package of problems is certainly real, whether or not it is labelled eugenics.

The third category is in effect the converse of the concern about indirect pressure (although they are not contradictory, and indeed are sometimes linked). This is the concern often described as 'back-door' eugenics. It

---

[18] Annas and Elias also note that only 40 per cent of the women in one study reported discussing the test with a physician.

[19] I am grateful to Karen Rothenberg for making this point.

[20] One such policy would be explicit application of the Americans with Disabilities Act of 1990 to cases of genetic discrimination. But the ADA exempts insurers. Thus an employer may not 'use the discriminatory practices of insurance companies as a pretext for refusing to hire, firing, or taking other adverse actions against an applicant or employee'. However, insurers themselves may discriminate on the basis of demonstrable risk, if this practice is compatible with state law (Natowicz *et al.* 1992: 471).

Faden R. R., *et al.* (1987), 'Prenatal Screening and Pregnant Women's Attitudes toward the Abortion of Defective Fetuses', *American Journal of Public Health*, 77 (Feb.): 1–3.

Galton, F. (1883), *Inquiries into the Human Faculty* (London: Macmillan).

Greely, H. T. (1992), 'Health Insurance, Employment Discrimination, and the Genetics Revolution', in D. J. Kevles, and L. Hood (eds.), *The Code of Codes: Scientific and Social Issues in the Human Genome Project* (Cambridge, Mass.: Harvard University Press), 264–80.

Holtzman, N. A. (1989), *Proceed with Caution* (Baltimore: Johns Hopkins University Press).

——and Rothstein, M. A. (1992), 'Eugenics and Genetic Discrimination', *American Journal of Human Genetics*, 50: 457–9.

International Huntington Association and World Federation of Neurology (1990), 'Ethical Issues Policy Statement on Huntington's Disease Molecular Genetics Predictive Test', *Journal of Medical Genetics*, 27: 34–8.

Karjala, D. J. (1992), 'A Legal Research Agenda for the Human Genome Initiative', *Jurimetrics*, 32 (Winter): 121–222.

Kevles, D. J. (1985), *In the Name of Eugenics: Genetics and the Uses of Human Heredity* (New York: Knopf).

Laudan, L. (1988), 'The Demise of the Demarcation Problem', in M. Ruse (ed.), *But Is It Science?* (New York: Prometheus Books), 337–50.

Ledley, F. D. (1991), 'Differentiating Genetics and Eugenics on the Basis of Fairness', poster 1818, Eighth International Congress of Human Genetics, Washington, DC, 6–11 Oct.

Lippman, A. (1991), 'Prenatal Genetic Testing and Screening: Constructing Needs and Reinforcing Inequities', *American Journal of Law and Medicine*, 17: 15–50.

Luria, S. (1989), letter, *Science*, 246: 873.

Modell, B. (1991), 'Cost/Benefit Considerations and Social Aspects of Genetic Services', paper given at Eighth International Congress of Human Genetics, Washington, DC, 6–11 Oct.

Muller, H. J. (1950), 'Our Load of Mutations', *American Journal of Human Genetics*, 2: 111–76.

Natowicz, M. R., *et al.* (1992), 'Genetic Discrimination and the Law', *American Journal of Human Genetics*, 50: 465–74.

Paul, D. B. (1987), '"Our Load of Mutations" Revisited', *Journal of the History of Biology*, 20: 321–35.

——(1992), 'Eugenic Anxieties, Social Realities, and Political Choices', *Social Research*, 59: 663–82.

Pauling, L. (1963), 'Our Hope for the Future', in M. Fishbein (ed.), *Birth Defects* (Philadelphia: J. P. Lippincott), 164–70.

——(1968), 'Reflections on the New Biology: Foreword', *UCLA Law Review*, 15: 267–72.

Proctor, R. (1992), 'Genomics and Eugenics: How Fair is the Comparison?', in G. J. Annas and S. Elias (eds.), *Gene Mapping; Using Law and Ethics as Guides* (New York: Oxford University Press), 57–93.

Reilly, P. R. (1991), *The Surgical Solution: A History of Involuntary Sterilization in the United States* (Baltimore: Johns Hopkins University Press).

Roberts, L. (1989), 'Ethical Questions Haunt New Genetic Technologies', *Science*, 243: 1134–6.

Russell, B. (1924), 'Eugenics', in *Marriage and Morals* (London: Liveright), 255–73.

Ryan, M. (1990), 'The Argument for Unlimited Procreative Liberty: A Feminist Critique', *Hastings Center Report* (July–Aug.): 6–12.

Saletan, W. (1989), 'Genes 'R Us', *New Republic* (17 July): 18–20.

Shaw, M. W. (1984a), 'Conditional Prospective Rights of the Fetus', *Journal of Legal Medicine*, 5: 63–116.

——(1984b), 'To Be or Not to Be? That is the Question', *American Journal of Human Genetics*, 36: 1–9.

——(1990), 'The Potential Plaintiff: Preconception and Prenatal Torts', in Λ Milunsky and G. Annas (eds.), *Genetics and the Law II* (New York: Plenum Press), 225–35.

Soloway, R. A. (1990), *Demography and Degeneration: Eugenics and the Declining Birthrate in Twentieth-Century Britain* (Chapel Hill, NC: University of North Carolina Press).

Wertz, D. C., and Fletcher, J. C. (1989), 'Fatal Knowledge? Prenatal Diagnosis and Sex Selection', *Hastings Center Report* (May–June): 21–7.

—— ——(1990), 'Ethical Decision Making in Medical Genetics: Women as Patients and Practitioners in Eighteen Nations', in K. S. Ratcliff *et al.* (eds.), *Healing Technology: Feminist Perspectives* (Ann Arbor: University of Michigan Press), 221–41.

——*et al.* (1991),'Attitudes toward Abortion among Parents of Children with Cystic Fibrosis', *American Journal of Public Health*, 81: 992–6.

Wilfond, B. S., and Fost, N. (1990), 'The Cystic Fibrosis Gene: Medical and Social Implications for Heterozygote Detection', *Journal of the American Medical Association*, 263: 2777–83.

Wright, R. (1990), 'Achilles' Helix', *New Republic* (9 and 16 July): 21–31.

28

NORMALITY AND VARIATION: THE HUMAN GENOME PROJECT AND THE IDEAL HUMAN TYPE

ELISABETH A. LLOYD

Certain issues involving science are widely regarded as ethical or social—the appropriate moral and medical responses to abnormal foetuses, for example. The 'concept of abnormality' *itself* is not usually one of these social or ethical issues. It is assumed that science tells us what is normal or abnormal, diseased or healthy, and that the social and moral issues begin where the science leaves off.

For many purposes, such an understanding of science is appropriate. In this chapter, however, I would like to challenge the 'givenness' of the categories of normality, health, and disease. By understanding the differences among various biological theories and the distinguishing features of their respective goals and approaches to explanation, we can analyse the way in which scientifically and socially controversial views are sometimes hidden inside apparently pure scientific judgements.

I am not suggesting that the misleading nature of some of the scientific conclusions I discuss implies some unsavoury *intention* on the part of the scientists involved. On the contrary, my point is that there are sincere scientists working among different theories and sub-fields, each with their own standards of explanation and evidence. The diversity of theories and models involved in implementing the Human Genome Project provides a unique challenge to both the producers and the consumers of the DNA-sequencing information. I contend that the problems arising from this diversity have not been recognized or addressed.

In drawing attention to some important differences among the biological theories involved in the Human Genome Project, I will explore several possible misunderstandings that may be arising from viewing biology as a monolithic and completely integrated science. Further, I raise some con-

First published in C. F. Cranor (ed.), *Are Genes Us?* (New Brunswick, NJ: Rutgers University Press, 1994), 99–112. Reprinted by permission.

cerns about the risks inherent in biological and medical reasoning about genetics.

HEALTH AND DISEASE

Judgements concerning health and disease inevitably involve questions of classification. 'Health' encompasses the thriving, fully functioning, or normal states of the organism, while 'disease' includes states of malfunction, disturbance, and abnormality. States of organisms do not announce themselves as desirable or undesirable, healthy or diseased, normal or abnormal; such classifications are inevitably applied by comparing the state of the organism to some *ideal* which serves a normative function. Where does this ideal come from?

Roughly speaking, our notions of the ideal state of an organism are informed by our understanding of the 'proper' or appropriate functions of various parts. The function of the kidneys is to clean the blood; if the blood is not cleaned thoroughly, and the organism loses the benefits of the 'proper functioning' of the kidneys, then the kidneys are 'diseased'. Overall, the organism does not function as well as it once did. But suppose instead that an organism with badly functioning kidneys never had kidneys that effectively cleaned the blood? Then the comparison must be made not to the prior state of that individual, but to a more abstract notion of 'proper functioning' of kidneys in people. In other words, the ideal is *normal* kidney functioning, where 'normal' signifies the function in a thriving person.

The difficulties of classifying diseases are well known, having been faced by every theory of medicine in human history. The range of definitions of 'normal functioning'—from proper balance of the four humours, to clear flow along the chi meridians, to freedom from cohabitation with microorganisms—has also received a great deal of attention. It seems, therefore, that it would be a major scientific advance, and a significant relief, to be able to understand disease and proper function on the molecular level, and it is just this that is promised by many proponents of the Human Genome Project.

Renato Dulbecco (1986), for example, argued that significant advances in cancer research would be made possible by knowledge of the exact human DNA sequences.[1] Similarly, Nobel Laureate Jean Dausset has defended the Human Genome Project by arguing that it will allow us to

---

[1] Cf. Fujimara 1988: 261.

predict when a person will develop an illness, or at least when he or she will have a predisposition to that illness. Early diagnosis and preventive measures are also emphasized. Dr Jerome Rotter, director of the Cedars–Sinai Disease Genetic Risk Assessment Center in Los Angeles, says: '[I]t's more than just knowing you're at risk. You can take steps to prevent coming down with the disease or be able to cure it at an early stage' (quoted in Roan 1991).

The descriptions of disease promised by the Human Genome Project are intended to be on the biochemical and, sometimes, even the molecular level. James Watson, co-discoverer of the structure of DNA, and later director of the National Institutes of Health segment of the Human Genome Project, asserted that genetic messages in DNA 'will not only help us understand how we function as healthy human beings, but will also explain, at the chemical level, the role of genetic factors in a multitude of diseases, such as cancer, Alzheimer's disease, and schizophrenia' (1990: 44). The geneticist Theodore Friedmann has proclaimed that 'molecular genetics is providing tools for an unprecedented new approach to disease treatment through an attack directly on mutant genes' (1989: 1275). Once diseases have been pin-pointed on the molecular level, treatment can begin: 'inherited diseases can be identified with biochemical as well as genetic precision, often detected *in utero*, and, in some cases, they can be treated effectively' (Culliton 1990: 211).[2]

The promise is that diseases will finally be subject to truly scientific classification, analysis, and treatment. Detailed descriptions of the molecular causes of disease will enable medical researchers to develop more precise preventive and therapeutic techniques. Once the human genome is sequenced, we will have a library of genes with which any potentially abnormal gene can be compared. Abnormal genes will be isolated, altered, replaced, or, in case they are present in implantable embryos, simply discarded.

In many ways, this picture of medical promise and possibility is undoubtedly positive. It is also inadequate and misleading. The primary problem is that the picture of disease most often presented in discussions of the Human Genome Project is oversimplified. Specifically, the presentation of genetic disease and abnormal gene function as self-announcing is unjustified, except in the most trivial sense. General physiological notions of normality, health, and disease are defined according to a *different* set of standards that go beyond molecular-level descriptions. Describing genes as 'causing' diseases is, on a basic scientific level, to confuse at least two

[2] See also Jukes 1988.

distinct levels of theory and description. While a genetic classification of disease may indeed be desirable and useful, it also involves a series of judgements about the ideal forms of human life. Moving the level of diagnosis down to the molecular level does *not* succeed in avoiding the fundamental value-judgements involved in defining health and disease, contrary to the suggestions of the genome researchers.[3]

So, while molecular techniques will certainly aid in the diagnosis, identification, and analysis of disease processes, they cannot replace the profoundly evaluative and essentially social decisions made in medicine about standards of health and disease. In fact, molecular techniques should be understood as offering an unprecedented amount of social power to label persons as diseased. Hence, it is more important than ever to gain insight into the normative components of judgements about health. The potential submersion of normative judgements under seas of DNA-sequence data should not persuade anyone that conclusions concerning health and disease have now, finally, become scientific. Any appearance to the contrary is a result, I will argue, of some rapid and illegitimate shifting between biological sub-theories. Once the structure of the theories involved has become clear, it will be easier to see how and where evaluative decisions are being made.

MOLECULAR DESCRIPTIONS

Let us begin by taking a closer look at the description and explanation of disease at the DNA level. The typical form of the model used in explanation is described by T. H. Jukes: '[I]nherited defects would be caused by changes in the sequence of DNA, perhaps a change in a single nucleotide. Such change might result in the replacement of one amino acid by another in a protein at a critical location, making the protein biologically useless' (1988: 16).[4]

. A paradigmatic case of genetic disease, sickle cell anaemia, fits this general model well, and is an early, compelling, and fairly complete case of medical genetics. The original mystery was why certain populations had a

---

[3] Another misleading aspect of the typical genetic presentations of disease concerns implicit assumptions of genetic determinism and disease. See the following for relevant discussion: Karjala 1992, Gollin *et al.* 1989.

[4] Friedmann describes the goals of molecular genetic explanations also: 'Predicting the exact structure and regulated expression of any gene, the tertiary and quaternary structures of its products, their interactions with other molecules, and finally, their exact functions, constitutes a problem of the highest priority in molecular genetics' (1990: 409).

high incidence of a type of red blood cell that seems to cause health problems. The red-cell abnormality stemmed from a difference in the haemoglobin molecule which impeded its ability to carry oxygen, hence damaging the bodily capacities of a person with these cells. An analysis of the biochemical causal pathway revealed that the genes that code for the haemoglobin molecule in these people were different from other people's haemoglobin genes in a particular way, thus affecting the ability of the haemoglobin to pick up oxygen. The differences in genes contributed to a difference in the proteins made according to the pattern on those genes, and these protein (haemoglobin) differences had systemic and detrimental effects on health.

This *biochemical causal pathway model* traces an isolated chain of events that yields the effect of interest. In the standard genetic/biochemical model for the production of haemoglobin, the haemoglobin gene codes for a protein that is included in the red blood cells, cells that, in turn, serve the function of carrying oxygen to the cells of the body. The model presents, in detail, a picture of 'normal or proper functioning'.

There is not much room for simple variation in this explanatory scheme using the standard model.[5] But why should there be? The purpose of the basic explanation is to explain *how it could be* that haemoglobin is produced and operates effectively in the body. The goal of the explanatory theory is to delineate at least one causal chain that could proceed from the initial state—DNA arranged on chromosomes in a zygote—to the final state—iron arranged in haemoglobin molecules carrying oxygen around the body. The final model of haemoglobin abnormality in sickle cell anaemia represents an astonishing piece of detective work, since it assumes an understanding of each chemical reaction involved in the 'normal' causal chain. The abnormality itself is explained through isolating the points in the sickle cell causal chain that are different from the model of normal functioning.

Watson, in motivating the genome projects, presents a similar picture: 'The working out of a bacterial genome will let us know for the first time the total set of proteins needed for a single cell to grow and multiply' (1990: 48). Such a goal is perfect for a biochemical causal model: what is desired is some complete set of causal steps yielding a living organism. But variation plays no role in this model. It is an uninteresting, and even distracting, feature of the processes on which the explanatory theory is focused.

---

[5] James Watson speaks ambiguously of 'the genetic diseases that result from *variations* in our genetic messages' (1990: 46, emphasis added). Not all variations result in genetic diseases.

In other words, under a biochemical causal model, there is no obvious approach to dealing with variation. One could classify all variations on this main scheme as 'abnormal'. Owsei Temkin, in ridiculing a definition of disease based on genetic origins, teased that 'there should also be as many hereditary diseases as there are different genes representing abnormal sub-molecular chemical structures' (1977: 444). Temkin argues against classifying diseases on causative principles in general, 'lest specific diseases be postulated which have no clinical reality' (ibid.). More sensibly, we would prefer to define a variant as abnormal or diseased only if it interfered with 'proper function'. Proper function, however, cannot be defined within the molecular genetic model itself: it must be defined in terms of the physiological functions of the resultant protein.[6] Here, proper function could involve simply the presence of an *effective* haemoglobin for carrying oxygen around. But there could be (and are) different degrees of effectiveness. How should we divide these up into abnormal and normal?[7] Simple variations and undesirable variations? Some differences in DNA lead to large changes in function, while other differences are imperceptible with regard to oxygen delivery in a normal person.

A friend was exposed to this problem personally. Having gone to a physician who was up to date on all the most recent screening techniques, my friend's slightly red eyes prompted the doctor to run a blood test. The results of this test were positive, and my friend was informed by his doctor that he had 'abnormal' haemoglobin and liver functioning, something called 'Gilbert's disease'. When my friend asked about the health consequences of this abnormality involving an essential protein in his body, he was told that the disease was 'nonfunctional', and the only known effect was a reddening of the whites of the eyes.

The point is this. If normality is defined at the level of the biochemical causal model, all variation in the DNA of the haemoglobin genes is abnormal. As such, however, the genetic abnormality tells us nothing about its effects on physiological function in the larger organism.

Another example emerges from recent research on human breast tissue. 'Fibrocystic breast disease' exists in approximately 45 per cent of the female population over age 35. In one sense, the relevant phenomena are called 'Fibrocystic disease', because they involve the process of

---

[6] Gollin *et al.* (1989) include in their list of uses 'an index of acceptable physiological function or behavior' (p. 50). Such views derive ultimately from Herman Boerhaave, who defined diseases as occurring when a person could not exercise a function.

[7] Theodore Friedmann admits that a great deal of sequence information and 'comparison' is needed in order to distinguish between 'polymorphisms' and 'disease-related mutations' (1990: 409).

encystation, which is additional to the usual physiological *functions* of breast tissue, such as milk production. On the medical and physiological level, however, it seems that the fibrocystic condition is completely 'normal' for the average adult woman; furthermore, the condition does not impede or alter the usual physiological functions of the breasts. In what sense, then, is the fibrocystic condition a 'disease'? This is a case in which refined understanding at the cellular level leads to the classification of a condition as an 'abnormality' or 'disease', while at the functional, medical, or physiological level it is unclear what sense can be made of labelling nearly half the women over the age of 35 'diseased'.

Again, if normality is defined according to some model of physiological function, molecular information alone cannot decide whether a certain person is normal or abnormal. The DNA information itself is potentially revealing about the functional state, but only potentially.[8] Abnormality in the DNA or in the causal chain may or may not have health consequences that we would consider significant. Carl Cranor, in his discussion of genetic causation and Hartnup disorder (1994), emphasizes the contingency of the emergence of disease on other factors. He also reviews the various types of confusion that can arise regarding genetic causation. I offer here a diagnosis of the underlying mechanism that produces the kinds of confusion discussed by Cranor.

PROPER FUNCTION

The second type of biological model involves reference to proper function on an organismic level. This level of model is usually considered most appropriate to medicine, and it could be called a '*medical model*'. Generally speaking, the *medical model* tends to be on the level of the whole organism, rather than on the cellular or molecular level. Take the example of the common cold. While part of the explanation of how a person contracts a cold is that they were exposed to a cold virus, the rest of the explanation requires taking account of the body as a whole: one does not get a cold simply from exposure to the virus; failure of the immune system to fight the virus invasion effectively is necessary, as is multiplication of the virus within the cells. Similarly, cancer cells are recognized as such on the microscopic level, but their undesirability is because of the damage they do to organ function. The presence of cells simply growing in the wrong place may not impair function—witness

---

[8] See Friedmann's discussion of 'normal, functional' genes in 1989: 1275.

the innumerable cases of benign tumours that lie undetected until death occurs from other causes.

The first important aspect of the medical model, then, is its *organismic basis*. A second feature, one that has far-reaching social and policy implications, is that the medical model must rely on *socially negotiated standards* of what counts as the proper functioning of a human being.[9] What range of functional performance is normal? Any answer involves a picture of what a human body should be like.[10] Probably the clearest recent demonstration of the social negotiation of categories of health and disease is the battle, in the past two decades, over the medical classification of homosexuality (Bayer 1981). Peter Sedgwick (1973), for example, argued persuasively that classifying homosexuality as a disease is clearly not just an empirical assessment of biological function.

To see the force of this argument, take the claims made in 1991 concerning differences in brain structure between homosexual and heterosexual men (LeVay 1991). Suppose, for the sake of argument, that these anatomical differences arise (in this environment) from genetic differences. (There is no evidence for this; the differences could just as well be caused *by* homosexual activity as be the causes of it.) Suppose, further, that we were able to isolate some genes whose functions included structuring this part of the brain, and that people with a particular sort of brain structure were more likely to be homosexual. Should this genetic character be considered an 'abnormality'?

To answer the question, we would need to assume some 'proper functioning' of the brain structure, and we must also align this proper function with a particular environment. (It could be that the same genes produce a different brain structure in a different environment, and that this brain structure is not correlated with a tendency to practise homosexuality.) What can we conclude about these genes? Only that they yield particular results in particular environments. But is this normal functioning? Clearly, the answer depends on something *outside the genetic causal story*: it depends on whether we think homosexual behaviour should count as normal functioning in human beings. In other words, this distinction between normality and disease depends on how we envision human life ought to be. This story introduces a further problem. If homosexuality is not seen as normal functioning, it is unclear which approach would most successfully move the population toward a higher incidence of normal functioning—

[9] Temkin 1977: 447. Cf. Caplan 1989.
[10] Caplan offers a useful summary of the debates about the extent of values in definitions of health and disease (1989: 55–9).

changing the environment in which this gene is expressed, or doing something on the genetic level to select out or replace this gene.[11]

The example of homosexuality brings out the profound value decisions involved in labelling certain functions as normal or abnormal. Without such evaluations, genes cannot be labelled as normal or abnormal in any but the most trivial respect—that is, in so far as they differ from the paradigmatic biochemical causal pathway currently accepted for that gene.[12] And such a weak classification system cannot do the work in medical genetics that has been advertised for the Human Genome Project.[13] The issue of defining the *standard* of health and disease is as open as it has ever been. Indeed, there is now the additional challenge of applying it to unimaginably fine biological differences.

POPULATION GENETICS MODELS

Very different types of description and explanation are used in population genetics, where the emphasis is on the analysis and maintenance of variation in populations. Population geneticists have posed persistent challenges to the grander claims made for the Human Genome Project, urging that a biological account of variation in human populations must accompany the DNA-sequence information that was originally targeted.[14] On average, any two human beings differ from each other in approximately 10 per cent of their nucleotides. Critics argue that any complete understanding of the functions of human DNA must be able to describe and account for this variation (Vicedo 1991).

Consider the meanings of 'normal' and 'abnormal' in the context of population genetics theory. Since the state of a population at a given time is given in terms of the distribution of different types of genes, a great deal of information is needed to delineate what is normal and abnormal, including (1) the range of the types of genes, (2) the range of phenotypes and functions associated with these genes, and (3) the range of environments

---

[11] See the discussion in Cranor 1994 and in Hubbard 1990.

[12] Gollin *et al.* describe this unjustified shift from developmental theory to the physiological level as follows: 'The observed regularities [in development] are mistaken for universals and the construed universals are regarded as indicators of health and developmental adequacy' (1989: 51).

[13] Even Theodore Friedmann, who is committed to the medical benefits of the Human Genome Project, discusses the distance between a DNA-level characterization of an 'abnormal' gene and a medical and physiological understanding of the disease. 'Most successes of medical genetics have begun with an understanding of an aberrant metabolic pathway; the genes responsible were identified and isolated through this physiological knowledge' (1990: 407).

[14] e.g. Council for Responsible Genetics 1990.

and the related norms of reaction. Finally, some decision about what will count as adequate functioning in a specific environment is also needed; only then can a specific type be categorized as *diseased*. Clearly, such information is not going to be provided by the biochemical causal models that are prominent in molecular genetics.

DEVELOPMENT AND EMBRYOLOGY

A fourth type of biological theory is needed to understand genetically based disease. The models of embryology, epigenesis, and developmental biology, though often confused with biochemical causal chain models, are distinct from them, since they are designed to describe different things and to answer different questions. Specifically, epigenetic models describe the process of what actually happens with genes in environments; ideally, the end result is a description of the emergence of a phenotype in an environment. In this context, 'normal' usually means that you get, at the end of the process of development, what is expected, given those genes in that environment. Something is labelled as 'abnormal' if it is not what is expected.[15]

The fundamental importance of environmental considerations in interpreting traits as normal or abnormal can be seen in the case of New Guinea highlanders, who often have urinary potassium/sodium ratios 400 to 1,000 times the 'normal' Western ratio. Daniel Carleton Gajdusek (1970) argues that this difference is a metabolic response to a sodium-scarce, water-poor environment. Notions of proper physiological function, then, depend fundamentally on the related environment.[16]

GENETICS AND DISEASE

Having considered these four types of biological theories, with their corresponding notions of normal and abnormal, we are ready to return to the specific issues surrounding the Human Genome Project. In the Genome Project, certain genes are labelled as abnormal, and the decision to do so is made by using as a comparison the DNA sequence of a gene that appears in an accepted model of the biochemical causal chain. What is

[15] The distinction between 'congenital' and 'genetic' birth disorders has long been recognized: 'congenital' refers only to disturbances from the normal pattern of development that are not believed to be genetically caused.

[16] See the discussion in Gollin *et al.* 1989.

abnormal under the biochemical model is not necessarily abnormal under a medical model. None the less, researchers interested in the Genome Project routinely slip from a DNA level of description to a medical usage of 'abnormal'.

P. A. Baird (1990), for example, promises that the Genome Project will yield a 'new model for disease', in which we will be able to diagnose on the basis of causes, and not simply treat symptoms. Carl Cranor (1994) points out that Baird's view exaggerates the role of genetic causes in disease. I would add that Baird is assuming the appropriateness of applying the biochemical causal chain model to all people carrying the gene; the implication is that the gene will produce disease 100 per cent of the time, which, as Cranor emphasizes, is very rarely true for genetic disease.

Victor McKusick, former head of the International Human Genome Organization, also tends to overstate the case for genetic causation: 'Mapping has proved that cancer is a somatic cell genetic disease. With the assignment of small cell lung cancer to chromosome 3, we know that a specific gene is as intimately connected to one form of the disease as are cigarettes' (1989; quoted in Culliton 1990: 212). Showing the genetic basis for one cancer, is, of course, not the same as showing that every cancer is best understood as a genetic disease.

A similar problem arises in the study of a gene region linked to liver cancer. This study found a region of the DNA where the gene is especially sensitive to exposures to toxins; the toxins induce mutations in that spot, which then prevent the gene from performing its usual physiological role (Angier 1991). This is a significant advance, especially since this gene has been implicated in many types of human cancer, including tumours of the breast, brain, bladder, and colon. This case seems to support Friedmann's claim that 'human cancer should be considered a genetic disease', because 'it is likely that most human cancer is caused by, or is associated with, aberrant gene expression' (1989: 1279). But the toxins appear to be a necessary condition here, in addition to the presence of the sensitive gene region. So in one sense, the disease is genetic; in another, it arises from environmental causes.

The differences in biological models I outlined above can help with this case. Under a biochemical causal model, the mutant gene is a necessary link in the biochemical causal chain of this liver cancer. Hence, the liver cancer has a genetic cause.[17] Under a physiological model, the usual presence of particular proteins is interrupted through the exposure to toxins in

---

[17] This supports Cranor's (1994) point against Hubbard that in cases where disease occurs, a single gene can be picked out as a cause.

the cellular environment. Hence, the abnormal functioning of the body is dependent on the interaction of environment and cells.

The implication in discussions of genetic bases of diseases is that an abnormal gene leads to abnormal functioning, which is itself deficient. But this, of course, is not shown from the strict biochemical description. Both a developmental and a functional model are necessary to support the identification of the genetic difference with what we traditionally identify as disease. One problem is that entities on the medical level, such as alcoholism, may be very difficult to pin down genetically. While a gene believed to be implicated in some severe cases of alcoholism has been found, researchers are more convinced than ever that (1) many genes are involved in the disease, and (2) they are *different genes* in different groups of individuals (Holden 1991). As Vicedo has argued, 'all the meaningful questions will *start* when all the sequencing is done' (1991: 19).

What about the presumption of genetic bases? Many researchers will cite Huntington disease, or cystic fibrosis, or Down's syndrome, as clear-cut cases of disease versus normality. But there are more than a hundred mutations catalogued that will produce cystic fibrosis as a clinical entity; and Down's syndrome is diagnosed as trisomy 21, although such a chromosomal arrangement can yield people with a very wide range of abilities. While some researchers are quick to cite the clear, deterministic cases, out of the total number of genetic screening tests available now, nearly all are for grey areas, where the genetic difference is a risk factor, or provides a vulnerability, to develop a specific physiological disease. How is this vulnerability to be understood?

Developmental models are crucial. Gene expression depends inextricably on environment, and environmental responses to knowledge of genetic predispositions can guide development away from dangerous outcomes.[18] Suppose that a person learns she has 'the gene for arteriosclerosis', and modifies her diet and exercise regime as a result. We cannot say she will develop arteriosclerosis; in fact, having changed her environmental circumstances, she may well have a *reduced* probability of developing arteriosclerosis in comparison to the population at large. The point is that having a gene for something does *not* imply having that phenotype. 'Abnormal' genes may or may not yield 'abnormal' or 'diseased' organisms. Dr Henry Lynch, director of a cancer genetics programme in Nebraska,

---

[18] A misleading over-reliance on biochemical causal models can lead to the complete disappearance of the influence of environment on phenotype, e.g. Watson, in arguing for the sequencing of the complete *Caenorhabditis elegans* genome, says: 'If we are to integrate and understand all the events that lead, for example, to the differentiation of a nervous system, we have to work from the whole set of genetic instructions' (1990: 48).

is worried that genetics programmes will emphasize genetics over simply life-style factors that are much more causally influential (see Roan 1991).

Professor Bernard Davis and his colleagues in the Department of Microbiology and Molecular Genetics at Harvard Medical School have been visible critics of the Human Genome Project. They argue that studies of specific physiological and biochemical functions and their abnormalities will be much more useful medically than the sequencing of the human genome (Davis *et al.* 1990). Furthermore, only through the refinement and application of developmental biology and population genetics studies of gene distributions can *susceptibility* be studied and interpreted scientifically. Public and scientific misperceptions of susceptibility are probably one of the most prominent problems facing those interested in the development of genetic medicine.

There is a tempting and widespread error in reasoning which is exacerbated by the slippage back and forth between distinct biological meanings of 'normal'. Under a biochemical causal model, let us suppose a person in whom arteriosclerosis is damaging their health and whose phenotype is clearly 'abnormal' and 'diseased' according to the medical model. The desired biological explanation traces a causal chain from the genes through the expression and development processes to the resulting pathological state. When asked, 'How does arteriosclerosis happen?', the answer is given: 'There's a gene for this, which, under these environmental circumstances, takes part in such-and-such a causal chain, resulting in buildup on the arterial walls.'

So far, so good. The problem arises when we attempt to understand what it means if a person tests positive for that gene. It is tempting to think that this means they either have or will have arteriosclerosis. But this would be to mistake a contributing cause for a sufficient condition.[19] Exposure to a cold virus is a contributing cause for coming down with a cold, but it is not sufficient; the immune system must also fail to control the spread of that virus in the body. Similarly, having a certain gene might contribute to getting arteriosclerosis, but it is not sufficient; the environmental conditions must also be right in order for arteriosclerosis to become a health problem.[20]

So, take persons with 'the arteriosclerosis gene'. Are they abnormal? If we define the standard biochemical causal model of fat metabolism as

---

[19] See the discussion in Hubbard 1990 and Cranor 1994. I believe that Hubbard's view can be interpreted as a rejection of the medical appropriateness of the biochemical causal model owing to the fact that it omits environment as a causal factor.

[20] Another problem with this model of genetic disease is that 'diseases caused by a malfunction in one gene tend to be rare', according to molecular geneticists at Harvard Medical School (Davis *et al.* 1990: 343).

'normal', then they are 'abnormal', and the cause of that abnormality is genetic. Are they abnormal on the phenotypic level? Are they diseased? Not ncessarily, if we are using the medical model. Inferences that slip from a discovery of genetic abnormality to conclusions of medical abnormality or disease are fundamentally mistaken and unjustified. It is not that the two levels are unrelated or irrelevant to each other; it is just that slipping from 'abnormal' in one to 'abnormal' in another without evidence is not defensible scientifically.

CONCLUSION

Claims that the Human Genome Project will give us 'the recipe to construct human beings' or the keys to understanding 'human nature' are misleading at best.[21] The usefulness of molecular- or DNA-level descriptions by themselves is extremely limited. Genes whose descriptions on the DNA level differ from an accepted paradigm of the biochemical causal model may or may not be physiologically significant. Regarding the medical uses of the Human Genome Project, then, the only relevant form of variation is determined by a medical or physiological model. It is important to understand that the medical model of health and disease is just as subject to value-judgements as it ever was. Its necessity has not diminished; it has simply gained a wider scope for use. Deciding how human beings *ought* to function is still a negotiated social decision. Proponents of the medical uses of the Human Genome Project have ignored the problems arising from the social nature of disease, but these will not disappear. On the contrary, biotechnology has new powers to implement and *enforce* codes of normality. Molecular biology cannot provide an objective and scientific code of health and normality. The scientific ability to make fine discriminations of variation, and the technological power to act on them, make it imperative that *variation itself* be the focus of a searching public debate and educational effort.

REFERENCES

Angier, N. (1991), 'Molecular "Hot Spot" Hits at a Cause of Liver Cancer', *New York Times* (14 Apr.).

Baird, P. A. (1990), 'Genetics and Health Care: A Paradigm Shift', *Perspectives in Biology and Medicine*, 33: 203–13.

Bayer, R. (1981), *Homosexuality and American Psychology: The Politics of Diagnosis* (New York: Basic Books).

[21] Zinder 1990: 96, Smith and Hood 1987; cf. Watson 1990 and Vicedo 1991.

Caplan, A. L. (1989), 'The Concepts of Health and Disease', in R. M. Veatch (ed.), *Medical Ethics* (Boston: Jones & Bartlett), 49–62.

Council for Responsible Genetics (1990), *Position Paper on Human Genome Initiative* (Boston: Committee for Responsible Genetics).

Cranor, C. F. (1994), 'Genetic Causation', in Cranor (ed.), *Are Genes Us?* (New Brunswick, NJ: Rutgers University Press), ch. 8.

Culliton, B. J. (1990), 'Mapping Terra Incognita (Humani Corporis)', *Science*, 250: 210–12.

Davis, B. D. *et al.* (1990), 'The Human Genome and Other Initiatives', *Science*, 249: 342–3.

Dulbecco, R. (1986), 'A Turning Point in Cancer Research: Sequencing the Human Genome', *Science*, 231: 1055–6.

Friedmann, T. (1989), 'Progress toward Human Gene Therapy', *Science*, 244: 1275–81.

——(1990), 'The Human Genome Project—Some Implications of Extensive "Reverse Genetic" Medicine' (opinion), *American Journal of Human Genetics*, 46: 407–14.

Fujimara, J. (1988), 'The Molecular Biological Bandwagon in Cancer Research: Where Social Worlds Meet', *Social Problems*, 35: 261.

Gajdusek, D. C. (1970), 'Physiological and Psychological Characteristics of Stone Age Man', *Engineering and Science*, 22: 56–62.

Gollin, E. S., Stahl, G., and Morgan, E. (1989), 'On the Uses of the Concept of Normality in Developmental Biology and Psychology', *Advances in Child Development and Behavior*, 21: 49–71.

Holden, C. (1991), 'Probing the Complex Genetics of Alcoholism', *Science*, 251: 163–4.

Hubbard, R. (1990), *The Politics of Women's Biology* (New Brunswick, NJ: Rutgers University Press).

Jukes, T. H. (1988), 'The Human Genome Project: Labeling Genes', *California Monthly* (Dec.): 15–17.

Karjala, D. S. (1992), 'A Legal Research Agenda for the Human Genome Initiative', *Jurimetrics*, 32: 121–222.

LeVay, S. (1991), 'A Difference in Hypothalamic Structure between Homosexual and Heterosexual Men', *Science*, 253: 1034–7.

McKusick, V. A. (1989), 'Mapping and Sequencing the Human Genome', *New England Journal of Medicine*, 320: 910–15.

Roan, S. (1991), 'Check It Out', *Oakland Tribune* (18 Apr.).

Sedgwick, P. (1973), 'Illness—Mental and Otherwise', *Hastings Center Report*, 1: 19–40.

Smith, L., and Hood, L. (1987), 'Mapping and Sequencing the Human Genome: How to Proceed', *BioTechnology*, 5: 933–9.

Temkin, O. (1977), *The Double Face of Janus* (Baltimore: Johns Hopkins University Press).

Vicedo, M. (1991), 'The History, Scientific Value, and Social Implications of the Human Genome Project' (MS).

Watson, J. D. (1990), 'The Human Genome Project: Past, Present, and Future', *Science*, 248: 44–9.

Zinder, N. D. (1990), 'The Genome Initiative: How to Spell Human', *Scientific American*, 263 (July): 128.

29

THE HUMAN GENOME PROJECT:
RESEARCH TACTICS AND
ECONOMIC STRATEGIES

ALEXANDER ROSENBERG

In the Museum of Science and Technology in San Jose, California, there is a display dedicated to advances in biotechnology. Most prominent in the display is a double helix of telephone books stacked in two staggered spirals from the floor to the ceiling twenty-five feet above. The books are said to represent the current state of our knowledge of the eukaryotic genome: the primary sequences of DNA polynucleotides for the gene products which have been discovered so far in the twenty years since cloning and sequencing the genome became possible.

I. THE ALLEGORY OF THE PHONE BOOKS

In order to grasp what is problematical about the Human Genome Project (HGP), I want you to hold on to this image of a stack of phone books, or rather two stacks, helical in shape. Imagine that each of the phone books is about the size of the Manhattan White Pages, and that the two stacks of phone books reach up a mile or so into the sky. Assume that the books are well glued together, and that there are no gusts of wind strong enough to blow the towers down. The next thing you are to imagine is that there are no names in these phone books, or on their covers—only numbers. We do know that each phone number is seven digits long, and we know that the numbers have been assigned to names listed alphabetically, but without the names, we can't tell to whom a number belongs. Moreover, the numbers are not printed in columns down the pages that will enable us to tell where one phone number ends and the next begins. Instead of being printed in columns down the page, the numbers begin at the top left and fill

First published in *Social Philosophy and Policy*, 13 (1996): 1–17. Reprinted by permission.

up the page like print, without any punctuation between them. They are grouped within area codes, of course, and we can tell when one area-code list stops and another begins, but we don't know the area codes, still less what geographical areas they cover.

Sounds like a set of phone books that would be pretty difficult to use, doesn't it? Well, let's make them harder to use. Of course, none of the individual phone books have names or any other identifying features on their covers. In fact, the books don't have covers, and, what's more, the binding of each directory was removed before the stack was constructed, and a random number of successive pages of adjacent phone books were rebound together. This rebinding maintains the order of pages, but it means that each volume begins and ends somewhere within each directory, and there is no indication of where these beginnings and endings are.

Can we make our mile-and-a-half-high double stack of phone books even harder to use? Sure. Imagine that somewhere between 90 and 95 per cent of all the phone numbers in all the phone books have been disconnected, or have never even been assigned to customers—and of course we don't know which ones they are. These unused numbers look just like sequences of assigned phone numbers; they even have area-code punctuation, though there is no geographical area assigned to these area codes. Remember, we can't tell which area codes represent a real area and which do not. We do know that between 5 and 10 per cent of the numbers are in area codes which have been assigned, and that within these assigned area codes there are long lists of phone numbers of real phone-company customers. Although we don't know which are the area codes that are real, or where they are in the directories, we do know some interesting things about these area-code phone number listings. First, sometimes area codes and their phone numbers are repeated one or more times rather close together in a single volume, and sometimes they are repeated in distant volumes in the stack. Second, sometimes there are sequences of phone numbers which are very similar in digits to the numbers in a real area code, but their area codes are unassigned and all the phone numbers are unused. Even within almost all of the real area codes, the lists of assigned phone numbers are interrupted by long sequences of digits which, when grouped into phone numbers, are unassigned; sometimes within a real area code there are several of these sequences, longer than the sequences of assigned phone numbers within the area code.

Perhaps you are tiring of all the bizarre details of this idea of telephone numbers impossible to read. So I will stop adding detail to our picture. But don't let go of the picture. Imagine that someone—a numerologist, say— now proposes to you that for 3 billion dollars of the US government's money, he will put together a team that will transcribe all the digits in the mile-and-a-half-high double stack of phone books into a computer. It's not the phone numbers he offers to transcribe, just the digits—one after the other—unsegmented into the phone numbers. The numerologist promises to make the list of digits available to anyone who asks for it, free. Assume further that 3 billion dollars is the cost of copying out the numbers, with no hidden profit for the numerologist.

I suppose one's first reaction to such a proposal would be to thank the numerologist for his offer, but to decline it on the grounds that the list of digits is of no immediate use to anyone, even if we were going to have a very large party and wanted to invite everyone who had a phone number. But since it is in our nature as philosophers to wonder, we ask our numerologist, 'What's in it for you? Why would you do this transcription at cost, without any profit?'

Imagine that our numerologist is candid, and comes clean as follows. He is not really a numerologist, but rather represents a relatively large number of privately held direct-sales companies, each of which has a potentially very useful product, which it can only sell over the telephone. The companies know that there are enough potential customers out there to go around, so that each of their shareholders can become rich through the sale of the very useful product they can manufacture, if only they had the phone numbers of the customers. So, our numerologist/direct-sales-marketing representative says that if the companies he represents had all the digits, they could sell the products, make every customer better off, and become rich themselves.

There are two responses one should make to the numerologist's admission. The first is: even if you can segment the 3 billion digits into phone numbers, why should the government pay for the phone lists, if you and the consumers are the ones to profit? The second and more important response is: surely putting all those digits from all those books into a large computer file, without being able to tell the meaningless ones from the meaningful ones, just for starters, is not the best way to get in touch with potential customers.

Consider the first question: Why should the government pay for the phone list? The answer given on behalf of the direct-sales-marketing companies is that their products are guaranteed to help people stay healthier and live longer. But in that case, if they think putting all these phone numbers in a computer memory is so valuable, why don't they arrange to fund the project themselves, and reap the rewards by selling the products? The answer we get is that their capital is tied up in even more valuable investments, and besides, there is a free-rider problem. If any of them get together to fund the transcription, the disks on which the transcription is recorded could easily fall into the hands of other members of the trade association without these other firms paying. So, we might reply, what's it to us? Why should the government get you out of this predicament? If it does, will you cut the government in on your profits from using the phone numbers? Oh no, comes the reply. That would be a disincentive to developing new products to sell to people on the phone list.

But wait a minute, let's consider the second question: What's the use of a transcription of all these digits? Aren't there far better ways to get the names of customers with telephones? In fact, is this any way of getting the names of customers at all? Well, comes the response, what if developing the technology to transcribe all these phone numbers will also enable us to identify the real area codes, to segment the digits into meaningful phone numbers, and to begin to tell which ones are actually in use? That might justify some investment in transcribing the phone numbers. Unfortunately, our numerologist can give us no such assurance. At most, he can promise that once we have the total list of real area codes, 5 to 10 per cent of the list of digits will become valuable. Are there ways of identifying these area codes? we ask. Certainly, says our interlocutor. Well, what are you waiting for, go and find them. When you have done so, you may or may not have any use for the transcription of all the digits. Until then, however, you would be wasting your time, or someone else's, along with a lot of money, government or private, to transcribe all these digits, including the ones in the unassigned area codes and the numbers with no customers.

This allegory, I suggest, pretty well matches the biochemical facts about the human genome, the molecular structure of the nucleic acids that compose it, the prospective pay-off to sequencing the 3 billion base pairs of the human genome, and the policy advocated by the proponents of the HGP. Here is a brief explanation of my simile between the human genome and the mile-and-a-half-high double stack of phone books. The human genome contains about 3 billion base pairs of purine and pyrimidine nucleic acids; these are the digits in our 'phone numbers'. It is hoped that the average cost of providing the whole sequence can be brought down to one dollar per base. Three of these bases constitute a phone number with three digits—a codon of three nucleic acid bases. Ninety to 95 per cent of these sequences code for DNA that has no role in gene transcription—this is the so-called junk DNA.[1] They compose phone numbers that are not in use.

---

[1] Molecular biologists are sensitive to the fact that calling 95 per cent of the human genome 'junk DNA' undercuts the rationale for sequencing the whole genome. As a result, some are suggesting that the scientific community was over-hasty in coming to the unanimous conclusion that DNA sequences with no known possible function are 'nonsense'.

for automated sequencing, etc. The HGP requires the joint problem-solving skills of biochemists, physicists, engineers, mathematicians, and computer scientists. The real value of the breakthroughs required to sequence the whole genome is that as 'spin-offs' they are crucial to answering currently pressing research questions that the scientists and commercial biotechnology firms have. These questions are not about the sequence of the human genome as a whole, but about functional units of it which have scientific and commercial interest. It may well turn out, of course, that for reasons no one can now identify, having the entire sequence of the human genome will be valuable information. But to act on the supposition that it will be is a very expensive gamble. Why, then, the strong pressure from the molecular biology community to establish and maintain the HGP?

We can shed considerable light on why leading figures in molecular genetics have committed us to the HGP, and perhaps draw some policy-relevant normative conclusions by applying a little of the economics of information to the project.

The first thing to know about information generally is that it is unlike most other commodities. Other things being equal, economically valuable information always benefits larger-scale producers over smaller ones—since the costs of discovery are the same for each, but can be spread over a larger production run by larger producers. This means that larger concerns can outbid smaller ones for information, and that concerns with exclusive rights to some valuable information will grow larger than others. The result will be a loss of general welfare well understood in the economics of monopoly and monopolistic competition.

The second thing of importance about information is its non-appropriability. An individual who has acquired some information cannot lose it by selling or giving away to others a non-exclusive right to use it. Moreover, information acquired at great cost in research, or purchased for non-exclusive use, can be sold very cheaply, or even given away. Consider what happens to a piece of software newly introduced in an office. The market price of a token of information is generally much smaller than the cost of research to develop the first token. The information cannot be fully appropriated by a purchaser since the seller can sell to others. Nor can a seller recapture its full commercial value in the price charged for selling a token of the information, because this value includes returns to purchasers who resell. Thus, the production of new information is always below the optimum level, *ceteris paribus*. According to established economic analyses, then, a competitive economy always under-invests in research and development. This is a form of market failure (Arrow 1984: 142–3).

Because it is non-appropriable, and not provided at optimum levels, information has an economic role rather like that of a public good. An economically rational individual will invest resources in the discovery of new information up to the level of the marginal expected value of the information to the discoverer. But the expected value of the new information to all researchers is the sum of the marginal expected value to each of them, and will be larger than the expected value to the individual discoverer or inventor alone. Unless some part of the marginal value to other potential users can be captured by the discoverer or inventor, he will lack the purely economic incentive to invest in research up to anywhere near the point at which the costs of research equal the benefits to the whole economy of the research. This leads to under-investment, because of the scope for 'free-riding'.

Solutions to problems of market failure in the provision of public or other non-appropriable goods usually involve governmental coercion. In the case of information, such coercion comes in the form of patent protection. The solution to the problem of undersupply is the establishment of governmentally enforced copyrights or patent rights in the discovery of the information. Patent and copyright protection enables the first discoverer to secure some of the marginal value that other users can produce by implementing the discoverer's information. Thus, it encourages individual investment at levels closer to the optimal level of investment for the whole economy. Patent protection is not a perfect solution to the underproduction problem; it cannot shift the market to a welfare-optimum equilibrium. No invention has a perfect substitute, and therefore an exclusive right to sell the invention allows its owners to charge a non-competitive price for their information. This reduces consumption of information by raising its cost. Thus, patents and copyrights alleviate the underproduction problem while fostering an under-utilization problem.[4]

Now, in the case of much biotechnological research, the resulting equilibrium does lead to investment in research. Witness the development of new products, the explosion of new firms, and their attraction of venture capital. In the case of the DNA sequence of *Homo sapiens*, however, patent protection is unlikely to have the same mitigating effects. Patents are unlikely to result in either the sequence or the spin-offs being produced at anything near optimum levels.

To begin with, there are reasons to think that neither the whole sequence nor large portions of it are open to patent protection. US patent law restricts patents to new or useful processes, machines, manufactures,

---

[4] For further discussion, see Hirshleifer and Riley 1992: ch. 7.

or compositions of matter, or useful improvements of them, including genetically engineered living organisms (Eisenberg 1992: 227; see also Sgaramella 1993: 299–302). But it excludes as unpatentable those scientific discoveries that cannot be immediately used for any human purpose. Absence of immediate utility would presumably exclude both the primary sequence of base pairs and most physical maps of the human chromosome until discovery of the functional units with which their landmarks are correlated. Patents are also unavailable for so-called obvious extensions or applications of prior information. Thus, attempts to patent large numbers of cDNAs produced by means well understood by all researchers have been challenged as failing the 'non-obviousness' test. Since these cDNAs are far more immediately useful than either the whole sequence, large parts of it, or a physical map of the whole chromosome, the likelihood of patent protection for any of these latter seems low.

However, usefulness and non-obviousness are tests which some of the spin-offs may satisfy. Technology useful for mapping and sequencing alone may not be any more patentable than the sequences are, because of either non-usefulness or obviousness (Eisenberg 1992: 227); and even if they were patentable, these innovations would probably be restricted in their foreseeable use to the HGP labs alone, and therefore would probably not be lucrative enough to net returns to their discoverers. However, some of the spin-off technologies will be of great value in producing useful things like gene products with significant pharmaceutical value. Thus, no single researcher or group has an economically measurable incentive to provide the whole sequence, because there is no patent protection available for it; nor do researchers have material incentives to produce technologies that do nothing but sequence. However, beyond the most obvious improvements in informatics and sequencing technologies, researchers cannot tell when spin-offs will turn out to be of great value, and will be patentable.

Eventually, the sequence will have value, once we have a great deal more functional information; but no one can predict how much value it will have, scientific or economic. Similarly, the spin-offs are sure to be harnessed to technological breakthroughs; but the probable time required for breakthroughs to be harnessed profitably is long, and the probabilities that any particular lab will secure returns from these breakthroughs is very small. Consider the transistor or the laser: the former was at first expected only to help produce better hearing aids; the latter was invented at Bell Labs, but was almost not patented, since it had no evident application in telephone technology, and would not until the advent of fibre optics twenty-five years later. No one anticipated its role in the household compact-disc player. No lab has an economic incentive to invest in the production of the sequence or the spin-offs alone, or a non-negligible rational expectation of inventing patentable and lucrative spin-offs.

How should an (economically) rational molecular genetics laboratory respond to this degree of uncertainty about information which may or may not be patentable, the well-known consequences in under-investment in research and development, and the potential public-goods effects of the sequencing project for the molecular genetics community? Let us raise the question for labs, and not for individual researchers, both because the unit of research nowadays is the lab, and because in the present context, the role of labs is rather like that of individual business firms—in fact, some of the participants in the HGP are private companies. How a rational molecular genetics lab might respond to this state of affairs depends on whether the lab is already a relatively large one or not.

The fact that sequence and spin-off information is a public good for molecular genetics labs is no incentive to the individual researcher to provide it; it is an incentive to free-ride on the willingness of others to provide the good. Could researchers in molecular genetics enter into an enforceable agreement among themselves to produce the sequence and the spin-offs? They could do so, provided they were willing to work together to secure a coercive agent to enforce the agreement, and to provide the resources to develop the technology to determine the sequences.

This last requirement is the most difficult one, of course. Not only do individual labs not have control over sufficient disposable research funds to pool together, but if they did, each laboratory would have an incentive to understate its resources. However, suppose they mutually agreed to propose that the government provide these resources, subject to each principal investigator's surrendering the patent right to information that might be generated. In return for this subvention, the larger labs surrender their right to patent sequences and some spin-offs.

If the analysis of patent rights sketched above is correct, the expected value of this right is very low: the sequence information that is predictably forthcoming cannot be patented, and the spin-off information which might be patentable is unlikely for any given lab. The expected value of patent rights is thus equally low for small labs and large ones. Accordingly, large labs will surrender patent rights 'cheaply', and labs too small to participate will not complain, even though they have not been made parties to the contract, and secure no governmental support from it. Note that federal support of the HGP does not prohibit the patenting by researchers of useful, non-obvious discoveries made while supported by HGP funds.

So, by offering the sequence of the human genome as a non-patentable public good to all labs in exchange for the funds to carry out HGP research, the large labs give up little in return for a great deal. And the larger the lab, the greater the economies of scale it will be able to apply for any unit of foreseeable spin-off information. Corporations pursuing research in biotechnology will happily support this strategy, since their immediate commercial interests are limited to foreseeable spin-off technology, and the expected value of this technology for each of them is sufficiently similar that none will have an incentive to oppose a no-patentability policy.

The HGP is a good deal for the molecular genetics community. Or rather, it is a good deal for those participants which can secure recognition as HGP research centres, or association with a centre. This will include the larger labs in universities, and of course the National Laboratories operated for the Department of Energy by leading universities. Moreover, to the extent that the allocation of funds for the HGP does not reduce support for other molecular-biological research, smaller laboratories which have no stake in the HGP will have no grounds to complain, and some interest in acquiescing: they may profit from the spin-offs—reagents, assays, informatics, etc. And the small labs have incentives not to criticize the programmes of established figures in the field who have advocated the HGP in public debate.

But is the trade-off—3 billion dollars in exchange for the sequence and the spin-offs that may not be patentable anyway—worth it to the governments and citizens whose taxes support the research? Is the exchange between the large labs and the government equitable? Beyond equity, does it provide incentives to pursue the most pressing questions, the most promising lines of enquiry, the highest-quality research, in molecular biology? Affirmative answers to these questions are doubtful. The multiplication of small, medium, and large biotechnology firms—none of them blindly sequencing the human genome, all of them seeking the sequences that code for gene products—suggests that the HGP is not the most promising line of enquiry for biomedical discoveries; interest in other sequences, like those of yeast, *E. coli, C. elegans,* the mouse, seems very great, but little research is devoted to sequencing for its own sake the entire genomes of these creatures.

If the government's intention in providing support for the HGP is its expected pay-off for health care, then the exchange cannot be a fair one. For were the expected pay-off high enough, the costs would have been internalized by the molecular genetics community, especially the biotech firms. If the government's intention in providing support for the HGP is simply to support the best research being done, then surely the bottom-up system which has worked well in the traditional grant-award process, and in the processes used by the National Science Foundation and the National Institutes of Health, would more fully ensure high-quality work on the most pressing problems. Of course, if sequencing the human genome were then the dimensions of the proper sphere of government it would make top-notch work more feasible. One argument for government support of large-scale scientific research projects—most arguments for government support for scientific research justify the support by appeal to its eventual instrumental value to the whole society, and its intrinsic value in expanding understanding of the way the world works. Even those who might doubt that such objectives are within the proper sphere of government can accept another argument for the public support of science. The regulatory burden—animal care and human subjects oversight committees, toxic waste prohibitions, fair employment practices, public disclosure requirements, as well as drug-free work-place-, conflict-of-interest-, anti-lobbying-, and other certifications—has so vastly increased the non-experimental costs of scientific research (in the USA at least) that government is obliged to fund these intrusive mandates, in science and elsewhere.

But everything the theory of public choice tells us about the behaviour of individuals and institutions suggests that the government must structure and administer the support of scientific research to maximize decentralization in decisions about the use of its funding. Only decentralizing the decision-making about research subjects and strategies to the lowest level possible will harness most effectively the distributed knowledge of the scientific disciplines. Like other human taste-goods, needs in science provides individuals with strong incentives to centralize decision making to themselves, direct interests, in the interests of institutions that enfeeble competition. The arguments of this section are examples of this tendency. The problem I have been at pains to point out is that the Human Genome proposals for the use of public funds from the defects of public resources in the other sphere, but is not effectively freed from these defects than the HGP is, at its centralizing mode?

I am grateful to James Fleischer for advice on several points of this chapter, and to Fleischer who has committed some sins of the grossest kind.

PART IX

PROGRESS

INTRODUCTION TO PART IX

MICHAEL RUSE

Progress is an idea of the eighteenth-century Enlightenment. Encouraged by advances in science and technology, people became increasingly convinced that virtually unlimited improvement in human knowledge and welfare is possible, if only we work long enough and hard enough. The idea was, nevertheless, judged heretical by Christians, because it was seen as opposed to the doctrine of Providence: that, save for God's grace, we humans are, and will ever remain, mired in sin, condemned to misery and pain. Naturally enough, more radical thinkers, outside the lines of authority, were thus stimulated to push progressivist doctrines to the extreme. Before the century's end there were those—notably Erasmus Darwin, the physician grandfather of Charles Darwin—who generalized from progress in human affairs, and became convinced that the whole of creation is in some sense progressive. In the world of organisms, where people were already used to thinking of everything as part of an ordered Chain of Being, from the simplest to the most complex, progress was taken to mean evolution: a natural process of development, from the most primitive life form, the 'monad', right up to the most complex and sophisticated and best, human beings, our own species.

Charles Darwin, in his *Origin of Species*, published in 1859, was much concerned to separate his theory of evolution from radical doctrines of the past, and with reason it is often thought that he broke the link between progress and evolution. After all, his central mechanism is natural selection, the survival of the fittest. Who in any particular circumstance is to say what is the fittest or best? In times of abundance, it may be the biggest and strongest. In times of starvation, it may be the very opposite. There is a relativism here which seems to negate any possibility of an upward climb, towards the best (however this may be defined).

As Robert J. Richards shows (Ch. 30), however, matters are somewhat more complex than this. Darwin himself, a good, solid, middle-class Victorian gentleman, was personally no less convinced of progress than was his grandfather. Richards argues that Darwin, like his contemporary, Herbert

Spencer (a passionate progressionist), was eager to see life's history as one of upward climb, from the primitive to the complex, ending with *Homo sapiens*. It is only our desire to see Darwin from a modern perspective that has hidden this fact from our view.

But how, then, is progress to be effected, given such a mechanism as selection? It is here that the chapter by Michael Ruse (31) picks up the theme, as it is shown that for Darwin and for his followers today a key evolutionary mechanism is a kind of competition, or 'arms race', producing a sort of comparative progress. Lines or groups of organisms compete against each other, and in the course of time this leads to improvement on both sides. The lion gets faster in order to catch its prey, but in turn the gazelle gets faster in order to escape its predator. There must be an upper limit on the speed achievable, but not before both sides have significantly strengthened their respective adaptations.

This is comparative progress. How can it lead to what Ruse calls 'absolute progress', where the end-point is humans, especially intelligent humans? In the *Origin*, this is a question which is left rather untouched. However, as Ruse's survey shows, it is a question which still needs answering today, for, although many evolutionists are no longer enthused by the idea, there remain many (including some very distinguished practitioners) who still believe that ultimately evolution has meaning. It is not a slow meandering process going nowhere. Rather, for all the qualifications one must make, it can be seen as a process that goes steadily from the simple to the complex, from the unimportant to the very significant, from the monad to the man.

The essay by Daniel McShea (Ch. 32) suggests that, if one continues in the Darwinian mode, one's chances of finding a happy answer to this question are minimal, if they exist at all. He argues that, impressions notwithstanding, the evidence for genuine progress is lacking. Indeed, the story is the other way, especially since the common practice of identifying complexity with progress is doomed to failure. Who dare say that the lion is 'better' than the whale, even though by any reasonable measure the backbone of the lion is significantly more complex than that of the whale? It is in its very simplicity that we find the success of the whale in its habitat, the sea. It would be a handicap if it had the complex vertebrate features of a land-based mammal like a lion.

Stephen Jay Gould (Ch. 33) is no more enthused by the notion of progress than is McShea. Indeed, he manages to be considerably ruder on the topic! Nevertheless, he does feel that there is something in nature which does need explaining, if only because the hold of the idea seems as strong today as it was two centuries ago. He feels that perhaps the notion of 'directionality' might capture the sense that we all have that the fossil record is not simply random, but has a definite pattern—a pattern that would be reversed were the record itself reversed. To this end, Gould points out that new kinds of life often seem 'bottom-heavy' when judged over time—they diversify and then start to scale down, so that the overall effect is pear-like, with the stem at the top. Although Gould himself does not incline to a Darwinian explanation, no doubt those who take selection as all-important would provide an account which stressed the initial emptiness of the ecological niche occupied by the new form (thus permitting maximum diversity), and then the consequent struggle as the existent forms battled for supremacy in the overcrowded space—with many forms which were able to survive in less stringent conditions now becoming extinct.

Whether Gould's suggestion is thought adequate is still not answered in a definitive manner. What is clear is that the question of progress is an ongoing debate. Is McShea right in denying a link between progress and simplicity? Is absolute progress an impossibility if Darwinism is true? Does Gould offer a genuine alternative? These are still open questions.

30

THE MORAL FOUNDATIONS OF THE IDEA OF EVOLUTIONARY PROGRESS: DARWIN, SPENCER, AND THE NEO-DARWINIANS

ROBERT J. RICHARDS

An eminent evolutionary thinker has interpreted natural selection as the muscle producing biological and social progress. He claims that 'as natural selection works solely by and for the good of each being, all corporeal and mental endowments will tend to progress towards perfection'. Most historians, philosophers, and biologists, however, would regard attaching the idea of progress to Darwin's theory as comparable to stitching a Victorian bustle on the nylon running shorts of a woman marathoner, a cultural atavism disguising the slim grace supplying the real power. Progress becomes Herbert Spencer's evolutionary theory, but that is a museum piece long ago shelved. Darwin's still vital conception, by contrast, threatens notions of progress embedded in mid-Victorian culture. 'Darwin's mechanism', Peter Bowler has recently insisted (1986: 41), 'challenged the most fundamental values of the Victorian era, by making natural development an essentially haphazard and undirectional process.' This historical assessment receives support from probably the most influential logical analysis of Darwinian theory, that of George Williams.

In *Adaptation and Natural Selection* (1966: 34–55), Williams isolates several simple features of the evolutionary process that would seem to defeat any possibility of long-term progressive development. First is the structure of the mechanism itself: natural selection is the substitution of alleles more favourable in a given environment for those less favourable. Since the selective environment constantly changes, what was adaptive in the past must become increasingly less so. Add to this the thermodynamical wheeze of entropy (see Wiley 1988)—indicated by genetic mutation and recombination—well, the situation would appear as Leigh Van Valen (1973) has characterized it. The Red Queen has to keep running on the

First published in M. H. Nitecki (ed.), *Evolutionary Progress* (Chicago: University of Chicago Press, 1988), 129–48. Reprinted by permission.

slipping treadmill of environmental change just to remain in the same place—no long-term progress appears possible. In addition to his a priori analysis of the possibility of evolutionary progress, Williams examines three candidates for empirically grounding the idea: progress as accumulation of genetic information, progress as morphological complexity, and progress as effectiveness of adaptation. All of these, at one time or another, have been proposed as measures of evolutionary advance. But these three candidates succumb to those structural features of the evolutionary process just mentioned. Further, Williams has little trouble in discovering obvious counter-examples to each. For instance, human beings may be more advanced than apes in their effective adaptation to a broad range of environments—so you might mark the transition from *Australopithecus afarensis* to *Homo sapiens* a progressive one. But where you find man, you also find the cockroach. And the cockroach will dwell happily in environments quite lethal to our species. According to the standard of effective adaptation, cockroaches may be the most important product of progress.

Let me mention one more recent examination of the idea of evolutionary progress—this by one who is, like Bowler, a historian, but also like Williams, a scientist. Ernst Mayr, in his *Growth of Biological Thought* (1982), has considered both the historical and the scientific questions: namely what was Darwin's view of progress, and what should we make of the idea of evolutionary progress? Not surprisingly, Mayr thinks we should believe what Darwin did. He maintains that 'Darwin, fully aware of the unpredictable and opportunistic aspects of evolution, merely denied the existence of a lawlike progression from "less perfect to more perfect"' (1982: 531). Darwin strongly objected to any notions of 'an intrinsic drive to perfection, controlled by "natural" laws', says Mayr (p. 532). Darwin might, as a cultural afterthought, have occasionally referred to evolutionary progress, but such short-term evolutionary innovations that every biologist must recognize were only the 'a posteriori results of variation and natural selection' (ibid.). Evolutionary laws leading to biological progress have been promulgated by Spencer and some minor Darwinians, but, it is believed, the master steered clear of that slope which leads to a yawning abyss of social Darwinism and sociobiology (but see Ospovat 1981: 210–28 for a different analysis of Darwin's notions about progress).

The quotation with which I began—that 'as natural selection works solely by and for the good of each being, all corporeal and mental endowments will tend to progress towards perfection'—of course, comes from the *Origin of Species* (1859: 489). Perhaps one might simply chalk it up to some carelessness on the part of Darwin, maybe even a sop to the theologically blinkered. But I don't think so. The delicate scroll-work and the emblems depicting cultural advance etched into the machinery of natural selection were not merely decorative motifs; they depicted functional parts of his device. Indeed, Darwin crafted natural selection as an instrument to manufacture biological progress and moral perfection. In this respect, his theory does not substantially differ from Spencer's, upon which much abuse is often heaped for making evolution necessarily progressive. Let me begin, then, by briefly considering the ways in which a moral vision structured Spencer's own theory of evolution. His case is instructive for understanding the historical logic of Darwin's theory.

THE MORAL FOUNDATIONS OF SPENCER'S THEORY OF EVOLUTION

Spencer's evolutionary theory is generally thought to have nasty moral and social consequences. Typically he is understood to have formed his moral conception around a razor-edged theory of survival of the fittest, an embrace most people think could only have deadly consequences for social philosophy. I believe, on the contrary, that the relation between Spencer's evolutionary theory and his moral theory is just the reverse of what is usually supposed. Spencer first formed a utopian social philosophy that would bear striking resemblance to that of his two contemporaries working at the British Museum, two Germans plotting revolution for their fatherland, Marx and Engels. Unlike their conception, though, Spencer's social theory was braced by distinctively English theological and moral principles. It was on this theological and moral loom that the fabric of his evolutionary theory was woven (Richards 1987: chs. 6 and 7; Peel 1971).

Spencer grew up in the English Midlands, the country of Erasmus Darwin, James Watt, and Josiah Wedgwood—those practical-minded men who founded the Derby Philosophical Society, which had Spencer's father as its recording secretary. The Spencer family counted itself among John Wesley's earliest followers. The father, though, dissented even from such heterodox association, and thereby set the son on the Nonconformist's path. In his youth, Spencer became involved in several Nonconformist agitation groups, often with his uncle Thomas Spencer serving as leader. It was his uncle, a curate with a parish near Bath, who arranged for the 21-year-old to contribute a series of letters to the newspaper *The Nonconformist* in 1842. These letters, bearing the title 'The Proper Sphere of Government', set down the theological, moral, and social considerations that served as templates for later developments of Spencer's theory of evolution.

In the letters (1842, 15 June to 23 November), Spencer argued that natural laws divinely designed for man's happiness, which regulated the physical, organic, and mental realms, governed the social realm as well. If God had established the laws of these dominions, only mischief would arise when human beings attempted to tinker with them. Government interference in the social sphere, Spencer believed, deformed the self-correcting natural forces controlling social development. If natural forces were left to play out their roles, then one could expect a divinely ordained consummation. Exactly what that terminus of social development might consist of, Spencer specified more precisely nine years later, in 1851, in his book *Social Statics*. Progressive development, both in the organic and the social realms, would lead to a classless society in which each person would freely exercise his or her talents, limited only by the freedom of others; it would be a society in which the state had withered away, land would be held in common, and women and children would have their freedom equally respected. This Godly conclusion and the natural laws leading to it demonstrated, according to Spencer, that:

Progress . . . is not an accident, but a necessity. Instead of civilization being artificial, it is a part of nature; all of a piece with the development of the embryo or the unfolding of a flower. The modifications mankind have undergone, and are still undergoing, result from a law underlying the whole organic creation; and provided the human race continues, and the constitution of things remains the same, those modifications must end in completeness. . . . So surely must the human faculties be moulded into complete fitness for the social state; so surely must the things we call evil and immorality disappear; so surely must man become perfect. (1851: 65)

Early in the 1840s, Spencer had read Lyell's (1830–3) description and refutation of Lamarckian transformation theory. In 1844, he enjoyed the speculative whimsy of Robert Chambers, whose *Vestiges of the Natural History of Creation* (1844) set forth an evolutionary conception of organic life—a conception that brought quick denunciation from British zoologists and gloomy regret from Charles Darwin, who feared Chambers's book might sink in ridicule his own evolutionary craft. Though Spencer's *Social Statics* resonated finely to these earlier evolutionary proposals, no one immediately felt the vibrations. Those few critics who took notice of the book were hardly disturbed by its faint evolutionary tremors, but they were quite agitated by the martial tempo of its anarchistic socialism. The author appeared as a British Proudhon holding up a moderately bloody fist.

As the result of attending lectures by Richard Owen, Britain's leading morphologist and student of Georges Cuvier, and of reading in physiology—especially the books of William Carpenter (1841) and Henri

Milne-Edwards (1841)—Spencer began consciously and more expressly to formulate his evolutionary ideas. In a series of essays in the early 1850s, in his *Principles of Psychology* (1855), and in his metaphysical work *First Principles* (1862), he laid down those evolutionary conceptions that led Alexander Bain to call him 'the philosopher of the Doctrine of Development, notwithstanding that Darwin has supplied a most important link in the chain' (1863).

The bio-social evolutionary theory that Spencer began to elaborate in the early 1850s was powered by a law that would produce for Spencer— and, I believe, for Darwin—the kind of evolutionary progress that their theological and ethical sentiments demanded. Spencer understood the law to govern 'an advance from homogeneity of structure to heterogeneity of structure'; and he believed that this 'law of organic progress is the law of all progress' (1857: 446). He derived his principle in part from von Baer's and Milne-Edwards's conception of a division of labour in organic structures. Spencer argued that continued adaptation of organisms to complex environments—chiefly through functional acquisitions of heritable characteristics, the Lamarckian device—would produce structures more adapted to deal successfully with those environments. It would create both organisms and societies of organisms that displayed greater specialization of parts within an overall integration of functions—more complex organisms and societies, that is, more progressive and perfect organisms and societies. As Spencer concluded his article 'Progress: Its Law and Cause', 'progress is not an accident, not a thing within human control, but a beneficent necessity' (1857: 484).

Spencer is usually supposed to have formulated his ethical ideas in light of evolution by natural selection, so that the leading principle of his moral philosophy is assumed to be the survival of the fittest. Even his friends— Thomas Huxley, for instance—later read Spencer as advocating a brutal individualism according to which survival would go to those most fit for the Hobbesian struggle, instead of to those who were most morally fit (Huxley 1902: 46–116). But this reading really does misrepresent Spencer's position. He supposed, except in one particular instance, that the leading principle accommodating man to harmonious social development was not competitive struggle, but Lamarckian absorption of the principles of justice and fair treatment. The one conspicuous use which Spencer made of Darwin's device of natural selection was in the form of community selection (ultimately group selection), by which the sentiments of altruism, that pivot of the moral life, might be established, so as to direct a society to ends dictated by the ideals of justice and greatest freedom. The passage to this end would be guaranteed by the unencumbered evolutionary process, which, as Spencer envisioned it in the last sentence of his *Principles of Psychology*, is a 'grand progression which is now bearing Humanity onwards to perfection' (1855: 620).

In his letters to the *Nonconformist* and in the first part of *Social Statics*, Spencer cast his eye to God as the promulgator of those providential laws designed to produce the New Jerusalem. Through the remaining parts of his book, however, God gradually receded into the wings, as a more impersonal nature stepped forward. Her actions revealed the same design, even though she relinquished the burning bush and tablets of stone. The nature that took centre stage in Spencer's conception would not bend to religious importuning—no special pleading, no friends at court, not even Laudian High Church incense could move her. Lawful nature replaced a provident God as disposer of progressive advance. The original theological goal of human development remained intact, but would now be ensured by fixed natural laws, by the principles of evolutionary progress. Spencer replaced divine providence with natural laws, ultimately to achieve the same moral end.

But that was Spencer, and we think his evolutionary ideas went the way of the dodo. What should we say of Darwin? Well, not, I think what has been suggested by authors like Bowler, Williams, Mayr, Gould (1988), and Provine (1988), for instance. I believe that Darwin's evolutionary ideas grew in an environment like to that of Spencer, and that the moral and theological pressures of that environment gave his theory a similar developmental curve. I will consider the history of Darwin's assumptions about progress in three parts: the first dealing with his early notebooks, composed prior to the *Origin of Species*; the second focused on the *Origin* itself and an early draft of it; and the last on the period embracing the composition of the *Descent of Man*. I will argue that from his earliest formulation of the idea of evolution through its mature form in the *Descent of Man*, Darwin conceived of the process as progressive. This progress was marked in several ways. From the early period to the late, Darwin thought that evolution gradually produced ever more complex creatures, that the embryo provided a living palaeontological deposit—an organic picture of this complex development—and that the most conspicuous instances of evolutionary progress were greater intelligence and a moral sense in the human species.

IDEAS OF PROGRESS IN DARWIN'S EARLY NOTEBOOKS

In the initial pages of his first transmutation notebook, opened the summer after returning from his five-year *Beagle* voyage, Darwin speculated that animal groups isolated in a new environment would progressively alter. In July 1837 he penned the following in his 'B Notebook':

> As I have before said, isolate species, especially with some change, probably vary quicker.—Unknown causes of change. Volcanic island.—Electricity. Each species changes. Does it progress. Man gains ideas. The simplest cannot help become more complicated; and if we look to first origin, there must be progress. (1960: 17–18)

From the context of his remarks, it is clear that the theory of biological progress over which Darwin's mind played was that of his French predecessor, Jean Baptiste de Lamarck. Shortly after, however, he also entertained Richard Owen's notions about progressive replacement of morphological types, as these jottings reveal: 'Every successive animal is branching upwards different types of organisation improving as Owen says simplest coming in and most perfect and others occasionally dying out' (ibid. 19). In these early pages of his first notebook, Darwin even mused that man was the goal of progressive evolution: 'Progressive development gives final cause for enormous periods anterior to man. Difficult for man to be unprejudiced about self, but considering power, extending range, reason and futurity, it does as yet appear' (ibid. 49). Unfortunately, the rest of this passage was cut from the notebook, but the gist remains: the final cause, the purpose of the long period prior to the human species' appearance on earth, was to allow time for maturation of those qualities of adaptability, reason, and promise that mark man the most progressive creature wrought by evolution. Ripeness is all.

Later, in his second transmutation notebook, his 'C Notebook', which he kept from February to July of 1838, Darwin reflected that the animal economy had conspired to produce human beings, and that if man were suddenly to die off, new, highly intellectual creatures, much like us, would evolve from current monkeys—though the transforming biological pressures would work no more swiftly nor more obviously than the geological forces Lyell described:

> The believing that monkey would breed (if mankind destroyed) some intellectual being though not MAN,—is as difficult to understand as Lyells doctrine of slow movements. . . . What circumstances may have been necessary to have made man! Seclusion want &c & perhaps a train of animals of hundred generations of species to produce contingents proper.—Present monkeys might not—but probably would,—the world now being fit, for such an animal—man, (rude uncivilized man) might not have lived when certain other animals were alive, which have perished. (Darwin 1960: 74, 79)

Darwin mused in these early jottings that evolving organic relations opened spaces in the economy of nature that could be filled only by those progressive qualities distinctive of our species; if human beings failed to occupy this garden, then another Adam and Eve would, even if they were more simian in demeanour.

Darwin did, however, allow that some more recently evolved species might have, in a sense, regressed; the individuals of such species might have adapted to new, but more simplified environments, and themselves have become less complex as a result. Mammals migrating back to the sea, for instance, might have lost no-longer-useful appendages. But despite the occasional regression of a species, the higher categories of animals, the genera and orders, would, he thought, continue to display progressive development:

My idea of propagation almost infers, what we call improvement. All mammalia from one stock, and now that one stock cannot be supposed to be most perfect (according to our ideas of perfection), but intermediate in character. The same reasoning will allow of decrease in character (which perhaps is case with fish, as some of the most perfect kinds the shark. Lived in remotest epochs). . . . It is another question whether whole scale of Zoology may not be perfecting by change of Mammalia for Reptiles which can only be adaptation to changing world. (1960: 204–5)

Though Darwin often expressed the assumption in his early notebooks that evolution was progressive, yet he handled the idea cautiously. In one often quoted and abused passage he reflected: 'It is absurd to talk of one animal being higher than another—We consider those where the cerebral structures intellectual faculties most developed, as highest.—A bee doubtless would where the instincts were' (ibid. 74). But this passage merely expresses an insight that is surely true: the standard for 'highness' is not a deliverance from nature, but one which we choose. Darwin simply recognized that any standard of measure must be relative to the reasons for selecting it. But he grew ever more certain that man's peculiar endowments—high intellect and a moral sense—were quite obviously ideals of perfection, against which other animals could reasonably be evaluated.

Perhaps a more troublesome kind of comment, found frequently enough in Darwin's notes and letters, is of the sort made to Alpheus Hyatt, an American neo-Lamarckian, with whom Darwin had corresponded in the early 1870s: 'After long reflection I cannot avoid the conviction that no innate tendency to progressive development exists, as is now held by so many able naturalists, and perhaps by yourself' (1902: ii. 344). This was a constant refrain in Darwin's work virtually from the beginning: no innate tendency toward perfection (see also 1909: 47). What Darwin objected to, though, was Lamarck's theory that organisms exhibited an *innate* drive toward complexity, toward greater perfection. But this objection to

Lamarck does not mean that Darwin rejected the idea that evolution was generally progressive. In his early notebooks, he portrayed progress as the inevitable outcome of the logic of the evolutionary process. Progress was the result, not of an internal drive pushing organisms to perfection, but of an external dynamic pulling them to perfection. He supposed that the environment against which a creature would be selected would be the living environment of other creatures, so that each increase of competitive efficiency, each augmentation of specialization, each new trait evolved to meet a new challenge—that all these alterations of one individual would call forth reciprocal development in others. The evolutionary situation had a built-in progressive dynamic, a belief in which Darwin frequently expressed, as in this passage from his fourth transmutation notebook, the 'E Notebook':

The enormous number of animals in the world depends on their varied structure & complexity.—hence as the forms became complicated, they opened *fresh* means of adding to their complexity.—but yet there is no *necessary* tendency in the simple animals to become complicated although all perhaps will have done so from the new relations caused by the advancing complexity of others.—It may be said, why should there not be at any time as many species tending to dis-development (some probably always have done so, as the simplest fish), my answer is because, if we begin with the simplest forms & suppose them to have changed, their very changes tend to give rise to others. (1960: 95)

Ever alive to the crafty ways of nature, Darwin did recognize, as I have already suggested, that some species seemed to have fallen back, to have been chiselled down to a simpler form. We might then ask, as he did in the continuation to the passage just cited: 'Why then has there been a retrograde movement in Cephalopods & fish & reptiles?' 'Supposing such be the case', he went on (ibid. 96), 'it proves the law of development in partial classes is far from true.' Thus, while the great classes of organisms would generally show improvement, some particular parts might not. But even this partial slippage was, he thought, to be expected—under the assumption that where it occurred, the underlying simpler animals had been destroyed. In which case, the apparent anomaly would disappear: natural selection inevitably would shave down some complex creatures and shove them into the gap. When the necessary back-filling was completed, however, natural selection would continue to stack species progressively higher in perfection.

THE IDEA OF PROGRESS IN THE ESSAY OF 1842 AND IN THE *ORIGIN OF SPECIES*

In the *Origin of Species*, Darwin retained his early view of the dynamic and progressive character of evolution. Through constant selection against an

tion exists, so must natural selection. Before historians had carefully scru-
tinized Darwin's early notebooks, it appeared that the structure of the
Origin mapped the path by which he had come to discover natural selection—that is, through analysis of artificial selection. We now think this not
to be the case. Only after he had initially formulated the principle of
natural selection, did Darwin recognize its analogy with artificial selection.
But after he perceived the similarity, I think he did begin to modify his
conception of natural selection according to its model, artificial selection.
An important feature of that modification was to implant in the mechanism a moral heart.

In an 1842 essay—a draft that would form the spine of his larger manuscript for the *Origin of Species*—Darwin first sketched a coherent outline
of his entire theory in which he pictured natural selection as comparable to
the work of an infinitely wise being that crafted nature according to a
divine plan for what is best. In sentences whose thought jumps about with
the energies of literary creation, he reflected:

If a being infinitely more sagacious than man (not an omniscient creator) during
thousands and thousands of years were to select all the variations which tended
towards certain ends . . . for instance, if he foresaw a canine animal would be better
off, owing to the country producing more hares, if he were longer legged and keener
sight—, greyhound produced. . . . Who, seeing how plants vary in garden, what blind
foolish man has done in a few years, will deny an all-seeing being in thousands of years
could effect (if the Creator chooses to do so), either by his own direct foresight or by
intermediate means,—which will represent the creator of this universe. (1909: 6)

In this essay, natural selection worked as a surrogate for the Creator. The
difference, as Darwin conceived it, between man's selection and divinely
guided natural selection was that man selected ineptly and for only those
qualities that pleased him, while natural selection worked with minute care
and for the good of the animal itself. This moral design for the evolutionary process was a direct consequence of Darwin's moulding natural selection in the image of the Creator, and of regarding natural selection to be a
secondary cause responsive to the primary cause of divine wisdom.
Though Darwin's faith in the biblical Creator waned during the decades
after he penned his essay—so that the *Origin of Species* retains its references to God as much from prudent protection against theologically zeal-
ous readers as from his own convictions about a designing Creator—yet
the moral pulse of natural selection remained strong. These moral rhythms
become palpable in this passage from the *Origin*, in which Darwin contrasts man's selection with nature's:

Man can act only on external and visible characters: nature cares nothing for
appearances, except in so far as they may be useful to any being. She can act on

every internal organ, on every shade of constitutional difference, on the whole machinery of life. Man selects only for his own good; Nature only for that of the being which she tends.... It may be said that natural selection is daily and hourly scrutinising, throughout the world, every variation, even the slightest; rejecting that which is bad, preserving and adding up all that is good; silently and insensibly working, whenever and wherever opportunity offers, at the improvement of each organic being in relation to its organic and inorganic conditions of life. (1859: 83–4)

Darwin's conception of the moral operations of natural selection— which watched over the fall of a sparrow and numbered the very hairs on the head of man—supported what was his general sentiment about nature, a sentiment rooted deeply in the Christian tradition: namely, that though suffering and death stalked the world, yet all somehow worked to the good. In that most lyrical passage of scientific prose, the final paragraph of the *Origin*, Darwin poignantly expressed his moral view of evolution:

Thus, from the war of nature, from famine and death, the most exalted object which we are capable of conceiving, namely, the production of the higher animals, directly follows. There is grandeur in this view of life, with its several powers, having been originally breathed into a few forms or into one; and that, whilst this planet has gone cycling on according to the fixed law of gravity, from so simple a beginning endless forms most beautiful and most wonderful have been, and are being, evolved. (1859: 490)

MAN'S INTELLECTUAL AND MORAL PROGRESS

The moral arpeggio of Darwin's theory of evolutionary progress shifted into another key in the *Descent of Man*. The central problem of the first part of the *Descent* was precisely evolutionary progress. Darwin's general theory required progress, particularly in the development of human intellectual and moral faculties, but several critics generally friendly to his theory suggested obstacles preventing such progress (Richards 1987: chs. 4 and 5).

Alfred Wallace, co-founder of the theory of evolution by natural selection, changed his mind concerning human evolution. He came to deny that natural selection could produce man's big brain and tender moral sentiments. He estimated that, after all, for sheer survival, man required a brain 'little superior to that of an ape' (1869: 392). For this reason, natural selection, which only operated to produce the sufficient and not the supererogatory, could not explain man's high intellect. Nor, according to Wallace, could it account for our acute moral sensitivity, our altruistic impulses. For moral behaviour usually benefits the recipient, not the agent: the fellow who jumps into a river to save a drowning child puts his own life in jeopardy—something natural selection would not countenance.

These difficulties for the theory were compounded for Darwin by the observations of William Rathbone Greg, a Scots political thinker who stopped to reflect on Darwin's theory. Greg pointed out that natural selection seemed to have a built-in governor, so that in the human line it should disengage itself. If natural selection produced the social and moral sentiments in man, Greg argued, such feelings would in proto-human groups prevent the beneficial culling of the morally and intellectually degenerate. Fuelled by the social instincts that Darwin deemed the foundations of human society, members of a tribe would prevent their dim-witted friend who wished to pet the sleeping sabre-tooth from meeting his natural end. The mentally inferior would live to procreate another day, so that the numbers of the least advantaged would increase, and thereby produce a regression within evolution. Greg, Scots gentleman that he was, took the case of the Irish as cautionary:

The careless, squalid, unaspiring Irishman multiplies like rabbits or ephemera:—the frugal, foreseeing, self-respecting, ambitious Scot, stern in his morality, spiritual in his faith, sagacious and disciplined in his intelligence, passes his best years in struggle and in celibacy, marries late, and leaves few behind.... In the eternal 'struggle for existence', it would be the inferior and less favoured race that had prevailed—and prevailed by virtue not of its good qualities but of its faults. (1868: 361)

In the *Descent of Man*, Darwin thus had two classes of problems bearing on evolutionary progress that threatened his theory: first, natural selection, as Wallace maintained, could not produce the high intellect and moral sentiments that human beings exhibited, since such faculties were in excess of what was required; and, second, retarding forces produced by natural selection itself appeared to prevent the further progressive development of the human species—our tender mercies would allow the inferior types to out-propagate the superior.

Darwin ingeniously resolved the first set of problems by applying a device that he originally developed to explain the traits of the social insects—the device of community selection. While altruistic impulse and even high intellect would little benefit individuals within those tribes of our ancestors—indeed, would even be damped down by the governor of individual selection—yet such traits would serve the tribe in its competition with other tribes. Communities themselves would re-establish the struggle for existence on a higher plane. Within a society, men would no longer struggle individually with each other for the resources of survival, but between societies, not yet bound by sentiments of universal brotherhood, the competition would continue just as effectively, as Darwin explained:

It must not be forgotten that although a high standard of morality gives but a slight or no advantage to each individual man and his children over the other men of the same tribe, yet that an advancement in the standard of morality and an increase in the number of well-endowed men will certainly give an immense advantage to one tribe over another. There can be no doubt that a tribe including many members who, from possessing in a high degree the spirit of patriotism, fidelity, obedience, courage, and sympathy, were always ready to give aid to each other and to sacrifice themselves for the common good, would be victorious over most other tribes; and this would be natural selection. At all times throughout the world tribes have supplanted other tribes; and as morality is one element in their success, the standard of morality and the number of well-endowed men will thus everywhere tend to rise and increase. (1871: ii. 166)

If one adds to the device of community selection that more venerable Lamarckian instrument of the inheritance of the effects of habitual practices—an instrument that Darwin never abandoned, even if he did not often deploy it—then the first set of problems dissolves.

The other set of difficulties, those suggested by Greg, would not yield so easily to Darwin's genius. Again, the problem was simply that qualities which identified the progressive attainments of the human race—high intellect and moral sense—appeared to be swamped out by the vicious, the criminals, and the imbeciles which inhabited the lower classes of Victorian society. Here natural selection seemed to be disengaged, since the depressingly worst appeared to be propagating at a faster rate than the obviously best. In his analysis, though, Darwin ventured that there were actually inherent checks on the possible increase among the inferior classes: the poor crowded into towns would die at a faster rate; the debauched would also suffer higher mortality; and gaoled criminals would not bear children. After careful reflection, it appeared a just world yet—fortune would finally snip the thread of the unworthy. And one could hope, as Darwin and Greg did, that enlightened social legislation and the impact of moral education would return the propagatory advantage to the more favourably endowed. None the less, it could be that civilized nations faced, after reaching a peak, a gradual decline. After all, as Darwin gloomily observed, 'progress is no invariable rule' (1871: i. 177). But he did think, it must be added, that progress was a general rule.

Darwin's worry that his own society might have been slipping into decline makes sense only under the supposition that he expected constant progress, a growth in complexity, finally of a florescence of intellect and moral sense, to be the natural outcome of natural selection. And though he recognized the possibility of devolution, he remained hopeful of continued progress among civilized men. He concluded his consideration of the problem in the *Descent* expressing just that sentiment:

It is apparently a truer and more cheerful view that progress has been much more general than retrogression; that man has risen, though by slow and interrupted steps, from a lowly condition to the highest standard as yet attained by him in knowledge, morals, and religion. (1871: i. 184)

CONCLUSION

Among contemporary evolutionary theorists, Darwin functions as an icon, an image against which theories may receive approbation or reprobation. To select from the historical Darwin those features that best comport with one's own predilections in the contemporary scientific debate is to have those predilections sanctioned by the master. So, for example, if one dislikes the political and social élitism that seems endorsed by the idea of evolutionary progress, one might claim, as one of our contributors has, 'To Darwin, improved meant only "better designed for an immediate, local environment"'. As Gould goes on:

Its [natural selection's] Victorian unpopularity, in my view, lay primarily in its denial of general progress as inherent in the workings of evolution. Natural selection is a theory of local adaptation to changing environments. It proposes no perfecting principles, no guarantee of general improvement; in short, no reason for general approbation in a political climate favoring innate progress in nature. (1977a: 45)

If we take Darwin whole, we see that his view of progress in evolution does not differ terribly from that of Spencer. Both had their vision of moral and intellectual development sharpened by the Christianity of their youth; and when the vision dimmed, so that the Creator seemed to recede into the vague interstices of nature, yet the moral aspects of the vision were retained. Both believed that evolutionary progress was to be expected, and it would be progress generally in the complexity of organization, and finally in the moral and intellectual faculties characterized by the higher races. The nineteenth-century version of evolution may not satisfy our own moral and political demands, but these demands should not obscure either Spencer's or Darwin's theories of evolutionary progress.[1]

REFERENCES

Bain, A. (1863), Letter to Herbert Spencer, 17 Nov., Athenaeum Collection of Spencer's Correspondence, MS 791, no. 67, University of London Library.

[1] I am extremely grateful to Richard Burkhardt and Phillip Sloan for aiding in what I suspect they regarded as an attempt to make the worse argument the better.

608   ROBERT J. RICHARDS

Bowler, P. J. (1986), *Theories of Human Evolution* (Baltimore: Johns Hopkins University Press).
Carpenter, W. (1841), *Principles of General and Comparative Physiology* (London: Churchill).
Chambers, R. (1844), *Vestiges of the Natural History of Creation* (London: Churchill).
Darwin, C. (1859), *On the Origin of Species* (London: Murray).
——(1871), *Descent of Man and Selection in Relation to Sex* (2 vols., London: Murray).
——(1902), *More Letters of Charles Darwin*, ed. F. Darwin (2 vols., Cambridge: Cambridge University Press).
——(1909), Essay of 1842, in F. Darwin (ed.), *Foundations of the Origin of Species* (Cambridge: Cambridge University Press), 1–53.
——(1960), *Darwin's Notebooks on Transmutation of Species*, ed. G. de Beer, Bulletin of the British Museum (Natural History), Historical Series, 2 (London: British Museum (Natural History)).
Gould, S. J. (1977*a*), *Ever since Darwin* (New York: Norton).
——(1977*b*), *Ontogeny and Phylogeny* (Cambridge, Mass.: Harvard University Press).
——(1988), 'On Replacing the Idea of Progress with an Operational Notion of Directionality', in M. H. Nitecki (ed.), *Evolutionary Progress* (Chicago: University of Chicago Press), 319–38; reproduced as Ch. 33.
Greg, W. R. (1868), 'On the Failure of "Natural Selection" in the Case of Man', *Fraser's Magazine*, 78: 353–62.
Huxley, T. H. (1902), *Evolution and Ethics and Other Essays* (New York: D. Appleton).
Lyell, C. (1830–3), *Principles of Geology* (3 vols., London: Murray).
Mayr, E. (1982), *The Growth of Biological Thought* (Cambridge: Cambridge University Press).
Milne-Edwards, H. (1841), *Outlines of Anatomy and Physiology* (Boston: Little and Brown).
Ospovat, D. (1981), *The Development of Darwin's Theory* (Cambridge: Cambridge University Press).
Peel, J. D. Y. (1971), *Herbert Spencer: The Evolution of a Sociologist* (New York: Basic Books).
Provine, W. B. (1988), 'Progress in Evolution and the Meaning of Life', in M. H. Nitecki (ed.), *Evolutionary Progress* (Chicago: University of Chicago Press), 49–74.
Richards, R. (1987), *Darwin and the Emergence of Evolutionary Theories of Mind and Behavior* (Chicago: University of Chicago Press).
Spencer, H. (1842), 'Letters on the Proper Sphere of Government', *Nonconformist*, 15 June–23 Nov.
——(1851), *Social Statics* (London: Chapman).
——(1855), *Principles of Psychology* (London: Longman, Brown, Green, and Longmans).
——(1857), 'Progress: Its Law and Cause', *Westminster Review*, NS 9: 445–85.
——(1862), *First Principles* (London: Williams and Norgate).
Van Valen, L. (1973), 'A New Evolutionary Law', *Evolutionary Theory*, 1: 1–30.
Wallace, A. R. (1869), Review of *Principles of Geology* and *Elements of Geology*, by Charles Lyell, *Quarterly Review*, 126: 359–94.

Wiley, E. O. (1988), 'Entropy, Evolution and Progress', in M. H. Nitecki (ed.), *Evolutionary Progress* (Chicago: University of Chicago Press), 275–91.

Williams, G. C. (1966), *Adaptation and Natural Selection* (Princeton: Princeton University Press).

31

EVOLUTION AND PROGRESS

MICHAEL RUSE

Evolution is the child of progress. Enlightenment hopes of ongoing social improvement were translated into beliefs about upward organic development through time since the Earth began (Richards 1992). As so often with family relationships, evolutionists tend to be ambivalent about their parentage. A critic says, 'Progress is a noxious, culturally embedded, untestable, nonoperational, intractable idea that must be replaced if we wish to understand the patterns of history' (Gould 1988). A defender responds: 'I do think that progress has happened, although I find it hard to define precisely what I mean' (Maynard Smith 1992).

Charles Darwin was torn on the subject, cautioning himself never to speak of 'higher' and 'lower', yet filling the *Origin* with flowery passages about the upward rise of life. Usefully, we find in his thinking a distinction between what we might label comparative progress, meaning the adaptive advance of one line of organisms over others, and absolute progress, meaning improvement up a scale of fixed value. This latter has been characterized as a directed change toward that which is better (Ayala 1988). Enthusiasts for absolute progress generally think that humans come out top.

COMPARATIVE PROGRESS

Comparative progress is a Darwinian notion, centring on selection. At the micro-level, all would agree that it occurs, although there is much debate about its precise nature and extent. Much attention has been paid recently to one particular form, the so-called arms race, in which organisms compete and evolve, throwing up methods of attack and defence in a way analogous to human weapon development (Dawkins 1986, Davies *et al.*

First published in *Tree*, 8/2 (1993): 55–9. Reprinted by permission of Elsevier Science Ltd., Oxford.

1989). Controversy arises when one tries to take the hypotheses and find-ings of micro-evolution and apply them to the long time-scale, the concern of the macro-evolutionist. There seem to be two particular points of dis-pute: namely, that over significant, new adaptations—'innovations'—and that over protracted shifts—'trends'.

Innovations

Innovations supposedly open the way to the occupation of new ecological niches or to the seizing of niches already occupied (Nitecki 1990). One has an 'adaptive breakthrough' (Fig. 31.1). Defining 'innovation' as something which has crossed a functional threshold, it has been claimed that they are the 'mainsprings' of macro-evolution (Jablonski and Bottjer 1990). Many evolutionists—particularly palaeontologists—simply take innovation as given, worrying more about its identification and implications. Endother-mal homoeothermy—the very essence of what it is to be a bird or a mammal—has been offered as a paradigmatic example of such a phenom-enon (Liem 1990). With such an adaptation, one can do many things barred to reptiles, particularly those involving nocturnal niches where there is no sunlight to warm the body.

The pressing concern of these enthusiasts is to find causal reasons why innovations should prove innovative. It is doubted that the phenomena could come about by pure chance, for some periods of the history of life seem to have been very much more prone to innovation than others. Opinion divides on two hypotheses, which may work more together than against each other. The ecological hypothesis suggests that an innova-tion is more likely to succeed in an empty niche, and that is why (for example) there was so much innovation in the early Palaeozoic (Erwin et al. 1987). The genomic hypothesis suggests that the genomes of earlier organisms were less canalized and had fewer epistatic interactions. Hence the organisms were more open to radical redesign. (Valentine and Erwin 1987).

Unfortunately, the fossil record yields few crucial predictions enabling one to make a decision between these hypotheses. Not that this is a matter of any great regret to some, who criticize the whole notion of innovation, arguing that the comparative evidence simply gives no support to the idea that some adaptations gave their possessors a major advantage over their rivals. Cracraft argues that close study shows that avian flight—something that a defender of innovation would see as a direct consequence of endo-thermal homoeothermy—crumbles into a long series of not-very-innova-tive parts (Cracraft 1990). Indeed, he believes that virtually all supposed

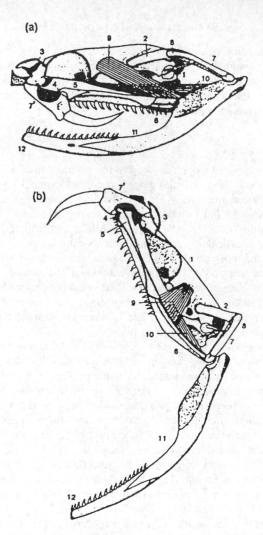

FIG. 31.1. The kinematics of the jaw mechanism of a solenoglyph snake, an example of an evolutionary innovation. Note how the fangs lock into an erect position when the jaw is open. This allows for a longer fang than otherwise. The snake is thereby more efficient at injecting venom into its victim, and consequently can attack larger prey, needing only (without struggle) to follow the stricken brute to its death-place. (a) Position of linkages, bones, and muscles with jaws closed and fangs folded. (b) Position of linkages, bones, contracted muscles, and erected fang with opened jaws. 1, brain case; 2, supratemporal; 3, ethmoid complex; 4, palatine; 5, ectopterygoid; 6, pterygoid; 7, quadrate; 7', maxilla; 8, supratemporal-quadrate joint; 9, levator pterygoideus; 10, protractor pterygoideus; 11, mandible; 12, mandibular tips of left and right sides. From Liem 1990: 149, with permission.

important innovations are without genuine 'ontological status' regarded as real evolutionary phenomena. As is so often the case with evolutionary disputes, one senses that much of the difference between Cracraft and those whom he criticizes rests on different definitions of what one would count as 'innovative', as well as the selective use of examples supportive of one's own case.

Trends

Trends likewise find evolutionists divided. Some, particularly students of marine invertebrates, state flatly that the fossil record shows trends, meaning paths up to improved adaptation (Jackson and McKinney 1990). Such trends include forms of growth, rates of growth, and potential for greater habitat choice. The favoured explanation of these trends—a phenomenon that has been referred to as 'escalation' (Vermeij 1987)—rests on some form of extended arms race. It is claimed that as predators get more efficient at gaining their prey—for marine invertebrate predators, an increased ability to break or cut through shells, and the like—so also there was the evolution of yet stronger shells, resisting being broken or otherwise torn apart (Fig. 31.2).

Critics of trends take a number of lines. One is to admit that there are phenomena which seem *prima facie* as if they are trends, but deny that they

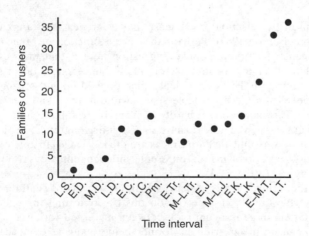

FIG. 31.2. An upward trend in numbers of marine families specializing in predation through shell breaking. From Vermeij 1987: 187, with permission.

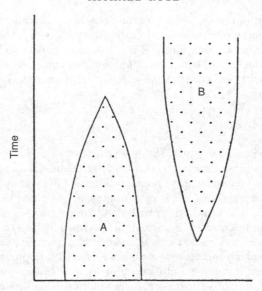

FIG. 31.3. One group replacing another through time. Have we the right to assume
that B is in some way superior to A, or that B's success is a function of its superiority
to A? From Benton 1987: 319, with permission.

are caused by selection, or at least that as trends they are caused by
selection (Benton 1987). Perhaps they are a function of other (possibly
non-organic) factors. A popular suggestion here is that mass extinction
might have been a key causal factor, a key chance causal factor from the
point of selection. Niches may have been opened up and organisms may
have moved into such available space, without any significant adaptive
pressure. This being so, it is hardly appropriate to talk of 'improvement'.
And in any case, even if we do have selection-driven trends, we should not
automatically assume that arms races are the key causal factors. We might
have two taxa evolving in different, independent ways (Gould and
Calloway 1980) (Fig. 31.3).

   Another critical line is to deny that the fossil record is all that trendy,
anyway (Benton 1987). One can do this by contesting the phenomenal
claim. Often, there is a confusion between increased variance in a group
and a changed mean, with the former being taken as evidence for the
latter. Again, one might feel that close examination shows all sorts of
fluctuations, incompatible with genuine trends. Or one can argue that the

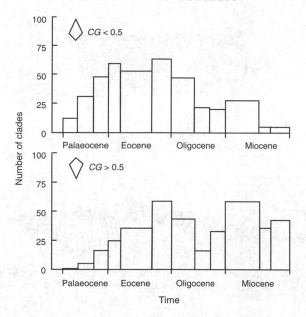

FIG. 31.5. Histograms showing bottom- and top-heavy mammalian clades (genera within families) during the Tertiary. 'CG' is the centre of gravity of the clade, determined as if one had a template of the clade cut from a flat sheet of uniform material. From Gould *et al.* 1987: 1440, with permission.

Complexity

Turning to the positive case for absolute progress, the most venerable criterion of improvement—with roots back to Aristotle's *De anima*—is that centring on complexity (Ruse 1988). It is not that people value complexity in itself, but that they regard it as a flag for other desirable qualities, like intelligence. Of course, everybody recognizes that you do not get a simple inevitable evolution of the more complex from the less complex. At least, every Darwinian recognizes that sometimes simplicity pays. Measuring variation in a single dimension along the vertebral column, it can be shown that in the return to the water, mammalian backbones underwent a significant simplification (Fig. 31.6) (McShea 1991). Yet this simplification was not only evolution, but was evolution in a sustained adaptive direction—an 'improvement' by any reasonable criterion.

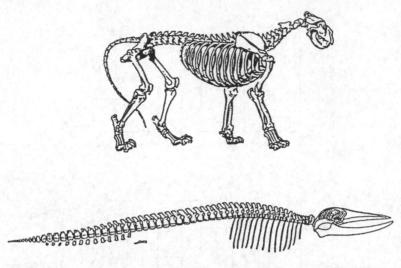

Fig. 31.6. The more complex lion vertebral column compared to the less complex fin-back whale vertebral column. From McShea 1991: 315, with permission.

To use complexity as a criterion, one must think of the overall picture of evolution. A recent extended case for this kind of progress has been made by Bonner (1988). Empirically, he shows that there is a general organic trend to increased size of the largest organisms (Fig. 31.7). This he explains simply by the fact that although niches at the bottom may be filled, there is always room at the top. But bigger organisms require a more complex support system; and, defining complexity as a simple function of the number of different cell types an organism has, Bonner concludes that there has indeed been such an evolution in the history of life.

There are many holes in this argument, starting with the fact that Bonner does nothing to show that the supposed evolution of size is other than an epiphenomenon of Markovian randomness. This apart, and ignoring the assumption that all differences of cell type have equal weight with respect to complexity, is it necessarily the case that great size spells complexity? The shrew is many orders of magnitude smaller than the big whales. Is it that much less complex? Or what about the shrew compared to the large dinosaurs? In any case, what about humans? Intuitively, we seem pretty complex. How do we rate, *vis-à-vis* the dinosaurs? Bonner himself seems to recognize that he has a problem here, for by the time he

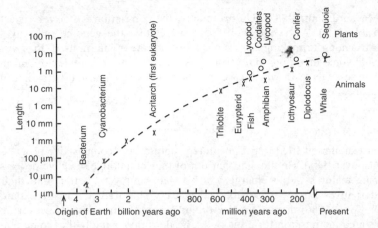

FIG. 31.7. The increase in maximum sizes of organisms over time. From Bonner 1988: 27, with permission.

TABLE 31.1 *DNA content per haplold genome*

	Genome size (pg)[a]	Percentage of genome coding for protein[b]	Coding DNA (pg)
Escherichia coli	0.004	100.0	0.004
Saccharomyces cerevisiae	0.009	69.0	0.006
Caenorhabditis elegans	0.088	25.0	0.022
Drosophila melanogaster	0.18	33.0	0.059
Homo sapiens	3.5	20.0	0.7
Triturus cristatus	19.0	3.0	0.57
Protopterus aethiopicus	142.0	0.8	1.136
Arabidopsis thaliana	0.2	31.0	0.062
Fritillaria assyriaca	127.0	0.02	0.025

[a] 1 pg of DNA corresponds to approximately 10^9 base pairs.
[b] The values of the per cent DNA coding for protein are the averages of different estimates, and are only approximate.

gets to humans he is talking about behaviour. Which may make sense, but is neither an obvious synonym for complexity, nor for size.

One may think that one could save the argument by suggesting that advance be grounded in terms of coding DNA present in the genome, which in some sense would give a measure of complexity/progress

(Maynard Smith 1988). Unfortunately, the correlation is not very exact (Table 31.1). Also, one should not forget that the genome only gives instructions for making an organism; it is not the organism itself. Progress could come in getting more from less, getting a more complex organism from less DNA. In which case, a reduction in DNA would be a sign of advance.

Darwinism

Darwin himself tried to get absolute progress from comparative progress (Ospovat 1981). He thought that out of the competitive selective process some features would emerge which would simply be better than their alternatives, on any reasonable value scale. This argument is still extant. Dawkins (1986) decries 'earlier prejudices' about progress. However, in his great enthusiasm for arms races, he does rather imply that some features, specifically intelligence, are more equal than others. Also he thinks in terms of adaptive breakthroughs, including the biggest breakthrough of them all: 'the evolution of evolvability (Dawkins 1989). Needless to say, humans look pretty good on this picture.

The most overt, living Darwinian enthusiast for absolute progress is E. O. Wilson. He is convinced that there has been such progress and that we won the race (Wilson 1978). Recently, in an attempt to clarify his position, he has taken to distinguishing between 'success' and 'dominance' (Wilson 1990). Success is to be defined in terms of the longevity of a species and of all of its descendants through geological time. Dominance, on the contrary, is to be measured both in terms of the abundance of a group compared to other groups and in terms of overall 'ecological and evolutionary impact' on all other organisms. By these measures, it would be too early yet to judge of human success, but we are clearly very dominant.

This suggestion does not in itself speak to causes, although Wilson would argue that the clarification introduced is a first step to a selection-based explanation of instances of success and dominance. One serious problem seems to be that, as presented, success simply goes to the oldest organisms which still have descendants. The reptiles are more successful than the mammals because they appeared first. One surely needs to modify the definition to take into account design potential for success? One might then say that the mammals were/are successful because their special adaptations made them adaptable in the face of change and disaster, and hence they could/did survive.

I suppose also one might say that humans do not promise to be a very successful organism, since they have the ability to produce twenty-first-century technology and yet are caught with Stone Age emotions (Wilson 1978). This does not seem much like progress. Likewise, although dominance may correspond to a recognizably intuitive concept, it is not necessarily that which we think is maximized when we speak of progress. The AIDS virus bids fair to be dominant, but we would hardly think its continued spread to be a matter of progress.

Yet, let us not end this discussion on an entirely negative note. A sophisticated approach to the fossil record may yield some positive support for some notion of progress, specifically in the Wilsonian sense of 'success'. If adaptations are getting better in some absolute sort of way, then perhaps taxa (species, genera, etc.) ought to last longer. With respect to genera, although a definitive case cannot yet be claimed, there are significant signs of improvement in survivorship—measured as a function of decrease in extinction rate—during the Phanerozoic (Raup 1988) (Fig 31.8). Indeed, we could get as much of a change in generic half-life from 7 myr to about 15 myr. Obviously, even if well-taken, this conclusion does not pick out one species as being superior to all others, nor does it relate success to other supposed marks of progress, such as complexity.

A BIOLOGICAL ANTHROPIC PRINCIPLE?

Concluding our survey, we see that the notion of progress continues to be of concern to evolutionists—especially those interested in macro-

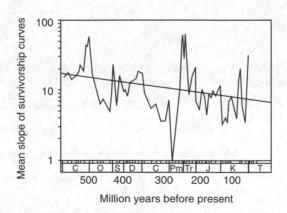

FIG. 31.8. Generic extinction rate plotted against time. Upward spikes are caused by mass extinction. From Raup 1988: 313, with permission.

evolution—as it continues to be a topic of controversy. Many evolutionists feel distinctly uncomfortable in discussing the very notion, and there is certainly a tendency to push such discussions into the semi-popular realm (Maynard Smith 1992, Dawkins 1986, Gould 1989). This is surely a function of the fact that absolute progress, certainly, is seen to have a blatant value component incompatible with the ideal of modern science. Note, however, that although in this discussion (by our very definition) we have allowed that there is such a component, this does not in itself say what value(s) are being endorsed. In flat opposition to just about every progressionist mentioned above, one prominent evolutionist (G. C. Williams) has argued recently that progress entails going in the face of that which is adaptively advantageous, because he feels (with T. H. Huxley) that that which is morally good is rarely if ever that which is biologically good (Williams 1989).

The question does remain why, for all its problems, progress of some kind remains so seductive a notion for so many evolutionists. No doubt there are many reasons; but my suspicion is that a major causal factor is some sort of biological version of the so-called anthropic principle: our understanding of the world is a function of our abilities to understand the world (Barrow and Tipler 1986). We are organisms, end-products of the evolutionary process, with the ability to ask questions about progress. Perhaps this alone is enough to turn us to favourable thoughts of progress (Ruse 1988).

REFERENCES

Ayala, F. J. (1988), 'Can "Progress" be Defined as a Biological Concept?', in M. Nitecki (ed.), *Evolutionary Progress* (Chicago: University of Chicago Press), 75–96.

Barrow, J. D., and Tipler, F. J. (1986), *The Anthropic Cosmological Principle* (Oxford: Oxford University Press).

Benton, M. J. (1987), 'Progress and Competition in Macroevolution', *Biology Review*, 62: 305–38.

Bonner, J. T. (1988), *The Evolution of Complexity by Means of Natural Selection* (Princeton: Princeton University Press).

Cracraft, J. (1990), 'The Origin of Evolutionary Novelties: Pattern and Process at Different Hierarchical Levels', in M. Nitecki (1990), 21–46.

Davies, N. B., Bourne, A. F. G., and Brooke, M. de L. (1989), 'Cuckoos and Parasitic Ants: Interspecific Brood Parasites as an Evolutionary Arms Race', *Trends in Ecology and Evolution*, 4: 274–8.

Dawkins, R. (1986), *The Blind Watchmaker* (New York: Norton).

——(1989), 'The Evolution of Evolvability', in C. Langton (ed.), *Artificial Life* (Reading, Mass.: Addison-Wesley), 201–20.

Erwin, D. H., Valentine, J. W., and Sepkoski, J. J. jun. (1987), 'A Comparative Study of Diversification Events: The Early Paleozoic Versus the Mesozoic', *Evolution*, 41: 1177–86.

Gould, S. J. (1988), 'On Replacing the Idea of Progress with an Operational Notion of Directionality', in M. Nitecki (ed.), *Evolutionary Progress* (Chicago: University of Chicago Press), 319–38; reproduced as Ch. 33.

——(1989), *Wonderful Life* (New York: Norton).

——and Calloway, C. B. (1980), 'Clams and Brachiopods—Ships that Pass in the Night', *Paleobiology*, 6: 383–96.

——Gilinsky, N. L., and German, R. Z. (1987), 'Asymmetry of Lineages and the Direction of Evolutionary Time', *Science*, 236: 1437–41.

Jablonski, D., and Bottjer, D. J. (1990), 'The Ecology of Evolutionary Innovation: The Fossil Record', in Nitecki (1990), 253–88.

Jackson, J. B. C., and McKinney, F. K. (1990), 'Ecological Processes and Progressive Macroevolution of Marine Clonal Benthos', in R. M. Ross and W. D. Allmon (eds.), *Causes of Evolution: A Paleontological Perspective* (Chicago: University of Chicago Press), 173–209.

Kitchell, J. A., and MacLeod, N. (1988), 'Macroevolutionary Interpretations of Symmetry and Synchroneity in the Fossil Record', *Science*, 240: 1190–3.

Liem, K. F. (1990), 'Plausibility and Testability: Assessing the Consequences of Evolutionary Innovation', in Nitecki (1990), 147–70.

McShea, D. W. (1991), 'Complexity and Evolution: What Everybody Knows', *Biology and Philosophy*, 6: 303–24; reproduced as Ch. 32.

Maynard Smith, J. (1988), 'Evolutionary Progress and Levels of Selection' in M. Nitecki (ed.), *Evolutionary Progress* (Chicago: University of Chicago Press), 219–30.

——(1992), 'Taking a Chance on Evolution', *New York Review of Books*, 39: 34–6.

Nitecki, M. (1990) (ed.), *Evolutionary Innovations* (Chicago: University of Chicago Press).

Ospovat, D. (1981), *The Development of Darwin's Theory* (Cambridge: Cambridge University Press).

Raup, D. M. (1988), 'Testing the Fossil Record for Evolutionary Progress', in M. Nitecki (ed.), *Evolutionary Progress* (Chicago: University of Chicago Press), 293–318.

Richards, R. (1992), *The Meaning of Evolution* (Chicago: University of Chicago Press).

Rosenzweig, M. L., and McCord, R. D. (1991), 'Incumbent Replacement Evidence for Evolutionary Progress', *Paleobiology*, 17: 202–13.

Ruse, M. (1988), "Molecules to Men: Evolutionary Biology and Thoughts of Progress', in M. Nitecki (ed.), *Evolutionary Progress* (Chicago: University of Chicago Press), 97–126.

Valentine, J. W., and Erwin, D. H. (1987), 'Interpreting Great Developmental Experiments: The Fossil Record', in R. A. Raff and E. C. Raff (eds.), *Development as an Evolutionary Process* (New York: Alan R. Liss), 71–107.

Vermeij, G. J. (1987), *Evolution and Escalation* (Princeton: Princeton University Press).

Williams, G. C. (1989), 'A Sociobiological Expansion of *Evolution and Ethics*', in J.

Paradis and G. C. Williams (eds.), *Evolution and Ethics* (Princeton: Princeton University Press), 179–214.

Wilson, E. O. (1978), *On Human Nature* (Cambridge, Mass.: Harvard University Press).

——(1990), *Success and Dominance in Ecosystems: The Case of the Social Insects* (Oldenorf/Luhe, Federal Republic of Germany: Ecology Institute).

32

COMPLEXITY AND EVOLUTION: WHAT EVERYBODY KNOWS

DANIEL W. MCSHEA

A CONSENSUS ON COMPLEXITY

Everybody seems to know that complexity increases in evolution. Darwin[1] thought it does, as did E. D. Cope, Herbert Spencer, and most of the Anglo-American palaeontological community from the last decade of the nineteenth century through the first three decades of this one (Swetlitz 1989). At mid-century, the consensus was still intact. Some of the major modern synthesis authors, notably Huxley (1953), Rensch (1960), and Simpson (1961),[2] said that complexity increases; and Goudge, in his 1961 book *The Ascent of Life*, included increasing complexity in a list of large-scale evolutionary patterns widely accepted among evolutionary biologists.

More recently, some have expressed doubts (Williams 1966, Lewontin 1968, Hinegardner and Engelberg 1983), but increasing complexity is still the conventional wisdom. Clear statements that complexity increases can be found in the work of Stebbins (1969), Denbigh (1975), Papentin (1980), Saunders and Ho (1976, 1981), Wake *et al.* (1986), Bonner (1988), and others. And lately the new thermodynamic school of thought has added its voice to the chorus: Wicken (1979, 1987), Brooks and Wiley (1988), and

First published in *Biology and Philosophy*, 6 (1991): 303–24. © 1991 Kluwer Academic Publishers. Reprinted with kind permission from Kluwer Academic Publishers.

[1] Darwin did not discuss his views on large-scale evolutionary patterns of complexity change in the *Origin*, but did so in his Notebook E (Darwin 1987: 422). His mechanism is similar to one proposed more formally by Waddington (see my discussion here under the heading 'Externalist Mechanisms').

[2] Simpson is known for his scepticism on complexity, especially for his apt remark that 'It would be a brave anatomist who would attempt to prove that Recent man is more complicated than a Devonian ostracoderm' (Simpson 1949: 252). However, he is forthright elsewhere: 'Perhaps the most common assumption about evolutionary sequences is that they tend to proceed from the simple to the complex. Such an over-all trend has certainly characterized the progression of evolution *as a whole*' (Simpson 1961: 97, emphasis original).

Maze and Scagel (1983) have all argued that complexity ought to, and does, increase in evolution. In my own experience, the consensus extends well beyond evolutionary biology and professional scientists. People seem to know that complexity increases as surely as they know that evolution has occurred. (For a recent declaration from the mainstream press, see Wright 1990.)

Does complexity *in fact* increase? Is the conventional wisdom true? Very little evidence exists. Empirical enquiries have been few; instead of gathering and evaluating data, students of complexity have been preoccupied with theorizing, with developing rationales for why escalating complexity is the expectation. Further, both empirical and theoretical studies have lacked rigour; for example, missing from most (until recently) have been clear discussions of what complexity means.

I will begin with a discussion of the modern solution to the question of meaning. Then, capitulating to the historical dominance of theory over evidence, I will discuss the various rationales that have been offered for why complexity ought to increase. Turning to the evidence, I will argue that not enough evidence exists to make an empirical case either for or against increase. Finally, guide-lines for future empirical studies will be suggested, and some explanations offered for what has sustained the consensus on complexity in the absence of evidence.

A preliminary word on progress in evolution: this discussion both is and is not about progress. It is not, because, despite the historical connection between progress and complexity, the two should be divorced, as I will argue later. For now, I simply ask the reader not to equate complexity with progress. However, progress *is* relevant here on account of its historical association with other apparent evolutionary trends, such as increasing adaptability and increasing control by organisms over their environment (discussed by Simpson 1949). On the basis of my limited investigation, I strongly suspect that the critique of complexity offered here could be applied to most other aspects of progress with little revision.

WHAT IS COMPLEXITY?

Until recently, the word 'complexity' has been used casually in the evolutionary literature, and interchangeably with 'order' and 'organization', to denote properties of organisms that seem to improve or progress in evolution. In the past thirty years, however, as students of information theory have begun to apply their concepts and formalisms to living systems, some essential clarifications have been made. There is some consensus now that the structural or morphological complexity of a system (biological or otherwise) is some function of the number of different parts it has and the irregularity of their arrangement (Kampis and Csányi 1987; Hinegardner and Engelberg 1983; Wicken 1979, 1987; Saunders and Ho 1976). Thus, heterogeneous, elaborate, or patternless systems are complex.

Order is the opposite of complexity (Wicken 1979). An ordered system has few different kinds of parts arranged in such a way that the pattern is easily specified. Homogeneous, redundant, or regular systems are ordered, such as the atoms in a crystal lattice or wallpaper patterns. Organisms have sometimes been described as well-ordered, and for some the central problem in biology has been explaining the generation of order (e.g. Needham 1936), meaning the origin and subsequent evolution of life. However, in the modern definitional scheme, organisms are not especially well-ordered (although they are well-organized, as discussed below).

Organization refers to the degree of structuring of a system for some function, independent of its complexity and order (Wicken 1979; although see Atlan 1974). Complex systems may be organized, as an automobile is, or disorganized, as a junk heap is (ordinarily). Both the automobile and the junk heap are complex, because they have many parts and the parts are irregularly arranged. This conceptual separation of complexity (a structural property) and organization (a functional property) is important, because we want to be able to evaluate structure even if we know little about function. The junk heap may be functional: for example, it may function as a work of art (perhaps one requiring a very specific arrangement of parts). On the other hand, it may be a haphazard pile of parts with no function. The point is that we do not have to know anything about its function in order to judge that its structure is complex.

In this definitional scheme, there is no necessary connection between complexity and organization. In evolution, however, complexity and organization probably *are* connected, because more complex organisms need more organization in order to survive (Saunders and Ho 1976).

Complexity has been discussed recently in a number of different contexts. For example, much has been written about the complexity of number sequences and geometric patterns (for recent commentaries, see Landauer 1988 and Maddox 1990). In biology, some current topics are the complexity of biological hierarchies (Salthe 1985), of evolutionary clades (Brooks and Wiley 1988), and of genetic systems (see Subba Rao *et al.* 1982 and Gatlin 1972 on genome 'information content', which is a kind of complexity).

The present discussion is concerned only with morphological complexity, which is a property of physical systems like organisms (as opposed to abstract systems like number sequences). More specifically, the focus is the morphological complexity of biological individuals (as opposed to, say, genomes or ecosystems). I leave it to others to discover the extent to which my remarks apply in other complexity domains.

MECHANISMS FOR INCREASING COMPLEXITY

Most theories of complexity increase can be classified as either internalist or externalist. (This division follows Gould's (1977a) classification of theories of directional evolutionary change.) Internalist theories conclude that complexity increase is driven by inherent properties of either complex systems generally or of organisms in particular—their design, genetics, and development. Externalist theories invoke natural selection or some aspect of the environment. A few remaining theories invoke no driving force at all.

Internalist Mechanisms

1. Invisible fluids   Lamarck (1809) believed that simple organisms arise spontaneously, and that their lineages transform over time in the direction of increasing complexity. Driving these transformations are invisible fluids, present initially in the environment and kept in constant motion by the sun's energy. Somehow these fluids become bottled up inside organisms, and once there, they act internally. Lamarck writes:

when reflecting upon the power of the movement of the fluids in the very supple parts which contain them, I soon became convinced that, according as this movement is accelerated, the fluids modify the cellular tissue in which they move, open passages in them, form various canals, and finally create different organs. (Lamarck 1809: 2)

The effect of adding new canals, organs, and so forth is to enhance complexity. The effect of the environment is mainly to deflect or retard the process.

2. The instability of the homogeneous   Spencer (1890) argues that dynamic systems tend to become more concentrated and heterogeneous as they evolve. He calls this tendency the Law of Evolution, although by 'evolution' he means directional change in a wide variety of dynamic systems, not just living ones. The abstract argument is this: given a homogeneous and diffuse collection of identical particles, let the collection begin to aggregate under the influence of various natural forces (such as gravity). As it does so, particles in different positions within the aggregate will find themselves in different environments (inside as opposed to outside, for example), and thus will experience different forces. Differences among the forces cause differentiation among the particles, thus increasing the heterogeneity or complexity of the whole system.

The law has a corollary: when fully aggregated and fairly homogeneous systems do manage to arise, they are unstable, because they are unable (in the long run) to maintain the identity and internal relations of their parts in the face of external perturbations. The instability is not exactly driven. It is not that of a stick balanced on its end, Spencer explains, but rather that of balanced scales, which eventually become unbalanced due to rust, abrasion, wind, and such.

Spencer's Law is clearly internalist. The corollary relies on external perturbations to trigger differentiation, but the instability which makes change possible is an internal condition. His law is supposed to capture a universal developmental principle, in the tradition of the *Naturphilosophen*, and its debt is to von Baer in particular (Gould 1977*b*).

3. Repetition and differentiation of parts Cope (1871) offers a mechanism derived from Haeckel's principle that evolution occurs by the acceleration of ontogenies (the speeding up of organismal development) and the terminal addition of parts (Richardson and Kane 1988, Gould 1977*b*). In Cope's mechanism, acceleration results in the repetition of existing parts, and their subsequent differentiation produces complexity, with both processes driven by an internal 'growth force'. Cope acknowledges another ontogenetic mechanism, retardation, which can (and frequently does) decrease complexity, but, on average, parts and differences accumulate and complexity increases.

Gregory (1934, 1935*a*, 1935*b*, 1951: i. 548–9) offers a similar mechanism: evolution occurs mainly by duplication of parts, producing a morphological condition he calls polyisomerism, and differentiation of parts, producing anisomerism. Gregory uses language that suggests his mechanism is internalist, consciously adopting Cope's phrase 'anteroposterior repetitive acceleration' to describe one kind of polyisomerism (Gregory 1934: 1), and referring at one point to polyisomerism as 'an inherent property of protoplasm' (Gregory 1935*a*: 289). He is cagier, however, about the forces of anisomerism; his remarks on the subject are few (Gregory 1935*a*, 1951: i. 548–9) and open to either internalist or externalist interpretation.

4. The path of least resistance Saunders and Ho (1976, 1981) suggest that component additions are easier to achieve in development than component deletions, because components already present will tend to have been integrated into developmental pathways, and thus will be hard to remove. The asymmetry is slight, but sufficient to drive an evolutionary trend in complexity, they aver (but see Castrodeza 1978).

5. *Complexity from entropy* There is currently much enthusiasm for the view that rising complexity has something to do with the Second Law of Thermodynamics. However, while much has been said about the role of non-equilibrium principles in complexity increase in pre-biotic evolution (Prigogine *et al.* 1972; Wicken 1979, 1987), organismal ontogeny (Robson *et al.* 1988), speciation and diversification (Brooks and Wiley 1988), and ecological succession (Salthe 1985; Wicken 1987), clear discussions explicitly linking the Second Law and *morphological* complexity are hard to find. From what *has* been said, I have pieced together two possible versions of a thermodynamic argument.

One version begins with the observation that dynamic systems far from thermodynamic equilibrium spontaneously develop complex structure, and that complexity increases as the systems grow and age (Salthe 1985; Wicken 1987). If evolutionary lineages are also far-from-equilibrium systems, then structural complexity might be expected to rise in all of them. Here the connection between the non-equilibrium condition and complexity is purely empirical. A mechanism is presumed to exist, but it is not known in any detail.

In a second version, entropic change would lead to morphological complexity by promoting what Wicken calls 'configurational disorder' (Wicken 1987: 179). This disorder would have two consequences in evolutionary lineages: the building up of some morphologies by the addition of components (along with the breakdown of some others) and the scrambling of arrangements of existing components (in all lineages). Most disorderings in most lineages would be disadvantageous, but some would be functional improvements, and these would tend to be preserved. Increasing complexity, in this version, consists of a raising of the upper level of functional disorder.

Externalist Mechanisms

1. *Selection for complexity* Rensch (1960) suggests that the addition of parts permits more division of labour among the parts, and that therefore complex organisms are more efficient. The superior efficiency of complex organisms gives them a selective advantage, and this advantage drives

complexity upward in evolution. Bonner (1988) makes this same argument and extends it: selection not only directly favours greater efficiency (which can be achieved by increasing complexity) but also favours large size, which may in some cases demand more efficiency, which in turn may require more complexity.

2. Selection for other features Complexity itself may not be advantageous, and yet may increase passively as a consequence of natural selection for other characters. In particular, selection for large size can permit increases (e.g. Rensch 1960, Katz 1987). Accompanying greater size may be greater cell numbers or tissue bulk beyond what is functionally necessary, thus freeing the redundant cells and tissue to vary. The resulting variation, or differentiation, is complexity. Rensch's version of this mechanism has a Spencerian spin: increase in body size makes possible a greater range of variation in environments within an organism. Different environments make different demands on local cells, leading (via natural selection) to cellular differentiation.

More generally, redundant parts, whether produced by selection for large size or any other process, are free to be modified by selection, and their modification has the incidental consequence of greater morphological complexity. Darwin expressed the principle nicely, in passing, in a discussion of morphology:

We have formerly seen that parts many times repeated are eminently liable to vary in number and structure; consequently it is quite probable that natural selection, during a long-continued course of modification, should have seized on a certain number of the primordially similar elements, many times repeated, and have adapted them to the most diverse purposes. (Darwin 1859: 437–8)

Here, selection drives the adaptive modification of parts, which results in complexity, but complexity itself is not directly favoured.

3. Niche partitioning Waddington (1969) suggests that as organismal diversity increases, niches become more complex. The more complex niches are then filled by more complex organisms, which further increases niche complexity (because niches are partly defined by these organisms), thus sustaining a cycle of ever-increasing complexity.

Undriven Mechanisms

1. Random walk Fisher (1986) proposes two mechanisms for producing any sort of evolutionary trend, both requiring no driving force at all. One is simple luck: most evolutionary lineages could, by chance alone, happen to wander in the direction of higher complexity.

2. *Diffusion* Fisher further suggests that if complexity in every lineage follows a random walk, decreasing as often as it increases on average, and if there is a complexity floor (a minimum below which no lineage can go), then the mean complexity of all lineages is expected to go up. Stanley (1973) offers the same logic as an explanation for the evolutionary trend in size described by Cope's Rule.

Maynard Smith (1970) makes the suggestion that if the first organisms were, and had to be, simple, then later ones could only have been more complex. The reasoning is the same as Fisher's, but considers only a limiting case: namely, the first moments of evolution when the only existing lineages sat right on the complexity floor and had nowhere to go but up.

3. *The ratchet* Stebbins (1969) proposes that major evolutionary jumps in complexity occur occasionally in the adaptive radiations accompanying the invasion of new habitats. The complexity increase in such adaptive radiations is not inevitable, not the consequence of relentless driving forces, but rather is contingent on the prior existence of promising morphological specializations in the radiating group. New levels of complex organization achieved in these occasionally successful radiations are conserved, however, and lay the foundation for future jumps. Over the long haul of evolution, the result is a ratcheting upward of the upper limit of complexity.

The main purpose of this list is to document the heavy emphasis on theorizing. The list recalls the characterization by Vogt and Holden (1979) of the superabundance of theorizing in the dinosaur extinction debate as a case of 'the multiple working hypotheses method gone mad'. As in the dinosaur case, few of the hypotheses or mechanisms have been refuted. (Lamarck's mechanism especially sounds false, but the major fault may be simply his language, invisible fluids now being unfashionable.) Unlike the dinosaur case, the main problem is a paucity of data, as I argue below.

The list is also a small data set for further study of the relationships among the mechanisms. In particular, the mechanisms share many similarities, some offering about the same product but in different packages. Consider Spencer's instability of the homogeneous, which predicts an increase in complexity in an aggregating system by the accumulation of perturbations. This sounds a lot like complexity from entropy, and also a lot like complexity permitted by size increase. Consider the Saunders and Ho argument that deletions are less likely on account of their developmental entrenchment. This sounds like selection against deletions, selection that acts at the embryo stage; they intend their mechanism to be internalist, but it has an externalist interpretation too. Even Lamarck's mechanism may be redundant: in some lights it looks vaguely thermodynamic. Finally, where the products seem to differ, they share many of the same ingredients. For example, most depend on some kind of ratcheting process, and most can be construed as variations on the repetition-and-differentiation-of-parts theme. Considering the large number of striking similarities among these mechanisms, the possibility seems worth investigating that they are all, at bottom, the same.

THE EVIDENCE ON COMPLEXITY

In what follows, I characterize and critique the empirical evidence that exists on complexity change, covering mainly the work of the principal students of complexity since Darwin: three internalists—Spencer, Cope, and (probably) Gregory—and two externalists—Rensch and Stebbins—and the several moderns. Broader coverage is unnecessary, because most of the remaining evidence in the literature consists of isolated and casually treated examples of increase. The literature is baroquely ornamented with these examples, and while together they add much to the apparent bulk of the evidence, they add little to its mass.

The purpose of critiquing the work of these serious students of complexity is not to diminish it. For one thing, the point is more to discover if enough evidence exists to make a case today, than to prove that a case has not been properly made in the past. For another, some of the work considered is only secondarily concerned with complexity; its authors had more general concerns, such as evolutionary progress, and can hardly be faulted if they failed to make a different point than the one they intended. Finally, the purpose of any such critique is to learn lessons from the inevitable mistakes attending all first efforts. We applaud those efforts for their courage, rather than damn them for their errors.

The Internalists

Spencer (1893) considers the evidence from the plant and animal kingdoms separately, but finds the same pattern of complexity change in both. His strategy is to arrange a number of existing plant and animal species along a scale according to their relative degrees of integration, which for Spencer encompasses degree of autonomy of wholes from their parts (organisms from organs, organs from cells, etc.), degree of physical cohesion among parts (mainly cells), depth of hierarchical organization (number of levels of sub-units within sub-units), and degree of interdependence of parts. Protozoans, for example, he describes as mere aggregates of protoplasm and assigns to the first order of integration, while sponges, coelenterates, and others are aggregates of first-order aggregates and thus belong to a second order. He assumes that the more highly integrated organisms arose from the less, and in accordance with his Law, these will also be the more highly differentiated and thus the more complex.

Spencer gives numerous other examples of evolutionary (biological sense) transformations that agree with his principle (Spencer 1890). Using worms as ancestors of crustaceans, insects, and arachnids, he points out that the transformation involved an aggregation of body segments to produce a smaller number of more integrated units. Using trilobites as ancestors of the crustaceans, he notes a progressive differentiation of the limbs; among the vertebrates, the transformation from fish to reptiles to mammals and birds is marked by increasing heterogeneity of the vertebral column. This is only a small sample of his biological examples: he also gives innumerable examples from other realms, invoking the same principle to explain the differentiation of the planets in the solar system from a homogeneous nebula, of social roles in advanced societies, of words in modern languages, and much more.

Cope's discussion centres of differentiation in the elements of homologous series, such as limb bones, teeth, and vertebrae (Cope 1871), while Gregory's major example is the reduction and differentiation of the skull bones in the evolution of the vertebrates (Gregory 1935b). For both, these are just focal cases among many; their discussions are shorter than Spencer's but no less densely packed with examples.

All three use examples of the same sort, and their evidence shares the same major weakness. That is, the examples were deliberately chosen in order to make a case for what we might call uniform complexity increase, and no finite list can make that case. Explaining this point first requires a distinction: Ayala (1974) distinguished uniform progress, or progress which occurs in all lineages at all times, from net progress, or progress occurring occasionally or only in some lineages (resulting in an increase in the mean). A similar distinction can usefully be made for complexity. Spencer and Cope both posit developmental laws or forces that act pervasively, and for Gregory the forces producing polyisomerism and anisomerism, whatever their origin, act in a wide variety of contexts. All three expect complexity to rise in all or most lineages, something more like uniform increase than net increase.

Now the point: instances chosen and marshalled in order to prove that a principle operates everywhere and always can do no better than to show that it has operated in those instances. Ten, a hundred, even a thousand photos of blue-eyed unicorns do not make the case that unicorns are uniformly (or even predominantly) blue-eyed if those animals were picked and photographed just to build that case.

For complexity, a long list of instances in which developmental forces have acted may *seem* to build a case; but really it cannot show that the forces have acted in all, most, or even a sizeable percentage of lineages. Good evidence would consist either of a complete list of all instances, or more practically, of an arbitrary or random sample. In a large random sample, robust patterns of change should emerge as statistical regularities (Gould *et al.* 1987).

The Externalists

Stebbins (1969) identifies eight grades of 'complexity of organization', the lowest being the earliest self-reproducing organic systems, followed by the prokaryotes, the single-celled eukaryotes, the simple multi-celled eukaryotes, organisms with differentiated tissues and organs, organisms with well-developed limbs and nervous systems, homeotherms, and finally human beings. He then observes that the order of levels corresponds well with time of appearance in the fossil record, suggesting a progressive increase in maximum complexity of organization. Rensch (1960) uses roughly the same strategy, but his grades are mainly levels of functional sophistication (e.g. in digestion or vision) rather than structural organization.

Stebbins and Rensch, unlike the internalists, use evidence appropriate to their cases. They seek to demonstrate only a net trend, a raising of the upper level, and they properly do so by trying to document the transitions in which net increase occurred. Nevertheless, their evidence does little to help support a case for increasing complexity. For Stebbins, the complexity of an organism seems to be its hierarchical depth (number of levels of nested sub-units), but he is not consistent. On this understanding, it seems clear enough that multi-celled organisms are more complex than single-celled, and that those with cells grouped into tissues and organs are more complex yet. However, the later transitions, to better nervous systems and then to homeothermy, do not seem to increase hierarchical depth. For these changes, Stebbins appears to have a different idea of complexity, and his ordering is strongly reminiscent of the old *scala naturae*.

Maynard Smith (1988) has assembled a similar list showing changes in hierarchical organization, a list which also seems reasonable up to the point of multi-cellularity. His next two organizational levels, demes and then species, however, do not correspond to evolutionary transformations in time.

Rensch is more consistent, but his evidence is no more helpful for present purposes. He is interested mainly in evolutionary progress, which he calls 'anagenesis', only one aspect of which is increasing complexity; other aspects are increasing rationalization of structure and function, increasing plasticity, and so on. A rise in any one of these constitutes anagenesis, although he clearly thinks they are somewhat correlated, and so the instances that constitute his evidence for anagenesis do not by themselves support a case for complexity increase. (Where he discusses complexity alone, his few examples do more to show what he means by complexity than to document major evolutionary transitions.)

The Sceptics

Williams (1966) questions whether recent animals are more complex than Palaeozoic members of the same taxa, noting that, for example, the skulls of Devonian fishes had far more bony elements, and thus were far more complex than the human skull. He also points out that some organisms typically considered 'low' or 'simple', such as the liver fluke, have enormously complex life cycles in which they take on a large diversity of forms.[3] His message is that decreases are common, complexity is difficult to measure, and the risk of gross error in making casual assessments is high. The more recent sceptics (e.g. McCoy 1977, Hinegardner and Engelberg 1983) give a few, somewhat different, examples, but make the same points.

The sceptics do a good job of illustrating the difficulties in many analyses of complexity, such as the lack of objective measures, but they provide us with no evidentiary case against the existence of a trend. A case against a *net* trend in complexity would require at a minimum a demonstration that the major transitions cited by Rensch, Stebbins, and others do not constitute increases; for that purpose, the sceptics examine the wrong examples and treat them too briefly. For a uniform trend, a series of deliberately

---

[3] Williams's example is salutary; it has become almost obligatory in discussions of complexity to make parenthetical mention of what is supposed to be an obvious decrease in complexity in the evolution of parasites.

chosen examples, such as the comparison of fish and human skulls, no more weighs against than the lists of the internalists weigh in favour.

The Moderns

Cisne (1974) studied evolutionary changes in limb tagmosis (complexity) of free-living aquatic arthropods. He defined tagmosis as the amount of information needed to specify the diversity of limb-pair types, and calculated it (using an expression from information theory) as a function of the total number of limb pairs and the number of each type. Cisne compared tagmosis values in a series of four malacostracan crustacean groups that may correspond roughly to an evolutionary-lineage, and found a pattern of continuous increase. He also plotted mean values for all orders world-wide against time, and found a trend in the shape of a logistic curve spanning the past 600 million years.

Cisne's result has a number of difficulties. For example, Van Valen (personal communication) has shown than if complexity is measured with a simple count of number of different leg-pair types (rather than the information-theory metric), complexity still increases, but the logistic curve disappears. And Raup (personal communication) has shown that the monotonic trend is an artefact. That is, the arthropod orders fall into two groups, an earlier one (trilobites) with low tagmosis values and a later one with higher values; the monotonic trend is a consequence of averaging in the time period of overlap, the later Palaeozoic. Cisne's data show, therefore, not a pattern of continuous increase, but a single increase in the later Palaeozoic.

The approach has considerable merit, however. First, Cisne provides a clear, operational definition of complexity. Second, he does not seem to have chosen arthropod limb types for the reason that they help make a case for complexity increase; more likely, his metric required easily differentiable, serial structures, and arthropod limbs were convenient. If so, then arthropod limb types are an arbitrary sample of all possible characters in all groups, and his study does provide some evidence for uniform increase. His sample size, a single character in a single group, is of course quite small.

Bonner (1988) measured the complexity of an organism as the number of different cell types it has. The measure has the drawback that cell types are difficult to distinguish without careful histological studies, and only very approximate counts are possible presently for organisms with more than about ten. The metric has the virtue, however, that it captures the complexity of a whole organism, rather than of just one or several

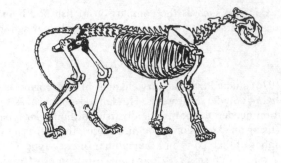

FIG. 32.1. Drawings (not to scale) of the skeletons of a lion (modified from Owen 1866: 493) and a fin-back whale (modified from Gregory 1951: ii. 899), showing their vertebral columns.

structures. Also, it will work with almost all taxa, thus allowing comparisons of organisms as different as fungi and frogs.

Bonner did not try directly to show ascending complexity with time; rather, he showed that the size of the largest organisms increases, and that complexity correlates with size. However, as he admitted, the correlation with size is weak, with some groups (e.g. vertebrates with 120 cell types and higher plants with 30 cell types) having size ranges spanning about ten orders of magnitude and broadly overlapping each other. The correlation between number of cell types and *maximum* size is better, but still quite rough.

Consider one final empirical study, this one my own. From among a number of different analyses,[4] I have deliberately chosen one that shows a decrease in complexity. This is hardly in keeping with the spirit of disinterested enquiry advocated earlier; I present it here mainly to offset the

---

[4] Most of my analyses have been ancestor–descendant comparisons in mammalian lineages. Specifically, I compared modern squirrels, ruminants, camels, whales, and pangolins with their ancestors in a number of vertebral dimensions, and found no tendency for complexity to increase. In fact, the number of complexity increases was about equal to the number of decreases for all three complexity metrics (see text).

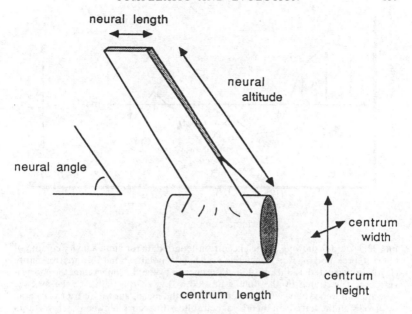

Fɪɢ. 32.2. A schematic drawing of a typical vertebra showing the vertebral dimensions relevant to the analysis.

impression doubtless left by the discussion of Cisne's and Bonner's work that the modern studies show nothing but rising complexity.

Following Cisne, complexity is understood in this analysis as the amount of differentiation within a homologous series of elements. A different metric has been adopted, however, in order to accommodate series in which change in the shape of elements along the series is continuous, and in which the elements therefore do not fall neatly into discrete types. The vertebral column is an example of such a series and the focus of my analysis. Figure 32.1 shows drawings of the vertebral columns of a lion and of a fin-back whale. Figure 32.2 shows a schematic drawing of a typical vertebra, indicating the vertebral dimensions relevant to the analysis.

Complexity is measured as a function of the amount of variation along a column in a single vertebral dimension. (Figure 32.3 shows artificial data for centrum length in a short imaginary series of ten vertebrae.) Actually, three different functions are used in order to capture three different aspects of that variation:

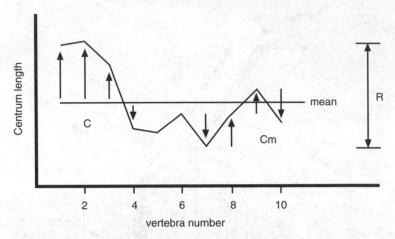

FIG. 32.3. A plot showing artificial centrum-length data for an imaginary column of ten vertebrae, and the way in which the range (R), polarization (C), and irregularity (Cm) are calculated. R: the absolute difference between the longest and the shortest vertebrae (illustrated by the double-headed arrow on the right); C: the average absolute difference between each vertebra and the mean (shown for four vertebrae by arrows on the left); Cm: the average absolute difference between each vertebra and the one before it (illustrated for four vertebrae by the arrows on the right). In real columns, averages are computed over all vertebrae, from the second cervical (the axis) to the last presacral (just ahead of the hip), inclusive. The formulae are as follows:

$$R = \log\left[X(\text{max}) - X(\text{min})\right];$$
$$C = \log\left[2\sum |X_i - \overline{X}|/N\right];$$
$$Cm = \log\left[\sum |X_{i+1} - X_i|/(N-1)\right],$$

where X_i is the raw measurement for the ith vertebra, $\overline{X}$ is the mean, and N is the number of vertebrae. Size correction is necessary for comparison among taxa: size-corrected R, C, and Cm values are estimated as residuals from a reduced-major-axis, best-fit line in a plot of C values against the log of a size factor (in this case, average posterior-centrum area). R, C, and Cm are strongly correlated, so to remove redundancies, R' and Cm' are substituted for R and Cm, computed respectively as residuals from reduced-major-axis, best-fit lines in plots of R and Cm against C.

1. Range of variation (R). R captures complexity as the maximum difference between elements, or as the difference between the two elements that are most different (Fig. 32.3).
2. Polarization, or overall complexity (C). C captures complexity as the average difference between the elements and their mean (Fig. 32.3). The word 'polarization' is appropriate, because C is maximal when half of the elements lie at each of the two extremes of the range: that is, when all the elements are as different as possible from the mean.
3. Irregularity (Cm). Cm captures complexity as the average difference between adjacent elements. Cm values reflect the number and magnitude of reversals in the direction of change along the column, or in other words, the bumpiness of the curve in Figure 32.3.

Precise formulae for computing R, C, and Cm from raw vertebral column measurements appear in the caption to Figure 32.3, along with a summary of the various transformations used to correct for size differences and to remove redundancies among the complexity measures.[5]

The analysis consists of a comparison between certain aquatic mammals and their terrestrial ancestors, or, more precisely, surrogates for those ancestors. Sufficiently complete vertebral columns for directly ancestral species have not been discovered, so their complexities are estimated using average values for a diverse assemblage of living terrestrial taxa. The substitution is reasonable, because what fossil vertebral material exists suggests that the ancestors of these aquatic species (among the early Carnivora for the otter and the pinnipeds and among the Condylarthra for the whale and the manatee) had vertebral columns generally similar to modern terrestrial mammals. The reason for the similarity is that terrestrial support and locomotion is associated with a characteristic pattern of variation in the direction and magnitude of forces acting along the column, and to accommodate these forces, the columns of most terrestrial mammals must be differentiated in broadly similar ways.

Table 32.1 shows the results of the comparisons. For all three aspects of complexity, most of the changes are decreases. Among the changes judged

---

[5] The metrics are based on a view of morphological differences as built up of small, standard units of difference. (These units are just heuristics, and their absolute size need not be specified.) Thus, a large difference is composed of proportionally more standard-difference units than a small one—it has more parts, in a sense—and thus it is proportionally more complex. A more important justification, however, is that the metrics correspond roughly to our intuitive idea of complexity. The correspondence is good, but not perfect: an alternating series, for example, has very high Cm, but strikes most people as fairly simple. Fortunately, alternating series (and other series that would fool the metrics) are rare in vertebral columns.

TABLE 32.1 *Results of comparisons of complexity values (R', C, and Cm') for aquatic mammals with those of their terrestrial ancestors (estimated using average values for an assemblage of 24 diverse, extant terrestrial taxa). Aquatic descendants considered are an otter, a fin-back whale, a group of three pinnipeds (seal, sea-lion, walrus; an average value for the three is used), and a manatee. Entries in the table indicate whether the complexity value for an aquatic mammal represents an increase (I) or a decrease (D) from the ancestral average. Significant differences (p ≤ 0.05, marked with an asterisk) were identified using a bootstrap of the mean of the 24 terrestrials (and in the case of the pinnipeds, a simultaneous bootstrap of the mean for the three).*

Descendants		Centrum height	Centrum length	Centrum width	Neural altitude	Neural length	Neural angle
Otter	R'	D*	D*	I*	D*	D	D*
	C	D*	D*	D*	D	D	I
	Cm'	I	D	I*	I	D*	D*
Finback	R'	D*	D*	D	D*	D*	no data
whale	C	D*	D	D*	D*	I*	no data
	Cm'	D*	D*	D*	D*	D*	no data
Pinnipeds	R'	I	D	D	D	D	I
	C	D	D*	D*	D*	D*	D*
	Cm'	I	D	D	D	D	D
Manatee	R'	D*	D*	D*	D*	D*	D*
	C	I*	I	I	D*	D*	D*
	Cm'	D	D*	D*	D*	D*	I

significant ($p \leq 0.05$), all but four are decreases. The table contains some redundancy: each vertebral dimension represents an independent test only to the extent that complexity values are uncorrelated, and while most pairwise correlation coefficients are low (only 6 out of 45 greater than 0.5), many are significant (14 out of 45, $p \leq 0.05$).

The decrease in complexity is almost certainly a consequence of adaptive modification of the vertebral column for life in the water. In terrestrial mammals, variation along the column in the forces associated with support and locomotion is quite high, so the columns of terrestrial mammals are highly differentiated—that is, quite complex. Aquatic mammals are weightless in water, thus eliminating gravity as a source of stress on the column, at least for most aquatics (the otter and the pinnipeds are partly terrestrial). Further, they propel themselves mainly with simple fish-like undulations of the body (or tail), which produce a more symmetrical distribution of forces than does quadrupedal walking or running, for example. The vertebral columns of these aquatic mammals apparently evolved to accommodate the new regime of forces, their columns becoming simpler and more fish-like, especially in the case of the whale (Fig. 32.1).

COMMENTARY AND RECOMMENDATIONS

To fill the (near) data vacuum, I encourage a shift in emphasis for the field of complexity studies from the theoretical to the empirical. Further, a high degree of rigour will be essential for future work, rigour which was largely absent in studies prior to Cisne's. For example, these studies do not define complexity operationally. That is, they give no careful explanation for how complexity is to be understood (or measured) that would enable anyone to repeat the analysis or to conduct a similar analysis independently.

Despite the recent consensus on meaning, specifying an operational metric remains a difficult problem. Useful metrics have been devised for whole organisms by Schopf et al. (1975) and Bonner (1988), for organic molecules by Papentin (1982) and Yagil (1985), for behaviour by Cole (1985), and for nucleotide sequences by Gatlin (1972). Others are needed, especially ones applicable to whole organisms across a wide range of taxa.

Further, most temporal studies have not examined complexity change in true ancestor–descendant lineages, mainly because few fossils of directly ancestral species have been discovered (or have been identified as such). Rather, they substitute either a complexity value for a related species or an average value for a more inclusive taxonomic group of which the ancestor was presumably a member. No general alternative solution to this problem has been found, but close attention to the phylogenetic relationships among the taxa compared will be desirable in future studies.

A few other lessons follow from past work. Tests for uniform or overall complexity trends should cover a random sample of lineages and characters. And documenting net increase will require a more-than-casual study of the major transitions, with serious attention to the precise *nature* of the morphological changes occurring as well as the *sense* in which they constitute complexity increase. Finally, as Fisher (1986) and Gould (1988) note in their discussions of trends generally, patterns of change may vary with temporal and physical scale, and may also have varied in absolute time, over the course of the history of life. As a consequence, a number of different kinds of empirical studies may be necessary before any generalizations can be made about patterns of complexity change. More detailed discussions of empirical approaches to the study of directionality in general, as well as the pitfalls of these approaches, can be found in Gould 1988, Hull 1988, Fisher 1986, and Raup 1977.

WHENCE THE CONSENSUS?

The case for increasing complexity in multi-cellular organisms is weak. Complexity must have increased early in the history of life, as Maynard Smith argues, if the first organisms were very simple, and the early transitions from uni-cellularity to multi-cellularity seem to be indisputable leaps in complexity. Since then, however, the only evidence is the trend Cisne found in a putative lineage of malacostracan free-living arthropods, and the difference he demonstrates between trilobites and later arthropods. On the other hand, the evidence against a trend is also pretty thin.

Even if the evidence is negligible, we must still account for the impression, the Gestalt, of increasing complexity that many experience in thinking about the history of life. To some extent, the conventional wisdom must be sustained by the Gestalt; I suggest (without exploring) a few of its possible sources.

One is that complexity does increase, and that we unconsciously compute complexity differences between earlier and later organisms with some innate, cognitive algorithm, or even perceive differences directly, in ways that we simply cannot yet articulate. If so, then our only project is to discover how to say what we already know.

On the other hand, properties other than complexity might cause the Gestalt. Comparing a cat with a clam, for example, many will get a vague impression that 'something more' is going on in the cat. Is the 'something more' greater complexity or is it greater intelligence, greater mobility, or greater similarity to us? Hard to say. Complexity has to do with number of different kinds of parts and the irregularity of their arrangement, and comparing parts and arrangements in cats and clams is not straightforward. They are anatomically *very* different animals. If cats seem to have more parts, it could be just that they are larger (on average), with parts that are easier to see and more familiar.[6] (Recall Williams's warning.) Possibly, as one reviewer suggested, organisms simply look more and more *different* from modern ones as we scan further and further back in time; if the moderns are assumed to be very complex, then less familiar might be mistaken for less complex. The human perspective, like any other, has its biases.

---

[6] Schopf *et al.* (1975) measure the complexity of a variety of taxa by counting the number of morphological terms used for each by anatomists. Mammals have about three times as many total terms as bivales (which include the clams), and about three times as many terms per genus, which suggests that cats really are more complex. The metric is ingenious, but I think it does not overcome the human bias. Mammals might have more terms just because many more people, with interests spanning a wider range, have spent more total time studying them.

Other possibilities: maybe the few spectacular (or even just clear-cut) cases of complexity increase, such as the transitions to multi-cellularity, so dominate our reflections on evolution as to create the impression of a pervasive, long-term trend. Or we may simply read into evolution the trend in technology toward increasing complexity of devices.[7]

Another cultural bias may also be at work. We have the double tendency to read progress into evolution (see discussions in Ruse 1988 and Gould 1988) and to connect complexity with progress. More complex organisms, like more complex machines, are commonly thought to have progressive qualities like efficiency or sophistication. I suspect that these connections enable us, by a kind of reciprocal reinforcement, to comfortably maintain (without close examination) pet notions of progressive evolution and complexity increase. On the one hand, complexity must increase because selection favours progressive qualities like efficiency. On the other, the 'observed' increase in complexity in evolution provides an objective basis for a belief in progress. These connections may seem so obvious as not to require demonstration, but in the absence of demonstration it is reasonable to ask if they are real.

The point has been made that progress is a poorly defined concept because it is heterogeneous, encompassing such diverse elements as organismal efficiency, intelligence, autonomy, and so forth (Fisher 1986), and that this fault, along with an inescapable evaluative or axiological component (Ayala 1974), makes it unsuitable for empirical enquiry. Complexity is more specific, more concrete, and therefore more tractable, but on account of its historical and still commonplace association with progress, it carries the axiological taint. This is unfortunate, partly because the taint will tend to spoil empirical analyses that are otherwise reasonably value-neutral. More importantly, even if a hint of the axiological were welcome, the connection is poorly thought out, and may well be baseless; the complex may not be better, in any sense.

The Gestalt of increasing complexity that emerges from an overview of the history of life is an unwavering one. We know, however, that Gestalts, even steady ones, can mislead, and thus a study of its possible sources in perceptual or cultural biases seems worthwhile. In the meantime, some

---

[7] This suggestion is a modification of an argument by Ruse (1988) that evolutionists tend to read into evolution the *progress* that seems evident in technology and science. The trend in technological *complexity* is almost as salient, although some counter-examples come to mind: ram-jet engines are at least superficially simpler than gasoline engines (in propeller-driven aircraft), and steel-beam arches seem simpler than the stone arches they succeed. The complexity trend in technology (like the one in evolution) might be worth closer examination. Has this been done?

evidence on complexity change from the natural world would do much to clarify matters.[8]

REFERENCES

Atlan, H. (1974), 'On a Formal Definition of Organization', *Journal of Theoretical Biology*, 45: 295–304.

Ayala, F. J. (1974), 'The Concept of Biological Progress', in F. J. Ayala and T. Dobzhansky (eds.), *Studies in the Philosophy of Biology* (London: Macmillan), 339–55.

Bonner, J. T. (1988), *The Evolution of Complexity* (Princeton: Princeton University Press).

Brooks, D. R., and Wiley, E. O. (1988), *Evolution as Entropy*, 2nd edn. (Chicago: University of Chicago Press).

Castrodeza, C. (1978), 'Evolution, Complexity, and Fitness', *Journal of Theoretical Biology*, 71: 469–71.

Cisne, J. L. (1974), 'Evolution of the World Fauna of Aquatic Free-Living Arthropods', *Evolution*, 28: 337–66.

Cole, B. J. (1985), 'Size and Behavior in Ants: Constraints on Complexity', *Proceedings of the National Academy of Sciences USA*, 82: 8548–51.

Cope, E. D. (1871), 'The Method of Creation of Organic Forms', *Proceedings of the American Philosophical Society*, 12: 229–63.

Darwin, C. (1859), *On the Origin of Species*, facsimile of the 1st edn. (Cambridge, Mass.: Harvard University Press, 1964).

——(1987), 'Notebook E', in P. H. Barrett *et al.* (eds.), *Charles Darwin's Notebooks* (Ithaca, NY: Cornell University Press).

Denbigh, K. G. (1975), 'A Non-Conserved Function for Organized Systems', in L. Kubat and J. Zeman (eds.), *Entropy and Information in Science and Philosophy* (New York: American Elsevier), 83–92.

Fisher, D. C. (1986), 'Progress in Organismal Design', in D. M. Raup and D. Jablonski (eds.), *Patterns and Processes in the History of Life* (Berlin: Springer-Verlag), 99–117.

Gatlin, L. L. (1972), *Information Theory and the Living System* (New York: Columbia University Press).

Goudge, T. A. (1961), *The Ascent of Life* (London: George Allen and Unwin Ltd.).

Gould, S. J. (1977a), 'Eternal Metaphors of Paleontology', in A. Hallam (ed.), *Patterns of Evolution* (Amsterdam: Elsevier), 1–26.

——(1977b), *Ontogeny and Phylogeny* (Cambridge, Mass.: Harvard University Press).

——(1988), 'On Replacing the Idea of Progress with an Operational Notion of Directionality', in N. H. Nitecki (ed.), *Evolutionary Progress* (Chicago: University of Chicago Press), 319–38; reproduced as Ch. 33.

[8] For their various contributions of ideas, helpful criticism, and encouragement, I thank D. M. Raup, L. Van Valen, M. LaBarbera, J. J. Sepkoski, Jun., A. A. Biewener, M. Ruse, M. Swetlitz, A. J. Dajer, R. J. Kunzig, S. Lyons, N. S. McShea, R. J. McShea, and S. D. McShea.

——Gilinsky, N. L., and German, R. Z. (1987), 'Asymmetry of Lineages and the Direction of Evolutionary Time', *Science*, 236: 1437–41.

Gregory, W. K. (1934), 'Polyisomerism and Anisomerism in Cranial and Dental Evolution among Vertebrates', *Proceedings of the National Academy of Sciences*, 20: 1–9.

——(1935a), 'Reduplication in Evolution', *Quarterly Review of Biology*, 10: 272–90.

——(1935b), '"Williston's Law" Relating to the Evolution of Skull Bones in the Vertebrates', *American Journal of Physical Anthropology*, 20: 123–52.

——(1951), *Evolution Emerging*, i (2 vols., New York: Macmillan).

Hinegardner, R., and Engelberg, J. (1983), 'Biological Complexity', *Journal of Theoretical Biology*, 104: 7–20.

Hull, D. L. (1988), 'Progress in Ideas of Progress', in M. H. Nitecki (ed.), *Evolutionary Progress* (Chicago: University of Chicago Press), 27–48.

Huxley, J. S. (1953), *Evolution in Action* (New York: Harper and Brothers).

Kampis, G., and Csányi, V. (1987), 'Notes on Order and Complexity', *Journal of Theoretical Biology*, 124: 111–21.

Katz, M. J. (1987), 'Is Evolution Random?', in R. A. Raff and E. C. Raff (eds.), *Development as an Evolutionary Process* (New York: Alan R. Liss), 285–315.

Lamarck, J. B. P. A. M. (1809), *Zoological Philosophy* (repr. Chicago: University of Chicago Press, 1984).

Landauer, R. (1988), 'A Simple Measure of Complexity', *Nature*, 336: 306–7.

Lewontin, R. C. (1968), 'Evolution', in D. L. Sills (ed.), *International Encyclopedia of the Social Sciences*, v (New York: Macmillan and the Free Press), 202–10.

Maddox, J. (1990), 'Complicated Measures of Complexity', *Nature*, 344: 705.

Maynard Smith, J. (1970), 'Time in the Evolutionary Process', *Studium Generale*, 23: 266–72.

——(1988), 'Evolutionary Progress and Levels of Selection', in M. H. Nitecki (ed.), *Evolutionary Progress* (Chicago: University of Chicago Press), 219–30.

Maze, J., and Scagel, R. K. (1983), 'A Different View of Plant Morphology and the Evolution of Form', *Systematic Botany*, 8: 469–72.

McCoy, J. W. (1977), 'Complexity in Organic Evolution', *Journal of Theoretical Biology*, 68: 457–8.

Needham, J. (1936), *Order and Life* (New Haven: Yale University Press).

Owen, R. (1866), *Anatomy of Vertebrates*, ii (London: Longmans, Green, and Co.).

Papentin, F. (1980), 'On Order and Complexity. I. General Considerations', *Journal of Theoretical Biology*, 87: 421–56.

——(1982), 'On Order and Complexity. II. Applications to Chemical and Biochemical Structures', *Journal of Theoretical Biology*, 95: 225–45.

Prigogine, I., Nicolis, G., and Babloyantz, A. (1972), 'Thermodynamics of Evolution', *Physics Today*, Dec., 38–44.

Raup, D. M. (1977), 'Stochastic Models in Evolutionary Palaeontology', in A. Hallam (ed.), *Patterns of Evolution* (Amsterdam: Elsevier), 59–78.

Rensch, B. (1960), *Evolution above the Species Level* (New York: Columbia University Press), 281–308.

Richardson, R. C., and Kane, T. C. (1988), 'Orthogenesis and Evolution in the Nineteenth Century', in M. H. Nitecki (ed.), *Evolutionary Progress* (Chicago: University of Chicago Press), 149–68.

Robson, K. A., Scagel, R. K., and Maze, J. (1988), 'Sources of Morphological Variation and Organization Within and Among Populations of *Balsamorhiza sagittata*', *Canadian Journal of Botany*, 66: 11–17.

Ruse, M. (1988), 'Molecules to Men: Evolutionary Biology and Thoughts of Progress', in M. H. Nitecki (ed.), *Evolutionary Progress* (Chicago: University of Chicago Press), 97–126.

Salthe, S. N. (1985), *Evolving Hierarchical Systems* (New York: Columbia University Press).

Saunders, P. T., and Ho, M.-W. (1976), 'On the Increase in Complexity in Evolution', *Journal of Theoretical Biology*, 63: 375–84.

———— (1981), 'On the Increase in Complexity in Evolution II. The Relativity of Complexity and the Principle of Minimum Increase', *Journal of Theoretical Biology*, 90: 515–30.

Schopf, T. J. M., Raup, D. M., Gould, S. J., and Simberloff, D. S. (1975), 'Genomic versus Morphologic Rates of Evolution: Influence of Morphologic Complexity', *Paleobiology*, 1: 63–70.

Simpson, G. G. (1949), *The Meaning of Evolution* (New Haven: Yale University Press).

—— (1961), *Principles of Animal Taxonomy* (New York: Columbia University Press).

Spencer, H. (1890), *First Principles*, 5th edn. (London: Williams and Norgate).

—— (1893), *The Principles of Biology*, ii (New York: D. Appleton and Company).

Stanley, S. M. (1973), 'An Explanation for Cope's Rule', *Evolution*, 27: 1–26.

Stebbins, G. L. (1969), *The Basis of Progresive Evolution* (Chapel Hill, NC: University of North Carolina Press).

Subba Rao, J., Geevan, C. P., and Subba Rao, G. (1982), 'Significance of the Information Content of DNA in Mutations and Evolution', *Journal of Theoretical Biology*, 96: 571–7.

Swetlitz, M. (1989), 'Facts, Values, and Progress: Julian Huxley's Use of the Fossil Record', a talk given to the Society for the History, Philosophy and Social Studies of Biology, University of Western Ontario, 25 June 1989.

Vogt, P. R., and Holden, J. C. (1979), 'The End-Cretaceous Extinctions: A Study of the Multiple Working Hypotheses Method Gone Mad', in W. K. Christensen and T. Birkelund (eds.), *Cretaceous/Tertiary Boundary Events Symposium Proceedings*, ii (Copenhagen: University of Copenhagen Press).

Waddington, C. H. (1969), 'Paradigm for an Evolutionary Process', in C. H. Waddington (ed.), *Towards a Theoretical Biology*, ii (Chicago: Aldine Publishing Company), 106–28.

Wake, D. B., Connor, E. F., de Ricqlès, A. J., Dzik, J., Fisher, D. C., Gould, S. J., LaBarbera, M., Meeter, D. A., Mosbrugger, V., Reif, W.-E., Rieger, R. M., Seilacher, A., and Wagner, G. P. (1986), 'Directions in the History of Life', in D. M. Raup and D. Jablonski (eds.), *Patterns and Processes in the History of Life* (Berlin: Springer-Verlag), 47–67.

Wicken, J. S. (1979), 'The Generation of Complexity in Evolution: A Thermodynamic and Information-Theoretical Discussion', *Journal of Theoretical Biology*, 77: 349–65.

—— (1987), *Evolution, Thermodynamics, and Information* (Oxford: Oxford University Press).

Williams, G. C. (1966), *Adaptation and Natural Selection* (Princeton: Princeton University Press).

Wright, R. (1990), 'The Intelligence Test', *New Republic*, 29 Jan., 29–36.

Yagil, G. (1985), 'On the Structural Complexity of Simple Biosystems', *Journal of Theoretical Biology*, 112: 1–23.

33

ON REPLACING THE IDEA OF PROGRESS WITH AN OPERATIONAL NOTION OF DIRECTIONALITY

STEPHEN JAY GOULD

PROGRESS AS HINDRANCE

Progress is a noxious, culturally embedded, untestable, non-operational, intractable idea that must be replaced if we wish to understand the patterns of history. Yet our obsession with progress records something larger, deeper, and vitally important in our search to understand the workings of time. Progress is a bad example of a crucial generality that we must pursue—the study of directional change in history.

Progress slots into the logic of our cultural hopes as a response to scientific discoveries that we, as geologists and palaeontologists, imposed upon an initially unwilling Western world. If life be stable on an Earth but a few thousand years old, then human domination pervades history, and we need no story of its gradual and inexorable development. But as soon as we learned that the Earth is billions of years old, and that human history occupies but a metaphorical microsecond at the very end, then our central notion of human superiority in a meaningful world met its strongest challenge from science. Our geological confinement to a moment at the very end of recorded time must engender suspicions that we are a lucky accident, an afterthought, rather than the goal of all creation. Progress is the doctrine that dispels this chilling thought—for if life moves inexorably forward, however fitfully, towards its ultimate embodiment in human consciousness, then the restriction of *Homo sapiens* to a final moment poses no challenge to the general hope; for all that came before may now be interpreted as part of a process scheduled to yield our form from the start. We may then continue to pervade time through a long chain of imperfect surrogates, mounting their steady course towards our exalted estate.

First published in M. H. Nitecki (ed.), *Evolutionary Progress* (Chicago: University of Chicago Press, 1988), 319–38. Reprinted by permission.

FIG. 33.1. The vernacular equation of the word 'evolution' with progress. Reproduced with permission of Granada TV Rental.

The continuing hold of progress upon our cultural perceptions may best be illustrated by a route too little exploited by scholars: the iconography of advertisements and cartoons, the leading visual styles of pop culture. I present just two examples from my large collection, anecdotes to be sure, but both well known to all of us as standard representations, not as idiosyncrasies.

The first (Fig. 33.1) demonstrates that the very word 'evolution' means progress in pop culture, for the ad makes no sense otherwise—that is, one must believe that Granada's theory of evolution is progress in communication from rock to television, not simply (as professionals would define the term) a series of changes that respond to shifting local environments. The second (Fig. 33.2) illustrates the confidence felt by cartoonists that

COMMENTARY

Fɪɢ. 33.2. The evolution of man, by Larry Johnson, *Boston Globe* (before a Patriots–Raiders football game). The humour depends absolutely on the cartoonist's confidence that we will all immediately grasp the iconography of linear progress. Reproduced with permission of Larry Johnson.

everyone understands the image of a line of progress as a representation of evolution. Otherwise, no laughs.

In this essay, I wish to make three major points, all supporting the argument that we can preserve the deeper (and essential) theme of direction in history, while abandoning the intractable notion of progress: (1) progress is a culturally conditioned, not an inevitable or necessarily true, account of history; (2) new themes in evolutionary theory have provided different interpretations of data conventionally treated as examples of progress; and (3) the larger issue of direction in history can be reformulated in a tractable manner.

THE POWER AND PERSISTENCE OF NON-PROGRESSIONIST VISIONS

Far from being either an obvious truth of nature or a psychological necessity of all cultures, the idea of progress is a late comer in one particular kind of society—our own. Many historians from Bury (1920) to Lovejoy (1936) to Tawney (1926), have located the origin of progress (viewed as a general feature of both nature and human life) in the seventeenth-century ferment of religious change, scientific discovery, and industrial innovation that signalled the spread and domination of Western culture throughout most of the world. Other scholars, including Eliade (1954) and Morris (1984), show that the opposite vision of an unchanging or strictly cycling Earth is not only far older but has been, and continues to be, the general view of time maintained by most peoples.

As Eliade argues in his classic treatise, the notion of directional history motivated by human action is viewed as terrifying, not uplifting, by most cultures—for such a concept of progress requires that direct human control over events be embraced, and we are then forced to admit that the pervasive tragedies of life, from plague to famine, are ordinary episodes under our influence and not (as the alternate view of stable immanence allows) anomalous moments in nature's constancy, and therefore subject to repeal or placation by prayer and sacrifice.

This restriction of progress may not impress readers, who might reply: since science is largely an invention of the culture that embraced progress, who cares about its unpopularity in other systems? But the limitation of progress goes further: many important Western scientific theories of time, life, and history have also placed the denial of progress at the centre of their conceptual structure. The 'uniformitarian' geology of Hutton and Lyell provides the most striking of all examples. Our failure to grasp this philosophical basis underlies the standard misinterpretation that textbook cardboard presents for the supposed founders of our science (see my book *Time's Arrow, Time's Cycle* (1987*b*), for an elaboration of this argument).

We view Hutton and Lyell as heroes of empirical science who transformed our view of the Earth by inductive field-work. In fact, the original and distinctive ideas, espoused by both and enshrined in our credo of uniformitarianism, arose prior to their field-work and had a foundation in their idiosyncratic rejection of progress for a strict belief in the cyclic and non-directional character of history. Hutton's cycles began so long ago that we see no vestige of beginning; they repeat with such precision that we can discern no prospect of an end so long as nature's current laws persist. Lyell anticipated the discovery of Palaeozoic mammals to prove the unchanging mean complexity of life through time, and he even predicted that extinct genera would reappear when the appropriate stage of the grand climatic cycle, or great year, came round again. The famous caricature drawn by De la Beche (Fig. 33.3) is not, as so long misinterpreted, a gentle satire on William Buckland's lectures, but a mordant dig from a classical progressionist at Lyell's view that ichthyosaurs would one day return (to lecture on a human skull of the 'previous' creation—see Rudwick 1975, Gould 1987*b*). For Lyell wrote (1930: 123):

AWFUL CHANGES.

MAN FOUND ONLY IN A FOSSIL STATE.——REAPPEARANCE OF ICHTHYOSAURI.

A Lecture.—" You will at once perceive," continued PROFESSOR ICHTHYOSAURUS, "that the skull before us belonged to some of the lower order of animals; the teeth are very insignificant, the power of the jaws trifling, and altogether it seems wonderful how the creature could have procured food.'

FIG. 33.3. De la Beche's mordant caricature of Lyell's belief in non-directional cyclicity as the pattern of life's history. The returned ichthyosaur lectures on a product of the last creation.

Then might those genera of animals return, of which the memorials are preserved in the ancient rocks of our continents. The huge iguanodon might reappear in the woods, and the ichthyosaur in the sea, while the pterodactyl might flit again through umbrageous groves of tree ferns.

WHY WE NEED DIFFERENT EXPLANATIONS FOR THE PHENOMENA OF PROGRESS

I do not deny that the fossil record contains legitimate cases of the primary phenomenon identified as progress—persistent trends within clades based on characters interpretable as structural improvements, and leading to increase in representation of taxa bearing these features (usually at the expense of assumed competitors who don't). Every word in this definition (excluding only articles and prepositions) can be challenged as ambiguous, hence the extraordinary difficulty and contentiousness of the concept. (And I have not even mentioned the primary issue of phylogenetics: shall we confine the idea to trends within strictly monophyletic clades, or shall we also admit parallel and iterative tendencies of several subclades within a monophyletic group?)

Among the many caveats, consider just two primary items:

A Question of Evidence

Our previous neo-Darwinian orthodoxies, now dispersed, often led us to assert a claim of progress based on ambiguous data that could yield no such firm interpretation without what my old logic teacher used to call the 'inarticulated major premises' (hidden assumptions to devotees of the vernacular) of both strict adaptationism and the competitive basis of faunal replacement. With these premises granted a priori, we falsely judged two classes of data as evidence *ipso facto* for conventional ideas of progress.

First, we read the simple documentation of a general trend as evidence for progress (what else, if it arose by anagenesis in a world dominated by natural selection?). With new ideas about the power of random processes and, particularly, the pervasiveness of side consequences (the 'spandrels' of Gould and Lewontin 1979) in a revised world with expanded causes (species selection with hitch-hiking morphology, to cite just one example), we now understand that simple documentation of a trend can only be the first step in asserting progress as its basis. I suspect that many of the most famous trends of the fossil record—stipe reduction in graptolites (Elles 1922, Bulman 1963), or increasing symmetry of the cup in crinoids (Moore and Laudon 1943), for example—will be candidates for reinterpretation, especially since their assumed basis in morphological improvement has been elusive for so long (and may never have existed in our new perspective).

Second, we have taken faunal replacement (by forms judged sufficiently similar to share a general mode of life) as evidence *ipso facto* for improvement by competitive replacement. Clearly, our revised view of the greater frequency, speed, effect, and causal distinctness of mass extinction must call this assumption into question, especially since so many of these classic replacements did not occur by gradual waning of one clade with co-ordinated waxing of its 'successor', but by 'relay' of one group to another at an extinction boundary (see Gould and Calloway 1980 on bivalves and brachiopods). Non-competitive replacement may be the rule rather than the exception in life's history. I need hardly add that our own existence is probably contingent on such a replacement of dinosaurs by mammals.

A Question of Definition

Do we consider a poker game progressive when players up the ante? Not usually, I think. The stakes are higher, but the rules don't change; a full house still beats a flush. I do not deny that co-ordinated change on this model does occur through time, especially when biological interactions are important (see Vermeij's elegant documentation of this theme (1973, 1987)). Is a snail with a thick shell 'better' than its thinner-shelled ancestor because an increase in the power of crushing predators requires this degree of strength to achieve the same adaptation that ancestors attained with thinner shells? The later world is different by virtue of such 'arms races', but in what usual sense of the term can we proclaim it better?

As a historical footnote, William Buckland invoked this phenomenon to resolve an apparent conflict between progress and God's timeless perfection (see his 1836 Bridgewater Treatise on the power, wisdom, and goodness of God, as manifested in the Creation). Progress might seem to imply past imperfection, a concept inconsistent with the notion of a benevolent and omnipotent deity. But if each prior state was optimal in itself, and change only ups the ante without modifying the rules, then a continuous perfection can escalate in time.

None the less, I do not deny that we have examples (probably many) of trends that meet our two major criteria for identification as progress in the usual vernacular sense: directional change fairly interpreted as biomechanical improvement (either in general design or in relation to a particular ecology continuously occupied by the group), and increasing relative representation in number, space, or taxa as an inferred consequence of the trend. Lidgard (1986), for example, has traced the gradual increase in representation of the more efficient zooidal budding mode (at the expense of intrazooidal budding) through 100 million years in the evolution of encrusting cheilostome bryozoans. And Stanley's (1975) views on mantle fusion and subsequent formation of siphons as key innovations in the history of bivalves gain support from both criteria of biomechanical improvement and increasing relative representation.

Yet even these properly winnowed cases may undergo a radical reinterpretation in the light of new ideas stemming from the most important contemporary reformulation of evolutionary theory—the hierarchical principle of interacting levels (Salthe 1985, Vrba and Eldredge 1984, Vrba and Gould 1986). The old notions of anagenesis and adaptation (both based on the assumption that trends are driven solely by natural selection operating on organisms) must be supplemented by new reasons for the phenomena of progress; these new reasons may be numerically predominant among the legitimate cases of progress.

I will present here only two revised notions of cause, one critique each for anagenesis and adaptation: (1) When we recognize that trends can be powered by the differential success of species, not only the struggles of organisms, we are led to view the lower level of selection within populations as part of an 'entity-making machine', with trends formed at the higher level of sorting among entities. This suggests a different perspective on trends as changes in variance rather than anagenesis of entities. (2) Classical adaptation may be an impediment to, not the basis of, most evolutionary trends. The chief ingredient of a trend is the size and flexibility of the 'exaptive pool', not the perfection of active and crucial adaptations.

Trends as Changes in Variance

Eldredge (1979) has argued that we must replace our traditional 'transformational' concept of evolution with a 'taxic' perspective. In this view, and adding the hierarchical theme, species (usually stable throughout their histories) are the entities that fashion trends by their differential births and deaths. A trend is the positive sorting of certain kinds of species, not the gradual transformation of a continuous population.

When we redefine trends as differential sorting of species, we may no longer interpret them as states (symmetry of a crinoid cup) with mean values gradually changing through time. The mean value of a clade is not a central adaptation, but only an abstraction, an average among many species each well adapted (by a particular state of the character in question) to its own environment. Means move because the distribution of variance changes. In ordinary directional selection within populations, we view an altered mean as an advantageous character, and therefore as the direct instigator of changes in variance that secure its state through time (if larger mean size be favoured by selection, then individuals so endowed leave more offspring on the average). But the story must be reversed for the taxic view: each species is an entity; they do not (for the most part) interact. The mean of a clade is an acausal abstraction, not the agent of anything. Means shift passively as a result of primary changes in variance produced by a differential sorting of species within the clade.

Trends, in other words, may be passive expressions of changes in variance—misperceptions based on our conventional look at the wrong thing: abstracted averages rather than distributions. Suppose, for example, that we plot average distance from shore for marine species in a clade that originated in near-shore waters (not an abstract or randomly chosen *Gedanken* experiment, given the impressive literature asserting just this as an empirical regularity—Jablonski *et al.* 1983 e.g.). We then plot the distribution of distances as the clade expands through time. Since new species are free to branch in only one direction (they are structurally precluded from invading land), the frequency distribution spreads out away from shore as time advances and the clade grows. The median value may never change—we may always find half the species between shore and 50 metres out, and half in deeper water. But the mean will continually increase because the range of deeper species may expand, while the scope of the shallower half remains limited by the nearby shore line.

I don't think that we would want to label this situation as a trend to increasing depth, yet we would so identify it by two standard and inappropriate measures: either by plotting an increase of the deepest species through time (and treating this abstraction as a continuous object, rather than an extreme variant of a system), or by calculating the mean value of the clade (and falsely thinking that we capture an unambiguous property of the whole). Instead, we witness an asymmetrical increase in variance, with a shifting mean as the skewness of the distribution expands, but no change in median. The asymmetry of increasing variation requires explanation, of course, but of a sort quite different from usual suggestions of the anagenetic view. Our traditional perspective requires that we view the shifting mean itself as a causal basis of change. But here, the asymmetry of the starting-point is the cause we seek—the curve was only free to expand in one direction. There never was an intrinsic advantage to deeper water (and no net movement in proportion of species in this direction), but only more room. The required explanatory principle is structural (a trivial point about geography of shore lines in this case), not adaptational. The trend is a change in variance, and needs to be so interpreted.

I first developed this argument in my work on the history of batting averages in baseball (Gould 1983, 1986); I shall set forth its full context for our concerns in my presidential address to the Palaeontological Society for 1987 ('Trends as Changes in Variance'). I shall, for now, only mention two examples to illustrate the extensive scope of this principle:

(a) Cope's Rule, as Stanley suggested (1973), is a prime candidate for such explanation. There may be no general advantage in larger size, only

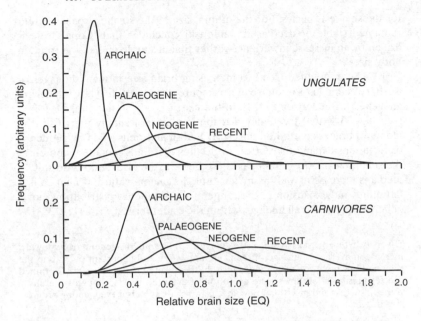

FIG. 33.4. Frequency distributions for EQ (encephalization quotient) for carnivores and ungulates during the Tertiary. Note how the variance expands, and how the region of small brains remains occupied throughout. From Jerison 1973: 315.

more room in this direction *if* ancestral species tend to be small (as another of Cope's principles—the so-called Law of the Unspecialized—suggests). Since animals can't attain negative size, origin near the lower limits would provide little room for decrease, but great scope for increase. Median size may remain constant, as mean size grows. We cannot interpret such a circumstance as a meaningful trend to larger size in the usual sense, but only as movement into more space away from an asymmetrical starting-point near one limit.

Jablonski (1987) has recently provided an example. He calculated sizes of both smallest and largest species within bivalve genera at their initial appearance and for their latest Cretaceous record. Of fifty-eight genera, thirty-three did follow Cope's Rule in the broad sense that the largest species of the latest Cretaceous exceeded the largest species at the point of origin. But only eleven of these genera showed a corresponding increase for the smallest species. For the others, size of the smallest species either remained stable or decreased. Jablonski concludes that 'Cope's rule is driven by an increase in variance rather than a simple directional trend in body sizes'.

(b) The undoubted trend to increasing brain size in many lineages of Tertiary mammals is often treated as a pure tale of selective advantage and anagenesis. But Jerison (1973), in the classic work on this subject, found (see Fig. 33.4) that frequency distributions do not simply march up the ladder of brain size. Rather, ancestors had small brains, and the frequency distributions expand thereafter, while their left ends remain anchored at the small size of ancestral lineages. To be sure, the distributions do shift— medians increase as well as means; but the major feature of change is a flattening and expansion in the range of the frequency distribution, not a general increase for all lineages within the clade. Jerison writes (1973: 315– 16):

Diversity evolved just as average size evolved. . . . Despite the general trend toward increase in average brain size, there is an interesting and important overlap in the region of low brain size which indicates that there were at least some small-brained species present at all times. The evolution of enlarged brains, though generally a route to success and survival of new species, was not universal even among progressive orders.

I can't help adding, as a footnote to this section, that the primal observation of global progress in life's history—the 'march from monad to man'— rests upon a biased perception in this category. Life began in simplicity; this fact of structure provided the 'entity machine' with but one open direction, for complexity is a vast field, but little space exists between the first fossils and anything both simpler and conceivably preservable in the geological record. Where else but up for the right tail by asymmetrical expansion of variance? But the modal life form on Earth has probably always been a prokaryotic organism; each of us harbours more *E. coli* in our gut than the Earth contains people.

Size and Flexibility of the Exaptive Pool

We often attribute improvement of design to the conventional process of direct adaptation by natural selection; but this view cannot be correct in most cases. First of all, most complex adaptations are restrictive in their intricate match with environmental particulars; they correlate with low geological longevity in the light of changing climates and geographies. This, of course, is generally admitted. Instead, we usually seek the link of progress with adaptation in a subset of morphological novelties qualifying as 'general adaptations' or biomechanical improvements in overall function. I believe that a logical error underlies this standard view.

If we understand adaptation aright, then even these general improvements arise for particular reasons in definite environments (legs for the possibility of land excursions in the search for more water, according to one classical view). They do not usually evolve 'for' the general use that underlies their role in fostering the expansion of a large clade. The possibility of fostering is a question of structure, not immediate adaptation: how flexible is the new feature for the set of related uses required in a large clade? how appropriate is the rest of the *Bauplan* for adaptive expansion? how tight are the developmental ties between this new feature and other parts? and how dissociable for mosaic evolution?—to name just a few. The wings of insects could play such a fostering role; those of flying fishes could not. Yet I am not sure that the first winged insect was shaped any less than the first flying fish by narrow adaptation to a particular environmental role. We must focus our concern upon structural possibilities *following* the adaptation's original utility. Insects had evolved a new feature with flexibilities; flying fishes co-opted a pectoral fin with other functions, and therefore limited in evolutionary malleability.

Exaptations (Gould and Vrba 1982) are features now useful to organisms (of adaptive value in the static and ahistorical sense of the term), but that originated for other reasons (therefore not by direct adaptation in the causal sense of the term favoured from Darwin 1859 to Williams 1966). The broad domain of exaptation includes several categories, three most important for our altered understanding of the source of progress: (a) features evolved for one use but suited, by virtue of structure, for co-optation to another (this, and only this, phenomenon counts as 'pre-adaptation' in our usual terminology; thus, pre-adaptation—besides being a dreadful and confusing term that should be expunged—is not a synonym for exaptation); (b) the expansion required by notions of hierarchy: features evolved for reasons, perhaps adaptive, at one level, but producing effects at the level of phenotypes; and (c) the expansion required by critiques of adaptation: features that never were adaptations, but arose as structural by-products of other features—the spandrels of Gould and Lewontin (1979).

The central attributes of evolving complexity, including human consciousness as a prominent example, must arise primarily from the last two sources of exaptation—those furthest removed from any ordinary view of adaptation. Most evolutionists agree, for example, that gene duplication may be a prerequisite for evolutionary flexibility to build complexity

(Ohno 1970), for some copies may continue to make required products while others become free to change. But duplicate genes did not evolve in any particular organism 'for' a flexibility only advantageous to descendants—unless our basic ideas of causality are awry. If they often evolve as 'selfish DNA' by genic selection (an attractive modern theory: see Orgel and Crick 1980, Doolittle and Sapienza 1980), then they are exaptations for organisms by theme (b). If we assume, as I believe we must (Gould and Vrba 1982, Gould 1987a), that most basic features of human consciousness arise as spandrels in the construction of such computing power, then most distinctively human traits are exaptations by theme (c).

One of the most powerful intellectual ideas of our century, Chomsky's theory of language, is an explicit argument about spandrels, though varying terminologies and lack of communication across disciplines have obscured this fact (Chomsky 1986 and personal conversations). Chomsky holds that the deep structure of language is universal and innate as an organ of the brain. It arises in normal development without special nurturing, so long as environments are adequate, just as any organ of morphology does. It works well, but it is too quirky and non-optimal, too much a historical particular rather than a predictable mechanism, to be viewed as something gradually and directly evolved for function. Chomsky prefers to view deep structure as something co-opted *in toto* from other sources—as a spandrel, originally evolved for other cognitive roles (chimps, lacking language entirely so far as we know, are richly endowed with cognition and intelligence by any meaning of the term). Chomsky's argument often strikes evolutionists as absurd, if not heretical; we are too long trained to view language as the immediate source of human triumph. But, given Chomsky's remarkable track record in prediction and efficacy (his insistence, for example, now confirmed, that signing experiments on chimps had been misinterpreted, and were not teaching language), I believe that evolutionists must weigh his views seriously.

Thus, exaptive possibilities, not adaptive usages, form the ground of progress, the source of flexibility in change (including the potential to build upon existing structure towards a new state that we judge more complex). The possibility of progress is specified by the size and flexibility of the 'exaptive pool'—the set of co-optable features. Progress is not regulated by adaptation, especially in the light of new views that locate so much of the exaptive pool not in features that first arose as adaptations of phenotypes (classical pre-adaptations), but either at other levels of the hierarchy, or as non-adaptations at the phenotypic level. The rules of structure and development, not the workings of adaptation, set the boundaries and possibilities of progress.

THE DEEPER ISSUE, RESTRICTIVELY AND MISLEADINGLY ADDRESSED BY PROGRESS, CAN BE REFORMULATED IN A TRACTABLE WAY

We have already seen how Hutton's rigidly cyclical view of time precluded progress entirely. But it did more, and to regrettable effect. It also ruled out the larger subject that progress represents as a poor and limited example—history itself. Hutton's geology is a curiously restricted subject to modern readers, for Hutton, in trying to make the Earth cycle through time as it moves through Newton's space, denied history in any meaningful sense of the term (Gould 1987*b*). What, then, does history require?

Two central attributes mark an interpretable history. First, events must have temporal signatures; they must represent their moment distinctly. Otherwise, history is not a sequence of definable events, since any episode may recur precisely, and we cannot, therefore, know where we are. (Hutton's geology denied history in this most fundamental sense, since his precise cyclicity granted no uniqueness to any moment.)

But distinctness is not enough, at least in our culturally conditioned view of intelligibility. History must be more than a string of isolated, if distinctive, events strung together one after the other. We view history as its last five letters—a story, a skein unwinding in some particular, if complicated, direction. A temporal sequence without directionality, and without causal links among its events, is not inconceivable (the world could be so fashioned), but neither would we call such a thing history. History needs both distinctiveness and directionality to meet our definitions and pique our interest.

What the science of history requires above all is a tractable way to study directional processes. This, I believe, is the proper reason why we have been so obsessed with the idea of progress (I leave aside the improper reason of our cultural hopes and traditions). The concept of directionality, or temporal vectors, is the core of our legitimate concern, for historical science requires its validation. We have been led astray by sinking a proper concern with directionality entirely into one possible and limited manifestation as progress. Thus, we have compromised the deeper theme by relying upon a poor and biased example to carry the richness of the entire subject. What we need is a methodology for studying time's directions.

I and my colleagues have recently made this argument *in extenso* (Gould *et al.* 1987), and will only present a sketch and an example here. Physical scientists have confronted the paradox that so little in nature's laws imparts a directionality to time, despite our overwhelming impression that such vectors exist. Beyond the Second Law of Thermodynamics, 'time's

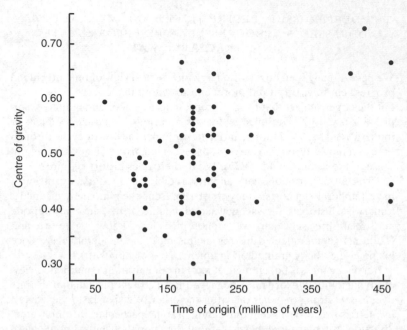

FIG. 33.5. Families within higher taxa. Centre of gravity plotted as a function of time of origin from the base of the Vendian for extinct groups. General rise in value shows that bottom-heavy clades are concentrated among those that originate early. From Gould *et al.* 1987: 1439.

arrow' in Eddington's phrase, we can point to little else without ambiguity or subjectivity (see Morris 1984). Morris suggests a base-level definition for temporal arrows: 'We mean only that the world has a different appearance in one direction of time than it does in the other.' In other words, we must look for temporal asymmetry. Expressed in our terms and by modern metaphor: if I hand you the tape of life's history but do not tell you which end is which, would you know, in watching the tape unfold, whether it was running properly forward or illegitimately backward? This may seem a risible small question compared with genera within families throughout the grandiose idea of progress, but it is definable, tractable, testable, and quantifiable. In other words, we may treat this approach to directionality as science rather than as an ambiguous instantiation of our hopes—a good trade.

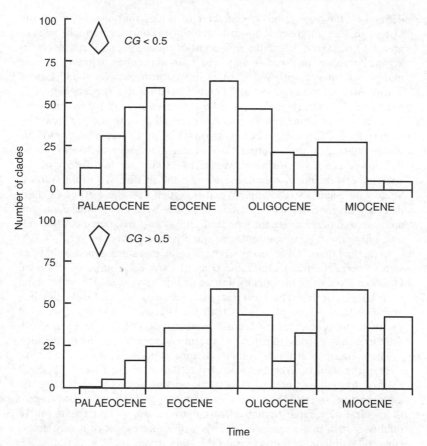

FIG. 33.6. Histograms for bottom-heavy (*above*) and top-heavy (*below*) clades of extinct Tertiary mammals. Note that early times featured a higher proportion of bottom-heavy clades. CG = centre of gravity. From Gould *et al.* 1987: 1440.

In our previous work (Gould *et al.* 1987), we looked for temporal asymmetry in the standard palaeontological representation of life's temporal structure: clade diversity diagrams. We asked the simple operational question: if I hand you a chart of clade diversity diagrams, but forget to label the vertical axis of time, would you know which end was Cambrian, which Recent? We found, by several criteria, that a temporal arrow could be defined by the bottom-heavy character of clades that arose early in the history of a larger group (bottom-heavy clades concentrate their times of greatest flourishing before the mid-point of their geological range; clades arising later show no directionality, and maintain centres of gravity at their mid-points, on average). We affirmed this pattern both for the history of marine invertebrates (genera and families) and for the Tertiary history of mammalian genera (Figs. 33.5 and 33.6). We also argued that its apparently fractal character indicated a possible generality, or true arrow of evolutionary time. (The bottom-heavy character of much larger monophyletic clades indicates a preservation across levels—consider the twenty or so Palaeozoic echinoderm classes, versus five today; or the history of the coelomate Metazoa, by comparing the astounding diversity of the Burgess Shale with current stereotypy.) We also argued that the method of clade diversity diagrams could be extended to any field that studies changing amounts and percentages through time (it has long been used in archaeology, but without appreciation for its quantitative power).

In reading these claims, many students trained in stereotyped definitions of a unitary scientific method and content may feel a sense of let-down. How can a vector of historically unique events be viewed as primary data, or as bearers of theoretical interest? Isn't science a search for the timeless and quantifiable laws of nature? Isn't the specification of a sequence of events no more than mere narrative, a description of uniquenesses worth little in an enterprise dedicated to experiment, repetition, and prediction?

These common attitudes embody a cultural prejudice almost as pervasive as the idea of progress—the restriction of science to an idealized method touted as canonical (but by no means always observed) in the high-prestige, so-called hard sciences of physics and chemistry. But science is a pluralistic search to understand nature's ways—and the narrative quality of historical sequences records a different aspect of nature accessible to legitimate methods beyond the stereotype.

Unfortunately, historical scientists often fall prey to the stereotype, and to the rank-ordering that places them low on the totem pole of reputation. We have often made the mistake of trying to ape the inappropriate methods, or bowing to the supposed data, of our colleagues in fields with greater prestige. Thus, Hutton denied history, and tried to formulate a geology that would emulate Newton's timeless cosmos. Kelvin proclaimed an impossibly young Earth, and most geologists accepted his authority, even though the data of sequential history pointed to times much longer. Charles Spearman invented factor analysis in order to misidentify intelligence as a single entity within the brain, and then touted his quantity as the key to a scientific psychology: 'This Cinderella among the sciences has made a bold bid for the level of triumphant physics itself' (see Gould 1981).

The sequential events of historical narrative provide the primary data for our palaeontological science of life's history and our evolutionary science of phylogeny. We can accord these data their due respect, while remaining true to the basic definition of science as a search for natural order, by insisting that the neglected subject of directionality in time become a focal point in the study of history.

REFERENCES

Buckland, W. (1836), *Geology and Mineralogy Considered with Reference to Natural Theology* (London: G. Routledge & Co.).

Bulman, O. M. B. (1963), 'The Evolution and Classification of the Graptoloidea', *Quarterly Journal of the Geological Society of London*, 119: 401–18.

Bury, J. B. (1920), *The Idea of Progress* (London: Macmillan).

Chomsky, N. (1986), *Knowledge of Language: Its Nature, Origin and Use* (New York: Praeger).

Darwin, C. (1859), *On the Origin of Species* (London: John Murray).

Doolittle, W. F., and Sapienza, C. (1980), 'Selfish Genes, the Phenotype Paradigm and Genome Evolution', *Nature*, 284: 601–3.

Eldredge, N. (1979), 'Alternative Approaches to Evolutionary Theory', in J. H. Schwartz and H. B. Rollins (eds.), *Models and Methodologies in Evolutionary Theory, Bulletin of the Carnegie Museum of Natural History*, 13: 7–19.

Eliade, M. (1954), *The Myth of the Eternal Return* (Princeton: Princeton University Press).

Elles, G. L. (1922), 'The Graptolite Faunas of the British Isles. A Study in Evolution', *Proceedings of the Geologists Association*, 33: 168–200.

Gould, S. J. (1981), *The Mismeasure of Man* (New York: W. W. Norton).

——(1983), Losing the Edge: The Extinction of the .400 Hitter', *Vanity Fair*, 120 (Mar.): 264–78.

——(1986), 'Entropic Homogeneity Isn't Why No One Hits .400 Anymore', *Discover*, Aug.: 60–6.

——(1987*a*), *An Urchin in the Storm* (New York: W. W. Norton).

——(1987*b*), *Time's Arrow, Time's Cycle* (Cambridge, Mass.: Harvard University Press).

——and Calloway, C. B. (1980), 'Clams and Brachiopods—Ships that Pass in the Night', *Paleobiology*, 6: 383–96.

——and Lewontin, R. C. (1979), 'The Spandrels of San Marco and the Panglossian Paradigm: A Critique of the Adaptationist Programme', *Proceedings of the Royal Society of London*, B205: 581–98.

——and Vrba, E. S. (1982), 'Exaptation—A Missing Term in the Science of Form', *Paleobiology*, 8: 4–15; reproduced as Ch. 4.

——Gilinsky, N. L., and German, R. Z. (1987), 'Asymmetry of Lineages and the Direction of Evolutionary Time', *Science*, 236: 1437–41.

Jablonski, D. (1987), 'How Pervasive is Cope's Rule? A Test Using Late Cretaceous Mollusks', *Geological Society of America. Abstracts with Programs*, 19: 713–14.

——Sepkoski, J. J., Jun., Bottjer, D. J., and Sheehan, P. M. (1983), 'Onshore–Offshore Patterns in the Evolution of Phanerozoic Shelf Communities', *Science*, 222: 1123–5.

Jerison, H. J. (1973), *Evolution of the Brain and Intelligence* (New York: Academic Press).

Lidgard, S. (1986), 'Ontogeny in Animal Colonies: A Persistent Trend in the Bryozoan Fossil Record', *Science*, 232: 230–2.

Lovejoy, A. O. (1936), *The Great Chain of Being* (Cambridge, Mass.: Harvard University Press).

Lyell, C. (1830), *The Principles of Geology*, i (London: John Murray).

Moore, R. C., and Laudon, L. R. (1943), *Evolution and Classification of Paleozoic Crinoids*, Geological Society of America Special Papers, 46.

Morris, R. (1984), *Time's Arrows* (New York: Simon and Schuster).

Ohno, S. (1970), *Evolution by Gene Duplication* (New York: Springer).

Orgel, L. E., and Crick, F. H. C. (1980), 'Selfish DNA: The Ultimate Parasite', *Nature*, 284: 604–7.

Rudwick, M. J. S. (1975), 'Caricature as a Source for the History of Science: De la Beche's Anti-Lyellian Sketches of 1831', *Isis*, 66: 534–60.

Salthe, S. V. (1985), *Evolving Hierarchical Systems* (New York: Columbia University Press).

Stanley, S. M. (1973), 'An Explanation for Cope's Rule', *Evolution*, 27: 1–26.

——(1975), 'Adaptive Themes in the Evolution of the Bivalvia (Mollusca)', *Annual Reviews of Earth Planetary Science*, 3: 361–85.

Tawney, R. H. (1926), *Religion and the Rise of Capitalism* (New York: Harcourt, Brace and Co.).

Vermeij, G. J. (1973), 'Adaptation, Versatility, and Evolution', *Systematic Zoology*, 22: 466–77.

——(1987), *Evolution and Escalation. An Ecological History of Life* (Princeton: Princeton University Press).

Vrba, E. S., and Eldredge, N. (1984), 'Individuals, Hierarchies and Processes: Towards a More Complete Evolutionary Theory', *Paleobiology*, 10: 146–71.

——and Gould, S. J. (1986), 'The Hierarchical Expansion of Sorting and Selection: Sorting and Selection Cannot be Equated', *Paleobiology*, 12: 217–28.

Williams, G. C. (1966), *Adaptation and Natural Selection* (Princeton: Princeton University Press).

PART X

CREATIONISM

INTRODUCTION TO PART X

MICHAEL RUSE

Most people think that science and religion are, and necessarily must be, in conflict. In fact, this 'warfare' metaphor, so beloved of nineteenth-century rationalists, has only a tenuous application to reality. For most of the history of Christianity, it was the Church that was the home of science. Until the sixteenth century, Christians took seriously the warning of St Augustine (in the fourth century) that one should beware of setting the truths of revelation, especially as given in the Bible, against the truths of reason, especially as applied to our world of the senses. Conflicts are never genuine, and only give support to the infidel.

The coming of Protestantism, with its emphasis on the Bible as the Word of God, changed things somewhat, although it was not until the seventeenth century, at the time of the Counter-Reformation, that the Catholic Church showed true hostility to science, when it condemned Galileo for his promulgation of Copernican heliocentrism. (Copernicus himself had been not merely a good Catholic, but a priest.) By the nineteenth century, the Catholic Church had reverted to its traditional role, being, if not friendly towards science, far from hostile to it. Most Protestants were also coming to see that science and religion can be reasonably good neighbours—it is true that the arrival of evolution, particularly in the form of Charles Darwin's *Origin of Species*, put this tolerance to severe test. But without denying that there were strong opinions held on both sides, the truth seems to be that much of the supposed controversy was a function of the imagination of non-believers (especially Thomas Henry Huxley and his friends), who were determined to slay theological dragons whether they existed or not.

In this secular century of ours, it seems fairest to say that most scientists are more indifferent to religion than strongly supportive or critical. There are exceptions. For instance, the pioneering statistician and population geneticist Ronald Fisher was a practising Anglican, seeing God's purpose in the workings of natural selection, and the Russian-born American evolutionist Theodosius Dobzhansky felt much the same way. Conversely, biologist and popular writer Richard Dawkins sees in Darwinian evolution the final nail in the coffin of Christianity. He believes that natural selection makes God redundant—there is no need of the supernatural to explain the design-like features of organisms—and that the cruelty of the struggle for existence negates a Being (as is supposed of the Christian God) who is simultaneously all-loving and all-powerful. The pain and destruction of life's processes point to atheism.

On the other side, most Christians (and members of other faiths) have learnt to bend to the winds of science. Some indeed—notably, the French-born Jesuit palaeontologist Pierre Teilhard de Chardin—have attempted full-blown syntheses of science and religion. (Although, given that the Church forbade him to publish his work in his own lifetime, it does suggest that the bending has its firm limits.) The major exception to this accommodation, if not harmony, has been American evangelical Protestantism, rooted as it is in a firm belief in the literal truth of every word of the Bible. Known as 'fundamentalists', such Christians have been carrying on a campaign right through this century against those aspects of science which they find opposed by God's Word, and evolution has always been at the top of the list. Since they believe in a short time span since earth's creation (about 6,000 years, as opposed to the conventional 4.5 billion for the Earth and 20 billion since the Big Bang), six literal days of creation, man and woman created miraculously from dust right at the end of the process, and shortly thereafter a universal destroying flood, the biblical literalists see evolution—particularly Darwinian evolution (that is, something supposing natural selection as the major mechanism)—as a totally false heresy. It is a product of the materialism or 'naturalism' which infects our culture, and must be opposed at every step. Above all, it must be countered in the class-room.

Notorious in the history of the fundamentalist campaign against Darwinism was the so-called Scopes Monkey Trial in Tennessee in the 1920s, when a young schoolteacher was convicted of teaching evolution (in defiance of a state law prohibiting such a practice). Although the conviction was overturned on appeal, the trial had a chilling effect on the teaching of evolution, which was essentially gutted from school biology texts—a state of affairs which persisted until around 1960, when (in response to the perceived threat from Russia's superiority as evinced by Sputnik), major efforts were made to upgrade American science teaching. These included the production of new texts—texts which included evolution.

The fundamentalists—or, as they now call themselves, 'Creationists'—responded with vigour. In order to get around the US Constitutional prohibition against the teaching of religion in state-supported schools, they developed arguments (in a tradition going back to a turn-of-the-century Canadian Seventh-Day Adventist, George McCready Price) that in fact the whole of Genesis taken absolutely literally can be proved by the best scientific methods. The supposed great age of the Earth is an artefact of inadequate or misunderstood measurement, the fossil record gives no support to continuous development, and so forth. Again the evolutionists and biblical literalists met in court, this time in 1981 in the state of Arkansas, and on this occasion it was the evangelicals who lost, as the judge ruled that Creationism has no place in school curricula—especially not in biology classes.

But again the literalists have fought on, and recently they have recruited very respectable academics to their cause. One is the distinguished philosopher of religion Alvin Plantinga, who wrote the first of our chapters (34). Although a professor at the Catholic University of Notre Dame, Plantinga is a Protestant, a Calvinist or 'Reformed' Christian, and as such stresses his respect for the power of reason. He claims not to be against science, but committed as a Christian to take the Bible as the Word of God, something to be interpreted metaphorically only if the science simply demands this. In the case of an Earth of great age, Plantinga seems prepared to accept that science may be right. But in the case of evolution, Plantinga finds no such demand: he does not argue absolutely and completely that evolution is false, but does conclude that evolution is not sufficiently well-established as an idea to merit the overthrow of the story of Genesis. He believes that the idea of evolution owes more to metaphysical commitments to the non-divine nature of reality than to genuine and compelling evidence in its favour.

Ernan McMullin, also a professor at Notre Dame, a Catholic priest and a philosopher of science, disagrees strongly (Ch. 35). He believes that the evidence in favour of evolution is overwhelming. At the same time, he finds no tension with his Christian commitment. The miraculous events of the Bible relate to (and only to) that specific drama leading from the sin of Adam and Eve to the sacrificial death of Jesus. As far as evolution is concerned, McMullin, referring back to Augustine, finds support for the idea that God's creation is natural and sequential. For him, therefore, the stance of Plantinga is scientifically inadequate and theologically unnecessary. There is no need for the warfare metaphor to rear its ugly head yet one more time.

34

WHEN FAITH AND REASON CLASH: EVOLUTION AND THE BIBLE

ALVIN PLANTINGA

My question is simple: How shall we Christians deal with apparent conflicts between faith and reason, between what we know as Christians and what we know in other ways, between teaching of the Bible and the teachings of science? As a special case, how shall we deal with apparent conflicts between what the Bible initially seems to tell us about the origin and development of life, and what contemporary science seems to tell us about it? Taken at face value, the Bible seems to teach that God created the world relatively recently, that he created life by way of several separate acts of creation, that in another separate act of creation, he created an original human pair, Adam and Eve, and that these our original parents disobeyed God, thereby bringing ruinous calamity on themselves, their posterity, and the rest of creation.

According to contemporary science, on the other hand, the universe is exceedingly old—some 10–15 billion years or so, give or take a billion or two. The Earth is much younger, maybe 4.5 billion years old, but still hardly a spring chicken. Primitive life arose on earth perhaps 3.5 billion years ago, by virtue of processes that are completely natural if, so far, not well understood; and subsequent forms of life developed from these aboriginal forms by way of natural processes, the most popular candidates being perhaps random genetic mutation and natural selection.

Now we Reformed Christians are wholly in earnest about the Bible. We are people of the Word; *sola Scriptura* is our cry; we take Scripture to be a special revelation from God himself, demanding our absolute trust and allegiance. But we are equally enthusiastic about *reason*, a God-given power by virtue of which we have knowledge of ourselves, our world, our past, logic and mathematics, right and wrong, and God himself; reason is one of the chief features of the image of God in us. And if we are enthusiastic about reason, we must also be enthusiastic about contemporary

First published in *Christian Scholar's Review*, 21 (1991): 8–32. Reprinted by permission.

natural science, which is a powerful and vastly impressive manifestation of reason. So this is my question: Given our Reformed proclivities and this apparent conflict, what are we to do? How shall we think about this matter?

I. WHEN FAITH AND REASON CLASH

If the question is simple, the answer is enormously difficult. To think about it properly, one must obviously know a great deal of science. On the other hand, the question crucially involves both philosophy and theology: one must have a serious and penetrating grasp of the relevant theological and philosophical issues. And who among us can fill a bill like that? Certainly I can't. (And that, as my colleague Ralph McInerny once said in another connection, is no idle boast.) The scientists among us don't ordinarily have a sufficient grasp of the relevant philosophy and theology; the philosophers and theologians don't know enough science; consequently, hardly anyone is qualified to speak here with real authority. This must be one of those areas where fools rush in and angels fear to tread. Whether or not it is an area where angels fear to tread, it is obviously an area where fools rush in. I hope this essay isn't just one more confirmation of that dismal fact.

But first, a quick gesture towards the history of our problem. Our specific problem—faith and evolution—has of course been with the Church since Darwinian evolution started to achieve wide acceptance, a little more than a hundred years ago. And this question is only a special case of two more general questions, questions that the Christian Church has faced since its beginnings nearly two millennia ago: first, what shall we do when there appears to be a conflict between the deliverances of faith and the deliverances of reason? And another question, related but distinct: How shall we evaluate and react to the dominant teachings, the dominant intellectual motifs, the dominant commitments of the society in which we find ourselves? These two questions, not always clearly distinguished, dominate the writings of the early Church Fathers from the second century on.

Naturally enough, there has been a variety of responses. There is a temptation, first of all, to declare that there really can't be any conflict between faith and reason. The no-conflict view comes in two quite different versions. According to the first, there is no such thing as truth *simpliciter*, truth just as such: there is only truth from one or another perspective. An extreme version of this view is the medieval two-truth theory associated with Averroës and some of his followers: some of these thinkers apparently held that the same proposition can be true according to philosophy or reason, but false according to theology or faith; true as science but false as theology. Thinking hard about this view can easily induce vertigo: the idea, apparently, is that one ought to affirm and believe the proposition as science, but deny it as theology. How you are supposed to do that isn't clear. But the main problem is simply that truth isn't merely truth with respect to some standpoint. Indeed, any attempt to explain what *truth from a standpoint* might mean inevitably involves the notion of truth *simpliciter*.

A more contemporary version of this way of thinking—the truth-from-a-standpoint way of thinking—takes its inspiration from contemporary physics. To over-simplify shamelessly, there is a problem: light seems to display both the properties of a wave in a medium and also the properties of something that comes in particles. And of course the problem is that these properties are not like, say, *being green* and *being square*, which can easily be exemplified by the same object, the problem is that it looks for all the world as if light *can't* be both a particle and a wave. According to Nils Bohr, the father of the Copenhagen interpretation of quantum mechanics, the solution is to be found in the idea of *complementarity*. We must recognize that there can be descriptions of the same object or phenomenon which are both true, and relevantly complete, but none the less such that we can't see how they could both hold. From one point of view light displays the particle set of properties; from another point of view, it displays the wave properties. We can't see how both these descriptions can be true, but in fact they are. Of course, the theological application is obvious: there is the broadly scientific view of things and the broadly religious view of things; both are perfectly acceptable, perfectly correct, even though they appear to contradict one another.[1] And the point of the doctrine is that we must learn to live with, and love, this situation.

But this view itself is not easy to learn to love. Is the idea that the properties in question *really are* inconsistent with each other, so that it isn't possible that the same thing have both sets of properties? Then clearly enough they *can't* both be correct descriptions of the matter, and the view is simply false. Is the idea instead that while the properties are *apparently* inconsistent, they aren't really inconsistent? Then the view might be correct, but wouldn't be much by way of a *view*, being instead nothing but a redescription of the problem.

---

[1] Perhaps the shrewdest contemporary spokesman for this view is the late Donald MacKay (1974*a*, *b*).

Let's look a little deeper. As everyone knows, there are various intellectual or cognitive powers, belief-producing mechanisms or powers, various sources of belief and knowledge. For example, there are perception, memory, induction, and testimony, or what we learn from others. There is also reason, taken narrowly as the source of logic and mathematics, and reason taken more broadly as including perception, testimony, and both inductive and deductive processes; it is reason taken this broader way that is the source of science. But the serious Christian will also take our grasp of Scripture to be a proper source of knowledge and justified belief. Just how does Scripture work as a source of proper belief? An answer as good as any I know was given by John Calvin and endorsed by the Belgic Confession: this is Calvin's doctrine of the internal testimony of the Holy Spirit. This is a fascinating and important contribution that doesn't get nearly the attention it deserves; but here I don't have time to go into the matter. Whatever the mechanism, the Lord speaks to us in Scripture.

And of course what the Lord proposes for our belief is indeed what we should believe. Here there will be enthusiastic agreement on all sides. Some conclude, however, that when there is a conflict between Scripture (or our grasp of it) and science, we must reject science; such conflict automatically shows science to be wrong, at least on the point in question. In the immortal words of the inspired Scottish bard William E. McGonagall, poet and tragedian,

> When faith and reason clash,
> Let reason go to smash.

But clearly this conclusion doesn't follow. The *Lord* can't make a mistake: fair enough; but *we* can. Our grasp of what the Lord proposes to teach us can be faulty and flawed in a thousand ways. This is obvious, if only because of the widespread disagreement among serious Christians as to just what it is the Lord *does* propose for our belief in one or another portion of Scripture. Scripture is indeed perspicuous: what it teaches with respect to the way of salvation is indeed such that she who runs may read. It is also clear, however, that serious, well-intentioned Christians can disagree as to what the teaching of Scripture, at one point or another, really is. Scripture is inerrant: the Lord makes no mistakes; what he proposes for our belief is what we ought to believe. Sadly enough, however, our grasp of what he proposes to teach is fallible. Hence we cannot simply identify the teaching of Scripture with our grasp of that teaching; we must ruefully bear in mind the possibility that we are mistaken. 'He sets the earth on its foundations; it can never be moved,' says the Psalmist (Ps. 104:5). Some sixteenth-century Christians took the Lord to be teaching here that the

Earth neither rotates on its axis nor goes around the Sun; and they were mistaken.

So we can't identify our understanding or grasp of the teaching of Scripture with the teaching of Scripture; hence we can't automatically assume that conflict between what we see as the teaching of Scripture and what we seem to have learned in some other way must always be resolved in favour of the former. Sadly enough, we have no guarantee that on every point our grasp of what Scripture teaches is correct; hence it is possible that our grasp of the teaching of Scripture be corrected or improved by what we learn in some other way—by way of science, for example.

But neither, of course, can we identify either the current deliverances of reason or our best contemporary science (or philosophy, or history, or literary criticism, or intellectual efforts of any kind) with the truth. No doubt what reason, taken broadly, teaches is by and large reliable; this is, I should think, a consequence of the fact that we have been created in the image of God. Of course, we must reckon with the Fall and its noetic effects; but the sensible view here, overall, is that the deliverances of reason are for the most part reliable. Perhaps they are most reliable with respect to such common everyday judgements as that there are people here, that it is cold outside, that the pointer points to 4, that I had breakfast this morning, that $2 + 1 = 3$, and so on; perhaps they are less reliable when it comes to matters near the limits of our abilities, as with certain questions in set theory, or in areas for which our faculties don't seem to be primarily designed, as perhaps in the world of quantum mechanics. By and large, however, and over enormous swatches of cognitive territory, reason is reliable.

Still, we can't simply embrace current science (or current anything else either) as the truth. We can't identify the teaching of Scripture with our grasp of it, because serious and sensible Christians disagree as to what Scripture teaches; we can't identify the current teachings of science with truth, because the current teachings of science change. And they don't change just by the accumulation of new facts. A few years back, the dominant view among astronomers and cosmologists was that the universe is infinitely old; at present the prevailing opinion is that the universe began some 10–15 billion years ago; but now there are straws in the wind suggesting a step back towards the idea that there was no beginning (see Hawking 1988: 115ff.). Or think of the enormous changes from nineteenth- to twentieth-century physics. A prevailing attitude at the end of the nineteenth century was that physics was pretty well accomplished; there were a few loose ends here and there to tie up and a few mopping-up operations left to do, but the fundamental lineaments and characteristics of physical reality had been described. And we all know what happened next.

As I said above, we can't automatically assume that when there is a conflict between science and our grasp of the teaching of Scripture, it is science that is wrong and must give way. But the same holds vice versa; when there is a conflict between our grasp of the teaching of Scripture and current science, we can't assume that it is our interpretation of Scripture that is at fault. It *could* be that, but it doesn't *have* to be; it could be because of some mistake or flaw in current science. The attitude I mean to reject was expressed by a group of serious Christians as far back as 1832, when deep time was first being discovered. 'If sound science appears to contradict the Bible,' they said, 'we may be sure that it is our interpretation of the Bible that is at fault'.[2] To return to the great poet McGonagall,

> When faith and reason clash,
> 'Tis faith must go to smash.

This attitude—the belief that when there is a conflict, the problem must inevitably lie with our interpretation of Scripture, so that the correct course is always to modify that understanding in such a way as to accommodate current science—is every bit as deplorable as the opposite error. No doubt science can correct our grasp of Scripture; but Scripture can also correct current science. If, for example, current science were to return to the view that the world has no beginning, and is infinitely old, then current science would be wrong. — why? Scientifically, surely...

So what, precisely, must we do in such a situation? Which do we go with: faith or reason? More exactly, which do we go with, our grasp of Scripture or current science? I don't know of any infallible rule, or even any pretty reliable general recipe. All we can do is weigh and evaluate the relative warrant, the relative backing or strength, of the conflicting teachings. We must do our best to apprehend both the teachings of Scripture and the deliverances of reason; in either case we will have much more warrant for some apparent teachings than for others. It may be hard to see just what the Lord proposes to teach us in the Song of Solomon or Old Testament genealogies; it is vastly easier to see what he proposes to teach us in the Gospel accounts of Christ's resurrection from the dead. On the other side, it is clear that among the deliverances of reason is the proposition that the earth is round rather than flat; it is enormously harder to be sure, however, that contemporary quantum mechanics, taken realistically, has things

[begs the question in favour of Scripture]

---

[2] *Christian Observer*, 1832, p. 437.

right.[3] We must make as careful an estimate as we can of the degrees of warrant of the conflicting doctrines; we may then make a judgement as to where the balance of probability lies, or alternatively, we may suspend judgement. After all, we don't *have* to have a view on all these matters.

Let me illustrate from the topic under discussion. Consider that list of apparent teachings of Genesis: that God has created the world, that the Earth is young, that human beings and many different kinds of plants and animals were separately created, and that there was an original human pair whose sin has afflicted both human nature and some of the rest of the world. At least one of these claims—the claim that the universe is young— is very hard to square with a variety of types of scientific evidence: geological, palaeontological, cosmological, and so on. None the less, a sensible person might be convinced, after careful and prayerful study of the Scriptures, that what the Lord teaches there implies that this evidence is misleading and that as a matter of fact the Earth really *is* very young. So far as I can see, there is nothing to rule this out as automatically pathological or irrational or irresponsible or stupid.

And of course this sort of view can be developed in more subtle and nuanced detail. For example, the above teachings may be graded with respect to the probability that they really are what the Lord intends us to learn from early Genesis. Most clear, perhaps, is that God created the world, so that it and everything in it depend upon him, and neither it nor anything in it has existed for an infinite stretch of time. Next clearest, perhaps, is that there was an original human pair who sinned and through whose sinning disaster befell both man and nature; for this is attested to not only here but in many other places in Scripture. That humankind was separately created is perhaps less clearly taught; that many other kinds of living beings were separately created might be still less clearly taught; that the Earth is young, still less clearly taught. One who accepted all of these theses ought to be much more confident of some than of others—both because of the scientific evidence against some of them, and because some are much more clearly the teachings of Scripture than others. I do not mean to endorse the view that all of these propositions are true: but it isn't just silly or irrational to do so. One need not be a fanatic, or a Flat Earther, or an ignorant fundamentalist, in order to hold it. In my judgement the view is mistaken, because I take the evidence for an old Earth to be strong and the warrant for the view that the Lord teaches that the Earth is young to be relatively weak. But these judgements are not simply *obvious*, or

[3] Here the work of Bas van Fraassen is particularly instructive—see his 1991.

inevitable, or such that anyone with any sense will automatically be obliged to agree.

II. FAITH AND EVOLUTION

So I can properly correct my view as to what reason teaches by appealing to my understanding of Scripture; and I can properly correct my understanding of Scripture by appealing to the teachings of reason. It is of the first importance, however, that we correctly *identify* the relevant teachings of reason. Here I want to turn directly to the present problem, the apparent disparity between what Scripture and science teach us about the origin and development of life.

First, I shall argue that the theory of evolution is by no means religiously or theologically neutral. Second, I want to ask how we Christians should in fact think about evolution: How probable is it, all things considered, that the Grand Evolutionary Hypothesis is true?

A. *Evolution Religiously Neutral?*

To turn to the bit of science in question, the theory of evolution plays a fascinating and crucial role in contemporary Western culture. The enormous controversy about it is what is most striking, a controversy that goes back to Darwin, and continues full force today. Evolution is the regular subject of court room drama; one such trial—the spectacular Scopes trial of 1925—has been made the subject of an extremely popular film. Fundamentalists regard evolution as the work of the Devil. In academia, on the other hand, it is an idol of the contemporary tribe; it serves as a shibboleth, a litmus test distinguishing the ignorant and bigoted fundamentalist goats from the properly acculturated and scientifically receptive sheep. Apparently this litmus test extends far beyond the confines of this terrestrial globe: according to the Oxford biologist Richard Dawkins, 'If superior creatures from space ever visit earth, the first question they will ask, in order to assess the level of our civilization, is: "Have they discovered evolution yet?"' Indeed, many of the experts—for example, Dawkins, William Provine, Stephen Gould—display a sort of revulsion at the very idea of special creation by God, as if this idea is not merely not good science, but somehow a bit obscene, or at least unseemly; it borders on the immoral; it is worthy of disdain and contempt. In some circles, confessing to finding evolution attractive will get you disapproval and ostracism, and may lose you your job; in others, confessing doubts about evolution will

Even if evolution is bad science it can still be good religion.

ALVIN PLANTINGA

have the same doleful effect. In Darwin's day, some suggested that it was all well and good to discuss evolution in the universities and among the *cognoscenti*; they thought *public* discussion unwise, however; for it would be a shame if the lower classes found out about it. Now, ironically enough, the shoe is sometimes on the other foot; it is the devotees of evolution who sometimes express the fear that public discussion of doubts and difficulties with evolution could have harmful political effects.[4]

So why all the furor? The answer is obvious: evolution has deep religious connections; deep connections with how we understand ourselves at the most fundamental level. Many evangelicals and fundamentalists see in it a threat to the Faith; they don't want it taught to their children, at any rate as scientifically established fact, and they see acceptance of it as corroding proper acceptance of the Bible. On the other side, among the secularists, • evolution functions as a *myth*, in a technical sense of that term: a shared way of understanding ourselves at the deep level of religion, a deep interpretation of ourselves to ourselves, a way of telling us why we are here, where we come from, and where we are going.

It was serving in this capacity when Richard Dawkins (according to Peter Medawar, 'One of the most brilliant of the rising generation of biologists') leaned over and remarked to A. J. Ayer at one of those elegant, candle-lit, bibulous Oxford dinners that he couldn't imagine being an atheist before 1859 (the year Darwin's *Origin of Species* was published); 'although atheism might have been logically tenable before Darwin,' said he, 'Darwin made it possible to be an intellectually fulfilled atheist' (Dawkins 1986: 6, 7). (Let me recommend Dawkins's (1986) book to you: it is brilliantly written, unfailingly fascinating, and utterly wrong-headed. It was second on the British best-seller list for some considerable time, second only to Mamie Jenkins's *Hip and Thigh Diet*.) Dawkins writes:

All appearances to the contrary, the only watchmaker in nature is the blind forces of physics, albeit deployed in a very special way. A true watchmaker has foresight: he designs his cogs and springs, and plans their interconnections, with a future purpose in his mind's eye. Natural selection, the blind, unconscious automatic process which Darwin discovered, and which we now know is the explanation for the existence and apparently purposeful form of all life, has no purpose in mind. It has no mind and no mind's eye. It does not plan for the future. It has no vision, no foresight, no sight at all. If it can be said to play the role of watchmaker in nature, it is the *blind* watchmaker. (Ibid. 5)

Evolution was functioning in that same mythic capacity in the remark of the famous zoologist G. G. Simpson: after posing the question 'What is man?' he answers: 'The point I want to make now is that all attempts to

---

[4] Thus, according to Anthony Flew, to suggest that there is real doubt about evolution is to corrupt the youth.

answer that question before 1859 are worthless and that we will be better off if we ignore them completely' (quoted in Dawkins 1976: 1). Of course, it also functions in that capacity in serving as a litmus test to distinguish the ignorant fundamentalists from the properly enlightened *cognoscenti*; it functions in the same way in many of the debates, in and out of the courts, as to whether it should be taught in the schools, whether other views should be given equal time, and the like. Thus Michael Ruse (1982: 326–7): 'The fight against creationism is a fight for all knowledge, and that battle can be won if we all work to see that Darwinism, which has had a great past, has an even greater future.'

The essential point here is really Dawkins's point: Darwinism, the Grand Evolutionary Story, makes it possible to be an intellectually fulfilled atheist. What he means is simple enough. If you are Christian, or a theist of some other kind, you have a ready answer to the question, How did it all happen? How is it that there are all the kinds of flora and fauna we behold; how did they all get here? The answer, of course, is that they have been created by the Lord. But if you are not a believer in God, things are enormously more difficult. How did all these things get here? How did life get started, and how did it come to assume its present multifarious forms? It seems monumentally implausible to think these forms just popped into existence; that goes contrary to all our experience. So how did it happen? Atheism and secularism need an answer to this question. And the Grand Evolutionary Story gives the answer: somehow life arose from non-living matter by way of purely natural means and in accord with the fundamental laws of physics; and once life started, all the vast profusion of contemporary plant and animal life arose from those early ancestors by way of common descent, driven by random variation and natural selection. I said earlier that we can't automatically identify the deliverances of reason with the teaching of current science because the teaching of current science keeps changing. Here we have another reason for resisting that identification: a good deal more than reason goes into the acceptance of such a theory as the Grand Evolutionary Story. For the non-theist, evolution is the only game in town; it is an essential part of any reasonably complete non-theistic way of thinking; hence the devotion to it, the suggestions that it shouldn't be discussed in public, and the venom, the theological odium, with which dissent is greeted.

B. The Likelihood of Evolution

Of course, the fact that evolution makes it possible to be a fulfilled atheist doesn't show either that the theory isn't true or that there isn't powerful evidence for it. Well then, how likely is it that this theory is true? Suppose we think about the question from an explicitly theistic and Christian perspective; but suppose we temporarily set to one side the evidence, whatever exactly it is, from early Genesis. From this perspective, how good is the evidence for the theory of evolution?

The first thing to see is that a number of *different* large-scale claims fall under this general rubric of evolution. First, there is the claim that the Earth is very old, perhaps some 4.5 billion years old: the *Ancient Earth Thesis*, as we may call it. Second, there is the claim that life has progressed from relatively simple to relatively complex forms of life. In the beginning there was relatively simple uni-cellular life, perhaps of the sort represented by bacteria and blue-green algae, or perhaps still simpler unknown forms of life. (Although bacteria are simple compared to some other living beings, they are in fact enormously complex creatures.) Then more complex uni-cellular life, then relatively simple multi-cellular life such as sea-going worms, coral, and jelly fish, then fish, then amphibia, then reptiles, birds, mammals, and finally, as the culmination of the whole process, human beings: the *Progress Thesis*, as we humans may like to call it (jelly fish might have a different view as to where the whole process culminates). Third, there is the *Common Ancestry Thesis*: that life originated at only one place on Earth, all subsequent life being related by descent to those original living creatures—the claim that, as Stephen Gould puts it, there is a 'tree of evolutionary descent linking all organisms by ties of genealogy' (1983: 256). According to the Common Ancestry Thesis, we are literally cousins of all living things—horses, oak-trees, and even poison ivy— distant cousins, no doubt, but still cousins. (This is much easier to imagine for some of us than for others.) Fourth, there is the claim that there is a (naturalistic) *explanation* of this development of life from simple to complex forms; call this thesis *Darwinism*, because according to the most popular and well-known suggestions, the evolutionary mechanism would be natural selection operating on random genetic mutation (due to copy error or ultraviolet radiation or other causes); and this is similar to Darwin's proposals. Finally, there is the claim that life itself developed from non-living matter without any special creative activity of God, but just by virtue of the ordinary laws of physics and chemistry: call this the *Naturalistic Origins Thesis*. These five theses are of course importantly different from each other. They are also logically independent in pairs, except for the third and fourth theses: the fourth entails the third, in that you can't sensibly propose a mechanism or an explanation for evolution without agreeing that evolution has indeed occurred. The combination of all five of these theses is what I have been calling the 'Grand Evolutionary Story'; the Common Ancestry Thesis together with Darwinism (remember, Darwinism isn't the view that the mechanism driving evolution is just what Darwin says it is) is what one most naturally thinks of as the Theory of Evolution.

So how shall we think of these five theses? First, let me remind you once more that I am no expert in this area. And second, let me say that, as I see it, the empirical or scientific evidence for these five different claims differs enormously in quality and quantity. There is excellent evidence for an ancient Earth: a whole series of interlocking different kinds of evidence, some of which is marshalled by Howard van Till in *The Fourth Day* (1986). Given the strength of this evidence, one would need powerful evidence on the other side—from scriptural considerations, say—in order to hold sensibly that the Earth is young. There is less evidence, but still good evidence in the fossil record for the Progress Thesis, the claim that there were bacteria before fish, fish before reptiles, reptiles before mammals, and mice before men (or wombats before women). The third and fourth theses, the Common Ancestry and Darwinian Theses, are what are commonly and popularly identified with evolution; I shall return to them in a moment. The fourth thesis, of course, is no more likely than the third, since it includes the third and proposes a mechanism to account for it. Finally, there is the fifth thesis, the Naturalistic Origins Thesis, the claim that life arose by naturalistic means. This seems to me to be for the most part mere arrogant bluster; given our present state of knowledge, I believe it is vastly less probable, on our present evidence, than is its denial. Darwin thought this claim very chancy; discoveries since Darwin, and in particular recent discoveries in molecular biology, make it much less likely than it was in Darwin's day. I can't summarize the evidence and the difficulties here.[5]

Now return to evolution more narrowly so-called: the Common Ancestry Thesis and the Darwinian Thesis. Contemporary intellectual orthodoxy is summarized by the 1979 edition of the *New Encyclopaedia Britannica*, according to which 'evolution is accepted by all biologists and natural selection is recognized as its cause. . . . Objections . . . have come from theological and, for a time, from political standpoints' (vii. 23). It goes on to add that 'Darwin did two things: he showed that evolution was in fact contradicting Scriptural legends of creation and that its cause, natural selection, was automatic, with no room for divine guidance or design'. According to most of the experts, furthermore, evolution, taken as the Thesis of Common Ancestry, is not something about which there can be

---

[5] Let me refer you to the following books: Thaxton *et al.* 1984; Shapiro 1986; Wicken 1987; Cairns-Smith 1982, 1985; Dyson 1985; see also the relevant chapters of Denton 1985. Thaxton *et al.* believe that God created life specially; the other authors cited do not.

sensible difference of opinion. Here is a random selection of claims of certainty on the part of the experts. Evolution is certain, says Francisco J. Ayala, as certain as 'the roundness of the earth, the motions of the planets, and the molecular constitution of matter' (1985: 60). According to Stephen J. Gould (1983: 254–5), evolution is an established fact, not a mere theory; and no sensible person who was acquainted with the evidence could demur. According to Richard Dawkins, the theory of evolution is as certainly true as that the Earth goes around the Sun. This comparison with Copernicus apparently suggests itself to many; according to Philip Spieth, 'A century and a quarter after the publication of the *Origin of Species*, biologists can say with confidence that universal genealogical relatedness is a conclusion of science that is as firmly established as the revolution of the earth about the sun' (1987). Michael Ruse trumpets, or perhaps screams, that 'evolution is Fact, FACT, **FACT!**' (1982: 58). If you venture to suggest doubts about evolution, you are likely to be called ignorant or stupid or worse. In fact, this isn't merely *likely*; you have already *been* so-called: in a recent review in the *New York Times*, Richard Dawkins claims that 'It is absolutely safe to say that if you meet someone who claims not to believe in evolution, that person is ignorant, stupid or insane (or wicked, but I'd rather not consider that)'. (Dawkins indulgently adds that 'You are probably not stupid, insane or wicked, and ignorance is not a crime'.)

Well then, how should a serious Christian think about the Common Ancestry and Darwinian Theses? The first and most obvious thing, of course, is that a Christian holds that all plants and animals, past as well as present, have been created by the Lord. Now suppose we set to one side what we take to be the best understanding of early Genesis. Then the next thing to see is that God could have accomplished this creating in a thousand different ways. It was entirely within his power to create life in a way corresponding to the Grand Evolutionary scenario: it was within his power to create matter and energy, as in the Big Bang, together with laws for their behaviour, in such a way that the outcome would be, first, life's coming into existence three or four billion years ago, and then the various higher forms of life, culminating, as we like to think, in humankind. This is a semi-deistic view of God and his workings: he starts everything off and sits back to watch it develop. (One who held this view could also hold that God constantly *sustains* the world in existence—hence the view is only *semi*-deistic—and even that any given causal transaction in the universe requires specific divine concurrent activity.[6]) On the other hand, of course, God

---

[6] The issues here are complicated and subtle, and I can't go into them; instead I should like to recommend Freddoso 1988.

could have done things very differently. He has created matter and energy with their tendencies to behave in certain ways—ways summed up in the laws of physics—but perhaps these laws are *not* such that given enough time, life would automatically arise. Perhaps he did something different and special in the creation of life. Perhaps he did something different and special in creating the various kinds of animals and plants. Perhaps he did something different and special in the creation of human beings. Perhaps in these cases his action with respect to what he has created was different from the ways in which he ordinarily treats them.

How shall we decide which of these is initially the more likely? That is not an easy question. It is important to remember, however, that the Lord has not merely left the cosmos to develop according to an initial creation and an initial set of physical laws. According to Scripture, he has often intervened in the working of his cosmos. This isn't a good way of putting the matter (because of its deistic suggestions); it is better to say that he has often treated what he has created in a way different from the way in which he ordinarily treats it. There are miracles reported in Scripture, for example; and, towering above all, there is the unthinkable gift of salvation for humankind by way of the life, death, and resurrection of Jesus Christ, his son. According to Scripture, God has often treated what he has made in a way different from the way in which he ordinarily treats it; there is therefore no initial edge to the idea that he would be more likely to have created life in all its variety in the broadly deistic way. In fact, it looks to me as if there is an initial probability on the other side; it is a bit more probable, before we look at the scientific evidence, that the Lord created life and some of its forms—in particular, human life—specially.

From this perspective, then, how shall we evaluate the evidence for evolution? Despite the claims of Ayala, Dawkins, Gould, Simpson, and the other experts, I think the evidence here has to be rated as ambiguous and inconclusive. The two hypotheses to be compared are (1) the claim that God has created us in such a way that (a) all of contemporary plants and animals are related by common ancestry, and (b) the mechanism driving evolution is natural selection working on random genetic variation and (2) the claim that God created mankind as well as many kinds of plants and animals separately and specially, in such a way that the thesis of common ancestry is false. Which of these is the more probable, given the empirical evidence and the theistic context? I think the second, the special creation thesis, is somewhat more probable with respect to the evidence (given theism) than the first.

There isn't the space, here, for more than the merest hand waving with respect to marshalling and evaluating the evidence. But, according to Stephen Jay Gould, certainly a leading contemporary spokesman,

our confidence that evolution occurred centers upon three general arguments. First, we have abundant, direct observational evidence of evolution in action, from both field and laboratory. This evidence ranges from countless experiments on change in nearly everything about fruit flies subjected to artificial selection in the laboratory to the famous populations of British moths that became black when industrial soot darkened the trees upon which the moths rest. (1983: 257)

Second, Gould mentions homologies: 'Why should a rat run, a bat fly, a porpoise swim, and I type this essay with structures built of the same bones', he asks, 'unless we all inherited them from a common ancestor?' Third, he says, there is the fossil record:

transitions are often found in the fossil record. Preserved transitions are not common, . . . but they are not entirely wanting. . . . For that matter, what better transitional form could we expect to find than the oldest human, *Australopithecus afarensis*, with its apelike palate, its human upright stance, and a cranial capacity larger than any ape's of the same body size but a full 1,000 cubic centimeters below ours? If God made each of the half-dozen human species discovered in ancient rocks, why did he create in an unbroken temporal sequence of progressively more modern features, increasing cranial capacity, reduced face and teeth, larger body size? Did he create to mimic evolution and test our faith thereby? (ibid. 258–9)

Here we could add a couple of other commonly cited kinds of evidence. (a) We, along with other animals, display vestigial organs (appendix, coccyx, muscles that move ears and nose); it is suggested that the best explanation is evolution. (b) There is alleged evidence from biochemistry: according to the authors of a popular college textbook, 'All organisms . . . employ DNA, and most use the citric acid cycle, cytochromes, and so forth. It seems inconceivable that the biochemistry of living things would be so similar if all life did not develop from a single common ancestral group' (Villee *et al.* 1985: 1012).[7] There is also (c) the fact that human embryos during their development display some of the characteristics of simpler forms of life (for example, at a certain stage they display gill-like structures). Finally, (d) there is the fact that certain patterns of geographical distribution—that there are orchids and alligators only in the American south and in China, for example—are susceptible to a nice evolutionary explanation.

Suppose we briefly consider the last four first. The arguments from

---

[7] Similarly, Ridley (1985) takes the fact that the genetic code is universal across all forms of life as proof that life originated only once; it would be extremely improbable that life should have stumbled upon the same code more than once.

vestigial organs, geographical distribution, and embryology are suggestive, but of course nowhere near conclusive. As for the similarity in biochemistry of all life, this is reasonably probable on the hypothesis of special creation, hence not much by way of evidence against it, hence not much by way of evidence for evolution.

Turning to the evidence Gould develops, it too is suggestive, but far from conclusive; some of it, furthermore, is seriously flawed. First, those famous British moths didn't produce a new species; there were both dark and light moths around before, the dark ones coming to predominate when the Industrial Revolution deposited a layer of soot on trees, making the light moths more visible to predators. More broadly, while there is wide agreement that there is such a thing as micro-evolution, the question is whether we can extrapolate to macro-evolution, with the claim that enough micro-evolution can account for the enormous differences between, say, bacteria and human beings. There is some experiential reason to think not; there seems to be a sort of envelope of limited variability surrounding a species and its near relatives. Artificial selection can produce several different kinds of fruit-flies and several different kinds of dogs, but, starting with fruit-flies, what it produces are only more fruit-flies. As plants or animals are bred in a certain direction, a sort of barrier is encountered; further selective breeding brings about sterility or a reversion to earlier forms. Partisans of evolution suggest that, in nature, genetic mutation of one sort or another can appropriately augment the reservoir of genetic variation. That it can do so sufficiently, however, is not known; and the assertion that it does is a sort of Ptolemaic epicycle attaching to the theory.

Next, there is the argument from the fossil record; but, as Gould himself points out, the fossil record shows very few transitional forms. 'The extreme rarity of transitional forms in the fossil record', he says, 'persists as the trade secret of palaeontology. The evolutionary trees that adorn our textbooks have data only at the tips and nodes of their branches; the rest is inference, however reasonable, not the evidence of fossils' (1980: 181).[8] Nearly all species appear for the first time in the fossil record fully formed, without the vast chains of intermediary forms evolution would suggest. Gradualistic evolutionists claim that the fossil record is woefully incomplete. Gould, Eldredge, and others have a different response to this difficulty: punctuated equilibriumism, according to which long periods of evolutionary stasis are interrupted by relatively brief periods of very rapid

---

[8] According to Simpson (1953), 'Nearly all categories above the level of families appear in the record suddenly and are not led up to by known, gradual, completely continuous transitional sequences.'

evolution. This response helps the theory accommodate some of the fossil data, but at the cost of another Ptolemaic epicycle.[9] And still more epicycles are required to account for puzzling discoveries in molecular biology during the last twenty years (see Denton 1985: ch. 12). And as for the argument from homologies, this too is suggestive, but far from decisive. First, there are of course many examples of architectural similarity that are not attributed to common ancestry, as in the case of the Tasmanian wolf and the European wolf; the anatomical givens are by no means conclusive proof of common ancestry. And secondly, God created several different kinds of animals; what would prevent him from using similar structures?

But perhaps the most important difficulty lies in a slightly different direction. Consider the mammalian eye: a marvellous and highly complex instrument, resembling a telescope of the highest quality, with a lens, an adjustable focus, a variable diaphragm for controlling the amount of light, and optical corrections for spherical and chromatic aberration. And here is the problem: how does the lens, for example, get developed by the proposed means—random genetic variation and natural selection—when at the same time there has to be development of the optic nerve, the relevant muscles, the retina, the rods and cones, and many other delicate and complicated structures, all of which have to be adjusted to each other in such a way that they can work together? Indeed, what is involved isn't, of course, just the eye; it is the whole visual system, including the relevant parts of the brain. Many different organs and sub-organs have to be developed together, and it is hard to envisage a series of mutations which is such that each member of the series has adaptive value, is also a step on the way to the eye, and is such that the last member is an animal with such an eye.

We can consider the problem a bit more abstractly. Think of a sort of space, in which the points are organic forms (possible organisms), and in which neighbouring forms are so related that one could have originated from the other with some minimum probability by way of random genetic mutation. Imagine starting with a population of animals without eyes, and trace through the space in question all the paths that lead from this form to forms with eyes. The chief problem is that the vast majority of these paths contain long sections with adjacent points such that there would be no

---

[9] And even so it helps much less than you might think. It does offer an explanation of the absence of fossil forms intermediate with respect to closely related or adjoining species; the real problem, though, is what Simpson refers to in the quote in the previous footnote: the fact that nearly all categories above the level of families appear in the record suddenly, without the gradual and continuous sequences we should expect. Punctuated equilibriumism does nothing to explain the nearly complete absence, in the fossil record, of intermediates between such major divisions as, say, reptiles and birds, or fish and reptiles, or reptiles and mammals.

adaptive advantage in going from one point to the next, so that, on Darwinian assumptions, none of them could be the path in fact taken. How could the eye have evolved in this way, so that each point on its path through that space would be adaptive and a step on the way to the eye? (Perhaps it is possible that some of these sections could be traversed by way of steps that were not adaptive and were fixed by genetic drift; but the probability of the population's crossing such stretches will be much less than that of its crossing a similar stretch where natural selection is operative.) Darwin himself wrote, 'To suppose that the eye, with all its inimitable contrivances . . . could have been formed by natural selection seems absurd in the highest degree.' 'When I think of the eye, I shudder' he said (1859: 3–4). And the complexity of the eye is enormously greater than was known in Darwin's time.

We are never, of course, given the *actual* explanation of the evolution of the eye, the actual evolutionary history of the eye (or brain or hand or whatever). That would take the form: in that original population of eyeless life forms, genes A_1–A_n mutated (due to some perhaps unspecified cause), leading to some structural and functional change which was adaptively beneficial; the bearers of A_1–A_n thus had an advantage, and came to dominate the population. Then genes B_1–B_n mutated in an individual or two, and the same thing happened again; then gene C_1–C_n, etc. Nor are we even given any possibilities of these sorts. (We couldn't be, since, for most genes, we don't know enough about their functions.) We are instead treated to broad-brush scenarios at the macroscopic level: perhaps reptiles gradually developed feathers, and wings, and warm-bloodedness, and the other features of birds. We are given possible evolutionary histories, not of the detailed genetic sort mentioned above, but broad macroscopic scenarios: what Gould calls 'just-so stories'.

And the real problem is that we don't know how to evaluate these suggestions. To know how to do *that* (in the case of the eye, say), we should have to start with some population of animals without eyes; and then we should have to know the rate at which mutations occur for that population; the proportion of those mutations that are on one of those paths through that space to the condition of having eyes; the proportion of *those* that are adaptive, and, at each stage, given the sort of environment enjoyed by the organisms at that stage, the rate at which such adaptive modifications would have spread through the population in question. Then we'd have to compare our results with the time available to evaluate the probability of the suggestion in question. But we don't know what these rates and proportions are. No doubt we *can't* know what they are, given the scarcity of operable time-machines; still, the fact is we don't know them. And hence

1. Evolution is bad scientific theory
2. ∴ we should believe in 'special creation'
3. ∴ God is involved
⇒ God of gaps

692   ALVIN PLANTINGA

we don't really know whether evolution is so much as biologically possible: maybe there is no path through that space. It is *epistemically* possible that evolution has occurred: that is, we don't know that it hasn't; for all we know, it has. But it doesn't follow that it is *biologically* possible. (Whether every even number is the sum of two primes is an open question; hence it is epistemically possible that every even number is the sum of two primes, and also epistemically possible that some even numbers are not the sum of two primes; but one or the other of those epistemic possibilities is in fact mathematically impossible.) Assuming that it *is* biologically possible, furthermore, we don't know that it is not prohibitively improbable (in the statistical sense), given the time available. But then (given the Christian faith and leaving to one side our evaluation of the evidence from early Genesis) the right attitude towards the claim of universal common descent is, I think, one of a certain interested but wary scepticism. It is *possible* (epistemically possible) that this is how things happened; God could have done it that way, but the evidence is ambiguous. That it is *possible* is clear; that it *happened* is doubtful; that it is *certain*, however, is ridiculous.

But then what about all those exuberant cries of certainty from Gould, Ayala, Dawkins, Simpson, and the other experts? What about those claims that evolution, universal common ancestry, is a rock-ribbed certainty, to be compared with the fact that the Earth is round and goes around the Sun? What we have here is at best enormous exaggeration. But then what accounts for the fact that these claims are made by such intelligent luminaries as the above? There are at least two reasons. First, there is the cultural and religious, the mythic function of the doctrine; evolution helps make it possible to be an intellectually fulfilled atheist. From a naturalistic point of view, this is the only answer in sight to the question 'How did it all happen? How did all this amazing profusion of life get here?' From a non-theistic point of view, the evolutionary hypothesis is the only game in town. According to the thesis of universal common descent, life arose in just one place; then there was constant development by way of evolutionary mechanisms from that time to the present, this resulting in the profusion of life we presently see. On the alternative hypothesis, different forms of life arose independently of each other; on that suggestion there would be many different genetic trees, the creatures adorning one of these trees genetically unrelated to those on another. From a non-theistic perspective, the first hypothesis will be by far the more probable, if only because of the extraordinary difficulty in seeing how life could arise even once by any ordinary mechanisms which operate today. That it should arise many different times, and at different levels of complexity in this way, is quite incredible.

From a naturalist perspective, furthermore, many of the arguments for evolution are much more powerful than from a theistic perspective. (For example, *given* that life arose naturalistically, it is indeed significant that all life employs the same genetic code.) So from a naturalistic, non-theistic perspective, the evolutionary hypothesis will be vastly more probable than alternatives. Many leaders in the field of evolutionary biology, of course, *are* naturalists—Gould, Dawkins, and Stebbins, for example; and according to William Provine, 'very few truly religious evolutionary biologists remain. Most are atheists, and many have been driven there by their understanding of the evolutionary process and other science' (1988: 68). If Provine is right or nearly right, it becomes easier to see why we hear this insistence that the evolutionary hypothesis is certain. It is also easy to see how this attitude is passed on to graduate students, and, indeed, how accepting the view that evolution is certain is itself adaptive for life in graduate school and academia generally.

There is a second and related circumstance at work here. We are sometimes told that natural science is *natural* science. So far it is hard to object: but how shall we take the term 'natural' here? It could mean that natural science is science devoted to the study of nature. Fair enough. But it is also taken to mean that natural science involves a *methodological naturalism* or provisional atheism:[10] no hypothesis according to which God has done this or that can qualify as a *scientific* hypothesis. It would be interesting to look into this matter. Is there really any compelling, or even decent, reason for thus restricting our study of nature? But suppose we ironically concede, for the moment, that natural science doesn't or shouldn't invoke hypotheses essentially involving God. Suppose we restrict our explanatory materials to the ordinary laws of physics and chemistry; suppose we reject divine special creation or other hypotheses about God as *scientific* hypotheses. Perhaps, indeed, the Lord has engaged in special creation, so we say, but that he has (if he has) is not something with which natural science can deal. So far as natural science goes, therefore, an acceptable hypothesis must appeal only to the laws that govern the ordinary, day-to-day working of the cosmos. As natural scientists we must eschew the supernatural—although, of course, we don't mean for a moment to embrace naturalism.

Well, suppose we adopt this attitude. Then perhaps it looks as if by far the most probable of all the properly scientific hypotheses is that of evolution by common ancestry: it is hard to think of any other real possibility. The only alternatives, apparently, would be creatures popping into existence fully formed; and that is wholly contrary to our experience. Of all the

---

[10] '[Science] must be provisionally atheistic or cease to be itself' (Willey 1961: 15).

scientifically acceptable explanatory hypotheses, therefore, evolution seems by far the most probable. But if this hypothesis is vastly more probable than any of its rivals, then it must be certain, or nearly so.

But to reason this way is to fall into confusion compounded. In the first place, we aren't just given that one or another of these hypotheses is in fact correct. Granted, if we *knew* that one or another of those scientifically acceptable hypotheses were in fact correct, then perhaps this one would be certain; but of course we don't know that. One real possibility is that we don't have a very good idea how it all happened, just as we may not have a very good idea as to what terrorist organization has perpetrated a particular bombing. And secondly, this reasoning involves a confusion between the claim that of all of those *scientifically* acceptable hypotheses, that of common ancestry is by far the most plausible, with the vastly more contentious claim that of all the acceptable hypotheses *whatever* (now placing no restrictions on their kind) this hypothesis is by far the most probable. Christians in particular ought to be alive to the vast difference between these claims; confounding them leads to nothing but confusion. My main point, however, is that Ayala, Gould, Simpson, Stebbins, and their coterie are wildly mistaken in claiming that the Grand Evolutionary Hypothesis is *certain*. And hence the source of this claim has to be looked for elsewhere than in sober scientific evidence.

So it could be that the best scientific hypothesis was evolution by common descent—that is, of all the hypotheses that conform to methodological naturalism, it is the best. But of course what we really want to know is not which hypothesis is the best from some artificially adopted standpoint of naturalism, but what the best hypothesis is *overall*. We want to know what the *best* hypothesis is, not which of some limited class is best—particularly if the class in question specifically excludes what we hold to be the basic truth of the matter. It could be that the best scientific hypothesis (again supposing that a scientific hypothesis must be naturalistic in the above sense) isn't even a strong competitor in *that* Derby.

Judgements here, of course, may differ widely between believers in God and non-believers in God. What for the former is at best a methodological restriction is for the latter the sober metaphysical truth; her naturalism is not merely provisional and methodological, but, as she sees it, settled and fundamental. But believers in God can see the matter differently. The believer in God, unlike her naturalistic counterpart, is free to look at the evidence for the Grand Evolutionary Scheme, and follow it where it leads, rejecting that scheme if the evidence is insufficient. She has a freedom not available to the naturalist. The latter accepts the Grand Evolutionary Scheme because from a naturalistic point of view this scheme is the only

[handwritten: Is motivation of scientific arguments relevant?]

[handwritten: Is Plantinga aiming to persuade a supposedly neutral rational observer or not?]

visible answer to the question *What is the explanation of the presence of all these marvellously multifarious forms of life?* The Christian, on the other hand, knows that creation is the Lord's; and she isn't blinkered by a priori dogmas as to how the Lord must have accomplished it. Perhaps it was by broadly evolutionary means, but then again perhaps not. At the moment, 'perhaps not' seems the better answer.

Returning to methodological naturalism, if indeed natural science is essentially restricted in this way, if such a restriction is a part of the very essence of science, then what we need here, of course, is not natural science, but a broader enquiry that can include *all* that we know, including the truths that God has created life on earth and could have done it in many different ways. 'Unnatural Science', 'Creation Science', 'Theistic Science'—call it what you will: what we need when we want to know how to think about the origin and development of contemporary life is what is most plausible from a Christian point of view. What we need is a scientific account of life that isn't restricted by that methodological naturalism.

[handwritten right margin: status of this knowledge & these truths? evidence?]

[handwritten: ≠ ?]

Finally, in all areas of academic endeavour, we Christians must think about the matter at hand from a Christian perspective; we need Theistic Science. Perhaps the discipline in question, as ordinarily practised, involves a methodological naturalism; if so, then what we need, finally, is not answers to our questions from *that* perspective, valuable in some ways as it may be. What we really need are answers to our questions from the perspective of *all* that we know—what we know about God, and what we know by faith, by way of revelation, as well as what we know in other ways. In many areas, this means that Christians must rework, rethink the area in question from this perspective. This idea may be shocking, but it is not new. Reformed Christians have long recognized that science and scholarship are by no means religiously neutral. In a way, this is our distinctive thread in the tapestry of Christianity, our instrument in the great symphony of Christianity. *[handwritten: .. blah, blah, blah...]*

[handwritten:
① This can't be science: public, potentially catholic etc.
② No critical analysis of Genesis in whole article.]

REFERENCES

Ayala, F. J. (1980), 'The Theory of Evolution: Recent Successes and Challenges', in E. McMullin (ed.), *Evolution and Creation* (Notre Dame, Ind.: University of Notre Dame Press).

Cairns-Smith, A. G. (1982), *Genetic Takeover and the Mineral Origins of Life* (Cambridge: Cambridge University Press).

——(1985), *Seven Clues to the Origin of Life* (Cambridge: Cambridge University Press).

Darwin, C. (1859), *The Origin of Species* (London: John Murray).
Dawkins, R. (1976), *The Selfish Gene* (Oxford: Oxford University Press).
——(1986), *The Blind Watchmaker* (London and New York: W. W. Norton).
Denton, M. (1985), *Evolution: A Theory in Crisis* (London: Burnet Books).
Dyson, F. (1985), *Origins of Life* (Cambridge: Cambridge University Press).
Eldredge, N. (1985), *Time Frames* (New York: Simon and Schuster).
Freddoso, A. (1988), 'Medieval Aristotelianism and the Case against Secondary Causation in Nature', in T. Morris (ed.), *Divine and Human Action* (Ithaca, NY: Cornell University Press).
Gould, S. J. (1980), *The Panda's Thumb* (New York: Oxford University Press).
——(1983), 'Evolution as Fact and Theory', in *Hen's Teeth and Horse's Toes* (New York: Norton), 253–62.
Hawking, S. (1988), *A Brief History of Time* (New York: Bantam Books).
Johnson, P. (MS), Science and Scientific Naturalism in the Evolution Controversy (unpub.).
Kitcher, P. (1985), *Vaulting Ambition* (Cambridge, Mass: MIT Press).
MacKay, D. (1974*a*), *The Clockwork Image: A Christian Perspective on Science* (London: Intervarsity Press).
——(1974*b*), '"Complementarity" in Scientific and Theological Thinking', *Zygon*, 9/3: 225–44.
Neill, S. (1958), *Anglicanism* (Harmondsworth: Penguin).
Provine, W. B. (1988), 'Progress in Evolution and Meaning of Life', in M. H. Nitecki (ed.), *Evolutionary Progress* (Chicago: The University of Chicago Press), 49–74.
Ridley, M. (1985), *The Problems of Evolution* (Oxford: Oxford University Press).
Ruse, M. (1982), *Darwinism Defended* (Reading, Mass.: Addison-Wesley Publishing Co.).
Shapiro, R. (1986), *Origins* (New York: Summit Books).
Simpson, G. G. (1949), *The Meaning of Evolution* (New Haven: Yale University Press).
——(1953), *The Major Features of Evolution* (New York: Columbia University Press).
——(1964), *This View of Life* (New York: Harcourt Brace and World).
——(1983), *Fossil and the History of Life* (New York: Scientific American Books and W. H. Freeman and Co.).
Spieth, P. (1987), 'Evolutionary Biology and the Study of Human Nature', presented at a consultation on Cosmology and Theology sponsored by the Presbyterian (USA) Church.
Stanley, S. (1981), *The New Evolutionary Timetable* (New York: Basic Books).
Stebbins, G. L. (1982), *Darwin to DNA, Molecules to Humanity* (San Francisco: W. H. Freeman and Co.).
Thaxton, C., Bradley, W., and Olsen, R. (1984), *The Mystery of Life's Origins* (New York: Philosophical Library).
van Fraassen, B. (1980), *The Scientific Image* (Oxford: Clarendon Press).
——(1991), *Quantum Mechanics* (Oxford: Clarendon Press).
Van Till, H. (1986), *The Fourth Day: What the Bible and the Heavens are Telling Us about the Creation* (Grand Rapids, Mich.: W. B. Eerdmans).
Villee, C. A., Solomon, E. P., and Davis, P. W. (1985), *Biology* (Philadelphia: Saunders College Publishing).
Wicken, J. S. (1987), *Evolution, Thermodynamics, and Information: Extending the Darwinian Program* (New York: Oxford University Press).

Willey, B. (1961), 'Darwin's Place in the History of Thought', in M. Banton (ed.), *Darwinism and the Study of Society* (London: Tavistock Publications, and Chicago: Quadrangle Books), 1–16.

35

EVOLUTION AND SPECIAL CREATION

ERNAN McMULLIN

How did God bring the ancestral living things to be? Two broadly different sorts of answer have found favour with believers in a Creator. One is to suppose that God brings the universe into existence already containing the potentialities that are required in order that the complexities of the world we know should 'naturally' develop within it. The other is to say that for some of these complexities to develop, God had to 'supplement' nature in certain respects, to act in a special way, special not only in the sense of being different from God's ordinary sustaining of the order accessible to us through natural science, but also in the sense that the interruption of that order is aimed at bringing about results that could not otherwise come to be. The first answer is the evolutionary one. What precise *theories* of evolution one chooses to defend is another matter. 'Evolution' is a generic label for the natural process whereby potentialities already present are actualized. The second alternative has the somewhat clumsy title of 'special creation'.

One who defends the hypothesis of special creation to account for the origin of a particular sort of being (like the first living cells or the first humans) may be quite content to allow an evolutionary account in other contexts. And one who argues, in principle, for the sufficiency of evolutionary models may (if a theist) insist that the natural order itself is created, dependent on God for its very existence. What separates the two is not the general admissibility of the notions of evolution and creation, but the need for 'special' episodes in the story of cosmic development. According to one account, they were needed; according to the other, they were not. On the face of it, both sides need to exercise logical caution. How can those who invoke special creation to account for a particular cosmic transition exclude the possibility that an as-yet-unthought-of evolutionary explanation might later be found for it? Short of providing an already completed

First published in *Zygon*, 28/3 (September 1993): 299–335. Reprinted by permission.

evolutionary account, how could defenders of evolution exclude the possibility that special creation might have occurred at some juncture? The evolutionist is not required to hold (and if a theist, will not hold) that special creation is in principle *impossible*, only that it is in general unlikely, or unneeded in specific contexts.

The vigorously negative reaction to the claims of 'creation science' in recent decades might easily lead one to overlook the logical and epistemological complexities of the underlying disagreement between proponents of evolution and proponents of special creation. What came to be called 'creation science' was an aberrant solution forced on defenders of the special creation alternative by the constraints imposed on public(=state) school education due to the accepted interpretation of the Constitution of the United States. Its manifest logical inadequacy led ultimately to the legal findings in the celebrated Overton judgment (Arkansas, 1981) striking down the mandatory teaching of creation science as an alternative to evolution[1] and might easily mislead one into supposing that special creation can at this point be dismissed out of hand in discussions of the origins of life. But creation science is only one of the many variant versions of special creation, and assuredly one of the more vulnerable.

It seems worthwhile, then, to look closely at a very different and much more sophisticated sort of defence of special creation. Alvin Plantinga is a well-known philosopher of religion whose work in epistemology, metaphysics, and modal logic is widely known and justly respected. In a recent essay, 'When Faith and Reason Clash: Evolution and the Bible' (reproduced as Ch. 34 above), he proclaims the merits of special creation in the light of what he perceives as inadequacies in the current evolutionary account of origins, and he proposes the antecedent likelihood, in a general way, of special creation from the theological standpoint of the Christian.[2] His principal targets are those evolutionists who, he believes, covertly rely on an anti-theistic premiss in order to make inflated claims for the certainty of what he calls the 'Grand Evolutionary Scheme'. His essay is an extended exercise in the epistemology of scientific theory from the perspective of a religious believer; though I disagree with some of its main conclusions, I shall not, I hope, underrate their force.

---

[1] For the text of the judgment, *McLean* v. *Arkansas*, see Gilkey 1985: 266–301. The judgment is not itself without some logical difficulties; see Quinn 1984.

[2] Plantinga's essay was featured in a special issue of *Christian Scholar's Review* (1991b) and is reproduced as Ch. 34. The issue carried critical responses by Howard Van Till and myself, as well as a detailed reply by Plantinga (here 1991a). The present essay is a revised and considerably augmented version of my paper in that volume. I am grateful to Dr Plantinga for our discussions of these issues, and for the characteristic care he took in responding to my original criticisms.

THEISTIC SCIENCE

Plantinga's thesis in regard to evolution is that, for the Christian, the claim that God created humankind, as well as many kinds of plants and animals, separately and specially, is more probable than the Thesis of Common Ancestry (TCA) that is central to the theory of evolution (Plantinga 1991*b*: 22, 28; pp. 687, 693 above). His larger context is that of an exhortation to Christian intellectuals to join battle against 'the forces of unbelief', particularly in academia, instead of always yielding to 'the word of the experts'. These intellectuals must be brought to 'discern the religious and ideological connections . . . [they must not] automatically take the word of the experts, because their word might be dead wrong from a Christian standpoint' (ibid. 30). The implication many would take from this is that Christian intellectuals should ally themselves with the critics of evolution, that it may somehow be to their *advantage* to find flaws in the case for evolution.

The 'science' these Christian intellectuals profess will not be of the usual naturalist sort. Their account of the origin of species, for instance, will be at odds with that given by Darwin, on grounds that are distinctively Christian in content. Despite the fact that claims such as these on the part of the Christian depend on what he or she knows 'by faith, by way of revelation', Plantinga believes that they can appropriately be called science, and he suggests as a label for them 'Theistic Science' (ibid. 29; p. 695 above). An important function of this broader knowledge would be revisionary; he reminds us that 'Scripture can correct current science'. His theistic science bears some similarity to the creation science that has commanded the headlines in the United States so often in recent decades. Like the creation scientists, he maintains that in the present state of knowledge the best explanation of the origin of many kinds of plants and animals is an interruption in the ordinary course of natural process, a moment when God treats 'what he has created in a way different from the way in which he ordinarily treats it' (ibid. 22; p. 687 above). Like them, he relies on a critique of the theory of evolution, pointing to what he regards as fundamental shortcomings in the Darwinian project of explaining new species by means of natural selection and emphasizing recent criticisms of one or other facet of the synthetic theory from within the scientific community itself. Like them, he calls for a struggle against prevailing scientific orthodoxy, one that may pit the teachers of Christian youth against the 'experts'.

But the differences between Plantinga and the creation scientists are even more basic. Most of the latter believe in a 'young Earth' dating back only a few thousand years, and they attempt to undermine the many arguments that can be brought against this view. Plantinga allows 'the evidence for an old Earth to be strong and the warrant for the view that the Lord teaches that the Earth is young to be relatively weak' (ibid. 15; p. 680 above). The creation scientists argue for a whole series of related cosmological theses (that stars and galaxies do not change, that the history of the Earth is dominated by the occurrence of catastrophe, and so forth); Plantinga focuses on the single issue of the origins of living things, especially of humankind. And he is in the end more concerned with combating the claims of certainty made by many evolutionists than he is with arguing that the Christian is irrevocably committed against a full evolutionary account of origins. He allows, as the creation scientists, I suspect, would not, that as evolutionary science advances, his own present estimate that special creation is more likely to account for some of the major transitions in the story of life on Earth might have to give way.

In the debates regarding the teaching of creation science in the state schools, its defenders attempted to detach their arguments from any sort of reliance on Scripture, or, more generally, from theological considerations, whereas Plantinga appeals explicitly to the scriptural understanding of the manner of God's action in the world. The former make a heroic attempt to qualify their creationism as scientific, in what they take to be the conventional sense of the term 'scientific'. Their effort, I think it is fair to say, was hopeless right from the start. They would undoubtedly have preferred to defend a view more explicitly based on Genesis, but the exigencies of the Constitutional restrictions on the state school curriculum prevented this. The scientists among them attempted to shore up their case by citing various consonances between the catastrophism of their young Earth account and the geological record. But the inspiration for their account lay, and clearly *had* to lie, in the Bible. Trying to fudge this, though understandable under the circumstances, proved a disastrous strategy.

Plantinga offers a far more consistent theme. True, his 'theistic science' will not pass Constitutional muster, so it will not serve the purposes for which creation science was originally advanced. But that is not an argument against it; it is merely a consequence of the unique situation of state education in the United States, a situation that imposes losses as well as gains. I do not think, however, that theistic science should be described *as* science. It lacks the universality of science, as that term has been understood in the later Western tradition.[3] It also lacks the sort of warrant that

---

[3] In defence of his usage, Plantinga notes that theology at an earlier time was called a science (1991*a*: 98). But this usage was recognized to be problematic from the Aristotelian viewpoint of that time. To the objection that theology cannot be regarded as a science because it proceeds from premises not admitted by all, Aquinas responds that because these premises are revealed

has gradually come to characterize a properly 'scientific' knowledge of nature, one that favours systematic observation, generalization, and the testing of explanatory hypotheses. Theistic science appeals to a specifically Christian belief, one that lays no claim to assent from a Hindu or an agnostic. It requires faith, and faith (we are told) is a gift, a grace, from God. To use the term 'science' in this context seems dangerously misleading; it encourages expectations that cannot be fulfilled.

Plantinga objects to the sort of methodological naturalism that would deny the label 'science' to any explanation of natural process that invokes the special action of God; indeed, he characterizes it, in Basil Willey's phrase, as 'provisional atheism'. 'Is there really any compelling or even decent reason for thus restricting our study of nature?', he asks (Plantinga 1991*b*: 27; p. 693 above). But, of course, methodological naturalism does not restrict our study of nature; it just lays down which sort of study qualifies as 'scientific'. Calling on the special action of God to explain the origins of the major phyla in the way Plantinga does transcends the boundaries of science.[4] This is not primarily because God is involved (Aristotle's argument for a First Mover, for example, could be counted a broadly naturalistic one), but because the action is a special one inaccessible to any sort of test on our part, and because of the sort of evidence that has to be invoked, evidence that does not lend itself to evaluation by the standard techniques of natural science, however loosely these be defined.[5]

If someone wants to pursue another approach to nature—and there are many others—the methodological naturalist has no reason to object. Sci-

by God, they can be accepted on authority, just as optics takes its principles from geometry (*Summa Theologica*, vol. I, q.1, a.2). But this does not really answer the objection adequately, since the revealed character of these premisses is not admitted by all. And the Aristotelian distinction between what is better known to us and what is better known 'in itself' will not do the work. When the Aristotelian conception of science (deduction from self-evident premisses) was gradually abandoned in the seventeenth century, the new conceptions that succeeded it made the extension of the term 'science' to theology even more problematic, particularly in the present context of the knowledge of nature.

[4] Calling it God's 'direct' action would leave matters ambiguous, since it could be said that God's action in sustaining the world in existence is direct action; this sort of action is, of course, not in dispute here. What makes God's 'special' action inaccessible to the methods of natural science is that it lies, as medieval philosophers put it, 'outside nature', outside the pattern of regularities that afford a foothold for later enquirers. The most that science could do where 'special' action is claimed, as in the case of miracles, would be to exclude, as far as possible, alternative 'natural' explanations. But when special creation is supposed to have occurred in the early history of life on earth, this (as we shall see) is *very* difficult to do.

[5] This argument does not depend on an ability to draw a sharp demarcation between science and non-science. Scientists often rely on principles of natural order of a broadly metaphysical sort, but these are in principle accessible to all; they are over the long run at least partially adjudicable in terms of the 'success' (in a fairly specific sense) of the theories employing them. (See McMullin, in press.) Reliance on Scripture is another matter entirely.

entists *have* to proceed in this way; the methodology of natural science gives no purchase on the claim that a particular event or type of event is to be explained by invoking God's 'special' action or by calling on the testimony of Scripture. Calling this *methodological* naturalism is simply a way of drawing attention to the fact that it is a way of characterizing a particular *methodology*, no more. In particular, it is not an ontological claim about what sort of agency is or is not possible. Dubbing it 'provisional atheism' is objectionable; the scientist who does not include among the alternatives to be tested, when attempting to explain some phenomenon, an action that would not lend itself to such test is surely not to be accused of atheism, even of a provisional sort. 'What we need', Plantinga tells us, 'is a scientific account of life that isn't restricted by methodological naturalism' (1991*b*: 29; p. 695 above). But, of course, if it is not so restricted, it is simply improper to call it 'scientific', in the light of long and unequivocal contrary usage.

Let me make myself clear. I do not object (as the concluding section of this essay makes clear) to the use of theological considerations in the service of a larger and more comprehensive world-view in which natural science is only one factor. I would be willing to use the term 'knowledge' in an extended sense here, though I am well aware of some old and intricate issues about how faith and knowledge are to be related. (See e.g., Kellenberger 1972: ch. 10.) But I would not be willing to use the term 'science' in this context. Nor do I think it necessary to do so in order to convey the respectability of the claim being made: the theology may appropriately modulate other parts of a person's belief system, including those deriving from science. I would be much more restrictive than Plantinga is, however, in allowing for the situation he describes as 'Scripture correcting current science'.[6] But before I analyse our differences, it may be useful first to lay out the large areas where we agree.

[6] As an illustration of how Scripture could 'correct current science', Plantinga remarks: 'If, for example, current science were to return to the view that the world has no beginning, and is infinitely old, then current science would be wrong' (Plantinga 1991*b*: 14; p. 679 above). I do not believe that Scripture *does* prescribe that the universe had a beginning in time, in some specific technical sense of the term 'time'; the point of the creation narratives is the dependence of the world on God's creative act, to my mind, not that it all began at a finite time in the past. A world that has always existed would still (as Aquinas emphasized) require a creator. As an illustration of how complex the notion of temporal beginnings has become, the Hawking model of cosmic origins mentioned by Plantinga does not imply that the universe is infinitely old (as that phrase would ordinarily be understood), but rather that, as we trace time backwards to the Big Bang, the normal concept of time may break down as we approach the initial singularity some 15 billion years ago. The history of 'real time' (as Hawking calls it) would still be finite in the same terms as before, as he explicitly points out (Hawking 1988: 138). The question of whether or not the time elapsed in cosmic history is finite or infinite depends, in part, on the choice of physical process on which to base the time-scale, particularly on whether it is cyclic or continuous. The

POINTS OF AGREEMENT

What really gall Plantinga are the views of people like Richard Dawkins and William Provine who not only insist that evolution is a proved 'fact', but who suppose that this somehow undercuts the reasonableness of any sort of belief in a Creator. Their argument hinges on the notion of design. The role of the Creator in traditional religious belief (they claim) was that of designer; the success of the theory of evolution has shown that design is unnecessary. Hence, there is no longer any valid reason to be a theist. In a recent review of a history of the creationist debate in the United States, Provine lays out this case, and concludes that Christian belief can be made compatible with evolutionary biology only by supposing that God 'works through the laws of nature' instead of actively steering biological process by way of miraculous intervention. But this view of God, he says, is 'worthless', and 'equivalent to atheism' (Provine 1987) (On this last point, Plantinga and he might not be so far apart.). He chides scientists for publicly denying, presumably on pragmatic grounds, that evolution and Christian belief are incompatible; they *must*, he says, know this to be nonsense.

Plantinga puts his finger on an important point when he notes that for someone who does not believe in God, evolution in some form or other is the only *possible* answer to the question of origins. Prior to the publication of the *Origin of Species* in 1859, the argument from design for a Creator was widely regarded as resting directly on biological science. The founders of physico-theology two centuries earlier (naturalists like John Ray and William Derham) had shown the pervasive presence in nature of means–end relationships, the apparently purposive adjusting of structure and instinctive behaviour to the welfare of each kind of organism. Someone who rejected the idea of a designer, therefore, had to face some awkward problems in explaining some of the most obvious features of the living world; it seemed to many as though science itself testified to the existence of God (McMullin 1988).[7]

Darwin changed all this. By undermining the classical arguments from design, he showed that atheism was not, after all, inconsistent with biolog-

---

question of the finitude or infinity of past time, so much debated by medieval philosophers and theologians, cannot straightforwardly be answered in absolute terms. The notion of time measurement is far more complex and theory-dependent than earlier discussions allowed. But the theological *point* of the biblical account of creation remains untouched by technical developments such as these (McMullin 1981: 35).

[7] The exponents of physico-theology were not entirely sure how to classify their arguments from design concerning origins. These could not be directly tested in the normal empirical ways, but it did seem as though 'naturalist' explanations could be systematically excluded.

ical science; from then on, the fortunes of atheism as a form of intellectual belief would, to some extent at least, be perceived as depending upon the fortunes of the theory of evolution. No wonder, then, that evolution became a crucial myth of our secular culture (as Plantinga puts it), replacing for many the Christian myth as 'a shared way of understanding ourselves at the deep level of religion' (Plantinga 1991b: 17; p. 682 above). No wonder also that an attack on the credentials of evolutionary theory would so often evoke from its defenders a reaction reminiscent in its ferocity of the response to heresy in other days.

Is evolution fact or theory? No other question has divided the two sides in the creation–science controversy as sharply. Plantinga argues that someone who denies the existence of a Creator is left with no other option for explaining the origin of living things than an evolutionary-type account. The account thus becomes 'fact' not just because of the strength of the scientific evidence in its favour, but because, for the atheist, no other explanation is available. Plantinga objects to the use of the word 'fact' in this context because it seems to exclude in principle the possibility of divine intervention, and hence by implication, the possibility of the existence of a Creator. 'Fact' seems to convey not just the assurance of a well-supported theory, but the certainty that no other explanation is open.

The debate may often, therefore, be something other than it seems. Instead of being just a disagreement about the weight to be accorded to a particularly complex scientific theory in the light of the evidence available, the debate may conceal a far more fundamental religious difference, each side appearing to the other to call into question an article of faith. Religious believers point out that calling the thesis of common ancestry a 'fact' violates good scientific usage; no matter how well-supported a theory may be (they argue), it remains a theory. To non-believers, the phrase 'merely a theory' comes as a provocation, because it suggests a substantial doubt about a claim that appears to them as being beyond question, a doubt prompted, furthermore, in their view by an illegitimate intrusion of religious belief.

At one level, then, Plantinga's essay can be read as a plea for a more informed understanding of the real nature of the creation–science debate, and a more sympathetic appreciation of what led the proponents of creation science to take the stand they did. Even their defence of a 'young' Earth (a major point of disagreement between his view and theirs) ought not (he says) to be regarded as 'silly or irrational'; a 'sensible person' might well subscribe to it after a careful study of the Scriptures. One need not be 'a fanatic, or a Flat Earther, or an ignorant fundamentalist' to hold such a view (Plantinga 1991b: 15; p. 680 above). The claim that the Earth is ancient is neither obvious nor inevitable; it has to be argued for, and disagreement may, therefore, legitimately occur.

Plantinga is right, to my mind, to see more in the creation–science debate than evolutionary scientists (or the media) have been wont to allow. And the sort of challenge he offers to the defenders of evolution, though it is not new, could serve the purposes of science in the long run if it forces a clarification and strengthening of argument on the other side, or if it punctures the sometimes troubling smugness that experts tend to display when dealing with outsiders. Plantinga leans too far in the other direction, however. In the first place, those who affirm that 'evolution is a fact' are not necessarily committed to a covert denial of God's existence. The affirmation itself is, of course, an ambiguous one. A plausible construal of it in this context might run as follows. The belief that the relationships attested to by the fossil record, by comparative morphology, and by molecular biology are best explained in broadly evolutionary terms is true. Calling a theoretical belief 'true' customarily means that the cumulative evidence in its favour is so strong that it is safe to affirm it without qualification, just as a geologist might, for example, affirm that the continents of Africa and South America, once in physical contact, have gradually separated from one another. This ought *not* be taken to mean that the alternative can be logically excluded in a completely conclusive way; nothing more than overwhelming likelihood is what scientists normally intend by this sort of usage. One may *object* to this usage, but one cannot impute an implicitly atheistic premiss to those who follow it. Such a premiss *may* be playing a covert role, but it is equally possible that it may not.

In the second place, the reading of creation science that he urges is rather too charitable. A claim does not have to be obvious or inevitable for its rejection to connote fanaticism or ignorance. If the indirect evidence for the great age of the Earth is overwhelming (Plantinga himself allows that it is 'strong'), if its denial would call into question some of the best-supported theoretical findings of an array of natural sciences (cosmology, astrophysics, geology, biology), then one is entitled to issue a severe judgement on the legitimacy of the challenge. Perusal of some of the standard works in creation science would lead one to suspect that no matter *how* strong the scientific case were in favour of an ancient Earth, it would make no difference to their authors. Their implicit commitment to a literalist interpretation of Genesis is such that (to my mind, at least) it appears to block a genuinely rational assessment of the alternatives.

What bothers Plantinga, I suspect, about the use of terms like 'fanaticism' here is that from *his* point of view the creation scientist's heart is in the right place. Anyone who stands up for the maxim of *sola Scriptura*

('Scripture alone') in the modern world, even in contexts as unpromising as the debate about the age of the Earth, ought not (he suggests) simply to be dismissed as irrational. Creation scientists may be wrong in holding that the Earth is only a few thousand years old, but their intellectual commitment to Scripture ought to be regarded with sympathy by their fellow Christians. I am much less sympathetic to them, in part because of a deeper disagreement about the merits of the *sola Scriptura* premiss, as well as of the remaining major theses of creation science. Though I would not be as harsh on creation scientists as leading evolutionists have been, I would, as a Christian, want to register disapproval of creation science at least as strong as the latter's, though for reasons that differ in part from theirs. These reasons will become clear, I hope, in what I have to say about Plantinga's analysis of what happens when 'faith and reason clash'.

GALILEO AND THE BIBLE

In his *Letter to the Grand Duchess Christina* (1615), Galileo provided the most extended account that anyone perhaps had given up to that time of how the Christian should proceed when an apparent conflict between science and Scripture arises.[8] Aided, doubtless, by some of his theologian-friends, he drew upon Augustine, Jerome, Aquinas, and an impressive array of other theological authorities, in order to show that the use made of Scripture by those who opposed the Copernican theory was illegitimate. There may be some lessons to be drawn from this historic document in the context of the more recent debate about evolution, apart from the obvious one of the embarrassment that the Church would later suffer because of its ill-advised attempt to make the geocentric cosmology of the Old Testament authors a matter, equivalently, of Christian faith.

What, then, did Galileo hold about the bearing of the Scriptures on our knowledge of the natural world? It does not take long for the reader to discover that several different hermeneutic principles are proposed in different parts of the *Letter*, and to realize that Galileo almost certainly was not aware of the resulting incoherence. On the one hand, he cites the traditional view, traced back to Augustine in his influential *De Genesi ad litteram*, that in cases of apparent conflict, the literal interpretation of Scripture is to be maintained, unless the opposing scientific claim can be *demonstrated*. In that case, theologians must look for an alternative reading of the scriptural passage(s), since it is a first principle that faith and

---

[8] Maurice Finocchiaro provides a new translation of the *Letter* in *The Galileo Affair* (1989).

natural reason cannot really be in contradiction. However, the straightforward interpretation of Scripture is to be preferred in cases where the scientific claim has something less than 'necessary demonstration' in its support, because of the inherently greater authority to be attached to the Word of God (Finocchiaro 1989: 94). Let us call this the 'literalist principle' because it maintains a presumption (though not, to be sure, an absolute one) in favour of the literal reading in cases of apparent conflict.

On the other hand, Galileo also argues that one should not look to Scripture for knowledge of the natural world in the first place. The function of the Bible is to teach us how to go to heaven, not how the heavens go, in the aphorism attributed to Baronius. God has given us reason and the senses to enable us to come to understand the world around us. Had the biblical authors attempted to describe the underlying structures of natural process, they would have baffled their readers and defeated the obvious purpose of Scripture. Galileo produces a number of convergent lines of argument to the effect that Scripture is simply not relevant to the concerns of the natural sciences to begin with.[9] This might be called the 'neutrality principle', since it proposes that the Scriptures are neutral in regard to natural science.[10]

The implications of these two principles were, of course, quite different for the resolution of the debate about the orthodoxy of the Copernican position (McMullin 1967: 33–5). But that is not our concern here. The requirement that the claims of 'reason' ought to be demonstrative in order to count is, of course, an echo of the classical Greek notion of science (equivalently, knowledge) that Augustine inherited. It is worth noting that, in practice, Augustine himself seems often to have been guided by a less strict norm, even in the *De Genesi ad litteram* itself. He did not require a *conclusive* demonstration on the side of natural reason before abandoning the literal reading of the narrative of the six days of creation and espousing a highly metaphorical alternative. And he constantly stressed the antecedent importance of literary norms in determining how biblical texts should be interpreted. The strong presumption in favour of literalism that has been the main source of conflict in the debates over Scripture and science

---

[9] It would be tempting to call this the 'Galilean principle', since it was Galileo's most distinctive contribution to the discussion, and fairly clearly the principle he favoured. But since he did, after all, allude to several others, it could be misleading to attach his name to one of them rather than to the others.

[10] Galileo introduced one further way of dealing with tensions between Scripture and natural science, suggesting that the biblical authors accommodated themselves to their hearers. This does not, in practice, reduce to either of the principles above. The notion of accommodation had already been hinted at by theologians as diverse as Thomas Aquinas and John Calvin. But this is not the place for an exhaustive analysis of the logical complexities of the famous *Letter*. See McMullin 1983 and Moss 1983.

('the text is to be interpreted literally unless a contrary reading can be established from an extrinsic source such as natural science') is much more characteristic of post-Reformation theology.

A troublesome feature of the literalist principle, even when interpreted quite broadly, is that it sets theologians evaluating the validity of the arguments of the natural philosophers, and natural philosophers defending themselves by composing theological tracts. Either way, immediate charges of trespass result. The theologian challenges the force of technical scientific argument; scientists urge their own readings of Scripture or their own theories as to how Scripture, in general, *should* be read. In both cases, the professionals are going to respond, quite predictably: What right have you to intrude in a domain where you lack the credentials to speak with authority? The techniques of the lawyer or of the logician are inappropriate in such a context. It is not a matter of persuading a jury of the inexpert that a particular assertion is supported beyond all reasonable doubt by the evidence at hand. Nor is it a matter of laying out an abstract argument that carries weight by force of logical rule alone. The assessment of theory strength is not a simple matter of logic and rule, but requires a long familiarity with the procedures, presuppositions, and prior successes of a network of connected domains, and a trained skill in the assessment of particular types of argument.

What, then, is to be done when tensions arise between a science-based assertion and a claim inspired by Scripture?[11] Can trespass be avoided? A first answer might simply rely on the neutrality principle. The Bible, it could be said, was not intended to convey insights about the underlying physical structure of the world around us. The biblical writers simply made use of the language and the cosmological beliefs of their own day while recounting the story of the covenant between Creator and creature. In particular, the creation narratives in the first two chapters of Genesis are not to be read as literal or quasi-literal history. Their meaning lies deeper; to discover it, one must take into account the wider literary context of that earlier day and the later theological appropriations of those texts, as well as the larger theological bearings of the biblical narrative as a whole (Anderson 1983, Bergant and Stuhlmueller 1985, Clifford 1988).

It is not, therefore, as though the creation stories are to be taken quasi-literally except where an opposing scientific claim can be strongly supported. If no likelihood is attached in the first place to the separate and special creation, say, of the ancestors of the major phyla of living things on the basis of a quasi-literal construal of the Genesis narrative, then the delicate

---

[11] I am expressing this question, of course, from the perspective of someone who takes the Bible seriously as an authoritative source.

balancing of opposing probabilities is not necessary. The majority of contemporary biblical scholars would, I think, favour the neutrality principle over the literalist principle in this particular context; this assessment on my part would, of course, require something more than an expression of opinion to carry any weight. The matter is, in the first instance at least, one for theologians and biblical scholars, not philosophers or biologists, to debate and resolve. This is the proper function of expertise, and the proper function of expertise is in part what is at issue in disagreements of this sort.

Does this mean that the two domains, scientific and biblical, are so disparate that real conflict *cannot* arise, that the appearance of conflict necessarily implies that one side or other is straying outside its proper boundary? Unfortunately, matters are not quite so simple; the neutrality principle only reaches so far. Even if agreement can be reached that the biblical writers are not communicating insights about the workings of nature that were specially revealed to them by God, there are still some common presuppositions about human nature that are integral to the biblical narrative as a whole: the reality of human free choice and the consequent moral responsibility for actions performed, for example (see p. 740 below). Were a psychological or psychoanalytical theory to call one or other of these presuppositions into question, real conflict *could* still arise.

At that point one would inevitably have to draw on a larger perspective, where the credentials of the scientific theory would be set in the balance with the claim that the disputed assertion is indeed an integral presupposition of biblical religion. Someone whose life is guided by that religion might, then, render a different assessment of a particular psychoanalytic theory than another would. Of course, this would not, as we have seen, constitute a new level of *science*. Were an analogue of the original Augustinian version of the literalism principle to hold true, one could say that if a 'necessary demonstration' were available on the scientific side, one could be assured that the disputed presupposition would have to be modified in some way. But, of course, if there is one thing that philosophers of science agree on, it is that such demonstration is in principle out of reach in the domain of large-scale theory. The underdetermination of theory by the available evidence is the fulcrum of much recent discussion in sociology of science and in feminist theories of science. In our context it explains how theological considerations can play a role for some in theory acceptance (or, more likely, rejection). Can such considerations, though not scientific, still count as *epistemic* (in the sense defined, for example, in McMullin 1984)? From the perspective of the religious believer, they would be held to be so, though this is treacherous ground indeed, and would require a far more extended discussion than can be given here (see p. 740 below).

The context where differences of this kind might properly occur seems restricted to issues concerning human nature. Does the theory of evolution conflict with any presuppositions that might be held to be essential to biblical theology? Human uniqueness? The promise of resurrection? Certainly it has been held to do so by some. I would argue that the apparent conflict in these cases is only apparent. Our topic here, however, is the more restricted one of special creation. Does the integrity of the biblical account of sin and salvation suggest that some plants and animals were specially created? Obviously not. Does it require a form of special creation of the first human pair that would be incompatible with the evolutionary account of human origins? It is not clear that it does, although many have held the two accounts to be incompatible on the point. Defenders of a dualistic account of human nature might come forward with a philosophical argument for the impossibility of the soul's coming to be from matter. But no such autonomous argument is available where the coming to be of the first cells or the major phyla (Plantinga's other candidates for special creation) are concerned.

These are large issues, requiring sensitive treatment from the epistemological standpoint because of the possibly 'mixed' character of the assessments involved. My intention here has been simply to draw attention to the various possible sources of tension between the Bible and the sciences, and some of the principles that have been proposed for dealing with such tensions.[12] This done, however schematically, I can now return to Plantinga's proposal of special creation in one form or another as a likely alternative explanation wherever the evolutionary account seems to him to be flawed.

THE ANTECEDENT LIKELIHOOD OF SPECIAL CREATION

The most distinctive feature of Plantinga's argument is that he makes a point of *not* calling explicitly upon the two creation narratives in Genesis.

[12] There is, of course, the larger issue of deciding on the proper approach for the Christian to take to Scripture generally. Plantinga characterizes the Reformed Christian as one who takes 'Scripture to be a special revelation from God himself'. Thus, for example, the story of Abraham, including the details of where he lived and journeyed and how he came to father a son, becomes a matter of history in the modern sense of that term, to be construed (in Plantinga's view) as having the standing of science. There is an implicit literalist presumption here that an Unreformed Christian like myself—someone unsympathetic, that is, to the constraints of the *sola Scriptura* maxim—would want to question.

Historically, these narratives have provided the main warrant for the traditional Christian belief that God intervened in a special way in the origins of the living world. Defenders of that belief have tended to rely on Genesis, unless they were prevented from doing so, as the recent advocates of creation science were, by extrinsic constraints. Plantinga is, however, under no such constraints. His reason for eschewing the reference to Genesis that one might have expected to find is, rather, an awareness of the problematic character of the literalist approach to the Genesis story of creation (Plantinga 1991a: 81). Instead, he rests his case not on specific scriptural passages, but on a central defining theme in the biblical account of God's dealings with the people of Israel. In this context, at least, God evidently 'intervened' or 'interrupted' normal human routines in all sorts of ways. (Words like 'intervene' are inadequate to convey the action of a Creator with the created universe, Plantinga reminds us, but we do not have any better ones.) Since the God of Abraham brought about God's ends in 'special' ways throughout the long history of Israel, it is to be expected (Plantinga suggests) that the same may very well be true at some moments in the much longer story of the development of life on Earth.

The issue, be it noted, is not whether God *could* have intervened in the natural order; it is presumably within the power of the Being who holds the universe at every moment in existence to shape that existence freely. The issue, is, rather, whether it is antecedently *likely* that God would do so, and more specifically whether such intervention would have taken the form of special creation of ancestral living kinds. Attaching a degree of *likelihood* to this requires a reason; despite the avowed intention not to call on Genesis, there might appear to be some sort of residual linkage here. In the absence of the Genesis narrative, would it appear likely that the God of the salvation story would also act in a special way to bring the ancestral living kinds into existence? It hardly seems to be the case.

Might it be that the supposed likelihood of special creation in given cases (e.g. for the 'founders' of the major phyla) derives directly from the unlikelihood of there being a scientific explanation in such cases? If there are only two possible types of explanation, and one can be shown to be highly improbable on present evidence, the other automatically gains in likelihood. In this event, a reference to God's dealings with Israel would not be needed. But Plantinga made it clear that this was not his strategy: 'It is a bit more probable, *before we look at the scientific evidence*, that the Lord created life and some of its forms—in particular human life—specially' (1992b: 22, my emphasis; p. 687 above).

It is this casting of special creation and evolution as rivals in the domain of cosmological explanation that I find so troubling. If one assumes that there is a presumption in favour of some sort of special creation at the critical moments in the historical development of life (a presumption whose plausibility wanes in regard to specific transitions as the strength of the evolutionary explanation of those transitions increases), one inevitably transforms the field of prehistory into a battleground where the religious believer is engaged in constant skirmishes with the protagonists of evolutionary-type theories, skirmishes that most often end in forced retreat for the religious believer.

Plantinga claims that the Christian believer 'has a freedom not available to the naturalist', because the believer is 'free to look at the evidence . . . and follow where it leads' (ibid. 28; p. 694 above). This would be more persuasive if he were to hold only that the believer holds an extra alternative that allows him or her to be more critical of the shortcomings of the scientific theory. But he proposes something much stronger than that. There is an antecedent *likelihood*, he says, of 'special' intervention of this kind at some points in the cosmic process, and hence where the scientific case is weak, the hypothesis of divine intervention has to be allowed the higher likelihood. I am not sure that this *does* in the end allow the Christian believer more freedom than the naturalist (see p. 740 below). But whatever one makes of that, it certainly ensures conflict; it is likely to maximize the strain between faith and reason, as the believer searches for the expected gaps in the scientific account.

In his 1991a, Plantinga appears to change ground somewhat. On the one hand, he says: 'I remain confident that TCA is relatively unlikely given a Christian or theistic perspective and the empirical evidence' (1991a: 108). But now the warrant for claiming the antecedent likelihood of special creation appears to shift from the salvation story to the 'empirical evidence'. Quoting Francis Crick and Harold Kein on the difficulty of explaining how the first cells originated, he concludes that 'we have every reason to doubt that life arose simply by the workings of the laws of physics' (ibid. 102). He goes on:

It therefore looks as if God did something special in the creation of life. (Of course, things may change; that is how things look *now*.) And if he did something special in creating life, what would prevent him from doing something special at other points, in creating human life, for example, or other forms of life? . . . I am therefore inclined to maintain my suggestion that the antecedent probability, from a theistic point of view, is somewhat against the idea that all the kinds of plants and animals, as well as humankind, would arise just by the workings of the laws of physics and chemistry. (ibid.)

The antecedent probability (no longer strictly antecedent) now seems to depend on the current lack of plausible scientific accounts of how the first cells could have originated. (Crick, who is notably unsympathetic to theistic belief, would surely not agree with the inference being drawn from this!) In his 1991*a*, Plantinga is more intent on shifting the burden of proof, and on combating claims for the antecedent probability, on theological grounds, of a naturalist account favouring TCA. If TCA were correct, 'we should expect much stronger evidence than we actually have. . . . The actual empirical evidence must be allowed to speak more loudly than speculative theological assumptions' (ibid.). So much for his original claim that the story of God's dealings with Israel spoke loudly in favour of special creation over TCA!

THE THESIS OF COMMON ANCESTRY

Though my disagreement with Plantinga centres especially on the conclusion he draws from Christian faith in regard to the antecedent likelihood of special creation, it may be worthwhile to say something very briefly about the scientific issues also. He dismisses the evidence ordinarily presented in support of the Thesis of Common Ancestry (TCA) as inconclusive, after a brief review. His conclusion is as follows: 'It isn't particularly likely, given the Christian faith and the biological evidence, that God created all the flora and fauna by way of some mechanism involving common ancestry' (1991*b*: 28; p. 694 above). The credentials of a thesis encompassing as much of past and present as TCA does, cannot, of course, be dealt with satisfactorily in a few pages. This is particularly true when these credentials are being *denied*, contrary to the firm conviction of the great majority of those professionally engaged in the many scientific fields involved.

Though a full-scale defence of TCA will not be attempted here, and would in any event be beyond my competence, it may nevertheless be worthwhile to indicate some of the lines along which a defence might proceed.[13] First, one should note an important distinction, one to which Plantinga alludes. TCA is a *historical* claim that the kinds of living things originated somehow from one another. The various theories of evolution, on the other hand, are an attempt to *explain* how that could have occurred. The dominant theory of evolution at the present time is the so-called modern synthesis, associated with such figures as Simpson, Dobzhansky, and Mayr. It has its critics: Goldschmidt and Schindewolf a generation ago,

---

[13] I would like to acknowledge at this point my debt to the many who in discussions past have helped me overcome the bafflement that evolutionary theory induces in the non-expert. In particular, my thanks go to Francisco Ayala, John Beatty, Bill Charlesworth, Ernst Mayr, Bob Richards, and Phil Sloan.

for example; Gould and Kimura today. Though all of these have found fault with the Darwinism of the modern synthesis and proposed alternatives to it, none would for a moment question TCA. Their confidence in TCA does not depend, then, on a similar degree of confidence in the explanatory adequacy of a specifically Darwinian account of the origin of species. Is it, perhaps, that they implicitly reject God's existence, and thus TCA is for them (in Plantinga's phrase) 'the only game in town'? I do not think it is nearly as simple as this.

Much of the evidence for TCA functions independently of the *detail* of any specific evolutionary theory. Plantinga mentions three such categories of evidence, so I will confine myself to those. There is the fossil record, which has already yielded innumerable sequences of extinct forms, where the development of specific anatomical features can be traced in detail through the rock layers. Palaeontologists have traced the development of eyes in no fewer than forty *independent* animal lineages, lineages being determined by overall morphological similarities (von Salvini-Plawen and Mayr 1977). As new fossil evidence is uncovered, palaeontologists continue to define stage after stage in crucial 'linking' forms, such as the therapsids, the forms that relate reptiles with the earliest mammals, gradually bridging troubling gaps. In cases like these (and there are a *lot* of them), palaeontologists can point to a variety of morphological features that gradually shift over time, retaining a basic likeness (a so-called *Bauplan*) throughout.

Gould's objection regarding the rarity of transitional forms (quoted by Plantinga) has to be taken in context. Gould would not deny the morphological continuities of the fossil record; like thousands of other researchers, he has given too much of his time to tracing these continuities for him to underrate their significance. What he *would* say (and what many defenders of the modern synthesis would now be disposed to admit) is that species often make their appearance in the record without the prior gradual sequence of modifications one would have expected from the traditional gradualist Darwinian standpoint. But this leaves untouched the implications, overall, of the fossil record for TCA. It *does*, of course, affect the sort of theory that could account for the sequence found in the record.

In a recent discussion of the relation between micro-evolution and macro-evolution, Mayr writes:

Almost every careful analysis of fossil sequences has revealed that a multiplication of species does not take place through a gradual splitting of single lineages into two and their subsequent divergence but rather through the sudden appearance of a new species. Early palaeontologists interpreted this as evidence for instantaneous sympatric speciation [speciation over a single area], but it is now rather generally recognized that the new species had originated somewhere in a peripheral isolate and had subsequently spread to the area where it is suddenly found in the fossil record. The parental species which had budded off the neospecies showed virtually no change during this period. The punctuation is thus caused by a localized event in an isolated founder population, while the main species displays no significant change. (1988: 415)

This theory of allopatric speciation (speciation involving a second—in this case a geographically isolated but adjoining—territory) allows Mayr to modify the gradualism of the original Darwinian proposal, while retaining the basic Darwinian mode of explanation and avoiding the 'punctual' events of the Gould–Eldredge scenario (events that in his view are objectionable). But the debate is by no means closed.

Instead of scrutinizing the fossil record, we might look to the living forms around us, and there discover all sorts of homologies and peculiar features of geographical distribution, which are best understood in terms of TCA. The arguments here are long familiar to the readers of *Zygon*, so I will not dally with them. But there is another category of evidence which has taken on a great deal of importance in the last twenty years: namely, that deriving from molecular biology. Comparison of the DNA, as well as of the proteins which DNA encodes, among different types of organisms shows that there are striking similarities in chemical composition between them. These similarities are just of the kind one would expect from the hypothesis of common ancestry. By now many of these similarities have been charted in great detail. They yield information of a quality that the fossil record, with its many limitations, could never hope to give; they point to branchings that occurred more than 2 billion years ago, when *Archaea*, a minute organism found in some hot springs, seems to have separated from its bacterial cousins. To recall one standard example, cytochrome C is found in all animals and is involved in cell respiration (Ayala 1985). It contains 104 amino acids, in a sequence which is invariable for any given species. For humans and rhesus monkeys, the sequence is identical except in one position; for horses and donkeys, the sequence also differs in only one position. But for humans and horses, the difference is twelve; for monkeys and horses, the difference is eleven. If, instead of cytochrome C another homologous protein is chosen, similar (though not necessarily identical) results are found. These very numerous resemblances and differences between the macromolecules carrying hereditary information can be explained by supposing a very slow rate of change in the chemical sequences constituting these molecules, and therefore a relationship of common descent among the organisms themselves.[14] Thus, the molecular-level dif-

---

[14] The rate of change depends on a variety of factors, including environmental ones, so that it is quite variable (with a variance two or three times the mean, in technical terms). By contrast,

ferences between species give an indication of the relative order of branching between the species; with three species, for example, one can infer whether A branched from B before C did. What is impressive here is the *coherence* of the results given by examining many different macromolecules in this light. Without common descent, this intricate network of resemblances would make no sense.[15]

What is even more impressive is that these results conform reasonably well with the findings of both palaeontology and comparative anatomy with regard to the ancestral relations between species, the postulated tree of descent that had already been worked out in some detail in these other disciplines.[16] The fit, as one would expect, is not exact in each case with regard to the 'closeness' between the species, but it is nevertheless quite

the rate of radioactive decay, also used for probing the distant past, is relatively uniform. Cytochrome C, a small molecule, changes relatively slowly, so that it would not serve to separate 'recent' events like the splitting of the hominid from the chimpanzee and gorilla lineages. (All three of these exhibit the same cytochrome C sequence.) Other molecules change much more rapidly, especially those 'silent' segments of DNA that do not seem to affect the development or functioning of the organism, and thus may not be subject to negative selection when changes in them occur. Because of the variability in the rates of change of particular proteins or segments of DNA, these rates must be used with caution to time branching events in the past. The 'molecular clock' allows at best only a rough estimate for any particular molecule, as Ayala and other geneticists have emphasized. Since, however, literally thousands of different chemical sequences are available for scrutiny, each with its own history, cross-comparison can enable a gradual convergence to occur. Whether it *does*, in fact, occur is challenged by Scherer (1990), quoted by Plantinga. The extent to which molecular change can be relied on to furnish a chronology of past branching events is debated; most workers in this very active field of research agree, however, that it does furnish a rough clock, whose accuracy will improve as more and more sequences are compared.

[15] Plantinga argues that because of the numerous gaps, 'the fossil record fits versions of special creation considerably better than it fits TCA: it suggests the independent appearance of the major *Bauplans* . . . with substantial evolution proceeding out from these *Ur* forms. The enormous gaps between the major forms would be much better accommodated on such a view than on TCA' (Plantinga 1991a: 104). Here the intricate molecular relationships between the different phyla loom large: they are much more easily intelligible in the TCA scenario than on the supposition of an independent 'special' origin for each phylum.

[16] Against this line of argument, Plantinga objects that many species, like the lamprey and the horseshoe crab, remain morphologically unchanged over tens of millions of years (Plantinga 1991a: 106). How is this possible if a steady change is going on at the molecular level on which heredity depends? He notes that the standard response to this is to say that the molecular and the morphological levels must be decoupled, so that change can go on in the one without substantially affecting the other, but he regards this suggestion as a mere 'epicycle' meant to save the theory. There is, however, a great deal of independent evidence for this sort of decoupling. Kimura and others have shown that many changes at the molecular level are neutral as far as the phenotype is concerned, and it is, of course, at the level of the phenotype that selection goes on. Mayr notes: 'DNA sequences believed to be functionless, such as pseudo-genes and certain introns, behave as if selectively neutral and may thus be subject to rapid change, owing to genetic drift and to their being immune to stabilizing selection' (Mayr 1988: 102). Even among the 'active' genes, most code for 'housekeeping' functions, like metabolism, and do not affect morphology directly. In the 1940s, Dobzhansky studied 'sibling' species which did not interbreed, though morphologically almost identical. In the 1960s, it was discovered that these species can be genetically very different. In some cases, where the difference amounts to upwards of half of the total gene content, the species must have diverged several million years

good. When a single explanatory hypothesis (TCA) underlies the binding together of domains so diverse in character, we have the sort of consilience that carries more weight with scientists than does, perhaps, any other virtue of a theory.

It should be emphasized that specific theories of evolution are not yet involved here. The support given TCA by these diverse types of evidence does not depend on any particular explanatory account of *how* species change takes place. One could reject natural selection as the primary agent of evolutionary change, for example, and still find this argument for TCA convincing.[17] Of course, a satisfactory explanatory account of how evolutionary change occurred would greatly strengthen the case for TCA. But in the light of the continuing debates about the adequacy of this or that feature of the neo-Darwinian model, it is important to stress that there is a vast body of evidence for common descent that does not depend for its logical force on the further issue of why the transitions from one life-form to another came about as they did.

Plantinga raises one objection that bears on TCA directly. Does there not seem to be an 'envelope of limited variability' surrounding each species, so that a departure of more than a small degree from the central species norm leads to reversion or sterility? Would one not expect to find evidence of new species now and then appearing in the present (or perhaps being deliberately produced) if indeed TCA is true? The first and simplest response is to note that in the plant world (in the forest, for example) new species have indeed been observed. And the production of fertile hybrids is an important part of agricultural research. The ability of populations of micro-organisms to alter their structures quite basically over relatively

ago, while their morphology remained substantially the same because of strong selection pressures against change.

[17] In this regard, the position adopted by Michael Denton, one of the most sweeping recent critics of evolutionary theory, is quite puzzling. On the one hand, he finds the sort of consilience described above altogether remarkable: 'It became increasingly apparent as more and more sequences accumulated that the differences between organisms at a molecular level corresponded to a large extent with their differences at a morphological level; and that all the classes traditionally identified by morphological criteria could also be detected by comparing their protein sequences. . . . The divisions turned out to be more mathematically perfect than even the most die-hard typologists would have predicted' (1986: 276, 278). But the distances between the molecular sequences characteristic of different species can only be explained (he argues) by postulating a remarkably uniform 'molecular clock' marking the rate of change in the constituents of particular kinds of molecules (and varying from one kind to another), and such a 'clock' (he maintains) is impossible to understand on neo-Darwinian principles. What would seem, at most, to follow from this is that neo-Darwinian theory cannot explain the uniformity of the postulated 'clock'. But he assumes that he has also refuted TCA, while providing no hint himself as to how the correspondences he finds so remarkable might be explained by something *other* than common ancestry. (Whether differences at the molecular level correspond as closely as he claims to differences at the morphological level is open to question; see n. 16.)

short times under the challenge of antibiotics is all too well known. But defenders of the modern synthesis themselves insist on the extraordinary stability of the genotype, in the animal realm particularly; this stability is essential to the maintenance of species differences, and some progress has been made toward an understanding of its molecular basis in the constellations of genes.

TCA does not require rapid change. The presumption is that the kind of species changes that would sustain TCA could take thousands of generations to accomplish. The rate of change required (as has been shown in detail in recent studies in population genetics) is far too slow for the sort of direct evidence to accumulate that Plantinga is asking for. There are also serious problems with the species concept itself, the concept underlying this objection. Should it, for example, be based on morphological differences (of the kind that palaeontologists or comparative anatomists can attest to), or should it be based on interbreeding boundaries (as naturalists have long preferred to maintain)? These are only two of the many possibilities (Sober 1984: sect. 7; Mayr 1963: 400–23). If we were to find the fossil remains of animals as different as a St Bernard and a chihuahua in the rock strata, we should assuredly label them different species. If we adopt the biological species concept according to which 'species are groups of interbreeding natural populations that are reproductively isolated from other such groups' (Mayr 1988: 318), how are we to apply this to populations that are widely separated in space or time? Mayr emphasizes that such application always involves complex and indirect forms of inference.[18] The moral is not that the species concept is so ambiguous as to be unusable, but only that such notions as species change are far more difficult to handle than at first sight they seem to be. And, more specifically, the claim that an 'envelope of limited variability' surrounds each species has no precise empirical foundation.

I suspect that in the end, this claim simply begs the question against TCA. It asserts that the sort of change TCA would require does not occur. But this is just the issue, and this is what is challenged by the three kinds of evidence described above, all of them pointing to TCA as the most reasonable explanation. Plantinga's way of dealing with this evidence is unconvincing: 'As for the similarity in the biochemistry of all life, this is reasonably probable on the hypothesis of special creation' (Plantinga

[18] A further problem is suggested by the notion of a 'natural' population. Reproductive isolation in the animal world is due, in the first instance, to *behavioural* barriers, which are the main isolating mechanisms (Mayr 1988: 320). Under artificial circumstances, such barriers can be overcome, but this will not necessarily give rise to new biological species. Likewise, deliberate interbreeding to produce new varieties of domestic dog, for example, will not produce a natural population with its own behavioural barriers to outbreeding.

1991*b*: 23; p. 689 above). But why *should* this be probable on the hypothesis of special creation?[19] Would *this* hypothesis have been able to predict in advance that such biochemical similarities would be found? Why would God, if 'specially' creating a new kind, give it the sort of biochemical constitution that would be likely to suggest that it shared a common ancestry with other organisms? Again, in regard to significant homologies between organisms, Plantinga remarks: 'Well, what would prevent [God] from using similar structures?' (ibid. 24; p. 690 above). But this is not the issue. Nothing would *prevent* this; that is, it would have been *possible* for the Creator to use similar structures. But is the finding of homologies a positive reason to suppose special creation has in fact occurred? (It *is* a reason to suspect common ancestry.) Homologies would have to be antecedently *likely* (not just possible) on the hypothesis of special creation for the finding of homologies *not* to give reason to prefer the evolutionary hypothesis.

Let me stress once again the criterion of consilience. Evidence from three quite disparate domains supports a single coherent view of the sequence of branchings and extinctions that underlie TCA. If TCA is *false*, if in fact the different kinds of organisms do not share a common ancestry, this consilience goes unexplained. It is all very well to say: 'but God *could* have. . .'. This hypothesis treats the consilience exhibited by TCA as a coincidence; it does not explain it. So it is not as though allowing the theistic alternative into the range of possible explanations alters the balance of probability drastically, as Plantinga supposes. TCA is, of course, a *hypothesis*, as any reconstruction of the past must be. But it remains by far the best-supported response, for the theist as for others, to the fast-multiplying evidence available to us.

THEORIES OF EVOLUTION

What about the objections to the neo-Darwinian theory of evolution, as such, as distinct from TCA? Plantinga outlines a familiar objection to any

---

[19] Plantinga responds without elaboration in 1991*a* that the molecular evidence 'fits particularly well' with those versions of special creation 'that involve typology, the idea that God created ancestors of the main types of animal and plant life, with subsequent evolution' (p. 105). But *is* there some antecedent reason why we should expect God to restrict the first members of each type to a narrow range of structures at the molecular level? Nineteenth-century critics of evolutionary theory, like Owen and Agassiz, claimed that the evidence from morphology and palaeontology points to the existence of discrete 'types'. These types were then taken to represent both ideas in the mind of God and immanent principles of living growth. The idealist assumption of 'ideas' in God's mind that would antecedently favour discreteness over continuity is obviously open to question.

theory which relies on natural selection as the primary mechanism of evolutionary change. There is no plausible evolutionary pathway (he argues) linking an eyeless organism, say, with an organism possessing the complex structures of the mammalian eye, such that every single stage along the way can be shown to be adaptively advantageous. This is the oldest of objections to Darwin's theory; it was the primary criticism raised by Mivart in his *Genesis of Species* (1871). Darwin's own first response was to emphasize that his theory did not rely on natural selection alone.[20]

Among the other processes that he proposed, one in particular is still emphasized: change of function, where a structure that originally developed because of the adaptive advantage offered by a particular function takes on a new function (especially under the impact of change of habitat or the like). Another process whose importance has only recently come to be recognized is genetic drift. In the isolated and often small populations that furnish the likeliest starting-point for the speciation process, there can be a sort of genetic random sampling error that eventually marks off the smaller population from the parent population. Additionally, there can be 'hitch-hiker' effects of all sorts due to genetic linkage. These processes do not operate independently of natural selection, but they can easily bring about results not possible with the model of evolutionary change that requires an adaptive advantage at every step (Mayr 1960). Defenders of the modern synthesis are as quick as Darwin was to insist that they are *not* limited in their explanatory strategies to the selectionist model only. Mayr, for instance, repudiates what he calls selectionist extremism:

Much of the phenotype is a byproduct of the evolutionary past, tolerated by natural selection but not necessarily produced under current conditions. . . . The mere fact of the vast reproductive surplus in each generation, together with the genetic uniqueness of each individual in sexually reproducing species, makes the importance of selection inescapable. This conclusion, however, does not in the least exclude the probability that random events also affect chances of survival and of the successful reproduction of an individual. The modern theory thus permits the inclusion of random events among the causes of evolutionary change. Such a pluralistic approach is surely more realistic than any one-sided extremism. (Mayr 1988: 136, 140)

Still, he also wants to say that the modern synthesis of which he is perhaps the leading representative 'was a reaffirmation of the Darwinian

[20] Indeed, he showed some uncharacteristic indignation in his comment in the last edition of the *Origin of Species* (1872): 'As my conclusions have lately been much misrepresented, and it has been stated that I attribute the modification of species exclusively to natural selection, I may be permitted to remark that in the first edition of that work, and subsequently, I placed in a most conspicuous position—namely, at the close of the Introduction—the following words: "I am convinced that natural selection has been the main but not exclusive means of modification." This has been of no avail. Great is the power of misrepresentation' (p. 395).

formulation that all *adaptive* evolutionary change is due to the directing force of natural selection on abundantly available variation' (ibid. 527, my emphasis).

Nevertheless, to some critics of the modern synthesis, these concessions are not enough. Gould, for example, has criticized what he calls the 'adaptationist programme' for its failure to take seriously the many alternatives to trait-by-trait selection on the basis of adaptive advantage. Instead, he notes the constraints that the integrity of the structure of the organism as a whole sets on possible pathways of change, so that the outcome is explicable rather more by the nature of the constraints than by the application of selectionist norms to individual traits (Gould and Lewontin 1979). Kimura has developed a controversial molecular-level theory according to which most changes in gene frequencies are 'neutral': that is, carry no selective advantage. More radical challenges come from those who rely on macro-mutations (saltations) to bridge major discontinuities in the fossil record; theories of this sort, it is generally thought, face intractable problems.[21]

Where does all this leave us? The defenders of the modern synthesis base their confidence on the substantial explanatory successes of their model. They have no illusions about having explained everything; in particular, they concede that the processes responsible for the origins of the main phyla are not well understood. In the early stages of life's development on Earth, sixty or seventy different phyla (morphological types) developed, most of which became extinct. Not a single new phylum, apparently, has originated since the Cambrian period, more than 400 million years ago. It would seem that the genetic structures of this early period were not as fixed as they later became. Thus, selection then may have had fewer constraints than later on, when highly cohesive genotypes developed; the rate of species change might thus have been quite rapid, lowering the chances of an adequate fossil record of the changes.

The Darwinian model has already been substantially reshaped over the last fifty years, while retaining the original emphasis on the transformative powers of selection operating on individual differences. Undoubtedly, more such reshaping lies ahead. Like any other active scientific theory, the

---

[21] One such problem is that a mutation affecting the phenotype in a major way would be likely to require co-ordinated change in hundreds of genes; another is that a macro-mutation in a single individual would not be enough, in a sexually reproducing species, to establish a new kind right away. The role of mutations in evolutionary change is much less dramatic than is often conveyed in popular accounts; they serve mainly to augment the stock of variations in a population upon which recombination can work. (Recombination is the blending of paternal and maternal DNA in each new biological individual in a sexually reproducing species; it is responsible for the fact that each such individual is different from all others.)

modern synthesis is incomplete, but its exponents argue that there are no *in-principle* barriers to its continued successful extension to the difficult cases. A minority has proposed that a more radical transformation is needed, one which abandons either the gradualism or the heavy reliance on selection that have marked the Darwinian approach.[22] The most extreme view is represented by Michael Denton, who argues that *all* current theories of evolution are in principle inadequate to handle macro-evolution, and that we have to await another quite different sort of theory.

Where does the burden of proof lie in a matter of this sort? The claim that principles of a broadly Darwinian sort are capable of explaining the origins of the diversity of the living world rests on the successes of the theory to date. These are very considerable; they span many fields, and have shown intricate linkages between those fields. In particular, the theory has shown an extraordinary fertility as it has been extended into new domains; even when it has encountered anomalies, it has shown the capacity to overcome these in creative ways that are clearly not *ad hoc*.[23] This is the sort of thing that impresses those who are actually in touch with the detail of this research. And it gives a prima-facie case for supposing that the theory can be further extended to contexts not yet successfully treated. But, of course, this cannot in the strong sense be *proved*; it can only be made to seem more (or less) plausible.

On the other side is the claim that theories of a Darwinian type are incapable of entirely overcoming certain kinds of problems: gaps in the fossil record, the origin of complex organs like the eye, the origin of the broad divisions of the living world (the phyla), or the like. Claims of this sort are hard to establish, because they cannot anticipate the trajectory that the theory itself may follow as it is reworked in the light of new challenge. (Could the changes of the last century leading up to the modern synthesis have been foreseen?) This is not to say that such claims can *never* be established, or at least shown to be strongly supported. So it is not that the burden of proof falls on one side exclusively. Adjudicating between modern Darwinists and their critics is a matter of weighing up the merits of the case on each side, and then making some kind of comparative

[22] The differences between the punctuated equilibrium model of Gould and Eldredge and the standard one of the modern synthesis are not nearly as great as was originally claimed. In particular, Gould's original assertion that only a 'non-Darwinian' theory could handle the evidence from the fossil record was quite clearly based on a very narrow construal of what ought to count as 'Darwinian'. Mayr has to my mind convincingly shown that Gould's own model is compatible with Darwinian principles (Mayr 1988: ch. 26).

[23] Denton's comparison of the modern synthesis to late Ptolemaic astronomy with its profusion of epicycles, and his conclusion that it is a paradigm in crisis (1986: ch. 15) cannot, I think, be sustained. The crucial question in this context would be what constitutes an *ad hoc* modification (what he oddly calls a 'tautology').

assessment, informed by parallels from the earlier history of science and a very detailed knowledge of the history and contemporary situation of the various fields where the neo-Darwinian paradigm is applied.

Concerning theories of evolution in general, Plantinga remarks that they can never tell the *whole* story of the genetic changes involved, the rates of mutation, the links between gene adaptation, and so forth: 'Hence we don't really know whether evolution is so much as biologically possible' (Plantinga 1991*b*: 26; pp. 691–2 above). But first of all, evolutionary explanation begins at the level of the biological individual and the population, not the gene; natural selection operates on adaptations of whose genetic basis we may be (and usually are) entirely unaware. And the explanation is none the less real for that. But, more important, evolutionary explanation is of its nature *historical*, and historical explanation is not like explanation in physics or chemistry. It deals with the singular and the unrepeatable; it is thus *necessarily* incomplete. One must be careful to apply the appropriate criteria when assessing the merits of a particular explanation. An evolutionary explanation can never be better than plausible; the real problem lies in discriminating between different degrees of plausibility. The dangers of settling for a very weak sort of plausibility are real (recall Gould's 'just so' stories). But the dangers of requiring too strong a degree of confirmation before allowing *any* standing to an evolutionary explanation ('Hence we don't really *know* . . .') are just as great.

The presumed inadequacy of current theories of evolution is part of what leads Plantinga to propose his own alternative: 'God created mankind, as well as many kinds of plants and animals, separately and specially' (Plantinga 1991*b*: 22; p. 687 above). Which kinds? More than 99.99 per cent of the species that existed since life first appeared on Earth are now extinct. (These have a part to play in the evolutionary story, but ought to be puzzling for defenders of special creation.) Plantinga's response is that he does not have to specify the points at which special creation is supposed to have occurred, since his aim is only to call TCA into question, not to propose an alternative explanation (Plantinga 1991*a*: 88–9). But surely his claim that for the theist TCA 'is less likely than not' depends essentially on the theist's producing an alternative explanation (i.e. special creation by some means) for those newly appearing forms for which an adequate evolutionary account is held to be lacking? His critique of TCA is aimed at establishing 'enormous gaps among the major forms' (ibid. 104), gaps which evolution cannot account for. When he holds that it is more probable than not that God specially created 'some forms' of prehuman life, he is presumably alluding to those forms which evolution cannot in his view explain. It is their supposed inexplicability in evolutionary terms that furnishes the warrant for his claim; there do not appear to be any independent *theological* grounds for it.[24]

Establishing the presence of gaps in the evolutionary account is thus essential to his case. This stress on gaps is reminiscent in one respect of eighteenth-century natural theology. Plantinga's intention is not, of course, to make of the gaps an argument for God's existence; his faith needs no such support. But he *needs* the gaps to sustain his argument, just as the natural theologians did for theirs. And he fills the gaps with God's special action, just as they did, while also emphasizing that God is at all moments sustaining the entire process as Creator. Should one use the unflattering label, 'God of the gaps', to describe this approach? Only in the sense that it has God operate 'specially' within the process of life's origins at just those points where gaps can be claimed to exist in the evolutionary account. Plantinga is open to the possibility that at some point in the future such gaps may close; his claim that there is, nevertheless, an antecedent probability that God must have intervened in the coming to be of life rests presumably on his belief that it is highly unlikely that all of the gaps will vanish.

THE INTEGRITY OF GOD'S NATURAL WORLD

Plantinga's original argument relied on the premiss that God's special intervention in the cosmic process is antecedently probable. Here is where he and I really part ways. My view would be that from the theological and philosophical standpoints, such intervention is, if anything, antecedently *improbable*. Plantinga builds his case by recalling that 'according to Scripture, [God] has often intervened in the working of his cosmos' (Plantinga 1991*b*: 22; p. 687 above). And the examples he gives are the miracles recounted in Scripture and the life, death, and resurrection of Jesus Christ. I want to recall here a set of old and valuable distinctions between nature and supernature, between the order of nature and the order of grace, between cosmic history and salvation history. The train of events linking Abraham to Christ is not to be considered an analogue for God's relationship to creation generally. The Incarnation and what led up to it were unique in their manifestation of God's creative power and a loving concern for the created universe. To overcome the consequences of human

---

[24] In his original 1991*b* (Ch. 34), as we have seen, he invoked a theological premiss (the salvation story reveals God as one who constantly 'intervenes'). To the extent that he has in his 1991*a* laid aside the idea of basing the antecedent likelihood of special creation on such a premiss, he is forced to rely exclusively on the 'gaps' strand of the argument.

freedom, a different sort of action on God's part was required, a transformative action culminating in the promise of resurrection for the children of God, something that (despite the immortality claims of the Greek
philosophers) lies altogether outside the bounds of nature.

The story of salvation is a story about men and women, about the
burden and the promise of being human. It is about free beings who sinned
and who therefore *needed* God's intervention. Dealing with the human
predicament 'naturally', so to speak, would not have been sufficient on
God's part. But no such argument can be used with regard to the origins of
the first living cells or of plants and animals. The biblical account of God's
dealings with humankind provides no warrant whatever for supposing
that God would have brought the ancestors of the various kinds of plants
and animals to be outside the ordinary order of nature. The story of
salvation *does* bear on the origin of the first humans. If Plantinga were
merely to say that God somehow leaned into cosmic history at the advent
of the human, Scripture would clearly be on his side. How this 'leaning' is
to be interpreted is, of course, another matter.[25] But his claim is a much
stronger one.

To carry the argument a stage further, what would the eloquent texts of
Genesis, Job, Isaiah, and the Psalms lead one to expect? What have theologians made of these texts? This is obviously a theme that far transcends
the compass of an essay such as this one. I can make a couple of simple
points. The Creator whose powers are gradually revealed in these texts is
omnipotent and all-wise, far beyond the reach of human reckoning. God's
providence extends to all creatures; they are all part of a single plan, only
a fragment of which we know, and that darkly. Would such a being be
likely to 'intervene' in the cosmic process: that is, deal in two different
manners with it? (Let me emphasize that I am uncomfortable with this
language of 'likelihood' with regard to God's action, as though we were
somehow capable of catching the Creator of the galactic universe in the
nets of our calculations.) Why should an omnipotent God not create a

---

[25] 'God fashioned Adam from the dust of the earth and breathed into his nostrils the breath
of life' (Gen. 2: 7). The 'fashioning' here could be that of a billion years of evolutionary
preparation of that 'dust' to form beings that for the first time could freely affirm or freely deny
their maker. Pope Pius XII in his encyclical *Humani Generis* (1950) allowed that such an
evolutionary origin of the human body was an acceptable reading of the Genesis text. But he
added that the human *soul* could not be so understood; souls must be 'immediately created' by
God (1950: 181). The Platonic-sounding dualism underlying this distinction requires further
scrutiny. The uniqueness of God's covenant with men and women and of the promise of
resurrection does not require that there be a naturally immortal soul, distinct in its genesis and
history from its 'attendant' body. But it is unnecessary to develop this issue here, since Plantinga's challenge extends to the evolutionary account of the plant and animal worlds, not simply of
the human alone.

universe in which God's ends with regard to all creatures except humans would be achieved in a *natural* way? Ought one not to expect a fundamental integrity in the work of such a Creator?[26] If one may use the language of antecedent probability at all here—and I am not at all sure that one may—it surely must point away from special creation.

St Augustine may help us, perhaps, to formulate the most persuasive theological response to this question. He was the first to weave from biblical texts and his own best understanding of the Church's tradition the full doctrine of creation *ex nihilo* as Christians understand it today. And in the *De Genesi ad litteram*, his commentary on the very texts in Genesis where the writer speaks of the coming to be of the plant and animal world on the fifth and sixth 'days' of creation, he enunciated the famous theory of the *rationes seminales*, the seed-principles which God brings into being in the first moment of creation, and out of which the kinds of living things will, each in its own time, appear (McMullin 1985: sect. 4). The 'days', said Augustine, must be interpreted metaphorically as indefinite periods of time. And instead of inserting new kinds of plants and animals ready-made, as it were, into a pre-existent world, God must be thought of as creating in that very first moment the potencies for all the kinds of living things that would come later, including the human body itself:

In the seed, then, there was invisibly present all that would develop in time into a tree. And in this same way we must picture the world, when God made all things together, as having had all things which were made in it and with it when day was made. This includes not only the heavens with sun, moon, and stars . . . but also the beings which water and earth contained in potency and in their causes, before they came forth in the course of time. (Augustine 1982: i. 175)

This is, of course, not an evolutionary theory; the species do not come from one another, so there is no common ancestry. But Augustine would not have attributed an antecedent probability to God's 'intervening' in the midst of the cosmic process to bring the first kinds of plants and animals abruptly to be, rather than having them develop in the gradual way that seeds do.

TOO MUCH AUTONOMY?

But what are we to make of Plantinga's objection that having life coming gradually to be according to the normal regularities of natural process is

[26] Van Till, in his contribution to this discussion (1991) and more fully in his 1986 work, also stresses this theme of the integrity of the natural order under the supposition that it is the work of an omnipotent Creator.

'semi-deistic': that is, that it attributes too much autonomy to the natural world? He says:

God could have accomplished this creating in a thousand different ways. It was entirely within his power to create life in a way corresponding to the Grand Evolutionary Scenario . . . to create matter . . . together with laws for its behavior, in such a way that the inevitable[27] outcome of matter's working according to these laws would be, first, life's coming into existence three or four billion years ago, and then the various higher forms of life, culminating as we like to think, in humankind. This is a semi-deistic view of God and his workings. (Plantinga 1991*b*: 21; p. 686 above)

He contrasts this alternative with the one he favours:

Perhaps these laws are *not* such that given enough time, life would automatically emerge. Perhaps he did something different and special in the creation of life. Perhaps be did something different and special in creating the various kinds of animals and plants. (ibid. 22; p. 687 above).

Plantinga's characterization of the first alternative as semi-deistic is intended to validate the second alternative as the appropriate one for the Christian to choose. But why should the first alternative be regarded as semi-deistic? He allows that it was within God's power to bring about cosmic evolution, but then asserts that to say God *did* in fact fashion the world in this way would be semi-deistic. This is puzzling. It would be semi-deistic, perhaps, if we *already* knew that God had intervened in bringing into existence some kinds of plants and animals, in which case the 'Grand Evolutionary Scenario' would attribute a greater degree of autonomy to the natural world than would be warranted. But this is exactly what we do *not* know. And to assume that we *do* know it would beg the question.

The problem may lie in the use of the label 'semi-deistic'.[28] A semi-deist, Plantinga remarks, could go so far as to allow that God 'starts everything off' and 'constantly sustains the world in existence', and could even maintain that 'any given causal transaction in the universe requires specific divine concurrent activity'. All this would, apparently, not be enough to make such a view acceptable. What more could be needed? Defining God's relationship with the natural order in terms of creation, conservation, and *concursus* has been standard, after all, among Christian theologians since the Middle Ages. Perhaps what still needs to be made explicit

---

[27] 'Inevitable' is a word that defenders of evolution, whether theists or not, would be inclined to challenge. It suggests that the evolutionary process is, at least in a general way, deterministic or predictable. But this is just what nearly all theorists of evolution would deny.

[28] In the entry under 'deism' in *The Encyclopedia of Religion*, Allen Wood remarks that the term 'deism' tended over time to become 'a vague term of abuse' when used by Christian writers with regard to hypotheses that in their view attributed an undue degree of autonomy to the universe.

is that God *could* also, if God so chose, relate to the created world in a different way, either by way of special creation, or in the dramatic mode of a grace that overcomes nature and of wonders that draw attention to the covenant with Israel and ultimately to the person of Jesus. The possibility of such an 'intrusion' on God's part into human history, of a mode of action that lies *beyond* nature, must not be excluded in advance, must indeed be affirmed. I take it that the denial that such a mode of action *is* possible on the part of the Being who creates and conserves and concurs is what constitutes semi-deism, in Plantinga's sense of that term.

But someone who asserts that the evolutionary account of origins is the best-supported *one* is *not* necessarily a semi-deist in this sense. Some defenders of evolution—notably those who deny the existence of a Creator and are, therefore, not deists of *any* sort—would, of course, exclude special creation in this way, in principle. But there is no intrinsic connection whatever between the claim that God did, in fact, choose to work through evolutionary means and the far stronger claim that God *could* not have done otherwise. Nor, of course, is there any reason why someone who defends the evolutionary account of origins should go on to deny that God might intervene in the later human story in the way that Christians believe God to have done.

In sum, then, at least *four* alternatives would have to be taken into account here. There are those who defend the evolutionary account of origins, and also rejecting the existence of God, would (if pressed) say that life could not *possibly* have come to be except through evolution. There may be those who maintain that God created, conserves, and concurs in the activity of the universe but *could* not 'intervene' in a special way in its history to bring new kinds of animals and plants to be, for example. These (if they exist) are the semi-deists Plantinga describes. Then there are those who prefer the evolutionary account of origins on the grounds of evidence that this is in fact most probably the way it happened, but who are perfectly willing to allow that it was within the Creator's power to speed up the story by special creation of ancestral kinds of plants and animals, even though (in their view) this was not what God did. This is a view that a great many Christians from Darwin's day to our own have defended; it is the view I am proposing here. It is *not* semi-deistic. And finally, there is the option of special creation: that God *did*, in fact, intervene by bringing various kinds of living things to be in a 'special' way.

When Plantinga presents two alternatives only, the second being that God might 'perhaps' have intervened as defenders of special creation believe occurred, he must be supposing that the other alternative, the 'Grand Evolutionary Scenario', is one that excludes such a 'perhaps': that

is, that excludes, *in principle*, the possibility that God could have intervened in a special way in the natural order. What I am challenging is this
supposition. The Thesis of Common Ancestry can claim, as we have seen,
an impressive body of evidence in its own right. It need not rely on, nor
does it entail, any in-principle claim about what God could or could not
do.[29]

CONCLUSION

So, finally, how *should* the Christian regard this thesis? Perhaps better,
since there are evidently 'distinctive thread[s] in the tapestry of Christianity' in Plantinga's evocative metaphor (1991*b*: 29; p. 695 above), how might
someone respond who sees in the Christian doctrine of creation an affirmation of the integrity of the natural order? TCA implies a cousinship extending across the entire living world, the sort of coherence (as Leibniz
once argued) that one might expect in the work of an all-powerful and all-
wise Creator. The 'seeds', in Augustine's happy metaphor, have been there
from the beginning; the universe has in itself the capacity to become what
God destined it to be from the beginning, as a human abode, and for all we
know, much else.

When Augustine proposed a developmental cosmology long ago, there
was little in the natural science of his day to support such a venture. Now
that has changed. What was speculative and not quite coherent has been

---

[29] There is one further perspective on this matter of semi-deism that I have set aside above.
The occasionalists of the fourteenth century maintained that God is the *only* cause, strictly
speaking, of what happens in the world. What appears to be causal action within the world is for
them no more than temporal succession. Things do not have natures that specify their actions;
rather, the fact that they act according to certain norms must be directly attributed to God's
intentions. There is no reason in this view why God should not, for example, suddenly make new
kinds of plants and animals appear, if God so wishes; since there is no order of *nature*, God is
committed only to the reasonable stability of (more or less) regular succession on which human
life depends. (The issue that separated the nominalists from the Aristotelian defenders of real
causation in nature is brought out very well in the essay by Alfred Freddoso (1988) cited by
Plantinga.) In this perspective, the issue of special creation comes to be posed in a quite different
way. Any view which affirms the sufficiency of the natural order for bringing about the origins
of life might be dubbed by the occasionalist as semi-deist. When I read the paragraph where
Plantinga says that someone who maintains that God creates, conserves, and concurs in the
activity of the universe can still be semi-deistic, my first reaction was to assume that this
committed him to occasionalism, since it would seem that it is *only* from the occasionalist
perspective that this view of God's relationship with the natural order would be classed as semi-
deist. But Plantinga is quite evidently not an occasionalist; his treatment of natural science
implies that he believes in the operation of secondary causation in nature. Thus, I have assumed
in the discussion above that he must have had something else in mind when speaking of semi-
deism: namely, the openness of creation to the supernatural order of grace and miracle. Incidentally, the occasionalist *would* be likely to believe that special creation is antecedently more
probable, and (in Berkeley's version, at least) might tend to question a theory, like the theory
of evolution, which depends on the reality of such causes as genetic mutation.

transformed, thanks to the labours of countless workers in a variety of different scientific fields. TCA allows the Christian to fill out the metaphysics of creation in a way that (I am persuaded) Augustine and Aquinas would have welcomed. No longer need one suppose that God must have added plants here and animals there. Though God *could* have done so, the evidence is mounting that the resources of the original creation were sufficient for the generation of the successive orders of complexity that make up our world.

Thus, common ancestry gives a meaning to the history of life that it previously lacked. From another perspective, this history now appears as preparation. The uncountable species that flourished and vanished have left a trace of themselves in us. The vast stretches of evolutionary time no longer seem quite so terrifying. Scripture traces the preparation for the coming of Christ back through Abraham to Adam. Is it too fanciful to suggest that natural science now allows us to extend the story indefinitely farther back? When Christ took on human form, the DNA that made him son of Mary may have linked him to a more ancient heritage stretching far beyond Adam to the shallows of unimaginably ancient seas. And so, in the Incarnation, it would not have been just human nature that was joined to the Divine, but in a less direct but no less real sense all those myriad organisms that over the aeons had unknowingly shaped the way for the coming of humanity.[30]

Anthropocentric? But of course: the story of the Incarnation *is* anthropocentric. Reconcilable with the evolutionary story as that is told in terms of chance events and blind alleys? I believe so, but to argue it would require another essay. Unique? Quite possibly not: other stories may be unfolding in very different ways in other parts of this capacious universe of ours. Terminal? Not necessarily: we have no idea what lies ahead for humankind. The transformations that made us what we are may not yet be ended. Antecedently probable from a Christian perspective? I will have to leave that to the reader.

REFERENCES

Anderson, B. W. (1983), 'The Earth is the Lord's: An Essay on the Biblical Doctrine of Creation', in R. M. Frye (ed.), *Is God a Creationist?* (New York: Scribner), 176–96.

[30] Though the alert reader will have caught echoes of the theology (not the biology) of Teilhard de Chardin, the affinities with the Christology of Karl Rahner are, perhaps, more immediate. See e.g. Rahner 1961.

Augustine, St (1982), *The Literal Meaning of Genesis*, trans. J. H. Taylor (2 vols., New York: Newman).
Ayala, F. J. (1985), 'The Theory of Evolution: Recent Successes', in E. McMullin (ed.), *Evolution and Creation* (Notre Dame, Ind.: University of Notre Dame Press), 59–90.
Bergant, D., and Stuhlmueller, C. (1985), 'Creation According to the Old Testament', in E. McMullin (ed.), *Evolution and Creation* (Notre Dame, Ind.: University of Notre Dame Press), 153–75.
Clifford, R. (1988), 'Creation in the Hebrew Bible', in Russell *et al.* (1988), 151–70.
Darwin, C. (1872), *On the Origin of Species*, 6th edn. (London: Murray).
Dawkins, R. (1986), *The Blind Watchmaker* (New York: Norton).
Denton, M. (1986), *Evolution: A Theory in Crisis* (Bethesda, Md.: Adler and Adler).
Finocchiaro, M. (1989), *The Galileo Affair* (Berkeley: University of California Press).
Freddoso, A. (1988), 'Medieval Aristotelianism and the Case against Secondary Causation in Nature', in T. V. Morris (ed.), *Divine and Human Action* (Ithaca, NY: Cornell University Press), 74–118.
Gilkey, L. (1985), *Creationism on Trial* (Minneapolis: Winston).
Gould, S. J., and Lewontin, R. (1979), 'The Spandrels of San Marco and the Panglossian Paradigm: A Critique of the Adaptationist Programme', *Proceedings of the Royal Society of London*, B205: 581–98.
Hawking, S. (1988), *A Brief History of Time* (New York: Bantam Books).
Kellenberger, J. (1972), *Religious Discovery, Faith, and Knowledge* (Englewood Cliffs, NJ: Prentice-Hall).
McMullin, E. (1967), *Galileo, Man of Science* (New York: Basic Books).
——(1981), 'How Should Cosmology Relate to Theology?', in A. R. Peacocke (ed.), *The Sciences and Theology in the Twentieth Century* (Notre Dame, Ind.: University of Notre Dame Press), 17–57.
——(1983), 'Galileo as a Theologian', Fremantle Lecture presented at Oxford University.
——(1984), 'The Rational and the Social in the History of Science', in J. R. Brown (ed.), *Scientific Rationality: The Sociological Turn* (Dordrecht: Reidel), 127–63.
——(1985), Introduction to E. McMullin (ed.), *Evolution and Creation*, (Notre Dame, Ind.: University of Notre Dame Press), 1–56.
——(1988), 'Natural Science and Belief in a Creator', in Russell *et al.* (1988), 49–79.
——(in press), 'Indifference Principle and Anthropic Principle in Cosmology', *Studies in the History and Philosophy of Science*.
Mayr, E. (1960), 'The Emergence of Evolutionary Novelties', in S. Tax (ed.), *The Evolution of Life* (Chicago: University of Chicago Press), 349–80.
——(1963), *Animal Species and Evolution* (Cambridge, Mass.: Harvard University Press).
——(1988), *Toward a New Philosophy of Biology* (Cambridge, Mass.: Harvard University Press).
Mivart, St. George Jackson (1871), *On the Genesis of Species* (London: Macmillan).
Moss, J. D. (1983), 'Galileo's *Letter to Christina*: Some Rhetorical Considerations', *Renaissance Quarterly*, 37: 547–76.
Pius XII (1950), 1981. *Humani Generis*, in Claudia Carlen (ed.), *The Papal Encyclicals 1939–1958* (Raleigh, NC: McGrath, 1981), 175–84.
Plantinga, A. (1991a), 'Evolution, Neutrality, and Antecedent Probability: A Reply to Van Till and McMullin', *Christian Scholar's Review*, 21: 80–109.

—— (1991*b*), 'When Faith and Reason Clash: Evolution and the Bible', *Christian Scholar's Review*, 21: 8–32; reproduced as Ch. 34.

Provine, W. B. (1987), review of *Trial and Error. The American Controversy over Creation and Evolution*, by E. J. Larson, *Academe*, 73/1: 50–2.

Quinn, P. (1984), 'The Philosopher of Science as Expert Witness', in J. T. Cushing, C. F. Delaney, and G. M. Gutting (eds.), *Science and Reality* (Notre Dame, Ind.: University of Notre Dame Press), 32–53.

Rahner, K. (1961), 'Christology within an Evolutionary View of the World', in *Theological Investigations*, (Baltimore: Helicon), 157–92.

Russell, R. J., Stoeger, W. R., and Coyne, G. V. (1988) (eds.), *Physics, Philosophy, and Theology* (Notre Dame, Ind.: University of Notre Dame Press, and Rome: Vatican Observatory Press).

Scherer, S. (1990), 'The Protein Molecular Clock: Time for Re-evaluation', *Evolutionary Biology*, 24: 83–103.

Sober, E. (1984) (ed.), *Conceptual Issues in Evolutionary Biology* (Cambridge, Mass.: MIT Press).

Van Till, H. (1986), *The Fourth Day* (Grand Rapids, Mich.: Eerdmans).

—— (1991), 'When Faith and Reason Cooperate', *Christian Scholar's Review*, 21: 33–45.

Von Salvini-Plawen, L., and Mayr, E. (1977), 'On the Evolution of Photo-Receptors and Eyes', *Evolutionary Biology*, 10: 207–63.

36

REPLY TO McMULLIN

ALVIN PLANTINGA

I note with pleasure the deep underlying agreement between McMullin and myself; nevertheless there remain some points of equally deep difference.

I. AUGUSTINIAN SCIENCE

In 1991*b* (Ch. 34 above) I argued that the Christian community ought to think about the subject-matter of the various sciences—again, in particular the human sciences, but also to some degree the so-called natural sciences—from an explicitly theistic or Christian point of view. It should do so, of course, only where that is relevant: only where it looks as if thinking about the matter at hand from that point of view might lead to conclusions or emphases different from those ordinarily to be found. I suggested calling the result 'Unnatural Science', or 'Creation Science', or 'Theistic Science'. A better name, I think, is 'Augustinian Science', which recalls Augustine's suggestion that serious intellectual activity in general is ordinarily in the service of a broadly religious vision of the world.

A. Why Do We Need Augustinian Science?

Fundamentally, because much of what goes on in the sciences is quite unsatisfactory, seriously flawed from the perspective of Christian theism. There are many examples, especially from psychology, sociology, sociobiology, political science, and other areas of the human sciences. Here I give one from sociobiology and a couple from evolutionary biology, the area where the disagreement between McMullin and I has been focused.

First published as 'Science: Augustinian or Duhemian?' in *Faith and Philosophy*, 13/3 (1996): 368–94. Reprinted by permission.

1. Simon and rationality According to Herbert Simon (1990),[1] there is a problem with *altruistic* behaviour, the sort characteristic of Mother Teresa, or the Little Sisters of the Poor, or the Jesuit missionaries of the seventeenth century, or the Methodist missionaries of the nineteenth. The *rational* way to behave, says Simon, is to act or try to act in such a way as to increase one's personal fitness: that is, to act so as to increase the probability that one's genes will be widely disseminated in the next and subsequent generations, thus doing well in the evolutionary Derby.[2] Mother Teresa and the Little Sisters, however, show very little interest in the propagation of their genes; this behaviour clearly requires explanation; so what is its explanation? Simon proposes two mechanisms: 'bounded rationality' and 'docility':

Docile persons tend to learn and believe what they perceive others in the society want them to learn and believe. Thus the content of what is learned will not be fully screened for its contribution to personal fitness. (ibid. 1666)

Because of bounded rationality, the docile individual will often be unable to distinguish socially prescribed behavior that contributes to fitness from altruistic behavior. In fact, docility will reduce the inclination to evaluate independently the contributions of behavior to fitness. . . . By virtue of bounded rationality, the docile person cannot acquire the personally advantageous learning that provides the increment, *d*, of fitness without acquiring also the altruistic behaviors that cost the decrement, *c*. (ibid. 1667)

The idea is that a Mother Teresa displays 'bounded rationality'; she adopts those culturally transmitted altruistic behaviours without making an independent evaluation of their contribution to her personal fitness. If she *did* make such an independent evaluation (and was clever enough to do it properly), she would see that this sort of behaviour does not contribute to her personal fitness, drop it like a hot potato, and get to work on increasing her fitness (perhaps by sponsoring a contest, among her younger relatives, to see who can have the most children).

But isn't this in clear conflict with Christian teachings about what it is rational for human beings to do? Behaving like Mother Teresa is not at all a manifestation of 'bounded rationality'—as if, if she thought about the matter with greater clarity and penetration, she would instead act so as to

---

[1] Simon won a Nobel Prize in economics, but also works in computer studies and psychology; he is currently Professor of Computer Studies and Psychology in the Department of Psychology at Carnegie-Mellon.

[2] More simply, says Simon, 'Fitness simply means expected number of progeny' (1990: 1665). *Why* he thinks this is the rational way to behave he doesn't explain; and could it really be part of science to make a normative claim of that sort? Simon apparently thinks it is simply given in our evolutionary origin that this way of behaving is the rational way. But couldn't I sensibly say that while my having lots of progeny might be best for my genes, *I'm* interested in *my* welfare, not theirs?

increase her personal fitness. Behaving as she does is instead a manifestation of a Christ-like spirit; she is reflecting in her limited human way the
splendid glory of Christ's sacrificial action in the Atonement. Indeed, is
there any sense of 'rational' in which, from a Christian perspective, there
is anything at all a human being can do that is *more* rational than what she
does?

Of course, we might be tempted to claim that Simon's project really isn't
science; but can we sensibly make that claim in these post-Kuhnian days?
If the scientists call it 'science' and get grants from the National Science
Foundation for doing it, if it is published in scientific journals and written
in that stiff, impersonal style characteristic of them, can we sensibly claim
that it really isn't science? So here we have an example of a scientific
project that, from a Christian perspective, is wholly misguided. It is, perhaps, a particularly flagrant example, but there are many others in the
same neighbourhood (see below, p. 739).

2. Randomness and design The next examples are taken from evolutionary biology, the specific area under dispute between McMullin and myself.
One of the most conspicuous examples of serious confusion on the part of
some of the experts is the claim that current evolutionary theory demonstrates, or at any rate supports, the claim that human beings are not the
product of intelligent design; they have not been designed by God or
anyone else. A number of the most prominent writers on evolution unite in
declaring that evolutionary biology reveals a substantial element of *randomness* or *chance* in the origin and development of the human species;
therefore, human beings (so they claim) have not been designed. Stephen
Gould writes: 'Before Darwin, we thought that a benevolent God had
created us' (1977: 267). Gould's sentiments are expressed less tersely by
Douglas Futuyma:

By coupling undirected, purposeless variation to the blind, uncaring process of
natural selection Darwin made theological or spiritual explanations of the life
processes superfluous. Together with Marx's materialistic theory of history and
society and Freud's attribution of human behavior to processes over which we have
little control, Darwin's theory of evolution was a crucial plank in the platform of
mechanism and materialism—of much of science, in short—that has since been the
stage of most Western thought. (1986: 3)

Clearer yet, perhaps, is George Gaylord Simpson:

Although many details remain to be worked out, it is already evident that all the
objective phenomena of the history of life can be explained by purely naturalistic or,
in a proper sense of the sometimes abused word, materialistic factors. They are
readily explicable on the basis of differential reproduction in populations (the main

factor in the modern conception of natural selection) and of the mainly random interplay of the known processes of heredity. . . , Man is the result of a purposeless and natural process that did not have him in mind. (1967: 344–5)

The same claim is made by Richard Dawkins:

All appearances to the contrary, the only watchmaker in nature is the blind forces of physics, albeit deployed in a very special way. A true watchmaker has foresight: he designs his cogs and springs, and plans their interconnections, with a future purpose in his mind's eye. Natural selection, the blind, unconscious automatic process which Darwin discovered, and which we now know is the explanation for the existence and apparently purposeful form of all life, has no purpose in mind. It has no mind and no mind's eye. It does not plan for the future. It has no vision, no foresight, no sight at all. If it can be said to play the role of watchmaker in nature, it is the *blind* watchmaker. (1986: 5)[3]

Modern evolutionary science has given us powerful reason to believe that human beings are, in an important way, merely *accidental*; that there are such creatures as human beings (creatures with the properties human beings display) is fortuitious, a matter of chance. There wasn't any plan, any foresight, any mind, any mind's eye involved in their coming into being or displaying the properties they have.

Now this is initially surprising: how would an empirical science show something like that? Could there be empirical evidence for it? But in fact there is confusion here. Evolutionary science speaks of 'randomness': random genetic mutation, for example. Now these events are 'random' in something like the sense of not arising from the proper function of the organism; more specifically, they are not a result of the organism's functioning in accord with any part of its design plan aimed at promoting or preserving its welfare. Thus Ernst Mayr: 'The term, when applied to variation, means that it is not in a response to the needs of the organism.' But the conclusion the above writers draw depends upon taking 'random' in a much stronger sense, a sense entailing *not supervised, orchestrated, caused, or planned by God*. As far as I can see, they simply confuse these two senses, leaping lightly from one to the other. This simple confusion, obviously, has enormous capacity for mischief; it can lead the unwary to think that science has somehow shown that human beings were not designed by God, and that their most crucial and characteristic capacities have arisen, not by way of divine design, but by way of chance or accident.

Someone might reply that the evidence for a theory lies in its success, and evolutionary theory, taken with the stronger sense of 'random', is a highly successful theory with much empirical confirmation. By way of response, note that there are two versions of the relevant scientific theory

---

[3] This claim is repeated by Dennett (1995).

here. The first, and stronger, version includes the claim that random events in the strong sense (the sense that entails being unplanned by God) play a crucial role in evolution; the second, and weaker, makes the same claim with respect to random events in the weaker sense of 'random'. Now consider the conjunction of the weak theory with the denial of the strong; and note that this conjunction is supported by the evidence at least as firmly as is the strong theory itself. Hence the evidence supports the strong theory no more firmly than its denial; hence it doesn't give us a reason to believe that theory as opposed to its denial. You might as well argue for theism by conjoining it with, say, relativity theory or quantum mechanics, pointing out that the resulting conjunction is empirically adequate; few, I take it, would regard that as much of a reason for starting to go to church. But the above confusion is no better reason for staying home.

3. Is TCA 'certain'?   Many of the experts tell us that evolution—TCA, at the least—is *certain* with respect to the empirical evidence, as certain as that the Earth revolves around the Sun rather than vice versa. But (as I argued in 1991*b* (Ch. 34 above)) this seems to be at best wild exaggeration. There are the problems with the fossil record: the great gaps and the fact that there aren't any documented or uncontroversial examples of macro-evolution. There is also Mivart's old objection: as Mivart (1871)[4] saw, the mammalian eye (for example) is an extraordinarily complex and functionally integrated structure, its various parts intimately dependent upon each other for their function. He pointed out how difficult it is to envision a series of organic forms leading up to the eye from creatures without eyes, where each step on the path through that space must be both close enough to the preceding step to be plausibly reachable in a single step, and also adaptive, or at any rate not unduly maladaptive. (Although this objection was one of the first, it has never really been answered; people have just grown accustomed to living with it.) Indeed, it isn't really known that such a series is so much as biologically possible.

Here McMullin (1993; Ch. 35 above) reminds me that the evidence for TCA is *necessarily* incomplete: 'evolutionary explanation is of its nature *historical*, and historical explanation is not like explanation in physics or chemistry. It deals with the singular and the unrepeatable; it is thus *necessarily* incomplete' (p. 322; p. 724 above). This is true, and important; but of course a hypothesis for which the evidence is *necessarily* weak is still one for which the evidence is weak. It is also part of my point; it is (partly) for

---

[4] See also Behe 1996.

attacking Gould, Dawkins, Ayala, etc. because
they have a "philosophical or religious axe to grind"
— somewhat hypocritical...?

REPLY TO McMULLIN   739

this reason that it is absurd to claim that TCA is certain; those strident declarations of certainty must come from some source other than a cool, reasoned, dispassionate look at the evidence. Perhaps these writers have a philosophical or religious axe to grind, or perhaps they confuse TCA's being the best available hypothesis (or the best available hypothesis that conforms to the demands of methodological naturalism) with its being certain; more likely, perhaps, they confuse the epistemic probability of TCA on the empirical evidence with its probability on that evidence together with naturalism. Whatever the problem, this assessment of the evidence is again wholly unsatisfactory from the standpoint of Christian theism; it is another reason why Christians must make their own estimates here, rather than blindly following the experts.

endorsing biased approach to evidence

There are plenty of other examples from this area. For example, the famous zoologist G. G. Simpson poses the question 'What is man?', and answers: 'The point I want to make now is that all attempts to answer that question before 1859 are worthless and that we will be better off if we ignore them completely' (quoted in Dawkins 1976: 1).[5] And of course there are many examples from the social sciences. The great psychologist Jean Piaget (1930) asserts that a 7-year-old child whose cognitive faculties are functioning properly will believe that everything in the universe has a purpose in some grand overarching plan or design; a mature person whose faculties are functioning properly, however, will learn to 'think scientifically' and realize that everything has either a natural cause or happens by chance. There is also the assumption, widely current in scientific (sociological, psychological) study of religion, that serious religious belief must be a manifestation of pathology, stupidity, backwardness, or invincible ignorance. And my point is that in these areas the Christian intellectual community has a stake in noting that these claims—that human beings are not designed, that the common genetic code decisively confirms TCA, that TCA is certain, and so on—are not at all established by the empirical evidence. It has a stake in noting the role that naturalism or other broadly religious views play in the acceptance and dissemination of these claims. And it should work at some of these areas—particularly in the human sciences, but in evolutionary biology as well—from the perspective of Christian theism. That is, it should pursue these sciences by *starting from*

---

[5] Speaking of Dawkins, he provides another example, this one perhaps resulting less from philosophical confusion than from eagerness to establish a point. In 1986: 81 ff. he states Mivart's objection as 'what good is 5% of an eye?' His answer, in essence, is to confuse 5% of an eye with 5% vision. But of course Mivart's point is that not just any 5% of an eye will lead to 5% vision, or indeed any vision at all.

the basic tenets of Christianity, taking them as part of the constant contextual background with respect to which the plausibility and probability of scientific hypotheses and claims are to be evaluated.

B. Objections to Augustinian Science

Now as far as I can see, McMullin agrees that the Christian community should pursue *something* like Augustinian science. He recognizes that there could indeed be conflict between scientific theories and the deliverances of the Christian faith, what Christians learn from the Bible; he thinks that 'the context where differences of this kind might properly occur seems restricted to issues concerning human nature' (p. 711 above); and he mentions psychological or psychoanalytical theories that deny 'human free choice and the consequent moral responsibility for actions performed' (p. 710 above). So perhaps McMullin would agree that the Christian community needs Augustinian science in these areas. He is deeply suspicious of this approach, however, in evolutionary biology and allied areas:

> It is this casting of special creation and evolution as rivals in the domain of cosmological explanation that I find so troubling. If one assumes that there is a presumption in favour of some sort of special creation at the critical moments in the historical development of life (a presumption whose plausibility wanes in regard to specific transitions as the strength of the evolutionary explanation of those transitions increases), one inevitably transforms the field of prehistory into a battleground where the religious believer is engaged in constant skirmishes with the protagonists of evolutionary-type theories, skirmishes that most often end in forced retreat for the religious believer. (1993: 313; pp. 712–3 above)

Augustinian science, he says, 'certainly ensures conflict; it is likely to maximize the strain between faith and reason, as the believer searches for the expected gaps in the scientific account' (ibid.) Here, I think, there is both misapprehension and error.

1. Faith and reason First, the failure of communication: McMullin believes that pursuing Augustinian science here 'is likely to maximize the strain between faith and reason' (p. 713 above). But why so? A strain between faith and reason is a possibility, of course, only for someone (or some community) that accepts the Christian faith, and then only if some deliverance of faith is in tension with some deliverance of reason. So far as I can see, however, the Christian faith doesn't teach us that TCA is false, and reason doesn't teach us that it is true. (Maybe reason plus *naturalism* does, but that is another matter entirely.) The believer need not be anxious about TCA, or desperately eager to refute it. What is clear, from the point of view of Christian theism, is that the Lord has created the heavens and the earth and all that is in them; there is no particular way of doing so, however, such that it is clear that he did it in that way. It is also not clear that he didn't do so by way of TCA. (Perhaps he did something different and special in creating our first parents; that is quite compatible with their having descended from non-human forms of life.) As far as I can see, the proper attitude for Christians to take towards TCA is a sort of genial scepticism; *maybe* things happened that way, but then again maybe not. Indeed, as I argued above, it is the *naturalist* who has a real stake here. Evolution is the only answer anyone can think of to what would otherwise be a very embarrassing question; it is this that calls forth all those declarations of certainty.

And hence it is not the case that the believer need spend a lot of time 'searching for the expected gaps in the scientific account', frantically looking for holes in the latest evolutionary theories. First, of course, there is no need to go *searching* for those gaps; the absence of transitional forms in the fossil record is one of its salient features. But more important, the believer, so far as I can see, has no particular stake in the outcome here. The Augustinian or theistic scientist has a certain freedom denied her naturalistic compeer: she can follow the evidence where it leads. If it leads towards TCA, no problem; God could surely have done things that way if he wished. If it leads away from it, again, no problem; God could also have done things by way of episodes of special creation, or in still other ways.[6] It is the naturalist who has a real stake in evolution, not the theist. My point is only that in deciding on the right epistemic attitude to take to TCA, Darwinism, and the rest, the Christian scientific community should use *all* that it knows, including what it knows by faith. In particular, it should use the idea that it is God who in one way or another has created life, and he certainly could have done so by way of episodes of special creation.

So far as I can see, therefore, Augustinian science doesn't at all 'maximize the strain between faith and reason'. Indeed, it should do just the reverse. In doing Augustinian science, you start by assuming the deliverances of the Faith, employing them as well as anything else you know in dealing with a given scientific problem or project. Conflict between faith and reason can certainly occur; but it is less likely to occur, I think, than if the scientific investigation is insulated from the deliverances of the Faith and left to develop of its own accord.

[6] For example, he could have done it by way of something like Augustine's 'seeds'; see McMullin 1993: 311; p. 711 above.

2. *Academic trespass?*   McMullin points out that Augustinian science crosses contemporary academic boundaries; the theologian and scientist may find themselves at loggerheads, bandying about charges of academic trespass. Given the present organization of the disciplines, this seems correct: Augustinian science, at least at first, will certainly involve people's making pronouncements or claims outside their areas of special competence. This is a real point, and has to be taken with real seriousness; in writing on these areas, I myself certainly feel acute discomfort at venturing beyond the areas where I might be thought to know something about what I am talking about.

But, as things presently stand, this kind of trespass is inevitable, at least if we propose to think about the broader involvements and significance of a theory like the Grand Evolutionary Story, or Darwinism, or TCA. We will inevitably stray outside our areas of competence, as Futuyma, Gould, and others do in claiming that contemporary evolutionary theory shows that human beings are not designed, or as G. G. Simpson does in declaring that nothing on the nature of man written before 1859 is worth reading. There is no way to address the important questions here without getting outside our areas of competence. The alternative would be not to think about these matters—or, if we think about them, not to speak or write about them. That seems to me a counsel that is dangerous as well as unduly diffident.

3. *Is Augustinian science 'science'?*   McMullin displays a certain sympathy with a project somewhere in the neighbourhood of Augustinian science; however, he doesn't think the result should be *called* 'science' (or perhaps what he thinks is that the result really *isn't* science, whatever we call it):

> I do not think, however, that theistic science [Augustinian science] should be described *as* science. It lacks the universality of science, as that term has been understood in the later Western tradition. It also lacks the sort of warrant that has gradually come to characterize a properly 'scientific' knowledge of nature, one that favours systematic observation, generalization, and the testing of explanatory hypotheses. Theistic science appeals to a specifically Christian belief, one that lays no claim to assent from a Hindu or an agnostic. . . . To use the term 'science' in this connection seems dangerously misleading; it encourages expectations that cannot be fulfilled. (1993: 303; pp. 701–2 above)

Now in a way it doesn't matter what we *call* this enterprise that (as McMullin and I agree) ought to be undertaken by the Christian community. And no doubt we would agree further that it is the scientists of the Christian community, the practitioners of the disciplines in question, who ought to undertake it. But there is *something* (research grants, for example) in a name, and I thoroughly disagree with McMullin's reasons for denying the name 'science' to this enterprise. He makes two points. First, Augustinian science, he says, lacks the *universality* of what is nowadays called 'science'; it couldn't be practised by an agnostic or a Hindu. And second, it lacks 'the sort of warrant that points to systematic observation, generalization, and the testing of explanatory hypotheses' (ibid.).

But is it really true that what is nowadays called 'science' is universal, in McMullin's sense? Certainly not. Remember Herbert Simon's account of rationality and his treatment of altruism (p. 735 above); no Christian theist could either accept that account of rationality or (initially) acquiesce in the conclusion that altruistic behaviour is a result of unusual docility and 'limited' rationality.[7] Simon's project is surely not universal; it doesn't start from and admit as premises only propositions everyone—Jew, Christian, Hindu, or agnostic—already accepts or is prepared to accept. Not by a long shot. Similarly for the claims that a common genetic code is decisive evidence for TCA, for the claim that current evolutionary science supports the conclusion that human beings are not designed by God, for the Piagetian claim that the mature person realizes that everything has either a natural cause or else happens by chance, for the assumption that serious religious belief is pathological or a result of stupidity, unusual backwardness, or social disorder, and the like. All these claims are assumed or accepted in one scientific project or another—that is, in one or another project to which the term 'science' is commonly applied; but none of them is universal in McMullin's sense. I conclude that McMullin is mistaken here: the term 'science' is *not* currently used in such a way that it applies only to projects universal in that sense; hence we don't have here an objection to applying the term 'science' to Augustinian science.

As for the second point (that Augustinian science wouldn't involve observation, generalization, and the testing of hypotheses characteristic of science), here there is misunderstanding. The way to try to understand, from a theistic perspective, how God created plants and animals and human beings is to take account of all that you know: what you know by faith, what you know as a Christian, as well as what you know in other ways. In the case at hand, the relevant considerations would be what, if anything, Scripture teaches or suggests on the matter, together with the antecedent probability of, for example, TCA from a theistic perspective,

[7] 'Initially': that conclusion is antecedently unlikely from the standpoint of Christian theism, but of course it is conceivable that it is none the less true. To test it, we would presumably have to administer altruism, docility, and intelligence tests to a properly chosen sample of the population, to see whether altruism really is accompanied by increased docility and lessened intelligence—carefully avoiding, of course, the tendency on the part of many social scientists to take serious religious belief and accompanying altruism as a *criterion* of low intelligence.

The whole enterprise of science relies on priority being given to empirical observation rather than dogma & authority.

744   ALVIN PLANTINGA

together with the 'empirical evidence': the fossil record, the molecular evidence, homologies, and the like. Clearly this involves precisely the sort of systematic observation, generalization, and testing of explanatory hypotheses that McMullin cites as the hallmark of science. It may involve *more*; but it certainly involves this much. To establish his point, McMullin would have to argue something else: that science (properly so-called) somehow *couldn't* involve those other matters, the looking to see what (if anything) Scripture says on the matter, the consideration of the antecedent probability of a theory on theism, and so on. And I haven't the faintest idea how that could be argued. Where is it laid down that anything that does that is not science? — EVERYWHERE ① methodologically ② sociologically

The answer, McMullin thinks, lies in the *methodological naturalism* he thinks essential to natural science: the idea, to put it crudely, that in science we ought not to appeal to what we know about God, or his activity, or to what we know by way of the testimony of Scripture (see Plantinga 1995). Speaking of methodological naturalism, he writes: 'Scientists *have* to proceed in this way; the methodology of natural science gives no purchase on the claim that a particular event or type of event is to be explained by invoking God's "special" action or by calling on the testimony of Scripture' (1993: 303; pp. 702–3 above). But where does this embargo come from? It is ordinarily supported only by bad arguments of the type 'God is not part of the universe; in science we can only refer to parts of the universe; therefore . . .'; or even 'To refer to God in science is to treat God as an object, which is idolatry; therefore . . .'. Why believe that scientists have to proceed the way McMullin says they have to?

Consider, for example, the question of how life originated: theists know that God created it in one way or another, and now the question is: How did he do it? Did he do it by way of the ordinary regularities or laws of physics and chemistry (the ordinary beaviour of matter, so far as we understand it) or did he do something special? If, after considerable study, we can't see how it could possible have happened by way of those regularities—if, as is in fact the case, after many decades of study, the enormous complexity and functional connectedness and integrity of even the simplest forms of life make it look increasingly unlikely that they could have originated in that way—the natural thing to think, from the perspective of Christian theism, is that probably God did something different and special here. (Such a conclusion, of course, would not be written in stone. All we can say is that it is likely with respect to our *present* evidence; perhaps things will change; the enquiry is never closed.) And why couldn't one draw this conclusion precisely as a scientist? Where is it written that such

God of gaps

a conclusion can't be part of science? Why should we accept methodological naturalism?

II. PROBABILITIES: ANTECEDENT AND CONSEQUENT

A. God: Classicist or Romantic?

The antecedent probability of a thesis or hypothesis, for you, is its epistemic probability[8] on your background information, prior to or independent of consideration of the particular evidence at hand. I argued that the antecedent probability of evolution with respect to naturalism is very different from its antecedent probability with respect to Christian theism; in fact, I said, the antecedent probability of TCA, for the Christian theist, is less than that of its denial. Here McMullin digs in his heels; here, he says, is where he and I 'really part ways'.

In order to address this issue properly, we need a couple of distinctions. First, suppose we use the phrase 'the Grand Evolutionary Scenario' (GES) to denote the conjunction of four theses: (a) the Naturalistic Origins Thesis, according to which life arose from non-life just by way of the regularities of physics and chemistry; (b) the Progress Thesis, according to which life has progressed from relatively simple uni-cellular forms[9] to relatively complex forms, culminating, as we human beings like to think, in us; (c) TCA, according to which life originated at one place on Earth, all subsequent living creatures being related by descent to those aboriginal creatures; and (d) Darwinism, the claim that these enormous changes occurred by way of the accretion of many small steps and that mechanisms driving the enormous changes occurring since the origin of life are broadly Darwinian (for starters, natural selection working on some source of variation such as random genetic mutation).

Now the epistemic probability of a theory such as GES or TCA is of course relative to a body of background belief or knowledge. I argued that the antecedent probability of GES on naturalism is high. For the naturalist, evolution is the only game in town, the only answer anyone can think of to the questions, Where did this teeming variety of flora and fauna come from? How did it get here? And what accounts for that appearance of design it displays? The theist has an easy answer: in one way or another, the

---

[8] For an account of epistemic probability (more specifically, conditional epistemic probability) see Plantinga 1993: ch. 9.

[9] *Relatively* simple: in fact, even prokaryotes (e.g. bacteria) are immensely complex.

Lord created all these creatures (why else are they called 'creatures'?). But that answer is not available to the naturalist; it is GES that gives an answer to this otherwise embarrassing question. Hence Richard Dawkins's remark to A. J. Ayer at one of those elegantly candle-lit and splendidly bibulous Oxford college dinners: he said he couldn't imagine being an atheist before 1859 (the year Darwin's *Origin of Species* was published); 'although atheism might have been logically tenable before Darwin', said he, 'Darwin made it possible to be an intellectually fulfilled atheist' (1986: 6–7). I doubt that it is possible to be an intellectually fulfilled atheist, but Dawkins's essential point is right: GES is a reasonably plausible answer to the above question, and the only reasonably plausible answer anyone can think of. For the serious naturalist, therefore, evolution is an absolutely essential plank in his platform; and hence the antecedent probability of evolution, given philosophical naturalism, is high. It is this, I take it, that is in part responsible for those triumphal cries of certainty (pp. 738–40 above).

Then I went on to argue that (because of the improbability of the Naturalistic Origins Thesis) the probability of GES with respect to Christian theism and the empirical evidence is very low, and the probability of TCA with respect to that same body is perhaps somewhat less than that of its denial—that is less than 1/2. I argued this by claiming, first, that the antecedent probability (its probability independent of the empirical evidence) of TCA is perhaps (judgements of this kind are of course necessarily infirm) less than 1/2; I then went on to say that when we add the empirical evidence, things don't change; on theism and the empirical evidence, TCA is still somewhat less probable than its denial.

Now why should we think that the antecedent probability of TCA on Christian theism is less than 1/2? First, according to Christian theism, God is constantly at work in his universe. He is in constant, close, intimate causal contact with his creation, supporting and upholding it in being: were it not for this constant upholding activity, the cosmos would disappear like a candle flame in a high wind. Second, most Christians hold that God has frequently treated the things he has made in unusual and special ways: water turns into wine, human beings emerge unhurt from a fiery furnace, and are miraculously cured of disease; and above all, there is the wondrous splendour of the life, death, and resurrection of Jesus Christ. Apparently, therefore, God is not averse to working in his creation in special ways. But then, so I said, there is no particular antecedent probability in favour of the idea that he wouldn't do anything different or special in the creation of life, say, or in the creation of special kinds of life. So it is hard to see how there is any antecedent probability in favour of GES, and hard to see that there is any antecedent probability in favour of TCA. Indeed, if God acts spe-

cially over *one* large and important range of his interaction with his creatures, isn't it a bit more probable than not that he would act specially over
other large and important ranges?

So I'm inclined to think that the antecedent probability of TCA with
respect to Christian theism is a bit less than 1/2. But we still have to factor
in what we know about the origin of life; and with respect to Christian
theism and the present evidence, the Naturalistic Origins Thesis seems
extremely improbable. As modern biochemistry reveals, the simplest
forms of life display an astonishing, stunning complexity, a complicated
interrelatedness and functional integrity that boggle the mind. According
to Francis Crick, life must be regarded as the next thing to a miracle;
according to Harold P. Kein of Santa Clara University, chairman of a
National Academy of Sciences committee that recently reviewed origin-of-
life research, 'The simplest bacterium is so damn complicated from the
point of view of a chemist that it is almost impossible to imagine how it
happened' (Horgan 1991: 120). It therefore looks as if God did something
different and special in the creation of life.[10] (Of course, things may
change; that is how things look *now*.) These things taken together suggest
that the Lord might very well have done something different and special,
not only in creating life in the first instance, but also in creating certain
subsequent forms of life. If he did something special in creating life, what
would prevent him from doing something special at other points in his
great creation drama, perhaps creating specially the original representatives of some of the phyla, or human beings, or still other forms of life? It
would seem entirely in character. I am therefore inclined to maintain my
suggestion that the antecedent probability of TCA, from a theistic point of
view, is perhaps a bit less than 1/2.

Now McMullin is especially inclined to dispute my claims about the
antecedent probability of TCA. He makes initial heavy weather over the
very asking of the question as to what God is likely to do; if I understand
him, however, he goes on to claim that in fact it is unlikely, indeed *very*
unlikely, that God would do something special and different, or create
something specially in bringing it about that there are human beings, or
certain kinds of plants and animals, or even, presumably, life itself:

To carry the argument a stage further: what would the eloquent texts of Genesis,
Job, Isaiah and the Psalms lead one to expect? What have theologians made of these

---

[10] McMullin comments: 'Crick, who is notably unsympathetic to theistic belief, would surely
not agree with the inference being drawn from this!' (1993: 314). But of course we are here
enquiring into the likelihood of God's creating life specially *given theistic belief*, not into its
likelihood given Crick's naturalism. With respect to the latter, its probability is presumably zero;
but nothing at all follows about its probability with respect to the former.

texts? This is obviously a theme that far transcends the compass of an essay such as this one. I can make a couple of simple points. The Creator whose powers are gradually revealed in these texts is omnipotent and all-wise, far beyond the reach of human reckoning. God's providence extends to all creatures; they are all part of a single plan, only a fragment of which we know, and that darkly. Would such a being be likely to 'intervene' in the cosmic process: that is, deal in two different manners with it? (Let me emphasize that I am uncomfortable with this language of 'likelihood' with regard to God's action, as though we were somehow capable of catching the Creator of the galactic universe in the nets of our calculations.) Why should an omnipotent God not create a universe in which God's ends with regard to all creatures except humans would be achieved in a *natural* way? . . . If one may use the language of antecedent probability at all here—and I am not at all certain that one may—it surely must point away from special creation. (1993: 324–5; p. 726 above)[11]

A couple of comments on this passage: first, the issue is not, of course, whether there is some way to *calculate* the probability that God would do this or that; at best, on a topic like that, we have little more than crude guesses. And I applaud McMullin's implicit suggestion that any ideas we might have about the antecedent likelihood of God's doing this or that should be at best tentative. I think it a bit more probable that God would do something different and special in the creation of life, and human beings, and perhaps some other forms of life; but any such views, surely, should be tentative and held with appropriate diffidence. It certainly befits no one to be at all cocksure here.

This said, however, I fail to see any force in the considerations McMullin puts forward. God is indeed omnipotent and all-wise; his providence does indeed extend to all creatures; and indeed we know but a fragment of his total plan. These things are true; but how do they bear on the question of whether God would or would not, for example, create in stages: first creating inanimate material, say, then later doing something special in creating life, perhaps, and then still later in creating human life? (It is part of the major theistic religions to think that God has created humankind *in his own image*; might he not have thought it appropriate to create human life in a special way, by way of an act of special creation?) We know, after all, that God is not averse to acting in special ways, as the many miracles recorded in the Bible attest.

Bible as source of empirical evidence.

---

[11] McMullin's word 'intervene' isn't a good one in the theistic context; God constantly supports every created thing in existence by way of activity without which the creature would disappear like your breath on a frosty morning; furthermore, according to the bulk of the theistic community, no creaturely action is so much as possible without his further concurrent activity. So he necessarily and constantly takes a hand in the operation of his creatures; he necessarily and constantly 'intervenes' in his creation; were he to leave it to its own devices, even for an instant, it would vanish like a dream upon awakening. Creating something new—life, for example, or human life—isn't really in any sensible sense an 'intervention' for him.

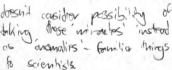

doesn't consider possibility of taking those 'miracles' instead as 'anomalies' – familiar things to scientists [margin annotation]

McMullin seems to think of God as like a classical artist, devoted to ideals of simplicity and elegance, economy and restraint. But why think of him like *that*? Perhaps God is more like a romantic artist with limitless resources, extravagant, prolific, fecund, overflowing with uproarious creative activity, disdaining restraint and economy of action. (The millions of species that have become extinct would be examples of this exuberant fertility.) After all, what are the attractions of economy for him? Creatures limited in energy, power, and time have need for economy; God suffers from no such limitations. Is it instead the idea that God's interest in economy of effort is a matter of aesthetic preference? But is there even the slightest reason to think so? The Lord constantly acts in his creation; apart from his upholding activity, it would disappear like a puff of smoke; why would he think it beneath his dignity, or aesthetically unpleasant, or otherwise disagreeable, to take a hand in his creation in other ways? Perhaps he is very much a hands-on God. Perhaps he marks various important transitions and junctures in the history of his creation by special celebratory or ceremonial activity of some sort. Perhaps an example of this sort of activity is his creating certain kinds of life specially, thus symbolically marking the importance of the transition. 'Are not sparrows two for a penny? Yet without your Father's leave', says Jesus, 'not one of them can fall to the ground.' If the Lord's creation scheme is such that his leave is needed for a sparrow to fall, might he not have created various forms of life—perhaps even sparrows—specially? Why not?

So my claim is that the antecedent probability of episodes of special creation with respect to Christian theism but prior to the empirical evidence—that is, the evidence bearing on TCA—is a bit greater than 1/2. McMullin, by contrast, insists that salvation history tells us nothing about natural history. What God does with respect to salvation gives us no probabilities with respect to special creation:

The story of salvation is a story about men and women, about the burden of being human. . . . The biblical account of God's dealings with humankind provides no warrant whatever for supposing that God would have brought the ancestors of the various kinds of plants and animals to be outside the ordinary order of nature. (1993: 324; p. 726 above)

The reason seems to be that the story of salvation is 'about free beings who sinned and who therefore *needed* God's intervention. Dealing with the human predicament 'naturally', so to speak, would not have been sufficient' (ibid.). Here McMullin suggests that God would perhaps have *preferred* to deal with the human predicament without doing something unusual, but that wasn't possible; he therefore had to act specially. Since he wasn't thus constrained when it came to creation, however, his acting specially in salvation history doesn't make it any more probable that he would do so in creation.

But how do we know that God was somehow *obliged* to act specially in salvation history? This seems to be a new theological idea; like most original theological ideas, it warrants suspicion. What is our source of information as to God's constraints here? In any event, he certainly wasn't obliged to act specially so *often* in salvation history. According to Catholic doctrine, a miracle occurs whenever the mass is properly celebrated; on nearly all Christian views, God regularly guides and directs his people individually and his Church collectively by virtue of the work of the Holy Spirit in the believer's heart;[12] each of these acts is special in the relevant sense. And so far as the 'ordinary order of nature' is concerned, on the views both of John Calvin and Pope Pius XII (and in the face of fire-power like that, who am I to demur?), God creates specially a new human soul or person whenever a human being comes into being. If so, the order of nature regularly and ordinarily involves very many acts of special divine creation; at present, the rate would be about three such acts per second. Some Christians might reject Pius's and Calvin's claim, but presumably not on the grounds that God has an aversion to acts of special creation.

Of course these are deep and difficult waters. I am inclined to think the antecedent probabilities slightly favour episodes of special creation, but I can certainly see the attractions of agnosticism here. Perhaps the most reasonable attitude is one of agnosticism: one just doesn't know what these antecedent probabilities are. What seems to me *un*reasonable, however, is to be confident that antecedent probability favours TCA or GES. And if I am right, then we must rely most heavily, here, on the empirical evidence.[13]   *So why the nonsense of the last 5 pages?*

---

[12] In fact, of course, the Holy Spirit does much more than guide and direct; for a genuinely powerful account, see Edwards 1746.

[13] I said that the antecedent probability of TCA just on Christian theism is a bit less than 1/2, and that when we factor in the evidence having to do with the origin of life, the antecedent probability of TCA remains less than 1/2; I added that we should rely more heavily, here, on the empirical evidence than on estimates of antecedent probability, which are bound to be a bit shaky. McMullin makes heavy weather of this; suggesting that 'Plantinga appears to change ground somewhat . . . now the warrant for claiming the antecedent likelihood of special creation appears to shift from the salvation story to the "empirical evidence"' (1993: 313; p. 713 above). He adds that the antecedent probability is 'no longer strictly antecedent' (ibid. 314; p. 713 above), notes that I say that 'actual empirical evidence must be allowed to speak more loudly than speculative theological assumptions', and comments: 'So much for his original claim that the story of God's dealings with Israel spoke loudly in favour of special creation over TCA!'

But here McMullin has uncharacteristically erred. First, the antecedent probability of a hypothesis is its probability with respect to some body of belief antecedent to the evidence under current consideration, whatever that is. My claim in 1991*b* (Ch. 34 above) was that, given

B. The Empirical Evidence

Turning briefly and finally to that empirical evidence, there are a couple of points that need to be made. (I must caution you that neither McMullin nor I can claim any expertise with respect to the empirical evidence; all we know is what we read in the papers. Think of each of the following sentences as prefaced by 'I'm no expert, but . . .'.) McMullin writes as if the theistic sceptic with respect to evolution is fighting a sort of desperate rearguard action, with a succession of new and powerful pieces of evidence ever compelling further retreat, as one major gap after another is closed, one major transition after another definitively nailed down. And it isn't only McMullin who thinks this; there is a sort of widespread impression, a kind of widely shared but uncritical assumption, among academics who view the subject from a certain distance, that these major gaps and transitions are in fact steadily closing, or at least narrowing.

But where *are* all these gaps being closed by further discoveries? They are not easy to find. As McMullin points out, the fossil record contains many sequences of extinct forms (e.g. trilobites) 'where the development of specific anatomical features can be traced in detail through the rock layers' (1993: 315; p. 715 above). This is indeed so, but does not bear on the main problems for TCA with the fossil record, which have to do with the lack, in that record, of sequences of intermediary forms between the really major taxa—phyla and classes, for example (Simon 1990). 'As new fossil evidence is uncovered,' he says (ibid.), 'palaeontologists continue to uncover stage after stage in crucial "linking" forms such as the therapsids, the forms that relate reptiles with the earliest mammals'. This is misleading exaggeration. It suggests that palaeontologists have discovered and continue to discover many forms that link, say, fish with amphibia, amphibia with reptiles, reptiles with birds, or reptiles with mammals. But so far as I

Christian theism and what most Christians believe, but antecedent to *any* of the empirical evidence bearing on GES, the probability of TCA is perhaps a bit less than 1/2. My claim in 1991a is that the probability of TCA on Christian theism *together with the empirical evidence bearing on the origin of life* is also a bit less than 1/2—and maybe also a bit less than the first probability. There is, of course, no inconsistency or tension here, or any shifting of ground. Furthermore, it is inaccurate to say, as McMullin does, that I originally claimed that the story of God's dealings with Israel 'spoke loudly in favour of special creation over TCA!' (1993: 314; p. 714 above). What I said was that 'it is a bit more probable, before we look at the scientific evidence, that the Lord created life and some of its forms—in particular, human life—specially' (1991b: 22; p. 687 above). 'A *bit* more probable', I said; nothing here about the story of God's dealings with Israel speaking *loudly* in favour of TCA. And in fact I think they speak *softly* here; the empirical evidence has much more weight, and should be allowed to speak more loudly than the antecedent probabilities. It should also be allowed to speak more loudly, a great deal more loudly, than semi-deistic theological assumptions according to which God would find it beneath his dignity or be somehow reluctant to act specially in the world he has created and lovingly sustains from moment to moment.

but empirical evidence can never lead to anything more than anomalies or puzzles. It cannot lead to 'God', nor Xtian God.

know, this is not so. So far as I know, therapsids are the only candidates for a link between reptiles and mammals (and they have been known for a long time). Although there is some controversy about the therapsids, perhaps they really could be thought of as something like a linking form between reptiles and mammals;[14] but if TCA were true, one would expect vastly many more such forms. Furthermore, *Archaeopteryx*, known since shortly after Darwin's death and formerly the only serious candidate for a similar post linking reptiles and birds, has, according to some, been demoted by the discovery of modern birds ante-dating it. And things stand no better with respect to those other major gaps.

In fact, it looks as if the shoe is on the other foot. 'When I think of the eye,' Darwin said, 'I shudder.' He was thinking of the enormous complexity of the eye, and the strain involved in believing that an instrument of that delicate interrelatedness and functional integrity could have evolved by anything like the mechanisms he suggested. Perhaps he was also thinking of Mivart's specific objections (above, p. 738). But Darwin had no idea at all of the true complexity of the visual system (Behe 1996: ch. 2); if he shuddered at what he knew then, he would have quaked uncontrollably had he known what we know now.

Secondly, Mivart's worry returns in spades when we think of 'irreducible complexity'. According to Michael Behe (1996: 59 ff.), the cilium (used by many kinds of cells for swimming) is composed of some six molecules. All six are required for the cilium's function; if any is absent, no ciliar function is possible. Cilia, says Behe, are 'irreducibly complex'—in that we can't envision any simpler forms that will carry out the cilium's function. In the case of Mivart's eye, we can certainly envision *some* intermediate forms; the problem is that it is hard to see how there could be the required complete series. (There is also the fact that it isn't really known that such a series is even biologically possible.) With the cilium, however, we can't envisage any members of a series of functional precursors at all. Still further, says Behe, this example of irreducible complexity is only one of several; there is also, for example, the system that targets proteins for delivery to sub-cellular compartments, as well as aspects of blood clotting, closed circular DNA, electron transport, the bacterial flagellum, telomeres, photosynthesis, and still other structures.

Thirdly, in Darwin's day it was possible to attribute the failure to find intermediate forms between the major taxa to the fact that the fossil record

---

[14] Of course they could also be thought of as outrigger forms on a typological classification. In general, the gaps in the fossil record are better accommodated by the successive but independent appearance of the major *Bauplans* with substantial evolution radiating out from these *Ur* forms.

was largely unexplored. Since Darwin's day, however, the number of fossils discovered and catalogued has increased a hundredfold; it is no longer possible (or at any rate plausible) to make that excuse, and the gaps are at least as great as ever. Lots of series with some modification have been found, as with trilobites; the great gaps, however, remain.

Fourth, there is the gap between life and non-life; as we have seen, this gap has greatly widened since Darwin's day.

Fifth, there is the Cambrian explosion. The fossil record displays unicellular life going all the way back, so they tell us, to 3 or 3.5 billion years ago—only a billion years or so after the formation of the Earth itself, and much less than a billion years after the Earth cooled sufficiently to permit life. There is no fossil record of skeletal animals until about 530 million years ago, 2.5 or 3 billion years after the appearance of uni-cellular life. Then there is a veritable explosion of invertebrate life, a riot of shapes and anatomical designs, with ancestors of the major contemporary forms and all the marine invertebrate phyla represented, together with a lot of forms wholly alien in the contemporary context.[15] None of this was known in Darwin's day, and would surely have given him pause. And now in a recent issue of *Science* we learn that the time during which this explosion took place was much shorter than previously thought; it all happened during a period of no more than 5 or 10 million years (Kerr 1993: 1274 and Bowring *et al.* 1993: 1293), a period that seems much too short to accommodate such furious evolutionary creativity, at least with respect to any known mechanisms.[16] On balance, it is likely that if Darwin knew what we now know about the complexity of such organs as the mammalian eye and the human brain, the enormous intricacy revealed by biochemistry and molecular biology (including the astonishing complexity of the simplest forms of life), the Cambrian explosion, the lack of closure in the fossil record, and so on, he would have been neither a Darwinian nor a devotee of TCA.

For a Christian, therefore, one who is not shackled by the demands of naturalism, the right attitude towards TCA is one of a certain cordial scepticism. TCA is a very pretty theory with many of the so-called theoretical virtues; it has been a fine source of research projects; incorporated into God's great drama of Creation and Incarnation, as in McMullin's concluding peroration, it is attractive. It doesn't follow, however, that it is true, or even that it is more likely than not.[17]

---

[15] 'The real shocker for me is the worm that looks like it has kneecaps,' said Dr Ellis L. Yochelson (1991), a palaeontologist at the Smithsonian Institution. He was referring to an animal known as *Microdictyon*.

[16] The initial response of Darwinists has been something like, 'Wow! Evolution can move much faster than we thought!'

[17] My thanks to Del Ratzsch, Gordon van Harn, and Ralph Stearley.

[handwritten note:] most weight in Plantinga's case is problems with TCA as a scientific hypothesis — he is just opening 'gaps'.

REFERENCES

Behe, M. (1996), *Darwin's Black Box* (New York: Free Press).
Bowring, S. A., Grotzinger, J. P., Isachsen, C. E., Knoll, A. H., Pelechaty, S. M., Kolosov, P., (1993), 'Calibrating Rates of Early Cambrian Evolution', *Science*, 261: 1293–8.
Dawkins, R. (1976), *The Selfish Gene* (Oxford: Oxford University Press).
——(1986), *The Blind Watchmaker* (New York: Norton).
Dennett, D. (1995), *Darwin's Dangerous Idea* (New York: Simon and Schuster).
Edwards, J. (1746), *Religious Affections* (Boston).
Futuyma, D. (1986), *Evolutionary Biology*, 2nd edn. (Sunderland, Mass.: Sinauer Associates).
Gould, S. J. (1977), 'So Cleverly Kind an Animal', in *Ever Since Darwin* (New York: Norton), 260–71.
Horgan, J. (1991), 'In the Beginning . . .', *Scientific American*, 264/2: 116–25.
Kerr, R. A. (1993), 'Evolution's Big Bang Gets, Even More Explosive', *Science*, 261: 1274–5.
McMullin, E. (1993), 'Evolution and Special Creation', *Zygon*, 28: 299–335; reproduced as Ch. 35.
Mivart, St G. J. (1871), *On the Genesis of Species* (New York: D. Appleton and Co.).
Piaget, J. (1930), *The Child's Conception of Physical Causality* (London: Kegan Paul).
Plantinga, A. (1991*a*), 'Evolution, Neutrality and Antecedent Probability: A Reply to Van Till and McMullin', *Christian Scholar's Review*, 21: 80–109.
——(1991*b*), 'When Faith and Reason Clash: Evolution and the Bible', *Christian Scholar's Review*, 21: 8–32; reproduced as Ch. 34.
——(1993), *Warrant and Proper Function* (Oxford: Oxford University Press).
——(1995), 'Methodological Naturalism?', in J. van der Meer (ed.), *Facets of Faith and Science* (Lanham, Md.: University Press of America).
Simon, H. (1990), 'A Mechanism for Social Selection and Successful Altruism', *Science*, 250: 1665–8.
Simpson, G. G. (1967), *The Meaning of Evolution*, rev. edn. (New Haven: Yale University Press).
Yochelson, E. L. (1991), 'Spectacular Fossils Record Early Riot of Creation', *New York Times* (23 Apr.).

NOTES ON THE CONTRIBUTORS

RON AMUNDSON is Associate Professor of Philosophy at the University Hawaii, Hilo. He is working on the reasons for development and embryology's exclusion from the Modern Syntheses, and on current efforts to integrate them into a more inclusive evolutionary theory.

ROBERT BRANDON is Professor of Philosophy at Duke University in Durham NC. He publishes on evolutionary theory with special attention to those aspects of the evolutionary process that tend to be ignored, in particular the environment. He has published two books, *Adaptation and Environment* (1990) and *Concepts and Methods in Evolutionary Biology* (1996), as well as an anthology with Richard M. Burian, *Genes, Organisms, Populations* (1984).

RICHARD DAWKINS is Lecturer in Zoology at Oxford University and a Fellow of New College. He is one of the half-dozen most influential and controversial biologists working on evolutionary theory today. His *The Selfish Gene* (1976) was followed by a series of books and papers amplifying and extending his gene-selections perspective.

DANIEL C. DENNETT is the Distinguished Professor of Arts and Sciences and Director of the Center for Cognitive Studies at Tufts University in Massachusetts. He is on a very short list of highly respected philosophers who have examined the assumptions of evolutionary theory as well as its implications for broader philosophical issues such as naturalism and reduction.

KEVIN DE QUEIROZ is a zoologist at the National Museum of Natural History of the Smithsonian Institution. He has published in both systematics and the philosophy of biology. His chief empirical publications deal with the phylogeny of aquamate reptiles, while his philosophical publications address the relationship between systematics and evolution. He is currently working on the phylogeny of the Holbrookia and Anolis lizards as well as a phylogenetic approach to biological nomenclature.

MICHAEL J. DONOGHUE is Professor at the Harvard University Herbarium. He publishes in the philosophy of taxonomy, evolutionary theory, and the phylogenetic analyses of the angiosperms, especially the genus Viburnum. Recently, while president of the Society of Systematic Biology, he urged the abandonment of the Linnaean hierarchy in classifying plants and animals.

MARC ERESHEFSKY is Associate Professor of Philosophy at the University of Calgary. His primary area of research is the philosophy of biology, especially philosophical problems concerning biological taxonomy. He is editor of *The Units of Evolution: Essays on the Nature of Species* (1992); and is writing a book on species and the Linnaean hierarchy.

PETER GODFREY-SMITH is Assistant Professor of Philosophy at Stanford University. He is author of *Complexity and the Function of Mind in Nature* (1996) and numerous articles on the philosophy of biology and the philosophy of mind. He is currently working on adaptationism and the evolution of behavioural complexity in individuals and populations.

STEPHEN JAY GOULD is Alexander Professor of Zoology and Professor of Geology at Harvard University and the Curator for Invertebrate Paleontology in the Museum of Comparative Zoology at Harvard. He has published fifteen books including a series of collections of the papers he has published, primarily in the *Natural History Magazine*. Not only does he contribute to science proper but he is also one of the most influential commentators on science writing today.

RUSSELL D. GRAY is Senior Lecturer in Psychology at the University of Auckland. He has published extensively on animal behavior, avian phylogenetics, and evolutionary theory. For some time he has been urging that the gene be dethroned and replaced by or at least supplemented with developmental systems theory.

PAUL GRIFFITHS is Associate Professor of Philosophy at the University of Auckland. He works in the philosophy of biology and psychology. He has edited *Trees of Life: Essays in Philosophy of Biology* (1992) and recently published *What Emotions Really Are: The Problem of Psychological Categories* (1997).

DAVID L. HULL is Dressler Professor in the Humanities in the Department of Philosophy at Northwestern University. He has published in the philosophy of biology for over thirty years. Among his books are *Darwin and his Critics* (1973), *The Philosophy of Biological Science* (1974), *Science as a Process* (1988), and the *Metaphysics of Evolution* (1989). He is currently working on science as a selection process.

EVELYN FOX KELLER is Professor of History and Philosophy of Science at the Massachusetts Institute of Technology. Although she began her career as a physicist, she is best known for her work on women in science and feminist critiques of science. With Elisabeth Lloyd she edited *Keywords in Evolutionary Biology* (1992).

PHILIP KITCHER is Presidential Professor of Philosophy at the University of California, San Diego, editor-in-chief of *Philosophy of Science*, and most recently philosophical adviser to the United States Congress on the Human Genome Project. His publications include *Abusing Science: The*

Case against Creationism, Vaulting Ambition: Sociobiology and the Quest for Human Nature, and *The Lives to Come: The Genetic Revolution and Human Possibilities*, as well as articles on many issues in the philosophy of biology and philosophy of science. He is currently working on the ethical implications of contemporary biology and on the connections among evolution, altruism, and ethics.

GEORGE LAUDER is Professor of Ecology and Evolutionary Biology at the University of California, Davis. He works in the area of vertebrate functional morphology and evolution as well as how mechanical systems in organisms evolve. His research involves both comparative (interspecific) and ontogenetic analyses of muscle and bone function.

ELISABETH LLOYD is Associate Professor of Philosophy at the University of California, Berkeley. She published an important book on *The Structure and Confirmation of Evolutionary Theory* in 1988. In addition to her work in the philosophy of biology, she has also contributed to the feminist literature, especially where biology impinges on gender. With Evelyn Fox Keller, she edited *Keywords in Evolutionary Biology* (1992).

ERNAN MCMULLIN is Director Emeritus of the Graduate Program in History and Philosophy of Science at the University of Notre Dame. He has written or edited a dozen books, *including The Concept of Matter* (1963), *Galileo, Man of Science* (1967), *Newton on Matter and Activity* (1978), and *The Inference that Makes Science* (1992).

DANIEL W. MCSHEA is Assistant Professor in the Department of Zoology at Duke University. His major area of work is evolutionary biology, but he has also worked in the areas of evolutionary trends, especially trends in complexity, developmental constraints and evolution, and evolutionary biology.

JOHN MAYNARD SMITH, one of the most innovative biologists working in evolutionary biology today, is Professor of Biology at the University of Sussex. He is best known for the application of the principles of game theory to evolution, in particular the introduction and development of evolutionary stable strategies.

BRENT D. MISHLER is Director of the University and Jopson Herbaria at the University of California, Berkeley. He works in taxonomic philosophy, evolutionary biology, and the classification and natural history of clonal organisms, especially the bryophytes.

SUSAN OYAMA is Professor of Psychology at the City University of New York's John Jay College and at the City University of New York Graduate Center. Trained in the Department of Social Relations of Harvard University, she has retained her early commitment to interdisciplinary scholarship. In addition to her best-known book, *The Ontogeny of*

Information, she has also collaborated on *Aggression: The Myth of the Beast Within*, published under the group pen name John Klama. She has written extensively on the nature/nurture opposition, on the relation between development and evolution, and on the concept of development itself.

DIANE B. PAUL is Co-director of the Program in Science, Technology and Values, and Professor of Political Science at the Boston campus of the University of Massachusetts. A selection of her papers on genetics and eugenics has been collected in *Evolutionary Biology, Social History and Political Power*. She is currently working on two books on the history of eugenics and human eugenics, one for the general public and the other for a scholarly audience.

ALVIN PLANTINGA is John A. O'Brien Professor of Philosophy at the University of Notre Dame and Director of the Center for Philosophy of Religion. Principal publications include: *God and Other Minds, The Nature of Necessity, God, Freedom and Evil, Does God Have a Nature?*, *Warrant: The Current Debate, Warrant and Proper Function, Warranted Christian Belief* (forthcoming), and *Advice to Christian Philosophers*.

ROBERT J. RICHARDS is Professor in the Departments of History and Philosophy and in the Committee on Evolutionary Biology at the University of Chicago. He is Director of the Fishbein Center for the History of Science. He has written extensively on the history and philosophical implications of evolutionary theory, including 'Darwin and the Emergence of Evolutionary Theories of Mind and Behavior' (1987) and 'The Meaning of Evolution: The Morphological Construction and Ideological Reconstruction of Darwin's Theory' (1992). Currently he is working on a book with the tentative title *Romantic Biology: From Goethe to the Last Romantic, Ernst Haeckel*.

ALEXANDER ROSENBERG is Director of the Honors Program at the University of Georgia. He is the author of eight books, including *The Structure of Biological Science* and *Instrumental Biology or the Disunity of Science*. He has been an ACLS Fellow, a Guggenheim Fellow, and is the recipient of the Lakatos Prize for distinguished contribution to the philosophy of science.

MICHAEL RUSE is Professor of Philosophy and Zoology at the University of Guelph in Ontario. He has written a dozen books and numerous papers on topics in history and philosophy of biology. He has edited *Biology and Philosophy* from its inception in 1986 to the present. His latest book is *Monad to Man: The Concept of Progress in Evolutionary Naturalism*.

ELLIOT SOBER is Hans Reichenbach Professor of Philosophy at the University of Wisconsin, Madison. He has published *The Nature of Selection* (1984), *Reconstructing the Past* (1988), *Philosophy of Biology* (1993), and *From a Biological Point of View* as well as the highly influential anthology *Conceptual Issues in Evolutionary Biology* (1984, 1989). He also is co-author, with David Sloan Wilson, of *Unto Others: The Evolution of Altruism.*

EDWARD STEIN teaches philosophy and gay studies at Yale University. He specializes in philosophy of science, philosophy of mind, epistemology, and ethics. He is the author of *Without Good Reason: The Rationality Debate in Philosophy and Cognitive Science* (1996) and *Sexual Desires: Science, Theory and Ethics* (forthcoming).

KIM STERELNY is Professor in the Department of Philosophy at Victoria University of Wellington in Wellington, New Zealand. He published "The Return of the Gene" with Philip Kitcher, a paper that has come to be a classic in the philosophy of biology. He is currently working on an introductory text in philosophy of biology, *Sex and Death*, with Paul Griffiths.

MARGA VICEDO is Assistant Professor of Philosophy at Arizona State University (West). Spanish born and educated, she is particularly interested in epistemological and ethical issues of genetics. Because of her sensitivity to questions of history, she is at the moment working on a book on the key players active in the science of heredity in America at the beginning of this century.

ELISABETH VRBA is Professor in the Department of Geology and Geophysics at Yale University. She is author of the effect hypothesis and helped to introduce and popularize such concepts as sorting and exaptation in connection with a hierarchical view of selection. In her early work she provided fossil data in support of speciation via punctuated equilibria.

MARY-JANE WEST-EBERHARD is a Senior Scientist of the Smithsonian Tropical Research Institute, resident in Costa Rica. She does field work in tropical social wasps and is at present writing on development and phenotypic flexibility in relation to selection and evolution.

DAVID SLOAN WILSON is Professor of Biology at the State University of New York, Binghamton. He is best known as a proponent of multi-level selection theory, which predicts that the properties of adaptation can evolve at all levels of the biological hierarchy. In addition, he has worked on a number of fundamental problems at the interface between ecology, evolution, and behaviour, including the nature of individual differences and the study of human behaviour from an evolutionary perspective.

FURTHER READING

The philosophy of biology is one of the hottest areas in philosophy today. There is a very large quantity of new, good material appearing each year. It is therefore quite impossible to cover all of the pertinent writings. What follows is a brief survey of some of the publications that anyone who is interested in the various topics would find useful. It is in no sense a definitive list. If you find yourself excited about something which we do not mention, do not think that we would disparage your choice.

Without a doubt, the place to start on the topic of *adaptation* is Charles Darwin's *Origin of Species*. For a 'classic', it is a remarkably readable text. Try to get hold of a reprint of the first edition, which reads very much more freshly than the later editions, where Darwin added answers to critics. There is a facsimile edition published by Harvard University Press (Cambridge, Mass., 1959), as well as a cheap reprint by Penguin (Harmondsworth, 1968) with a very good introduction by the historian John Burrow. You must at a minimum read the first four or five chapters, where Darwin introduces his causal mechanisms and explains why he thinks adaptation is so important an issue. For those who want to dig a little more deeply into the historical background—and philosophers of biology are in the forefront of those philosophers of science who stress the importance of history—start with Michael Ruse's *The Darwinian Revolution: Science Red in Tooth and Claw* (Chicago: University of Chicago Press, 1979), an overall account of the great revolution in biology, together with his 'The Darwin Industry: A Guide' (*Victorian Studies*, 39/2 (1996): 217–35), which covers all of the recent literature. David Hull (ed.), *Darwin and his Critics* (Cambridge, Mass.: Harvard University Press, 1973), a reprint of nineteenth-century reactions to the *Origin*, will prepare you well for today's debates about adaptation. Supplement this with 'Historical Development of the Concept of Adaptation', by one of our authors, Ron Amundson (in M. R. Rose and G. V. Lauder (eds.), *Adaptation* (New York: Academic Press, 1996), 11–53).

Moving to contemporary writings, begin with Richard Dawkins's *The Blind Watchmaker* (New York: Norton, 1986). This is a first-class survey of modern evolutionary thinking by a committed Darwinian. You may not agree with everything he has to say, but you will certainly be impressed by the skill with which Dawkins plays with his ideas. Another good source, directly on the subject of adaptation, is an article entitled 'Adaptation' (*Scientific American*, 239/3 (1978): 212–30) by the leading population geneticist Richard Lewontin. He explains brilliantly the importance of adaptation for evolutionary biology, while not fudging the difficulties faced in handling the subject. Lewontin has also been one of the strongest critics of the Darwinian enthusiasm for unbridled attributions of adaptation. Essential reading is 'The Spandrels of San Marco and the Panglossian Paradigm: A Critique of the Adaptationist Programme', by palaeontologist Stephen Jay Gould and Lewontin (*Proceedings of the Royal Society of London*, ser. B: *Biological Sciences*, 205 (1979): 581–98). A most interesting collection of essays directly on this particular article, edited by the student of rhetoric Jack Selzer, is *Understanding Scientific Prose* (Madison: University of Wisconsin Press, 1993). A vigorous Darwinian response to the Gould and Lewontin paper can be found in Daniel Dennett's *Darwin's Dangerous Idea* (New York: Simon and Schuster, 1995), a chapter of which is included in our collection (Ch. 3).

In recent years, philosophers have written comparatively little on the topic of *development*. In the past, it tended to be the domain of the 'vitalist'—that is to say, of the person who wanted to argue that living things require some sort of life force of their own. (The classic work of this ilk is *Creative Evolution* by the French philosopher Henri Bergson (London: Macmillan, 1911).) Recently, however, a number of philosophers have been turning their attention towards development, considering matters from a perspective more in tune with modern biology. An excellent overview of the history of development as it bears on evolutionary questions is Stephen Jay Gould's *Ontogeny and Phylogeny* (Cambridge, Mass.: Belknap Press, 1977). This book is part science and part history. Although it is long, it is very readable. It gives a huge amount of information from one who, although no philosopher himself, is sensitive to the sorts of issues which excite philosophers.

A recent work looking at development philosophically, one which has attracted much praise, is by the psychologist Susan Oyama: *The Ontogeny of Information* (Cambridge: Cambridge University Press, 1985). She is concerned with long-standing troublesome dichotomies, such as that between nature and nurture, and she argues that the time has come for fresh thinking, moving beyond the categories of the past. Paul Griffiths, who has a chapter in our collection with R. D. Gray (Ch. 7), and who has been much influenced by Oyama, has edited a collection of writings on the philosophy of biology entitled *Trees of Life* (Dordrecht: Kluwer, 1992). This includes a section on development, centred on a paper by Oyama putting forth her ideas in a more condensed form.

In the past several decades, the *units of selection* controversy has given rise to a large philosophical literature. It is worth going back to the historical roots: Charles Darwin and Alfred Russel Wallace, the co-discoverers of natural selection, thought long and hard about the units of selection issue, coming down on opposite sides of the fence. This is discussed by Michael Ruse in 'Charles Darwin and Group Selection' (*Annals of Science*, 37 (1980): 615–30). This article is included in a collection on the topic edited by Robert Brandon and Richard Burian: *Genes, Organisms, Populations: Controversies over the Units of Selection* (Cambridge, Mass.: MIT Press, 1984). Also not to be missed is biologist George Williams's *Adaptation and Natural Selection* (Princeton: Princeton University Press, 1966), which really convinced people that there is a real problem here in need of solution. (Williams should also be read by those interested in adaptation.)

The seminal single-author work on the subject, from a philosophical perspective, is that by Elliott Sober: *The Nature of Selection* (Cambridge, Mass.: MIT Press, 1984). He shows in some detail the conceptual issues at stake, teasing out important questions of causation. Follow up Sober with important works, from somewhat different perspectives, by Elisabeth Lloyd, *The Structure and Confirmation of Evolutionary Theory* (New York: Greenwood, 1988), and Robert Brandon, *Adaptation and Environment* (Princeton: Princeton University Press, 1990). Both of these books have scopes beyond the units of selection issue. Lloyd is worth reading as an excellent introduction to some of the logical issues connected with the structure of evolutionary theory. Brandon can be read for a very detailed discussion of adaptation, as well as a careful discussion of the ways in which organisms interact with their environments. People interested in development will find this useful. Finally, for this section, we recommend a work of a somewhat different pace by the palaeontologist Niles Eldredge and the philosopher Marjorie Grene: *Interactions: The Biological Context of Social Systems* (New York: Columbia University Press, 1992). These authors do not restrict themselves to questions about biological evolution: importantly, they move on to broader issues reaching up into culture.

Function is a topic which has a long history. Post-Second World War treatments of the subject were much influenced by technological advances in that conflict, and even today the classic treatment of the subject—by the logical empiricist Ernest Nagel, in his *The Structure of Science* (London: Routledge and Kegan Paul, 1961)—can be read with profit. Follow this with two works devoted exclusively to the topic: Larry Wright's *Teleological Explanations* (Berkeley: University of California Press, 1976) and Andrew Woodfield's *Teleology* (Cambridge: Cambridge University Press,

1976). Wright's basic ideas can also be found in an often-reprinted article entitled 'Functions' (*Philosophical Review*, 82 (1973): 139–68); this should be complemented by a discussion which takes a somewhat different perspective: Robert Cummins's 'Functional Analysis', (*Journal of Philosophy*, 72 (1975): 741–64). Both the Wright article and the Cummins article have been reprinted in an excellent collection on issues in the philosophy of evolutionary biology: Elliott Sober (ed.), *Conceptual Issues in Evolutionary Biology* (Cambridge, Mass.: MIT Press, 1984; 2nd edn. 1993). Back this reading with a new book by Peter Godfrey-Smith: *Complexity and the Function of Mind in Nature* (Cambridge: Cambridge University Press, 1996). As the title hints, it is much broader than simply a discussion of teleology, and touches on many issues covered in other parts of our collection.

One last point. Anyone who is going to delve in detail into questions of function or teleology ought to look at some of the standard analyses of teleology by the great philosophers, particularly Aristotle and Kant. Good places to start are S. Gotthelf and J. D. Lennox (eds.), *Philosophical Issues in Aristotle's Biology* (Cambridge: Cambridge University Press, 1987), and J. D. McFarland, *Kant's Concept of Teleology* (Edinburgh: University of Edinburgh Press, 1970). One might also want to look at some more historical material on teleology in biology. Works mentioned earlier on the historical aspects of adaptation will prove useful here. The definitive work on function in history is E. S. Russell's *Form and Function: A Contribution to the History of Animal Morphology* (London: John Murray, 1916), recently reprinted in paperback (Chicago: University of Chicago Press, 1982).

The *species* question has been a topic of intense interest to philosophers. A good collection, including most of the seminal papers from the first part of this century, is C. N. Slobodchikoff (ed.), *Concepts of Species* (Stroudsburg, Pa.: Dowden, Hutchinson and Ross, 1978). This should then be supplemented with a more recent collection edited by the philosopher Marc Ereshefsky: *The Units of Evolution: Essays on the Nature of Speciation* (Cambridge, Mass.: MIT Press, 1992). This latter, a nice mix of writings by biologists and philosophers, will bring the reader right up to date. More scientifically oriented are two collections: M. Claridge, H. Darwah, and M. Wilson (eds.), *Species: The Units of Biodiversity* (New York: Chapman and Hall, 1977), and Q. Wheeler and R. Meier (eds.), *Species Concepts and Phylogenetic Theory: A Debate* (New York: Columbia University Press, forthcoming). Of somewhat broader interest, but essential reading for all who are interested in questions of classification, is Elliott Sober's *Reconstructing the Past: Parsimony, Evolution, and Inference* (Cambridge,

Mass.: MIT Press, 1988), dealing with problems to do with inference of historical lineages (phylogenies) and their connections to classification. Supplement your philosophical reading with a work which looks at contemporary history of taxonomy: David Hull's *Science as a Process* (Chicago: University of Chicago Press, 1988). This work is especially recommended to those readers who have never laboured as real scientists. It will give them a good idea of some of the issues which are raised by the practical problems faced in the laboratory and the field, not to mention the ways in which the personalities of the participants can influence outcomes.

Moving now to the more social issues included in this collection, *human nature* has attracted a vast literature: philosophical, biological, and social. An excellent place to start is with the Pulitzer Prize-winning work *On Human Nature*, by the Harvard entomologist and sociobiologist Edward O. Wilson (Cambridge, Mass.: Harvard University Press, 1978). This very controversial work sees humankind as having a nature rooted strongly in our evolutionary background. For a vigorous critical attack of this approach, turn to philosopher Philip Kitcher's *Vaulting Ambition* (Cambridge, Mass.: MIT Press, 1985). This is a brilliant case for the prosecution, even if (as is the nature of these things) it is not always quite as sensitive to its targets as justice demands. For the defence, read Michael Ruse's *Taking Darwin Seriously: A Naturalistic Approach to Philosophy* (Oxford: Blackwell, 1986). This work covers both epistemology (theory of knowledge) and ethics (theory of morality). Of a somewhat different pace is a recent work, Steve Jones's *In the Blood: God, Genes and Destiny* (London: HarperCollins, 1996). You will be overwhelmed by the huge number of stimulating ideas, albeit coming often in a rather chaotic and disconnected order, revealing the origins of the work in a television series.

Feminist philosophers have authored many works, some of which emphasize the positive, others of which focus on problems with more traditional viewpoints. Especially recommended is Helen Longino's *Science as Social Knowledge* (Princeton: Princeton University Press, 1990), which, although very critical of much which has gone before, is nevertheless sympathetic to the strengths of traditional philosophical analyses. If you want a taste of how savage the critiques can get, look at Fiona Erskine's 'The Origin of Species and the Science of Female Inferiority' (in D. Amigoni and J. Wallace (eds.), *Charles Darwin's "The Origin of Species": New Interdisciplinary Essays* (Manchester: Manchester University Press, 1996), 95–121). Complement your reading of feminist literature with some of the many available works on human nature considered from the gay perspective. The starting-point for this genre is the first volume of the (unfinished) *History of Sexuality* by Michel Foucault (New York: Pantheon, 1978). A good recent collection, edited by one of our contributors, is Edward Stein (ed.), *Forms of Desire: Sexual Orientation and the Social Constructivist Controversy* (New York: Routledge, 1992). Finally, let us recommend a recent controversial work on human nature by the historian-psychologist Frank Sulloway: *Born to Rebel: Birth Order, Family Dynamics, and Creative Lives* (New York: Pantheon, 1996). His claim is that birth order is an absolutely crucial determinant in forming of human nature. No doubt this is a thesis which will be subjected to stringent analysis in years to come; but Sulloway's is one of those works such that, even if it is refuted, we will all move forward in our understanding.

Altruism is, in a sense, a spin-off from more traditional discussions about evolution and ethics. A good historical introduction to some of the issues is Michael Bradie's *The Secret Chain: Evolution and Ethics* (Albany, NY: SUNY Press, 1994). This should be complemented by a collection on the connections between evolution and ethics, edited by Paul Thompson, *Issues in Evolutionary Ethics* (Albany, NY: SUNY Press, 1995), as well as a historical-philosophical work by Robert J. Richards: *Darwin and the Emergence of Evolutionary Theories of Mind and Behavior* (Chicago: University of Chicago Press, 1987). From a more scientific perspective, Richard Dawkins's *The Selfish Gene* (Oxford: Oxford University Press, 1976) is a deservedly well-established source: still highly readable and pertinent, even though now over twenty years old. (There is now a second edition, with a new introduction.) An excellent account of recent attempts to put morality on a firm biological basis, written by a science journalist, is Robert Wright's *The Moral Animal: Evolutionary Psychology and Everyday Life* (New York: Pantheon, 1994). Also well worth considering, although of a much sterner philosophical nature, is a new work jointly authored by Elliott Sober and David S. Wilson: *Unto Others: The Evolution of Altruism* (Cambridge, Mass.: Harvard University Press, 1997). Very new is a collection of essays on biology and ethics, looking at matters both historically and from a contemporary perspective, edited by Jane Maienschein and Michael Ruse: *Biology and Ethics* (Cambridge: Cambridge University Press, 1998).

The Human Genome Project has stimulated the production of several good books. Before one plunges into the philosophy, one should certainly master the scientific and technological material. Especially recommended here is the book by the science historian Dan Kevles and molecular biologist Leroy Hood: *The Code of Codes: Scientific and Social Issues in the Human Genome Project* (Cambridge, Mass.: Harvard University Press, 1992). From a philosophical perspective, an absolutely first-class work is

766 FURTHER READING

Philip Kitcher's *The Lives to Come: The Genetic Revolution and Human Possibilities* (New York: Simon and Schuster, 1996). He deals first with the technical material, and then goes on to lay out the pertinent philosophical issues. This work should be complemented by a couple of collections from conferences on the Human Genome Project: a somewhat older one edited by R. F. Weir, S. C. Lawrence, and E. Fales, *Genes and Self-Knowledge: Historical and Philosophical Reflections on Modern Genetics* (Iowa City: University of Iowa Press, 1994), and an excellent new one edited by Notre Dame professor Phillip Sloan, *Controlling Our Destinies: Historical, Social, Philosophical, and Ethical Perspectives on the Human Genome Project* (Notre Dame, Ind.: University of Notre Dame Press, 1997). This is a fast-moving field, and readers will probably find much new material appearing all the time.

The topic of *progress* has recently been the focus of much attention. You should start with the definitive critique of biological progressionism in the already mentioned *Adaptation and Natural Selection*, by George Williams. A useful collection based on a 1986 conference in Chicago is Matthew Nitecki (ed.), *Evolutionary Progress* (Chicago: University of Chicago Press, 1988). This volume presents a spectrum of views, written by scientists, philosophers, and historians. Complementing this work is a major new historical analysis of the concept of progress as it has played itself out in the history of evolutionary thought: Michael Ruse's *Monad to Man: The Concept of Progress in Evolutionary Biology* (Cambridge, Mass.: Harvard University Press, 1996).

You should look at the article entitled 'Arms Races Between and Within Species' by Richard Dawkins and John Krebs (Proceedings of the Royal Society, B 205 (1979): 489–511). Note that this was given at the same symposium as the Gould and Lewontin critique of adaptationism. Is this a reflection of a crucial difference between English and American evolutionism? You might think so if you turn to later writings by Stephen Jay Gould, for whom progress has become something of an obsession. Critical of views of progress as applied to humankind is Gould's *The Mismeasure of Man* (New York: Norton, 1981). A work which shows how ideas of progress have influenced people's perspectives on the past, as they try to analyse the palaeontological record, is his *Wonderful Life: The Burgess Shale and the Nature of History* (New York: Norton, 1989). Gould's most recent book, *Full House: The Spread of Excellence from Plato to Darwin* (New York: Random House, 1996), takes up some of these issues from a broader perspective, attempting to show exactly why ideas of progress have such a magnetic hold on so many people.

Finally, coming to *Creationism*, one should start with two excellent works which were written in response to the explosion of interest resulting from the 1981 Arkansas Creationism Trial, when a pro-Creationism law was overthrown following a constitutional challenge by the American Civil Liberties Union. Still very pertinent are Philip Kitcher's *Abusing Science: The Case Against Creationism* (Cambridge, Mass.: MIT Press, 1962) and biologist Douglas Futuyma's *Science on Trial: The Case for Evolution* (New York: Pantheon, 1983). Complement these single-author works with a collection edited by Arkansas ACLU witness Michael Ruse: *But is it Science? The Philosophical Question in the Creation/Evolution Controversy* (Buffalo: Prometheus Books, 1988). This work, just reissued in paperback, deals directly with philosophical questions emerging from the clash of Creationism and evolution, stressing the historical background to the whole issue.

The definitive history of Creationism is by the historian of science Ronald Numbers: *The Creationists* (New York: Knopf, 1992). This is an exceptionally balanced account, with a full and fair exposition of Creationist ideas. However, if you are going to really dig into this matter, then you really must look into writings by the Creationists themselves. Probably the best place to start is with the work by lawyer Phillip Johnson, *Darwin on Trial* (Washington: Regnery Gateway, 1991). Another work favourable to the cause, pretending to be more objectively disinterested than it really is, is *Darwin's Black Box* by biochemist Michael Behe (New York: Free Press, 1996). A new work looking at the Creationism controversy critically, but fairly, is by a philosopher who is also an active Quaker, Robert Pennock's *Tower of Babel: Scientific Evidence and the New Creationism* (Cambridge, Mass.: MIT Press, 1998).

Finally, if you are going to study this subject in any detail, then you really ought to learn something of the history of evolutionary biology. An excellent place to start is the recently reissued *The Death of Adam*, by the distinguished historian of science John C. Greene (Iowa City: University of Iowa Press, 1959). Also recently reissued is the single most important work on the relationship between science and religion in the years leading up to Darwin: *Genesis and Geology*, by Charles C. Gillespie (Cambridge, Mass.: Harvard University Press, 1950). Similarly excellent reading is the volume edited by Ernan McMullin: *Evolution and Creation* (Notre Dame, Ind.: University of Notre Dame Press, 1985). After reading these books, you will know how shallow are claims that biblical literalism is the definitive Christian tradition.

Bringing to an end these suggestions for 'Further Reading', we note that there are several other works which do not fit neatly into compartments, but which should be on everybody's reading list. Especially important are a number of collections of individual people's writings on the philosophy of biology. These include Robert Brandon, *Concepts and Methods in Evolutionary Biology* (Cambridge: Cambridge University Press); David Hull, *The Metaphysics of Evolution* (Albany, NY: SUNY Press); Michael Ruse, *The Darwinian Paradigm: Essays on its History, Philosophy and Religious Implications* (London: Routledge, 1989), and *Evolutionary Naturalism: Selected Essays* (London: Routledge, 1995); and Elliott Sober, *From a Biological Point of View* (Cambridge: Cambridge University Press, 1994). Also, the interested student will want to consult the journal in the field, *Biology and Philosophy*, published quarterly. This carries a very wide range of articles on the philosophy of biology, both from the point of view of epistemology and from the point of view of ethics. Finally, your attention is directed to Michael Ruse's *Philosophy of Biology Today* (Albany, NY: SUNY Press, 1988), a handbook on the subject. It provides a detailed catalogue of pertinent writings. Although now ten years old and somewhat dated, it will be a useful complement to the material presented in this survey. Taken together with the references at the end of the various articles in our collection, there will be little that you have missed.

INDEX OF NAMES